The Ecology of Kalimantan

The Ecology of Indonesia Series

Volume III

The Ecology of Indonesia Series

Volume III: The Ecology of Kalimantan

Other titles in the Series

Volume I: The Ecology of Sumatra
Volume II: The Ecology of Java and Bali
Volume IV: The Ecology of Sulawesi
Volume V: The Ecology of Maluku and Nusa Tenggara
Volume VI: The Ecology of Irian Jaya
Volume VII: The Ecology of the Indonesian Seas

Produced by
Environmental Management Development in
Indonesia Project, a cooperative project of the
Indonesian Ministry of the Environment
and
Dalhousie University, Halifax, Nova Scotia
under the sponsorship of the
Canadian International Development Agency

The Ecology of Kalimantan

Dr. Kathy MacKinnon
Ir. Gusti Hatta
Dr. Hakimah Halim
Dr. Arthur Mangalik

PERIPLUS
EDITIONS

Copyright © 1996 Dalhousie University
All maps in this edition copyright © 1996 Periplus Editions (HK) Ltd
All rights reserved

Published by Periplus Editions (HK) Ltd.

ISBN 0-945971-73-7

Publisher: Eric Oey
Design: Pete Ivey
Prepress production: JWD Communications Ltd.
Copy editing: Sean Johannesen and Kathy McVittie

Distributors:
Indonesia:
C.V. Java Books, P.O. Box 55 JKCP
Jakarta 10510

Singapore and Malaysia:
Berkeley Books Private Ltd.
5 Little Road #08-01
Singapore 536983

The Netherlands:
Nilsson & Lamm B.V.
Postbus 195, 1380 AD Weesp,

Printed in the Republic of Singapore

Table of Contents

EMDI *xvii*

Foreword *xix*

Acknowledgements *xxi*

Abbreviations and Acronyms *xxiii*

Introduction *1*

Chapter 1 The Island of Borneo *9*
    Geography *9*
    Geological History *12*
    Geology *22*
    Soils *25*
    Climate *30*
    Flora *35*
    Fauna *41*
    Biogeography *47*
        Biogeographical Units in Borneo *51*
    Human History in Borneo *55*
        Current Land Use *62*

Chapter 2 Coastal Habitats *71*
    Offshore Islands and Rocky Islets *73*
    Rocky Shores *74*
    Beaches and Beach Forests *75*
    Estuaries and Mudflats *80*
    Seagrass Beds *84*
    Coral Reefs *85*
    Mangrove Forest *94*
        Mangrove Zonation *96*
        Mangrove Ecology *101*
        Mangrove Fauna *106*
        Mangrove Habitats as a Valuable Resource *114*

Chapter 3 Freshwater Habitats *117*
    Swamp Forests *117*
    Peat Swamp Forest *119*
        Formation of Ombrogenous Peat Swamps *120*
        Ecology of Peat Swamp Forest *121*
        The Fauna of Peat Swamp Forests *126*

Freshwater Swamp Forest *127*
  Fauna of Freshwater Swamp Forests *130*
Rivers *131*
  Physical Patterns in Rivers *133*
  Water Chemistry *134*
  River Communities *139*
  Energy Flow in Rivers *141*
  Invertebrate Drift *143*
  Freshwater Fishes and Vertebrate Predators *143*
  Riverbank Communities *148*
Kalimantan Lakes *151*
  Weedbeds and Swamps *155*
  The Kapuas Lakes - A Wetlands Ecosystem *159*
Human-Made Lakes *163*
Human Impacts on Freshwater Ecosystems *169*

Chapter 4 The Lowland Rainforest of Borneo *175*
  Lowland Dipterocarp Forest *177*
  Forest Structure *179*
    Stratification *179*
    Tree Form *180*
    Ground Layer *182*
    Climbers, Stranglers and Epiphytes *183*
  Forest Succession *186*
  Biomass and Productivity *190*
  Nutrient Cycling *193*
  Flowering, Fruiting and Leaf Production *196*
  Plant-Animal Interactions *201*
    Herbivory *202*
    Pollination *204*
    Seed Dispersal *209*
  Animal Communities *213*
    Stratification of Animal Communities *214*
    Competition and Niche Separation *216*
    Soil and Litter Communities *222*
    Forest Floor Community *226*
    Larger Herbivores *230*
    The Forest at Night *231*

Chapter 5 Other Lowland Forest Formations *241*
  Heath Forest or Kerangas *241*
    Forest Structure and Composition *243*
    Heath Forest Regeneration *256*

Fauna *257*
Ironwood Forests *262*
Forest over Ultrabasic Rocks *265*
    Gunung Silam: an Ultrabasic Mountain *266*
    Pleihari-Martapura Wildlife Reserve *267*
    Land Use on Ultrabasic Soils *271*

Chapter 6 Limestone Habitats *273*

Soils and Drainage *275*
Limestone Flora *277*
Limestone Forests *280*
    Lowland Scree Forest *280*
    Lowland Limestone Cliff Communities *281*
    Lowland Limestone Forest *281*
    Lower Montane Limestone Forest *282*
    Upper Montane Forest *282*
The Fauna of Limestone Hills *284*
Caves *287*
Cave Communities *290*
    Bats *292*
    Swiftlets *294*
    Roof Community *297*
    Floor Community *298*
The Mulu Caves *299*
Caves and Human History *303*
Caves as a Resource *308*
Conservation of Caves and Limestone Formations *310*

Chapter 7 Mountain Habitats *315*

Climate *317*
Mountain Soils *318*
Productivity and Nutrient Cycling *319*
Zonation of Mountain Forests *320*
Zonation of Animals on Tropical Mountains *329*
Mount Kinabalu - Summit of Borneo *339*
    The Fauna of Mount Kinabalu *343*
Bukit Raya *348*
Effects of Disturbance on Montane Habitats *350*
Mountains as Centres for Biological Diversity *352*

Chapter 8 Borneo Peoples - Migrations and Land Use *355*

Penan: Harvesters of the Forest *357*
Dayak Groups *358*

Dayak Migrations *360*
Coastal Communities *361*
Shifting Agriculture: Subsistence Farming *363*
   Shifting Agriculture: Ecologically Sound or
     Environmentally Damaging? *369*
Harvesting the Forest *375*
Omens and Augury *384*
Transmigration and Resettlement *387*
The Future *392*

Chapter 9 Forest Resources *395*

Forests for Timber *395*
   Effects of Logging on Forest Structure and Dynamics *403*
   The Sustainability of Timber Harvests *407*
Other Forest Products *410*
   Rattans *410*
   Resins and Incense Wood *413*
   Illipe Nuts *415*
Forests for Water and Soil Conservation *417*
Forest Disturbance and Wildlife *421*
Reforestation *432*
Forest Fires on Borneo *435*
   Fire and Forest Ecosystems *438*
Conservation of Forests *442*

Chapter 10 Wetland Resources *445*

The Ecological Importance of Wetland Habitats *445*
   Water Flow and Water Quality *448*
Plant Harvests From Wetlands *450*
   Timber From Swamp Forests *454*
Food Chains and Animal Resources *456*
Fisheries in the Middle Mahakam Lakes *460*
Aquaculture *464*
Reclaiming Tidal Swamplands for Agriculture *465*
   Acid Sulphate Soils *467*
   Success and Failure in the Tidal Swamps *469*
   Peat Swamps for Agriculture *471*
Peat for Fuel *472*
Creation of New Wetland Habitats *474*
   Rice Fields *474*
   Introducing Exotic Species *479*
The Sungai Negara Swamps - A Converted Wetland Habitat *479*
Wetlands for Conservation *483*

Chapter 11 Coastal Resources *491*
    Estuarine and Coastal Fisheries *491*
      Harvests From the Sea *494*
    The Value of Coral Reefs *500*
    Mangroves as a Resource *503*
      Mangrove Fisheries *504*
      Mangroves for Commercial Timber and Chipwood *505*
      Other Mangrove Products *510*
      Mangrove Services *513*
    Mangrove Management *514*
      Conversion to Tambak *516*
    Oil Spills and Coastal Ecosystems *520*
    Development and Coastal Resources *520*

Chapter 12 Agriculture and Plantations *531*
    Agricultural Potential *531*
    Agroecosystems *538*
      Rice *540*
      Cassava *543*
      Fruit Trees *544*
      Pepper *547*
      Other Crops of Ladang and Home Gardens *548*
    Plantations *550*
      Coconuts *550*
      Rubber *554*
      Oil Palm *555*
      Cocoa *555*
      Timber as a Plantation Crop *556*
      Plantation Ecology *558*
      Pest Ecology and Control *562*
    Livestock *565*
    Genetic Resources *568*
    Future Trends in Agriculture in Kalimantan *569*

Chapter 13 Development and the Environment *573*
    Mineral Resources *573*
      Gold *576*
      Coal *578*
      Limestone *580*
      Diamonds *580*
    Mining and the Environment *581*

Petroleum *582*
  Oil Pollution *586*
Industrial Developments *589*
  Industrial Effluents and Pollution *592*
Forest Conversion *598*
Forest Loss and Climatic Change *598*
  The Greenhouse Effect *599*
Rehabilitating Critical Lands *602*
Agroforestry on Critical Soils *608*
The Urban Environment *610*
  Urban Ditches and Ponds *615*
  Houses *616*
Indonesia's Expanding Population *618*
Ecology of Diseases *620*
The Future *624*

Chapter 14 Conservation: Protecting Natural Resources for the Future *631*

Biodiversity *631*
  The Biological Importance of Borneo *632*
Species Extinctions *636*
  Endangered Species in Borneo *639*
The Need to Conserve Biological Resources *642*
Habitat Protection *643*
  Stabilisation of Hydrological Functions *644*
  Soil Protection *644*
  Stabilisation of Climate *644*
  Conservation of Renewable Harvestable Resources *645*
  Protection of Genetic Resources *645*
  Preservation of Breeding Stocks, Population Reservoirs and Biological Diversity *645*
  Maintenance of the Natural Balance of the Environment *646*
  Economic Benefits from Tourism *646*
  Environmental Monitors *647*
  Future Benefits *647*
Genetic Resources *649*
The Protected Areas System within Borneo *652*
  Species Protection in Reserves *663*
  Will the Reserve System Alone be Adequate to Protect Borneo's Biodiversity? *663*
Conservation Outside Protected Areas *670*
  Buffer Zones *672*
  The Value of Secondary Forests *672*

Production Forests *673*
  Captive Breeding and Reintroductions *676*
 Research Needs *680*
 The Future *681*

Appendices

Appendix 1   Plant genera and families mentioned in text *685*
Appendix 2   Land birds found on offshore islands of Borneo *695*
Appendix 3   Distribution of Bornean montane birds *697*
Appendix 4   Distribution of Bornean snakes *699*
Appendix 5   Simple methods for inventory and monitoring of species *703*
Appendix 6   Useful forest plants used by Iban people *710*
Appendix 7   Field-sketch landscape profile, based on Field Rapid Rural Appraisal, for reforestation planning in South Kalimantan. *715*
Appendix 8   Species databases *716*

Bibliography *718*

Bibliography Addendum *771*

Index *775*

List of Boxes

Box 2.1. The dugong: mermaid or endangered marine mammal.
Box 2.2. Simplified classification of the major invertebrates occurring in the benthos of coral reefs.
Box 2.3. Feeding guilds of coral reef fishes.
Box 2.4. Mangrove species in Kalimantan.
Box 2.5. Mangrove forests in the Barito River estuary.
Box 3.1. Riverine habitats in Borneo: a classification of flowing waters.
Box 3.2. Ecomorphology of fish.
Box 3.3. Feeding guilds of fish.
Box 3.4. Rheophytes - riverine plant specialists.
Box 3.5. Watchful mothers - parental care in crocodiles.
Box 3.6. The freshwater fish of the Kapuas River system.
Box 4.1. Forest regeneration.
Box 4.2. Mass flowering and mast fruiting in dipterocarps.
Box 4.3. Dioecism and cross-pollination among scattered rainforest trees.
Box 4.4. Species richness in lowland rainforest.
Box 4.5. Fig eating and seed dispersal by rainforest birds.
Box 4.6. Niche separation in tropical squirrels.

Box 4.7. Niche separation in Sundaland primates.
Box 4.8. Gliding – a rainforest adaptation.
Box 5.1. Mandor Nature Reserve.
Box 5.2. Myrmecotrophy: plants fed by ants.
Box 5.3. Plants as predators: pitcher plants.
Box 5.4. *Agathis*, a tropical conifer.
Box 5.5. White sands and blackwater rivers.
Box 5.6. Monospecific dominance in tropical rainforests.
Box 6.1. The Sangkulirang limestone formations.
Box 7.1. Rhododendrons.
Box 7.2. Orchids.
Box 7.3. The trig-oak, a missing link.
Box 8.1. Shifting agriculture.
Box 8.2. People and forests in East Kalimantan.
Box 8.3. Forest management by local villagers in West Kalimantan.
Box 8.4. Wild game meat in Sarawak.
Box 8.5. Traditional medicines of the Kenyah of East Kalimantan.
Box 8.6. Uses of forest plants in Sarawak.
Box 8.7. The cows of Long Segar: a lesson in development aid.
Box 9.1. Commercial forest management systems.
Box 9.2. The economic values of some forest species.
Box 9.3. Gaharu (aloe wood or incense wood).
Box 9.4. Valuation of a rainforest.
Box 9.5. The effects of logging.
Box 9.6. Forest functions.
Box 9.7. Logging and wildlife in the Sungai Tekam Forestry Concession, Pahang, West Malaysia.
Box 9.8. Watershed management: Riam Kanan, South Kalimantan.
Box 9.9. Some possible effects of burning on five ecological processes.
Box 9.10. Goods and services provided by tropical forests.
Box 10.1. Making aquatic weeds useful.
Box 10.2. Sago orchards of the Melanau.
Box 10.3. *Melaleuca* - the paper bark tree (gelam).
Box 10.4. *Ikan siluk*, the dragonfish of West Kalimantan.
Box 10.5. Wetland resources of economic value.
Box 10.6. Alabio ducks.
Box 10.7. Acid sulphate soils.
Box 10.8. Tidal swamplands and the Banjarese system of agriculture.
Box 10.9. Rice harvesting at Gambut and Kertak Hanyar, South Kalimantan.
Box 10.10 Threats to wetlands.
Box 11.1. Management implications of harvest, hunting and mining activities in estuaries.
Box 11.2. Harvesting turtle eggs in the Berau turtle islands, East Kalimantan.

Box 11.3. Connections between coral reefs and neighbouring linked habitats.
Box 11.4. Products of mangrove ecosystems.
Box 11.5. Commercial and traditional uses of some Bornean mangrove species.
Box 11.6. Pollutants and their effects on coastal ecosystems.
Box 11.7. Coastal ecosystems in Kalimantan.
Box 12.1. Good agricultural practices.
Box 12.2. Irrigated rice fields in the Kerayan.
Box 12.3. Buginese pepper farmers.
Box 12.4. Cultivation of rattans.
Box 12.5. Some effects of grazing by domestic stock on five ecological processes.
Box 13.1. AMDAL: a guide to environmental assessment in Indonesia.
Box 13.2. Protection and wise utilisation of valuable ecosystems during development.
Box 13.3. The ozone layer - a global issue.
Box 13.4. Riam Kiwa - agriculture in a degraded valley.
Box 13.5. Simplified comparison of some system properties between a natural ecosystem and a man-made structure.
Box 13.6. Habitat manipulations that may be ecologically beneficial.
Box 13.7. Two visions for the future of Indonesia.
Box 14.1. Within- and between-habitat diversity.
Box 14.2. Characteristics of species affecting their survival.
Box 14.3. Nonconsumptive benefits of conserving biological resources.
Box 14.4. Simplified scheme for assessing suitable protection category for protected habitats.
Box 14.5. Different categories of protected areas within Kalimantan.
Box 14.6. Linking conservation with development.
Box 14.7. Habitat fragmentation and loss of species: an Indonesian example.
Box 14.8. Benefits of buffer zones.
Box 14.9. Selection of crops for buffer zones.
Box 14.10. Habitat conservation within a large timber concession: Danum Valley, Sabah.

EMDI

The Environmental Management Development Project (EMDI) was designed to upgrade environmental management capabilities through institutional strengthening and human resource development. A joint project of the Ministry of State for Environment (LH), Jakarta, and the School for Resource and Environmental Studies, Dalhousie University, Halifax, Nova Scotia, EMDI supported LH's mandate to provide guidance and leadership to Indonesian agencies and organizations responsible for implementing environmental management and sustainable development. Linkages between Indonesian and Canadian organizations and individuals in the area of environmental management are also fostered.

EMDI received generous funding from the Canadian International Development Agency (CIDA). CIDA provided Cdn$2.5 million to EMDI-1 (1983-86), Cdn$7.7 million to EMDI-2 (1986-89), and contributed Cdn$37.3 million to EMDI-3 (1989-95). Significant contributions, direct and in kind, were made by LH and Dalhousie University.

EMDI-3 emphasized spatial planning and regional environmental management, environmental impact assessment, environmental standards, hazardous and toxic substance management, marine and coastal environmental management, environmental information systems, and environmental law. The opportunity for further studies was offered through fellowships and internships for qualified individuals. The books in the Ecology of Indonesia series form a major part of the publications programme. Linkages with NGOs and the private sector were encouraged.

EMDI supported the University Consortium on the Environment comprising Gadjah Mada University, University of Indonesia, Bandung Institute of Technology, the University of Waterloo, and York University. Included in EMDI activities at Dalhousie University were research fellowships and exchanges for senior professionals in Indonesia and Canada, and assistance for Dalhousie graduate students undertaking thesis research in Indonesia.

For further information about the EMDI project, please contact:

Director
School for Resource and Environmental Studies
Dalhousie University
1312 Robie Street
Halifax, Nova Scotia
Canada B3H 3E2
Tel. 1-902-494-3632
Fax. 1-902-494-3728

Foreword

Indonesia covers only 1.3% of the earth's surface, yet it harbours some of the world's richest biodiversity: 10% of all flowering plants, 12% of the world's mammal species, 16% of all reptiles and amphibian species, 17% of the world's birds and a quarter of all marine and freshwater fishes. This incredible biological richness is not only an important part of our natural heritage, it provides the life support systems, natural harvests and environmental services on which the nation's health, livelihood and well-being depend. An estimated 40 million people (almost one-quarter) of Indonesia's population are dependent on biodiversity for subsistence through harvesting of coastal, freshwater and marine fisheries, collection of non-timber forest products or cultivation of indigenous fruits, vegetables, cereals and spices. Conservation of the country's natural ecosystems and the biodiversity and environmental services they provide is crucial to the sustainability of economic development in sectors as diverse as forestry, agriculture and fisheries; health care; science; industry and tourism.

Protection and wise utilisation of biological resources requires that we understand how ecosystems work and the interactions and interdependence between natural habitats and a healthy environment. The Ecology of Indonesia series, part of the CIDA-funded Environmental Management Development in Indonesia Project attempts to explain the principles of ecology in the Indonesian context. This third volume in the series focuses on Kalimantan, part of Borneo, the world's fourth-largest island. Studies in Kalimantan provide valuable insights into ecological processes and the complex interactions between plants and animals in natural ecosystems. The book also considers human dependence on this island's biological resources and how human actions are changing and threatening the natural habitats on which so many Kalimantan peoples depend. As Kalimantan's vast tropical rainforests and coastal wetlands come under increasing pressure, this book reminds us that sustainable development depends on the maintenance of ecological processes and environmental functions.

The Ecology of Kalimantan was prepared as a collaborative effort between western and Indonesian scientists based at the environmental study centres in Kalimantan. The project provided training opportunities and resources for Kalimantan scientists and students to undertake ecological research in wetland and forest habitats and also established important literature collections and databases at KPSL-UNLAM, Banjarbaru, and PSL-UNTAN, Pontianak. I congratulate the authors and contributors on publication of this book and welcome this useful contribution to environmental literature of Indonesia.

Jakarta, September 1994

Sarwono Kusumaatmadja
State Minister of Environment
Republic of Indonesia

Acknowledgements

The Ecology of Kalimantan is the third volume in The *Ecology of Indonesia* series prepared as part of the Environmental Management Development in Indonesia (EMDI) Project. EMDI is a joint project implemented by the Indonesian Ministry of State for Environment (LH) and the School for Resources and Environmental Studies (SRES) at Dalhousie University, Halifax, Nova Scotia, Canada. The project is funded by the Canadian International Development Agency (CIDA).

This book was prepared originally at the environmental study centre, KPSL-UNLAM, at Banjarbaru, South Kalimantan between 1988 and 1990. After delays in publication Kathy MacKinnon undertook to revise and update the text prior to this publication. The authors are grateful to the Rector of UNLAM, Dr. Supardi, and to Ir Hazairin Noor and other staff at the KPSL for their assistance and support with fieldwork and research. Kathy and Richard Eaton, then CUSO volunteers at the KPSL, helped with research, fieldwork and reference collection, and established and catalogued the KPSL library. Kevin Teather and the staffs of PSL-UNTAN, Pontianak, PSL-UNPAR, Palangkaraya and PSL-UNMUL, Samarinda undertook fieldwork and data collection for the book. Wim Giesen and Olivier Klepper of the Asian Wetland Bureau collaborated with the KPSL staff to survey the Sungai Negara wetlands.

The authors also wish to acknowledge the support of Dr. Koesnadi Hardjosoemantri (former Rector of Gadjah Mada University, Yogyakarta) Arthur Hanson, Shirley Conover, George Greene, Gerry Glazier, John Patterson, Barbara Patton and Diane Blachford (SRES, Halifax and EMDI, Jakarta). Special thanks are due to Ani Kartikasari, who helped with all stages of the book from compiling the original references to final revisions. Ani took full responsibility for the production of the Indonesian version, which was translated by Professor Gembong Tjitrosoepomo (Yogyakarta) Agus Widyantoro and herself. The whole book was reviewed by Kuswata Kartawinata, Mien Rifai, John MacKinnon and Tony Whitten. S.C. Chin, George Greene, Lesley Potter, Carla Konsten, Bob Maher, Kevin Teather and Andrew Vayda all read and commented on one or more chapters. Great thanks are due to these people for their constructive criticisms; any faults that remain are the responsibility of the authors.

A volume like this is necessarily a compilation of much information drawn from other sources. The authors are grateful to the many scientists and researchers who generously shared their research data and provided copies of publications. We would also like to thank all those who provided us with hospitality, reference materials, encouragement and assistance with field visits. Special thanks are due to Abdurraman, Pak Acon of

Camera Foto, Lamri Ali and the staff of Sabah Parks, Wim Andriesse, Patrick Andau, Ir Asfihani, Max van Balgooy, Sonny Barnas, Chip Barber, Liz Bennett, Dr. Boeadi, Jean-Marie Bompard, Henk van Bremen, Nick Brown, Arie Budiman, Jules Caldecott, Ir Chaeruddin, Leo Chai, Paul Chai, Cecilia and Danny Chew, Lucas Chin, Carol Colfer, Mark Collins, Alain Compost, Rokhmin Dahuri, Rene Dekker, Michael Dove, Djasmani Hisbi, John Dransfield, Julian Dring, Ros and Ian Edwards, Mochtar Effendi, Clara van Eyck-Bas, Birute Galdikas, Wim Giesen, Ron Greenberg of USAID, Colin Groves, Jerry Harrison, Heruyono, Paul and Coby Hillegers, Jeremy Holloway, Derek Holmes, L.B. Holthuis, Indra, Robert Inger, Alan Irving and staff from Kaltim Primacoal, Clive Jermy, Tim Jessup, Andy Johns, Kuswata Kartawinata, Mike Kavanagh, Victor King, Olivier Klepper, KOMPAS Borneo, Carla Konsten, Jan de Korte, the late Doc Kostermans, Maurice Kottelat, Drs. Koesoemanto of GAMA Press, Jan Krikken, Tony and Anthea Lamb, Danna Leaman, Mark Leighton, Cornelius Lintu, Richard Luxmoore, John MacKinnon, Clive Marsh, Ian McKelvie, Jeff McNeely, Willem Meijer, John Mitani, Margaret Mockler, Peter Neame, Philip Ngau, Hazairin Noor, Riwai Noor, Christine Padoch, Jim Paine, Junaidi Payne, Nancy Peluso, Ron Petocz, Roger and Miranda Pratt, Herman Prayitno, John Proctor, Karen Phillipps, Gusti Rahmat, R. Rajanathan, Ir. Rifani, Ans and Herman Rijksen, Widodo S. Ramono, Anoma and Charles Santiapillai, Toga Siallagan, Marcel Silvius and other staff of AWB, Karta Sirang, Willy Smits, Dr. Soetikno, Belinda Stewart-Cox, Nigel Stork, Anwar Sullivan, Effendy Sumardja and staff from PHPA both in Bogor and Kalimantan, Sunardi, Stephen Sutton, Akira Suzuki, Syamsuni Arman, Tan Fui Lian, Andrew Vayda, Jan Vermeulen, Noel Vietmeyer, Ed de Vogel, David Wall, Dick Watling, Tim Whitmore, Tony and Jane Whitten, Kumpiadi Widen, Jan Wind, Richard and Cory Wink, Nengah Wirawan, Eric Wong and the staff of Mount Kinabalu National Park, WWF Bogor, WWF Malaysia, Yayasan Sabah and the staff of Danum Valley research centre. The authors also wish to record their thanks to the staff and librarians of Bogor Herbarium, Bogor Zoological Museum, Brunei Museum, the British Museum (Natural History), Kew Gardens, Leiden Herbarium and Zoological Museum, Sarawak Museum, Sabah Museum and Sabah Forestry Research Centre and the University Malaysia campus in Kota Kinabalu.

Katarina Panji and Rosalind Edwards typed and revised the text with care and much patience. Ismed Inono and Zain Noktah in Kalimantan and David MacKinnon in England prepared the artwork. James and Andrew MacKinnon assisted with the bibliography and proof reading. Kathy McVittie copy edited the English text and prepared the index. Ani Kartikasari and Anoma Santiapillai patiently checked the Indonesian translation. This has been a truly collaborative effort; the authors are grateful for the friendship, support, collaboration, encouragement and sheer hard work of the many people who made it possible.

Abbreviations and Acronyms

AWB	Asian Wetland Bureau
BAPPENAS	Badan Perencanaan Pembangunan Nasional (National Planning Agency)
BOD	Biochemical oxygen demand
B.P.	Before Present
Bt.	*Bukit* (hill)
CFC	Chlorofluorocarbons
CIDA	Canadian International Development Agency
CITES	Convention on International Trade in Endangered Species of Wild Flora and Fauna
CPVD	Citrus Phloem Virus Disease
D.	*Danau* (lake)
DVCA	Danum Valley Conservation Area
FACE	Forest Absorbing Carbon Dioxide Emissions
FAO	Food and Agriculture Organisation, UNDP
Gn.	*Gunung* (mountain)
ha	hectare
HPH	*Hak Pengusahaan Hutan* (Forest Consession Right)
HPT	*Hutan Produksi Terbatas* (Limited Production Forest)
ICBP	International Council for Bird Preservation, now Birdlife International
ITCI	International Timber Corporation Indonesia
IUCN	World Conservation Union
KEPAS	*Kelompok Penelitian Agro-ekosistem* (Research Group on Agroecosystems)
KLH	*Kantor Menteri Negara Kependudukan dan Lingkungan Hidup* (Ministry of State for Population and Environment, now Ministry of State for Environment LH)
KPC	Kaltim Primacoal
KPSL-UNLAM	*Kelompok Program Studi Lingkungan - Universitas Lambung Mangkurat* (Environmental Study Centre - Lambung Mangkurat University)
LNG	Liquefied natural gas
MAB	Man and the Biosphere
M$	Malaysian dollar
Malaya	Peninsular Malaysia
Mt.	Mountain
m	metre
cm	centimetre

ODA	Overseas Development Administration (U.K.)
P.	*Pulau* (island)
PIR	*Perkebunan Inti Rakyat*
PHPA	*Perlindungan Hutan dan Pelestarian Alam*
PROSEA	Plant Resources of South East Asia
PSL	*Pusat Studi Lingkungan* (Environmental Study Centre)
PSL-UNTAN	*Pusat Studi Lingkungan Universitas Tanjungpura* (Environmental Study Centre - Tanjungpura University)
RePPProT	Regional Physical Planning Project for Transmigration
S.	*Sungai* (river)
$	US $
TAD	Transmigration Area Development
TGHK	*Tata Guna Hutan Kesepakatan* (Forest Land Use)
Tj.	*Tanjung* (peninsula)
TPI	*Tebang Pilih Indonesia*
UNDP	United Nations Development Program
UNEP	United Nations Environment Program
UNESCO	United Nations Education, Social and Cultural Organisation
UNIDO	United Nations Industrial Development Organisation
USAID	United States Agency for International Development
WWF	World Wide Fund for Nature

Introduction

Kalimantan is the name given to the Indonesian portion of the great island of Borneo, the third largest island in the world after Greenland and New Guinea. Kalimantan covers 73% of Borneo's land mass. The four provinces, Kalimantan Barat (West Kalimantan), Kalimantan Tengah (Central Kalimantan), Kalimantan Selatan (South Kalimantan) and Kalimantan Timur (East Kalimantan) together have a total area of 539,460 km^2. This is 28% of the total land mass of Indonesia, with East Kalimantan alone accounting for 10% of the republic. The north of Borneo comprises the Malaysian states of Sarawak and Sabah and the small independent Sultanate of Brunei Darussalam (fig. I.1). Present-day political boundaries are a reflection of past colonial interests.

Despite its large area, Kalimantan supports just 5% of the total population of Indonesia, 9.1 million people (in 1990) at an average density of 17 people per square kilometre. Nevertheless, Kalimantan has played a key role in Indonesia's economic development and is a major earner of foreign revenue. In 1987 just one province, East Kalimantan, accounted for 21% of Indonesia's export revenues. This wealth is not due to industrial production nor to Kalimantan's agriculture or plantations but to the island's rich reserves of natural resources: forests, oil, gas, coal and other minerals. The exploitation of these natural gifts is the major development on Kalimantan. Although this exploitation began much earlier, the pace of development increased rapidly during the 1970s and 1980s to make Kalimantan a key area in the national development process. As development continues and gains momentum, several questions need to be asked. Is exploitation of Kalimantan's natural resources wise in the long-term? Is the exploitation sustainable? Or will it lead to environmental damage, degradation of the soils and forests, and pollution of waterways? What can be done to mitigate such potential damage?

Why is such a large area so underpopulated? Kalimantan's vast open spaces would appear to have massive potential for agricultural development and industrial plantations. The island seems to have great potential for large-scale human resettlement or transmigration from overcrowded Java. For a government planner from Java, this vast green island of low-lying undulating land, with no dry season and plenty of sunshine, must seem full of agricultural promise. Yet agronomists agree that the soils of Kalimantan are very poor, very fragile and very difficult to develop for agricultural use.

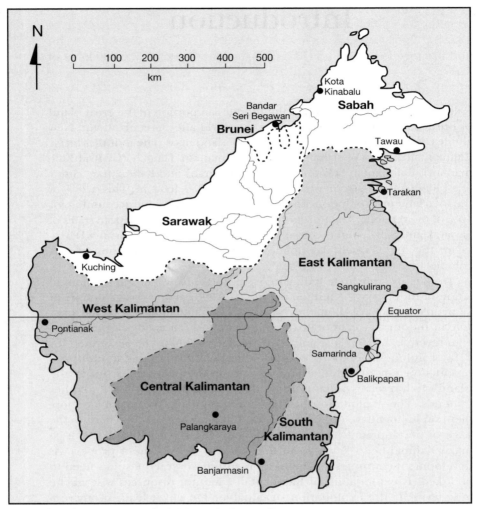

Figure I.1. State and provincial boundaries on Borneo.

Kalimantan can be developed, but only within rather tight ecological constraints and only with great care. Already large areas of degraded lands, poorly managed logging operations and failed agricultural schemes scar the Kalimantan landscape. Extensive white sandy *padang* and red laterised heaths are abandoned where great forests once grew. Each year a great sea of *alang-alang* grassland curls dry and burns. The forests are not allowed time to regenerate, and the sea of grassland expands, providing little other than fodder for cattle. Fires are set to clear *ladang* (agricultural fields) and to stimulate new grazing for cattle. Each year, like a ravenous herbivore, the fires nibble away at the forest boundaries. In the drought year of 1982-83 huge fires swept across the forests of East Kalimantan, and drought and fire together caused damage to 3.6 million hectares (36,000 km^2) of forest, an area the size of Belgium. At the same time fires set by agriculturalists raged out of control in Sabah, affecting another million hectares of forest and leading to considerable financial loss. Again the cause for such massive resource loss was poor primary development and careless forestry practice.

River pollution, from opening up the forests, from untreated industrial discharges, from domestic sewage and from illegal methods of gold extraction, is making many waterways dangerous for human use and is causing loss of fish stocks. Clearance of mangroves for *tambak* (fish pond) development can result in consequent loss of valuable offshore fisheries of shrimp and milkfish. Without careful planning, development of Kalimantan may bring short-term economic gain that creates long-term environmental damage.

Global climate is warming, a result of the "greenhouse effect" caused by an increase in the amount of carbon dioxide in the upper atmosphere. Indonesia would be one of the countries most affected if sea level were to rise by even a few metres. It would lose a huge area of prime agricultural land, and millions of coastal families would have to be moved to higher ground. Most of the "greenhouse gases" result from the burning of fossil fuels, but destruction of rainforests also plays a key role, as burning *ladang* and forests release carbon dioxide into the atmosphere. At the same time, as the area of forest shrinks, the forests become less effective as carbon "sinks" that remove carbon dioxide from the atmosphere and bind the carbon into productive biomass. As the largest expanse of rainforest in Southeast Asia, Kalimantan's forests are of environmental significance on a global as well as a national scale.

With its great mineral and forest reserves, Kalimantan does have great development potential, but development must be carefully planned with full understanding of the ecology of the land units being developed and of their inter-relationships with surrounding habitats. These constraints can be ignored by developers only at great peril.

Humankind generally has two options: to harvest products from a natural or modified natural productive system, or to replace a natural system with an artificial one. In either case environmental factors such as soil type, slope and climate place strict limits on what types of development are possible. This book aims to outline the ecology, and potential for development, of the various natural land systems of Kalimantan: mountains, lowland forests, special forest types, peat swamps, mangroves, coastal habitats and other landforms.

The book takes a long-term view of development. The rapid harvesting of timber, the stripping of land for underlying minerals, the clearing of hills for two fast crops while the soils are still fertile, these are all highly profitable activities in the short-term, but can leave the land stripped, useless and unproductive. This type of development is hardly desirable. If development is to be useful to the nation it must be sustainable. Development should be seen as an investment for the future, establishing long-term productivity by adoption of harvesting techniques that can continue to give good yields, year after year, without depleting the resource base on which they depend.

The need for sustainable development was the message emphasised in the World Conservation Strategy (UNEP/IUCN/WWF 1980). This message is the theme of this book. The World Conservation Strategy (WCS) identified three main global objectives, all highly relevant to the development of Kalimantan:

- maintenance of essential ecological processes and life-support systems;
- preservation of genetic diversity; and
- sustainable utilisation of species and ecosystems.

Essential ecological processes are those that are governed or moderated by ecosystems and are essential for food production, health and other aspects of human survival and sustained development. Life-support systems are the main ecosystems involved. The most threatened life-support systems are agricultural systems, forests, wetlands and coastal systems.

Kalimantan lacks Java's rich volcanic soils, the source of that island's fertility and agricultural productivity. Agricultural systems in Kalimantan are constrained by low fertility; soil erosion; mineral stress; pernicious weed invasion; failure of irrigation systems; increased flood hazard; salinisation; chemical pollution; loss of natural control agents for insects and other pests; and industrial and urban expansion. The traditional style of shifting agriculture, with short periods of cultivation and long fallow, is well suited to Kalimantan's poor soils, but is being replaced by less appropriate forms of land use.

Forests are being threatened by overexploitation, poor harvesting methods, limited regeneration, lack of reforestation technology, and losses to shifting cultivation and to fire. Kalimantan's important wetlands are threatened by changed rates of water flow, increased erosion and sedimentation, pollution, and drainage for other developments. The coastal areas are threatened by clearance of mangrove and other coastal vegetation, causing loss of fisheries and increasing the danger of marine erosion and storm damage.

The conservation of natural habitats and the preservation of genetic diversity are both an insurance and an investment. They are necessary to sustain and improve agriculture, forestry and fisheries production; to keep open future options; as a buffer against harmful environmental change; and as the raw material for much scientific and industrial innovation. Conservation and protection of biodiversity are matters of prudent management as well as of moral principle.

Much of Borneo is not suitable for *sawah* (wet rice cultivation), yet the island's productive forests prove that it has high potential for plant growth. Making the best use of the land in the future will certainly depend on finding more suitable crops and appropriate combinations of species to grow. Species that occur naturally in Kalimantan-timber trees, fruits, rattans, medicinal plants-should all be considered for their potential as plantation and agricultural crops. Only a small number of the world's plants have been developed on a commercial scale, yet many more are useful to human communities. The native peoples of Kalimantan use many species for food, medicines and as handicraft and construction materials. It is vital that the widest possible range of plant and animal species is protected as a reservoir for potential new crops and breeding material.

The sustainable utilisation of species and ecosystems is particularly important in an island where such a large proportion of the population depends heavily on direct harvesting from nature (hunting, fishing, collecting firewood and other plant materials) and where forestry is such an important sector of the local and national economy. Development should follow the goal of achieving maximum sustainable yield rather than over-harvesting natural resources to extinction.

Sustainable utilisation is analogous to spending interest whilst keeping the capital. A society that insists that all utilisation of living resources be sustainable ensures that it will benefit from those resources virtually indefinitely. Unfortunately, current levels of utilisation of many natural resources in Kalimantan are not sustainable, and some controls will need to be introduced.

This book starts by examining each of the major ecosystems in Kalimantan and the inter-relationships of some of the component species. An attempt has been made to identify the limitations on direct harvesting of products and the vulnerability of different ecosystems to overexploitation and modification. A chapter on Bornean peoples describes their traditional attitudes to, and relationships with, their forest environment. There are many lessons to be learned about sustainable use of natural resources from the island's indigenous peoples. The chapter traces the changing patterns of land use, emphasising some of the problems and dilemmas that now confront the developer and land-use planner.

The book examines the potential of different land units for the development of artificial production systems such as agricultural cropping, plantations, and agroforestry, and the development options open for forests, wetlands and coastal resources. Later chapters outline the ecological consequences that should be considered when planning mineral exploitation and industrial development, and the need to conserve habitats and species on Borneo, which is a global centre for biodiversity.

Although the book is entitled *The Ecology of Kalimantan*, political boundaries on Borneo bear no relation to biogeographical boundaries and plant and animal distributions. For this reason, many topics such as climate, vegetation and faunal characteristics cover the whole island. Throughout the book considerable reference is made to studies conducted outside Indonesian Borneo, especially where findings are directly relevant to the Kalimantan situation. For the sake of completeness, special Bornean features such as the biology of Sabah's Mount Kinabalu (Borneo's highest mountain) are included. Nevertheless, the text is generally oriented towards Kalimantan, and Indonesian research is quoted wherever appropriate.

This book is a companion volume to *The Ecology of Sumatra* and *The Ecology of Sulawesi* (Whitten et al. 1987a, 1987b) which are readily available within Indonesia. Since these volumes are comprehensive in their coverage of basic ecological principles and case studies, coverage of similar topics in this book has been condensed, to avoid duplication and to allow greater emphasis on the wise use of natural resources for sustainable development.

This book was prepared originally between 1988 and 1990 as part of the EMDI project to promote institutional development in the Kalimantan environmental study centres (PSL). The authors worked at the environmental study centre (KPSL-UNLAM) of Universitas Lambung Mangkurat at Banjarbaru, South Kalimantan, and were assisted by collaborative teams at the environmental study centres at Pontianak, Palangkaraya and Samarinda. Changes in project design and objectives have delayed publication, but the manuscript has been revised and updated for this edition.

Chapter One

The Island of Borneo

Borneo straddles the equator between latitudes 7° N and 4° S. It is the third largest island in the world (after Greenland and New Guinea) and the largest land mass in the Sundaic Region, more than five times the size of Java (fig. 1.1). Borneo lies in a region of constant rainfall and of high temperatures throughout the year, ideal conditions for maximum plant growth. As a result, the island has some of the most luxuriant tropical habitats on earth, and contains the largest expanse of tropical rainforests in the Indomalayan realm. The island is rich in biodiversity and is a fascinating place to study natural ecosystems.

The bulk of Borneo (539,460 km^2 or 73%) lies within Indonesian territory and is known as Kalimantan; the rest of the island consists of the states of Sarawak and Sabah (together forming East Malaysia) and the oil-rich, independent sultanate of Brunei Darussalam. Together these political states form one geographic unit and share a wealth of biological resources and diverse tropical habitats. Although this book focuses primarily on Kalimantan it begins with a general description of the whole island of Borneo. Geography, geology, climate and human land use have shaped the island's natural ecosystems to create the present Bornean landscape.

GEOGRAPHY

The first-time visitor to Borneo is often surprised at the flatness of the island, with vast areas of low coastal and river plains, especially in the south. Over half the island lies below 150 m in altitude, and water can be tidal up to 100 km inland. Borneo has no active volcanoes, but its main mountain ranges are igneous in origin. The mountain chains run down the island's centre like an inverted trident from north to south, with the three spurs diverging in the south (fig. 1.2). Borneo's highest mountain is Mount Kinabalu in Sabah. At 4,101 m Kinabalu is the highest peak in Southeast Asia and the highest mountain between the Himalayas and the snow-capped peaks of Irian Jaya. It consists of a granitic plug forced up by volcanic pressures, and it is still rising. Few of the other Bornean mountain peaks exceed 2,000 m.

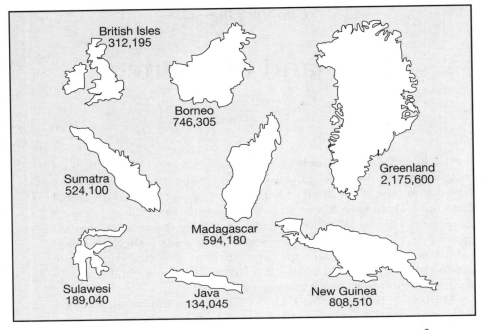

Figure 1.1. Borneo compared with other major islands: areas are given in km^2.

The Iran Mountains, between East Kalimantan and East Malaysia, rise in the north to 2,160 m at Mount Harden (Harun) near the Sabah border. A western spur of the central Iran mountain chain forms the Kapuas Hulu range along the border between Sarawak and West Kalimantan, rising to Mount Lawit (1,767 m) and Mount Cemaru (1,681 m). From the central mountains around Mount Cemaru, the Müller Mountains (highest point Mount Liangpran 2,240 m) and Schwaner Mountains (Bukit Raya 2,278 m) run southwest along the border between West and Central Kalimantan. To the southeast a lower offshoot, the Meratus Mountains, (highest point Gunung Besar 1,892 m) separates Central and East Kalimantan and extends southwards along the coast. These are all secondary mountain chains with an average height of 1,000-1,500 m with only occasional peaks rising above 2,000 m. In the Iran range, Gunung Makita (2,987 m) near Longnawan and Gunung Siho (2,550 m) near Longsaan, both on the Sarawak border, are the highest peaks in Indonesian Borneo, followed by the massif of Gunung Mantam (2,467 m) west of Tanjung Redeb, East Kalimantan (fig. 1.2). None of the higher Kalimantan mountains are volcanic in origin, in contrast to the volcanic cones that dominate the other Sunda islands and Sulawesi to the east.

Borneo is dissected by great rivers which run from the interior heart-

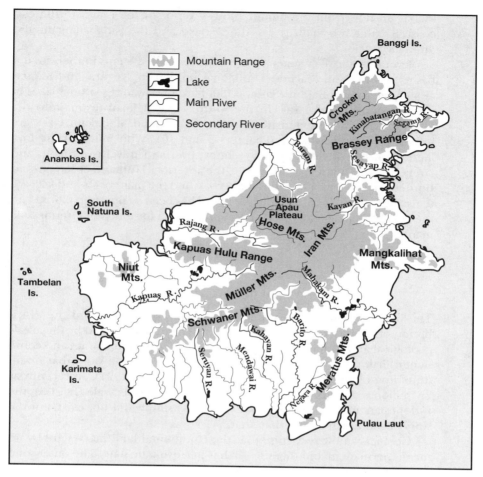

Figure 1.2. Map of Borneo showing main mountain ranges, main rivers and lake systems.

lands to the coast and provide the main routes of transportation and communication. The island boasts Indonesia's three longest rivers: the Kapuas (1,143 km), the Barito (900 km) and the Mahakam (775 km). The Kapuas (as long as the Rhine in Europe) flows west from Mount Cemaru, drains the greater part of West Kalimantan, and is navigable by small steamers 500-600 km upriver from Pontianak. Also flowing westwards are the Rajang (Sarawak's longest river) and the Baram. The great Barito originates in the Müller Mountains and flows southwards, draining the southern swamplands and emptying into the sea near Banjarmasin. Smaller, but historically important, the Kahayan also drains to the south coast. The Kayan and the

Mahakam flow from the mountainous interior of the island to the east coast. Sabah's two main rivers, the Segama and the Kinabatangan, also drain eastwards.

Several major river systems have extensive lake systems in their inland basins and lowland reaches. The Mahakam, Barito, Negara, and Kapuas (Kalimantan) and Baram (Sarawak) all have oxbow and seasonal lakes in their lowland floodplains. In the south, Lake Belajau drains into the Seruyan River. The extensive Kapuas, Negara and Mahakam lakes are important inland fisheries (chapters 3 and 10). With much of the lowland plain poorly drained and swampy, overland travel is difficult, and rivers are the main highways into the interior. Human settlements in Borneo are concentrated around the coast and the main rivers and lake systems. Perhaps this explains why the island has been so poorly explored and the lack of reliable maps and topographical data for much of interior Kalimantan.

GEOLOGICAL HISTORY

The biogeography of the whole Indonesian archipelago and the distribution of its soils, plants and animals have been determined by the area's geological and climatic history. This is a story of plate tectonics and continental drift, climatic events and changing sea levels, a story that is still unfolding (Audley-Charles 1981, 1987; Burrett et al. 1991). As recently as 25 million years ago, very recent on the geological time scale (fig. 1.3), the Indonesian archipelago as we know it today simply did not exist, but the story began much earlier than that.

Geologists now recognise that the continental land masses are by no means permanent and that the earth is in a dynamic state. The outer solid part of the earth, the crust, is quite thin, like the rind of an orange. There are two kinds of crust: oceanic and continental. Oceanic crust is usually young (0-200 million years), thin (5-15 km) and composed mostly of dense volcanic rock. Continental crust often has a core of older rocks (200-3,500 million years), is thicker (20-50 km) and is less dense than oceanic crust, composed of rocks such as sandstones and granites (RePPProT 1990). Western Indonesia, comprised of much of Kalimantan, Sumatra and west and central Java, is composed predominantly of continental crust, as is much of the shallow sea floor between these islands. Below the earth's crust is a zone where the rock is hotter and more plastic. Continental and oceanic plates float on the fluid, underlying material.

Turbulent convection currents rising from the planet's molten core carry the continents and oceanic plates about, creating zones of weakness and disturbance. Where two plates move apart, liquid rock wells up to fill

GEOLOGICAL HISTORY 13

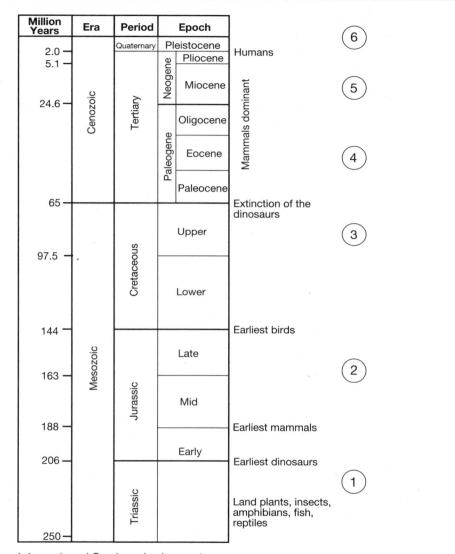

Figure 1.3. Geological time scale showing the appearance of major life forms and the occurrence of major geological events.

the gap; where two plates collide, one subducts to dive beneath the other, creating deep trenches and crumpled mountain ranges.

On a human time scale, plate movements are very slow, only a few centimetres per year, but over a period of 60 million years (the time since the dinosaurs became extinct) a plate drifting only 1 cm per year would have moved 600 km. The Indonesian region is dominated by three major plates, the southeast Asian plate, the Indo-Australian plate and the Pacific plate, as well as several smaller platelets that have sheared off Irian Jaya (Katili 1989).

The process is still continuing today. Some of the Indonesian plates move as much as 13 cm per year (RePPProT 1990). Australia is moving northwards and at its present rate of progress will crash through most of Indonesia during the next 50 million years. At the same time the Pacific plates are moving south and west to meet the Australian plate. New Guinea, Java, Sumatra and the islands of Nusa Tenggara lie on the points of impact and are regularly shaken by earth tremors and volcanic eruptions caused by the colliding plates.

Until about 280 million years ago (280 Ma), the earth's land mass formed a single continent, Pangea. During the Triassic, about 250 Ma, Pangea rifted into two supercontinents, Laurasia (including present- day North America, Europe and much of Asia) and Gondwanaland (including present-day South America, Africa, India, Australia, Antarctica and the rest of Asia). The two supercontinents were separated by the Tethys Ocean.

From this stage the early geological history of Indonesia is unclear. Until recently it was believed that the two halves of the Malesian archipelago were derived totally from different supercontinents and had been separate from the Triassic (250 Ma) until a mid-Miocene collision (15 Ma). It was thought that the western half (the Malay Peninsula, Sumatra, Java, Borneo and western Sulawesi) was derived from Laurasia, while the eastern islands, including the rest of Sulawesi, were derived from Gondwanaland much later (Audley-Charles 1981). According to this model, much of what is now mainland Southeast Asia, including Kalimantan, was on the southern side of the large continental mass of Laurasia (Sengor 1988). By 260 Ma this southern margin was the site of a subduction zone and volcanic arc (RePPProT 1990). Old volcanic rocks in Kalimantan date from this period and are associated with limestones in South Kalimantan, indicating that fringing coral reefs were present as they are today around volcanic islands in Nusa Tenggara.

To the south, separated from Laurasia by the Tethys Ocean, lay the continental mass of Gondwanaland. The northern flank of Gondwanaland was unstable, and fragments split off and were transported northwards, where they eventually collided with the southern flank of Laurasia between 230 Ma and 200 Ma. The older or basement rocks of much of Sumatra and Peninsular Malaysia and the surrounding continental crust were welded onto Borneo to create the block of continental crust known as Sundaland, about 180 Ma. As a result of the collision a large amount of granite was intruded in the Thailand-Malaysia-Bangka-Belitung-West Kalimantan zone; economic tin deposits are associated with much of this granite (RePPProT 1990).

Between 200 Ma and 150 Ma, further fragmentation of Gondwanaland occurred, and Africa, India, Australia and Antarctica (until then a single land mass) started to separate. During the Mesozoic, tectonic movements caused the Tethys Ocean to move northwards accompanied by subduction and arc formation at the southern side of Sundaland. The late Mesozoic granites of Kalimantan date from this time (RePPProT 1990). The most dramatic result of the breakup was the separation and rapid movement northward of India, which collided with the Laurasian plate. This massive collision destroyed the Tethys Ocean and caused the uplift of the Himalayas.

In the light of more recent paleontological and geological discoveries, an alternative theory has been proposed. This suggests that most of Southeast Asia, Sumatra and Borneo were not part of Laurasia but separated from Gondwanaland much later, in the mid-Jurassic and Cretaceous (Audley-Charles 1987; Burrett et al. 1991). Veevers (1988) recognised three main stages of rifting on Australia's northwest shelf: Cambrian (equated with the rifting of North China), late Carboniferous to early Permian (rifting of Shan Thai), and the late Jurassic. The Jurassic-Cretaceous event rifted the many Gondwana fragments now found in Indonesia (Burrett et al. 1990). These continental fragments rifted from Australia-New Guinea (the northern margin of east Gondwanaland) and provided "stepping-stones" of dry land during the Tertiary. Parts of Sumatra and southern Kalimantan probably rifted from Australia in the early Cretaceous. Northern Borneo had probably rifted even earlier. These and other Southeast Asian blocks became relatively isolated within the Tethys Ocean between Gondwanaland and the Asian mainland for perhaps as long as 60 million years, but from the late Cretaceous (100 Ma) onwards they provided an archipelago of islands that could have permitted land plant dispersal between the Australasian and Asian land masses (fig. 1.4).

About 90 Ma the continental plate of Australia and New Guinea broke away from Antarctica and began its rapid drift northwards. During the Paleocene (60 Ma) northern Borneo and southern Kalimantan were quite separate, but they probably collided in the Eocene, 50 Ma (Pieters and Supriatna 1990). By the Oligocene (30 Ma) Borneo was part of a land mass that also included western Sulawesi (Burrett et al. 1991). By 20 Ma the Australian plate was beginning to interfere with the volcanic arc on the southern edge of Sundaland. Fragments of the continental plate sheared off and were forced westward, accommodated by subduction of the ocean floor beneath Sundaland. The plate fragments derived from the area of

16 THE ISLAND OF BORNEO

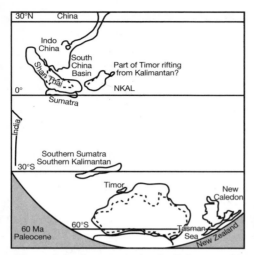

Figure 1.4a. Reconstruction for Paleocene 60 Ma. NKAL, North Kalimantan.

Dotted lines are present coastlines.

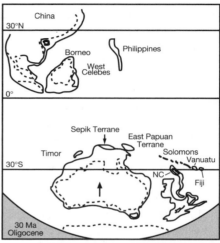

Figure 1.4b. Reconstruction for Oligocene 30 Ma. South Kalimantan and northern Borneo already joined.

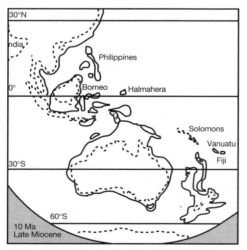

Figure 1.4c. Reconstruction for Late Miocene 10 Ma.

Figure 1.4. The changing locations of the components of Southeast Asia since the first rifting from Gondwanaland.

Source: Burrett et al. 1991

Irian Jaya have collided progressively with eastern Sundaland to cause trapping and then obduction of the sea floor to form ophiolites in the Meratus Mountains of Kalimantan, in east Sulawesi and in Halmahera (Moluccas).

The last 15 million years have been a time of dramatic geological activity in the Indonesian archipelago, marked by the uplift of sedimentary rocks from the ocean bed and the creation of new volcanic islands. Tectonic movements, resulting from the northwards and westwards movements of the major plates, led to the uplift of new land from the sea bed and the upthrust of the Bornean mountain blocks. During the later Tertiary, erosion of these mountain chains created thick sedimentary deposits. In Borneo the thickest and most economically important of these Tertiary sedimentary sequences were deposited in deltas, such as the Mahakam delta in East Kalimantan. Biological material washed into the Mahakam basin was changed by heat and pressure into coal and oil deposits.

During the Tertiary and Quaternary, the Indonesian archipelago experienced periods of lowered sea levels and more seasonal climates than those recorded today. Sea level changes during the Tertiary resulted from tectonic activity. The expansion of the oceans due to sea floor spreading resulted in marked changes in ocean volumes. The mid-Oligocene onset of Antarctic glaciation also caused a marked lowering of global sea levels (fig. 1.5).

The late Cenozoic was a period of repeated alternations of glacial and interglacial periods. The major climatic cycle, with a period of about 100,000 years, was probably initiated by astronomical forces, originating from small changes in the ellipticity of the earth's orbit. During the Quaternary the resulting changes in insolation were enhanced by changes in the carbon dioxide (CO_2) concentration in the atmosphere. Since concentration of atmospheric CO_2 causes the "greenhouse effect" (chapter 13), lowered CO_2 at glacial maxima may have had a profound influence on world temperatures (Morley and Flenley 1987). Simultaneously the increase in albedo (whiteness) of ice-covered surfaces causes decreased absorption of sunlight in glacial regions, explaining why temperate climates fluctuated more than tropical ones in the Quaternary.

Changes in sea level during the Quaternary can be calculated with some precision from oxygen isotope data. The maximum lowering of the sea level in the Quaternary was about 200 m, which would have exposed the Sunda platform and the Sahul Shelf. Submarine channels on both shelves indicate the courses of rivers when these continental platforms were above sea level (Verstappen 1975). In the Malesian region sea level reached its minimum, and land areas were most exposed, in the middle Pleistocene. Later, in the Holocene, sea levels were slightly above present levels (fig. 1.6; Tjia et al. 1984). Reduction in sea levels provided land linkages from mainland Southeast Asia to the islands of the Sunda Shelf.

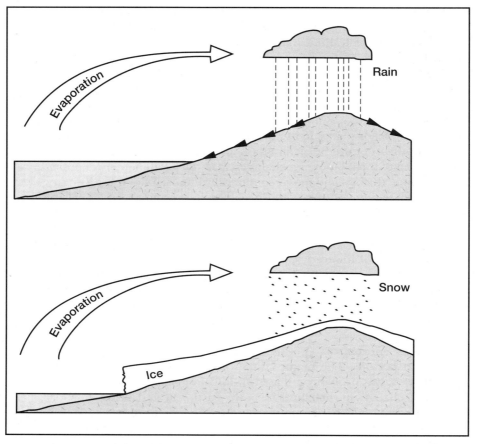

Figure 1.5. Hydrological cycles (a) in warm conditions, and (b) in cold conditions, when sea level falls as water is retained in the ice cap.

There is considerable evidence for more strongly seasonal climates in both the late Tertiary and the Quaternary. Studies of peat swamps and analysis of the pollen record indicate that the climate of South Kalimantan was much more seasonal then than today (Morley 1982).

The more seasonal climates of the late Tertiary and mid-Pleistocene created a savanna corridor between Southeast Asia and Borneo which allowed migration of savanna plants and animals across the area from Thailand to Java. Cooler climatic conditions in Asia during the glacial periods encouraged animals to move southwards along the land bridges over a period of half a million years, and successive waves of animal, and human, immi-

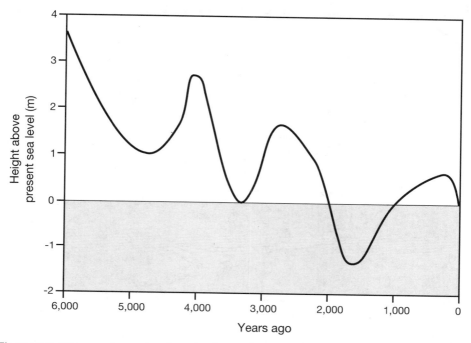

Figure 1.6. Changes in sea level on the Sunda Shelf during the last 6,000 years. (After Tjia 1980.)

grants were able to reach the Greater Sunda islands from Asia. During glacial maxima, cooler temperatures lowered montane vegetation zones, providing "stepping-stones" for the migration of mountain and temperate taxa. During the Pleistocene, glacial periods were longer than the interglacial periods, and for the greater part of the last two million years most of the Sunda Shelf was exposed above sea level (fig. 1.7) but everwet habitats were less extensive than today. Late Tertiary and Quaternary paleoclimates thus help to explain many disjunct plant and animal distributions in the Indonesian archipelago (Morley and Flenley 1987).

At the times of lowest sea level during the Pleistocene, all the islands on the Sunda Shelf—Java, Borneo, Sumatra, Bali and Palawan—were thus connected by land bridges. Connections were formed and broken several times. Java and Palawan, at the extremities of the shelf, were connected less often and less recently than either Borneo or Sumatra, and this is reflected in their respective distinctiveness.

Even at the times of lowest sea levels in the late Cenozoic, Borneo was

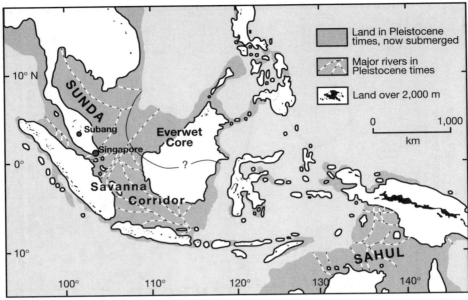

Figure 1.7. A generalised paleogeographical reconstruction of the Sunda-Sahul region during one of the Quaternary glacial maxima. A savanna corridor may have extended across Sundaland in the middle Pleistocene. Everwet refuges may have occurred in montane areas. (After Morley and Flenley 1987.)

not connected to Sulawesi. Thick sediments in the Makassar straits indicate that Borneo and Sulawesi have been separate for at least 25 Ma (Audley-Charles 1981) and possibly longer. This long isolation is reflected in differences in flora (Balgooy 1987), lepidopterans (Holloway 1987) and mammals (Musser 1987) between Borneo and Sulawesi. Sulawesi's flora and fauna show much stronger similarities to those of the Philippines and Lesser Sundas than to those of Borneo and the other Sunda islands.

The shallow continental shelves of the South China Sea and Java Sea are incised with several ancient river channels, including three major rivers, between Borneo and Sumatra: the Anambas, the North Sunda (with the proto-Kapuas as a tributary), and the proto-Lupar valleys (Haile 1973; fig. 1.8). Two large, parallel rivers run along the bed of the Java Sea between Java and Borneo, towards the Straits of Makassar (Verstappen 1975). Similarities in freshwater fish between western Borneo and Sumatra indicate that the rivers of these islands were once linked; the Musi of Sumatra and the Kapuas of Borneo were once part of the North Sunda river system. Some of these large rivers probably served as barriers to floral and faunal distribution; dipterocarps show some breaks in distrib-

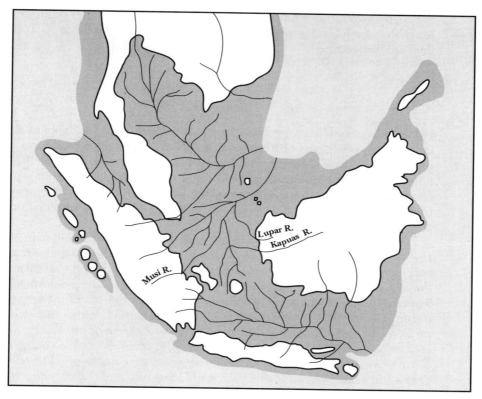

Figure 1.8. The Sunda Shelf showing present coastlines (unshaded), the area of Sundaland exposed at times of lowest sea level (dark shade) during the last glacial about 12,000 years ago, and past and present river systems. (After Tjia 1980.)

ution at the Lupar River (Ashton 1972). Similarly, the major rivers between Borneo and Java probably slowed faunal dispersals between the two islands.

The climate of the region was probably only 2°–4°C cooler in the Quaternary and Pleistocene than it is today, but conditions were often more arid (Morley and Flenley 1987). Elsewhere climatic changes have been cited as a cause of species extinctions, but of 200 large mammal extinctions worldwide in the Pleistocene only 11 occurred in Southeast Asia (Medway 1972b). Of the 32,000-year-old fauna recorded from excavations at Niah only one species, the giant pangolin, is truly extinct (Harrisson 1961), although other mammals, such as the tapir and Javan rhino, have been locally extirpated in Borneo, probably by overhunting. The lesser gymnure *Hylomys suillus* and ferret badger *Melogale orientalis* recorded from the Niah excavations are now found only on the higher slopes of Mount Kinabalu, which lends support to the theory of a cooler climate at Niah in the Late Pleistocene.

Geology

Publication of systematic geological maps is less advanced for Kalimantan than for any other part of Indonesia (RePPProT 1990). Table 1.1 gives a simplified overview of major geological events in Borneo. Figure 1.9 illustrates the geology of Kalimantan. There are four main geological units represented in Kalimantan: rocks associated with plate margins; basement rocks; younger consolidated and nonconsolidated rocks; and alluvium and young superficial deposits.

The Kalimantan basement complex in west and central Kalimantan (including the Schwaner Mountains) represents the largest continental basement outcropping in Indonesia. Basement rocks, at the bottom of the stratigraphic sequence, are generally older than the overlying rocks. They are usually metamorphosed by heat. Typical products of metamorphism are marbles (from alteration of limestones), green schists from volcanic rocks, and gneiss from sandstone or granite. Areas of metamorphic rock or basement are typical of continental crust, often invaded by younger intrusive rocks. The Kalimantan basement complex consists of schist and gneiss intruded by granites of Paleozoic, Mesozoic and Tertiary age to form a widespread crystalline terrain.

Rocks associated with plate margins in Kalimantan include ophiolites (derived from oceanic crust) and melange. Fragments of oceanic floor are found on land in several parts of Borneo. Characteristically they are composed of dark, dense igneous rocks of basic and ultrabasic type with a granite component. Thin oceanic siliceous sediments (chert) and carbonate sediments may also be present. This suite of rocks is called **ophiolite**. Such ophiolites are created by plate collisions when oceanic crust, instead of subducting, is trapped by plate tectonic movements and forced onto the adjacent plate margin where it remains exposed. This obduction process is often accompanied by crumpling and fracture of the rocks. The ophiolite complexes of Pulau Laut and the Meratus Mountains originated in this way.

Melange is a mixture of broken fragments of different rock types and sizes in a sheared clayey matrix, indicating very strong compression. Fragments can be very small (a few centimetres) to large (hundreds of metres or more). Melange is often associated with the subduction process. It is a combination of material scraped off the downgoing oceanic plate with sediments derived from the nearby land mass or volcanic arc. The whole mass is squeezed and sheared due to the underthrust of the downgoing

Table 1.1. Summary of geological time scale and events relevant to Borneo over the last 350 million years.

Era	Period	Epoch	Began (millions of years ago)	Geological events	Biological events
Cenozoic	Quaternary	Holocene	0.01		Modern humans.
		Pleistocene	2	Low sea level; Sundaland connections.	Earliest hominids.
	Tertiary	Pliocene	10	Latest possible connections with Sulawesi either via Doang-doang shoals or a reduced Makassar Straits.	Large carnivores.
		Miocene	25	Plate collisions and volcanic activity create Indonesian archipelago.	Abundant grazing animals.
		Oligocene	30	Borneo in approximately present position.	Large running animals.
		Eocene	50	Fragments rift from Australian plate and drift north.	Many modern types of mammals evolve.
		Paleocene	65	Australia broken away from Antarctica.	Dinosaurs extinct. First placental mammals.
Mesozoic	Cretaceous		145		First flowering plants; extinction of ammonites at end of period.
	Jurassic		215	Western Indonesia, Tibet, Burma, Thailand, Malaysia and Kalimantan break away from Gondwanaland.	First birds & mammals; dinosaurs and ammonites abundant.
	Triassic		250	Pangea rifts into two: Laurasia and Gondwanaland; insular and some mainland parts of Southeast Asia part of Gondwanaland.	First dinosaurs; abundant cycads and conifers.
Paleozoic	Permian		280	All land together as one continent, Pangea.	Extinction of many forms of marine animals including trilobites.
	Carboniferous		350		Great coal-forming conifer forests; first reptiles, sharks and amphibians abundant.

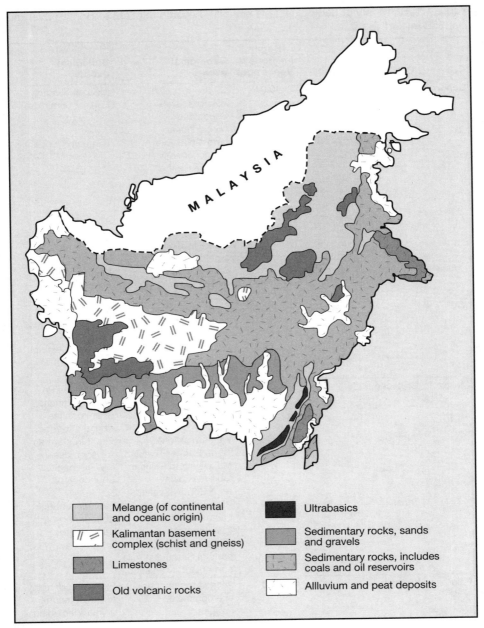

Figure 1.9. Geology of Kalimantan.
Source: RePPProT 1990

plate. Rocks formed in this way are associated with the large upthrust slabs of ophiolite in the Meratus Mountains.

The large area of melange in the centre of Borneo, straddling the border between Kalimantan and Malaysia, is less well understood. This is a zone of broken rocks often including fragments of ophiolite, but its width and extended geological age (late Mesozoic to older Tertiary) are difficult to explain in terms of simple plate tectonics (Williams et al. 1989).

Much of Kalimantan consists of consolidated and semiconsolidated rocks, including Quaternary limestones (in the Sangkulirang peninsula and Meratus range), volcanic rocks, and Tertiary sediments. Although Borneo has no active volcanoes, as do Sumatra and Java, it does have substantial areas of old volcanic rocks in southwest and eastern Kalimantan. These are a legacy of Indonesia's geological history, which has included many periods of volcanic activity from 300 Ma to the present day. Volcanic rocks are formed when magma from deep in the earth reaches the surface. Where magma cools and consolidates below the earth's surface, intrusive rocks such as granodiorite are formed. Where the old Kalimantan volcanic rocks have been eroded, stocks of gold-bearing intrusives, originally below the volcanoes, have been exposed. The interaction of magma with groundwater beneath volcanoes is an important part of the process leading to the deposition of minerals such as gold.

Large areas of Central, East and South Kalimantan are composed of sedimentary rocks such as sandstone and shales. Apart from older formations in West Kalimantan, most sedimentary formations are relatively young, and include coal and oil source rocks. The southern part of Borneo consists mainly of loosely consolidated sand and gravel terraces, often overlain by young, superficial deposits of peat and alluvial fans deposited by flooding rivers.

Soils

Soil conditions are very important in affecting the distribution of vegetation. There are five key factors in soil formation: lithology, climate, topography, biological organisms, and time. Knowledge concerning soil distribution in Kalimantan is generally limited: 90% of soil survey reports produced by the Centre for Soil Research have been for specific project sites for transmigration, tree crop estates or irrigation schemes (Sudjadi 1988). Figure 1.10 shows the distribution of major soil types in Kalimantan; table 1.2 lists the main soils and their properties.

The majority of Kalimantan soils have developed on rolling plains and dissected hills on sedimentary and old igneous rocks. These soils range from strongly weathered and acid ultisols to young inceptisols. In the

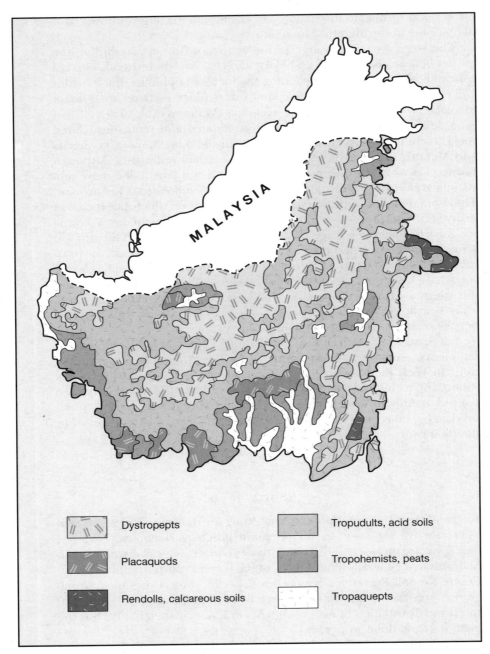

Figure 1.10. Major soil types of Kalimantan.
Source: RePPProT 1990

south, extensive alluvial plains and peat soils extend into the Java Sea. Accretion is still occurring on the shallow shelf of southern Kalimantan, with alluvial sediments building up behind coastal mangroves.

Weathering is strong in the humid tropics, favoured by both warmth and moisture. Because of the high rainfall, soils are constantly wet, and their soluble constituents are removed; this process is called **leaching**. High levels of weathering, leaching and biological activity (degradation of organic matter) are characteristic of many Bornean soils (Burnham 1984). The island's rocks are poor in metal bases, and Bornean soils are generally much less fertile than the rich volcanic soils of neighbouring Java.

Table 1.2. Main soil types of Kalimantan, ranging from young, undeveloped histosols to heavily weathered spodosols and oxisols.

Order	Character	Suborder	Group	Character
Histosol	mainly organic	Hemist	Tropohemist	swampy, half-decomposed
Entisol	weakly developed	Aquent	Hydraquent Fluvaquent	poorly drained/ acid sulphate soils, in alluvial lowlands
		Psamment	Tropopsamment Quartzipsamment	undeveloped, infertile sands
		Fluvent	Tropofluvent	more fertile alluvial soil
Inceptisol	distinct profile, moderate weathering	Aquept Tropept	Tropaquept Dystropept	poorly drained, less fertile more fertile soil on steep slopes
Mollisol	thick base-rich A horizon	Rendoll		shallow soil over calcareous parent material
Alfisol	clay-enriched, B horizon relatively base-rich	Udalf	Tropudalf	more fertile soil, less leached
Ultisol	clay-enriched, B horizon base poor	Udult	Tropudult	old, infertile, loam and clay soil
Spodosol	heavily weathered with 'podzol' B horizon	Aquod	Placaquod	infertile sands
Oxisols	highly weathered			old infertile soils over basic igneous and ultra basic rocks

Nomenclature according to USDA soil groupings.

Source: RePPProT 1990

Deep complete weathering, combined with leaching, results in soils of low fertility in many lowland areas. Steeper slopes may be more fertile as erosion and landslips constantly expose new parent material.

Soils over the main part of central and northeastern Borneo are ultisols (acrisols). These strongly weathered soils form a high proportion of the red-yellow podsolic soils typical of the rolling plains of Kalimantan. In Kalimantan ultisols are mainly represented by tropudults. Udults are difficult to utilise intensively because of low nutrient levels beneath the topsoil and the combination of high aluminium levels and strong acidity (RePPProT 1990). Traditionally local people have worked these soils by shifting cultivation, with a short cropping regime and a longer fallow to allow fertility to recover. This allows the topsoil to regain some humus and organic matter, which are important as stores of nutrients and for regulating soil moisture and temperature.

The commonest soils found in Kalimantan are the inceptisols, moderately weathered soils with a distinct profile. These are some of the more fertile of the Bornean soils. Within this order, aquepts and tropepts are recognised in Kalimantan (RePPProT 1990). The poorly drained tropaquepts formed in river sediments eroded from Tertiary siliceous sandstones and shales in Kalimantan are some of the least fertile aquepts. The more fertile tropepts are widespread, particularly in strongly dissected hills and mountain areas where slopes are steep and erosion is active. Some old tropepts are associated with plains landscapes. Reddish brown dystropepts form over the more acid and siliceous rocks such as conglomerates, sandstones, shales and siltstones, and are common in Kalimantan.

Histosols, nonmineral or predominantly organic soils called peats, cover vast areas of lowland Kalimantan (RePPProT 1990). These soils develop initially in depressions in marshy alluvial plains, where organic litter and debris accumulate rapidly, up to 4.5 mm/yr. (Anderson 1964), because of the permanently saturated and anaerobic conditions. In tropohemists the organic material is only partly decomposed. Histosols also occur in Borneo as relatively thin (50-150 cm) layers of organic material that accumulate on high altitude plateaux and ridges under conditions of high cloud and humidity. These are ombrogenous peats closely associated with moss forests (chapter 3). Nearly all histosols are very acidic, with low levels of major and minor plant nutrients, making them difficult and expensive to cultivate (RePPProT 1990).

Spodosols are normally found in cool temperate climes. In Indonesia they only occur where there are strongly acid parent materials, impeded subsoil drainage and minimal erosion (RePPProT 1990). Such conditions are found on quartz sand terraces and on the gentle dips of sandstone cuestas in Central Kalimantan, and on unusually siliceous rocks such as those of the Usun Apau plateau in Sarawak. Spodosols are very strongly weathered and medium to coarse in texture, with an accumulation of iron and humus compounds in the subsoil. These sandy soils are strongly acid, poor in plant nutrients, and will not retain added fertiliser without the liberal mixing of organic material; they are particularly poor for agriculture. These soils are naturally associated with heath forest (*kerangas*) vegetation (chapter 5).

Alfisols occur where rocks, such as calcareous marls and limestones in East Kalimantan, yield large amounts of bases on weathering. Mollisols are confined to lime-rich landscapes in Kalimantan. They are dark, because of a high humus content, and rich in bases, especially calcium (chapter 6). They are commonly deficient in the major nutrient potassium. Lime-induced deficiencies of minor nutrients pose a problem for the growth of many crops on these soils, which are weakly acidic to weakly alkaline. Well-drained rendolls are locally common in eastern Kalimantan, especially in the Sangkulirang peninsula.

The most weathered soils are oxisols, dominated by clays which have few weatherable minerals and yield few plant nutrients. They occur over ultrabasic rocks in Ranau and Tawau, Sabah, and in the Meratus Mountains in South Kalimantan (chapter 5). Even though these soils have a high magnesium/calcium ratio and high levels of nickel, chromium and cobalt, the vegetation is not distinct from surrounding forests. In contrast, the organic-rich soils of high altitudes with peat over ultrabasic rocks, as on Mount Kinabalu at 2,000-2,800 m, do support a distinct vegetation (Burnham 1984).

Entisols are weakly developed soils of recent origin. Fluvents and aquents (alluvial soils) are found in the floodplains of river valleys, and on the coastal plains, which receive new deposits of alluvium at frequent intervals. Aquents are saturated with water for long periods of the year and are characteristically deep, grey and mottled. Fluvaquents are associated with flat and poorly drained alluvial plains; their fertility depends on the mineralogy and organic matter of the alluvial deposits where they form. Hydraquents occur in tidal swamps in Kalimantan; they are young, soft, muddy and "unripe". Sulfaquents commonly occur with hydraquents. These poorly drained soils have only limited potential for agriculture as they contain pyrites which, if drained, can produce strongly acid conditions with toxic levels of iron and aluminium sulphates (chapter 10). Such acid sulphate soils are encountered in the Pulau Petak area, South Kalimantan.

Fluvents are important soils of the larger river and lake floodplains of Kalimantan. They are common where rivers carrying large sediment loads are liable to flood and shift course. The mineral content and fertility of the tropofluvents found in Kalimantan depend very much on the geological formations of the upper catchment and the surrounding topography. The two main alluvial soil environments are the levee and the back swamp. Recent freshwater alluvial soils in Borneo mostly support freshwater swamp forest. They are usually more fertile than soils on the surrounding slopes but not as fertile as soils on recent marine alluvium or volcanic ash (Burnham 1984). The riverine alluvial flats of Kalimantan are some of the most fertile and accessible habitats for agriculturalists. In contrast, the fertility of all psamments-distinctive young soils normally found on young and old beaches-is characteristically low. Large areas of psamments occur in Central Kalimantan.

Each plant has its own ideal range of soil requirements, both physical and chemical. Different soil types support different vegetation types and have different potentials for agriculture. Kalimantan soils range from highly fertile to virtually sterile. The most fertile parts of Kalimantan have deep, well-drained soils of moderately fine texture, with a well-balanced nutrient status; these include alfisols, and small areas of vertisols (over volcanic rocks), and hapludolls and haplustolls (over limestone). These soils are already used intensively for non-irrigated cultivation. Young riverine and estuarine plains on Kalimantan also support intensive irrigated agriculture on poorly drained, but relatively fertile, tropaquepts (chapter 10). The poorest soils for agriculture are the very strongly weathered and leached, waterlogged placaquods and quartzipsamments of Kalimantan quartz sand terraces and the rendolls of limestone karst. These are still largely under forest.

In Bornean soils most of the nutrient-holding capacity is a function of the humus content and is very low where humus content is low, as on degraded *kerangas* sands (chapter 5). Often more than half of the adsorbed bases in 1 m of soil are in the top 25 cm (Nye and Greenland 1960). This explains the short duration of fertility of cleared *ladangs*, since burning of the original cover and erosion of topsoil leads to loss of this fertile top layer. For sustained agricultural use many Bornean soils require conservation of topsoil organic matter, by erosion control, balanced application of fertiliser and good management.

CLIMATE

The distribution of plants and animals is not determined solely by geological events, soils and the geography of mountain and water barriers. Other physical factors also play their part, including climate (especially rainfall) and altitude.

Borneo lies on the equator and enjoys a moist, tropical climate. Temperatures are relatively constant throughout the year, between 25°C and 35°C in lowland areas. In Borneo the main climatic variable is rainfall. The type of vegetation is determined not only by the total annual rainfall but by its distribution through the year. Equatorial lowland areas that receive at least 60 mm of rain every month can support evergreen rainforest

(Holdridge 1967). All parts of Borneo lie within this everwet zone.

The pattern of rainfall in Indonesia is determined by two main monsoons, a southeast or "dry" monsoon (May-October) and a northwest or "wet" monsoon (November-April). From May to October the sun passes directly over northern Indochina and southern China, and a belt of low pressure builds up over the hot Asian mainland (fig. 1.11a). Rain-bearing winds blow north from high pressure zones over Australia and the Indian Ocean. These winds pick up moisture as they pass over wide areas of sea. When they reach the Greater Sunda islands and the Asian mainland, they must rise over hills and mountain ranges. As they climb, the air cools, and moisture drops as rain. Heavy monsoon rains fall on India and South China, and more moderate rains fall on the islands of the Sunda Shelf, including Borneo.

From October to March the sun lies south of the equator. Central Asia is very cold, and the hot low pressure zones are now over the southern continent of Australia. The monsoon winds blow southwards but are met by other winds blowing east from high pressure zones over the southern Indian Ocean (fig. 1.11b). Where the hot and cold air masses meet, heavy rain is shed over all of the Sunda Shelf, as well as Sulawesi, Nusa Tenggara (Lesser Sundas) and New Guinea.

Borneo has very few months with rainfall of less than 200 mm. The northwest monsoon (November to April) is generally wetter than the southeast monsoon, but some coastal areas show a bimodal rainfall pattern. Kalimantan can be divided broadly into five agroclimatic zones (fig. 1.12; table 1.3). Most of the hilly inland areas receive between 2,000 and 4,000 mm of rain each year. Most of Kalimantan falls into the wettest agroclimatic regimes (Oldeman et al. 1980). In contrast to Sumatra, there is no dominating coastal mountain range influencing rainfall, but some lower mountain ranges influence climate locally, especially in eastern Borneo. West and central Borneo are the wettest areas while parts of the east coast are considerably drier.

The northwest monsoon reaches West Kalimantan in August-September, and the wet season continues to May. Rainfall is especially heavy in November with a second peak in April. The climate is relatively drier from June to August, but no month has less than 100 mm rainfall. The annual rainfall in Putussibau is over 4,000 mm with no month having less than 200 mm. In South and Central Kalimantan rainfall generally increases northwards from the coast. The influence of the southeast monsoon is much more pronounced than in West Kalimantan, and dry months occur between July and September, particularly in rainshadow areas west of the Meratus Mountains, for example, at Martapura. Nevertheless, the dry season is not as pronounced as in Java and Nusa Tenggara. The southeast coast and Pulau Laut are generally much wetter than the south coast because of the influence of the Meratus Mountains (Oldeman et al. 1980).

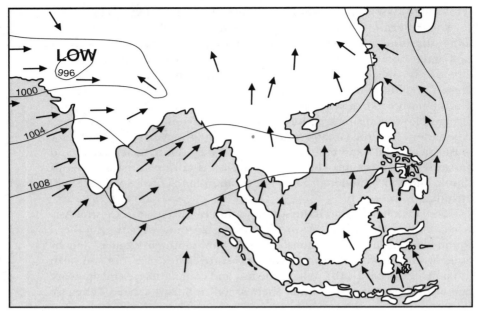

Figure 1.11a. Isobars and wind direction over Asia in the dry monsoon (May-October).

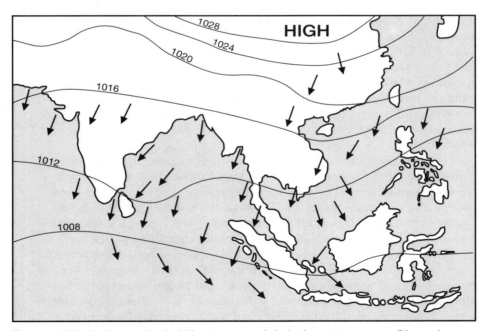

Figure 1.11b. Isobars and wind direction over Asia in the wet monsoon (November-April).

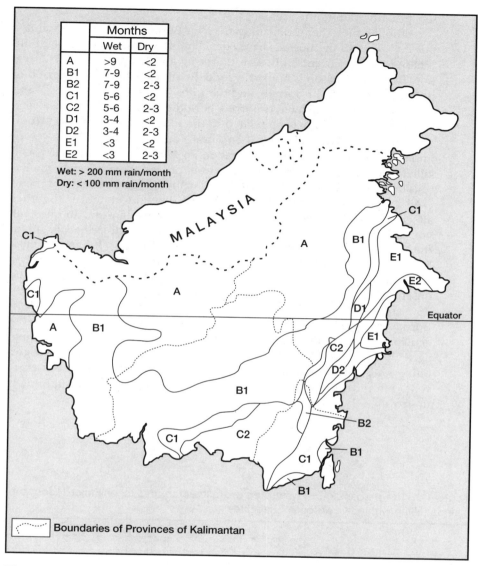

Figure 1.12. Agroclimatic map of Kalimantan. (After Oldeman et al. 1980.)

The coastal areas of East Kalimantan and eastern Sabah are considerably drier than other parts of Borneo. The influence of the northwest monsoon is weak, as most of the rain has fallen in the central mountains. Even during the wet season, rainfall is relatively low and often less than 200 mm per month, especially in the Sangkulirang peninsula. There is no marked dry season, as the southeast monsoons have travelled over open sea, bringing rain to the area.

Although the general climatic pattern for Borneo is one of high rainfall, the short annual dry periods have considerable significance for plant life (chapters 4 and 5) and influence flowering and fruiting patterns. Occasionally the dry season is unusually long. In 1982-83 there was a period of prolonged drought in Borneo, and the rains were late again in 1987 and 1991. Such droughts have occurred periodically throughout historical times in Borneo and may be related to the El Niño Southern Oscillation (Leighton and Wirawan 1986). Drought conditions can have a severe impact on the island's natural vegetation. In 1972-73 a prolonged dry spell killed all woody plants in some pockets of upper montane forest on shallow soils on Mount Kinabalu; five years later many species had still not recovered (Whitmore 1984a). During 1982-83 drought and fires together affected 3.6 million hectares of forest in East Kalimantan, and another million hectares in Sabah (chapter 9), in what has been described as one of the greatest natural catastrophes in historical times (Johnson 1984). Nevertheless, prolonged dry periods play an important role in shaping Bornean forest ecosystems, since dry conditions seem to stimulate the onset of mass flowering and fruiting of dipterocarp species (Ashton 1988).

Natural vegetation is closely correlated with rainfall, but the picture in Borneo, as elsewhere throughout the tropics, is further complicated by the activities of the human population. Forest clearance related to logging, agriculture and resettlement has been extensive and continues at a frightening rate. Kalimantan alone loses more than 500,000 ha of forest every year. Plant and animal species are lost when their forest habitat is destroyed.

Table 1.3. The percentage of land area of Kalimantan and other major Indonesian islands falling within five agroclimatic zones.

	Agroclimatic zones				
	A	B	C	D	E
Kalimantan	43	32	15	4	6
Java	4	23	39	25	9
Sumatra	24	47	15	12	2
Sulawesi	1	25	25	22	27
Lesser Sundas and Bali	<1	<1	3	69	26
Moluccas*	5	11	40	12	20
Irian Jaya	48	23	17	8	4

* note the figures in the text do not add up to 100%.

Source: Oldeman et al. 1980

Flora

The natural vegetation of any area is dictated by a combination of several factors: topography, altitude, geology, soils, climate and water supply, especially rainfall. Borneo lies on the equator in a region experiencing high temperatures throughout the year and within the wettest part of the Indonesian archipelago. These conditions and the island's geological and climatic history have encouraged speciation and high species diversity. As a consequence Borneo supports some of the largest expanses of tropical rainforests in Southeast Asia, providing some of the most species-rich habitats on earth.

Forest types include mangrove, peat swamp and freshwater swamp forests, the most extensive heath forests (*kerangas*) in Southeast Asia, lowland dipterocarp forests, ironwood (*ulin*) forests, forests on limestone and ultrabasic soils, hill dipterocarp forest and various montane formations (fig. 1.13). Coastal habitats include beach vegetation, sea grass beds and coral reefs. Areas of original, remaining and protected habitat types are given in table 1.4.

Borneo is the richest of the Sunda islands floristically both in terms of total species richness and diversity, with small plot tree diversity as high as found anywhere in New Guinea or South America (table 1.5). The island is a major centre for plant diversity, with 10,000 to 15,000 species of flowering plants, a flora as rich as that recorded for the whole African continent, which is 40 times larger. It includes both Asian and Australasian elements (fig. 1.14).

Borneo has at least 3,000 species of trees including 267 species of dipterocarps, the most important group of commercial timber trees in Southeast Asia; of these dipterocarps 58% are endemic to the island (Ashton 1982). Borneo harbours more than 2,000 species of orchids and 1,000 species of ferns, and is the centre of distribution for the remarkable carnivorous pitcher plants *Nepenthes*. Endemism levels are high through the whole flora with about 34% of all plants, but only 59 genera (out of 1,500), unique to the island. In comparison Sumatra has only 12% endemism at species level and 17 endemic genera (MacKinnon and MacKinnon 1986). Only one plant family, the Scyphostegiaceae, is endemic to Borneo (Ashton 1989).

Areas of plant richness can be associated with soil types (fig. 1.15). The diversity of habitat types and local endemism associated with soils over geologically young rocks, especially in northwest Borneo, both contribute to this richness (Ashton 1989). Borneo's diverse forest habitats range from mature climax dipterocarp forest, with high canopy, clear stratification and lofty leguminous and dipterocarp emergents (chapter 4), to less species-rich heath forests of lower stature, found on impoverished white sand soils (chapter 5). Local conditions favour the growth of floristic

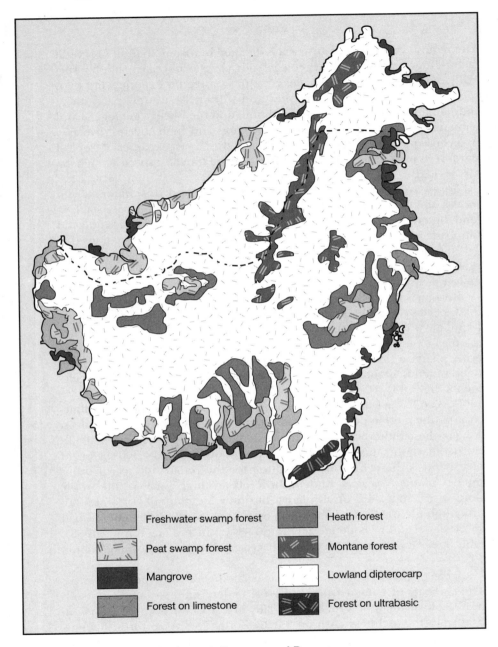

Figure 1.13. Original natural vegetation cover of Borneo.
Source: MacKinnon and MacKinnon 1986

Table 1.4. Original, remaining and reserve areas, and habitat product, of the different habitats in Kalimantan (in ha $\times 10^3$). Habitat product is an index of the rarity of a habitat, its rate of loss and its priority for conservation, with high indices denoting rarest habitats. "Wet", "moist" and "dry" refer respectively to the A and B, C and D agroclimates.

	Original area ha $\times 10^3$	Remaining area (and % original area) ha $\times 10^3$	Area including in existing reserves (and % of original and remaining areas) ha $\times 10^3$	Habitat product
Moss forest	50	50	24	0.35
Montane on other rocks/soils	2,004	2,004	985	0.18
Wet hill forest on other rocks/soils	7,456	7,175	1,899	0.23
Moist hill forest on other soils	222	210	13	0.64
Lowland forest:				
wet areas:				
alluvium	2,201	265	16	1.42
limestone	17	17	–	2.67
dipterocarp	4,271	10,936	970	0.40
lowland plains	3,437	1,362	63	0.78
moist areas:				
alluvium	87	25	–	2.57
limestone	48	41	–	2.14
dipterocarp	1,983	850	89	0.67
lowland plains	1,614	85	–	2.35
dry areas:				
alluvium	21	–	–	10.03
limestone	162	87	–	1.93
dipterocarp	970	225	190	0.67
lowland plains	360	50	–	2.45
Ironwood dipterocarp forest	210	73	20	0.84
Heath forest	8,076	2,475	210	0.74
Freshwater swamps	3,895	1,717	362	0.51
Peat swamps	4,403	3,531	257	0.45
Mangrove forest	1,560	920	78	0.69
Beach vegetation	8	3	2	1.39
Coral reefs	(110)	(110)	(c.15)	0.59
Freshwater lakes	(92)	(92)	(25)	0.46

(Based on MacKinnon and Artha 1981.)

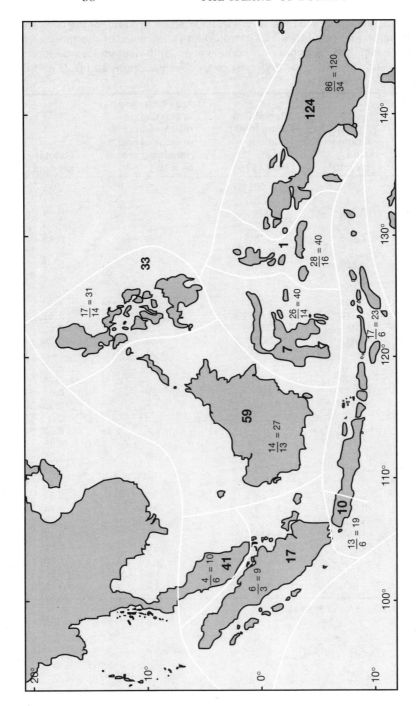

Figure 1.14. Phytogeographical divisions of the Malesian archipelago based on the distribution of flowering plant genera. Bold numbers indicate the number of endemic genera in each division. Small numbers indicate the number of eastern-centred genera in each division, with Australian genera above the hyphen and Pacific-Subantarctic genera below the hyphen.

Source: van Steenis 1950

facies such as ironwood forests on alluvial terraces, while Borneo's coasts are fringed with mangrove forests (chapter 2), some of the most extensive stands in Indonesia. Differences in vegetation types and forest structure provide a wide variety of niches for other plants and animals. Thus many of Borneo's 146 rattans (climbing palms) are associated with specific forest types (Dransfield 1992).

As well as the commercially valuable dipterocarps and other timber trees, Bornean forests are rich in fruit trees important both to wildlife and local peoples. These include mangoes *Mangifera*, durians *Durio*,

Table 1.5. Matrix of floral plot richness in different habitat types on different Indonesian islands.

Habitat type	New Guinea	Borneo	Sumatra	Java	Sulawesi	Moluccas	Lesser Sundas	Habitat Mean
Upper montane	15 (1)	22 (2)	18	10 (1)	10 (3)	11	7	14.3
Fertile lower montane	62	61 (2)	66 (3)	49 (2)	49 (4)	32 (1)	25	51.4
Infertile lower montane	41 (1)	43	35 (1)	30	23 (2)	25	16	33.0
Fertile hill forests	63	64 (6)	75 (1)	45 (2)	45 (3)	35 (1)	–	52.0
Infertile hill forests	40 (2)	44 (1)	38 (2)	37 (1)	34 (3)	19 (1)	17	34.5
Fertile lowland rainforests	75 (4)	39 (6)	80 (5)	51 (3)	58 (3)	37 (1)	–	65.0
Infertile lowland rainforests	44	50 (3)	51 (4)	35 (3)	34 (3)	28 (2)	21 (2)	36.5
Monsoon forest	32 (1)	–	–	15 (1)	19	30 (2)	14 (2)	22.7
Heath forest	–	40 (3)	50	–	–	–	–	40.0
Beach forest	20	25 (1)	21	20 (1)	18 (2)	13	4 (1)	16.8
Volcanic scrub	17	–	20 (1)	12 (1)	10 (2)	11	7	14.0
Complex mangrove	13	15 (1)	14	12	7 (1)	8	5	11.0
Simple mangrove	6	6 (1)	6	5 (1)	4 (1)	4 (1)	2	4.8
Peat or freshwater swamp forest	46	45 (2)	46 (4)	35	25 (1)	29	19	38.6
Island mean richness index	1.20	1.30	1.26	0.91	0.82	0.76	0.48	

Data from 250 forest samples in 102 different localities. Figures represent mean number of tree species greater than 15 cm diameter found in 0.5 hectare plots; figures in brackets indicate number of localities.

Source: MacKinnon and Artha 1982

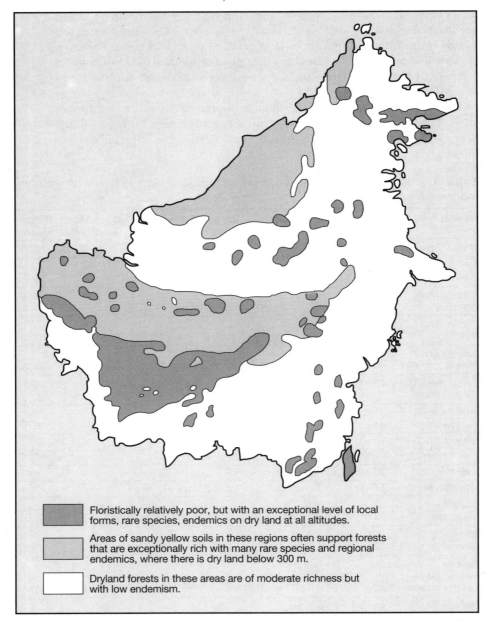

Figure 1.15. Generalised distribution of plant richness categories in Borneo, based on lithology.

Source: MacKinnon and Artha 1981

Baccaurea (Euphorbiaceae), breadfruits and jackfruits *Artocarpus*, and rambutans *Nephelium*. Some of these species are domesticated in home gardens. Among the Bornean palms are several genera that produce fruit, food and other products widely used by local communities. Sugar palms include *Arenga pinnata, Nypa fructicans* and *Borassodendron borneensis*. The indigenous sago palm *Eugeissona utilis* is also harvested for food, as is the introduced *Metroxylon sagu*. Rattan palms *Calamus* spp. are commercially valuable, and four or five species are grown in traditional home gardens in East Kalimantan. Rattans, bamboos, ferns and other plants are widely utilised in local economies.

Many forest products are harvested by local Dayak communities and used for food, handicrafts, building materials, medicines, fish poisons, wrappings and ritual purposes (chapter 8). Treatments for a wide range of illnesses and diseases are still obtained from plant extracts. Many are widely believed to be efficacious, and some plants (e.g., *Pityogramma, Blechum, Nephrolepis, Urena, Celrodendrum*) may repay phytochemical and pharmaceutical investigation (Pearce et al. 1987). Other plant products with known physiological effects include the root of *Derris elliptica* (an insecticide and fish poison), *Goniothalamus* (an insect repellent with antimicrobial action), fish poisons such as *Diospyros*, and the poison present in *Parartocarpus latex*.

With increasing exposure to modern lifestyles and the availability of alternatives, communities are utilising natural plant products less frequently. It is important to document their ethnobotanical knowledge before it is lost.

Fauna

Like its flora, Borneo's fauna reflects its geological history and ancient land connections. The island's rich fauna is characteristically Asian in origin, with families such as deer, wild cattle, pigs, cats, monkeys and apes, squirrels and many families of oriental birds. Borneo shares much of its fauna with the Asian mainland and the other Sunda islands, but shares few species with Sulawesi and the eastern islands, which have a somewhat different faunal composition (fig. 1.16).

For vertebrate groups, Borneo has similar numbers of species to the smaller island of Sumatra. Borneo has 222 mammals compared with 196 on Sumatra and its offshore islands, 183 on Java, and only 127 on Sulawesi. In relation to its size, Borneo is less rich in mammals than the smaller island of Sumatra, although it shares many of the same species (fig. 1.17). This impoverishment can be explained by the fact that Borneo lies further offshore from mainland Asia and probably was separated earlier from the mainland by rising sea levels. Thus the leopard *Panthera pardus* and wild

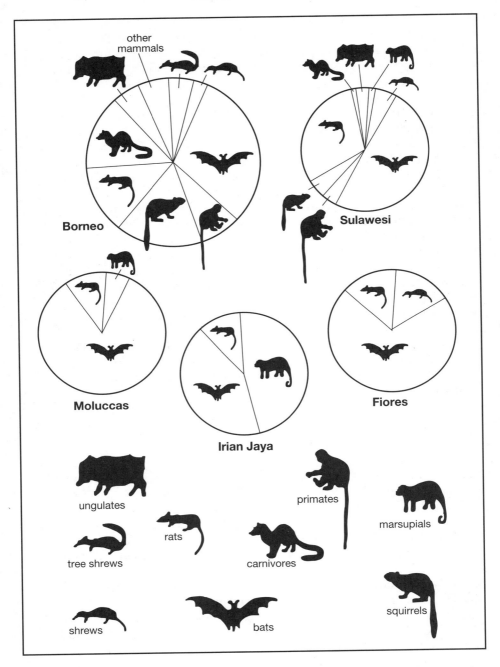

Figure 1.16. Composition of the mammal fauna on Borneo and other Indonesian islands. (After Whitten et al. 1987b.)

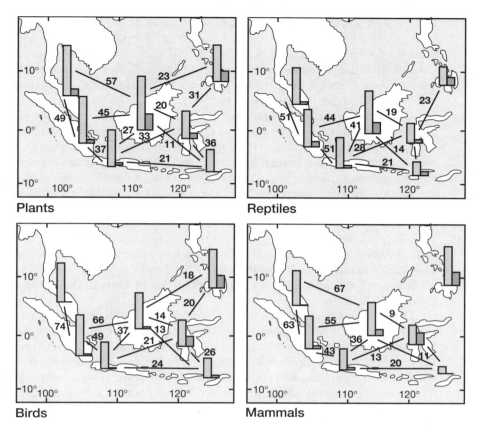

Figure 1.17. Comparison of species richness, endemism and species overlap between islands in the Malay archipelago. (Numbers refer to percentage of total species list shared between neighbouring islands.)

Source: MacKinnon and MacKinnon 1986

dog *Cuon alpinus* probably never reached Borneo, even though both occur on Java and the wild dog (but not the leopard) occurs in Sumatra. Nevertheless, Borneo has more endemic species than Sumatra, with 44 endemic land mammals (Payne et al. 1985) compared with only 23 endemic mammals on mainland Sumatra (MacKinnon 1990a).

Borneo has 13 species of primates and 10 species of tree shrews (more than any other Asian mainland or island area). Large mammals include banteng *Bos javanicus*, Sumatran rhinoceros *Dicerorhinus sumatrensis* and the elephant *Elephas maximus*. Elephants are confined to a small area in Sabah and the very northern regions of East Kalimantan. It has been suggested that this is a feral population descended from introduced captive elephants kept by the Sultan of Sulu. It is just as likely that the elephant is native to the island. Fossil elephant teeth are known from Borneo, and it is simpler to postulate a range reduction than local extinction followed by reintroduction (Hooijer 1972; Groves 1992).

Although the two largest Sundaic predators – the tiger *Panthera tigris* and the leopard – are not now found on Borneo, there is some evidence that the tiger may have occurred on the island during the Stone Age (see table 6.8). Borneo has several species of medium-sized and small carnivores, ranging from the beautiful clouded leopard *Neofelis nebulosa* and the sunbear *Helarctos malayanus* to numerous civets and mustelids. Most of the endemic mammals are small rodents and bats which, though inconspicuous, play an important and major role in forest ecosystems as seed predators and seed dispersers. Many of the endemic species are montane, with 21 endemic mammals living in montane and submontane habitats and only 15 restricted to the lowlands (table 1.6). Ten endemic mammals are recorded only from the high mountains of Sabah and Sarawak and may not occur in Kalimantan.

The bird fauna of Borneo is typically Asian in origin and similar to that of Peninsular Malaysia and Sumatra, with rich representation of the hornbills (8 species), woodpeckers (18 species), pittas (13 species) and other forest families. Borneo has 420 species of resident birds compared with 465 on Sumatra, 340 on Java and 240 bird species on Sulawesi. There are 37 endemic birds known for Borneo (table 1.7), and again many are montane species. At least 28 Bornean birds (24 endemics including four endemic genera *Haematortyx*, *Chlamydochaera*, *Chlorocharis* and *Oculocincta*) are restricted to the Bornean mountains (appendix 3). Eight of these are not yet recorded for Kalimantan, though it is probable that more faunal surveys of the central mountains will add new species to the Kalimantan list. The range of the black-breasted triller *Chlamydochaera jefferyi*, previously known only from Mount Kinabalu and other Sabah mountains, was recently extended to Bukit Baka in the Schwaner range (Rice 1989).

Borneo is probably one of the richest islands of the Sunda Shelf for fishes, amphibians, reptiles and invertebrates, but figures are not so accurate for these less well-known groups. Borneo has at least 166 species of snakes (appendix 4) compared with 136 species on the Malay peninsula, 150 species on Sumatra and only 64 species on Sulawesi (Medway 1981). The island has at least 100 species of amphibians compared with 36 species known for Java and only 29 species for Sulawesi (table 1.8). Borneo has 394 species (149 endemic) of freshwater fish compared with 272 species (30 endemic) for Sumatra, 132 for Java and 68 for Sulawesi (Kottelat et al. 1993). Borneo has far more endemic fish species than Sumatra, but this

Table 1.6. Endemic mammals of Borneo.

*Suncus ater (m)	black shrew
Tupaia splendidula	ruddy treeshrew
Tupaia montana (m)	mountain treeshrew
Tupaia gracilis	slender treeshrew
Tupaia picta (m)	painted treeshrew
Tupaia dorsalis	striped treeshrew
Dendrogale melanura (m)	smooth-tailed treeshrew
Hipposideros dyacorum	Dayak roundleaf bat
*Hipposideros coxi (m)	Cox's roundleaf bat
Pipistrellus kitcheneri	red-brown pipistrelle
*Pipistrellus cuprosus	coppery pipistrelle
Presbytis hosei	Hose's langur/grey leaf monkey
Presbytis rubicunda	maroon langur/red leaf monkey
Presbytis frontata	white-fronted langur
Nasalis larvatus	proboscis monkey
Hylobates muelleri	Bornean gibbon
*Callosciurus baluensis (m)	Kinabalu squirrel
*Callosciurus adamsi (m)	ear-spot squirrel
Callosciurus orestes	Bornean black-banded squirrel
Sundasciurus jentinki (m)	Jentink's squirrel
Sundasciurus brookei (m)	Brooke's squirrel
*Glyphotes canalvus (m)	grey-bellied sculptor squirrel
*Glyphotes simus (m)	red-bellied sculptor squirrel
*Lariscus hosei (m)	four-striped ground squirrel
Dremomys everetti (m)	Bornean mountain ground squirrel
Exilisciurus exilis	plain pygmy squirrel
Exilisciurus whiteheadi (m)	Whitehead's pygmy squirrel
Rheithrosciurus macrotis	tufted ground squirrel
Petaurillus hosei	Hose's pygmy flying squirrel
*Petaurillus emiliae	lesser pygmy flying squirrel
Aeromys thomasi	Thomas' flying squirrel
*Rattus baluensis (m)	summit rat
*Maxomys alticola (m)	mountain spiny rat
Maxomys ochraceiventer (m)	chestnut-bellied spiny rat
*Maxomys baeodon	small spiny rat
Chiropodomys major	large pencil-tailed tree mouse
Chiropodomys muroides (m)	grey-bellied pencil-tailed mouse
Haeromys margarettae (m)	ranee mouse
Haeromys pusillus (m)	lesser ranee mouse
Hystrix crassispinis	thick-spined porcupine
*Hemigalus hosei (m)	Hose's civet
Herpestes hosei	Hose's mongoose
Felis badia	bay cat
Muntiacus atherodes	Bornean yellow muntjak

(m) denotes montane species
* species not recorded for Kalimantan

may reflect differences in methods and extent of collections from the two islands. Borneo has a high number of endemic Balitoridae (hill stream species) and Belontiidae. Among invertebrates, swallowtail butterflies are one of the few well-known groups. Borneo has 40 species of swallowtails including four endemic species; three of these are likely to occur in the Kayan-Mentarang reserve in East Kalimantan (table 1.8).

Table 1.7. Endemic birds of Borneo.

Species	Common name
Microhierax latifrons	white-fronted falconet
*Spilornis kinabaluensis	mountain serpent eagle
Arborophila hyperythra (m)	red-breasted partridge
Haematortyx sanguiniceps (m)	crimson-headed partridge
Lophura bulweri (m)	Bulwer's pheasant
Polyplectron schleiermacheri	Bornean peacock pheasant
*Batrachostomus harterti (m)	Dulit frogmouth
Harpactes whiteheadi (m)	Whitehead's trogon
*Megalaima eximia (m)	black-throated barbet
Megalaima monticola (m)	mountain barbet
Megalaima pulcherrima (m)	golden-naped barbet
Calyptomena hosei (m)	Hose's broadbill
Calyptomena whiteheadi (m)	Whitehead's broadbill
Pitta arquata (m)	blue-banded pitta
Pitta baudi	blue-headed pitta
Oriolus hosei (m)	black oriole
Dendrocitta cinerascens (m)	Bornean treepie
Chlamydochaera jefferyi (m)	black-breasted triller
Malocincla perspicillata	black-browed babbler
*Zoothera everetti (m)	Everett's thrush
Ptilocichla leucogrammica	Bornean wren-babbler
Napothera atrigularis	black-throated wren-babbler
*Napothera crassa (m)	mountain wren-babbler
Yuhina everetti (m)	chestnut-crested yuhina
Copsychus stricklandi	white-browed shama
*Urosphena whiteheadi (m)	short-tailed bush warbler
*Bradypterus accentor (m)	Kinabalu friendly warbler
Cyornis superba (m)	Bornean blue flycatcher
*Pachycephala hypoxanthe (m)	Bornean mountain whistler
Prionochilus xanthopygius	yellow-rumped flowerpecker
Dicaeum monticolum (m)	black-sided flowerpecker
Arachnothera juliae (m)	Whitehead's spiderhunter
Arachnothera everetti	Kinabalu spiderhunter
*Oculocincta squamifrons (m)	pygmy white-eye
*Chlorocharis emiliae (m)	mountain blackeye
Pityriasis gymnocephala	Bornean bristlehead
Lonchura fuscans	dusky munia

(m) denotes montane species
* species not recorded for Kalimantan

Table 1.8. Species richness on Borneo and other Indonesian islands.

	Borneo	Sumatra *	Java	Sulawesi	New Guinea
Plants	10,000-15,000	9,000	4,500	5,000	15,000-20,000
Mammals	222 (44)	196 (9)	183 (19)	127 (79)	220 (124)
Resident birds	420 (37)	465 (18)	340 (31)	240 (88)	578 (324)
Snakes	166	150 8	7 (4)	64 (15)	98
Lizards			42 (1)	40 (13)	184 (59)
Freshwater turtles		8			8
Amphibians	100	70	36 (10)	29 (19)	197 (115)
Fish	394 (149)	272 (30)	132 (12)	68 (52)	282 (55)
Swallowtail butterflies	40 (4)	49 (4)	35 (2)	38 (11)	26 (2)

Numbers in brackets are island endemics.
* Sumatra mainland

BIOGEOGRAPHY

Studies of the biogeography of islands have shown that small islands support fewer species than do larger islands of similar habitat, and that isolated islands support fewer species than islands close to the mainland (Diamond 1975; Simberloff 1974). According to the Theory of Island Biogeography (MacArthur and Wilson 1967), a given island can support only a limited number of species, and species numbers will return to this equilibrium point, even after artificial attempts to enrich or impoverish the island fauna by adding or removing species. Equilibrium is maintained by a balance between species immigration and extinction, these rates being determined by island size and distance from a colonising source (fig. 1.18). This holds true even for the largest island. Total species lists for both plants and animals on different Indonesian islands fit the predicted pattern closely (figs. 1.19 and 1.20).

For both flora and fauna, Borneo shows much closer relationships to the Asian mainland and other Sunda islands (ancient Sundaland), than to its eastern neighbour Sulawesi. Although separated only by the Makassar straits (200 km at their widest point), Borneo and Sumatra have been separated for many millennia, probably at least since the Pliocene (10 Ma).

The faunal break between Borneo and Sulawesi was first recognised by the great nineteenth century explorer and naturalist Alfred Russel Wallace, who spent nine years exploring the Indonesian archipelago. He proposed a faunal boundary between Bali and Lombok running northwards through the Makassar straits between Borneo and Sulawesi, and included the Philip-

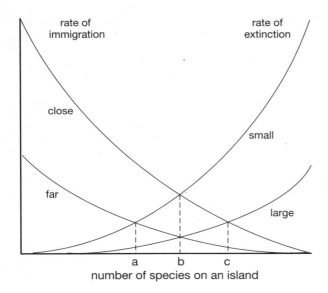

Figure 1.18. The relative number of species on small, distant islands (a) and large, close islands (c). The number of species on large, distant islands or small, close islands (b) is intermediate.

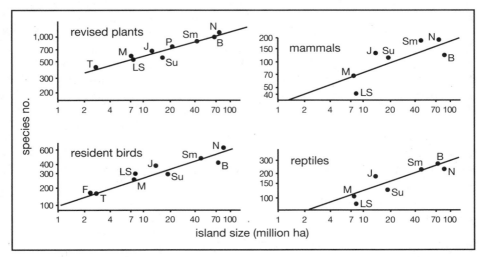

Figure 1.19. Relationship between number of species and island size for revised plants (reviewed in Flora Malesiana), resident birds, mammals and reptiles. B - Borneo, F - Flores, J - Java, LS - Lesser Sundas, M - Moluccas, N - New Guinea, Sm - Sumatra, Su - Sulawesi, T - Timor.

Source: MacKinnon 1982a

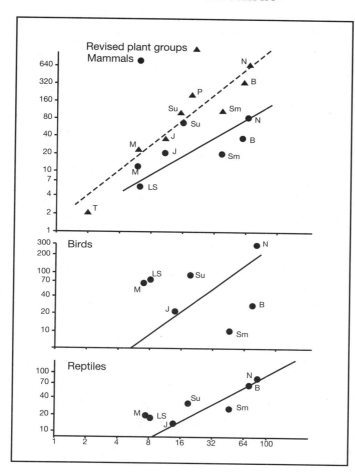

Figure 1.20.
Relationship between number of endemic species and island size.
B = Borneo,
J = Java,
LS = Lesser Sundas,
M = Moluccas,
N = New Guinea,
P = Philippines,
Sm = Sumatra,
Su = Sulawesi

Source: MacKinnon 1982a

pines on the Asian side of the line. Later his friend, the Victorian biologist Thomas Huxley, modified Wallace's Line to exclude all the Philippines except Palawan, thus following the limits of the Sunda Shelf and including all of ancient Sundaland (fig. 1.21). Although there has been much debate over the validity of Wallace's Line, it now seems that it does indeed hold good as a useful limit to the distribution of many plant and animal genera (Balgooy 1987; George 1987).

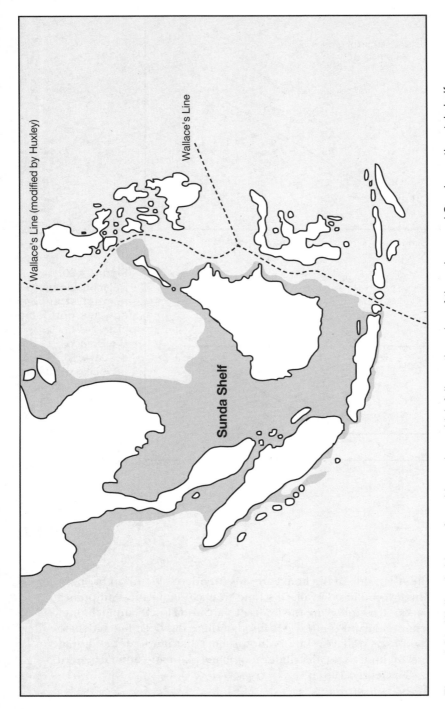

Figure 1.21. Wallace's Line, a faunal boundary which follows the edge of the submerged Sunda continental shelf.

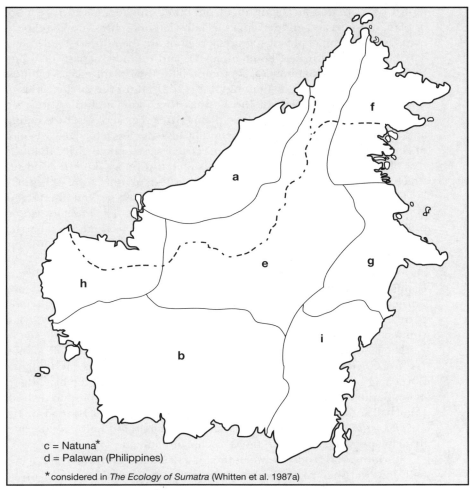

Figure 1.22. Biogeographical divisions of Borneo.
Source: MacKinnon and MacKinnon 1986

Biogeographical Units in Borneo

Within Borneo the distribution of animal and plant species is far from uniform. Apart from obvious altitudinal and habitat limits on distribution, there are several different phytogeographical and zoogeographical divisions which presumably reflect differences in geological history, Pleistocene land connections and geographical barriers to dispersal of species. Borneo and its offshore islands can be divided into nine biogeographical units (MacKinnon and MacKinnon 1986), with seven biounits on the island of Borneo itself (fig. 1.22).

The Meratus Mountains of South Kalimantan, marked (i) in

figure 1.22, are floristically distinct from other hilly regions and are a site of plant richness and endemism, especially rich in orchids (de Vogel pers. com.). Other distinct phytogeographic regions are the northeast, including Sabah and northern East Kalimantan (f), and the north coast including Brunei and eastern Sarawak (a), both noted for their plant species richness (Myers 1988). A fourth distinct biounit is centred on the ancient hills of northwest Borneo north of the Kapuas River and including western Sarawak (h). The southern lowland plains (b), a region of vast peat swamp and freshwater swamp forests, is generally less species-rich. The east coast of Borneo (g) is somewhat distinct because it is seasonally drier than the rest of the island. Finally there is a distinct and poorly known zone (e) which includes the hilly and mountainous, little-explored centre of Borneo.

Faunal distribution is related not only to habitat types but also to geographical barriers such as mountain ranges and rivers. The lowland regions of Borneo are drained by major rivers, whose width has acted as a barrier to animal movements. The headwaters, where these rivers and their tributaries are narrower and fast-flowing but generally easier to cross, may lie above the altitudinal limits of many lowland species. Thus the broad Kapuas and Barito rivers have acted as barriers to separate the endemic Bornean fauna from Sumatran invaders which reached Borneo at the end of the last Pleistocene glacial, when Sumatra and West Kalimantan were last connected by land (fig. 1.23).

The land between the Kapuas and the Barito is home to the agile gibbon *Hylobates agilis* (also found in southern Sumatra), whereas the endemic Bornean gibbon *H. muelleri* occurs elsewhere on Borneo. In the headwaters of the Barito where the two species meet, they interbreed to produce hybrids (Marshall and Sugardjito 1986). The southwest also lacks the banded langur *Presbytis melalophos*, while the orangutans *Pongo pygmaeus* here are intermediate in appearance between the Bornean and Sumatran races.

In southeast Borneo, between the Mahakam and Barito rivers, is a faunally impoverished region with no orangutans and a distinctive race of the Bornean gibbon. The northern lowlands are faunally, as well as floristically, much richer in species than the south, including two endemic squirrels *Petaurillus hosei* and *P. emiliae*, a rat *Chiropodomys major* and the mongoose *Herpestes hosei*, which are all confined to this unit, as are Borneo's only elephants.

The central mountain unit including Mount Kinabalu and Kalimantan's highest ranges contains the distinct montane fauna of the island with 28 restricted-range birds (24 endemic; see table 1.9.) and 21 endemic mammals. The Kinabalu massif is much higher than any other Bornean mountain and is the only known locality for several Bornean mammals and birds, but their known ranges may extend as other Bornean mountains are better explored.

The offshore islands of Natuna and Anambas and the Tambean islands

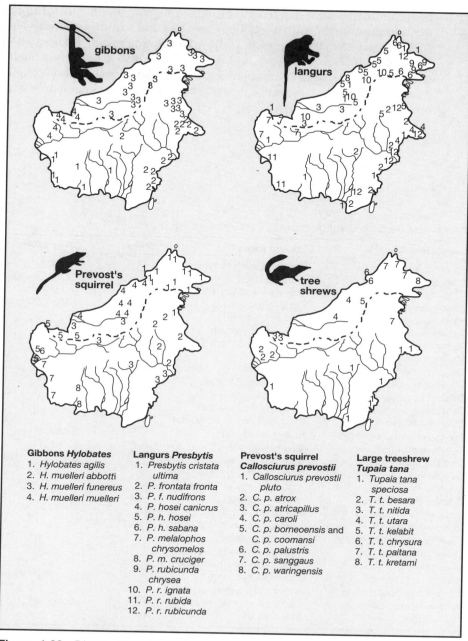

Figure 1.23. Distribution of species and subspecies of gibbons, langurs, Prevost's squirrels and tree shrews on mainland Borneo, related to past geographical separation.

Sources: Payne et al. 1985; Swindler and Erwin 1986

lie within the Bornean biogeographical province but are included within the administrative province of Riau, Sumatra and are described in *The Ecology of Sumatra* (Whitten et al. 1987a). Zoogeographically Palawan also falls within the Bornean province. Like Borneo, it lies on the Sunda Shelf and derived much of its fauna and flora from the Asian mainland, although it is less species-rich. However, since Palawan is part of the Philippines it lies outside the scope of this book.

Table 1.9. Restricted-range birds of the Bornean mountains.

Latin name	English name	Status	Other EBAs
Spilornis kinabaluensis	mountain serpent-eagle	T	
Arborophila hyperythra	red-breasted partridge		
Haematortyx sanguiniceps	crimson-headed partridge	N	
Otus brookii	rajah scops-owl	N	Sumatra, Java
Batrachostomus harterti	Dulit frogmouth	T	
Batrachostomus mixtus	Bornean frogmouth		
Harpactes whiteheadi	Whitehead's trogon		
Megalaima eximia	Bornean barbet		
Megalaima monticola	mountain barbet		
Megalaima pulcherrima	golden-naped barbet		
Calyptomena hosii	Hose's broadbill		
Calyptomena whiteheadi	Whitehead's broadbill		
Chlamydochaera jefferyi	black-breasted fruit-hunter	N	
Pycnonotus nieuwenhuisii	blue-wattled bulbul	T	Sumatra
Zoothera everetti	Everett's thrush	T	
Garrulax calvus	bare-headed laughing-thrush		
Garrulax palliatus	Sunda laughing-thrush		Sumatra
Napothera crassa	mountain wren-babbler		
Bradypterus accentor	friendly warbler	N	
Urosphena whiteheadi	Bornean stubtail		
Rhinomyias gularis	eyebrowed jungle flycatcher		
Pachycephala hypoxantha	Bornean whistler		
Dicaeum monticolum	black-sided flowerpecker		
Arachnothera juliae	Whitehead's spiderhunter		
Chlorocharis emiliae	mountain blackeye		
Oculocincta squamifrons	pygmy white-eye	N	
Zosterops atricapillus	black-capped white-eye		Sumatra
Oriolus hosii	black oriole		

T = threatened
N = insufficient information
EBA = Endemic Bird Area

Source: Bibby et al. 1993

Human History in Borneo

In 1891 a young Dutchman, Eugene Dubois, found part of a skull of human form near Trinil in Java; he called this find *Pithecanthropus (Homo) erectus*. This and subsequent *Homo erectus* finds in Java are extremely important because they provide some of the earliest evidence for the radiation of hominids from Africa through the tropical zones of the Old World. The earliest remains of *Homo erectus* in Africa date from between 1.7 and 1.5 million years ago (Wolpoff 1980). *H. erectus* teeth and associated stone tools from Yunnan (China) have been dated to about 1.7 million years ago. The Jetis and Trinil fossils from Java fall between 1.7 million years (at the earliest) and 500,000 years ago.

The earliest known evidence for human presence in Borneo is a skull of *Homo sapiens* found in the West Mouth of the Great Cave at Niah, Sarawak (Harrisson 1959; Majid 1982). The cave mouth was excavated under the direction of the late Tom Harrisson between 1954 and 1967. The skull was dated at more than 35,000 years old, using carbon-14 dating methods of the soils in which the skull was buried. Although some doubts have been expressed about the age of the skull, Niah is an important site, because it contains the longest stratified record of human occupation in Southeast Asia (Bellwood 1985). Niah is a late Pleistocene site revealing much about the lifestyle of tool-using Paleolithic peoples.

Early humans in Borneo 40,000-20,000 years ago hunted wildlife, fished and gathered forest products. Among the broken and burned bones at Niah are those of some animals now extinct in Borneo including the tapir *Tapirus indicus*, the giant pangolin *Manis palaeojavanica* and the white-toothed shrew *Crocidura fuliginosa*. Early humans also hunted mousedeer *Tragulus* spp., the orangutan, deer *Cervus unicolor* and bovids, the Sumatran rhinoceros and the sunbear. Apart from mammals, early humans also brought fish, birds, monitor lizards and crocodiles to the caves. The cave middens at Niah reveal an increasingly sophisticated stone tool culture from about 20,000 years ago; these stone tools were probably used both to kill and to prepare food, and to model and manufacture other artifacts from wood, bamboo and bone.

More recent excavations at Madai, Sabah (fig. 1.24) have provided further evidence of the early migrations and human settlements throughout the Indonesian archipelago (Bellwood 1989). Since 1980 the Sabah Museum staff have carried out excavations in the Madai and Baturong limestone massifs, at cave and open sites with deposits dating back 30,000 years (Bellwood 1984). Although these sites are now near the coast, during the low sea levels of the late Pleistocene they may have lain up to 150 km inland. Baturong is surrounded by a large area of alluvial deposits, formed by the damming of the Tingkayu River by a lava flow. Several open sites lie directly on the shoreline of the old lake and can be dated to between

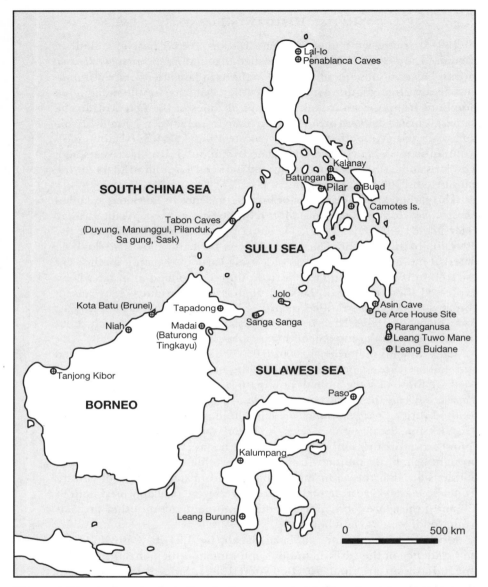

Figure 1.24. Archaeological sites in Borneo and neighbouring islands that have yielded information on human prehistory. (After Bellwood 1988.)

28,000 and 17,000 years old (Bellwood 1985). The Tingkayu stone industry shows a unique level of skill for its period. In the shelter of Hagop Bilo, at the foot of the Baturong massif, excavations have revealed shell middens of lake gastropods, dating back to 17,000 to 12,000 B.P. The remains of many mammals, snakes and tortoises have also been found, all food items collected by early occupants of the rock shelters (Bellwood 1989).

During the land connections of the Pleistocene, waves of ancestral humans swept through the islands of the Sunda Shelf from Asia. Negrito peoples, ancestral to Australians and Melanesians, probably inhabited Niah about 50,000 years ago but were supplanted by later waves of southern Mongoloids. As successive waves of migrants swept through the archipelago, they intermixed and interbred with the original inhabitants.

Some Southeast Asian tribes such as the Negritos of Malaysia are primitive hunters and gatherers, and it is tempting to believe that the Penan, too, may be derived from the original Negrito inhabitants of Borneo. There is, however, some speculation that the Penan may have returned to a hunting-gathering way of life from a more agricultural society (Bellwood 1985; Hoffman 1981). Penan occupy many forested areas of Sarawak and Kalimantan, dwelling in temporary camps of a few families, hunting with blowpipes, exploiting stands of wild sago *Eugeissona utilis*, collecting wild fruits such as rambutans, durians and mangosteens, and trading forest produce with surrounding agricultural societies such as the Kayan (Hose and McDougall 1912; Kedit 1978). Whether or not the Penan are Negrito in origin or belong to a later Mongoloid wave of immigrants like the Dayaks, their lifestyle most closely reflects that of early humans. They are discussed further in chapter 8.

The pre-Austronesian inhabitants of the archipelago may have used edge-ground axes as at Niah but did not use pottery. Although they exploited fruit trees, sago and tubers, they did not systematically cultivate these species (Bellwood 1985). Probably the earliest form of settled agriculture was associated with the introduction of the sago palm *Metroxylon sagu* from eastern Indonesia; this palm grows prolifically in the wet coastal swamps. Early peoples probably harvested the pith of the sago palm then tended the palms, much as the Melanau people, in the Rajang delta of Sarawak, do today. Coastal and river communities began to fish and collect freshwater molluscs; with the ability to harvest regularly from sago, permanent settlements were formed (Avé and King 1986).

The Austronesian-speakers who later expanded into the Indomalayan archipelago from mainland Asia carried a full agricultural economy, which was initially focussed on cereals, and introduced pottery and a new range of unibevelled stone adzes. At the beginning of the Holocene, about 7,000 years ago, wild rice and millets were being cultivated in the Yangtze basin in southern China in areas of seasonal wetlands. Rice was probably introduced to Indonesia with the southern Mongoloid immigrants. Rice culture may have received early setbacks in Borneo, as there is no evidence for it at Niah or Madai (Bellwood 1985), but it is likely to have thrived in the fertile lands of Java.

Artifacts from burial sites in the Niah caves, dated between 2000 and 600 B.P. (0 - 1400 A.D.) show that by the end of the Neolithic period Borneo peoples had quite an advanced culture, with beautiful stone adzes, earthenware pottery, bone and shell ornaments, bamboo caskets, wooden coffins, pandan mats and cotton textiles. The people practised ritual burials including cremation and secondary burial. In the Painted Cave at Niah wall paintings in haematite depict "ships of the dead" or "soul boats", illustrating a belief that is widespread among other Bornean cultures including the Ngaju of Central Kalimantan. Carbon-14 dates for the wooden coffins at the Painted Cave range from 460 B.C. to 1000 A.D.; glass, pottery and metal remains are associated with the coffins.

A significant change in lifestyle occurred with the discovery of iron ore deposits and the means for extracting and working iron. There were ample deposits of iron ore in Borneo, and the native inhabitants were already working it at Kuching, in the Sarawak river delta, in 1000 A.D. (Avé and King 1986). Skills in iron working and the manufacture of iron tools may have emerged even earlier, associated with the introduction of copper-bronze and iron artifacts and the appropriate technologies from Vietnamese, Chinese and Indian sources between the fifth to tenth centuries A.D. (Bellwood 1985). The Agop Atas cave in the Madai massif was inhabited from about 200 to 500 A.D. and includes fragments of pottery, bronze and iron. Jar burials related to this period have also been found at Madai and in the Tapadong caves on the Segama River; jar burials were also practised in the late Neolithic at Niah. The earliest jar burials date from 200 B.C.; this tradition may have derived from India since there is known to have been Indian contact with Southeast Asia from the first millennium B.C.

The use of iron brought fundamental changes to the lives of the local people. With iron tools the forest could be more easily cleared, and these clearings were planted with rice and taro. The Dayaks changed from gathering natural sago to planting and cultivating rice actively. Shifting cultivation of rice is still practised widely in Borneo. Traditional systems require short cropping periods interspersed with long fallows (chapter 8). Traditionally the Kayan of central Borneo moved villages about every fifteen years as fields lost their fertility (Hose and McDougall 1912).

Iron was used for making knives and agricultural implements and for hollowing the bores in the hard ironwood blowpipes, which are characteristic to Borneo. Early hunters already had access to bows and arrows, but the wooden blowpipes were far superior weapons, more accurate, and capable of killing prey over longer distances. The fine darts were dipped in natural poisons derived from the sap of forest trees.

The Apo Kayan was an area of plentiful iron ore, as were Mantalat (upper Barito), Mantikai (a tributary of the Sambas) and the Tayan in West Kalimantan. *Parang* and *mandau* (special fighting blades) from these areas were much sought after by other Dayak peoples (Avé and King 1986).

The stone megaliths found in the headwaters of the Baram around Mount Murud and elsewhere in the Kelabit highlands and central Kalimantan may date from this period (Harrisson 1962; Chin 1980). The rock carvings of human figures at Sungai Jaong, Sarawak, an early iron-working site, date back to 1000 A.D. or even earlier (Chin 1980). The upland megaliths include massive dolmens (three or more upright stones supporting a large roof stone), menhirs (pairs of tall natural stone slabs), urns and relief carvings on natural boulders. The Kelabits continued to create megaliths until 1950 when they were converted to Christianity. The megaliths were associated with the funeral rituals of important persons, such as tribal chiefs. The upper Bahau region of East Kalimantan offers probably the highest concentration of archaeological remains in Kalimantan. Fifty sites of so-called "Ngorek" settlements and graveyards have stone funerary monuments and stone tools.

Archaeological evidence from burial sites shows that Borneo has had a long history of trade with outsiders. Indian traders began to visit Indonesia in the first few centuries A.D. The Hindu kingdom of Kutai was established at this time, and Brahmin sacrificial posts at Muara Kaman and Hindu statues in caves on Gunung Kombeng in East Kalimantan date back to the fifth century (Boyce 1986).

Diplomatic links between Chinese and Bornean coastal settlements are recorded in Chinese dynastic histories from the seventh to the sixteenth centuries A.D. Brunei was an important trading port known to the Chinese since the Song dynasty. The Chinese exchanged ceramics, wine jars and coins for natural products such as incense wood (gaharu), bezoar stones (from the gall bladders of langurs), hornbill "ivory", rhinoceros horn and edible birds' nests. Already the trade in wildlife and forest products was substantial. Siamese traders brought the highly prized stone jars still used for marriage dowries and burials. There was also considerable trade within the archipelago, especially with the Bugis and Javanese.

Reliefs depicting a hunter with a blowpipe, on the galleries at the ninth century Buddhist temple of Borobudur in Central Java, suggest there were trade connections between the Dayaks and Java by this time (Avé and King 1986). During the fourteenth and fifteenth centuries the southern, eastern and western parts of Borneo came under the suzerainty of the East Javan kingdom of Majapahit. Even before this, Borneo had contact with other Hindu-Buddhist states, as evidenced by a number of Hindu and Buddhist remains, including the Hindu temple at Amuntai in South Kalimantan.

As the power of Majapahit declined, Islam spread throughout the archipelago, following the trade routes. There is evidence that Muslim traders were already visiting Borneo's ports during the thirteenth century.

The Muslim state of Brunei may have been founded as early as the fifteenth century. The sultan acquired great wealth by taxation of riverine districts from Pontianak to the southern Philippines. Other important Muslim states included Sambas, Sukadana and Landak on the west coast and Banjarmasin in the south.

As Muslim traders settled around the coasts, the Dayaks penetrated further inland in several waves of migration (chapter 8). One group of more recent immigrants, the Iban, are renowned for their phenomenal expansion from the Kapuas basin of western Kalimantan through most of Sarawak in the last 400 years. By 1850 they had colonised most of the Rajang basin (St. John 1974), and during the late nineteenth century they continued northwards into Brunei, clearing large areas of rainforest along river valleys for shifting cultivation. This expansion was probably not triggered entirely by population pressure and the need to grow more food, but was also related to the cultural and ritual need to collect heads.

Dayak settlements focused on nuclear villages for social reasons and defence, with communities living in longhouses. Most communities either made, or had access to through trade, pottery and iron, with barkcloth for clothing. The blowpipe was developed in or around Borneo (Jett 1970). Another Dayak practice, tattooing, is found throughout the Austronesian world. Murut men traditionally tattooed stars on their shoulders to denote captured heads (Rutter 1929). Mourning rituals were a major stimulus for headhunting. Many groups such as the Kelabits (Harrisson 1962) and the Benuaq of Kalimantan (Massing 1981) practised secondary forms of burial of defleshed bones, in log coffins or large stoneware jars.

When Europeans first began to arrive on the coasts of Borneo, in the early sixteenth century, they found a scatter of Muslim states, positioned at strategic points to control the forest products coming down Borneo's great rivers. The first European visitor to Borneo was the Italian Ludovico da Varthema, who arrived between 1504 and 1507. Later the Portuguese visited Borneo, especially Brunei. Anton Pigafetta, chronicler of Magellan's last voyage, visited the sultan's splendid court in 1521. Later the Spanish launched occasional forays against Brunei from their bases in the Philippines.

From these early European contacts the whole island came to be known as Borneo, a corruption of Brunei. The name Kalimantan originates from the word *lamanta* (raw sago) but later the Javanese changed this to Kalimantan, meaning "rivers of precious stones", a reference to the island's wealth in gold and diamonds *intan* (Avé and King 1986). The earlier Europeans were concerned only with trade with coastal settlements, and it was not until the nineteenth century that the interior came under European control, when the Dutch and the British staked out their colonial claims.

Dutch trading ships arrived in the Indonesian archipelago in the late sixteenth century, and people began to map the islands (fig. 1.25). For the next two hundred years the Dutch East India Company made several unsuc-

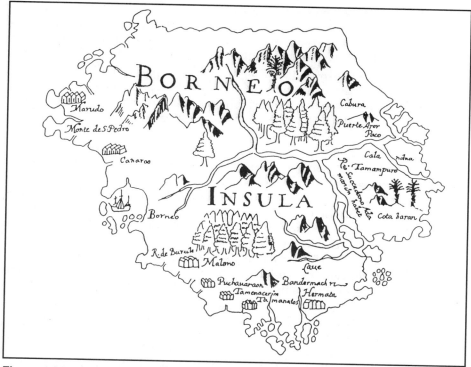

Figure 1.25. One of the earliest maps of Borneo, published in 1607. It was drawn by Olivier van Noort, the first Dutchman to circumnavigate the globe.

cessful attempts to establish a permanent presence in Borneo to control the trade in gold, diamonds and pepper. In the early 1800s the Dutch concluded treaties with the sultans of Banjarmasin, Pontianak and Sambas and finally, in 1840, began to assert their sovereignty in Borneo. This action was prompted by the establishment of a British presence in north Borneo, and especially the cession of Sarawak to the British adventurer James Brooke by the Sultan of Brunei in 1842, which began the hundred-year rule of the "White Rajahs of Sarawak" until the Japanese Occupation in 1942. The remaining areas of north and northeast Borneo eventually came under the control of a British trading company. Colonialism encouraged forest clearance for the establishment of plantations.

From the mid-1800s the Dutch increasingly established their claims to southern Borneo, sending military expeditions to subdue the conflicts between rival groups of Chinese gold miners who had settled in West Kalimantan from the seventeenth century onwards (chapter 13). The descendants of these Chinese, now mostly farmers, account for the large Chinese presence in West Kalimantan today.

In 1859 the Dutch became involved in the bloody Banjarmasin war when their candidate for the sultanate met with much local opposition. They established control after the death of the rightful heir, Antasari, and the execution in 1862 of Demang Lehman, the principal resistance leader, but the struggle continued, sustained by Surapati, leader of the Siang Dayak, and the sons of Antasari. The resistance finally ended in 1905 with the death of Mohamad Seman, the last pretender to the throne (Avé and King 1986).

By the late nineteenth century the Dutch had established a strong presence in the coastal districts of the south and west, and from here they began to extend their authority inland. With establishment of colonial rule in southern and northern Borneo came an influx of Christian missionaries and a succession of naturalists and explorers. The latter have left some fascinating accounts of the natural and cultural histories of the island and its peoples (Beccari 1904; Wallace 1869; Bock 1881; Lumholz 1920; Mallinckrodt 1926; Tillema 1938; St. John 1974). Colonial administrators throughout the twentieth century continued to document in writings and pictures the island's natural and cultural wealth (Hose 1926; Harrisson 1959; Nieuwenhuis 1904; Macdonald 1956; Morrison 1957, 1972).

The Second World War and the Japanese occupation of Borneo and other islands of the Indonesian archipelago spelled the end to the colonial period. In 1945 Indonesia, including Kalimantan, declared independence; in 1949 the Dutch acknowledged the new republic. In the north, Sarawak and British North Borneo (now Sabah) were ceded to the British crown in 1946 and became independent as part of the new Federation of Malaysia in 1963. The state of Brunei remained a British protectorate until it became fully independent in 1984 as Negara Brunei Darussalam, although it had been self-governing since 1959.

Current Land Use

Today the island of Borneo supports a total of 12.5 million people, three-quarters of them based in Kalimantan (table 1.10). The four provinces of Kalimantan cover an area of almost 54,000 km^2 and sustain a human population of 9.1 million, an average density of 17 people/km^2 (fig. 1.26; table 1.11). East Kalimantan (Indonesia's second largest province with an area of 202,440 km^2) has a population of only 1.9 million people, mainly distributed in the coastal lowlands, fertile Mahakam valley and upland river valleys. When the above density is compared with the population density of more than 800 people/km^2 in Java, one can understand why Kalimantan is considered to be underpopulated, and is designated a prime target for development by the Indonesian government.

Although the human population of Borneo is small, human activities have had dramatic and far-reaching effects on the natural habitats of the island. In the last twenty years development has accelerated rapidly to exploit the island's rich tropical forests, oil and gas resources, and other mineral wealth. Extensive areas of the original forest have been cleared by logging and agriculture, and now only 60% of the island remains forested (fig. 1.27). Development is continuing apace, with rapid deforestation as a result of logging (chapter 9) and land clearance for agriculture, plantations, human settlement and transmigration schemes (chapter 12). Although the natural rainforest habitat gives an impression of luxuriant plenty, Borneo's soils are generally poor and unsuitable for intensive agriculture (table 1.12). It is no coincidence that until recently the island was sparsely populated, with settlements clustered on the more fertile alluvial river plains and around the coast.

The traditional peoples of Borneo have long practised slash-and-burn agriculture, an agricultural regime that can be sustained on poor soils when practised at low densities and with long fallow periods. Even so, in the Kelabit highlands along the Sarawak-Kalimantan border there are extensive, man-created grasslands, already recorded a century ago (Blower et al. 1980). Expanding populations, recent immigrants, logging concessions and government-sponsored transmigration programmes have opened up new areas of forest to agriculture, often exposing poor soils and growing inappropriate crops. Without long fallows the soils lose their fertility, and the farmers eventually abandon their degraded fields. Almost one-quarter

Table 1.10. The provinces and states of Borneo, and their populations in 1990.

	Area km²	Inhabitants x 10³	Population density/km²	Percent of Indonesia	Percent Indonesian population	Capital
West Kalimantan	146,760	3,228	22	7.65	1.81	Pontianak
Central Kalimantan	152,600	1,396	9	7.95	0.78	Palangkaraya
South Kalimantan	37,660	2,597	69	1.96	1.45	Banjarmasin
East Kalimantan	202,440	1,875	9	10.55	1.05	Samarinda
Total Kalimantan	539,460	9,096	17			
Sarawak	124,449	1,600	13			Kuching
Sabah	73,700	1,400	19			Kota Kinabalu
Brunei Darussalam	5,765	300	52			Bandar Seri Begawan

Source: *Buku Statistik 1991*; Collins et al. 1991

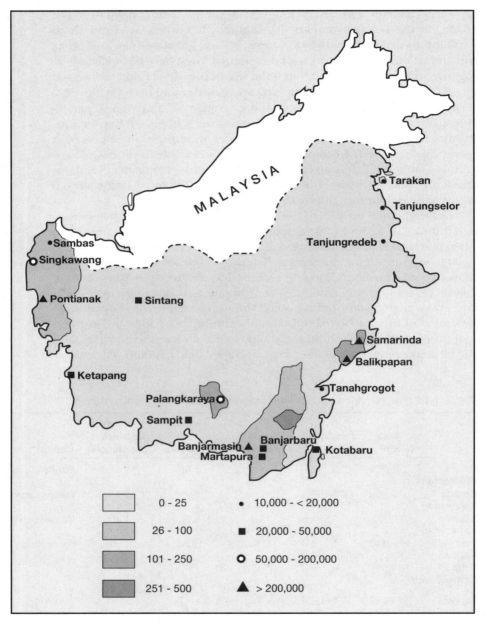

Figure 1.26. Population density in Kalimantan.
Source: RePPProT 1990

of the land area of South Kalimantan, 900,000 ha, is now covered in unproductive *alang-alang* grasslands of *Imperata cylindrica*. This area is growing by 10,000 ha each year because of poor agricultural practices.

As development and agricultural expansion continue, it is timely to look at the ecology of the island, as many of Borneo's habitats are critical lands unsuited to random development and wet rice agriculture. This is not to say that habitats such as the tidal wetlands cannot be converted and cultivated, but they need to be developed in an appropriate and sustainable manner, taking into consideration the special nature of their fragile ecosystems.

Borneo has rich natural and mineral resources. Kalimantan's forests

Table 1.11. Rural and urban populations of Kalimantan, 1986.

Province	Kabupaten	Area km^2 (x 10^3)	Total pop. 1986 (x 10^3)	Urban pop. (x 10^3)	Rural pop. (x 10^3)	Rural density /km^2	Rate of pop. increase 1980 to 1986 (%)
West Kalimantan	Sambas	12.3	678.8	98.1	580.7	47	
	Pontianak	18.2	699.4	52.5	646.8	36	
	Sanggau	18.3	395.8	12.4	383.3	21	
	Ketapang	35.8	295.9	23.6	272.0	8	
	Sintang	32.3	337.4	21.3	316.1	10	
	Kapuas Hulu	29.8	155.9	2.9	153.0	5	
	Total	146.7	2,563.2	210.8	2351.9	16	2.73
Central Kalimantan	Kota Waringin Barat	21.0	142.7	18.8	123.9	6	
	Kota Waringin Timur	50.7	329.1	34.6	294.8	6	
	Kapuas	34.8	418.0	22.9	395.1	11	
	Barito Selatan	12.9	145.0	14.3	130.8	10	
	Barito Utara	32.0	133.4	5.4	127.9	4	
	Total	151.4	1,168.2	96.0	1,072.5	7	4.57
South Kalimantan	Tanah Laut	2.2	158.9	5.0	153.9	72	
	Kotabaru	13.5	284.9	23.5	262.4	20	
	Banjar	6.3	390.7	57.7	333.6	54	
	Barito Kuala	3.3	201.1	9.4	191.7	58	
	Tapin	2.3	119.8	2.9	116.9	50	
	Hulu Sungei Selatan	1.7	180.2	14.5	165.7	97	
	Hulu Sei Tengah	1.5	208.2	16.7	191.5	130	
	Hulu Sei Utara	2.8	252.5	5.3	247.2	89	
	Tabalong	3.9	135.3	6.6	128.8	33	
	Total	37.5	1,931.6	141.6	1791.7	49	2.33
East Kalimantan	Pasir	20.0	112.0	18.3	93.7	5	
	Kutai	91.0	481.4	39.2	443.2	5	
	Berau	32.7	52.7	14.9	37.8	1	
	Bulungan	64.0	209.7	86.4	123.3	2	
	Total	207.7	855.8	158.8	698.0	3	4.42

Source: RePPProT 1990

66 THE ISLAND OF BORNEO

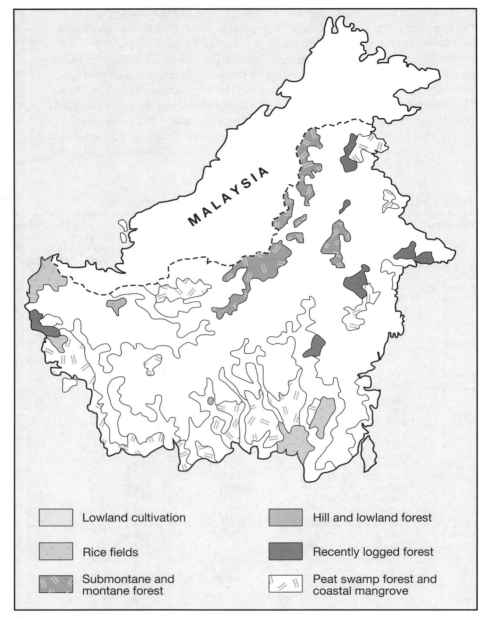

Figure 1.27. Land use and forest cover in Kalimantan.

and the timber boom of the last twenty years have fuelled the rapid growth of the Indonesian economy. Kalimantan gold has been traded for centuries, and the rising price of gold on world markets is leading to a new gold rush (chapter 13). East Kalimantan has extensive oil and natural gas fields, centred around Balikpapan and Bontang, which are an important source of revenue for the nation. Since 1984 attention has turned increasingly to other mineral deposits, including coal and bauxite, as well as to the traditional panning for diamonds and semiprecious stones. Most of lowland Kalimantan has already been given out in timber concessions (HPH); many are overlapped by mineral exploration concessions for gold and coal. Some of the preliminary surveys have proved promising, and mining will begin soon, with further destruction of the natural environment.

Kalimantan's diverse tropical habitats and rich biodiversity already contribute substantially to local and national economies and human welfare. Borneo's forests and natural wetlands also provide numerous environmental benefits: protecting watersheds and the integrity of water supplies; preventing soil erosion and conserving soil fertility; and acting as reservoirs

Table 1.12. Current land use and land with potential for agriculture in Kalimantan.

Province	Area km²	Area forest (H)	Area used (NH)	Area suitable (S)	Area unsuitable (NS)	"suitable" (H)	"suitable" (NH)	"unsuitable" (H)	"unsuitable" (NH)
West Kalimantan	47,516	87,010 59%	60,500 41%	72,270 49%	75,260 51%	19,550 27%	52,720 73%	67,480 90%	7,780 10%
Central Kalimantan	145,290	111,125 76%	34,165 24%	77,772 54%	67,518 46%	50,992 66%	26,780 34%	60,133 89%	7,385 11%
South Kalimantan	37,480	7,960 48%	19,500 52%	23,690 63%	13,800 37%	6,910 29%	16,780 71%	11,080 80%	2,720 20%
East Kalimantan	197,110	180,000 9%	17,110 9%	79,330 40%	117,880 60%	63,360 80%	15,970 20%	116,640 99%	1,230 1%
Total	527,406	396,095 75%	131,275 25%	253,062 48%	274,458 52%	140,812 56%	112,250 44%	255,333 93%	19,115 7%

H - Forest NH - Nonforest
"suitable" (S) is gross suitability for agricultural use.

Note that 56% of land assessed as suitable for agriculture and 93% of unsuitable land remain under forest. This proportion varies with province.

Source: RePPProT 1990

for genetic material that may have future potential for crops, drugs and commercial projects. It is essential that we learn to understand the ecology of the island's rich but fragile ecosystems, and the role they play in the livelihoods of local communities, so that development can be planned in a rational way for sustainable use.

Chapter Two

Coastal Habitats

The coastal zone is generally defined as the band of land and sea straddling the coast (Salm, and Clark 1984). This is a very small part of the total area of Kalimantan but its ecology is influenced by events in a much broader area, from the inland watersheds to the offshore waters of the coastal shelf. The coastal zone includes some of the most biologically productive habitats on the island, including estuaries, tidal wetlands, mangroves and coral reefs. It is also the area where most of the people of Kalimantan live and where most development is occurring.

Environmental disturbance within the coastal zone affects offshore ecosystems which rely on coastal habitats for export of nutrients (Du Bois et al. 1984). At the same time, a coastal habitat can be damaged by activities far outside its immediate boundaries, whether these be an offshore oil slick, mismanagement of a watershed many kilometres inland, or pollution from urban, agricultural and industrial activities. In addition to the environmental impacts of development, the coastal zone is threatened by natural hazards such as sea storms and tidal waves (Salm and Clark 1984). Probably more than for any other ecosystem, wise development and utilisation of the coastal zone requires a good understanding of the ecology of natural habitats.

The coastline of Kalimantan extends 8,054 km from the Sambas peninsula in the west to the island of Nunakan on the Sabah border. Most of the Kalimantan coastline is fronted by a shallow, shelving shore, backed by mangrove and mudflats or extensive sandy beaches fringed with *Casuarina* trees. Major habitats around Kalimantan, and the rest of Borneo, include islands and rocky islets, coral formations, rocky coastlines including promontories and headlands, sandy beaches, mangrove/nipa associations and mudflats, and estuaries (fig. 2.1). This chapter will deal only with these coastal habitats. The ecology of inshore waters and the open seas will be covered in *The Ecology of the Indonesian Seas*, a later volume in this series.

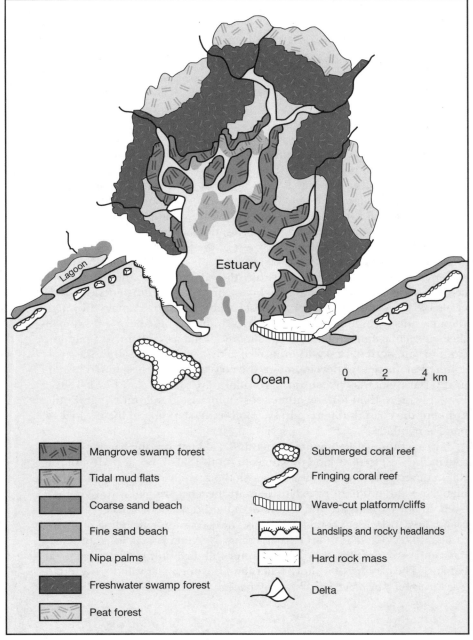

Figure 2.1. A simplified sketch of natural ecosystem types and their zonal distribution in a typical portion of coastal Kalimantan. (After Knox 1980.)

Offshore Islands and Rocky Islets

Several large islands lie close to Kalimantan: Pulau Padangtikar and Pulau Maya off West Kalimantan and Pulau Laut, with its satellite island Pulau Sebuku (together 13,043.5 km^2), off the coast of South Kalimantan. They support fine stands of lowland dipterocarp forest and fringing mangrove. On the east coast there are several major islands of deltaic origin, including Tarakan and Nunakan. Here, too, the natural vegetation is dipterocarp and mangrove forest, much of it now being logged. Many of the smaller island groups off Kalimantan are coral islets such as Karimata in the west and Birah-Birahan and Pulau Maratua off the east coast.

A few small rocky islets occur off Borneo's coasts. They form a graded series from small rocks just breaking the sea surface to those with sufficient soil to support low vegetation and bushes, and larger islets where the original vegetation has often been cleared to create coconut plantations and fruit-tree groves. Several of the small rocky islets are the sites of nesting colonies of terns, including black-naped tern *Sterna sumatrana*, bridled tern *Sterna anaetheta*, brown noddy *Anous stolidus*, Bulwer's petrel *Bulweria bulweri*, masked and brown boobies *Sula dactylatra* and *S. leucogaster*, and the lesser frigate bird *Fregata ariel* (Salm and Halim 1984; Chan et al. 1985). Some tern populations are exploited by fishermen who collect eggs and young birds during the breeding season.

Other island-adapted birds around Borneo's coasts include the pied imperial pigeon *Ducula bicolor* and the Nicobar pigeon *Caloenas nicobarica*. The pigeons are fruit eaters which help to disperse fig seeds so that young figs are commonly found growing on rocky islets, their roots struggling for purchase and nutrients among the rocks. Grey-rumped edible swiftlets *Collocalia fuciphaga* are also often found on islands, where they nest in caves carved out by the sea in cliffs and headlands.

Megapodes *Megapodius freycineti (cumingii)* were formerly common on most of the small islands around Borneo (Balambangan, Maratua, Banggi and Labuan) and at a few sites on the Sabah mainland, but they are now scarce due to excessive collecting of their eggs (MacKinnon and Phillips 1993). These moundbuilders are related to the maleo birds of Sulawesi and the brush turkeys of Australia. The moundbuilders are unusual among birds since the hen does not brood her eggs. Instead, the eggs are buried in a heap of soil and rotting vegetation, and heat from the decomposing material keeps the eggs warm until they hatch (Crome and Brown 1979). The moundbuilders represent an Australian element in the Bornean fauna. Megapodes are found from Borneo eastwards to Australia and the islands of the Pacific.

One type of fruit bat, the island flying fox *Pteropus hypomelanus*, is often found on small islets. Several other Kalimantan mammals have island races, usually found on the larger islands. Island races of the ruddy and

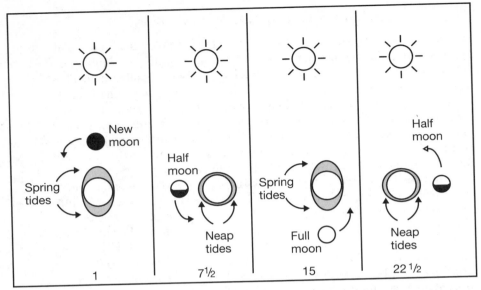

Figure 2.2. The tidal cycle. The ebb and flow of the tides determines the characteristic features of the sea shore. Tides are greatest when the sun, moon and earth are in a straight line (i.e., on days of the full and new moon). High tides are 'spring tides'. (After Pethik 1984.)

slender treeshrews *Tupaia splendidula carimatae* and *T. gracilis edarata* are restricted to the Karimata islands. Natuna (within the Bornean biogeographical region but administratively part of Sumatra) has a subspecies of flying lemur different from the one on mainland Borneo; the island race is also found on Pulau Banggi (Payne et al. 1985). Since their isolation from mainland Borneo, the island populations have gradually evolved new characteristics better adapted to the island habitat. This is how new species evolve.

ROCKY SHORES

The ebb and flow of the tides determine the characteristics of the seashore (fig. 2.2). Rocky shores are found where hard and resistant rock formations are washed by the sea, but the products of this weathering are swept away rather than deposited to form wide beaches. Such shores are often steep, with the rock face continuing below the sea surface. Commonly there is a narrow shingle beach in upper tidal regions. Such steep cliffs are usually formed of old limestone (as on Sangkulirang) or granites and Tertiary sandstones (as at Bako, Sarawak, and on the Natuna archipelago). In the Anambas, Tambelan and Natuna island groups, steep rock-strewn slopes plunge into the sea. There is no single type of vegetation associated with rocky shores, but the trees *Barringtonia, Casuarina* and *Calophyllum*, the figs *Ficus*, and the pandans *Pandanus* may be found clinging to the upper rock face, above the level of extreme high tides but still affected by sea spray.

Rocky shores are of biological interest for their associated fauna, especially certain species of molluscs that are not found on sandy or muddy beaches. Barnacles are common, as well as small oysters, limpets and *Nerita* snails. Long-legged rock crabs *Grapsus grapsus* scuttle beside the water's edge, and small blennies jump from rock to rock in the spray zone, grazing the algae (MacKinnon 1986). The animals of rocky shores are adapted to withstand the force of waves, periodic desiccation, high temperatures and variable salinity. They have efficient means of retaining a grip. Some, like barnacles, are permanently attached; others can move around to forage or graze. Animals and plants are found in distinct zones up a rocky shore (Purchon and Enoch 1954). Many shellfish found on rocky shores are edible and some, such as rock oysters, are much prized.

Many of the limestone and granite cliffs have clefts or small wave-washed caves that support colonies of edible-nest swiftlets *Collocalia fuciphaga*, whose white nests attract collectors (see chapter 6). Crevices in sea caves also provide shelter for venomous sea snakes, 19 species of which are common in Borneo's coastal waters (Stuebing 1991). Sea snakes are lung breathers that frequent shallow inshore waters where they can quickly come up for air. They weave through the water using their flattened tails as paddles. They feed on fish and strike at their prey just as land snakes do; the broader tail acts as an anchor, weighting the snake as it strikes.

BEACHES AND BEACH FORESTS

Beaches and sand dunes occur around Kalimantan's coastline; offshore they function as cays and barrier beaches adjacent to mangroves, coral reefs and other marine ecosystems (Burbridge and Maragos 1985). Compared with other tropical ecosystems, beaches are neither organically productive nor biologically diverse. Exposed beaches are often inhospitable environments for marine or strand species but perform several important ecological functions by dissipating wave action and protecting other terrestrial and aquatic ecosystems.

As the boundary between land and sea, beaches cover a wide spectrum of physical and chemical conditions (fig. 2.3). Plants and animals occur in distinct zones, according to where each species can survive on the beach profile (Morton 1990). In the upper beach zones the dominant organisms

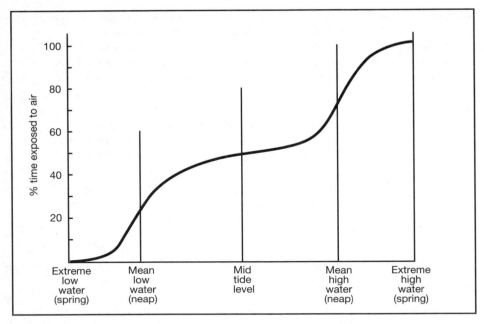

Figure 2.3. General pattern of tidal exposure up a beach. According to the tides, areas of beach are inundated with seawater for varying amounts of time. At the mean high water level for neap tides, the beach is exposed for about 70% of the time. The exposure gradient determines the occurrence of plants and animals up the shore. (After Brehaut 1982.)

are strand plants. Where the beach tends to build, the outer fringe of vegetation usually comprises a ***pes-caprae* community**, named after the characteristic bindweed *Ipomoea pes-caprae* (table 2.1). The plants in this community are low, sand-binding herbs, spiky *Spinifex* grass, other grasses, and sedges. To survive here, plants must be deep rooted, tolerant of salt, wind and high temperatures, and capable of producing floating seeds (Paijmans 1976). They must also be able to survive being buried periodically under wind-blown sand. The plant community exhibits low species diversity and, because of their specialised nature, most species are restricted to this habitat (van Steenis 1958a).

Many sandy beaches have almost pure stands of *Casuarina equisetifolia*, an important pioneer species which helps to stabilise beach soils. The golden sand beaches of Pasir Panjang and Sambas in West Kalimantan, for instance, are fringed by a belt of *Casuarina* trees. *Casuarina* is able to grow in this nutrient-poor environment because it is able to fix nitrogen in nodules on its roots. Other common beach trees include *Terminalia catappa*, *Calophyllum inophyllum* and *Hibiscus tiliaceus* (Whitmore 1984a). Several strand species are widely utilised by coastal villagers: hibiscus for making ropes and cordage and various pandans for making mats, baskets and sails (Soegiarto and Polunin 1980).

Stable soils behind the beach (or rocky soils where there is no beach) are generally colonised by trees and shrubs which form a **Barringtonia community** (table 2.1), named after the pioneer tree *Barringtonia asiatica*. Spreading *Barringtonia* trees line the shore above the high tide line and are easily recognised by their "shaving brush" flowers, which are bat pollinated, and heavy fruits that disperse with the tide. In undisturbed areas the tree canopy can be dense with little ground vegetation, but in open clearings ferns, grasses, gingers and herbs grow. This coastal formation is rarely more than 50 m wide. Only a few species occur in this habitat, though many are widespread throughout the tropics (Whitmore 1984a). Some members of

Table 2.1. Some component plant species of the *pes-caprae* and *Barringtonia* communities in the vegetation of Indonesian beaches.

Community	Species in the flora	Community	Species in the flora
pes-caprae	*Canavalia cathartica*	Barringtonia	*Ardisia elliptica*
	C. microcarpa		*Barringtonia asiatica*
	Cyperus stoloniferus		*Caesalpinia bonduc*
	C. pedunculatus		*Casuarina equisetifolia*
	Euphorbia atoto		*Calophyllum inophyllum*
	Fimbristylis sericea		*Cocos nucifera*
	Ipomoea pes-caprae		*Colubrina asiaticum*
	Ipomoea gracilis		*Crinum asiaticum*
	Ischaemum muticum		*Cycas circinalis*
	Launea sarmentosa (rare)		*C. rumphii*
	Lepturus repens		*Desmodium umbellatum*
	Spinifex littoreus		*Erythrina variegata*
	Thuarea involuta		*Guettarda speciosa*
	Triumfetta repens		*Hernandia peltata*
	Vigna marina		*Hibiscus tiliaceus*
			Mammea odorata
			Messerschmidia argentea
			Morinda citrifolia
			Pandanus tectorius
			Pemphis acidula
			Pisonia grandis
			Pluchea indica
			Pongamia pinnata
			Premna obtusifolia
			Scaevola taccada
			Sophora tomentosa
			Terminalia catappa
			Thespesia populnea
			Wedelia biflora
			Ximenia americana

Source: van Steenis 1958a; Soegiarto and Polunin 1980

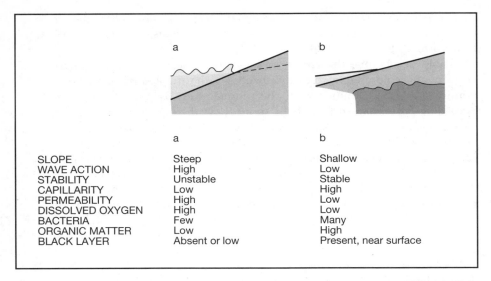

Figure 2.4. Comparison between (a) a steep coarse-grained beach and (b) a shallow fine-grained beach. The different characteristics provide living conditions for different animals. (After Bradfield 1978.)

the *Barringtonia* community may also occur behind mangrove formations (van Steenis 1958a). Beach vegetation has been locally modified or replaced around most human settlements, with *Barringtonia* being felled for wood or cleared for coconut plantations, rice and other crops.

Exposed beaches are a hostile environment. They are physically unstable, with wide variations in temperature, salinity and humidity, and they are exposed to wind and waves (fig. 2.4). A number of animals have adapted to these conditions. The beach fauna shows well-developed zonation (Soegiarto and Polunin 1980). Ghost crabs *Ocypode* and burrowing beach fleas live in the upper zones, whereas the bivalve *Donax* and isopod crustaceans occur lower down the beach. Hippid crabs, sea urchins and sand dollars are found next to the sea, where they burrow beneath the sand to avoid the breaking waves (Soegiarto and Polunin 1980). Zones regularly inundated by tides support a specific fauna of burrowing marine invertebrates such as sand dollars, razor shells and small clams. These are a major food resource for fish and waterbirds; some species are also collected for food by local people.

Under the beach surface, in air pockets between the sand grains, live nematodes, copepods and flatworms. These invertebrates are generally more abundant closer to the surface (Soegiarto and Polunin 1980), and they are a source of food for shorebirds which range the beaches, probing the sand with their bills. Many of the less disturbed Kalimantan beaches provide important resting and feeding habitats for resident and migratory waders and seabirds (Soegiarto and Polunin 1980; Salm 1984a; NPWO/ Interwader 1985; Eve and Guigue 1989).

The animals of the lower shore feed mainly on marine organisms, whereas the food chain of the intermediate zone is based on flotsam left behind by the receding tide (Soeigarto and Polunin 1980). Sea urchins feed on all organic material, plant or animal, living or dead, and have a highly developed chewing apparatus, the "Aristotle's lantern", which projects through the mouth. Though the lantern is efficient at chewing, feeding is slow, and it can take a sea urchin several days to munch its way through a small bunch of seaweed.

Several sandy beaches are important nesting habitats for the five species of marine turtles which breed in Indonesian waters (fig. 2.5). Major green turtle rookeries are found on the Anambas, South Natuna and Tambelan islands (Schulz 1987; Greenpeace 1989). Green turtles *Chelonia mydas*, hawksbills *Eretmochelys imbricata* and occasional leatherbacks *Dermochelys coriacea* are known to nest on the beaches and islands at Paloh (West Kalimantan). Loggerhead turtles *Caretta caretta* nest occasionally on Pulau Lemukutan off West Kalimantan (Salm 1984b; Salm and Halim 1984). The Berau islands (East Kalimantan) are important green turtle rookeries, especially the islands of Sangalaki, Bilang-Bilangan, Belambangan, Sambit and Mataha (fig. 2.6). Approximately twelve thousand female green turtles are believed to nest in the Berau islands each year (see box 11.2). The Talang-Talang islands and Satang Besar of Sarawak and the Turtle Islands off northern Sabah are also important nesting sites for marine turtles. Although turtles are legally protected in Indonesia, there is a continuing licensed trade in turtles and turtle eggs for domestic consumption and export, with two to three million eggs collected annually in the Berau islands alone (Schulz 1984). Excessive harvesting of eggs is affecting every major sea turtle rookery in Kalimantan, and there has been a noticeable decline of nests on many beaches over the last few years (Schulz 1984, 1987). Even on the closely monitored turtle islands of Sarawak, where there are traditional injunctions against taking the adult turtles themselves, turtle populations are declining (Chan et al. 1985).

Beaches of golden sand are comparatively rare in Kalimantan and elsewhere in Borneo. Most beaches are graded from slightly muddy sands to sandy muds. Beaches with golden sands, at Balikpapan, Takesung and Pagetan in South Kalimantan and along parts of the west coast, have important recreational and tourism value. The digging of sand for building purposes, the construction of access roads, and other developments destroy this potential value and can affect drainage patterns and lead to beach erosion.

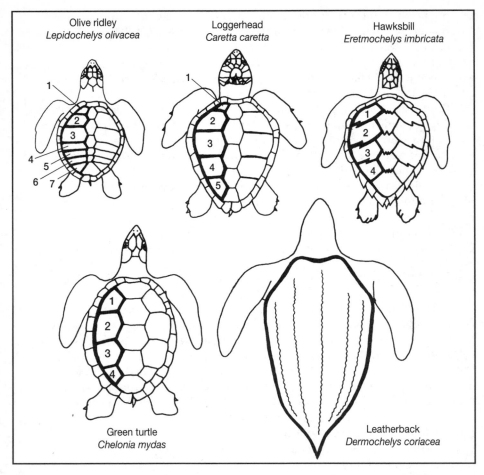

Figure 2.5. Carapace patterns of the five species of marine turtles which feed and nest in Indonesian waters. Three of these species (green, loggerhead and hawksbill) nest on Kalimantan beaches. Leatherbacks occasionally nest on the turtle beach at Paloh, West Kalimantan.

ESTUARIES AND MUDFLATS

Estuaries are areas where inland and marine influences meet; their ecology exemplifies the interdependence of terrestrial and marine systems (Salm 1984a). The silt-laden fresh waters of Kalimantan's great rivers pour into the sea substantial inputs of nutrients from terrestrial systems, enriching coastal waters and the associated mudflats. Estuaries provide the filtering system and settling basin for river silt. This process has created the large deltas of the

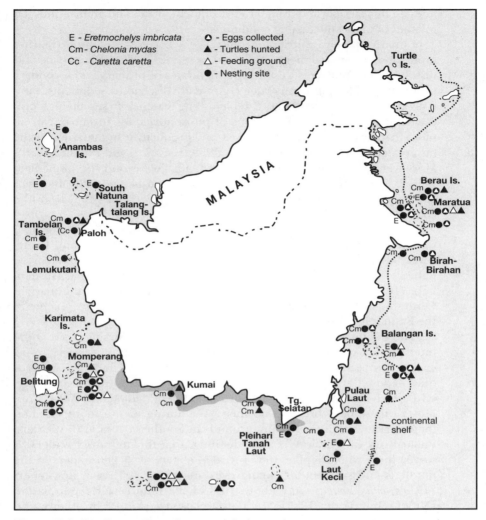

Figure 2.6. Feeding and nesting grounds for turtles.

Kapuas, Barito and Mahakam rivers, where series of islands are interspersed with radiating patterns of major and minor tributaries whose patterns may vary over time (Voss 1979). In the Mahakam delta, for instance, land is accreting in some areas while other parts of the intertidal mangroves are being eroded (Eve and Guigue 1989). Chemical conditions in river estuaries also vary as significant volumes of fresh and salt water meet and mix, usually creating two-layered water environments with the salt "wedge" retreating and advancing with the tides. In this variable and dynamic environment, plants and animals must be tolerant of widely fluctuating conditions.

Estuaries receive nutrients from three sources: river inputs, marine inputs and bottom sediments. Tidal and river flows facilitate the mixing and distribution of nutrients to stimulate high rates of primary and secondary productivity. Flowing fresh water continually discharges sediments, minerals and nutrients into estuaries, replenishing materials to sustain high productivity. Estuarine concentrations of phosphorus, for instance, can be more than twice those in the open sea, and productivity in estuaries may be far greater than offshore (Doty et al. 1963; Soegiarto and Polunin 1980). Moreover, the influence of estuaries extends far beyond the immediate vicinity of river mouths: in the Java Sea plankton is especially abundant around estuaries, but the number of individuals decreases away from the coast (Hahude et al. 1979).

Estuaries support a rich variety of plant and animal life. Associated plant communities include tidally influenced swamp forests and mangroves, all of which export organic matter and nutrients into the estuarine habitat. The nutrient-rich waters support a diverse fish fauna. Hardenberg (1936, 1937) reported 80 species of fish from the Kumai estuary in Central Kalimantan and 222 species from the estuary and lower reaches of the Kapuas in West Kalimantan.

The infauna of the sediment shows considerable variation across the mudflats. It is generally poor in numbers of species, but some species, especially gastropod molluscs, are locally common (figs. 2.7 and 2.8). Areas of mud exposed at low tide offer rich feeding grounds to migratory shorebirds which feed on the invertebrate communities (Medway and Wells 1976). Most of these birds breed in northern Asia during the northern summer, then migrate to winter feeding grounds in Southeast Asia and Australia. Pulau Bruit off Sarawak is a key feeding area for migratory waders. A survey there in November 1985 revealed 50,000 to 80,000 waders in the mudflats and surrounding mangroves, including 400 Asian dowitchers *Limnodromus semipalmatus*, about 10% of the known world population (NPWO/Interwader 1985). Mudflats are also frequented by other waterbirds, including Chinese egrets *Egretta eulophotes* and lesser adjutant storks *Leptoptilus javanicus*. Surveys of the birds of the Mahakam River delta recorded a total of 146 species, including two new records for Borneo and 22 for Kalimantan (Eve and Guigue 1989).

Estuaries are among the most productive of all ecosystems and are important in the life cycles of many marine animals. Dolphins and dugongs are often found in estuarine waters where they come to mate and give birth. Estuaries and adjacent wetlands provide spawning and nursery grounds for many marine fish and crustaceans, thereby supporting valuable fisheries both within their confines and offshore (see chapter 11). Estuaries provide a variety of products useful to people: fish, crabs and prawns, other

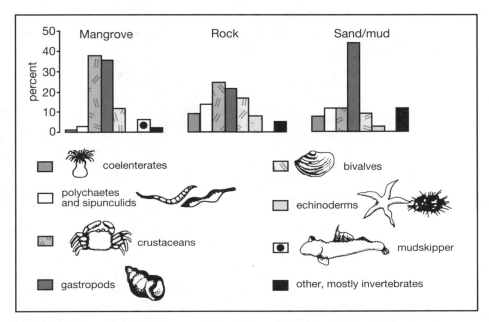

Figure 2.7. Aquatic animals recorded on three different shore habitats. Crustaceans and gastropod molluscs account for 74% of the total on the mangrove shore. (After Berry 1972.)

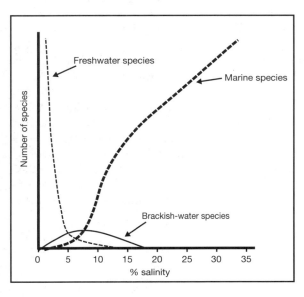

Figure 2.8. Changes in the number of freshwater, brackish-water and marine water species in different salinities. (After Barnes 1984.)

shellfish, marine mammals, reptiles and reptile skins, timber and wood products, fodder, oystershell, clay and sand for building, and water for industrial cooling (Salm and Clark 1984; Salm 1984a).

Seagrass Beds

The Indonesian term *rumput laut* refers collectively to both submerged marine flowering plants (seagrasses) and macroscopic benthic algae (seaweeds). The former are the dominant plants in seagrass beds, though seaweeds may also occur, including the nitrogen-fixing green algae *Halimeda* and *Neomeris*. Seagrass diversity in Indonesia is high (Soegiarto and Polunin 1980; Salm and Halim 1984). Common genera include *Halodule, Halophila, Enhalus* and *Thalassia*, all recorded off Kalimantan (table 2.2). Seagrasses grow on sandy substrates in shallow, well-lit waters.

Productivity in seagrass beds is high. Net primary productivity was measured at 16.4 t/ha for *Thalassodendron ciliatum* meadows off Sulawesi (Whitten et al. 1987b); this is higher than the rate for many lowland forests. Seagrasses and their epiphytic algae provide grazing for dugongs, sea turtles, certain species of fish and some species of sea urchins which possess cellulose-digesting bacteria. Only 5% of seagrass production is consumed directly; the rest enters the offshore food chain as decomposing material eaten by detritivores.

Many adult and juvenile stages of fish and invertebrates spend parts of their life cycles in seagrass beds, feeding or taking shelter. Important commercial and subsistence species include rabbit fish, goat fish and mullets (Polunin 1983; Burbridge and Maragos 1985), and edible invertebrates such as crabs, shrimps, clams and sea cucumbers. Other common invertebrates include sea stars, gastropods and sometimes corals. Some of the common seaweeds in the beds are also edible. Since seagrass beds are in shallow, nearshore, and often accessible areas, they are popular fishing grounds in Indonesia.

Seagrass beds serve several other useful functions, including stabilisation of offshore sand reservoirs and regular transport of carbonate sand to dynamic beach systems nearer to shore. They are a source of commercially valuable algae and may be particularly favourable for seaweed mariculture (Burbridge and Maragos 1985). As coastal ecosystems in shallow, well-lit waters, seagrass beds are susceptible to damage from increased sedimentation levels, dredging, and thermal and chemical pollution (Zieman 1975) as well as overexploitation.

CORAL REEFS

Coral reefs are tropical, shallow-water ecosystems that flourish best at temperatures between 25°C and 29°C. Since reef-building corals need sufficient light to grow well, significant reef development occurs only in water shallower than 30 m in clear seas and much shallower than that in turbid areas (Salm and Clark 1984). As a result, fringing reefs form close to shore, but this makes them particularly vulnerable to pollutants and silt. On most areas of the Sunda Shelf, and especially around the Kalimantan coasts where large rivers flow into the sea, the water is too muddy for growth of corals, and reefs are absent or poorly developed. Figure 2.10 shows the distribution of coral reefs around Kalimantan. The finest corals, and greatest species richness and productivity, are found on the fringing reefs around the Karimata islands off West Kalimantan and the Berau islands off East Kalimantan, both areas proposed as marine national parks. All three of the main forms of coral reef, fringing reefs, barrier reefs and atolls, are found off the east coast of Kalimantan (Salm and Halim 1984). Just a little farther north, in the proposed Semporna marine park on the Sabah/Kalimantan border, the shallow reef at Pulau Bodgaya plunges dra-

Table 2.2. Seagrass species recorded from Indonesia.

Family	Species	Distribution Kalimantan				
		1	2	3	4	5
Potamoge-	*Halodule uninervis*	X	X	X		X
tonaceae	*H. pinifolia*	X	X	X	X	X
	Cymodocea rotundata	X	X	X	X	X
	C. serrulata	X	?	X		X
	Syringodium isoetifolium	X	?	X		X
	Thalassodendron ciliatum			X	X	X
Hydrochari-	*Enhalus acoroides*	X	X	X	X	X
taceae	*Thalassia hemprichii*	X	X	X	X	
	Halophila ovalis	X	X	X	X	X
	H. minor	X	?	X	X	X
	H. decipiens		X		X	X
	H. spinulosa	X	X			X
	H. beccarii	?	X			
		Kalimantan total: 12 species				

1. Malaya, Singapore, Sumatra
2. Java, Bali, Kalimantan
3. Sulawesi, Philippines
4. Maluku, Nusa Tenggara
5. New Guinea

Sources: Soegiarto and Polunin 1980; den Hartog 1957

> **Box 2.1**. The dugong: mermaid or endangered marine mammal.
>
> Largest of the grazers in the seagrass meadows off south Kalimantan, dugongs *Dugon dugon* (fig. 2.9) are one of four surviving species in the order Sirenia. With their cousins, the manatees, they are the only vegetarian sea mammals. Found in shallow coastal areas in the southwestern Pacific and Indian oceans, dugongs are allegedly the inspiration for ancient mariners' tales of beautiful mermaids. One cannot help feeling, however, that the sailors must have been suffering badly from their long voyages to be captivated by large, lethargic sea cows, well padded with layers of fat. Dugongs, with their insulated bodies and fishlike tails, are totally adapted for a simple aquatic life.
>
> Dugongs are believed to share a common ancestor with elephants and, like elephants and horses, are nonruminant herbivores (Anderson 1981). They grow to 3 m in length and may weigh up to 300 kg. Dugongs are often caught accidentally in the nets of local fishermen, but are not freed as their oily meat and tusks fetch a good price at local markets. Sometimes they are kept for exhibition at sideshows, but such ventures are doomed to end in tragedy since the dugongs refuse to eat and gradually pine away and die. Often the unhappy creatures seem to be crying, and these salty secretions known as *air mata duyung* (dugong tears) are sold as an aphrodisiac.
>
> Dugongs graze on seagrasses and macro-algae, which they tear from the shallow sea bed, shake free of sand and devour completely. They dig into the sediment to obtain the rhizomes of the seagrasses since it is here that most carbohydrates are concentrated (Johnstone and Hudson 1981). The other large grazer of the seagrass beds, the green turtle *Chelonia mydas*, eats only the leaf blades, so there is little competition for food between the two species. Much can be learned from the study of dugongs since they can exploit a food source that humans cannot. They are able to convert the higher marine plants like seaweed and seagrass into protein fit for human consumption. Properly farmed, these marine herbivores might make a useful addition to the world's food supply. Their South American relatives, the riverine manatees, have already proved their usefulness at keeping river systems free from waterweeds such as water hyacinth (Anderson 1981).

matically, producing a spectacular underwater coral cliff.

In spite of their floral appearance, corals are not plants but colonies of cnidarians (Coelenterata), animals which collectively deposit calcium carbonate exoskeletons to build ornate formations. Each reef consists of colonies of living corals growing on the accumulated calcium carbonate skeletons of dead corals beneath (fig. 2.11). Coral polyps and colonies come in all shapes and sizes and many colours: convoluted brain corals, fast-growing staghorn *Acropora*, scarlet fan corals, and soft corals with feathery tentacles such as *Dendronephthya* and *Xenia* (Morton 1990). The polyps of most species are retracted by day, but at night their tentacles protrude to capture passing plankton.

Coral reefs are diverse ecosystems, supporting as many as three thousand species of plants and animals (Henrey 1982). Yet tropical seas are generally poor in life-supporting nutrients such as nitrates and phosphates (Salm and Clark 1984). The high productivity of coral reefs results principally

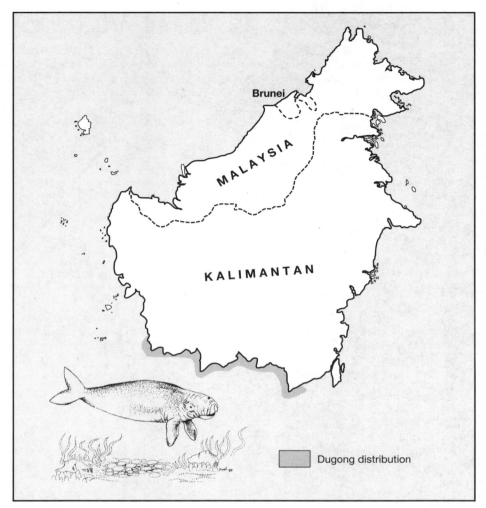

Figure. 2.9. Dugong distribution around Kalimantan.

from the import of nutrients by flowing water, efficient biological recycling, cling and high retention of nutrients. Within their tissues coral polyps harbour tiny one-celled plants, symbiotic algae called zooxanthellae. These green algae utilise waste products and carbon dioxide produced by the polyps, thus retaining vital nutrients such as phosphates (Muscatine 1973). Using nitrates, phosphates and carbon dioxide produced in the polyps, the zooxanthellae photosynthesise to produce oxygen and organic compounds that the coral polyp then absorbs.

Figure 2.10. Distribution of coral reefs off Kalimantan. (After Salm and Halim 1984.)

Flowing waters carry nutrients across the reefs, and they also receive usable nitrogen essential for photosynthesis from nitrogen-fixing blue-green algae which flourish on the adjacent reef flats. The algal beds are grazed by surgeon fish and parrot fish, which carry nutrients back to the reef with them (Wiebe et al. 1975). Similarly fixed nitrogen produced by bacteria in the sediments of seagrass beds is carried by feeding fishes back to the reefs. Often overlooked, the reef flats and seagrass beds are an important part of the coral ecosystem (Salm and Clark 1984).

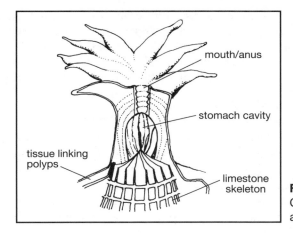

Figure 2.11. Cross section through a coral polyp.

Box 2.2. Simplified classification of the major invertebrates occurring in the benthos of coral reefs.

Phylum Porifera	- sponges
Phylum Cnidaria (Coelenterata)	
Class Hydrozoa	- sea firs or hydroids
Class Scyphozoa	
Order Rhizostomeae	- jellyfish
Class Cubozoa	
Order Cubomedusae	- sea wasps
Class Alcyonaria	
Order Alcyinacea	- soft corals
Order Gorgonacea	- sea whips/sea fans/sea feathers
Class Zoantharia	
Order Actinaria	- sea anemones
Order Scleractinia	- true corals
Class Ceriantipatharia	
Order Antipatharia	- black or thorn corals
Phylum Annelida	
Class Polychaeta	- marine worms
Phylum Mollusca	
Class Gastropoda	- limpets, snails, sea slugs
Class Bivalvia	- clams, cockles, mussels
Phylum Bryozoa	
Class Gymnolaemata	- marine bryozoans
Phylum Echinodermata	
Class Asteroidea	- starfishes
Class Ophiuroidea	- brittlestars and basketstars
Class Echinoidea	- sea urchins
Class Holothuroidea	- sea cucumbers
Class Crinoidea	- sea lilies and featherstars
Phylum Chordata	
Class Ascidiacea	- sea squirts or tunicates

The phylum Chordata is the one to which all animals with backbones also belong.

Source: Barnes 1984

Living corals grow as a veneer on a porous limestone base which forms a structure for colonisation by other animals. After corals, sponges are probably the most conspicuous reef colonisers. More than a hundred species have been recorded on the Semporna reefs, Sabah. The great number of holes in a reef provide abundant shelter for fish and invertebrates and are important fish nurseries. Lobsters, octopus and fragile fanworms lurk in the crevices. Some rock holes provide shelter to different fish by day and night (Smith and Tyler 1972; Collette and Talbot 1972). Snappers, grunt fish and moray eels shelter in crevices during the day but move away to feed at night, when their homes are taken over by "tenants" who do their foraging during the daylight hours. In effect, the reef supports two fish communities: a daytime community and a nocturnal one. In this respect coral reefs are similar to tropical rainforests, which also have high species richness and distinct day and night communities of animals (see chapter 4).

There is a greater diversity of fish species on coral reefs than anywhere else in the sea, with as many as two hundred species per hectare (Salm and Clark 1984). Reef fish display an amazing array of shapes, patterns, colours and habits. Most of the herbivores and invertebrate feeders are diurnal (White 1987). Colourful butterfly fish, rabbit fish and trigger fish prowl among the branching coral stems, browsing on algal fronds, while parrot fish graze the coral polyps. Frilled scorpion fish, bat fish, puffed-up box fish and orange-and-white banded clownfish are all part of the community of brightly coloured reef fishes. Many of these fish are caught for food or for sale as ornamental fish. There are several useful field guides to coral reef fishes of Borneo and the rest of Indonesia (Carcasson 1977; Schroeder 1980; White 1987).

Despite their high diversity, coral reefs show little endemism. Most reef species reproduce by dispersing thousands of eggs or larvae which float freely in the plankton. Dispersed on ocean currents, the tiny larvae are carried to new areas so that many species have a wide distribution in tropical seas (Salm and Clark 1984).

Gause's Theory of Competitive Exclusion states that species can coexist only if each is exploiting a different set of limiting resources, or **niche**. Coral reefs provide a wide range of niches occupied by representatives of many animal groups: sessile detritus feeders like the sponges; rock-boring clams; crawling predators such as the sea slugs and brittle stars; mobile shrimps and lobsters; clinging sea anemones; and a host of fishes (Morton 1990). Giant clams *Tridacna* gape on the reef ledges, filtering detritus from the sea water. Contrary to popular myth, these clams close their shells slowly enough that no underwater swimmer need fear that his foot will be caught in the closing gape. The clam's feeding behaviour is remarkable; it is both a filter feeder and a "farmer" of one-celled green algae which live in its fleshy siphon (Mash 1975).

Box 2.3. Feeding guilds of coral reef fishes.

I. **Pelagic and midwater guilds**
   A. **Strainers and siphoners**
      herrings Clupeidae, whale-shark *Rhinchodon*, manta ray *Manta birostris*, sea horses and pipefish Syngnathidae
   B. **Fast predators**
      sharks Carcharinidae and Isuridae
      mackerels Scombridae
      tuna and tunnies Thunnidae
      horse mackerels Carangidae

II. **Benthic guilds**
   A. **Omnivores and mixed carnivores**
      coral browsers and grazers
         demoiselles Pomacentridae
      coral probers and fossickers
         butterfly and angel fishes Chaetodontidae
         file fishes Monacanthidae
         some wrasses Labridae
      coral crunchers
         trigger fishes Balistidae
      bottom fossickers
         wrasses Labridae
         trunk fish Ostracionidae
         goat fish or red mullet Mullidae
         trigger fishes Balistidae
         spiny puffers Didontidae
         puffers and globe fish Tetrodontidae
      fine foragers
         gobies Gobiidae
         blennies Blenniidae

   B. **Carnivores**
      crushing carnivores
         rays Myliobatidae
      roving carnivores
         snappers and sea perch Lutjanidae
      lurking carnivores
         groupers Serranidae
         scorpion fish Scorpaenidae
         moray eels Muraenidae

   C. **Herbivores**
      algal scrapers
         parrot fish Scaridae
      bottom grazing herbivores
         surgeon fish Acanthuridae
      sand-sifting herbivores
         grey mullets Mugilidae
         sweepers Eliotridae

Source: Morton 1990

The food web of the coral reef is just as complex as that found in the rainforests on land. The primary source of energy is sunlight, which is fixed during photosynthesis by the tiny green algae of the reef and reef flats. Numerous animals graze on seaweeds, seagrasses and algal films; these herbivores include small shrimps and molluscs, sea hares and sea urchins, as well as fish. Yet other animals feed on microscopic plant cells which float free in the waters that swirl around the reef. Tubeworms, bivalves, sea lilies and sea squirts all have special tracts of whiplike cilia which create currents to wash detritus and plankton toward the animal's mouth. Fanworms construct long tubes which project from, and help to strengthen and cement, the reef. Other filter feeders include bivalves like the oysters and mussels, and some of the sea snails.

These herbivores and detritivores are eaten by larger animals which may in turn become prey for other carnivores. Predators abound on the coral reefs, and include the individual corals themselves and their close relatives the sea anemones, which have special stinging cells (nematocysts) for catching prey. Each nematocyst consists of a tightly coiled and sometimes barbed line. When an invertebrate bumps against an outstretched tentacle, the nematocysts are triggered, and threads shoot out to harpoon and bind the prey and draw it into the mouth (fig. 2.12). These stinging cells, however, do not protect the corals from predators such as carnivorous starfish (Henrey 1982).

Other predators on the reef include the octopus, the spotted cowries, and cone shells which have become collectors' items because of their beautifully patterned shells. Some cone shells have an extremely venomous sting, lethal even to humans. That a cone shell, which preys on small fish, produces sufficient poison to kill a man seems astonishing. Different poisons vary in their toxicity to different organisms, and the dose of neurotoxin required to paralyse a fish can also be lethal to people (White 1987). In fact, several marine creatures are venomous and best avoided by swimmers. Lumpy stonefish lurk in the shallows off the beach, spiny sea urchins group on the sandy bottom, predatory jellyfish float with the currents, and even stinging corals can cause considerable pain.

In the fascinating undersea world of the coral reef there are many curious animal relationships. Reef fish and many of the larger, deep-sea fish are serviced by small "cleaner" wrasse *Labroides dimidiatus*, which clean their clients' gills, teeth and skin of parasites, algae and debris. Every small reef seems to have a cleaner fish with a regular cleaning station where clients queue to await their turn. Experiments have shown that if the resident cleaner fish is removed there is a drastic reduction in the numbers of fish on the reef, and those that do remain are often less healthy (Slobodkin and Fishelson 1974). Another smaller fish, the blenny *Aspidontus tractus*, has similar markings and behaviour, but this mimic bites the flesh of larger fish rather than cleaning them.

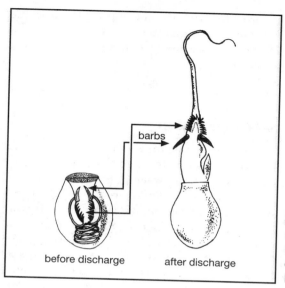

Figure 2.12.
Cnidarian nematocyst before and after firing of the harpoon. (After Henrey 1982.)

Clownfish (Amphiprionidae) lessen their chance of falling prey to other fish by sheltering among the stinging tentacles of large sea anemones. The clownfish produces a coat of mucus to protect itself from the anemone's venom (Brooks and Mariscal 1984). In this symbiotic relationship the clownfish gains protection, and the anemone benefits from scraps of food dropped by the fish. The fish may also act as a lure, attracting prey to the anemone. Similarly, the armoury of the black sea urchin *Echinothrix diadema* is exploited by shrimpfish (Centriscidae), which hover head-down between the long poisonous spines (White 1987). The shrimpfish's long, thin body and camouflaging stripe of contrasting colour make it difficult to detect among the spines.

The internal space (cloaca) at the end of the digestive tract of the sea cucumber is well supplied with circulating, oxygen-rich water and provides a secure home for smaller commensal animals, such as small fish, pea crabs and polychaete worms. One resident, the pearlfish, enters the sea cucumber tail-first through the anus but exits head-first in case of danger. Since the sea cucumber breathes through its anus, the aperture is never closed for long. A pearlfish impatient to enter merely nudges his host and, as the anus opens, the fish reverses quickly inside (Mash 1975).

With their complex plant and animal inter-relationships, coral reefs are fascinating and beautiful places (Mash 1975). Scientific studies of species-rich coral reefs provide information on community ecology, niche separation and species distributions (Connell 1978). Coral reefs are also an extremely valuable resource, providing rich fish harvests for local people and a solid substrate for bottom-living organisms, many of commercial value. Reefs also act as barriers between sea and land, protecting waveswept shores from coastal erosion (see chapter 11).

Mangrove Forest

Mangrove (or mangal) is the collective name for the tree vegetation that colonises muddy shores within the tidal zone from extreme high water to low tide level. Mangrove occurs only on shores where the vigour of the surf is broken by sand bars, coral reefs or islands. There are three main types of mangrove ecosystem: the coastal/deltaic form, the estuarine/lagoonal form and the island form. All three are represented in Kalimantan and elsewhere in Borneo (Ong 1982). Mangrove is most extensive on the deltas of major rivers, with particularly extensive stands at the mouths of the Kapuas, Mahakam and Sebuku. It also occurs in small bays and lagoons and can penetrate far inland along rivers; for example, mangrove extends 240 km up the Kapuas in West Kalimantan (Whitmore 1984a).

Mangrove forests fringe most of the Kalimantan coastline (fig. 2.13). Estimates for the extent of remaining mangrove habitat in Kalimantan vary from 275,000 ha (Burbridge and Koesiobono 1980), through 383,000 ha (FAO 1985) to 900,000 ha (MacKinnon and Artha 1981); the last figure includes large areas of nipa swamps as well as mangrove forests. The greatest extent of remaining undisturbed mangrove habitat is found in East Kalimantan, but much of this is allocated for logging or conversion to fish ponds *tambak* (see chapter 11). The original, remaining and protected areas of mangrove forests in Kalimantan are listed in table 2.3. Elsewhere in Borneo fine mangrove stands are found in the Rajang delta of Sarawak (Chan et al. 1985) and in Brunei Bay.

The mangrove association consists of a number of species tolerant of the saltwater-and-mud environment, together with the brackish-water forest at its inland edge, which consists of almost pure stands of nipa palm *Nypa fruticans*. The species composition and ecology of mangrove ecosystems is well described by van Steenis (1958b), Whitmore (1984a) and Whitten and associates (Whitten et al. 1987b). There have been few studies of mangrove ecosystems in Kalimantan, but the plant community has been studied in some detail in Sarawak (Chai 1975a, 1982), and elsewhere in Indonesia and Southeast Asia (Kartawinata et al. 1979; Whitmore 1984a). Chai (1975b) provides a useful key for mangrove trees and shrubs.

Mangrove shelters and feeds, directly or indirectly, a large mixed community of animals, including shorebirds and many marine organisms.

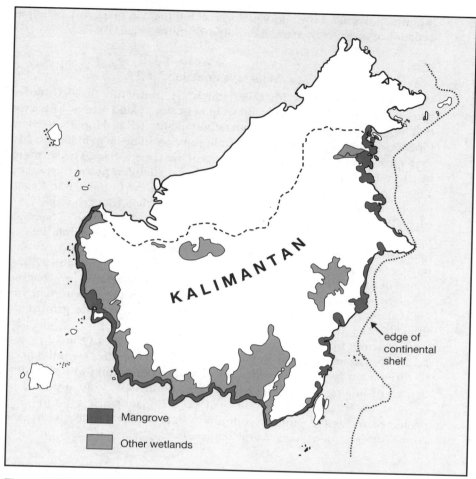

Figure 2.13. Mangrove habitat and other Kalimantan wetlands. (After Salm and Halim 1984.)

Mangrove forests have a rich fauna of large crustaceans and molluscs (Kartawinata et al. 1979) and are also important spawning grounds and nurseries for prawns and many pelagic fish species of commercial importance. Some of Borneo's interesting vertebrates associated with mangrove forests are the endemic *bekantan* or proboscis monkey *Nasalis larvatus*, the silvered langur *Presbytis cristata*, monitor lizards *Varanus salvator*, crocodiles *Crocodylus porosus*, the mangrove pitta *Pitta megarhyncha* and the mangrove blue flycatcher *Cyornis rufigastra*. Mangroves also provide important feeding grounds and high tide roosts for flocks of migrant Palaearctic waders. The

Indonesian Wetland Inventory lists several Kalimantan mangrove forests as wetlands of major conservation importance (Silvius et al. 1987).

Mangrove Zonation

Mangrove forests, especially those which are frequently flooded, differ markedly from dryland forests and from most swampland forests, with a virtual absence of climbing and understorey plants (Ding Hou 1958). The forest is one-storeyed, with trees generally not exceeding 25 m in height, and undergrowth is sparse, except for recruits of the same species. This seems to be a result of regular tidal inundations rather than different soil tolerances. There are about fifty species of mangrove trees in the Indomalayan Realm with more than thirty recorded for Indonesia and 26 for Borneo (table 2.4). Chai (1975a) lists 54 species of trees, shrubs and climbers found in Sarawak mangrove forests. While trees and young seedlings comprise the bulk of mangrove vegetation, a few lianas may occur (for example, *Derris*), and ferns, grasses and sedges sometimes colonise openings (Chapman 1977). A fringe of the holly-leaved shrub *Acanthus ilicifolius* sometimes grows along the edge of estuarine mangroves. Where mangrove has been disturbed, the giant mangrove fern *Acrostichum aureum* forms dense undergrowth, so thick that the mangrove trees cannot regenerate.

Mangroves frequently show a marked zonation of species (figs. 2.14 and 2.15). The occurrence and abundance of individual species are influenced by three main factors: frequency and duration of flooding with seawater; degree of mixing with freshwater at river mouths and concentration of brackish water; and the consistency of the soil (sandy or clay). On open protected coasts the pioneer community is generally dominated by *Avicennia marina*, *Avicennia alba* or *Sonneratia alba* (Chai 1975a), with *Avi-*

Table 2.3. Mangrove forests in Kalimantan.

	Mangrove and nipa[1]			Mangrove only[2]		
	Original area ha	Remaining ha	Protected ha	Proposed ha	Remaining ha	Timber concessions
West Kalimantan	425,000	60,000	-	26,000	40,000	39,500
Central Kalimantan	40,000	20,000	10,000	10,000	10,000	?
South Kalimantan	165,000	90,000	4,000*	80,000	66,650	67,650
East Kalimantan	950,000	750,000	1,000	170,000	266,800	143,000
Total	1,580,000	920,000	15,000	286,000	183,450	50,150

Mangrove is one of the most threatened habitats on Kalimantan. Areas granted as timber concessions overlap with many areas proposed for conservation.

* The Pleihari Tanah Laut Reserve, South Kalimantan, which includes mangroves, was degazetted in 1992 to become an industrial plantation.

Sources: [1]*MacKinnon and Artha 1981*; [2]*Bina Program 1982 in FAO 1985 cennia* occurring on firm sandy soil and *Sonneratia* associated with soft muds. Behind these colonisers and in small creeks, the plant community is dominated by *Rhizophora* trees, supported on characteristic branching stilt

Table 2.4. Mangrove tree and shrub species recorded from Borneo and Indonesia.

Family	Species	SW 1	Kal 2	Indonesia 3	4	5	6
Species exclusive to mangrove							
Avicenniaceae	Avicennia alba	X	X	X	X	X	X
	A. marina	X	X	X	X	X	X
	A. officinalis	X	X	X	X	X	X
Bombacaceae	Camptostemon schultzii					X	X
Combretaceae	Lumnitzera littorea	X	X	X	X	X	X
	L. racemosa		X	X	X	X	X
Euphorbiaceae	Excoecaria agallocha	X	X	X	X	X	X
Flacourtiaceae	Scolopia macrophylla	X	X	X			
Leguminosae	Cynometra ramiflora	X				X	X
Meliaceae	Xylocarpus granatum	X	X		X	X	X
	X. moluccensis		X	X	X	X	X
Myrsinaceae	Aegiceras corniculatum	X	X	X	X		X
Myrtaceae	Osbornia octodonta		X		X		X
Palmae	Nypa fruticans	X	X	X	X		X
Plumbaginaceae	Aegialitis annulata		?	?			
	A. rotundifolia					X	X
Rhizophoraceae	Bruguiera cylindrica	X	X	X	X		X
	B. exaristata						X
	B. gymnorrhiza	X	X	X	X	X	X
	B. hainesii						X
	B. parviflora	X		X	X	X	X
	B. sexangula	X	X	X	X	X	X
	Ceriops decandra	X	X	X	X		X
	C. tagal	X	X	X	X	X	X
	Kandelia candel	X		X			
	Rhizophora apiculata	X	X	X	X	X	X
	R. mucronata	X	X	X	X		X
	R. stylosa		X				X
Rubiaceae	Scyphiphora hydrophyllacea		X	X	X		X
Rutaceae	Paramignya ahgulata	X	X	X			
Sonnerataceae	Sonneratia alba	X	X	X	X	X	X
	S. caseolaris	X	X	X	X	X	X
	S. ovata	X	X	X	X	X	X
Sterculiaceae	Heritiera littoralis	X	X	X	X	X	X
Nonexclusive species							
Apocynaceae	Cerbera manghas	X	X	X	X	X	X
Bignoniaceae	Dolichandrone spathacea	X			X	X	
Lecythidaceae	Barringtonia acutangula	X	X	X	X		X
	B. racemosa	X	X	X	X		X
Malvaceae	Thespesia populnea	X	X	X	X	X	X
	Hibiscus tiliaceus	X	X	X	X	X	X
Palmae	Oncosperma tigillarium	X	X	X	X		
Tiliaceae	Brownlowia argentata	X	X			X	X

1. Sarawak 2. Kalimantan 3. Sumatra, Java 4. Sulawesi 5. Maluku, Nusa Tenggara 6. New Guinea

Sources: Chai 1975a, 1975b; Soegiarto and Polunin 1980; Saenger et al. 1983

Box 2.4. Mangrove species in Kalimantan.

RHIZOPHORACEAE
- *Rhizophora* — Medium to large trees with wide oval leaves. No pneumatophores, but root struts and descending aerial roots. Cigar-shaped hypocotyl. Flower: 4 sepals, 4 petals.
- *Bruguiera* — Mainly tall trees, with slenderer oval leaves. Long hypocotyl. Pneumatophores as kneelike hoops. No aerial roots. Flowers: up to 11 sepals and petals.
- *Ceriops* — Large shrubs, leaves round or notch-tipped. Small hypocotyls, clustered and up-pointing. Flowers with 5 petals, fringed and lobed.

AVICENNIACEAE
- *Avicennia* — Trees or shrubs, with leaves rounded or lanceolate, shining above, tomentose below. Pneumatophores like thin pencils. Fruit roundish, velvety coated; embryo with folded green cotyledons. Heads of small flowers.

SONNERATIACEAE
- *Sonneratia* — Bushy trees; leaves thick, dull green, round-tipped or notched. Pneumatophores strong conical stumps. Flowers with numerous long, white stamens. Fruits hard, green, depressed, with single style.

COMBRETACEAE
- *Lumnitzera* — Slender trees; leaves thick, bright green both sides, apex notched. Pneumatophores hoop-shaped. Flowers small, in clusters, red or white. Fruits small, flask-shaped.

MELIACEAE
- *Xylocarpus* — Large shrubs or trees; leaves compound with large oval leaflets. Roots with tall, thin flanges (pneumatophores). Flowers small, white, clustered. Fruits big and spherical, thick-skinned with large "nuts" dovetailed together.

MYRSINACEAE
- *Aegiceras* — Large bushes or trees, with leaves oval, edges slightly incurled. No pneumatophores. Flowers white, clustered in leaf axils. Fruits slender, crescentic.

MYRTACEAE
- *Osbornia* — Bushy shrubs; small leaves dull green with reddish stalk, aromatic glands. No pneumatophores. Flowers sessile with many stamens, calyx 8-lobed; fruits small and dry.

RUBIACEAE
- *Scyphiphora* — Bushy shrubs, with shiny green, oval, opposite leaves, sticky when young. No pneumatophores. Flowers white, clustered in axils. Fruit green when young, then drying out.

PLUMBAGINACEAE
- *Aegialitis* — Shrubs or small trees, with swollen, conical stem-bases, and surface-spreading roots. Leaves with white salt glands. Flowers small, blue, tubular with 5 petals, in terminal inflorescences.

EUPHORBIACEAE
- *Exocoecaria* — Small trees, with reddish, stripey bark; leaves pointed oval, glossy, keeled ("milky mangrove" with poison latex). Flowers in male and female catkins on separate trees. Fruits 3-locular capsules.

PALMAE
- *Nypa* — Small, erect palms, with trunk underground, horizontal. Pinnate leaves 3-9 m long. Fruit a spherical clump of carpels.

Source: Morton 1990

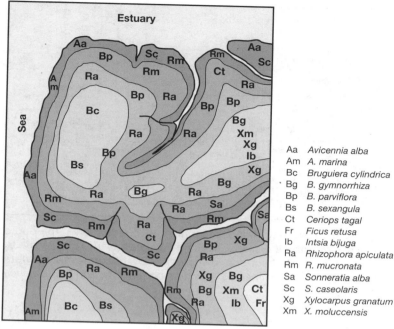

Figure 2.14. Typical distribution of mangrove tree species in undisturbed habitat near the mouth of a large river. Actual distribution of mangrove species depends on water salinity. (After Watson 1928.)

roots; *R. mucronata* gives way to *R. apiculata* farther up the shore (fig. 2.16). *Bruguiera gymnorrhiza* and *Xylocarpus granatum* are commonly associated with the *Rhizophora*, and in older stands there may be an understorey of *Ceriops tagal* (Soegiarto and Polunin 1980). Drier areas, higher up the shore and on hard clay soils, are colonised by *Bruguiera* trees, their knee roots covered only by the highest tides. A well-developed *Bruguiera cylindrica* association may occur in actively accreting forest, while *Bruguiera gymnorrhiza* is found on the landward side of the mangrove (Soegiarto and Polunin 1980). Along tidal river banks *Sonnerata caseolaris* grows as far as salt water penetrates, and this 20 m tree can survive in fresh water. Its conelike aerial roots project from the mud and are so light that fishermen use them as fish floats (Polunin 1988).

Stands of nipa palm flourish inland, along tidal creeks and estuaries. Nipa grows between the tidemarks where there is a considerable mixture with fresh water. The palm's thick stem creeps over the surface of the mud, and the leaves may reach heights of 10 m. Nipa fruits are dispersed by water, and the floating yellow fruit, often with a shoot emerging, can survive for many months before being washed up on a mudflat and taking root

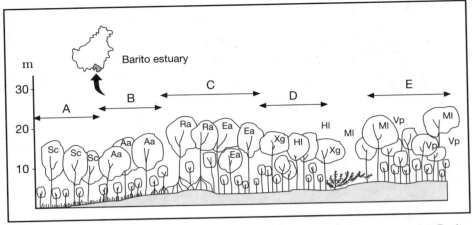

Figure 2.15. Diagram of community changes in a mangrove formation in the Barito estuary, from the beach towards the inland. **A.** *Sonneratia caseolaris* community; **B.** *Avicennia alba* community; **C.** *Excoecaria agallocha - Rhizophora apiculata* community; **D.** *Heriteria littoralis - Xylocarpus granatum* community; and **E.** *Melaleuca leucadendron* community.

Source: Mirmanto et al. 1989

Box 2.5. Mangrove forests in the Barito River estuary.

The mangrove forests in the Barito estuary are extensive, extending westwards along the coast for more than 30 km and inland along the river for about 20 km. The mangrove ecosystem on the Barito includes a spectrum of vegetation types from undisturbed primary forest to secondary communities dominated by herbs and shrubs in fallow rice fields. Inland, the Barito mangroves are bordered by peat and freshwater swamp forests which have been converted in most places to rice fields.

Of the 22 tree species identified in sampling plots, only seven were true mangrove species. This indicates the homogeneity and low species richness of the mangrove community. The Barito mangroves can be zoned into five community types, depending on soil and water conditions.

The *Sonneratia caseolaris* community type was always found close to the river mouth and extended inland up to the area not affected by tides. Near the river mouth *S. caseolaris* usually formed pure stands of large trees, 10-25 m high, with a dense canopy and good regeneration beneath. Upriver, *Sonneratia* occurred on shallow to deep sediments that were permanently flooded, and gradually became mixed with an increasing number of species.

The *Avicennia alba* community occurred on flooded habitats with high salinity and with shallow to deep sediments of sandy mud. Trees of *A. alba* with dense crowns and heights of 10-20 m occurred at high density in almost pure stands. A few saplings and seedlings of *Rhizophora apiculata* occurred throughout the *Avicennia* stand.

Further inland was the *Excoecaria agallocha-Rhizophora apiculata* community which extended inland about 300-500 m from the beach, where soils were drier and flooded only during high tides; *R. apiculata* was the dominant tree here. In disturbed areas the mangrove fern *Acrostichum aureum* and holly-leaved *Acanthus ilicifolius* could become dominant and prevent regeneration of the mangrove trees.

The communities comprising the *Heriteria littoralis-Xylocarpus granatum* community occurred farther inland and formed the transition between mangrove and terrestrial communities; this community type showed the greatest floristic richness. Still farther inland, almost pure stands of gelam *Melaleuca leucadendron* had developed in areas of formerly cleared peat swamp forests.

Source: Mirmanto et al. 1989

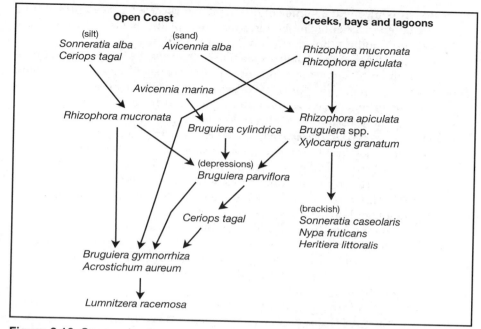

Figure 2.16. Succession in mangrove forests. (After Walsh 1974.)

(Polunin 1988).

The fringe of coastal mangrove may be only a few trees deep or may extend many kilometres inland. In established stands the mangrove root system slows the flow of silt-laden water and allows particles to settle, in the process building up the sediment on the landward side of the mangrove swamp. This succession allows mangrove "pioneers" to move steadily seawards, accelerates shore development and ensures coastal stability. The march of mangroves towards the sea can exceed 100 metres per year. Palembang, a thriving port when Marco Polo visited Sumatra in the thirteenth century, is today 50 kilometres inland (Maltby 1986). Mangrove and mangrove/nipa associations act as buffers to wave action, and protect stabilised muds and river banks from the effects of tidal currents. This stabilising function is so important that in many cases mangrove habitats should be protected for this function alone.

Mangrove Ecology

Many mangrove trees are only facultative halotypes in that they can survive and regenerate in freshwater conditions (van Steenis 1958a), but they flourish only in tidal saline conditions. Their success here can probably be attributed to lack of competition from other plants that are not tolerant of salt and flooding (Soegiarto and Polunin 1980).

The plants of the mangrove habitat have evolved several adaptations to allow them to survive in this saline and waterlogged environment. Most trees of the mangrove forest have developed peculiar root systems to allow gaseous exchange above the waterlogged and oxygen-poor soils (Mann 1982). These breathing roots are known as **pneumatophores** (fig. 2.17). The stilt roots of *Rhizophora* may also be effective in preventing the growth of seedlings too close to the parent tree. *Sonneratia* and *Avicennia* have horizontal cable roots held in place by anchor roots; the spiked pneumatophores grow upwards from the cable roots and put out new nutritive roots. In *Bruguiera* the cable roots loop in and out of the soil, and the "knee" roots act as pneumatophores. *Ceriops* has no special root adaptations, but its bark has many adaptations for gaseous exchange. Apart from enabling gaseous exchange, pneumatophores play an important role in protecting young trees and germinating seeds from wave action, trapping sediment and organic material, and providing a habitat for burrowing crabs which help to aerate the soil (Chambers 1980).

Mangrove roots take in abundant salt with the water they absorb. As the water evaporates from the leaves, the salt is left behind. The more seawater is absorbed, the more salt accumulates and the more water is needed by the plant. In this sense the environment is physiologically dry. This is probably why mangrove plants, like those of desert regions and temperate salt marshes, have thick fleshy leaves which reduce water loss (Polunin 1988). A few plants such as the holly-leaved *Acanthus* and mangrove *Avicennia* also have special glands on the leaves which excrete concentrated salt solution; during extended dry periods whitish patches of salt can be seen on the leaves (Polunin 1988).

Some mangrove tree species flower when very small, and several have unusual means of seed dispersal to ensure rapid establishment of the seedlings on the inundated mud banks. Mangrove stands may show some flowering and fruiting in all months of the year, but most species flower during the dry season and drop ripe fruits during periods of peak rainfall (Christensen and Wium-Anderson 1977). The feathery flowers of *Sonneratia* are pollinated by pollen-eating bats, while the Rhizophoraceae have a range of mechanisms for effecting pollination. The anthers of *Bruguiera* open explosively when the flowers are visited by sunbirds or by butterflies and other insects in search of nectar. *Ceriops tagal* is also pollinated by moths. *Rhizophora* species set only a few large and conspicuous flowers each day and are visited by the wide-ranging, cave-dwelling bat *Eonycteris*; other common pollinators are the bees *Apis* and *Trogona* (Appanah 1990; Tomlinson et al. 1979). Sunbirds are sometimes seen visiting *Rhizophora* trees, but this is to lick the sweet exudate from leaf buds or young flowers that have suffered

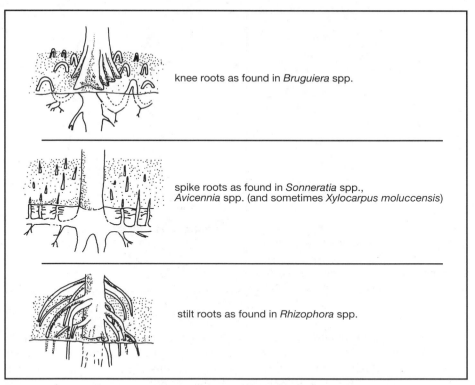

Figure 2.17. Different types of roots in mangrove trees.

insect damage (Wium-Anderson 1981).

Most species of mangrove trees exhibit viviparous reproduction. In most of the Rhizophoraceae, including *Rhizophora* and *Bruguiera*, the fruits ripen on the parent tree, and the seeds germinate inside the hanging fruit. The embryonic root of the seedling grows down through the wall of the fruit to form a spearlike plantlet, which is the resting and dispersal phase of the plant (fig. 2.18). When the seedling finally falls, it may stick in the mud like a javelin, put out a tap root and develop rapidly, or it may float and be carried by the tide away from the parent tree to lodge and grow elsewhere. Predation of these young seedlings is high. Experiments in Australian mangroves show that about 75% of all the propagules of five mangrove trees were consumed by predators, primarily grapsid crabs (Smith 1986). To minimise predation some species have developed high tannin levels, which make the propagules unpalatable to most herbivores.

Mangrove forests are highly productive natural ecosystems. The net primary productivity of the mangrove forest can be assessed from the rate at

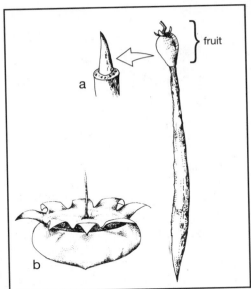

Figure 2.18.
a) Propagule of *Rhizophora mucronata* showing the root and top of the seedling after it is detached from the parent tree b) fruit of *Sonneratia alba*.

which litter is produced. Litter production (leaves, twigs, fruit and flowers) in Peninsular Malaysia and New Guinea is about 14 t/ha/year (Sasekumar and Loi 1983; Leach and Burgin 1985), and similar figures are likely for Kalimantan. These figures are similar or higher than those for lowland rainforest and support the contention that mangroves grow, reproduce and die relatively quickly (Jimenez et al. 1985).

Although mangrove forests are highly productive, only about 7% of live leaves are eaten by herbivores (Johnstone 1981). Most of the mangrove forest production enters the energy system as detritus or dead organic matter (fig. 2.19). As it decomposes, the detritus becomes rich in nitrogen and phosphorus because of the fungi, bacteria and algae growing on and within it; it is therefore an important food source for many **detritivore** (detritus-eating) animals such as zooplankton, other small invertebrates, and prawns, crabs and fish, which may depend on mangrove litter fall for their food. These detritivores are eaten in turn by carnivores, including people. Thus the mangrove litter and detritus play an extremely important role in the productivity of the mangrove ecosystem as a whole and of other coastal ecosystems (Lugo and Snedaker 1974; Ong et al. 1980; Saenger et al. 1983; Mann 1982).

In the Sungei Merbok mangrove forests in Malaysia, 40% to 90% of the litter disappeared within twenty days, eaten or buried in the mud by crabs. The importance of crabs in the decomposition process in the mangrove association is illustrated by the fact that when litter was protected from crabs,

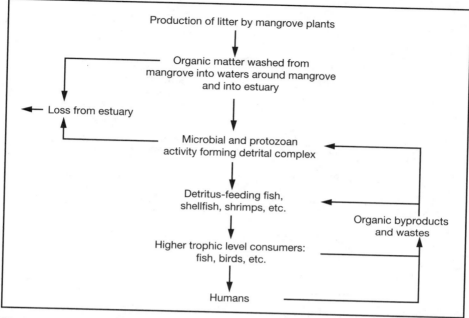

Figure 2.19. Major pathways of energy flow in a mangrove-fringed estuary. (After Saenger et al. 1983.)

decomposition took four to six months (Ong et al. 1980). Similar results were obtained at a study site in West Kalimantan (PSL-UNTAN 1989).

The high productivity of mangroves, and the physical structure and shading they provide, form a habitat for many organisms, several harvested for local consumption and export (see chapter 11). The influence of mangroves extends far beyond the coastal mangrove belt as they export nutrients on the tides to other coastal ecosystems (fig. 2.20). Probably most of the microfauna and macrofauna in the mangroves and surrounding coastal areas are dependent on litter production from mangrove forests (Ong et al. 1980). Carbon from mangrove trees has been found in the tissues of commercially important bivalves such as cockle *Anadara granosa*, oyster *Crassostrea*, shrimps *Acetes* (used to make belachan paste), crabs such as *Scylla serrata*, and many fish (Rodelli et al. 1984). Mangrove litter is the basis of the food chain of many commercial fish species including mullet *Mugil*, milkfish *Chanos* and barramundi or giant perch *Lates* (MacNae 1968; Moore 1982; Polunin 1983).

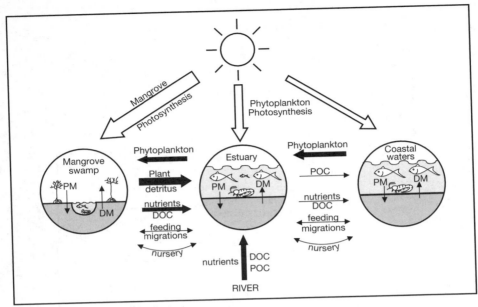

Figure 2.20. Processes of energy flow and interchange among mangrove swamps, estuaries and coastal waters. POC = particulate organic carbon; DOC = dissolved organic carbon; PM = particulate materials; DM = dissolved materials.

Mangrove Fauna

Many classes of the animal kingdom are represented in the mangrove swamps, and there is much to fascinate the biologist who decides to brave the mosquitoes and stinking mud of this fringe habitat where land and sea meet.

At high tide, banded archer fish *Toxotes jaculator* swim among the tangled roots, hunting for insects and grubs. Although the archer fish cannot survive out of water, it is able to "shoot down" insects from above the water surface by spitting droplets of water at its prey (fig. 2.21). So great is its skill that a fully grown fish can shoot down an insect from 1.5 metres. The fish reduces refraction of light by positioning itself almost vertically below its prey. Its aim is remarkably accurate. If the insect is only a few centimetres above the water, the archer fish will jump to snatch its prey rather than shoot it down (O'Toole 1987).

As the tide recedes, vast expanses of mudflats and mangrove roots are left uncovered. These mudflats support a rich fauna of large crustaceans and molluscs (Kartawinata et al. 1979). Barnacles encrust the mangrove trunks, with species showing vertical zonation according to their tolerance to desiccation (Whitten et al. 1987b). Small, colourful fiddler crabs *Uca* spp. scuttle

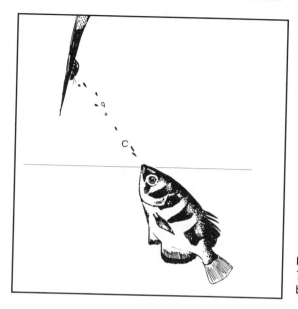

Figure 2.21. An archer fish *Toxotes jaculator* shooting a beetle off an over-hanging leaf.

over the mud, feeding on algae and detritus left by the receding tide. Male fiddler crabs march sideways from their holes, beckoning with one huge claw. Either the right or left claw is grossly enlarged and brightly coloured; it is waved vigorously in territorial disputes between males or as a signal to attract a female. The more mundane task of food gathering is left to the other claw, which is small and spoon-shaped to scrape algae from the mud surface (MacKinnon and MacKinnon 1974).

When the tide comes in, fiddler crabs withdraw into their holes, blocking the entrance with balls of mud and sand to retain a bubble of air to allow them to breathe beneath the waters. The ghost crab *Ocypode* adopts a different strategy, building an igloo of mudballs around itself, then burrowing vertically down into the mud (fig. 2.22). It plasters the excavated sand into the roof of its shelter and burrows even deeper. The finished burrow protects the crab's vital bubble of air until the crab can emerge when the tide ebbs (MacKinnon 1986).

One of the strangest creatures of the mudflats, the king crab or horseshoe crab, is not a true crab at all. It is a living fossil, an ancient marine relative of the spiders and scorpions, apparently unchanged in appearance from its ancestors which inhabited the oceans during the Paleozoic over 200 million years ago. At full moon and high tide, king crabs come up onto the muddy tidal flats to breed. The female carries a bundle of two to three

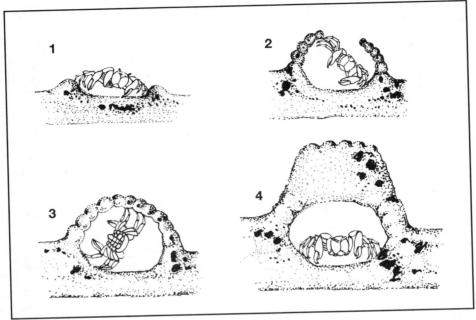

Figure 2.22. Ghost crabs *Ocypode* build an igloo of mud balls to escape the incoming tide.

thousand eggs between her front legs. She excavates a depression in the mud, and the male, clinging to her back, fertilises the eggs as they are deposited. A month later, again at high tide, the eggs hatch, and the small larvae are washed away on the tide to become part of the floating marine zooplankton.

Just as the mangrove vegetation changes in species composition with distance from the sea, so does the fauna (Berry 1972; Budiman 1985). This is well illustrated by the molluscs. Molluscs are common in mangrove swamps and accounted for 60% of the infauna biomass in an Australian mangrove (Wells 1984). Snails such as *Littoraria scabra*, *Neritina violacea* and the large, conical *Telescopium telescopium* crawl over the mudflats and climb the stilted roots of the mangrove trees (Sabar et al. 1979). Further inland air-breathing pulmonate snails occur (Budiman and Darnaedi 1982). The sea hare, a snail without a shell, rests on arched mangrove roots, the only sea slug to spend part of its life out of water. Molluscs at the seaward edge of the mangrove are mainly gastropods and bivalves, but further inland these carnivores and filter feeders are largely replaced by snails that graze algae growing on the mud surface. All of these snails can breathe efficiently in air, and the pulmonate snails, such as *Ellobium*, have

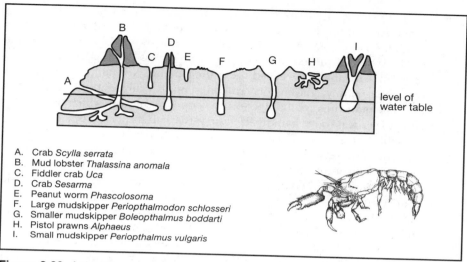

Figure 2.23. Animal burrows in mangrove forest. (After Berry 1972.)

A. Crab *Scylla serrata*
B. Mud lobster *Thalassina anomala*
C. Fiddler crab *Uca*
D. Crab *Sesarma*
E. Peanut worm *Phascolosoma*
F. Large mudskipper *Periopthalmodon schlosseri*
G. Smaller mudskipper *Boleopthalmus boddarti*
H. Pistol prawns *Alphaeus*
I. Small mudskipper *Periopthalmus vulgaris*

true lungs. As the tide comes in, these snails climb up the mangrove stems, away from the rising water.

In the landward areas of true mangrove forest, large mounds of displaced mud, as much as 1 m high, mark the burrows of the mud lobster *Thalassina anomala* (fig. 2.23). The burrow is up to 3 m long and extends down below the water surface; the entrance is usually plugged with earth. The mud lobster is active mainly by night and feeds on mud, digesting algae, protozoa and other organic material. The lobster's habit of living in a partly flooded burrow in oxygen-poor mud suggests that it may be capable of anaerobic respiration (Malley 1977).

Invertebrates are not the only animals adapted to the successive inundation and exposure of the mudflats. Mangrove swamps are rich in fish species, and two groups of fish, the mudskippers Periopthalminae and the climbing perch, are equally at home in or out of the water. The mudskippers must be very like the first fishlike creature that emerged onto dry land and eventually gave rise to all the land vertebrates. In the sea, mudskippers swim and breathe like other fish, but on land they behave more like amphibians, filling their gill chambers with a mixture of air and water. This water has to be renewed every five to six minutes (Burhannudin 1980). In addition, gas exchange through the skin and fins supplements this oxygen supply (Stebbins and Kalk 1961). The mudskipper's curious, stalked eyes, high on the head, are also an adaptation to the terrestrial habit, giving the fish a better field of view to detect prey, predators and territorial intruders.

In the water, mudskippers use their pectoral fins for balance just like any other fish, but on land these become efficient limbs for walking. This ability to walk ensures that the mudskippers are not trapped in shrinking tidal pools. When disturbed, mudskippers skim across the mud or water in a series of skips that has earned them their name. Normally, however, they move more slowly, crawling or "crutching" on specially adapted stiff pectoral fins. The pelvic fins are fused to form a ventral sucker with which the fish can cling to vertical roots, hauling itself upwards with its pectorals.

Different species of mudskipper are found on various parts of the mudflats and have very different diets, from detritivores like *Boleopthalmus boddarti* to carnivorous species that feed on small crabs, insects, snails and even other mudskippers (Burhanuddin and Martosewojo 1978; Martosewojo et al. 1982). Most species occupy deep burrows which they enter briefly during the day to replace water in the gill pouches. Males are territorial, and challenge and chase other males who approach their patch of mud. The male mudskipper excavates shallow craters which quickly fill to form small pools, then displays to attract a female to spawn in his pool. After mating the male guards the developing eggs (Nursall 1981).

The climbing perch *Anabas testudineus* is also a mud-dwelling fish, more commonly found in swamp habitats inland. The perch has gone one step further in the transition from sea to land. It has evolved an air-breathing lung and can survive out of water even longer than the mudskippers; in fact, it will drown in water if it cannot reach the surface to gulp in fresh air. Disappointingly, this fish's climbing abilities have been much over-rated. Although sometimes found high in mangrove trees, it has not reached the higher branches by its own efforts, but has been dropped there by predatory birds who have lost hold of a slippery meal (MacKinnon and MacKinnon 1974).

Ants are found in a wide range of habitats including mangrove swamps. Colonies of weaver ants *Oecophylla* live in untidy leaf nests made by binding five or six leaves together with fine silk thread. Within the nest lives the colony's strange, green queen, guarded by big, black soldier ants. If the nest is damaged or disturbed, swarms of worker ants hurry out to ward off the attacker and carry out immediate repairs. The workers cling to one edge of the tear and seize the other with their jaws, drawing the two sides together. The tear is then "sewn" by another corps of ants brandishing living shuttles. Each worker holds an ant grub in its jaws and, as the grub is passed back and forth, it secretes a sticky thread that binds the rent.

Weaver ants are fierce predators, but the small spider *Amyciaea* not only lives in ant nests but also preys on its hosts. Why the ants do not attack the spider is not clear; they probably mistake it for another ant. To aid this illusion, the spider mimics the coloration and movements of the ants and even bears a false ant head with two realistic "eyes" on its abdomen, a disguise that requires it to run backwards.

Beetles are the most successful group of animals on earth, including almost one-third of all described animal species and about two-fifths of all insects. In the mangrove swamps one group of beetles, the fireflies (*kelip-kelip*), are especially common and put on a spectacular light show. As dusk falls the first firefly starts to flash, and soon others in the same tree take up the rhythm until the whole bush is twinkling. Fireflies contain photoluminescent chemicals in the tip of the abdomen which glow clearly at night. The glow can be switched on and off to produce a regular series of synchronised flashes, each species having its own distinct pattern. The male fireflies display the light signal, and the wingless females signal back from the nearby forest; size and brightness of the flash may be as important as sequence for species recognition (O'Toole 1987). If the male receives an answering signal within two seconds of his own, he flies to join the female.

Male fireflies of the same species congregate in one tree; even so, there would be confusion if each male flashed at random, but the members of the group synchronise their rhythm so that they flash in unison. The group response ensures a more dramatic display, more likely to evoke a female response, and is also an antipredator device. About half the mangrove trees in any swamp are infested with predatory weaver ants. If one firefly is already settled in a tree and flashing, then there is a good chance it is safe for others to land there too. Firefly trees are so spectacular that in Perak, Malaysia, they were preserved as aids to river navigation at night (Polunin 1988).

The mangrove forest has a rich and varied bird fauna, including many fish-eating species, and waders which feed off the invertebrate community of the exposed mudflats. Studies in Malaysia showed that the mangrove avifauna was distinctive (table 2.5) and shared only 27 species with the inland rainforest (Wells 1976). The mainly frugivorous families of lowland dipterocarp forest are rare in mangrove. The inland boundary of the mangrove avifauna is sharp, with species replacement in some groups. Thus in West Malaysia the white-collared kingfisher *Halcyon chloris* occurs in mangrove but is replaced by *Halcyon smyrnensis* in dryland forest. In Borneo, where *H. smyrnensis* does not occur, *H. chloris* is found in both habitats. Interestingly, the bird faunas of villages and cleared lands are more similar to the avifauna of the mangrove and other fringe habitats than to that of the evergreen rainforest (Wells 1976). The 51 commonest mangrove species are all found in village lands. Few mangrove birds are restricted to this habitat; 85% of resident mangrove bird species also breed elsewhere in Peninsular Malaysia. Mangroves have a resident bird community but also provide temporary feeding for visiting migrants, as well as nocturnal roosts for species such as white egrets, which feed inland in the rice fields where they perform a valuable role by preying on insect pests.

The mangrove forest shares many species of mammals with riverine habitats, including three species of monkey, long-tailed macaque *Macaca fascicularis*, silvered langur *Presbytis cristata* and the proboscis monkey *Nasalis lar-*

vatus, a Bornean endemic. Mangrove leaves, particularly the young leaves of *Rhizophora* and *Sonneratia*, are an important part of the diet of proboscis monkeys in coastal areas. These large, reddish monkeys are found only along the coasts and rivers of Borneo and have been recorded hundreds of kilometres inland in riverine forest (MacKinnon 1974a, 1986; Giesen 1987; Bodmer et al. 1991; fig. 2.24). The male's large, red nose has earned the proboscis monkey the name of *kera Belanda* because of its alleged resemblance to a sunburnt Dutchman. In fact, the male's nose, the largest found among all the primates, is not just a respiratory organ but may act as a sound booster when he gives warning calls (MacKinnon and MacKinnon 1974). Recent research suggests that the male's nose may also be a sexual signal to attract females to join his harem, another example of the curious strategies animals adopt to attract a mate (Bennett 1988a). Always living close to water, the proboscis monkeys are excellent swimmers and can easily cross wide estuaries, sometimes staying submerged for several minutes at a time.

Table 2.5. Bird species dependent extensively or exclusively on mangroves.

Species	Borneo	Peninsular Malaysia	Sumatra	Java
Anhinga melanogaster	X			X
Ardea sumatrana	X	X	X	X
Ardea purpurea	X	X	X	X
Butorides striatus	X	X	X	X
Nycticorax nycticorax	X	X	?	X
Nycticorax caledonicus	X			X
Leptoptilos javanica	X	X	X	X
Haliastur indus	X	X	X	X
Halcyon coromanda minor	X	X		
Picoides moluccensis	X	X		X
Pitta megarhyncha	X	X	X	
Parus major	X	X		
Orthotomus ruficeps	X	X	X	?
Orthotomus sericeus	X	X	X	
Cyornis rufigastra	X	X	X	X
Pachycephala cinerea	X	X	X	
Nectarinia calcostetha	X	X		
Nectarinia jugularis	X	X		
Rhipidura javanica	X	X		
Zosterops palpebrosa	X	X		
Zosterops flava	X			X

Source: Wells 1985

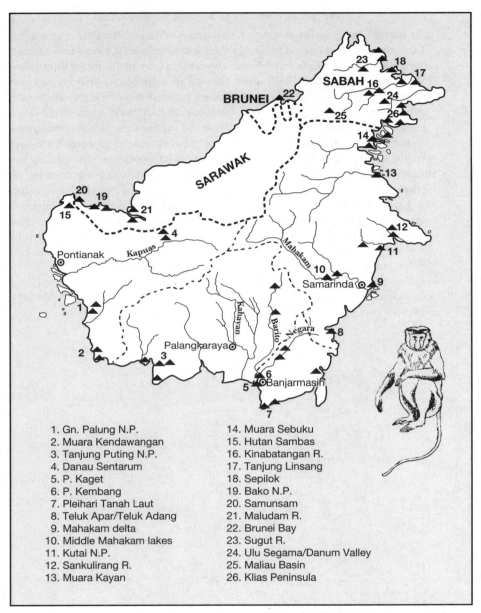

Figure 2.24. The distribution of proboscis monkeys *Nasalis larvatus* in Borneo. These monkeys occur only in Borneo, where they are found in coastal mangroves and in riverine forest hundreds of kilometres inland.

Mangrove Habitats as a Valuable Resource

The mangrove forests that fringe Kalimantan's coasts are the most extensive of the coastal habitats and probably the most threatened. Of all Kalimantan mangrove forests, 95% have been allocated as timber concessions (Burbridge and Koesobiono 1980). Less than 1% of Kalimantan's mangroves and mangrove/nipa associations lie within gazetted reserves (MacKinnon and Artha 1981), and many of these habitats are already badly disturbed.

Because of their high productivity and the shelter they provide, mangrove forests are key habitats in the life cycles of many marine organisms, including several species of commercial importance. Mangroves are also vital to the health and productivity of other coastal ecosystems, trapping river-borne sediments and exporting organic nutrients through the detrital food chain. As a peripheral habitat, mangroves are particularly sensitive to environmental change and pollution. Mangrove habitats are important for coastal protection, fisheries productivity, timber and numerous products utilised by local communites (Soegiarto and Polunin 1980; Saenger et al. 1983; Burbridge and Koesoebiono 1980; Salm and Halim 1984; Salm and Clark 1984). Figure 2.25 illustrates the complex inter–relationships of the mangrove ecosystem. The economic and environmental values of mangroves, and the need for their conservation, are discussed in chapter 11.

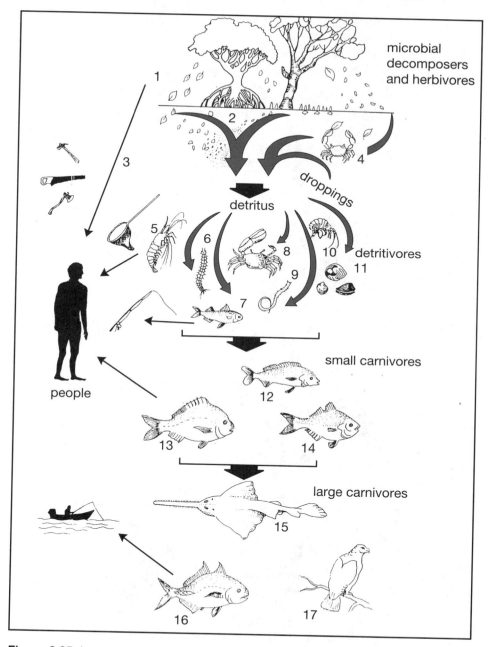

Figure 2.25. Inter-relationships of a mangrove ecosystem.
Source: Salm and Clark 1984

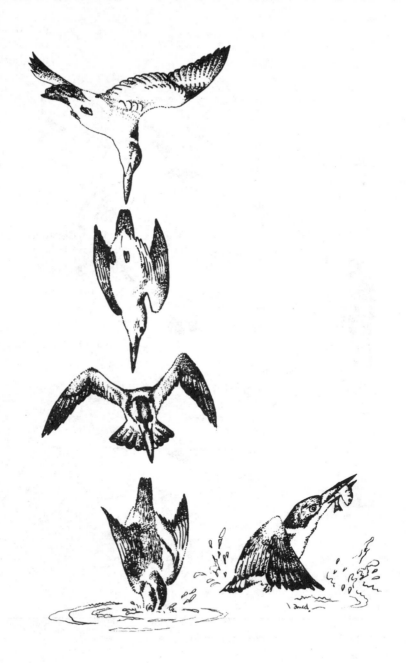

Chapter Three

Freshwater Habitats

Behind the coastal margin of mangrove and nipa, the waterlogged soils of the Kalimantan lowland plains support vast tracts of peat swamp and freshwater forests. These are intersected by the island's great rivers, meandering in broad ribbons to the sea. Hundreds of kilometres inland the land is still low-lying, poorly drained and swampy. In the basins of the Rivers Kapuas, Mahakam, Negara and Seruyan, lakes have formed, oxbow lakes from meanders cut off from the river and seasonal lakes which expand as the rivers flood during the wet season. Freshwater habitats (swamp forests, riverine forest, rivers and lakes) cover more than one-sixth of the land surface of Kalimantan (table 3.1). Ecologically and economically these habitats are of major importance to the peoples of Kalimantan.

Although it is convenient to discuss the various freshwater habitats as separate ecosystems with characteristic vegetation types, in fact all of these wetlands are closely interlinked, with many common species of plants and animals. Vegetation cover depends on soil types, drainage patterns and, often, the activities of people. Many areas of open swamp vegetation were originally covered in freshwater or peat swamp forests but have been cleared by fire and by logging and agricultural activities to create open swamp grasslands, as in the Sungai Negara wetlands and the Mahakam basin.

SWAMP FORESTS

Bornean swamp forests are of three major types: mangrove, which is under the influence of seawater (chapter 2), rain-fed peat swamp forest, and river-fed freshwater swamp forest. Freshwater swamps receive dissolved mineral nutrients in floodwaters from rivers and streams, whereas peat swamp forests receive moisture mainly from rainfall and are therefore nutrient-poor (Whitmore 1984a). These differences in nutrient input are reflected in vegetation composition, with peat swamp forests generally less species-rich than the adjacent freshwater swamp forests.

Table 3.1. Wetland habitats in Kalimantan.

WEST KALIMANTAN	Original area (ha)	Remaining area (ha)	Area included in reserves (ha)	Percent protected
Freshwater swamp	1,305,000	475,000	120,000	9
Peat swamp	2,201,000	1,805,000	125,000	5.6
Mangrove forest	425,000	60,000	17,000	4.0
Beach vegetation	5,000	2,000	1,000	20
Freshwater lakes	28,000	28,000	25,000	89
Total	3,964,000	2,370,000	288,000	
	100 %	59.7 %	7.2 %	

CENTRAL KALIMANTAN	Original area (ha)	Remaining area (ha)	Area included in reserves (ha)	Percent protected
Freshwater swamp	1,880,000	940,000	78,000	4.1
Peat swamp	996,000	872,000	98,000	9.8
Mangrove forest	120,000	100,000	11,000	9.1
Beach vegetation	?	?	-	-
Freshwater lakes	20,000	20,000	-	-
Total	3,016,000	1,932,000	187,000	
	100 %	64.0 %	6.2 %	

SOUTH KALIMANTAN	Original area (ha)	Remaining area (ha)	Area included in reserves (ha)	Percent protected
Freshwater swamp	288,000	107,000	20,000	6.9
Peat swamp	457,000	180,000	-	-
Mangrove forest	165,000	66,000	14,000	8.4
Beach vegetation	?	?	?	
Freshwater lakes	4,000	4,000	-	
Total	914,000	357,000	34,000	
	100 %	39 %	3.7 %	

EAST KALIMANTAN	Original area (ha)	Remaining area (ha)	Area included in reserves (ha)	Percent protected
Freshwater swamp	422,000	195,000	110,000	26
Peat swamp	749,000	594,000	42,000	5.6
Mangrove forest	950,000	750,000	-	-
Beach vegetation	3,000	1,000	-	-
Freshwater lakes	40,000	40,000	-	-
Total	2,164,000	1,580,000	152,000	
	100 %	73 %	7.0 %	

KALIMANTAN TOTAL	Original area (ha)	Remaining area (ha)	Area included in reserves (ha)	Percent protected
Freshwater swamp	3,895,000	1,717,000	328,000	8.4
Peat swamp	4,403,000	3,451,000	265,000	6.0
Mangrove forest	1,660,000	976,000	42,000	2.5
Beach vegetation	8,000	3,000	1,000	12.5
Freshwater lakes	92,000	92,000	25,000	27
Total	10,058,000	6,239,000	661,000	
	100%	62%	6.5%	

Source: Adapted from MacKinnon and Artha 1981

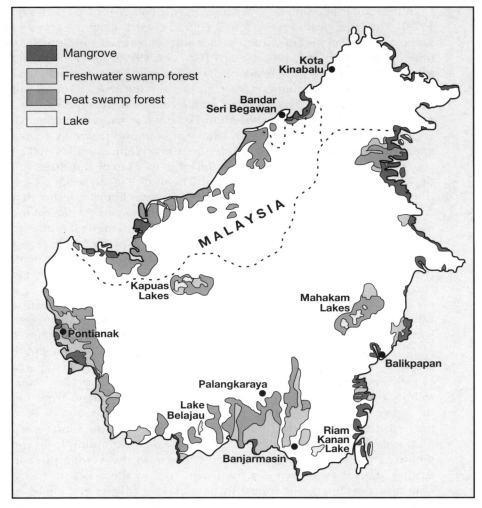

Figure 3.1. Sketch map of major areas of remaining wetland habitats in Borneo.

PEAT SWAMP FOREST

Peat swamps cover extensive areas of lowland Kalimantan, with estimates varying between 8% and 11% (MacKinnon and Artha 1981; Soepraptohardjo and Driessen 1976). The greatest extent of peat areas in Kalimantan is found over marine alluvium along the west and south coasts, and along the lower reaches of the Barito and other south-flowing rivers. Inland, peat swamps occur around the lakes of the Kapuas and Mahakam (MacKinnon and Artha 1981; RePPProT 1990). The coastal plains of Sarawak also support extensive areas of peat swamp forest (fig. 3.1).

A peat soil is one with 65% or more organic matter content (Driessen 1978). The large peat deposits found behind coastal mangrove forest in Kalimantan and in the coastal and deltaic areas of Sarawak and Brunei are **ombrogenous** (rain-fed) peat swamps (Driessen 1977; Morley 1981; and see fig. 1.9). Peats also form in poorly drained depressions at high altitudes, and are characteristic under upper montane rainforest, as on Mount Kinabalu (Whitmore 1984a). A thin peat layer is found also on the surface of the periodically waterlogged heath forest known as *kerapah* (chapter 5).

The surface of the extensive lowland peat swamps is markedly domed and is not subject to flooding (Anderson 1964; Whitmore 1984a). The peat deposits are usually at least 50 cm thick, but they can be very deep, and depths of up to 20 m have been recorded. The surface of the peat is a solid, fibrous and sometimes soft crust overlying a semi-liquid interior that contains large pieces of wood and other vegetable remains. As most of the incoming water is from rain, it is extremely deficient in mineral nutrients. The peat and its drainage water are very acid (with a pH of 4 or less) and poor in nutrients (oligotrophic), especially calcium. The many small rivulets that drain the peat swamp are tea-coloured, and are described as "blackwater" rivers.

The lowland peat soils support a distinctive forest formation with a rather restricted flora (Anderson 1972; Brunig 1973). In Kalimantan, peat swamp forests account for 14.6% of the total forest cover (RePPProT 1990). Elsewhere in Borneo, peat swamp forests occupy nearly the entire coastline of Sarawak and Brunei, covering 12.5% of Sarawak's land area (15,000 km^2) and 22.5% of Brunei (Anderson 1963). The biggest single swamp, on the Maludam peninsula, stretches 64 km inland from the mangrove forest fringe and covers 1,070 km^2. These forests, and those of northwest Kalimantan, are unique in the presence of the dipterocarp *Shorea albida*, a valuable timber tree with a range extending from the mouth of the Kapuas River south of Pontianak to the Tutong River in Brunei. This dipterocarp is a dominant tree in these swamp forest communities but is not found in peat swamp forests elsewhere in Kalimantan (Whitmore 1984a).

Formation of Ombrogenous Peat Swamps

The ombrogenous peat swamps of southern Kalimantan and Brunei and Sarawak are relatively recent in origin. The peat has formed in the few thousand years since the last glaciation of the Ice Age (Wilford 1960; Muller 1965; Morley 1981). As the ice melted, sea levels rose, and Sundaland flooded. By about 5,500 years ago the sea level had risen to its maximum height, and fine silt and sediments carried down by Borneo's great rivers were deposited to create levees and coastal plains (Kostermans 1958; van Meese 1982). Behind the silt-filled estuaries poorly drained swamps were created. This process continues to the present day.

Peat formation begins at the inland edge of the mangrove. Fine sediments carried down by the rivers are trapped behind the tangled mangrove roots to build up new land. As the coastline advances seawards, tidal inundation is less frequent, and the landward margin of the mangrove becomes less saline. Other plant communities develop in its wake (fig. 3.2). Because of tidal influence and the backing up of fresh water in estuaries, the water table is high, and the soil is permanently waterlogged. The micro-organisms that would normally decompose falling plant debris are unable to survive in the anaerobic conditions and high sulphide concentration, so that partly decomposed organic matter builds up to form a layer of peat over the mangrove clay soil (Anderson 1964). Rivers deposit alluvium along their banks, forming raised levees and creating backswamps. Cut off from the river water by the levees, the peat receives its only water input from rainfall. Peat swamp forest species replace species more characteristic of the mangrove formation.

In Sarawak and Brunei, a catena of six distinct forest types can be identified from the edge to the centre of the peat swamp (Anderson 1964). Pollen analysis of peat cores from the Baram delta in Sarawak has shown that this sequence represents a succession in time (Anderson 1964; Muller 1965, 1972). The peat swamps of the Baram delta have formed over the last 4,500 years, with the coastline advancing at a mean rate of 9 m per year (Wilford 1960). Peat was deposited more rapidly during the earlier stages and accumulated more slowly later. Decomposition occurs more slowly in the acid centre of the swamp so that organic material accumulates here to create the domed shape.

The ombrogenous peat swamps of southeast Kalimantan are exceptional since they developed under a marginally seasonal climate, whereas elsewhere in the Sunda region peat swamps are mostly limited to everwet conditions with minimal seasonality (Morley 1981). Six pollen samples taken from the peat swamp adjacent to the Sebangau River in Central Kalimantan show that ombrogenous peat formation began over a freshwater swamp, where grasses and the clubmoss *Lycopodium cernuum* were conspicuous elements. This peat formation at Sebangau may have been initiated by a change to a less seasonal climate during the mid-Holocene. The major vegetational differences between the Sebangau and the Marudi (Baram) peats relate to the nature of the substrate. The Marudi peats developed over mangrove clay (Anderson and Muller 1975), whereas the Sebangau peats developed over freshwater deposits (Morley 1981).

Ecology of Peat Swamp Forest

The ecology of peat swamp forests has been well studied in Sarawak and Brunei (Anderson 1958, 1961a, 1961b, 1963, 1964, 1983; Brunig 1969,

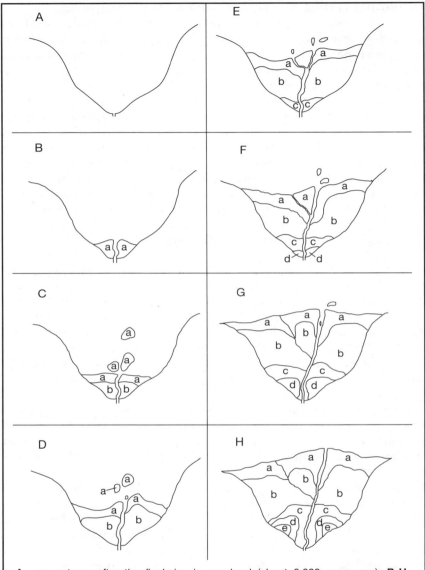

A - an estuary after the final rise in sea level (about 8,000 years ago); **B-H** - deposition of alluvium, colonisation by mangrove and peat swamp forests showing species succession: **a** = mangrove pioneers (e.g., *Avicennia*); **b** = late mangrove species (e.g., *Bruguiera*); **c** = peat swamp forest pioneers on thin peat and slightly brackish soil; **d** = mixed peat swamp forest on thicker soils above river level; **e** = dwarf padang forest on thick peat.

Figure 3.2. Hypothetical formation of a coastal peat swamp.

Source: Whitten et al. 1987a

1973) and, to a lesser extent, in Kalimantan (Dilmy 1965; Anderson 1976; Giesen 1987, 1990). Most peat swamps have concentric forest zones, changing from an outer uneven-canopied high forest, similar to lowland dipterocarp forest, to zones of lower height, decreased tree girth and less species richness towards the centre of the swamp (Anderson 1963; Whitmore 1984a; see figs. 3.3. and 3.4). Floral composition also changes across the catena. Several tree species have prominent aerial roots (**pneumatophores**) for obtaining oxygen in the waterlogged conditions of the peat swamp habitat.

Zonation is especially clear in the peat swamp forests of Sarawak and Brunei (Anderson 1961a, 1963), where six concentrically zoned communities can be distinguished, each with characteristic species and structure. All six communities are not necessarily present in every locality (fig. 3.3). The number of tree species declines across the catena (Anderson 1976). Whitmore (1984a) lists species for each zone, reducing from 34 species in the outer zone to only six species in the centre of the swamp. The sixth and innermost zone consists of stunted trees with marked adaptations, such as thick bark and thick leaf cuticles, for preventing water loss (Anderson 1963).

The Kalimantan peat swamp forests are less highly developed than those of Brunei and Sarawak, with fewer distinct zones (Anderson 1976). In Sarawak the more recent Lawas swamps are less developed than the older Baram swamps (Muller 1965), and it seems likely that the Kalimantan swamps are even more recent in origin. The swamp forests in West and South Kalimantan also show a marked decrease in number of species per unit area across the swamp (Anderson 1976), with high forest trees giving way to densely packed poles in the inner part of the swamp. Feeding roots form a dense mat above the water table, drawing nutrients from the top 15 cm of soil, which is richer in nutrients than the deeper peat (Muller 1972).

Although peat swamp forests are generally considered florally impoverished, Anderson (1963) recorded 927 species of flowering plants and ferns, from 224 genera and 70 families, from peat swamp forests in Brunei and Sarawak. Most of the tree families of lowland evergreen dipterocarp forest are found in swamp forest. Exceptions are Combretaceae, Lythraceae, Proteaceae and Styraceae (Whitmore 1984a). In the peripheral mixed swamp forest, where drainage is best, species composition is similar to that of lowland dipterocarp forest. The three outer zones (mixed swamp forest, high forest and high pole forest) contain the most important timber trees ramin *Gonystylus bancanus* and *Shorea albida* (Anderson 1964). Jelutong *Dyera costulata* is another commercially valuable species characteristic of peat swamp forest. Palms are few and usually occur in peripheral communities. The colourful sealing-wax palm *Cyrtostachys lakka* is commonly seen along rivers in peat swamp forests (Whitmore 1973). There are very few plant species endemic to peat swamp

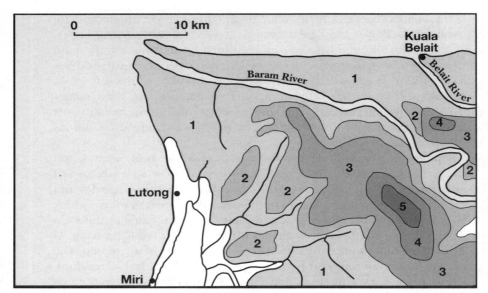

Figure 3.3. Peat swamp forest in the delta of the Baram River, western Borneo. Five vegetation zones can be distinguished, each with its own forest type. The poorest (type 6), found on the high central parts of raised bogs, occurs further inland. (After Whitmore 1984a.)

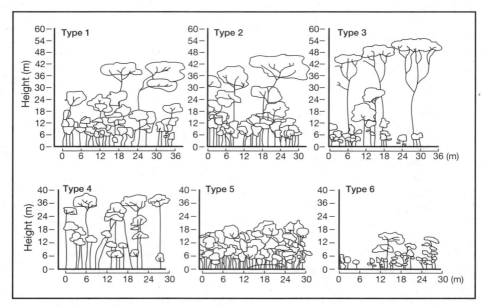

Figure 3.4. Tree profiles across transects of peat swamp forest. (After Whitmore 1984a.)

forests; this may be because of the relatively recent origin of this habitat, which is probably less than 11,000 years old (Muller 1965).

In Tanjung Puting National Park, Central Kalimantan, peat swamp forests are characterised by ramin *Gonystylus bancanus*, *Dyera*, *Tetramerista*, *Palaquium*, *Campnosperma*, *Ganua*, *Mesua*, *Dactylocladus* and *Alstonia* trees (MacKinnon 1983a). The dipterocarp *Shorea balangeran* is common on the edge of peat swamps and occurs in pure stands along the flooded margins of the reserve's rivers. This species probably replaces *Shorea albida* in the peat swamp forests of southern Kalimantan (Ashton 1964).

Towards the centre of the swamps the amounts of inorganic nutrients in the soil decline, especially phosphorus and potassium (Anderson 1958; Muller 1972). Forest structure from swamp edge to centre mirrors conditions of increasing infertility, characterised by decreasing canopy height (and hence decreasing total biomass per unit), smaller girth of individual trees, and thicker leaves (Whitmore 1984a). In the centre of the swamp there are also many plants with supplementary means of nutrition, such as ant plants *Hydnophytum* and *Myrmecodia* and pitcher plants (see table 7.2). Many of the plant species found at the centre of the swamp also occur on poor, podsolised heath forest soils (chapter 5); thus 146 species are common to both heath and peat swamp forests (Brunig 1973).

There are conspicuous parallels in structure and physiognomy between the more central peat swamp forest types and heath forest (chapter 5). There is also an increasing percentage of heath forest species along the catena (Whitmore 1984a). As with heath forest, the relative importance of periodic water stress and oligotrophy has not been fully determined. In the inner peat swamp communities, many of the tree roots form a platform which may periodically dry out above the water table. Moreover, water uptake may be slow from waterlogged peat soil, and low transpiration rates may be essential to avoid uptake of toxic solutes from the highly acidic, phenol-rich groundwater (Brunig 1971).

The low nutrient levels of soils in mature peat swamp forests almost certainly limit primary productivity (Whitten et al. 1987a). The energy costs of replacing leaves and other plant parts are therefore high. Many of the larger trees recorded in a Sarawak peat swamp forest belong to families rich in defence compounds such as latex, essential oils, resins, tannins and phenols (Anderson 1963). These compounds all make plant material less palatable, and their production has probably evolved as a mechanism against predation by herbivores. Many other plants growing in the inner part of the swamp — on the most nutrient-poor soils — also have very thick leaves with a resinous, acrid or aromatic taste; this is also probably a defence against animal predators. In nutrient-poor conditions it may be more advantageous for plants to invest in chemical and physical defences to protect leaves, fruit and stems than to replace herbivore damage with new growth (Janzen 1974a).

The Fauna of Peat Swamp Forests

Many species of mammals and birds occur in peat swamp forests, though none are specific to this habitat type (Medway 1977a; Davies and Payne 1982; MacKinnon 1983a; Wells 1985). Conspicuous species such as diurnal primates are good indicators of the faunal richness of a habitat. Monkeys, gibbons and even orangutans are found in peat swamp forests, but most species occur at lower densities than in dry lowland forests (Galdikas 1978; Marsh and Wilson 1981; K. MacKinnon 1983). Long-tailed macaques *Macaca fascicularis* and silvered langurs *Presbytis cristata* occur at higher densities in peat swamp forests than in other lowland forests, but only along the rivers (Wilson and Wilson 1975; Marsh and Wilson 1981; Davies and Payne 1982; K. MacKinnon 1983). Away from the rivers all primate densities are much lower. Studies in Peninsular Malaysia showed average primate densities reduced from 10 groups per square kilometre along the river to three groups per square kilometre inland (Marsh and Wilson 1981). This is to be expected in view of the lack of major fruit trees and of big trees for travel, especially towards the centre of the swamp.

At Tanjung Puting National Park primate studies have extended over more than twenty years (Galdikas 1979, 1982, 1985a; Supriatna et al. 1986; Yeager and Blondal 1992). All seven diurnal primates recorded in the park (orangutans *Pongo pygmaeus*, endemic maroon langurs *Presbytis rubicunda*, silvered langur, agile gibbon *Hylobates agilis*, long-tailed macaques, pig-tailed macaques *Macaca nemestrina* and proboscis monkeys *Nasalis larvatus*) are known to visit the mixed peat swamp forests, sometimes on the way to preferred food sources in other habitats, or to visit a fruiting tree.

Peat swamp forests are a key habitat for proboscis monkeys (Bennett 1988a, 1988b). These monkeys are common in Tanjung Puting, even in areas of previously logged swamp forest. They are easy to census when they return to the riverside in the late afternoon to sleep in tall lookout trees. A 1989 survey by the authors estimated a population of at least three hundred proboscis monkeys along the Sikonyer River and its two main branches. Proboscis monkeys are also common along other rivers within the park, for instance the Sungai Buluh Besar. Probably the total population within the park numbers several hundred proboscis monkeys. This compares favourably with the situation in Sarawak, where the total proboscis monkey population is estimated to be only one thousand animals (Bennett 1988b). Tanjung Puting is obviously a key site for the protection of this endemic Bornean primate.

Birds are easily recognised and are good indicators of species richness. More than two hundred bird species are recorded for Tanjung Puting reserve (MacKinnon 1983a; Nash and Nash 1986; Galdikas and Bohap 1986). Many of these species are seen in the peat swamp forests, especially along the river, though this may not be their preferred habitat. In the

Gunung Mulu National Park, Sarawak, a small area of peat swamp forest had only about half the species of birds found in nearby lowland evergreen rainforest (Wells et al. 1979).

The animal life of peat swamp rivers has not been studied in any detail for Kalimantan, but several of the larger rivers draining the swamps are known to be rich in fish species (Giesen 1987, 1990). The rare arowana fish *Scleropages formosus* is found in deep pools in peat swamp rivers in Tanjung Puting and the Kapuas lakes (MacKinnon 1983a; Giesen 1987). These rivers also support other typical riverine fauna such as otters, waterbirds, crocodiles and monitor lizards, though frogs are rare in the blackwater rivers (Giesen 1987). Towards the centre of the swamp, however, the blackwater streams are more acid and impoverished in fauna (see box 5.5).

FRESHWATER SWAMP FOREST

Freshwater swamp forests are widespread over alluvial soils that are flooded for long periods with fresh water. They are associated with coastal swamps, inland lakes and huge low-lying river basins, such as those of the Kapuas, the middle Mahakam and its tributaries, the Barito and Negara (South Kalimantan) and the Seruyan and Kahayan (Central Kalimantan). Freshwater swamp forests are the natural vegetation cover for about 7% of the land surface of Kalimantan (MacKinnon and Artha 1981). Much of this original vegetation cover has been cleared or modified by the activities of people, especially in South and West Kalimantan. Recent estimates of freshwater swamp forest for Kalimantan suggest that only 7,500 km^2 of these areas is still forested, about 1.4% of the land area (RePPProT 1985, 1987, 1988). Much of the rest has been cleared for agriculture, especially for *sawah* rice.

Whereas peat swamp forests are rain-fed, the soil surface of freshwater swamp forests is periodically inundated with mineral-rich river floodwaters of fairly high pH (above 6). The less acid conditions and greater nutrient input are reflected in greater productivity. Freshwater swamp forests are generally taller and more species-rich than those of peat swamps. The water level fluctuates, allowing periodic drying of the soil surface.

The freshwater swamp habitat is extremely heterogeneous in soils and vegetation. Although a few centimetres of peat may occur, this forest formation is quite different from the peat swamp forest which grows on deep peat (Whitmore 1984a). Floristic composition varies from floating grass mats, as seen in the Sungai Negara and Mahakam lake systems and believed to be a fire-climax (Endert 1925; van Steenis 1957; Giesen 1990), to pandan and palm swamp, scrub and forest. In Gunung Palung reserve, West Kalimantan, swamp forest varies from low scrub with trees 10 m tall to

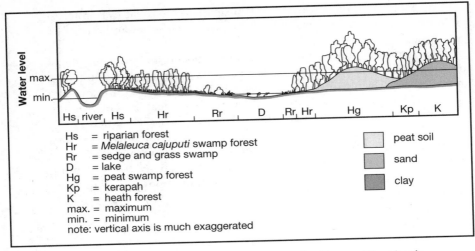

Figure 3.5. Vegetation catena (hypothetical) across the Negara River basin.

Source: Giesen 1990

forests similar in structure to mixed lowland forest (MacKinnon and Warsito 1982). If the water level is high for long periods, no forest develops at all. Instead, there is a herbaceous type of vegetation, consisting of swamp plants. Elsewhere, the forest is flooded during the wettest season, as in the Kapuas lakes area (Giesen 1987).

Swamp forests with constant water level can be differentiated into belts, which may reflect successional stages (Jacobs 1988). Corner (1978) studied freshwater swamp forests in Johore and Singapore and distinguished seven belts of swamp forest between sea and river: mangrove, nipa, then four belts of varying depth and composition of freshwater swamp forest and a final riverine belt dominated by the orange-flowered *Saraca* tree. The succession may develop further in either of two ways, to mixed "dry land" rainforest or to peat swamp forest, depending upon the height of the soil above sea level and the height of the water table (Jacobs 1988). Figure 3.5 illustrates the mosaic of vegetation types found on different landforms in the Sungai Negara basin, South Kalimantan.

Prime freshwater swamp forest has trees with an average height of 35 m, some lianas and many epiphytes. Tree roots may be under water for long periods when the forest is flooded. In this waterlogged habitat, where respiration is difficult, root systems are close to the surface and have developed various adaptations for gaseous exchange: stick roots, knee roots, stilt roots and flying buttresses (fig. 3.6).

This type of freshwater swamp forest, found along the coasts of west and south Borneo, is very mixed in species composition. Sometimes forest

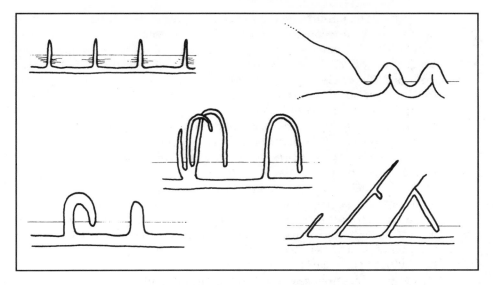

Figure 3.6. Types of pneumatophores found in freshwater swamp forest. (After Corner 1978.)

stands are species-poor, with gregarious associations of one genus (e.g., *Mallotus leucodermis* and *M. muticus*); other formations may be quite rich in tree species. Corner (1978) recorded 1,082 species in freshwater swamp forests in Johore, one-seventh of the entire Malayan flora. The flora of the swamp forest is not distinct, but includes many of the species that also occur in the neighbouring lowland forests. This has important implications for conservation. Areas of adjacent dryland forests must be protected as well as the swamp forest itself, otherwise the freshwater forest will be cut off from its most important source of seeds (Jacobs 1988).

Although freshwater swamp forests share many species with lowland dry forest, they are generally less rich in species. Forest structure is less layered, and the diameters and heights of trees remain smaller (Whitmore 1984a). The most important trees in this type of forest are *Campnosperma, Alstonia, Eugenia, Canarium,* the tall legume *Koompassia, Calophyllum* and *Melanorrhoea*. The swamp sago *Metroxylon sagu,* probably introduced to Borneo from eastern Indonesia by early settlers, also thrives in this habitat (chapter 10).

The distinctive riparian forests of the middle reaches of the Kalimantan and Sarawak rivers are seasonal swamp forests (called *empran* in Sarawak). These forest formations grow on the raised, storm-deposited levees that form where the fast-flowing hill torrents become slow-moving rivers. These riparian forests are conspicuous on aerial photographs from the very large, fluffy crowns of emergent trees. This is the habitat of *tengkawang* trees, *Shorea macrophylla*, *S. seminis* and *S. splendida*, and of the *ulin*, Bornean ironwood *Eusideroxylon zwageri*. Ulin is not restricted to these alluvial clay soils but is also found as scattered trees on clay hillsides up to 360 m (chapter 5). Other tree species commonly found in this habitat are *Dipterocarpus oblongifolia*, *D. apterus* and the tall legume *Intsia palembanica*. Yellow-flowered *Dillenia* trees occur locally where drainage is poor, as do *Cerbera manghas*, *Dracontomelon puberulum*, *Heritiera littoralis* and *Lagerstroemia speciosa* with its spectacular mauve blossoms. The riparian forests are frequently backed by swampy, alluvial, valley-floor forest, which is flooded occasionally when river levels are high.

Riparian forests may carry commercial timber at a productivity of up to 90 m^3/ha, a very heavy stocking level (Whitmore 1984a). Since the soil on these alluvial terraces is deep, fertile, friable and well-watered, much of this habitat has been cleared for cultivation.

Original freshwater swamp forests in south and southeast Kalimantan have been converted to single-species stands of the paperbark *Melaleuca cajuputi*, sedge and grass swamp, or wet-rice fields, succumbing to the same fate as most freshwater swamp forests in Southeast Asia (Whitmore 1984a). This process has been going on for at least the last sixty years in South Kalimantan (Schophuys 1936). *Melaleuca cajuputi*, with its striking, white papery bark, is an understorey tree in primary swamp forest but becomes gregarious after repeated burning, as it produces root suckers and coppice shoots. Big *Melaleuca* trees have thick, loose, corky bark which gives protection against fire. Sloughed bark, dry leaves and other litter accumulate on the forest floor and are highly inflammable in periods of drought such as those experienced in Kalimantan in 1982-83 and 1987. *Melaleuca* is a fire-adapted species and regenerates quickly after burning when competing species have been eradicated. The stands in South Kalimantan are fired deliberately to encourage new growth. Thus trees can be harvested on a sustainable basis for firewood and building materials (PHPH/AWB/KPSL-UNLAM 1990). Burning too frequently kills the *Melaleuca*, but a moderate burning regime favours the *Melaleuca* trees by eradicating competitors (see box 10.3).

Fauna of Freshwater Swamp Forests

Faunal diversity and abundance in freshwater swamp forests varies with the structure and floral diversity of the forest, but is usually higher than in peat swamp forests. Primate densities can be as high as at dryland forest sites in some areas. As in peat swamp forests, primates are commonest along rivers (Marsh and Wilson 1981). Long-tailed macaques *Macaca fascicularis* are the commonest primates in freshwater swamp forests.

Long-tailed macaques are some of the most adaptable of all primates and can live almost anywhere: in rainforest, mangrove swamps or even in temples, as at Banjarmasin, where they supplement their diet with food snatched from unwary visitors. These monkeys live in troops of twenty or more and wander along the beach, picking over the flotsam and catching small crustaceans. This habit has earned them the alternative name of crab-eating macaques.

During the day macaques move away from the river to forage in spread-out groups, searching out fruit, leaves and flowers (MacKinnon and MacKinnon 1980). Groups are large, with complex social hierarchies and much squabbling among the males. Young females spend their whole life with the group into which they were born, but young males leave the mother's group at the age of five or six years (Aldrich-Blake 1980). If these youngsters are accepted into another macaque group, they may reach high enough status to breed, but after a few years they move on again. These male migrations ensure outbreeding among macaque populations. This constant genetic exchange may be one reason why these are such versatile and adaptable monkeys (van Schaik 1985). The long-tailed macaque is one monkey species that can adapt to disturbed forest, achieving higher numbers there than in unlogged forest (Marsh and Wilson 1981).

RIVERS

Freshwater bodies can be flowing (rivers) or static (lakes). Borneo is a land of great rivers, the Kapuas, Barito, Kahayan, Kayan and Mahakam in Kalimantan, the Rajang and Baram in Sarawak and the Segama and Kinabatangan in Sabah. These rivers are the main access routes to the interior of the island and to the central highlands.

A trip up one of Kalimantan's great rivers from the estuary to its source takes the traveller through all the major habitats on the island, from the mangrove forests of the delta to the hill dipterocarp forests that shade the river headwaters. Nipa-fringed estuaries, where sea and fresh water meet, give way further inland to freshwater swamps and lowland dipterocarp forests. In the low-lying coastal plains the rivers are broad, winding ribbons of brown water, meandering through the forest. The Kapuas, for instance, drops only 50 m between Putussibau (900 km inland) and the sea, and the river's course is tortuous. Oxbow lakes along its length are evidence of an earlier course, where deposited sediment has cut off sections of the river. Lowland rivers carry heavy silt loads. The silt-laden waters of the Kapuas extend 50-60 km out to sea, and the mouth of the Kapuas Kecil,

Pontianak's link to the sea, must be dredged twice yearly to keep it open for freight boats (Giesen 1987).

Up in the headwaters the river is narrower. It flows through hill forest, running fast and clear. The canopies of giant trees extend over the flowing water, providing shade. Shallow rapids break the surface, and the traveller must haul his small canoe over the jutting rocks. Even to the casual observer, it is obvious that the various stages of the river provide very different physical and biological conditions for plant and animal life. The ecology of the river changes along its length, with species adapted to the peculiar chemical, physical and biological conditions of their habitats. The ecology of freshwater rivers and lakes is well described by Bishop (1973), Hutchinson (1975), Moss (1980), Payne (1986), and Whitten et al. (1987a).

Box 3.1. Riverine habitats in Borneo: a classification of flowing waters.

Montane streams above 1,000 m are cold-water torrents with few plants, little fringe vegetation and a small fauna, mostly of aquatic insects. Some of these insects have elaborate life cycles adapted to nutrient scarcity. Montane streams harbour some rare and primitive insects (e.g., corduliids *Procordulia* and cordulegasterids *Chlorogomphus*). Many fish have adaptations such as suckers to attach themselves to the substrate, and thus avoid being swept away.

Upland streams at 100-1,000 m are cool-water torrents with a diverse riparian fringe characterised by aroids (e.g., *Piptospatha*) in smaller tributaries and the leguminous tree *Saraca* along larger streams. These streams support a richer fauna, including a great diversity of aquatic insects. In unpolluted upland streams, groups such as dragonflies, mayflies and stoneflies may each be represented by more than twenty species.

Below 100 m, upland streams flow into **lowland rivers**. A riparian fringe of tall trees provides shade and a nutritive input of fallen leaves and fruit. The natural fauna is very rich and diverse. Many larger fish are restricted to larger rivers. Water conditions are often turbid; many fish have barbels to sense the movement of potential prey. Fish communities change downriver, with some families (Gobiidae, Hemiramphidae, Tetradontidae) more common in brackish and marine conditions near the river mouth. Lowland rivers in Borneo support important fisheries.

Lowland slow-flowing streams are found in flat country, where water levels vary in response to rainfall, and the direction of water flow may change according to whether a river is flooding or receding. In undisturbed environments, under closed canopy forest, these streams are extremely rich in animal life, with large numbers of fish species, prawns and crabs, aquatic insect larvae and water skaters. In acid and/or peaty conditions, the waters are tea-coloured, and known as "blackwater" streams; in conditions of very low pH, the fauna is much reduced. Removal of the natural tree cover, especially if coupled with increase in sediment loads and other polluting effects, severely affects aquatic life.

Freshwater tidal rivers naturally possess a rich diversity of riparian vegetation, characterised by screw pines *Pandanus*, wild jambu *Eugenia,* and *Palaquium* trees. These rivers support a rich fauna, especially of insects and fish. They are important fisheries for cyprinids, and provide breeding sites for estuarine fish.

Source: Adapted from Chan et al. 1985

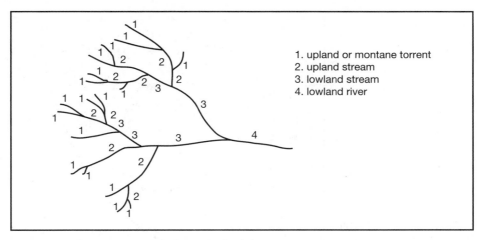

Figure 3.7. Stream order in a hypothetical river system.

Physical Patterns in Rivers

Most of Kalimantan's main rivers rise in the central mountain ranges. They increase in width and volume as they flow towards the sea, fed by tributaries to form a main river that drains a large watershed. The Kapuas, Indonesia's longest river, is estimated to drain two-thirds of the whole province of West Kalimantan and to have a watershed of 100,000 km^2 (Giesen 1987). Drainage patterns are influenced by geology, geography and human use of the catchment area.

Rain falling in the forested uplands percolates down to the forest floor and drains into small streams which join with other tributaries to feed the great rivers winding down to the sea (fig. 3.7). The rate at which this drainage occurs depends on the extent of vegetation cover in the catchment. Tropical forest acts like a sponge, slowing and regulating water discharge into the rivers and streams.

The discharge (water volume per unit time) increases as the river flows downhill and receives input from tributaries and runoff. The average rate of flow is estimated to be 6,000-7,000 m^3 per second at the Kapuas delta. Far upriver at the mouth of its tributary, the Tawang, average flow in the Kapuas is estimated at 2,000 m^3 per second (Giesen 1987). Discharge varies according to season. Heavy rains can produce dramatic changes in water level. Along much of the upper course of the Kapuas, water levels rise 10-12 m overnight. Headwater streams respond quickly to short, intense rainfall to produce flash floods. Floodwaters scour river beds and riverbanks and sweep away

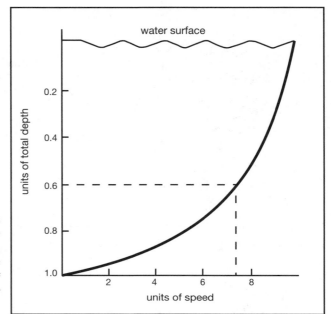

Figure 3.8. Relationship between current velocity and depth in an open channel. Average velocity is measured at 60% of the total depth. (After Townsend 1980.)

vegetation, fertile alluvial soils and even homesteads.

Current velocity, depth of water, and substrate composition vary along the length of the river and also across its width, influencing the biota that can live there (fig. 3.8). Surprising though it may seem, average velocity is greater in the lower reaches of a river than in the headwaters. Current velocity in rivers is greater in the main body of flow just under the surface than on the river bed or near the riverbank; this affects the distribution of plants and animals in the river. Turbulent flow near the surface may create eddies which dislodge rocks, pebbles and particles. The greater the shear stress on the river bed, the greater the probability that a plant or benthic (bottom-living) organism will be dislodged. Current velocity increases downstream, but shearwater stress is greatest in the headwaters.

Water Chemistry

Water enters the river systems from two main sources, either directly as rainwater or indirectly as groundwater, the water that has already percolated

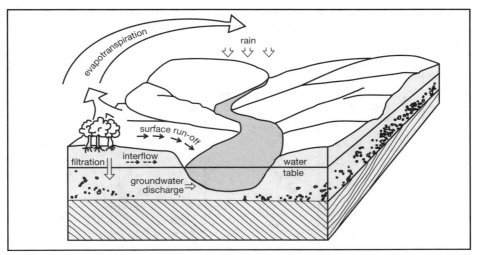

Figure 3.9. Diagram showing the various pathways of water in a catchment area. (After Townsend 1980.)

through the catchment rocks. The largest input to rivers and lakes comes from groundwater (fig. 3.9). Rain is not pure water; it contains significant amounts of dissolved gases, salts and trace elements. As it percolates through the catchment area, it dissolves more chemicals, by leaching them from soils and rocks, and picks up organic materials. The chemical composition of the groundwater will be related to the nature of the parent rock in the catchment area and its solubility. Rocks in the headwaters are more resistant to weathering. Thus the concentration of dissolved salts is generally greater downstream, with inputs from alluvial and sedimentary areas and human waste (Townsend 1980). Where groundwater has percolated through an acid medium such as peat, drainage waters are highly acid, as in blackwater rivers (see chapter 5).

The plant life found in rivers and lakes depends on nutrient content, pH, and concentrations of dissolved oxygen and carbon dioxide. All plants require a number of elements for healthy growth, especially nitrogen, phosphorus and potassium. Excesses of these elements in still water can make it unsuitable for plant and animal life. Where a large amount of organic waste enters a stream, near human habitation or at the effluent drain from a factory, its decomposition by micro-organisms causes severe depletion of dissolved oxygen.

The concentration of dissolved oxygen in rivers is determined by a

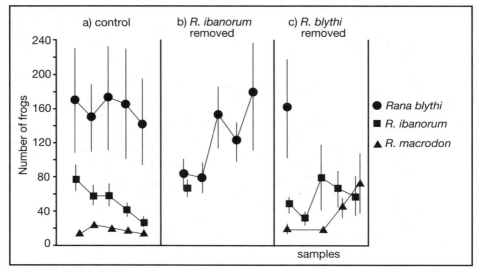

Figure 3.10. Ecological separation of frog populations in three Sarawak streams. The three species show considerable niche overlap in habitat requirements and food species. After the frog populations were censused, *R. ibanorum* was removed from the stream where *R. blythi* was least common (b), and *R. blythi* was removed from the stream where *R. ibanorum* was least common (c). In situation (b) the population of *R. blythi* increased; in (c) the population of *R. ibanorum* fluctuated but remained low. *R. ibanorum* seems to have been restricting the niche of *R. blythi* but not vice versa. How the competition was effected is not known.

Sources: Inger 1986; Inger and Greenberg 1966

combination of physical, chemical and biological factors. The headwaters of rivers tend to be rather turbulent; this mixing of the waters increases the amount of dissolved oxygen. A rise in temperature reduces the solubility of gases, leading to a reduction in the concentration of dissolved oxygen. River temperatures are strongly influenced by the environmental characteristics of the catchment (Crowther 1982a). Water temperatures are lower where the river is shaded by forest, and where there is a major input from groundwater, than in open plains where surface water is the main input, and there is no shading vegetation. Felling increases the rate of surface runoff and reduces shade, thus increasing the river's capacity to absorb heat (Crowther 1982a). With increased warming and reduced oxygen levels, certain species of plants and animals may be lost. With higher temperatures, less oxygen and more nutrients, eutrophication is likely to occur in stiller waters, especially in lowland lakes and swamps. This

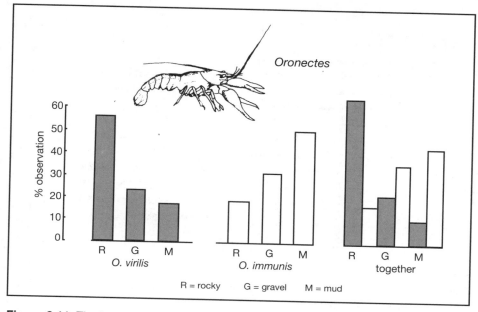

Figure 3.11. The importance of particle size in determining distribution of two species of the freshwater crayfish *Oronectes* in the same river in the U.S.A.

Source: Bovbjerg 1970

can lead to less diversity of plant and animal life.

All of these physical and chemical factors influence the distribution of animals and plants in river and lake communities. Nevertheless, the distribution of an organism is not determined simply by physical factors such as temperature and water flow. Every organism has its own preferred niche within a given set of variables, but there may be competition from other species with similar requirements. Like the animals of lowland forest communities (chapter 4), aquatic animals show niche separation and resource partitioning (Moyle and Senanayake 1984). Frog populations and tadpoles in rainforest streams in Borneo avoid competition, by different species exploiting different microhabitats within the stream and taking different foods (Inger 1986; fig. 3.10). The distribution of freshwater prawns and crayfish is determined by particle size of sediments (fig. 3.11). A detailed study of the ecology of a small river in Peninsular Malaysia showed a change in distribution of fish species along the river's course with a change in

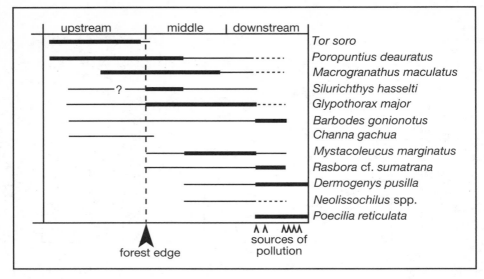

Figure 3.12. Change in distribution of fish species along the course of a river. (After Bishop 1973, with nomenclature according to Kottelat et al. 1993.)

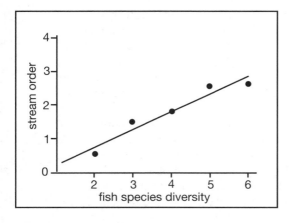

Figure 3.13. Relationship between fish species diversity and stream order. (For stream order see fig. 3.7.) (After Bishop 1973; KPSL-UNLAM 1989a.)

species at the forest edge and with increased pollution (fig. 3.12). There was also a significant correlation between fish diversity and the size of the stream, with more fish species in larger streams where there were more spatial and feeding niches (fig. 3.13).

River Communities

As one passes from the headwaters to the river mouth, there are obvious changes in water speed, total volume of water, turbidity, substrate and types of food available. These differences are reflected in the distribution of plant and animal communities along the course of the stream or river. The only plants found in very fast-flowing headwaters are encrusting algae, filamentous algae and mosses. Adaptations that enable higher plants to live in fast-flowing water include reduced resistance to water flow, good anchorage, and resistance to abrasion, as well as the ability to propagate from small fragments.

Many invertebrates have adapted to life in turbulent headwaters. Some live beneath boulders, and others have adaptations such as flattened body form, the presence of hooks or suckers, or streamlined shapes that offer little resistance to water flow. Different invertebrate communities occupy different stretches in the river. Thus the slow-flowing inner bend of a river in Java had a lakelike invertebrate fauna, whereas the invertebrate community in the fast-flowing outer bend supported species with adaptations for headwater torrents (Lieftinck 1950).

River headwaters are turbulent, with little sediment, but benthic (bottom-living) invertebrates such as caddis fly *Trichoptera* and dipteran larvae filter fine organic particles from the water. Aquatic mosses, liverworts and algae attached to rocks are grazed by other benthic invertebrates, including mayflies and freshwater molluscs. Freshwater shrimps feed on micro-organisms, especially fungi, that colonise pieces of litter lodged on the stream bed. Carnivorous invertebrates (including leeches and flatworms) and some insect larvae prey on these herbivores and detritivores. Fish feed on plant material, invertebrates and other fish (Moss 1980).

Fish living in the fast-flowing waters of Kalimantan rivers are often strong swimmers, and most have streamlined bodies. However, some species, such as the loaches Cobitidae, spend most of their time on the river bed, are flattened in shape, live among stones and rocks and generally avoid the strongest currents. Their eyes are close to the dorsal surface, and the mouth is located ventrally for browsing on the bottom. Many have a much-reduced swim-bladder to decrease their buoyancy. Some fish have developed suckers to cling on to rocks; the sucker fish *Gastromyzon borneensis* can maintain its hold even in fast-flowing rapids, where it moves slowly over the rocks, grazing on algae (Jacobson 1986). Mountain-stream fish such as *Gastromyzon* probably also have distinctive larval adaptations for life in headwater torrents.

In its lower reaches the river widens. Water flows more slowly at the river's edge, and sediment is deposited on the bottom, providing a medium for larger aquatic plants (**macrophytes**), including flowering plants and mosses. Those macrophytes that neither float nor have most of their leaves above water survive by being adapted to low oxygen levels, slow rates of oxygen diffusion and low light intensity. They often have much large-celled aerenchyma tissue through which gas can pass easily to facilitate gaseous exchange. Adaptations to promote photosynthesis under low light conditions include the absence of cuticle from leaves and stems, and high concentrations of chloroplasts in the epidermis. Many plants, such as water lilies *Nymphoides*, duckweed *Lemna*, and the smallest of all flowering plants, *Wolffia*, avoid these problems by having leaves that float on the water surface.

Box 3.2. Ecomorphology of fish.

Much can be determined about the ecology and behaviour of a fish from its shape. Fish with a flat ventral (lower) profile and belly are generally bottom-dwelling species. If the mouth is inferior, then the species probably feeds on detritus, on invertebrates living among the bottom substrate, or on algae scraped off rocks (e.g., Balitoridae and Cobitidae). Some species, however, have a flat, ventral profile and a mouth facing upwards. Such fish are generally bottom-dwelling, darting out at small, passing fish or other animals; these are known as sit-and-wait predators (e.g., *Chaca bankanensis*). Species such as *Scleropages formosus* also have an upwards-facing mouth. These typically swim at or just below the surface. Fish with a terminal mouth probably live in middle strata. A body that is laterally compressed, as in gouramis, allows fish to move easily through water with dense vegetation. Long, slim bodies, such as those of the snakeheads (*Channa*), suggest fast-swimming, predatory fish that lie in wait among vegetation to ambush suitable prey.

Fish with small mouths (e.g., Syngnathidae) tend to feed on plankton or on organisms attached to aquatic plants and other submerged structures. Scavengers often have medium-sized mouths to take in considerable quantities of mud or sand which they sift for small animals and plants. Many cyprinoid fish (barbs, catfish) have barbels, which are used to detect food in turbid or dark water. A large mouth generally indicates that the fish is a predator, and closer examination often reveals teeth.

Some fish have developed suction pads from their pelvic fins and other parts of their bodies (e.g., Gobiidae and *Gastromyzon*). This is an adaptation to fast-flowing waters (either in a river or in the tidal or wave zone) which allows the fish to attach to the substrate. Long fins are associated with fish that normally swim slowly but are capable of fast acceleration. These may be predators or fish living where predator attacks are frequent. Shorter fins indicate a capacity for sustained swimming.

Fish with large eyes tend to live in clear water where they can see. Fish living on the bottom of turbid waters of estuaries tend to have small eyes. Their lack of visual acuity may be offset by the possession of long barbels, as in many catfish. A few species, such as *Lepidocephalus spectrum*, have no eyes at all; they are generally associated with caves or with rapids.

Source: Kottelat et al. 1993

Emergent reed beds slow the water flow and cause the river to drop its sediment load. These reed beds and submerged macrophytes provide a habitat for a diverse array of invertebrates and fish which feed on epiphytic algae and bacteria. Mud-living invertebrates, including oligochaete worms, chironomid larvae (Diptera) and bivalve molluscs, burrow in deposited sediments or feed on the mud surface.

Organisms adapted to living suspended in water (**plankton**) are found only in the lower reaches of slow-moving rivers and in the still waters of lakes and ponds. Zooplankton includes mostly small crustaceans, such as the water flea *Daphnia*, protozoans and rotifers. Zooplankton feed on microscopic algae (phytoplankton) or, if carnivorous, on other species of zooplankton. Many Kalimantan fish feed on plankton and are in turn eaten by carnivorous fish, reptiles, birds and mammals.

Energy Flow in Rivers

In most ecosystems the energy base for successive trophic levels is provided by plants converting solar energy through photosynthesis into high-energy sugars. Exceptions to this rule are cave ecosystems (see chapter 6), where there is not enough light for green plants, and rivers, where much of the energy base is provided by dead organic matter. This organic matter consists of two components, allochthonous material, from outside the system (e.g., falling leaf litter, fruit) and autochthonous material, originating within the system (e.g., from photosynthesising plants). Most of the organic input to Bornean river systems is allochthonous. Organic material entering the river system can be divided into three classes: dissolved organic material (DOM) less than 0.45×10^{-3} mm in diameter, fine particulate organic matter less than 1 mm in diameter (FPOM) and coarse particulate organic matter (CPOM), including whole twigs and leaves (fig. 3.14). This plant and animal detritus forms the basis of the food chain and is processed by a host of riverine organisms. Bacteria and fungi decompose organic material, and a host of invertebrates, fish, reptiles and aquatic mammals are part of the food chain as algal grazers, shredders feeding on large plant parts, collectors (detritivores) and predators.

The majority of river headwaters in Kalimantan flow through forested catchments and receive a substantial allochthonous input from material falling from the forest canopy (fig. 3.14). Little light reaches the stream, thus hindering the growth of attached algae and macrophytes. As the river broadens, shading occurs only at the edges, and the autochthonous component increases (fig. 3.15). Both macrophytes and attached filamentous algae (such as *Spirogyra*) contribute significantly to energy input until the lower reaches, when depth and turbidity are such that light no longer reaches the river bed, and these plants are restricted to the river margins (Townsend 1980). Phytoplankton can only make a significant

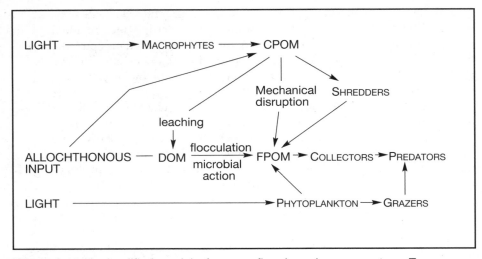

Figure 3.14. A simplified model of energy flow in a river ecosystem. To preserve clarity some arrows have been omitted. For example, all the animals contribute to FPOM in the form of faeces, dead bodies, etc.; some of the allochthonous input contributes directly to FPOM; the principal food of many fish in rivers consists of invertebrates, so the "predator" category includes fish. However, some fish feed on the macrophytes and detritus. (After Townsend 1980.)

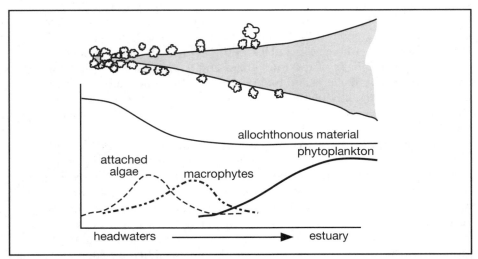

Figure 3.15. Hypothetical representation of the relative contributions of possible energy inputs to a river along its course. (After Townsend 1980.)

contribution where the river is long enough to allow this component to build up; this is true for the Kapuas. The generation time for phytoplankton is one to two days, but for zooplankton it is three to four days and frequently longer. Forest clearance in river headwaters can seriously disturb a river's energy input and therefore the life that depends on it (Townsend 1980).

Invertebrate Drift

After heavy rain, river flow increases, and the river bed is scoured, often sweeping away river organisms. Even in a river that is flowing normally, benthic organisms are being carried by the water; this is known as invertebrate drift. In a Malayan river 222,280 invertebrate specimens were estimated to drift past a transect in a major headwater stream in just 24 hours (Bishop 1973). This is equivalent to 160 individuals per 100 m^3 of river water. Drift varies with time of day as well as river flow. It is greatest just after sunset, when invertebrates emerge from beneath boulders to forage.

A study by Townsend and Hildrew (1976) showed that 2.6% of benthic invertebrates shifted their position each day, but 60% travelled less than 10 m before regaining a foothold (McLay 1970). Drifting is an energy-saving way of moving along the river bed to new food sources. This movement is not entirely passive; at least some invertebrates landing on an unsuitable substrate will re-enter the drift within five to thirty minutes (Walton 1978). Drifting is also an effective dispersal mechanism for eggs and larvae. Since drift always occurs downstream, how is a viable population maintained in river headwaters? Presumably there is a colonisation cycle whereby eggs are laid in the headwaters, larvae disperse downstream and adults move upstream again to breed. To date, however, there is very little evidence of how adult invertebrates recolonise headwaters. They may be airborne by the prevailing winds, or they may travel along the river edge in waters where stress and flow are least. Upstream movements represent only about 7% to 10% of the individuals that move downstream (Moss 1980; Williams 1981). Nevertheless, if only one female insect reaches the headwaters, she may lay hundreds or even thousands of eggs.

Freshwater Fishes and Vertebrate Predators

The **nekton** (free-swimming community) of freshwaters consists of fish and other vertebrates. The greatest number of freshwater fish species is found in tropical waters, and 394 species (149 endemic) are known for Borneo (Kottelat et al. 1993). More than 290 species are recorded for the Kapuas alone (Roberts 1989). The lower reaches of Kalimantan rivers are especially rich in species, more than one-third of them marine species such as rays and finfish. Kottelat et al. (1993) have prepared a useful field guide to the freshwater and estuarine fishes of western Indonesia, including Kalimantan. The ecology of tropical fishes is discussed by Lowe-McConnell (1977).

Like other freshwater organisms, fish show a series of intergrading but often distinct communities. The occurrence of a species in a given river depends on many environmental factors, levels of predation (including fishing) and interspecific competition. Species composition of fish communities varies between different river systems in South Kalimantan (KPSL-UNLAM 1989a). Similarly in Sarawak, studies of the fish communities of the Rivers Ai (31 species), Rajang (59 species) and Baram (43 species) showed only 9 species common to all three river systems, and 8, 10 and 11 species, respectively, were exclusive to a single river (Chan et al. 1985). In general as river size increases, so the number and diversity of fish species increase (Bishop 1973). At the same time the number of feeding guilds and the number of species in each guild also increase. A study in Sarawak found that while the number of bottom-dwelling species increased closer to the river mouth, there was a consequent reduction in the breadth of each species' niche as well as a decrease in mean life span and adult size of the individual species (Watson and Balon 1984).

Species composition of the fish fauna also changes along the course of a river (fig. 3.12). Factors that may influence this are the presence of plant material, the presence of a shading canopy (which affects water temperatures), the distribution of riffles and pools, and the invertebrate food supply. In the Kala'an River, South Kalimantan, the species community in shaded forest areas was different from that found downstream where the river flowed through cleared ladang lands (fig. 3.16). The adjacent lake community of Riam Kanan again showed different species, including introduced *Tilapia* (KPSL-UNLAM 1989a).

Within a river or lake community of fishes, different species occupy different levels or niches. A study in eastern Sabah revealed that most forest streams had one or two surface species, one or two living just below the surface, three to five midwater species, two to four living just above the bottom and three to ten living on or in the bottom (Inger and Chin 1962). Differences between sites in the proportions at each level could be accounted for partly by the nature of the bottom and by the current. For example, rivers with flat sandy or silty bottoms with few rocks tended to have of bottom-dwelling species than rivers with rocks, boulders and riffles. The ecological separation of freshwater fish communities has important implications for conservation since certain fish species, including some of commercial importance, may be represented only in one or a few river systems or only in certain parts of the riverine habitat.

The floodplains of several Kalimantan rivers are particularly important fisheries (Zehrfeld et al. 1985; Giesen 1987). As the river floods, fish move into small tributaries and onto the floodplains. Cyprinids, which are

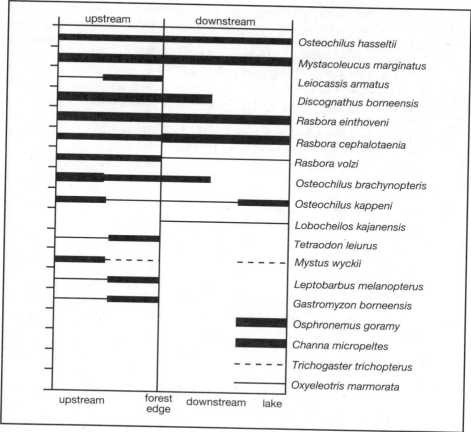

Figure 3.16. Distribution of some stream and river fish in the Kala'an River system, South Kalimantan.

Source: KPSL-UNLAM 1989a

important food fish, are dominant in Kalimantan rivers (Roberts 1989). They spawn on the floodplain almost as soon as they reach it; eggs ripen quickly and may hatch in a few days (Lowe-McConnell 1977). This means that the fry are produced just in time for the burst of algal, plant and invertebrate productivity that follows flooding.

During the period when the floodwaters are receding, the area of accessible aquatic plant beds is reduced. As the fish are concentrated into a smaller area, there is greater competition for food resources. Predation also increases from piscivorous fish, birds and other predators, including fishermen. This situation is well illustrated in the Kapuas lakes area in West Kalimantan (Giesen 1987). Only a few zooplankton feeders breed during this period (Moss 1980).

Fish and other free-swimming animals are wide ranging in their movements; they can exploit several parts of the aquatic ecosystem, modifying prey communities (both plant and animal) by selective feeding. Thus, when river levels rise, larger river fish may enter smaller side streams and, through competition and predation, may exert pressure on the resident communities. In a closed ecosystem like a fish pond the fish population literally may eat out their food source. Much evidence now exists for pronounced changes in zooplankton populations, even in large lakes, caused by fish predation on the larger members of the plankton in preference to smaller species (Moss 1980).

Little is known of the impact of vertebrate predators other than fish.

Box 3.3. Feeding guilds of fish.

Fish species can be divided into at least eight feeding guilds, although some species change their dietary preferences with age, season or the availability of alternative food items:

herbivore A (endogenous) - feeds on plant material, such as filamentous algae, fungal hyphae, diatoms and blue-green algae, growing in the water or in the mud;

herbivore B (exogenous) - feeds on plant material, such as fruits, seeds and leaves, that falls into the water;

predator 1 (endogenous) - feeds on small aquatic animals such as nematodes, rotifers, suspended plankton and other invertebrates, ingested as detritus in mud or sand;

predator 2 (endogenous) - feeds on insect larvae and other small aquatic animals;

predator 2 (exogenous) - feeds on aquatic animals such as insects and spiders that fall into the water;

predator 3 - feeds on larger aquatic animals such as shrimps, snails and small crabs, generally on or near the bottom;

predator 4 - feeds on other fish;

omnivore - feeds on both animal and plant material.

In tropical America fish have been shown to play an important role in actively dispersing seeds from fruits that drop into the water (Goulding 1980). In Southeast Asia some of the larger, *Puntius*-like species may play this role. Species such as the beaked puffers *Tetraodon* and *Chonerhinos* destroy any seeds they eat. Exogenous foods such as fruit and terrestrial invertebrates are important to many river fish, and the loss of waterside vegetation can have significant effects on fish communities.

The different feeding guilds relate to each other and to the food sources in various ways. Figure 3.17 illustrates these interactions in a forest stream in Sabah. Some of the most economically important fish species are predators, such as *Hampala macrolepidota*, *Clarias* spp., *Kryptopterus micronema*, *Channa* spp. and *Oxyeleotris marmorata*; these species are clearly dependent on their food webs remaining intact.

Sources: Inger and Chin 1962; Kottelat et al. 1993

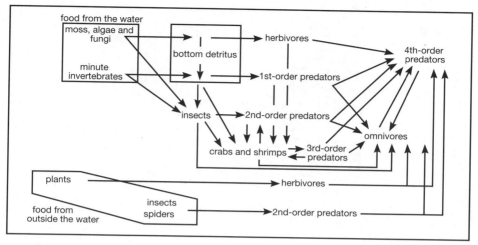

Figure 3.17. Generalised food web in a small forest stream in Sabah. "Herbivores", "predators" and "omnivores" refer to fishes only. (After Inger and Chin 1962.)

Riverine predators in Kalimantan include at least one endangered mammal, the rare Irrawaddy dolphin *Orcaella brevirostris* which feeds on fish and freshwater prawns. The most curious inhabitant of Kalimantan's rivers, the dolphin (pesut) is found in the Mahakam River and the freshwater lakes of Danau Semayang and Jempang in East Kalimantan (Wirawan 1986). Dolphins and whales are mainly marine creatures, but a few, like the pesut, are found in tropical rivers. Once found in several Bornean estuaries (Chan et al. 1985), the pesut is now rare except in the Mahakam River, where it is found hundreds of kilometres from the sea at Kota Bangun (fig. 3.18). Unlike other dolphins the pesut has a bulbous head and tiny eyes (probably both adaptations to muddy waters). These freshwater dolphins travel in small schools. In spite of their poor sight and the fact that they live in heavily silted waters, these river dolphins are expert at detecting and avoiding obstacles. It is likely that they use ultrasonics for echolocation as do their marine relatives (Kamminga et al. 1983).

Two species of crocodile, *Crocodylus porosus* and *Tomistoma schlegeli*, river turtles, monitor lizards and otters are all predators at the head of the riverine food chain. The curious otter civet *Cynogale bennetti* fishes the forest rivers; its dense hair, valvelike nostrils, thick whiskers and webbed toes equip it for the aquatic life, but it is also able to climb trees.

The river ecosystem is influenced by conditions and activities far beyond its banks. Wide-ranging water birds and amphibious mammals may import nutrients to a river or lake from outside the catchment area. Fishermen preying on fish populations can affect their species composition and

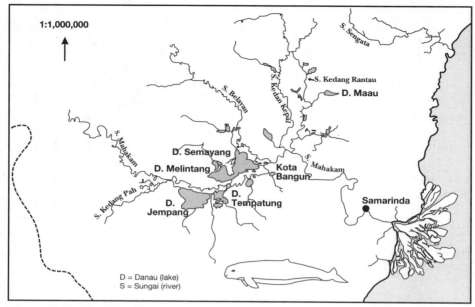

Figure 3.18. The Mahakam lakes system, freshwater habitat of the pesut *Orcaella brevirostris*.

demography. Human populations may also modify the riverine habitat by increasing organic inputs with human waste or by removing vegetation from the riverbank or shoreline.

Riverbank Communities

The open riverbanks receive far more sunlight than the surrounding forest floor and support a tangle of ground herbage and impenetrable vegetation. The tree species composition of these riverside communities generally reflects the composition of the inland forests. Nevertheless, certain species such as *Eugenia, Palaquium, Tristania*, with its peeling bark, *Pandanus* and colourful sealing-wax palms *Cyrtostachys lakka* are characteristic of the lowland riverbanks (Whitmore 1973). The riverbank habitat supports a high density of wildlife, including aquatic amphibians and snakes (tables 3.2 and 3.3). Large herbivores emerge from the forest to drink or to graze on the lush greenery at the river's edge. Wild pigs *Sus barbatus* root among the herbs, and sambar deer *Cervus unicolor* and muntjak *Muntiacus* spp. feed here at night. Banteng *Bos javanicus*, too, hide up in the forest fringe during the day but forage at night along the riverbanks.

The most visible animals of the riverbanks, especially in the early morning and late afternoon, are the monkeys: proboscis monkeys (chapter 2), long-tailed macaques and silvered langurs. The latter are silvery black as adults, but the infants are bright orange. This colouring makes the infant no more vulnerable to predators that lack colour vision but may act as a signal to other langurs to give the infant special care (MacKinnon and MacKinnon 1974). Although they often wander away from the river to forage during the day, monkey groups return to the riverbanks to sleep in tall trees, with many groups clumping close together. Whether this is a protection against predators or a means of maintaining social contact with neighbouring troops is not known. In any case, this behaviour is a great asset to the scientist wanting to census monkey populations in different swamp and riverine habitats (K. MacKinnon 1983). The monkeys may also play a valuable role in transporting nutrients from adjacent forests to the river ecosystem.

Table 3.2. Species distribution of Bornean amphibians and reptiles in broad habitat types.

	Frogs		Lizards		Snakes		Total	
	No.	%	No.	%	No.	%	No.	%
Sea and sea coast	0	0	1	1	17	12	18	5
Man-made habitats	8	9	9	9	0	0	17	5
Lowland forest	66	73	62	60	101	71	229	68
Montane forest	11	12	9	9	9	6	29	9
Unknown	6	7	23	22	16	11	45	13
Total	91		104		143		338	

Table 3.3. Species distribution of Bornean amphibians and reptiles in broad niche types.

	Frogs		Lizards		Snakes		Total	
	No.	%	No.	%	No.	%	No.	%
Fossorial (underground)	4	4	3	3	32	22	39	12
Cavernicolous (in caves)	0	0	1	1	0	0	1	0.3
Aquatic (in or close to water)	24	26	5	5	38	27	67	20
Arboreal (in trees)	20	22	58	56	25	17	103	30
Terrestrial (on ground)	22	24	7	7	48	34	17	23
Unknown	21	23	30	29	0	0	51	15
Total	91		104		143		338	

Source: Dun Committee 1982

The riverside is an important habitat for birds. Swallows and swiftlets swoop to catch insects that hover above the pools. Fish-eating raptors, the osprey *Pandion haliaetus,* sea eagle *Haliaetus leucogaster* and Brahminy kite *Haliastur indus,* patrol the skies above the river mouth. The Brahminy kite, with its chocolate body and white head, is a bird much revered by the Iban Dayaks who consider it a reincarnation of one of their foremost deities, Singalong Burong.

Many of the riverine birds prey on fish, and each has a different hunting strategy. Large herons stalk the shallows on stiltlike legs. The grey and purple herons stand boldly in the open, but green herons, night herons and bitterns are more secretive and keep in thick cover. The herons spear fish with the beak, using a sideways lunge of the neck. The eagles, osprey and kites snatch fish from just below the surface with curved talons. Brightly coloured kingfishers, most conspicuous of the riverine birds, dive from tree perches to stab their prey with heavy beaks. Not all kingfishers are fish eaters. The forest kingfishers prey on insects, but all adopt the same strategy, scanning their hunting ground from a perch, diving on the prey,

---

Box 3.4. Rheophytes - riverine plant specialists.

Plants especially adapted to living on the river beds and banks of fast-flowing streams are called rheophytes, and are typically shrubs with narrow leaves and brightly coloured fruits dispersed either by water or fish (van Steenis 1981). Many rheophytes belong to the families Euphorbiaceae, Melastomataceae, Meliaceae, Myrtaceae or Rubiaceae.

The ability of vegetation to grow in river beds is limited by the quantity of water that passes through at irregular intervals. Stream beds that are commonly dry or with just a trickle of water can become a raging torrent within hours of heavy rainfall upstream. Rheophytes typically possess an extensive root system and tough, flexible branches. The dipterocarp *Dipterocarpus oblongifolius,* for instance, grows on islets formed by boulders in shallow river beds. How rheophytes germinate, attach themselves to the river bed and disperse their seeds is mostly unknown.

The colourful *Saraca* and *D. oblongifolius* are rheophytes with fruits dispersed by water, but some seeds of rheophytic plants are fish-dispersed. Fish dispersal gets round the perennial problem of how plants can move upstream. Several species of riverine figs drop their fruit in the water, where they are eaten by catfish which disperse the seeds. Riverine species from Peninsular Malaysia that are fish-dispersed are *Aglaia ijzermannii, Disoxylum angustifolium* and *Ficus pyriformis* (Whitmore 1984a), and it is likely that some closely related Bornean species also employ this strategy. In Kalimantan the seeds of the commercially important ramin *Gonystylus bancanus* are dispersed by small catfish *Clarias batrachus* which inhabit the streams draining peat swamp forests (Anderson 1961b). At certain times of year the flesh of the fish is tainted from eating ramin fruits.

and returning to the perch to eat it.

Perhaps the most curious of all the fishing birds is the darter or snakebird, *Anhinga melanogaster*. Like other cormorants, the darter chases fish underwater. As it has little oil on its feathers, its buoyancy is reduced, and it can swim more easily beneath the surface. After hunting, these birds sit for hours on a riverside branch with wings outstretched, drying their feathers in the sun. At night these daytime predators are replaced by large fishing owls which patrol the river.

Several fish-eating birds, like the Brahminy kite, have white plumage on their undersides. This is an effective camouflage device. Fish looking up cannot distinguish a white outline from the sky, and the hunting bird has a better chance of surprising its prey. A small, predatory riverine fish *Eleoctris oxycephala* has adopted a similar strategy that confuses both would-be predators and its own intended prey. It positions itself beneath a floating leaf, then paddles the leaf where it wants to go. The fish cruises along beneath its protective shelter, safe from sharp-eyed birds and undetected by the smaller fish on which it preys, because they notice only the leaf outline and not the predator beneath (K. MacKinnon 1992).

River courses often serve as traffic arteries for birds, bats and other animals, which move through the forest fulfilling the roles of pollinators and seed dispersers. Ribbons of fruit bats *Pteropus vampyrus* follow the river's winding course when they leave their roosts for a night's foraging. Riverine forests often serve as corridors and refuges between remaining blocks of natural forest surrounded by open agricultural lands.

Kalimantan Lakes

Lakes can be formed in many ways, by landslides or glacial activity, in volcanic craters or rifts in the earth's surface, or as seasonal lakes and oxbows cut off large rivers. Most Bornean lakes are seasonal lakes or oxbows formed in river floodplains. Kalimantan's main lakes occur in the lowland river systems of the Mahakam, Barito, Negara, Kapuas and Seruyan rivers (table 3.4). Most lakes can be viewed as very slow-moving rivers in which the river bed has become very wide and deep. Many of the same species of plants and animals live both in rivers and lakes. This is especially true in Kalimantan where many of the lakes are ephemeral, formed when rivers flood. Where the water is less than 2 m deep (or deeper if it is very clear), aquatic macrophytes may be found. The fish include fast swimmers of open water, bottom feeders, species that skulk among the vegetation, and surface-layer fishes, which wait beneath overhanging vegetation and feed on falling detritus, fruits and small invertebrates (fig. 3.19).

Streams and rivers are relatively shallow, but lakes are of varying depth according to their formation. The depth determines much of how the lake ecosystem functions. The ecology of deepwater tropical lakes in Indonesia has been discussed thoroughly elsewhere (Green et al. 1976; Whitten et al. 1987a, 1987b). Most of the natural Kalimantan lakes are relatively shallow, and fed by the floodwaters of rivers. As a river widens and broadens, it becomes more and more like a shallow lake. Deeper lakes, whether natural such as Danau Empangau in West Kalimantan or man-made as at Riam Kanan in South Kalimantan, can be compared to rivers where the flow has been so reduced that the water body becomes stratified, either horizontally through the entry of different water sources or, more often, vertically. Water rapidly absorbs light and heat and this leads to a layering of warmer, illuminated water above deeper, darker, colder water. This creates a range of physical and chemical conditions, suitable for

Table 3.4. Major lakes of Kalimantan.

	size (ha)	depth* (m)
Kapuas floodplains		
Bekuan	1,268	-
Belida	600	-
Genali	2,000	-
Keleka Tangai	756	6.5
Luar	5,208	6.7
Pengembung	1,548	6.9
Sambor	673	-
Sekawi	672	-
Sentarum	2,324	7.2
Sependan	604	6.6
Seriang	1,412	6.6
Sumbai	800	-
Sumpa	664	7.3
Tekenang	1,564	-
Mahakam lakes		
Jempang	14,600	7-8
Semayang	10,300	-
Melintang	8,900	4-6
Sungai Negara wetlands		
Panggang	<1,000	1.8-2.4
Sambuyur	<1,000	-
Bangkau	c.1,000	2.0-3.2
Central Kalimantan		
Belajau (Sembuluh)	7,500	-

* average depth in wet season
All of these lakes are associated with backswamps which flood during the wet season to extend the area of open water.

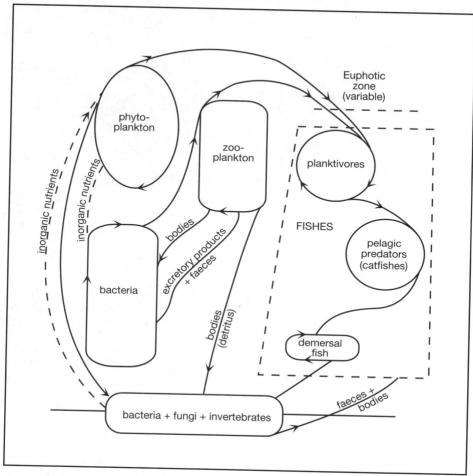

Figure 3.19. Spatial community structure in a tropical lake ecosystem.
Source: Payne 1986

different living organisms.

Only the deeper lakes are stratified (fig. 3.20). The upper layer (epilimnion) is the zone of greatest productivity. Dead and decomposing material, including phytoplankton, drifts downwards, using up oxygen as it decomposes. When nutrients are released back into the lake, there may be a bloom of algae. Eutrophic (nutrient-rich) habitats support good fisheries because of their high productivity. Naturally eutrophic sustems are usually well balanced, but the addition of artificial nutrients can upset this balance. The nutrient loadings of lakes can be increased significantly by wastes

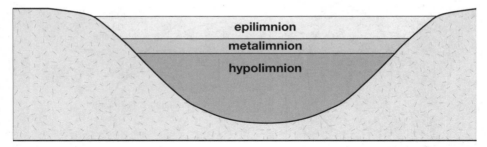

Figure 3.20. Hypothetical vertical section of a deep lake showing the layers.

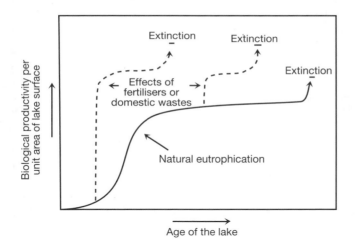

Figure 3.21. The pathways through which a lake may become extinct. The solid line indicates the process of natural eutrophication. Addition of nutrients from fertilisers or sewage speeds the process so that the extinction of the lake occurs sooner than under natural conditions.

Source: Simmons 1981

from human activities, as at Lake Toba in Sumatra. Excessive nutrient loading (**eutrophication**) leads first to increased production and then to stagnation as the algal blooms die and rot (fig. 3.21). The ensuing rapid decomposition of organic debris by bacteria robs the water of its oxygen, sometimes to the extent that fish and other organisms suffocate, and lake life becomes "extinct".

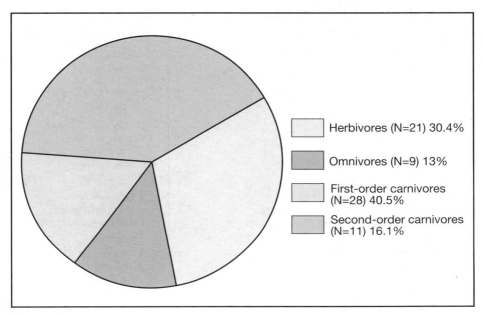

Figure 3.22. Feeding niches of fish species in the Sungai Negara wetlands, South Kalimantan. (After Chaeruddin 1990.)

Weedbeds and Swamps

Several of Kalimantan's major rivers, including the Mahakam, Negara and Kapuas, open out into great marshes and swampy lakes. Conditions within a well-developed swamp at the margin of a lake or river are different from those in adjacent open water. Swamp macrophytes usually have a very rich nutrient supply in the sediments where they are rooted. Emergent aquatic plant communities are among the most productive per unit area in the world (Maltby 1986). This is not surprising since they have ample supplies of nutrients and water, two of the main factors which limit growth on land. Aquatic plant communities can be expected, therefore, to support more secondary production, including fish, than open waters. The whole wetland ecosystem supports a diverse fish fauna, with species adapted to exploit different niches and food sources (table 3.5; fig. 3.22).

The shallow water conditions of swamps and floodplain lakes provide an environment that is physiologically more taxing than that found in fast-flowing or deeper waters. Shallow, still water can become very hot in dry weather. The hotter the water, the less oxygen it can hold. To survive in

Table 3.5. Fish feeding habitats in the Sungai Negara wetlands.

Diet	Number of species	Genera
Plankton	6	Dangila Helostoma Osteochilus Thynnichthys
Periphyton and leaves	9	Amblyrhynchichthys Epalzeorhynchus Osteochilus Puntius Trichopterus
Leaves, fruits and seeds of submerged, inundated vegetation	6	Leptobarbus Ctenopharyngodon Osphronemus Puntius
Zooplankton, water insects and larvae	6	Cryptopterus Macrognathus Polyacanthus Rasbora
Insects on the water surface or in aquatic vegetation	1	Toxotes
Organisms in sediments	10	Akysis, Anabas Clarias Macrones Mastacembelus Pangasius
Omnivorous	9	Arius Bagroides Ketengus Hemisilurus Macrones
Small fish, insects, shrimps and other small animals	11	Apodoglanis Callichrous Cryptopterus Datnioides Hampala Macrochirichthys Zenarchopterus
Small and larger invertebrates and fish	11	Belondontichthys Monopterus Notopterus Ophiocephalus Oxyeleotris Wallago Xenopterus

Source: Chaeruddin 1990

these conditions, swamp fish have developed additional ways to absorb oxygen. The swamp eels *Monopterus albus* and *Ophisternon bengalense* have many blood vessels in the walls of the gill cavity as well as a specialised oxygen-absorbing hind gut. The catfish *Clarias* and the anabantoids (gouramis, climbing perch *Anabas* and their relatives) have saclike chambers above the gill cavity (Kottelat et al. 1993). Some of these fish, such as *Monopterus, Anabas* and *Clarias,* are even able to travel overland, particularly on wet nights when there is little danger of drying out. Fishes of urban canals and ditches, typically the guppy *Poecilia reticulata,* the tinhead *Aplocheilus panchax,* and *Oryzias javanicus,* have superior mouths and flat heads, which allows them to take advantage of the oxygen-rich surface layer of the water without recourse to modified breathing organs.

Although a fringing swamp may appear to have a distinct boundary with the open water of its lake, the two habitats may have considerable influence on each other. Water moves from the swamp into the lake, flushed by catchment water, and carries organic and inorganic material with it. Swamps certainly alter the chemical composition of water passing through them; for instance, the Sudd swamps on the White Nile, Sudan, reduce the sulphate content of the water as it filters through (Talling 1957). Displacement of swamp water by heavy rain may cause fish kills in tropical lakes, as open-water fish are suddenly engulfed in deoxygenated water (Moss 1980). More gradual input, however, results in a zone of high animal production, as in the Mahakam and Kapuas lakes (Zehrfeld et al. 1985; Giesen 1987).

The flooded swamp forests and open swamps are important feeding, breeding and nursery sites for many freshwater fish which move freely between the swamp, lake and river habitats. These swampy areas are also feeding grounds for migrating waterbirds, including many species of ducks and waders. Some of the smaller lakes are important breeding sites for waterbirds such as herons, egrets and cormorants. The "Lake of a Thousand Birds" in the Tanjung Puting National Park is one such site.

Open swamp areas are also a favourite habitat for reptiles, including monitor lizards, turtles and tortoises. Freshwater swamps and rivers in Kalimantan once supported large populations of estuarine crocodiles *Crocodylus porosus* and false gavial *Tomistoma schlegeli,* but sadly these species have been hunted almost to extinction because of their valuable skins. Tanjung Puting National Park is one of the few sites in Kalimantan where both still occur.

Both crocodilians are predators at the head of the food chain, preying on fish. The false gavials are specialised fish eaters, but the crocodiles also take migrating pigs, deer, monkeys (Galdikas 1985b) or even the unwary villager who strays too close to the river's edge or is caught midstream. The crocodile dazes its prey with a swipe of the powerful tail, then tosses the victim in the air and catches it in vicious jaws; prey is usually drowned.

Crocodiles cache their food in underwater stores beneath riverbanks, a habit which has given rise to some lurid stories.

Estuarine crocodiles are found in a range of habitats from coastal mangrove swamps to freshwater swamps far inland. Like all animals that can live in both seawater and freshwater habitats, the estuarine crocodile has physiological adaptations that allow it to control the osmotic pressure of its plasma. In the sea the crocodile conserves water by reabsorbing it in its kidneys and from its faeces. Calcium, magnesium, potassium, ammonium, uric acid and bicarbonate are excreted in the urine. Sodium

---

Box 3.5. Watchful mothers - parental care in crocodiles.

Crocodiles are the last surviving members of the Archosaurs, the group of reptiles, including the dinosaurs, that inhabited the earth during the Mesozoic. Like all reptiles, crocodiles lay eggs. The female seeks a shady location on the riverbank where she builds a dome-shaped nest of leaves or tall grass. Fifty or more eggs are laid here and covered with vegetation, which soon becomes a rotting mass. The eggs are kept damp, protected from direct sunlight and at a constant temperature in the range 30°C to 32°C. Incubation temperature is important, as it determines the sex of the young crocodiles. Temperatures of 30°C or less produce all female offspring; high temperatures (34°C) produce all male young. Intermediate temperatures produce a batch of mixed offspring. Sometimes the mother sprays the nest with urine if she senses that the eggs are becoming too hot.

A crocodile's brood is vulnerable at three stages. Predation occurs on eggs, on hatchlings as they emerge and on young crocodiles in the water. Many mammals and reptiles are persistent egg hunters, including wild and feral pigs, monitor lizards and people. Monitor lizards, feral dogs, raptors and crows take newly emerged hatchlings before they reach the water. Once in the water young crocodiles are preyed upon by large fish and wading birds such as storks and herons. To minimise this predation, maternal care in many crocodile species begins as soon as the eggs are laid, and sometimes lasts long after hatching.

The female estuarine crocodile *Crocodylus porosus* excavates one or more wallows just beside her nest mound and lies here, submerged and unseen but on guard. False gavials *Tomistoma schlegeli* are also known to visit their nests regularly. When ready to emerge, the hatchlings break through the hard egg shell and, with just their snouts poking out, wait for their mother. They will respond to any stimulation, such as vibrations, caused by the mother crawling over the nest area, by calling with loud, high-pitched grunts that can be heard at considerable distance and are probably the trigger for the mother to excavate the nest. In many species the female picks up eggs that have not hatched and gently crushes them in her jaws to free the late hatchlings. In captivity, males have been observed to help.

The female crocodile gently picks up the young in her mouth to carry them to a safe nursery area in the water, where they will stay for one or two months feeding on insects, fish and frogs. The female carries the young in batches in her gular (throat) pouch. Hunters have reported approaching a female estuarine crocodile which "swallowed" its young before diving. Evidently this was a mother taking her offspring into her mouth for a ride to safety. The female guards her vulnerable young closely. This maternal care lasts for at least the first few months and often longer. American alligators are known to care for their young for as long as one or even two years.

Source: Halliday and Adler 1987

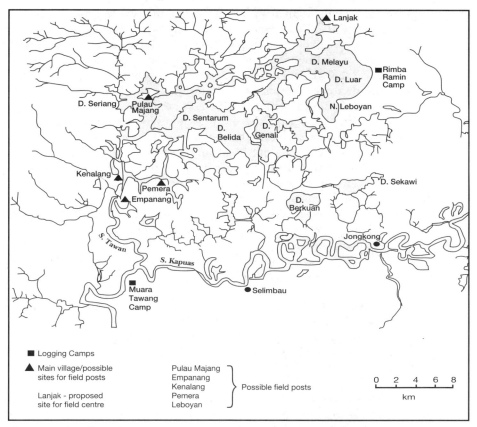

Figure 3.23. The Kapuas lakes and Danau Sentarum Wildlife Reserve, West Kalimantan.

chloride does not pass into the urine but is excreted through the crocodile's external nasal gland and lachrymal glands in the corner of the eye, giving rise to the legend of "crocodile tears".

The Kapuas Lakes - a Wetlands Ecosystem

The Kapuas lakes in West Kalimantan lie in an inland basin of about 6,500 km^2 hemmed in by mountain ranges, the Upper Kapuas Mountains to the north, Müller Mountains to the east, Madi Plateau to the south and Kelingkang Mountains in the west. The lakes are shallow and ephemeral and are best regarded as temporary floodplain extensions of the Kapuas River (fig. 3.23).

During periods of heavy rainfall during the wet season the Upper Kapuas rises by 10-12 m or more. As the Kapuas waters rise, floodwaters enter the lake area via the Tawang River to create a series of ephemeral lakes. During peak floods the water flow from the Kapuas to the Tawang is estimated at 1,000 m^3/s. The Tawang siphons off about one-quarter of the Kapuas floodwaters (Giesen 1987), thereby preventing major floods in the lower river basin (Molengraaff 1900; Polak 1949).

During the wet season the Kapuas lakes system is a complex of interconnecting waterways and flooded forests. Giesen (1987) mapped 83 lakes with a total area of about 27,500 ha. Of these 39 were less than 100 ha in size, and only seven were larger than 1,000 ha at the time of maximum depth at the end of the wet season. By August all the lakes were dry except for a few former Kapuas River channels and Danau Empangau, a deep oxbow lake reported to be 15-20 m deep during peak floods. The total water-holding capacity of the upper Kapuas basin in the open-water lakes and seasonally inundated forest was estimated to be about 3.8×10^9 m^3, about 6% of the total annual volume of the Kapuas (Giesen 1987). Thus the Kapuas lakes can be seen to play an important water-regulating role in the Kapuas basin.

The waters of the lake region are high in humic acid, giving them the colour of weak tea, and can be described as "tropical blackwaters" (Janzen 1974). Vegetation types in the Danau Sentarum area range from aquatic vegetation on the open lakes, through freshwater and peat swamp forests and heath forest, to lowland and hill dipterocarp forest on the surrounding hills.

Two factors are of primary importance in the ecology of the Kapuas floodplains: seasonality and the open nature of the ecosystem. The very existence of the floodplain lakes depends upon high levels of rainfall in the upper Kapuas basin. When monthly rainfall is about 300 mm or more, the Kapuas spills into the floodplain lakes. If rainfall is less, lake levels drop. If this situation persists for three months or more, many of the lakes dry out; this phenomenon probably occurs two years out of three.

During the early floods, fish enter the floodplain to spawn and feed, and migratory birds, bees and mammals move out. Peak flood time is a period of dormancy for the inundated vegetation, with little flowering and fruiting. Fish populations flourish, but densities are low; birds and mammals are scarce on the floodplains other than adapted species such as kingfishers and proboscis monkeys. As the floodwaters recede and the lakes shrink, fish populations become concentrated and attract fishermen from the lake and Kapuas River villages. Whitefish (mainly cyprinids) migrate to the Kapuas, while blackfish (catfish, channids and anabantids) remain in floodplain waters. Terrestrial mammals and migratory birds return to the area to feed. In the dry season (June to September), vegetation sprouts on the dry lake beds, and inundated forest begins to flower and set seed, attracting bees, other pollinators and seed predators.

The Kapuas floodplain lakes form an open ecosystem with energy flow in and out. Floodwaters move in and out; fish migrate in from the Kapuas, feed, breed and emigrate back to the river system, or are caught by fishermen. Vegetable matter from immersed shrubs and trees forms the base of the food web, in terms of both energy and nutrients. Phytoplankton and algal mats play a secondary role. Fish are the main herbivores and

---

Box 3.6. The freshwater fish of the Kapuas River system.

During the last Ice Age until about 6000 B.P. the Kapuas was connected with the rivers of south Sumatra, Java and the Malay peninsula via the Sunda river system, which is now flooded. The Sunda drainage system, the only great equatorial river system of Asia, was an important evolutionary centre for many fish groups, especially the cyprinids. As a result, the Kapuas probably supports the richest fish fauna of the Sundaland rivers.

The Kapuas drainage system supports 290 species of freshwater fish, from 120 genera and 40 families. The two largest groups are carp and catfish. The Kapuas fish fauna is typical of tropical freshwater systems, rich in species but with individuals of each species rare. Many species are small or minute. The food habits and modes of reproduction of the Kapuas fish are diverse. Kapuas fish feed on a range of foods from algae to fruit, insects and other fish. Some fish have remarkable feeding habits. The halfbeak *Hemiramphodon pogonognathus*, for instance, feeds almost exclusively on terrestrial insects, especially ants. Very little is known about fruit- and seed-eating fishes in the Kapuas, but the fact that many trees in the lakes area fruit just as the forests are submerged by floodwaters suggests that fish are important seed dispersers.

The way in which fish care for their young is also ecologically significant. Fish show virtually every reproductive adaptation known in vertebrates. Eggs can be fertilised either externally (*Chaca*) or internally (Phallostethidae). The Kapuas has a diverse assemblage of mouth-brooding fish involving at least 11 species from six of the 11 world families known to brood in the mouth. Mouth brooders in the Kapuas include *Channa orientalis, Betta pugnax, Sphaerichthys,* Luciocephalidae, arowana *Scleropages formosus* and Charidae. In four species of the freshwater pipefish *Doryichthys*, the male carries the developing young in an abdominal brood pouch. The Kapuas is also home to a viviparous freshwater elasmobranch and four species of viviparous halfbeaks (Hemiramphidae). Fish in the Kapuas breed throughout the year, though certain species, including *Mastocembelus erythrotaenia*, have restricted reproductive periods. The migrations of the large ikan tapah *Wallago leerii* are probably related to reproduction.

Kapuas fish also illustrate different strategies for providing food to their young. The Bornean catfish may secrete a milk-like substance from the pectoral glands to feed the fry. In several species, fry feed on eggs produced by the mother or on food regurgitated by the parent. Feeding on faeces is important among fish in general. Predatory fish with protracted parental care, for example, the snakehead *Channa micropeltes*, may kill excess prey to supply food for the young fry. By employing different feeding and reproductive strategies, the Kapuas fish achieve ecological separation and are able to exploit the wide range of riverine niches.

Source: Roberts 1989

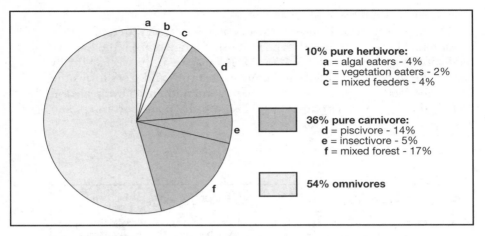

Figure 3.24. Feeding habits of Kapuas fish.
Source: Giesen 1987

detritivores, but zooplankton, hydrozoans and molluscs also exploit these niches. First- and second-order carnivores are mainly fish but also include crocodiles, birds, otters and people. Frogs are virtually absent on the floodplains. More than half the Kapuas fish are omnivores, and 36% are predators (fig. 3.24); in table 3.6 the fish of the Kapuas ecosystem are classified according to their feeding habits.

Both plant life and animal life in the lake system are adapted to the seasonal inundations of floodwaters. This is a relatively harsh environment, characterised by stunted vegetation which is dominated by just a few tree families: Anacardiaceae, Dipterocarpaceae, Ebenaceae, Euphorbiaceae, Guttiferae and Myrtaceae. Fish are the most abundant and diverse fauna of the lakes system with 113 species recorded (Giesen 1987; fig. 3.25).

The floodplains are major fisheries, benefiting both local villages and, via exports, the whole province. Fish yields are an estimated 37.5 kg/ha for the lakes and inundated forest together, or 75 kg/ha for the lakes alone. Two-thirds of all freshwater fish caught in West Kalimantan come from the Upper Kapuas basin, and about half of this total is caught in Danau Sentarum. This productivity depends on the seasonal flooding of the forests and the rise and fall of the Kapuas. Any changes to this hydrological regime could seriously affect these fisheries and the livelihood of the local communities that depend upon them. The areas have been extensively fished for centuries (Beccari 1904) with little impact on fish populations, but some species now seem to be showing the first signs of overexploitation (Giesen 1987). Recent developments in the Kapuas basin, including logging and the construction of a new road from Sintang to Putussibau, have opened up new lands, giving access to immigrant farmers who are clearing parts of the watershed. Plans for mining and construction of dams and irrigation works in the Upper Kapuas could all have adverse effects on the area's hydrology and on the lake fisheries, and should be assessed and monitored carefully.

The Danau Sentarum Wildlife Reserve preserves a unique ecosystem of interconnecting seasonal lakes, peat swamp and inundated freshwater swamp forests. It also protects the role of the lakes system in regulating water flow and preventing floods, a benefit that is felt all the way along the lower course of the river to the river mouth. The uniqueness of the vegetation/fish ecosystem justifies protection for conservation and rational utilisation of the resource. Danau Sentarum reserve has been identified as a major Kalimantan wetland (Silvius et al. 1987) and an important addition to the Kalimantan reserve network (MacKinnon and MacKinnon 1986). In 1992, with assistance from the U.K. government, the Asian Wetland Bureau (AWB) and the Indonesian Conservation Department (PHPA) began to develop a comprehensive management strategy for the reserve, involving local communities in protection and management of this important freshwater ecosystem.

HUMAN-MADE LAKES

Human activities exploit and affect natural wetlands but can also lead to the creation of new wetland habitats: lakes behind dams for reservoirs, extensive areas of *sawah* (fields for wet-rice cultivation), and fish ponds. Large dams for hydroelectric projects have been built at Riam Kanan in South Kalimantan and at Batang Ai in Sarawak. As the water rises behind a new dam, it tends to be turbid from river sediment, and phytoplankton growth is hindered. Flooding of the old river valley submerges vegetation which dies and begins to rot. This and the settling sediment release nutrients into the water, leading to an increase in plant and animal growth. Many of the original trees remain dead but standing for several years, and their submerged branches become covered with abundant periphyton organisms, stimulated by high nutrient levels. These algae and the high production of invertebrates in the littoral zone of the lake contribute to high fish production in the early stages. High nutrient levels in new lakes may also

Table 3.6. Fish fauna of the Kapuas and adjacent lakes and rivers, classified according to their feeding habits.

	a	b	c	d	e	f	g	h	i	j	k	l
Plankton feeders												
Helastoma temmincki	m					a						
Thynnichthys thynnoules	m	m			a	a						
Dungila ocellata	m	a			a	a						
Dungila festiva	m	a	a			a						
Periphyton and vegetable feeders												
Amblyrynchichthys truncatus	a	m		a		a						
Osteocheilus melanopleura	a	m	a	a	m							
O. brevicaudata	a	m	a	a	m							
O. waandersi	a	m	a	a	m							
O. vittatus	a	m	a	a	m							
Vegetable feeders on submerged higher plants (inundated land plants, fruits and seeds)												
Puntius waandersi				a		m		a				
P. nini	a		a	a	m	a	a	a				
P. bulu				a		m		a				
P. schwanefeldi				a		m		a				
Leptobarbus hoeveni				a		m						
L. melanotaema				a		m						
Pristolepis fasciatus				a		m						
Osphromenus goramy				a	a	m		a				
Omnivores feeding mainly on insects and larvae, zooplankton												
Balantiocheilus melanopterus	a							m	a			
Cyclocheilus repasson					a			m				
Luctosoma trinema	a				a	a	m					
Rasbora argyrotaema	a				a	a	m					
R. vaillanti	a				a	a	m	a	a			
Eaters of insects at surface												
Chela oxygastroides							a	m	a	a		
Toxotes chatareus							a	a	a	m	a	
Omnivorous bottom feeders												
Barynotus microlepis					a			m	a			
Pangasius polyuranodon					a			a			m	
Mastacembelus armatus					a			a			m	
M. argus					a			a			m	
Omnivorous predators												
Macrones nigriceps							a	m		m		a
M. nemurus							a	m		m		a
Hermisilurus chaperi							a					m
H. scleronema							a					m

m = main food   a = additional food

Key to headings:
a = Phytoplankton
b = Periphyton
c = Thread algae
d = Bottom algae
e = Submerged plants (inundated land plants, fruits, seeds)
f = Small zooplankton
g = Cladocera, copepods, rotifera
h = Insects and their larvae
i = Aerial insects
j = Shrimps
k = Insects, larvae, worms
l = Fish, prawns, crabs

Table 3.6. *(continued)* Fish fauna of the Kapuas and adjacent lakes and rivers, classified according to their feeding habits.

	a	b	c	d	e	f	g	h	i	j	k	l
Predators on small fish and small animals, insects, shrimps												
Lycothrissa crocodilus												
Cryptopterus cryptopterus								a		a		m
C. schilbeides								a		a		m
C. limpok								a		a		m
C. micronema								a		a		m
Macrochirichthys macrochirus								a		a		m
Setipinna melanochir								a		m		m
Datninoides microlepis												m
Hampala bimaculata										m		m
								a		a		m
Large predators eating fish of all sizes, shrimp, prawn and crabs												
Ophicepalus striatus												
O. micropeltis										a		m
O. pleurophthalmus										a		m
O. lucius										a		m
Notopterus chitala										a		m
Wallago leeri										a		m
Silurodes hypothalmus										a		m
										a		m

m = main food   a = additional food

Key to headings:

a = Phytoplankton
b = Periphyton
c = Thread algae
d = Bottom algae
e = Submerged plants (inundated land plants, fruits, seeds)
f = Small zooplankton
g = Cladocera, copepods, rotifera
h = Insects and their larvae
i = Aerial insects
j = Shrimps
k = Insects, larvae, worms
l = Fish, prawns, crabs

Source: Giesen 1987

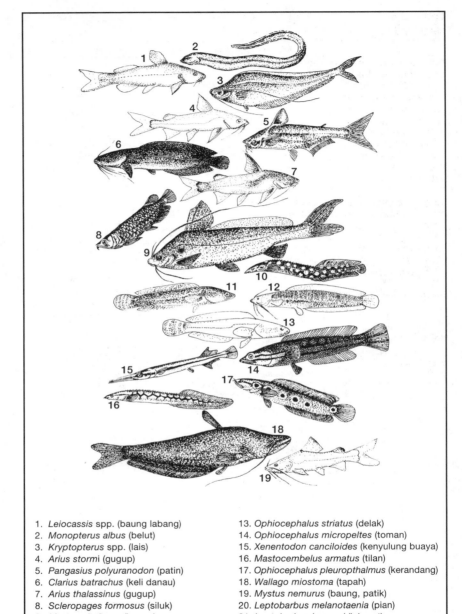

Figure 3.25. Some fish of the Kapuas lakes, illustrating diversity of form which can be related to feeding and habitat niches (not drawn to scale).

1. *Leiocassis* spp. (baung labang)
2. *Monopterus albus* (belut)
3. *Kryptopterus* spp. (lais)
4. *Arius stormi* (gugup)
5. *Pangasius polyuranodon* (patin)
6. *Clarius batrachus* (keli danau)
7. *Arius thalassinus* (gugup)
8. *Scleropages formosus* (siluk)
9. *Mystus nigriceps* (baung)
10. *Mastocembelus argus* (tilan)
11. *Ophiocephalus melanosoma* (runtuk)
12. *Clarias teysmanni* (keli sungai)
13. *Ophiocephalus striatus* (delak)
14. *Ophiocephalus micropeltes* (toman)
15. *Xenentodon canciloides* (kenyulung buaya)
16. *Mastocembelus armatus* (tilan)
17. *Ophiocephalus pleuropthalmus* (kerandang)
18. *Wallago miostoma* (tapah)
19. *Mystus nemurus* (baung, patik)
20. *Leptobarbus melanotaenia* (pian)
21. *Leptobarbus hoeveni* (jelawat)
22. *Puntius tetrazona* (berbaju)
23. *Osteochilus vittatus* (bantak)
24. *Luciosoma* spp. (kelekui)

25. *Chela oxygastroides* (kelampa)
26. *Pristolepis fasciatus* (patong)
27. *Tor douronensis* (kulir)
28. *Rasbora myersi* (jajau)
29. *Puntius collingwoodi* (tengadak)
30. *Epalzeorhynchus kalopterus* (tengibau)
31. *Puntius bulu* (ternengelang)
32. *Hampala macrolepidata* (langkung)
33. *Helostoma temmincki* (biawan)
34. *Tetraodon leiurus* (bantal lemas)
35. *Puntius schwanefeldi* (suain)
36. *Trichogaster leeri* (sepat)
37. *Trichogaster pectoralis* (sepat siam)
38. *Tetraodon fluviatilis* (bantal)
39. *Cylocheilichthys apogon* (buin)
40. *Rasbora argyrotaenia* (seluang maram)
41. *Oxyleotris marmorata* (ketutuk)
42. *Botia macrantha* (ulanguli)
43. *Cynoglossus wandersii* (kenilah)
44. *Notopterus chitala* (belida)
45. *Osphronemus goramy* (kaloi)
46. *Anabastes testudineus*

Figure 3.25. (*continued*)

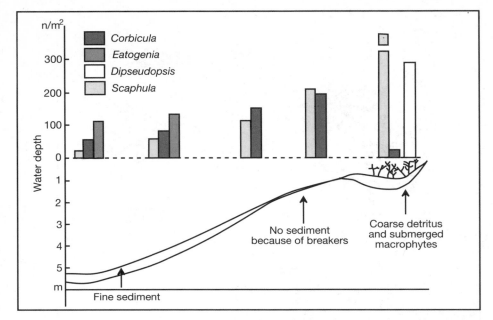

Figure 3.26. Distribution of benthic organisms in Bung Borapet reservoir, Thailand. (After Junk 1975.)

lead to a flourishing crop of the exotic water hyacinth *Eichhornia crassipes*. Mats of this pantropical weed, originating from South America, can become a problem, closing the open waters and clogging hydroelectric turbines and irrigation canals (Soerjani 1980).

As the lake stabilises and the internal nutrient loading declines, the lakes settle to a less fertile state, and the productivity of lake fisheries also declines. Settling sediment produces different substrates which are colonised by different benthic organisms (fig. 3.26). The stages of filling a man-made lake create an extended period of rising river level for the original river fish fauna, and most species flourish initially. As the lake level stabilises, some species disappear from the main waterbody but may persist near river mouths if they need flowing water for spawning (Moss 1980).

During the formation of a new lake, reduced water flow below the dam may change the downstream ecology with resulting loss of species.

Once a new lake has formed, the average river flow below the dam may be only a little less than before, but some fish species may not return. If water is removed from the lake for irrigation at certain times of year (as is planned at Riam Kanan) changes in the seasonal flow pattern may cause changes in the lower river ecosystem (Moss 1980). Once the river is dammed, the silt load is deposited in the lake itself. This is a major reason to protect the watershed area, to reduce the levels of siltation in the lake, to stop silting of the turbines, and to extend the life of irrigation schemes (Sumardja et al. 1984). Construction of a dam may also prevent or interfere with fish migration upriver for spawning, so damming a river may lead to loss of some fish species, with serious adverse effects on local fisheries.

The creation of new lakes can also have important impacts on local communities. River valleys in the tropics are relatively densely populated, and creation of the Riam Kanan dam flooded some of the best alluvial rice-growing lands and fruit groves (Schweithelm 1987). Villagers were forced onto less productive hillsides and began to deforest the watershed.

New lakes and other artificially created still-water bodies may harbour the hosts of waterborne diseases. Irrigation canals extend the breeding habitats of disease vectors such as mosquitoes (table 3.7; fig. 3.27). Fortunately schistosomiasis, common in the Lake Lindu area of Sulawesi, does not occur in Kalimantan. Mosquito-borne malaria and Bancroftian filariasis are both common in Kalimantan. Filariasis is caused by the nematode *Wuchereria bancrofti*, which has been recorded in both humans and the silvered langur. The adult nematode lives in dilated lymph vessels in the human host; blockage of these causes the swelling called elephantiasis. Microfilariae are ingested in blood meals by biting mosquitoes and transmitted to new human hosts when the mosquito bites again. The filarial parasite may be carried by *Anopheles, Culex* and *Mansonia* mosquitoes. *Mansonia* can be controlled by weed clearance since it lays its eggs on the undersurface of floating water plants such as water hyacinth. By understanding the ecology of the vector, it is possible to take appropriate action to control the disease (chapter 13).

Human Impacts on Freshwater Ecosystems

Freshwater systems are important to human communities for drinking water, other domestic and industrial purposes, agriculture and aquaculture, hydroelectric power, commercial and subsistence fisheries, navigation, irrigation, scientific research, and recreation. The bulk of animal protein available to most of the world's population comes from fish, and freshwater fish are critically important. The Bornean river and lake fisheries are an important natural resource and a source of food and income to local

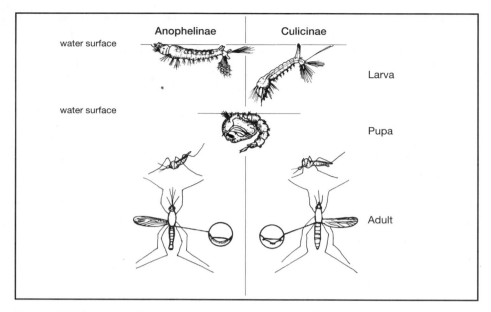

Figure 3.27. Some differences in body shape and lifestyle between Anophelinae (*Anopheles*) and Culicinae (*Aedes*, *Culex*, *Mansonia*) mosquitoes.
Source: Whitten et al. 1987b

Table 3.7. Disease-carrying mosquitoes of Kalimantan and their major breeding habitats.

	Breeding sites	Diseases carried
Anopheles	fresh or brackish water	malaria
Culex	clean water	malaria, filariasis
Aedes	clean water	dengue fever
Mansonia	clean water	filariasis

Source: Whitten et al. 1987b

communities (Zehrfeld et al. 1985; Giesen 1987; Watson 1982; Chan et al. 1985).

The complex food webs and inter-relationships of freshwater communities are easily disrupted by environmental changes. Forest loss is by far the most serious threat to freshwater ecosystems. Aquatic communities are dependent on forests as animal and plant detritus, from overhanging vegetation and riparian forests, forms the basis of the food chain for many invertebrates and fish.

When forests are destroyed, water temperatures rise because of decreased shading. As water temperature increases, the concentration of dissolved oxygen in that water decreases. The higher the water temperature, the greater the metabolic rate of the fish (and of some other aquatic animals) and the greater their demand for oxygen, but the lower the affinity of their haemoglobin for oxygen. The effect is aggravated where there is decaying organic matter present, which uses oxygen in the decomposition process. During the night aquatic plants do not photosynthesise, so oxygen concentrations can fall rapidly, perhaps below the threshold required by certain species. This can lead to mass fish mortality as observed in the deforested lower reaches of the Kanan River in Sumatra when water levels were very low during the dry season of 1984 (Santiapillai and Suprahman 1984).

In inland areas of Kalimantan, deforestation, due to logging and shifting agriculture, leads to increased sediment loads in rivers, upsets hydrological regimes, and may cause flooding. Siltation and turbidity, caused by soil washed down from cleared and eroding hillsides, limit light penetration with serious implications for aquatic plant life. Silt encrusts and smothers photosynthetic algae at the base of the freshwater food chain (Ang 1984). Increased sediment loads also impact on aquatic animal communities.

Suspended sediment and flocculated iron salts can accumulate on the gills of certain fish, causing death from suffocation. As river flow slows, the silt settles on the bottom and can smother food resources, eggs and spawning grounds, as well as reducing the depth and width of the channel. This can have serious effects on bottom-dwelling species such as loaches. Exposed silt can be colonised by plants; this gives permanence to the change. All these effects of deforestation can be reduced by maintenance of buffer strips of forest, of perhaps 30 m or wider, along rivers (Newbold et al. 1980)

Human activities modify aquatic ecosystems through predation (by subsistence or commercial fishing); through changing nutrient loadings by input of effluents or by altering sediment loads through land-use changes; and by use of waterways. The most common pollutants of water are: high levels of organic matter, plant nutrients, suspended mineral particles and deoxygenating substances; heat; and small quantities of poisons such as heavy metals, pesticides, chemicals, acids and radioactive substances. Artisan gold miners, using cyanide and arsenic in the extraction process, are polluting the upper Barito (chapter 13). Inland lakes may become polluted with human waste and fertiliser runoff, while urban and industrial effluents threaten river life in the lower reaches.

Plants and animals associated with freshwaters are closely dependent on the quality of the water and the adjoining riparian or lake environment. Microscopic algae, especially desmids and red algae and, more importantly, decomposer bacteria and fungi (which can survive without light), all process dead and waste organic matter and thereby purify the water. The most numerous animals are detritus-eating and predatory insects (especially bugs, beetles and dragonflies) which also help to break down organic matter.

Conditions created by human activities tend to be more extreme variations of situations that may occur naturally (Moss 1980). Thus the discharge of raw sewage creates conditions paralleled by the decomposition of an animal corpse in a small stream. Leaching of agricultural fertilisers is an extension of conditions in naturally fertile catchments. Natural hot volcanic springs are often acid, with high concentrations of toxic heavy metals. This does not mean that uncontrolled effluent is acceptable, but indicates that often there are natural organisms that can cope with the polluted conditions. Sometimes these organisms can be harnessed to treat pollutants biologically, as happens in sewage treatment (Moss 1980).

Living organisms can be useful biological indicators of environmental conditions. The resident biota of upland streams, for instance, require waters that are naturally clear and well aerated; they are poorly adapted to survive in oxygen-depleted or otherwise biologically polluted waters. The presence of invertebrates such as dragonflies (Odonata), anyclid snails, mayflies (Ephemeroptera) and midges (Chironomidae) are good indicators of the ecological health of upland streams. Unpolluted waters also support a variety of fishes, predominantly of the carp family (Cyprinidae) which breed in upland streams. In lowland rivers, dragonflies and bugs (Hemiptera) are useful indicators of water quality since they are sensitive to disturbance and pollution (Chan et al. 1985). Certain fish and waterfowl are also good indicators of the health and environmental quality of freshwater systems; they are at or near the top of freshwater food chains and are highly susceptible to wetland contamination and disturbance (Kottelat et al. 1993; Scott 1989).

Chapter Four

The Lowland Rainforest of Borneo

Evergreen rainforest is the natural vegetation type of the hot, wet tropics and one of the most favourable environments for plant growth on earth. Here the two main prerequisites for life, sunshine and water, are plentiful. The constant high temperatures and humidity, the heavy year-round rainfall, and the long hours of sunshine in equatorial regions together provide optimal conditions for maximum plant growth.

The rainforests of Borneo have had a long and relatively stable history. Earliest evidence of the occurrence of dipterocarps in Borneo is from fossil pollen in Sarawak from more than 30 million years ago (Muller 1970). Although the climate of Malesia cooled appreciably during the Ice Ages of the Pleistocene, and altitudinal zones were lowered, the rainforests below 1,000 m did not suffer periods of prolonged drought (Jacobs 1988). The area covered by rainforests may have expanded and contracted several times during the Pleistocene, but they remained essentially unchanged in character and composition.

The rainforest's long history has allowed a great diversity of plants to evolve. There may be as many as 240 different species of tree growing within one hectare of lowland forest in Kalimantan (fig. 4.1), and the neighbouring hectare may add half as many again to the list (Kartawinata et al. 1981; Ashton 1989). Tree species abundance and diversity is greatest in the valleys, and declines with increasing altitude (fig. 4.2). The flora of the whole of Malesia (Malaysia, Indonesia, the Philippines and New Guinea) comprises between 25,000 and 30,000 plant species (van Steenis 1950). The flora of Borneo alone consists of about 10,000 to 15,000 species (Burley 1991), compared to 8,500 for the Malay Peninsula, 8,000 for Sumatra, 7,000 for the Philippines and 9,000 for New Guinea (Jacobs 1988). The flora of Borneo is richer than that of the whole continent of Africa, which is 40 times larger. Small areas of Bornean rainforest are richer than similar areas in Africa and most similar areas in tropical America (Whitmore 1984a). Small plot tree diversity is as high as found anywhere in New Guinea. About two-thirds of all these species are found only in lowland forests. Borneo has more than 3,000 species of trees (Whitmore and Tantra 1987), 2,000 orchids and numerous ferns, palms, vines, mosses and fungi. It is the centre of distribution for dipterocarps, the

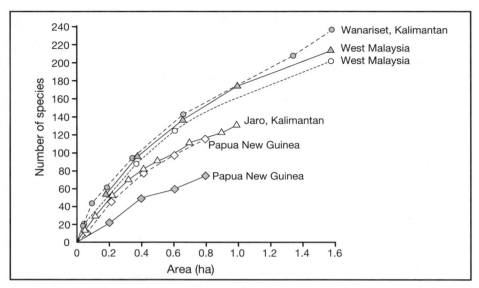

Figure 4.1. Species-area curves for small plots in tropical lowland evergreen rainforest.
Source: Whitmore 1984a

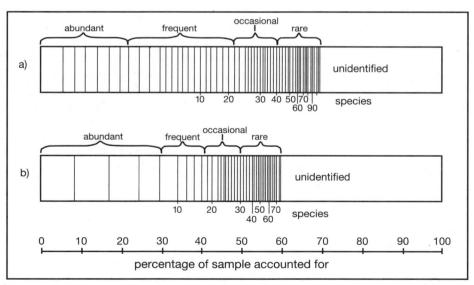

Figure 4.2. Tree species abundance and diversity in **a)** valleys and **b)** hills in rainforest, Ulu Segama, Sabah.
Source: MacKinnon 1974a

most commercially important timber trees of Southeast Asia, with 267 species, 60% of them endemic (Ashton 1982).

The term "lowland rainforest" includes both wet swamp forests (mangrove, freshwater swamp and peat swamp forest) as well as dry lowland forests. The best general account of all of these forest types for Borneo and the rest of Southeast Asia is Whitmore's classic book *Tropical Forests of the Far East* (Whitmore 1984a). The Malesian rainforests are the richest in species in the world. Good accounts of rainforest ecology are given by Jacobs (1988), and Whitten and associates (Whitten et al. 1987a, 1987b). Photographic essays celebrating some of this diversity are presented by Rubeli (1986), J. MacKinnon (1975) and K. MacKinnon (1986 and 1992). Vegetation maps showing rainforest types and their extent in Borneo are given by MacKinnon and Artha (1981), Whitmore (1984b) and MacKinnon and MacKinnon (1986); a more detailed map of Kalimantan forests is being prepared by a French project working with BIOTROP. In this chapter the focus is on dryland forests, especially the lowland dipterocarp forests, though many of the principles of forest dynamics apply equally to all rainforest habitats.

Lowland Dipterocarp Forest

The extensive lowland dipterocarp forests are ecologically and commercially the most important feature of Sundaland and reach their greatest species richness in Borneo. Dipterocarps (so named after their winged fruits) grow as very tall trees with canopy heights commonly reaching 45 m and sometimes 60 m or more. Typically, several species of several dipterocarp genera grow together. In the richest formations 10% of all trees and 80% of all emergents are dipterocarps (Ashton 1982; fig. 4.3). Figure 4.4 shows a section through mature Bornean dipterocarp forest in Brunei. Tropical rainforests of this stature and this density of top-of-canopy trees are unique to dipterocarp forest (Whitmore 1984a). The combination of very high stocking of trees with huge boles, commonly 20 m long or more, and of relatively light weight, has encouraged extensive exploitation of dipterocarp forests throughout Southeast Asia. Indeed, in the Philippines dipterocarp forests have been almost totally logged (Jacobs 1988; Collins et al. 1991). Borneo has the finest and most extensive remaining dipterocarp forests in the region but here, too, the habitat is threatened by logging and clearance for agriculture. Recent estimates suggest that more than 60% of the island's lowland rainforest remains, but only 2.9% is included within gazetted conservation areas; most of the rest is designated as timber concessions or conversion forests (MacKinnon and Artha 1981; MacKinnon and MacKinnon 1986).

178 THE LOWLAND RAINFOREST OF BORNEO

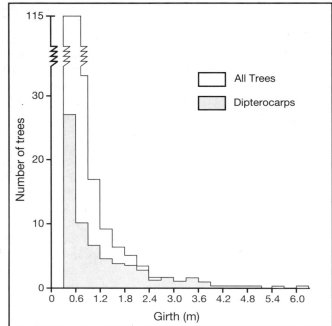

Figure 4.3. Predominance of dipterocarps among the big trees in Sabah lowland rainforest.
Source: Burgess 1961

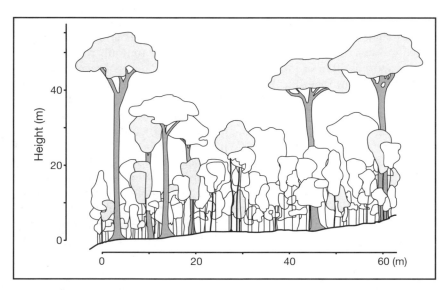

Figure 4.4. Mature and building phases in lowland dipterocarp forest at Belalong, Brunei. Dipterocarps are shaded; note that in the building phase these retain the tall, narrow crowns of youth.
Source: Ashton 1964

Forest Structure

The great richness of flora in lowland rainforests is to some extent a consequence of the very complex structure of vegetation; tall trees provide a framework and an environment within which smaller trees and other plants grow. Richards (1952) provided a useful classification for the plants of lowland evergreen rainforests:

Autotrophic plants (with chlorophyll)
1. mechanically independent plants
   (a) trees and treelets
   (b) herbs
2. mechanically dependent plants
   (a) climbers
   (b) stranglers
   (c) epiphytes

Heterotrophic plants (without chlorophyll)
1. saprophytes
2. parasites

Stratification

The canopy of a tropical rainforest is often considered to be layered (stratified) into three or more main strata: emergents and main canopy; lower storey trees; woody treelets; and forest floor herbs and seedlings (Richards 1952). This concept is a source of great controversy among ecologists (Whitmore 1984a; Jacobs 1988) since it takes no account of the dynamic nature of the rainforest canopy with different patches at various phases of the forest growth cycle. Strata are sometimes easy to recognise in the forest and in profile diagrams, but more often they are not. The concept of stratification, however, does have some usefulness, especially when considering animal use of the forest (MacKinnon 1978).

In Borneo the topmost or emergent layer is composed mostly of Dipterocarpaceae and Leguminoseae. Of the dipterocarps, *Dipterocarpus*, *Dryobalanops* and *Shorea* are emergents, while *Hopea* and *Vatica* are smaller trees of the lower layers. Among the legumes, *Dialium*, *Koompassia* and *Sindora* are common emergents; their fine, pinnate leaves offer little resistance to wind in these exposed situations. The tallest emergent of all is the distinctive *kempas* tree *Koompassia excelsa*. One specimen from Sarawak measured 83.82 m, the tallest rainforest broadleaf tree in the world (Whitmore 1984a). Wild bees choose the open, lofty boughs of the *kempas* to hang their pendulous honeycombs. While clearing the rainforest for *ladang* (agricultural fields), Dayaks traditionally leave one *kempas* standing for the bees. Honey is collected at night by a nerveless climber who mounts a flimsy peg ladder, brandishing a flaming torch to keep the furious bees at bay (MacKinnon 1974b).

The emergents and trees of the upper canopy trap most of the incident sunlight. Below them lies the middle storey, made up of trees which thrive in the comparative shade and the growing saplings of the taller dominant species. Burseraceae and Sapotaceae are common in the main canopy layer, with the lower tree layer including many species of Euphorbiaceae, Rubiaceae, Annonaceae, Lauraceae and Myristicaceae. Euphorbiaceae are the second major family in Bornean forests, sometimes commoner than dipterocarps (Newbery et al. 1992). Trees at the top of the canopy are exposed to intense sunlight, high temperatures and considerable wind. More species are adapted to live within the main canopy than reach the emergent layer (Whitmore 1984a).

Tree Form

There is a wide range of types of crown construction among tropical rainforest trees, but this diversity is of little use in species identification. While each species has a distinct crown form (Hallé and Oldeman 1970, 1975; Hallé et al. 1978), this architecture does not correlate well with taxonomy, so that some families are rich in models (e.g., Euphorbiaceae) while others exhibit only a few forms (e.g., Myristicaceae). A few trees, such as the dipterocarps *Hopea* and *Vatica*, retain the same structure throughout life. For most species, however, as a tree gets larger the original model is repeated on a smaller scale in the crown, a process called reiteration (fig. 4.5). Species show the same crown architectural model in both primary and secondary forest (Ashton 1978), but the height at which the canopy forms varies according to forest type (Richards 1983; Ng 1983). This has implications for commercial exploitation of forests. Intensive modern logging destroys much of the forest canopy so that in regenerating forests, after logging, canopy top trees reiterate at lower levels than in natural forest. The canopy layer becomes correspondingly lower, bole heights are reduced and the volume of commercial timber is less (Ng 1983).

Trunks of rainforest trees are variously coloured from black (*Diospyros*) to white (*Tristaniopsis*). Bark surface characteristics and the appearance and smell of the "slash" when the trunk is cut with a parang are important aids to species recognition. Some species bear flowers and fruits on their trunks. This phenomenon, known as **cauliflory**, is commonest in lowland forest and diminishes with altitude. Trunk-borne flowers may even occur at ground level, as in some wild durians and *Baccaurea parviflora*, or below ground, as in several gingers and the geocarpic figs where they appear on the end of string-like strands, sometimes 6 m long (Corner 1940).

Most roots of rainforest trees are found in the top 30 cm of soil, where nutrients are concentrated. Many trees have entirely superficial root systems

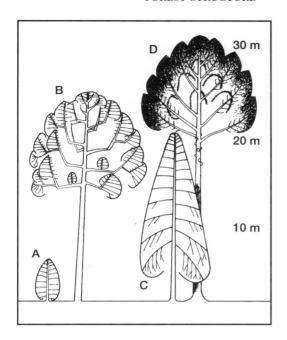

Figure 4.5. Crown construction of two dipterocarps *Shorea mecistopterix* (A and B) and *Dryobalanops aromatica* (C and D). The monopodial crown structure of the juvenile tree (A and C) is reiterated many times in the crown of the adult.

Source: Hallé and Ng 1981

with no deep penetrating roots at all, as can be seen at tree fall sites when the root mass is exposed. Since they lack long tap roots, many rainforest trees have evolved great buttress roots for added support for their tall, straight trunks. Buttress height, spread, thickness and surface form are fairly constant within species and a useful guide to tree identification (Wyatt-Smith 1954), but not all trees of any one species have buttresses (Whitmore 1984a). It seems likely that buttresses help support trees on surfaces that provide poor anchorage (Richards 1952) and act as tension members to reduce the strain on the roots. Buttresses develop on the windward side where there is a prevailing wind direction and on the uphill side on slopes. The incidence and size of buttresses diminish with altitude, presumably since smaller trees need less support.

Besides acting as anchors, the relatively few roots which do penetrate to depth must play an important role in bringing up mineral nutrients from weathering rocks (Nye and Greenland 1960). Most nutrients, however, are derived from the surface layers of the soil, products of the decomposition of fallen litter. Most lowland forest trees, including dipterocarps, also develop a mutualistic symbiotic relationship (mycorrhiza) between their roots and fungi to aid nutrient transfer. The hyphae of the fungi may either penetrate tree roots or, as with many dipterocarp species, form a closely woven mantle over the surface of the root. The fungi obtain carbohydrates from the tree roots, while some of the nitrogen and other inorganic nutrients assimilated by the fungi from the forest soils are returned to the host tree. Mycorrhizal associations thus play an important role in facilitating the uptake of phosphates and other nutrients by the trees (Janos 1980). There is a close correlation between time of fungal fruiting and flowering of dipterocarp trees (Smits et al. 1987). Dipterocarp seedlings must become infected with appropriate fungi if they are to flourish. This has important implications for forest regeneration and reforestation schemes (Smits 1989; Smits et al. 1992; Alexander et al. 1992; see also chapter 9).

Ground Layer

Where the canopy is closed, little sunlight reaches the forest floor and few plants grow. It is easy to move around on the carpet of dead leaves and rotting wood interspersed with occasional shade-tolerant palms, young seedlings (many short-lived), liana roots and wild gingers. Only where the forest is disturbed and there is a canopy opening or along the river banks where light can penetrate does one find the thick tangle of climbers and secondary vegetation that fits the popular idea of "jungle".

Many families contribute to the herb layer, including monocotyledons such as gingers and wild bananas *Musa*, as well as begonias, Gesneriaceae, Melastomataceae, Rubiaceae, various ferns and orchids. In spite of the shady conditions, many herbs regularly produce flowers and fruits, though vegetative reproduction is also common (Kiew 1978). One striking and common feature of many rainforest herbs is their red or silver colouring and variegation of the leaves. These conspicuous colorations reflect red light back onto the chlorophyll-containing tissues, an adaptation which increases the amount of photosynthetically useful light in very dark forest.

Also found on the forest floor are two groups of plants which need no sunlight: saprophytes (fungi) and parasites. The world's most spectacular parasitic plant *Rafflesia* is named after Sir Stamford Raffles, the founder of Singapore and one-time governor of Java. It has been described as "the greatest prodigy of the vegetable world". There are five, or possibly six, species known from Borneo, including the recently discovered *Rafflesia tunkuadlinensis* from Sabah (Meijer in litt.). *Rafflesia* has no leaves. Instead, it derives all its energy from the tissues of the ground-trailing vine *Tetrastigma*, which it parasitises. Cabbage-like buds burst out of the vine and eventually open as enormous *Rafflesia* flowers, coloured in vivid splashes of brown, red and white. The largest of these phenomenal blooms, *Rafflesia arnoldi*, found in West Kalimantan, can be more than 60 cm across and weigh as much as 9 kg (Meijer 1985). Five leathery petals surround a deep cup studded with spikes which exude a powerful stench like rotting meat. The smell attracts insects, which act as pollinating agents. How the seeds are dispersed and reach new vines is still a mystery. Foraging pigs, rooting on the forest floor, may pick up the seeds on their hooves and unwittingly disperse them. Treeshrews and squirrels also consume and probably disperse *Rafflesia* fruit and seeds (Emmons 1992).

Climbers, Stranglers and Epiphytes

Within the matrix of the forest trees, other plants have evolved different strategies to reach the life-giving sunshine. A multitude of climbing palms, vines and lianas cling to the trunks of the big trees, climbing to the canopy to expose their leaves. Thick ropes of hanging lianas are one of the most striking sights in the rainforest; they can reach lengths of 60 m or more (Putz and Chai 1987).

Many climbers that reach the canopy top have crowns of the form, and often the size, of a tree crown. Among the commonest big woody climbers are *Bauhinia*, the Annonaceae and the climbing palms, the rattans. Most big woody climbers are light-loving and grow prolifically in clearings and forest fringes. After logging operations they may also hinder the growth of new forest, as *Mezoneuron sumatranum* does in Sabah (Whitmore 1984a). The other group of climbers, the bole climbers, grow close to the tree, adhering to the trunk (Richards 1952; Jacobs 1976).

Strangling figs are a common sight in Bornean forests. Most stranglers start as epiphytes and send roots down to the ground; these roots increase in number and girth, ultimately encasing the host tree, which often dies. A strangling fig starts life as a seed dropped on a branch by a feeding bird, bat or monkey (fig. 4.6). The fig seedling drops a thin, dangling aerial root, which eventually anchors on the forest floor. The fig grows quickly and soon develops a sizeable crown that competes with that of its host. As it gets larger, the fig drops other roots, and these thicken and join to form a tough, woody network that eventually strangles the host tree. The fig takes over the lofty crown with its own busy leaves and shoots. The original host is reduced to a mere support and eventually rots away.

Other plants gain access to the light by lodging in the branches of the tallest trees. Tree-crown epiphytes include many orchids and ferns. Unlike the strangling figs, these epiphytes have no supply system from the ground, so they have the continuous problem of finding sufficient water and nutrients for growth. Different epiphytes have adopted different solutions to meet their needs. Many have roots and tubers adapted for collecting and storing moisture when it rains. Most epiphytic orchids have swellings at the base of the leaves to store water; in dry weather these pseudobulbs shrivel. The staghorn ferns *Platycerium* have special bracket leaves which lie against the tree trunk and form pockets to collect water and humus. The epiphytic fern *Drynaria* makes its own soil. In addition to its fernlike foliage, it has short bracket leaves which collect and hold decomposing debris. The

184   THE LOWLAND RAINFOREST OF BORNEO

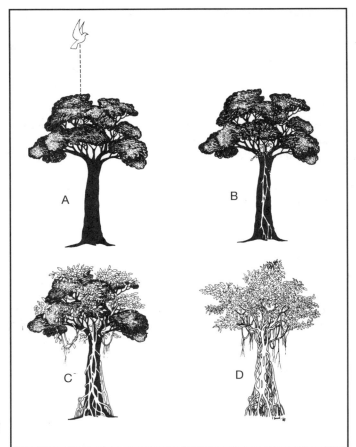

Figure 4.6. Establishment of a strangling fig.

creeper *Dischidia rafflesiana* is even more remarkable. It has two kinds of leaves, one for photosynthesis and another which grows to form an oval sac where dead animal and vegetable matter collects. Nutrients from this decomposing matter are absorbed by roots growing into the sac from the *Dischidia* stem. The mistletoes Loranthaceae are semiparasitic epiphytes, attached to a host tree from which they derive water and mineral nutrients. These mistletoes have long, conspicuous, red flowers and are bird pollinated (Proctor and Yeo 1973).

Bryophytes and other small epiphytes cling to damp substrates at all

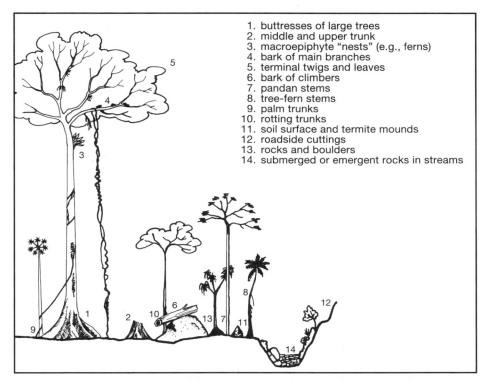

Figure 4.7. Bryophyte microhabitats in a lowland forest. Bryophytes (mosses and liverworts) are small but common epiphytes in humid lowland forests. (After Pocs 1982.)

levels of the rainforest (fig. 4.7). **Epiphylls** are mosses, liverworts, algae and lichens that grow on the surface of living leaves in very humid, shady situations. The leaf is colonised in a fixed sequence, first by nitrogen-fixing bacteria, then by fungi, yeasts, algae and lichens, followed by unicellular flagellates, slime moulds and even mosses (Jacobs 1988). Kiew (1982) studied epiphyll spread on the leaves of the low *Iguanura* palm. The "working life" of a leaf was estimated at three years. In an ever-wet climate, the first lichens appeared on the leaf surface after seven months, and after five years much of the leaf was covered, especially along the veins where water collects after rain. Epiphylls may be more common on those leaves which lack drip tips.

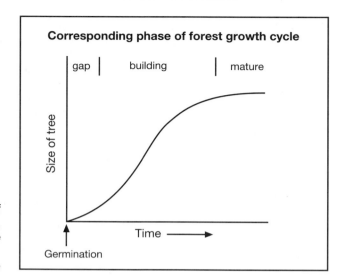

Figure 4.8. Growth of a freely growing tree, and its relation to the forest growth cycle.
Source: Whitmore 1984a

Forest Succession

There is almost nothing in nature that can be called a stable environment. All forest ecosystems are in a dynamic state. Climax forests age, and trees die or fall. The death of an individual or group produces a gap in the forest canopy into which other trees grow. These in turn mature, age and die; the forest canopy is continually changing. This growth cycle of the canopy can be classified into three phases: the gap phase, the building phase and the mature phase (Watt 1947; Cousens 1974; Whitmore 1984a; fig. 4.8). Lowland rainforest consists of a mosaic of patches at different stages of maturity: from gaps, to stands of small saplings or poles, to mature high forest, often topped by giant emergents (fig. 4.9). The cycle starts with the gap phase. Gap size depends on how the gap was formed – by treefall, tree death due to fungal or insect attack, landslip, gales, fire, or clearance for timber or ladangs. In lowland dipterocarp forest in Malaysia, where small gaps are the norm, gaps accounted for almost 10% of a 23 ha study area (Poore 1968). In a 10.5 ha plot in East Kalimantan, gaps covered 17% of the study area (Partomihardjo et al. 1987). The larger the gap the less the microclimate within the gap is like that of closed forest. The theory of gap regeneration dynamics proposes that the size of gap has an important influence on species composition and spatial arrangement, with different species successful in growing up in gaps of different size (fig. 4.10). Recent studies in Sabah, however, suggest the most important determinant of seedling survival and growth in gaps is seedling size at the time of gap

Figure 4.9. Forest in Sabah, showing how small dipterocarps (shaded) are growing up in a light canopy opening.
Source: Burgess 1961

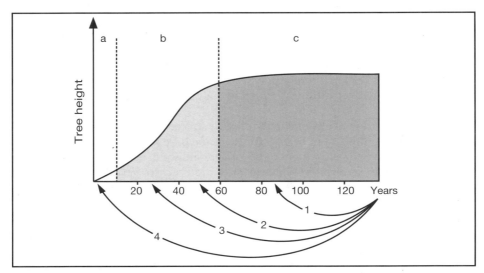

Figure 4.10. The forest growth cycle is influenced by the size of gaps in the canopy. 1, 2 and 3 are short cycles due to the replacement of dead trees by young trees growing up to fill the gap. Large gaps allow the full forest growth cycle to occur as in the long cycle 4. **a** = pioneer species invade large gap, **b** = building phase, **c** = mature phase forest. (After Huc and Rosalina 1981.)

creation, regardless of species (Brown and Whitmore 1992).

Established seedlings and saplings most often grow up to maturity in small gaps. The drastic change in microclimate caused by the formation of a large gap is commonly followed by severe damage or death of many of the young seedlings established in the cool, shady, humid microclimate under the closed canopy. Big gaps are therefore colonised by species that are rare or absent in the undergrowth of high forest but are adapted to the drier, hotter conditions of open sites (Whitmore 1978).

Rainforest trees can be divided into two groups in early life: shade-tolerant species, which regenerate in the shade of high forest, and light-demanding species, which regenerate in gaps (Swaine and Whitmore 1988). Species that are light-demanding cannot grow even in the shade of their own species and have been described as biological nomads (van Steenis 1958c) because they appear at new sites, then disappear when later growth shades them out. A more appropriate term is pioneer species (Whitmore 1984a).

Pioneer species characteristically grow rapidly in height and girth, since a pioneer tree that can overtop others in the same gap and shade them out will be at an advantage. Pioneer species become fertile at an early age and flower and fruit frequently. They produce large numbers of small seeds which are easily dispersed, often by wind, squirrels and birds. Many pioneer species have big leaves, as in *Macaranga gigantifolia* and *M. gigantea*, which colonise roadsides in recently cleared areas in Kalimantan. Seeds of pioneers have long dormancy and germinate when a gap forms (Whitmore 1984a). This is an obvious advantage as regeneration is from seeds already in the soil and not from seeds shed after the gap is created (Kennedy 1991). Pioneer species often occur in pure stands; adjacent gaps, created at different times in logging areas, often have different species.

In contrast, seeds of shade-tolerant species are able to germinate and establish in the gloom of high forest. Some have large seeds and substantial food reserves to allow them to become established. Seedlings can persist for several years, growing very slowly. Most dipterocarps germinate within a few days and persist as a "seedling bank", whereas many Leguminoseae seeds lie dormant and germinate over a period of time, surviving as a "seed bank" (Kennedy 1991). When a gap forms, young seedlings grow upward, perhaps stimulated by an increase in light, reduced root competition or related factors. The lack of dormancy of many rainforest seeds has important implications for silviculture. Natural forest regeneration for dipterocarps, for instance, depends on seedling survival in mass fruiting years as seeds cannot be stored and planted later (Whitmore 1984a).

The species that colonise gaps often vary from place to place. Variation among species in the degree of dependence upon gaps has been offered as one explanation for high tree species diversity in tropical rainforests

(Orians 1982). There have, however, been few long-term studies of gap regeneration. Artificial small gaps of 0.1 ha in primary forest on Gunung Gede in West Java were soon filled by surviving young individuals of primary forest trees, but in larger gaps of 0.2 to 0.3 ha primary forest trees were suppressed by the lush growth of secondary forest pioneers (Kramer 1926, 1933). These experiments are now being repeated at Danum Valley in Sabah. Results indicate that gap size is unlikely to influence the composition of colonising vegetation by controlling seed germination (Kennedy and Swaine 1992).

Riswan et al. (1985) studied regeneration and species succession in mixed dipterocarp forest in East Kalimantan and estimated that it took sixty to seventy years after the formation of a large gap for the number of growth phase species to reach a maximum, and as long again for mature phase species to dominate (fig. 4.11). It may therefore take several hundred years after logging for a dipterocarp forest to recover to something like its original mature state.

Dipterocarp seedlings are recruited in large populations after gregarious fruiting, and there is very high mortality in the first few months (Liew and Wong 1973; Fox 1973). Dipterocarp species that fruit infrequently,

---

Box 4.1. Forest regeneration.

A study of forest structure, floristic composition and vegetation dynamics was made in a 0.8 ha plot on an old secondary forest developed from a pepper plantation abandoned 35 years ago at Lempake, East Kalimantan. This secondary forest had high species diversity comparable to some primary tropical rainforests. Although 70% of the 121 tree species recorded were primary forest trees, the biggest trees, the emergents, and the most common species were secondary forest species, particularly *Macaranga* species, which appear to play an important role in nutrient conservation. The study recorded 22 species of Euphorbiaceae but only one dipterocarp *Hopea rudiformis,* although the plot was surrounded by lowland dipterocarp forest, where 12 species of dipterocarp were recorded (Riswan 1987). The dominant tree of the adjacent primary forest, ironwood *Eusideroxylon zwageri,* was represented by eight small trees but no seedlings or saplings. After 35 years the forest was still in an early stage of succession, with little recruitment of primary forest trees. If there was no further disturbance, and if succession proceeded normally, it would take 150 to 500 years for the forest to reach a stage similar to primary forest. Ashton (1981) has tried to estimate the age of the dominant ironwood in this forest using C_{14} dating and obtained an estimate of 250 years; Riswan (1982) calculated the age of primary forests on the basis of girths of *Shorea smithiana* as 200 to 488 years. The minimum possible regeneration for the forest as a whole would be about 150 years, but individual species may take from 200 to 500 years to regenerate. Clearly this has implications for the logging industry, which operates on a 35-year rotation cycle. It should be noted that even when a tree is mature (above 50 cm girth), it may still not have reproduced itself as many dipterocarps fruit only at intervals of several years. Without such fruiting, there is no seedbank to establish forests of the future.

Source: Riswan and Kartawinata 1988

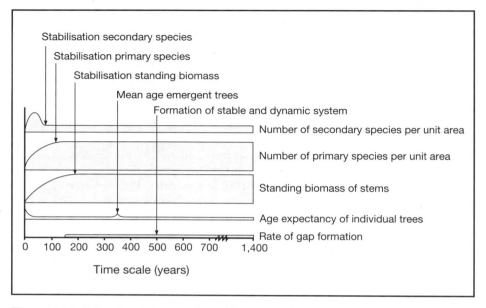

Figure 4.11. Estimates of composition, biomass and age expectancy of trees after the creation of a large gap in lowland forest in East Kalimantan. (After Riswan et al. 1985.)

such as *Hopea*, tend to have longer-lived seedlings than species which fruit more often, such as the red meranti group of *Shorea* (Fox 1973). After establishment dipterocarp seedlings grow only very slowly in closed forest but show a dramatic increase in height once light increases due to a gap (Whitmore 1984a). The differences in response to light between dipterocarp species can have important silvicultural implications, with quick-growing species such as *Parashorea* and *Dipterocarpus* able to compete with other species and even with smothering climbers, which rapidly invade big gaps (Wyatt-Smith 1963).

Biomass and Productivity

The total dry weight of the forest community, including leaves, branches, trunks and roots, is the **plant biomass** (fig. 4.12). Plant biomass increases as plants fix carbon from the atmosphere into organic matter during photosynthesis; the rate at which this occurs is the **gross primary productivity**. A large part of this production is lost during respiration; what is left is the **net**

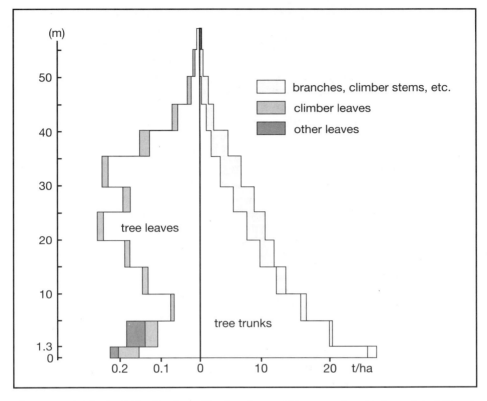

Figure 4.12. Vertical distribution of leaf and wood biomass density in a plot at Pasoh Forest Reserve, Peninsular Malaysia. (After Kato et al. 1978.)

primary production, which accumulates over a period of time. **Net primary productivity** is the rate of production of new plant matter per unit area over time. By estimating productivity from measurements of the standing crop and litterfall, ecologists can measure energy flow through ecosystems. Net primary productivity is lowest in the gap phase and greatest during the building phase, when growth rates are fastest. In mature forests growth rates slow down, and there are greater losses due to death. When a forester talks about production, he is referring only to above-ground net biomass.

A forest's above-ground biomass can be estimated crudely from the volume of trees (height x basal area x 0.5) multiplied by specific gravity, about 0.6 t/m^3. Above-ground biomass of lowland primary forest is usually about 400 t/ha (Whitmore 1990) and varies, according to forest type, from 210 to 650 t/ha in Gunung Mulu National Park, Sarawak (Proctor et al. 1983a). Lowland dipterocarp forests in West Malaysia and on a ridge crest at Mulu, Sarawak, had very high values for biomass (table 4.1). To obtain total biomass, it is necessary to add the below-ground (or root) biomass, which is difficult to measure but may be as much as 20% to 25% of biomass above ground. In tropical rainforests more than three-quarters of the carbon is in the wood, whereas in temperate coniferous woodland half is in the soil (Whitmore 1984a).

It is extremely difficult to estimate net primary production in tropical rainforests. To do so it is necessary to add litterfall, an allowance for grazing and browsing, and the weight of fine sloughed roots and root exudates (often impossible to estimate) to the final standing weight of the forest. Fine roots up to 2 mm thick form 20% to 50% of total root biomass, and their supposed rapid turnover is probably a significant part of ecosystem nutrient cycles. Litterfall can be divided into fine litter (fallen leaves, thin twigs, fruits, caterpillar frass, etc.) and coarse litter (larger twigs, branches and trunks). At Mulu fine litterfall was greatest in the wet season (Proctor et al. 1983b). Figures for biomass and litterfall for forests at Mulu and other Southeast Asian forests are shown in table 4.1.

There have been few attempts to study total forest production in Southeast Asia. At a site in Peninsular Malaysia, results suggested a net primary production of 30 t/ha/yr and a gross primary production of 80 t/ha (Whitmore 1984a). The net primary productivity is lower than for rubber and oil plantations (Kato et al. 1978), but mature forest would not be expected to have as high a productivity as a forest plantation, which is harvested during the building phase.

Table 4.1. Biomass and fine litterfall in Malesian rainforests.

Forest formation	Area sampled (ha)	Above-ground biomass (t/ha)	Below-ground biomass (t/ha)	Fine litterfall (t/ha/yr)
Lowland evergreen rainforest				
Pasoh, Malaysia	0.1	664	(20.5)	10.5
	0.2	475	(20.5)	
Mulu, Sarawak				
broad ridge crest	1.0	650		8.8
valley alluvium	1.0	250		11.5
forest over limestone	1.0	380		12.0
heath forest	1.0	470		9.2
Lower montane rainforest				
New Guinea		310	40	7.7

Sources: Proctor et al. 1983a and b; Kato et al. 1978; Edwards and Grubb 1977

Tropical rainforests are often described as the "lungs" of the world, and there is growing alarm that their destruction is removing an important oxygen-producing and carbon dioxide-absorbing system, thereby enhancing the "greenhouse effect" (chapter 13). The value of tropical rainforests as "carbon sinks" is, however, sometimes overemphasised. All forests in their building phase fix carbon, and temperate forests and plantations do this as well as tropical rainforests. Moreover, a fast-growing crop, whether herbs or forest plantation, is a much better carbon-fixing system than natural forest. The net production, and carbon-fixing role, of mature tropical forests is nil except for peat swamp forests (where peat is accumulating), which continue to remove carbon dioxide from the atmosphere and supply a net release of oxygen (Whitmore 1984a). There are many good reasons for saving tropical rainforests (see chapter 14), but their importance for gas exchange is not the best argument. Burning of extensive areas of tropical forests to clear agricultural fields, however, may enhance the greenhouse effect by raising levels of carbon dioxide in the atmosphere.

Nutrient Cycling

The mineral nutrient cycles of plants in tropical rainforests, both above and below ground, are extremely complex and difficult to study. Proctor (1989) provides a comprehensive review of recent studies of nutrient cycles. Because of their great height and luxuriant growth, tropical rainforests give the appearance of great fertility, yet the soils when cleared often give poor agricultural yields. It is often suggested that most of the plant nutrients are held in the above-ground biomass of the forest rather than in the soil and are lost when the forest is felled or burned. Forest studies in Mulu (Proctor et al. 1983a, 1983b) and elsewhere in Southeast Asia indicate that this is not the case. While a considerable fraction of the inorganic nutrient capital is in the boles of big trees, another considerable fraction is in the forest floor, roots and soil, not the vegetation itself.

Nutrient ions enter the forest system from rain and from weathering of rocks, and leave in streamwater in solution or as leaf litter and eroded soil in suspension. Nutrients from the canopy reach the forest floor in litterfall and in rain, percolating through the canopy, which is enriched by nutrients leached from the leaves. Dead organic matter (animal corpses, faeces, etc.) falls to the forest floor and is incorporated partly in the mineral soil as humus and partly in the forest floor litter. Leaf litter decomposes slowly over four to twelve months. In any forest there will also be some dead branches and a few standing dead trees. There is considerable internal cycling with nutrients, especially potassium, being leached from the canopy

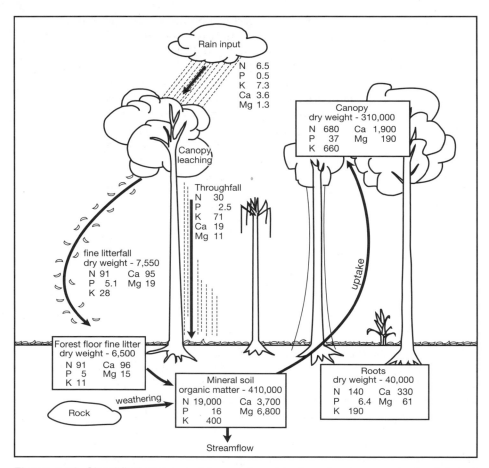

Figure 4.13. Simplified diagram of inorganic nutrient cycling in a montane rainforest (figures in kg/ha). (After Edwards 1982.)

and returned to the forest floor. These nutrients, and those reaching the soil by decay of fine litter, may be taken up quickly by the roots or absorbed on soil organic matter or clay minerals and then either released slowly or lost in streamflow. Nutrients are held in wood for many years and eventually form coarse litter which decomposes slowly. In most soils the main decomposers are litter-feeding invertebrates (Collins 1984; Burghouts et al. 1992). The main flows of nutrients in a rainforest are shown in figure 4.13. The amounts of different nutrients in the different compartments, and the amounts flowing between them, vary with space and time.

The organic matter (humus) in the mineral soil usually decreases at depth; at Mulu it also varied between different lowland forest formations (Proctor et al. 1983c). Organic matter is the main pool of soil nitrogen and phosphorus. The main chemical properties of the soils of the Mulu forests are shown in table 4.2. Humus provides much of the cation exchange capacity of most tropical soils and therefore retains a substantial fraction of the exchangeable cations K^+, Ca^{2+} and Mg^{2+}. Humus also contributes to soil structure and water-retaining capacity, both of which are important for development of root systems and plant growth. Most plant roots are in the top 10 to 30 cm of the soil and are important for uptake of nutrients that enter the soil. Most tree species have mycorrhiza, which enhances the uptake of nutrients, especially phosphorus. Disruption of the root systems in the top soil by activities such as logging can have serious impacts on nutrient cycling.

It was once widely believed that tropical rainforests have closed cycles of inorganic nutrients. It now seems, however, that there is an important distinction between rainforests with deep soils, which receive nutrients solely

Table 4.2. Inorganic nutrients in litterfall, litter layer and soil (kg/ha) in four rainforests at Mulu, Sarawak.

Forest formation	Dry weight	N	P	K	Ca	Mg
Lowland evergreen rainforest						
Broad ridge crest						
fine litterfall	8.8	81	1.2	33	13	8.9
floor fine litter	5.9	42	1.0	9.6	7.2	3.8
soil, 0-0.3 m		6,000*	360*	96	4.6	22
litterfall/soil		0.13%	0.3%	30%	110%	34%
Valley alluvium						
fine litterfall	11.0	110	4.1	26	290	20
floor fine litter	5.2	39	1.9	4.4	110	4.7
soil, 0-0.3 m		7,800*	420*	95	1,600	69
litterfall/soil		1.4%	1%	26%	17%	27%
Heath forest						
fine litterfall	9.2	55	1.6	18	83	12
floor fine litter	6.0	26	0.85	4.8	44	4.5
soil, 0-0.3 m		7,800*	190*	50	62	82
litterfall/soil		0.7%	1%	32%	78%	14%
Forest over limestone						
fine litterfall	12.0	140	4.5	16	370	33
floor fine litter	7.1	77	2.6	4.7	270	14
soil, 0-0.3 m		5,000*	120*	46	2,400	150
litterfall/soil		2.7%	4%	30%	14%	20%

* Much of this N and P is probably not available to plants.
N = nitrogen, P = phosphorus, K = potassium, Ca = calcium, Mg = magnesium

Source : Proctor et al. 1983 in Whitmore 1984a

in rainfall, and forests with soil parent material within the rooting zone (Whitmore 1990). In the former, or where the parent material is very low in nutrients (for example, the sedimentary soils of parts of Sarawak and West and Central Kalimantan), nutrient cycles are almost closed, and recycling is very important. In the latter, the nutrient cycles are more open. The young, shallow soils of Sabah, for instance, show high concentrations of certain major nutrients (Burnham in Proctor 1989). Comparison of nutrients in fine litterfall and in leaves still attached to trees show that lowland rainforests cycle little phosphorus in their litterfall. Similarly, montane rainforests cycle little nitrogen (Whitmore 1990). Thus phosphorus is a limiting nutrient in lowland ecosystems, while in the mountains nitrogen is more strongly limiting; this has implications for tropical farming (chapter 12).

Flowering, Fruiting and Leaf Production

Within the rainforest most of the trees are evergreen, and there are no true seasons such as occur in temperate deciduous forest. Throughout the year temperatures and rainfall are high, even though some months are wetter than others. As a result, at any time of year some trees are in flower, some are fruiting and some are putting out new leaves. A few trees seem to produce leaves almost continually, but most evergreens produce their leaves in flushes, with intervals varying from place to place and year to year. Some species flower at consistent but non-annual intervals. Some flower more or less annually, while others (e.g., dipterocarps) flower irregularly at intervals of several years. Irregular fruiting seasons are also the norm for wild fruit trees such as durians *Durio zibethinus*, rambutan *Nephelium*, mangosteen *Garcinia*, langsat *Lansium* and rambai *Baccaurea* (MacKinnon 1974a), although these trees fruit annually when cultivated.

Nevertheless, there is some evidence for seasonal cycles in plants, which may be correlated to changes in water supply and other variations such as hours of sunshine. In areas with two wet and two drier seasons, as at Ulu Gombak, Malaysia, there are two flushes of new leaves each year, with the major peak just after the driest time of year and a second, lesser peak just before, and extending into, the wettest time of year (Medway 1972b; fig. 4.14). Similarly, maximum new leaf production at Semengo, Sarawak, occurred at the wettest time of year, January, which there follows the driest period (Fogden 1972). While there is an overall pattern of some seasonality in leaf production, there are marked differences between species. Leaf production and fall is commonly associated with water stress, but the relationship is not simple (Whitmore 1984a).

Flowering and fruiting are also irregular, varying between species and even between trees of the same species in different valleys (MacKinnon 1974a). Some trees blossom every year, but most dipterocarps flower only

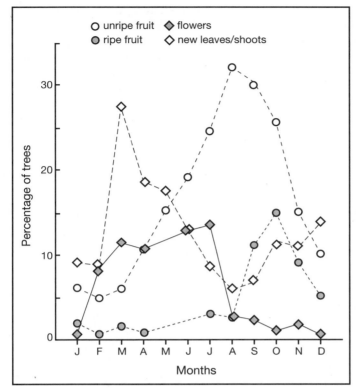

Figure 4.14. Monthly percentages of trees with fruits, flowers and new leaves at Ulu Gombak, Malaysia. (After Medway 1972.)

every four to five years in response to dry periods (Ashton 1988). The flowering of the delicate *merpati* (pigeon) orchid *Dendrobium crumentatum* can also be related to weather conditions (Richards 1970). It blooms eight to ten days after the marked drop in temperature associated with a thunderstorm. Many merpati orchids come into flower on the same day, but the flowers wither after only a day.

A four-year study of the phenology of trees in lowland rainforest in Peninsular Malaysia found no strong seasonality, though more species flowered after dry spells at whatever time of year these occurred (Putz 1979). Of the species studied, 85% did not flower or produce new leaves at regular intervals, and even within species there was a lack of synchrony. Similarly, in the Ulu Gombak study area flowering followed the early dry season, but the species which flowered varied from year to year (Medway 1972b). In more seasonal forests most flowering occurs at the end of the dry season (Mabberley 1983).

Flowering of dipterocarps is especially infrequent but noteworthy because several species flower at the same time, and there is often exceptional flowering and fruiting activity in other tree families as well (Wood 1956; Appanah 1985). Mass flowering and fruiting of dipterocarps seem to occur on a five-to seven-year cycle in both Peninsular Malaysia and Borneo and again can be related to a decrease in rainfall and prolonged periods of drought (Wood 1956; Baillie 1972; Ashton et al. 1988; box 4.2). Interestingly, dipterocarps in the peat swamp forest of Sarawak flower out of phase with those of the dryland forests, that is, not during the dry season (Anderson 1961a).

Box 4.2. Mass flowering and mast fruiting in dipterocarps.

A striking feature of dipterocarp forests in Borneo and western Malesia is mass flowering followed by mast fruiting (Cockburn 1975; Ashton 1982). At irregular intervals of two to ten years, several species of dipterocarps, as well as canopy members of the Burseraceae, Fagaceae, Myristicaceae, Polygalaceae and Sapotaceae, fruit almost simultaneously. Over a period of a few weeks or months, nearly all dipterocarps, and up to 88% of all canopy species, can flower after a long period of no reproductive activity (Appanah 1979, 1981). Enormous numbers of seeds ripen and fall after a mass flowering; an individual dipterocarp may have four million flowers and set 120,000 fruits.

Such regionalised, synchronous mast fruiting at irregular intervals can swamp seed predators, especially in habitats where such predators normally occur at low densities. Dipterocarp seeds are large, energy-rich and poorly protected chemically; they are eaten by beetles before maturation and by wild pigs after dispersal (Chan 1977; Ashton 1982). The seedlings are resinous and suffer little predation, but individuals fruiting out of synchrony suffer high levels of seed mortality (Burgess 1972). Dipterocarps have winged fruits and are wind-dispersed, but they are dependent on animals for pollination.

This irregular but synchronous fruiting poses several key questions:
 1. What pollinates species that flower at multi-year intervals, and how do they avoid competition for pollinators?
 2. What is the environmental cue for the beginning of flowering?
 3. What factors cause an aggregation of fruiting times?

Several mass flowering dipterocarps are pollinated by thrips, which persist at low levels between mass flowerings and can explode in density as dipterocarps come into flower. Thrips are ideal pollinators for mass flowering species, since they have a very short generation time. When flowers and pollen are plentiful, they can quickly multiply. Studies of six mass flowering species of *Shorea* revealed staggered flowering periods, thereby reducing the competition for pollinators and/or reducing the clogging of stigmas with foreign pollen. The time of flowering in *Shorea* is triggered by a drop of roughly 2°C or more in minimum night temperature for three or more nights. Flowering on such a cue does not appear to be directly adaptive to the aseasonal tropics and suggests that the centre of origin for dipterocarps may be the seasonal tropics. There dipterocarps flower at the beginning of the dry season and fruit at the beginning of the wet, when the seeds, which are viable for only a few days, have the best chance of germinating. Possibly mass flowerings in Borneo may be triggered by the climatic events of the El Niño Southern Oscillation in the western Pacific. Mast fruiting in tropical Malesia often occurs in El Niño years.

Source: Ashton 1988; Ashton et al., 1988

Mast fruiting is followed by simultaneous germination of the seeds. This could have several advantages for the fruiting species by "swamping" predators and ensuring that at least some fruits and seedlings escape predation. Natural selection therefore favours simultaneous maturation of fruit (Janzen 1976).

Seasonal variations in plant activities may be reflected by marked changes in animal numbers or behaviour. Twice-yearly seasonal production of fruit bodies in higher fungi in Malaya and Singapore seems to be due to revival of mycelial growth by rain after a dry spell (Corner 1978). The whole soil microflora increases at such times, and soil invertebrates then peak in numbers, feeding on the increased microflora and fungal fruit bodies (Murphy 1973).

Since many trees in lowland forest are pollinated by insects, it is not surprising that peaks in abundance of certain insect species correspond with peaks of leaf and flower production, usually just after the driest part of the year. Other insect species, especially those associated with dead wood, are most abundant in the wettest months of the year (McClure 1978). Insect abundance related to flowering and leaf production affects the breeding behaviour of birds. Among Malayan rainforest birds, insectivores and partial insectivores showed marked breeding and moulting peaks, depending on availability of food (fig. 4.15). Similarly, insectivorous birds in Sarawak forests showed a marked breeding season corresponding with time of maximum insect abundance (Fogden 1972).

Many rainforest insectivorous birds probably live closer to the food limit than species of more seasonal climates, and are especially sensitive to slight seasonal fluctuations in food. Birds that are only partly insectivorous in diet also show marked breeding seasons, probably because insects are an important high-protein part of the diet for feeding young nestlings (Wells 1976). The main annual breeding peak therefore corresponds to the time of greatest insect abundance. Breeding of insect-feeding bats Microchiroptera shows a similar seasonal pattern (Lim 1973; Gould 1978).

Animal feeding and reproductive behaviour can be related to food availability and seasonality (figs. 4.16 and 4.17). Primate feeding and foraging behaviour can be related to availability of fruits (MacKinnon 1974a; Chivers 1980; Rijksen 1978; Rodman 1978; Galdikas 1979). Forest rats in Peninsular Malaysia, which eat fallen fruits, show maximum breeding activity coinciding with the annual peak in fruiting (Medway 1972b). Several large mammals show seasonal ranging patterns related to fruiting patterns; for example, migrations of the wild pig *Sus barbatus* follow the harvests of illipe nuts *Shorea* in Sarawak and East Kalimantan (Pfeffer 1959; Caldecott 1988a), and variations in ranges of orangutans, pigs and parrots at Gunung Palung occur in mast fruiting years (Leighton in prep.). Studies in lowland rainforest at Kutai, East Kalimantan, showed that large hornbills and pigeons, which eat large, lipid-rich fruits, and seed-predating parrots

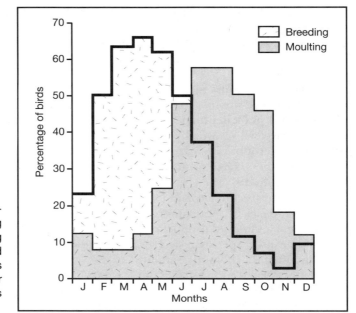

Figure 4.15. Incidence of moulting and breeding among insectivorous and partly insectivorous birds in Peninsular Malaysia. (After Wells 1976.)

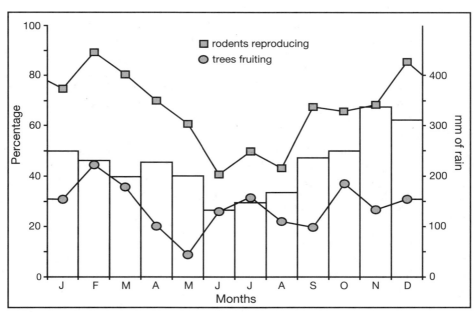

Figure 4.16. The relationship between rainfall (columns), reproductive activity among 13 species of rodents, and fruiting trees in a lowland forest in Zaire. (After Dieterlen 1982.)

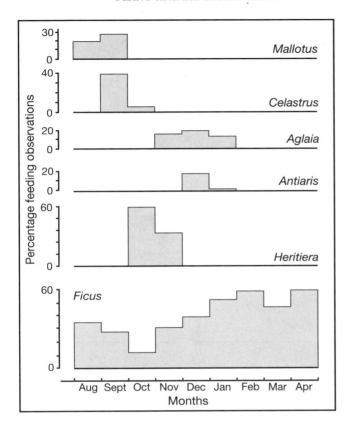

Figure 4.17. Main sources of food for orangutans in North Sumatra. Note the importance of figs *Ficus* and *Heritiera* as "keystone" resources when other fruits were scarce. (After Rijksen 1978.)

left the study area when these food sources became scarce (Leighton and Leighton 1983).

Frugivores with fixed home ranges, such as orangutans and long-tailed macaques, respond to reduced availability of favoured fruits by increasing the amount of non-fruit items, such as bark and insects, in their diet, and by increasing reliance on tree species that do fruit at times of low fruiting (MacKinnon 1974a; Rijksen 1978). Such "keystone" plant species are extremely important in maintaining sedentary populations of arboreal frugivores (Leighton and Leighton 1983; Howe 1984).

Plant-Animal Interactions

It is apparent that animals play key roles in plant life cycles and in the maintenance and evolution of tropical rainforests. Plant-animal interactions

Figure 4.18. A pied hornbill *Anthracoceros coronatus* inspects a potential nest hole in a hollow tree. Foresters contend that a rotten core in a tree is fatal or a disadvantage to the tree, but Janzen (1976) suggests that a hollow core may be an adaptation to poor soils. Hollow cores attract animals that deposit mineral-rich faeces and organic debris inside the trunk, thereby enriching the soil.

and their effects on forest composition and structure are being studied in Borneo at research sites at Gunung Palung, Barito Ulu, Bukit Baka and Danum Valley (chapter 14). Animals as predators, pollinators and seed dispersers may have stimulated species evolution among rainforest plants and helped to control plant species density. Animals also play a significant role in nutrient cycling (fig. 4.18).

Herbivory

Though there are many different tree species in tropical rainforest, any one species will be rare. Some tree species occur in clusters, especially those (such as dipterocarps) which have inefficient dispersal, but others occur as scattered individuals. This may be at least partly a protection against leaf-eating animals, especially insects, and may result from predation of dense seedling populations (Janzen 1971). Leaf predation is common in all forest types (fig. 4.19). Caterpillars are known sometimes to cause substantial damage in dipterocarp forests. Many hectares of *Shorea albida*

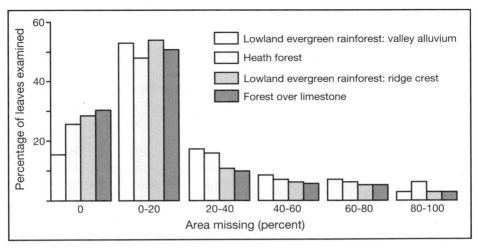

Figure 4.19. Leaf predation by invertebrates in four lowland rainforests at Mulu, Sarawak. During a 12-month study, leaf predation was similar in all four forest types. (After Proctor et al. 1983b.)

stands in peat swamp forests in Sarawak were defoliated and killed by tussock moth caterpillars Hymantridae (Anderson 1961b).

Plants have evolved secondary compounds such as gums, resins, latexes, tannins, and phenols to render their leaves and fruits unpalatable to animals (Waterman 1983). An animal which overcomes such a palatability barrier will be at an advantage because it has an exclusive food source. Such pest pressure may have encouraged speciation in the rainforest. The relationship between plants and animals may evolve even further so that an insect which is able to overcome chemical repellents and eat the "protected" plant parts may itself be distasteful to predators; such insects often have warning coloration. Leaf-eating insects may be far more restricted in the number of species they utilise than are pollinating insects. By specialising on breaking down specific chemical defences, the leaf eater narrows its range of available foods but also has less competition from other herbivores unable to cope with those chemicals. Most moth caterpillars eat only specific leaves and may even starve rather than eat the "wrong" leaves.

Not all plants have adopted or rely solely on chemical defences. Some have physical structures to deter predation, such as thorns, siliceous hairs (bamboos), leathery leaves and internal spicules (Araceae). A few species (e.g., the pioneer tree *Macaranga* and the climbing palm *Korthalsia*) have evolved special associations with ants which live in nodes in the plant and deter predation of plant parts. The ants may be rewarded with nectar secretions or other foods. The ants *Crematogaster borneensis* live in the hollow twig-tips of *Macaranga* trees, where they cultivate sap-sucking scale insects *Coccus* and eat starch grains (Khoo 1974).

There is probably no plant whose defences preclude being eaten by all herbivores, and there is no animal that can eat all kinds of leaves. Leaves are eaten by a wide range of animals, including mammals and larval and adult insects, but not by amphibians and by very few reptiles and birds. An estimated 7% to 12% of leaf production in tropical rainforests is eaten by insects and only 2% to 4% by vertebrates (Leigh 1975; Wint 1983). About 50% of a leaf consists of cellulose, a complex carbohydrate molecule which makes up outer cell walls. Most animals lack the necessary enzymes to break down cellulose into easily digestible molecules, but some, like termites, use certain bacteria, protozoa or fungi to conduct this first stage of digestion. In vertebrates there are two kinds of bacteria-assisted digestion: foregut fermentation, as in deer and langurs, and hindgut fermentation, as in rodents.

Pollination

Wind pollination is common in temperate forests but is of little significance in rainforests. Among 760 tree species in 40 hectares of rainforest in Brunei, Ashton (1969) found only one that was wind pollinated. Even though the crowns of the upper canopy and emergents are exposed to wind, most rainforest trees are animal pollinated. Animal-borne pollen is more likely to reach another flower of the same species than pollen scattered by the wind, especially as most rainforest trees are self-pollinated. Numerous insects and bats are available as pollinating agents throughout the year.

Animals do not pollinate flowers intentionally; they visit them to feed on energy-rich nectar and pollen. However, usually some pollen adheres to the visitor's body and is transported to the next flower, where it rubs off on the stigma and fertilises the ovules. These develop into seeds. Flowers that are pollinated by animals have evolved features to attract and reward them (nectar, pollen) but also may have deterrents to avoid overexploitation (Baker 1978; Faegri and van der Pijl 1979).

Different flowers are adapted to attract various different pollinators (table 4.3). Large, night-flying hawk-moths are common in rainforests and visit white, heavily-scented flowers. Day-flying butterflies visit brightly coloured, less-scented flowers, which are often trumpet-shaped. Beetle flowers are fragrant, simple, open dishes or bowls, and pollination occurs as beetles scramble to eat the pollen. Beetles and flies also pollinate flowers smelling of carrion or dung, such as the dark red *Rafflesia* and *Amorphophallus*. Bees are strongly attracted to blue and yellow flowers. Birds are not very responsive to scent; they prefer red, orange and yellow flowers.

Table 4.3. Some typical morphological and behavioural characteristics of flowers associated with common types of animal pollinators. Many variations from these generalisations occur.

Flower characteristics	Type of primary pollinator						
	unspecialised	bee	butterfly	moth	bird	bat	beetles
Size and shape	small, radial symmetry	large enough to admit bee, bilateral symmetry	small, tubular	small, deeply lobed or fringed	long, tubular	large, bowl or beaker shaped	simple, open, bowl shaped
Colour	various	yellow, white, blue	vivid red, white, blue	drab or white	red or strong contrasts	drab or white	white, (purple & red)
Perfume	various	fresh, weak	fresh, weak	sweet, strong	absent	foetid, strong	fragrant (carrion)
Nectar hidden when flower open	no	yes	yes	yes	yes	no	-
Time of nectar production and/or flower opening	most of time	day	day	night	day	night	day
Examples	*Syzigium lineatum*	orchids	*Mussaenda*	*Ceriops tagal* (mangrove) *Ipomoea alba*	*Hibiscus* Loranthaceae	*Maranthes Parkia* Banana Durian	Annonaceae *Magnolia* (*Rafflesia*, *Amorphophallus*)

(Adapted from Deshmukh 1986.)

> **Box 4.3.** Dioecism and cross-pollination among scattered rainforest trees.
>
> The large number of species per hectare in rainforest often means there is a considerable distance between individuals of one species, often 100 m or more (Jacobs 1988). This greatly diminishes the chances of cross-pollination between trees, yet cross-pollination is essential for long-term survival and 'fitness' of the species. To encourage such cross-pollination, many rainforest trees are dioecious (i.e., they produce flowers that are either male or female, with the two produced on different plants). Ashton (1969) found that 26% of trees in Sarawak rainforests were dioecious, compared to only 2% of the whole flora of the British Isles. Another 14% of Sarawak trees facilitate cross-pollination by stigmas and stamens maturing at different times, or by unisexuality or polygamy (male and female flowers as well as bisexual ones). Certain plant families are almost completely dioecious (e.g., Burseraceae, Myristicaceae and Rafflesiaceae). Dioecism encourages cross-pollination but reduces the chances of pollination by 50%, so it must have a strong selective advantage in maintaining genetic fitness. To maintain a viable population of dioecious species, the minimum area needed will be considerably greater than that necessary for monoecious species (where both male and female reproductive organs occur on one plant). This has important implications for conservation of lowland rainforests.
>
> *Source: Jacobs 1988*

Bats are important pollinators in the Asian tropics; many bats feed on nectar. Bat-pollinated flowers typically are open at night, are large and are dull white, greenish or brownish in colour, often with a heavy scent like that of the durian flower.

The relationship between flowers and their pollinators can be highly specialised. Throughout the world's tropics there are more than 900 different species of figs whose flowers are pollinated by very small wasps. These lay their eggs in the young flowers and spend most of their lives inside the developing fruits. Incredibly, every species of fig has its own unique species of fig wasp belonging to the family Agaonidae.

Borneo alone has more than 130 species of figs, from large, strangling figs to small, free-standing trees. As often as twice a year, a fig tree produces a crop of five hundred to a million fruits (the figs), each containing a hundred to a thousand florets (Janzen 1979). A few minute, female fig wasps enter the developing fig, pushing through the narrow mouth, which is partially closed by scales. The female pushes her ovipositor down the style of one of the female flowers and lays a single egg in the ovary; she leaves one egg in the ovaries of many florets. The ovipositor can only reach the ovary in florets with a short style; the long-styled florets do not receive a wasp egg and thus escape predation of the developing seed. As the female wasp moves around the fig looking for suitable female flowers, she deposits pollen from the fig in which she hatched and so effects cross-pollination

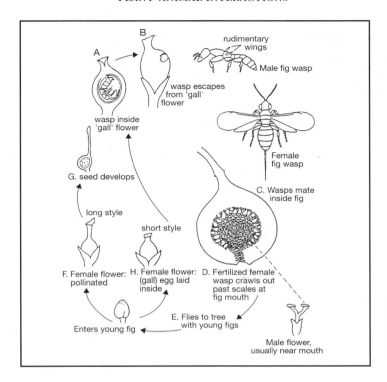

Figure 4.20. The pollination of fig flowers by fig wasps. (After Whitmore 1990.)

(fig. 4.20).

The wasp larvae eat the developing seed and pupate inside the seed coat. The wingless males emerge first and search for female pupae and fertilise them before they emerge. The males then tunnel through the fig wall to the outside, thus bringing about a reduction in carbon dioxide levels within the fig. This change seems to stimulate the development of male flowers, the emergence of the females and the process of ripening. The male dies within the fig, but the winged female flies off to continue the cycle. As the female crawls out of the fig, she gets pollen on her body from male flowers and carries this to another fruiting fig tree, which she probably locates by scent.

The fig pays a high price for this service. Although the wasps are essential to the figs for pollination, they also prey on the fig seeds. Of 160 figs collected from four species in Costa Rica, 98% of figs had more than 30% of their potential seeds killed by pollinating wasps (Janzen 1979). Other minute wasps also parasitise the fig seeds, fig wasps or both, and either oviposit in the same way as the pollinating wasps or through the fig wall. Various moth

> **Box 4.4.** Species richness in lowland rainforest.
>
> 1. There are between 80 and 200 tree species per hectare in tropical rainforests. The total number of vascular plant species per hectare is about three times as high.
> 2. No two hectares have exactly the same species composition. Many species occur only once, even in a large plot.
> 3. These large numbers of species, and the large proportion of woody plants, imply a long tenure in one place. Regeneration of the rainforest proceeds very slowly.
> 4. No single tree species is dominant. In primary forest the commonest species seldom makes up 15% of the total trees. Higher percentages than this indicate the presence of limiting conditions or damage.
> 5. The more species there are per hectare, the fewer individuals of one species there will be. Density of tree species varies between about 25 trees in one ha and one tree in 25 ha or more.
> 6. The fewer individuals of a species there are in a hectare, the greater the distance between them, on average 100 m.
> 7. A mixed forest with long distances between individuals of the same species prevents massive outbreaks of pests and diseases.
> 8. Long distances have to be bridged for cross-pollination to occur; this promotes dioecism.
> 9. The more species per hectare, the smaller the ecological niche available to each.
> 10. All rainforest trees are virtually dependent on animals for pollination, and for dispersal if they have large seeds.
> 11. Dispersal by wind and by animals results in different patterns of seeding. Difficulty in dispersal leads to concentrations of trees in clusters.
> 12. The longevity of many species, and a tendency to cluster, result in a forest mosaic. This makes it difficult to delimit minimum areas for species protection.
> 13. The number of species declines with increasing altitude. Most species are confined to altitudes below 300 to 500 m. Lowland rainforests are the richest in species.
> 14. Because of the high species richness, damage and destruction of rainforest will endanger more species per unit area than damage and destruction to other vegetation types.
> 15. Because of the small numbers of individuals per species, the elimination of a few individuals has a strong effect on the size of a population and biological balances.
> 16. The long-term minimum population with sufficient genetic diversity for survival is found only in very large areas.
>
> *Source: Jacobs 1988*

(Pyralidae) and weevil (Curculionidae) larvae also prey on the developing seeds in maturing figs. Other seeds are lost to fig-eating vertebrates, such as parrots and *Treron* pigeons, which prey on the ripe figs as well as dispersing them. The fig reproduction cycle is a fascinating story of the complex interrelationships between plants and animals.

Seed Dispersal

Mere fertilisation of the seed is not enough to ensure survival. The seed must mature and be dispersed. The success of a new seedling depends also on where the seed is dropped. Within tropical rainforest the two main agents of seed dispersal are animals and wind. Most epiphytes are wind dispersed. Tiny orchid seeds are carried great distances by the wind. Wind dispersal is more common in upper storey and emergent trees, such as the dipterocarps. Most dipterocarp fruits fall within 20 m of the parent tree (Burgess 1975; Fox 1972). The wings on the dipterocarp fruits are thought to reduce the rate at which the fruit falls by causing it to spin, but several dipterocarp species do not have winged seeds. Most (98%) of the fruits of *Shorea fallax*, a species with wingless seeds, fell within 10 m of the parent (Fox 1972).

Eighty percent of canopy and understorey tree species in tropical forest on Barro Colorado Island, Panama, are dispersed by animals (Foster 1982); animal dispersal is also very important in Bornean rainforests (fig. 4.21). As most rainforest trees germinate in the shade, it is advantageous to have a large fruit to provide a food store for the young seedling. Animals are more effective than wind at distributing larger seeds over greater distances. Many rainforest trees have fruits adapted to animal dispersal, with colours, flesh and scents attractive to animals, and seeds resistant to digestive juices.

Bat-dispersed fruits are mainly yellow or brown, with strong musty odours. They ripen on the tree, and as with bat flowers, are held away from the foliage, which facilitates visits by bats (van der Pijl 1957). Larger fruits which contain a single, large seed are carried to habitual roosting spots, where the flesh is eaten and the seed dropped; mangoes are dispersed in this way. Flying foxes *Pteropus vampyrus* can carry fruits that weigh as much as 200 g. Fruits such as figs, with small seeds, are squeezed to extract their juice, and the discarded pulp is dropped below feeding roosts. Bat-adapted fruits include *Annona, Artocarpus, Spondias, Baccaurea, Diospyros, Dracontomelum, Lansium, Mammea, Mangifera* and *Musa* (van der Pijl 1957). Many of these are also eaten by primates, including humans.

Birds are also important seed dispersers. Bird fruits are typically showy, coloured bright red, yellow, black, purple or blue, and often shiny. Frugivorous birds can be divided into fruit specialists, such as the hornbills, and opportunists or nonspecialists. In his survey on tropical frugivorous birds and their food plants, Snow (1981) concluded that fruits eaten by specialised frugivores are generally large (up to 70 mm x 40 mm), have relatively large seeds and are highly nutritious. Major Bornean families with fruits of this kind include Lauraceae, Burseraceae and Myristicaceae. Specialised frugivores feed on high-quality fruits, rich in fats and proteins, while unspecialised or opportunistic frugivores feed on small, many-seeded, less nutritious fruits such as *Melastoma* and *Trema*. These have watery flesh and contain mainly carbohydrates.

Figure 4.21. Some Bornean rainforest fruits and their animal dispersers.

Specialised frugivores void the seeds intact, either by regurgitation or defaecation. Some birds, such as pigeons and parrots, are seed predators feeding on seeds adapted for dispersal by other birds. Flowerpeckers are specialists, feeding almost entirely on a diet of mistletoe fruits Loranthaceae (Snow 1981).

Fruits are also an important part of the diet of fruit-eating carnivores, especially civets, mongooses and even bears. The largest civet, the binturong *Arctictis binturong*, is a fruit specialist. These mammals are mainly nocturnal and so presumably are attracted by smell, though many of their favoured fruits are also brightly coloured (Medway 1969; Leighton and Leighton 1983).

Monkeys and apes are important fruit dispersers. Primate fruits include mangosteen, rambutan and durian, notably many of the fruits that people also prize and have chosen to domesticate. The eminent tropical botanist Professor E.J.H. Corner believed the durian to be primitive, with its spines protecting the unripe fruit. He made it the cornerstone of the "durian theory", which suggests that bright, contrasting colours and edible, fleshy and sometimes strong-smelling arils are attractants to animal dispersers (Corner 1949). Certainly orangutans and people will go to great lengths to find durians and probably locate them at least partly by smell. Other mammals such as elephants, pigs and even tigers feed on the fallen fruit and disperse the undigested seeds through the forest, depositing them in their faeces.

Many fruit eaters can be regarded as both seed dispersers and seed predators (MacKinnon 1978; Payne 1980). An animal is a seed predator if it destroys the seed or eats the fruit before it is ripe. Plants have evolved various strategies to protect themselves from seed predation. Seeds may be too small to be worth exploiting. Seeds or fruit may contain toxic compounds (as in Lauraceae) or other chemical defences. Seeds or fruit may have physical protection, such as a hard shell around the seed (*Maranthes*), latex (*Lansium, Garcinia*), spines (*Durio*), irritant hairs (*Ficus, Gnetum*) or a thick pericarp (*Disoxylum*). In Bornean forests, the heavily protected dehiscent fruits of many species of Burseraceae and Meliaceae appear to be dispersed exclusively by hornbills, as only these birds possess sufficiently strong bills to open the thick husk and extract the large seeds (Leighton 1982). Fruit design can be quite subtle to attract seed dispersers but deter predators. Thus a fruit may be attractive to monkeys, which swallow the seeds whole, yet inedible to squirrels, which bite and destroy the seeds. A toxic or distasteful resin in a thick peel, such as in *Garcinia*, will deter a squirrel chewing through the rind but is no deterrent to monkeys, which remove the peel easily with their fingers, eat the juicy pulp and swallow the seeds which pass undamaged through the digestive tract. Yet another antipredator strategy is for the trees of the same or different species to produce vast numbers of seeds simultaneously (mast fruiting).

For successful regeneration, it is essential that mature seeds be dispersed, so they must not be eaten before they are ready for dispersal. In many unripe fruits the seeds are protected with various defence compounds, in concentrations that reduce as ripening progresses. These compounds make the fruit unpalatable to most potential predators, but at least one major frugivore, the orangutan, is very tolerant of bitter or sour, young fruits. This probably gives the orangutan a competitive advantage over other frugivores such as gibbons, monkeys and birds, which feed only

> **Box 4.5. Fig eating and seed dispersal by rainforest birds.**
>
> Kalimantan's lowland forests are rich in figs, which are an important food supply for birds and several arboreal mammals, including monkeys, squirrels and civets. Birds are important seed dispersers for many fig species. Of 38 *Ficus* species identified in a 2 km^2 study area of lowland forest in West Malaysia, 29 species, all stranglers or vines, possessed seeds eaten primarily by birds. Figs were available every month; figs are an important 'keystone' resource, providing food for frugivores when other fruits are scarce. Nevertheless, crops of large figs (> 25 mm mean dimension) were rare, with only 13-16 large-fruited fig patches per km^2 of forest per year.
>
> Sixty bird species, representing about 25% of the forest avifauna, ate figs in the Kuala Lompat study area. Several species are specialised fig-eaters. *Treron* pigeons are specialised fig-seed predators and ate large proportions of some fig crops, about 30% in one case. Figs of all sizes were eaten by birds, but the resource was partitioned according to fruit size. As a general rule, large birds took bigger fruits, while smaller birds fed on the small figs. Among the *Treron* pigeons (three species) and the *Megalaima* barbets (four species), partitioning was distinct, with large congeners preferentially taking larger figs. Although gape sizes may influence the fig-size preferences of smaller birds, data suggest that for large birds, bird weight in relation to the width or strength of branches bearing fig, is a more important constraint.
>
> All birds, except some pigeons and possibly the parrot *Loriculus galgulus*, rapidly defaecated ingested *Ficus* seeds. Radiotracking and observations of foraging frugivorous birds showed that many species stayed close to large fruit patches. As a consequence of this behaviour and rapid gut-passage rates for fig seeds, the seed shadows of most bird-dispersed *Ficus* are narrow, and most seeds will germinate close to the parent fig. The most important fig seed dispersers are predicted to be the larger, specialised frugivorous birds, such as hornbills, hill myna *Gracula religiosa* and fairy bluebird *Irena puella*, which forage over wide areas of forest and thereby carry fig seeds to new areas of forest.
>
> Bird-dispersed *Ficus* at Kuala Lompat were epiphytic species, growing predominantly on large commercial timber trees. Selective logging of such lowland forest is likely to deplete the density of bird-dispersed figs severely; only about 25% of the figs would remain at Kuala Lompat if the site were selectively logged. Large-fruited *Ficus* would become exceptionally rare, and birds particularly dependent on such figs, such as *Treron capellei* and some of the larger hornbills, are therefore threatened by such logging practices. Leaving small patches of forest containing important *Ficus* species, or post-logging enrichment planting of carefully chosen *Ficus* species, could help preserve these vulnerable birds in logging areas.
>
> *Sources : Lambert 1989a, 1991*

on the ripe fruits of the same species (MacKinnon 1974a).

Some fruits do not lose all their defence compounds when they are ripe. This is thought to be a means of protecting fruit from being eaten or destroyed by non-disperser species (Herrera 1982; Janzen 1983). High toxicity may be an extreme adaptation. One extremely toxic seed is that of *Antiaris toxicara,* which contains poisonous glucosides. The latex from this tree contains the same poisons and is used for poisoned darts in Borneo.

Macaques are able to eat *A. toxicara* fruit and disperse the seeds in their faeces; presumably the toxins deter other non-disperser predators such as hornbills.

The proportion of a crop likely to be taken by predators, and the proportion successfully dispersed, can be estimated from study of fruiting figs. One fig tree in Malaysia was estimated to have a crop of about 41,000 figs ripening over a period of eight days. Consumption by bats and other nocturnal mammals was not estimated, but among diurnal feeders macaques took about half the crop, birds 29% and gibbons about 13%; squirrels were observed to eat only a small proportion of the crop (MacKinnon and MacKinnon 1980). Not all of these seeds would have been dispersed. A detailed study in Costa Rica monitored the fate of 100,000 figs over five days and found that birds took about 65% of the figs (the rest were taken by bats and mammals), and most of these were eaten by parrots which are seed predators. Taking into account seed predation by both pollinator wasps and parrots, only about 6% of the total crop was dispersed by birds (Jordano 1983).

ANIMAL COMMUNITIES

The complexity of tropical forests, both in terms of plant species and structure, permits many animal species with different specialisations to coexist in lowland rainforest. This leads to species richness in many animal groups. Thus Borneo has more than 200 land mammals (Payne et al. 1985), more than 500 resident and visiting birds (Smythies 1960), 166 species of snakes (Stuebing 1991), 183 species of amphibians (Inger 1966) and countless invertebrates, including tens of thousands of species of beetles (Hammond pers. com.).

Small areas of lowland rainforest in Malaysia are richer in bird species than similar areas of rainforest in Africa; they are comparable in richness to South and Central America (Wells 1971). Rainforests in Borneo are just as species-rich. In a small study area (200 ha) at Semengo, Sarawak, 141 species of resident birds were recorded, although one-third of all species were represented by only one pair (Fogden 1976). The pattern is similar to that for plants, namely high species richness, with individual species occurring at low density. Similarly, the herpetofauna of lowland rainforest in Sarawak is very rich with 135 species (48 amphibians, 40 lizards and 47 snakes) recorded in an area of 52 km^2 (Lloyd et al. 1968).

Primate densities, however, are fairly low in Sundaland forests. In Malayan rainforest the average density of primates was 13.5 groups/km^2 (Marsh and Wilson 1981). This is considerably lower than for comparable forests in tropical Asia and Africa and perhaps can be attributed to the dominance of unpalatable dipterocarps in this forest. Biomass values for primates in Borneo, in Sabah and in Kutai, are even lower (Davies and Payne 1982).

Stratification of Animal Communities

During the course of a year, tropical rainforests receive twice as much sunshine as do forests in temperate regions. Most of this sunlight is trapped in the canopy. Here the photosynthesising leaves convert this energy into sugars and proteins, which the plants then store in primary foods such as new leaves, shoots, buds, seeds and fruits. Because of this rich food supply, many animals have taken to an arboreal way of life. Apart from flying insects, birds and bats, which are at an obvious advantage in the canopy, a fascinating array of other animals — monkeys, civets, squirrels, snakes, lizards and even frogs — live and feed in the trees. Some 45% of the non-flying mammals of Borneo are arboreal (Davis 1962).

Just as the trees and vegetation can be zoned horizontally, so can the animals. Of course animals can move between zones, but each forest layer from the forest floor to the top canopy tends to have its own characteristic complement of animals. Naturally, there is a close link between the kinds of food available at different levels and the feeding habits of the animals that live there. Young leaves and ripening fruits are much more plentiful in the middle canopy and treetops so that it is at these levels that most of the larger arboreal mammals and frugivorous birds feed.

Harrison (1962) defined six communities of birds and mammals in lowland rainforest in Peninsular Malaysia:

- **above the canopy:** insectivorous and carnivorous birds and bats;
- **top of canopy:** birds and mammals feeding mainly on leaves and fruit, and to a lesser extent on nectar and insects;
- **middle-of-canopy flying animals:** mainly insectivorous birds and bats;
- **middle-of-canopy seasonal animals:** omnivorous mammals which range up and down tree trunks from crown to the ground; also a few carnivores;
- **large ground animals:** herbivores and carnivores;
- **small ground or undergrowth animals:** mammals and birds, mainly insectivores and mixed feeders.

Within each layer community, there is a distinction between animals active by day or by night (Harrison 1962; MacKinnon 1974a).

Certain animal groups are well represented at all levels of the forest. The squirrels are a good example. Species of diurnal squirrels have been studied in some detail at Kuala Lompat, Malaysia (MacKinnon 1978; Payne 1979). Squirrel densities in Malaysia are only about four animals per hectare and are probably limited by food availability and competition with primates. Bornean squirrels show similar stratification and niche separation according

ANIMAL COMMUNITIES 215

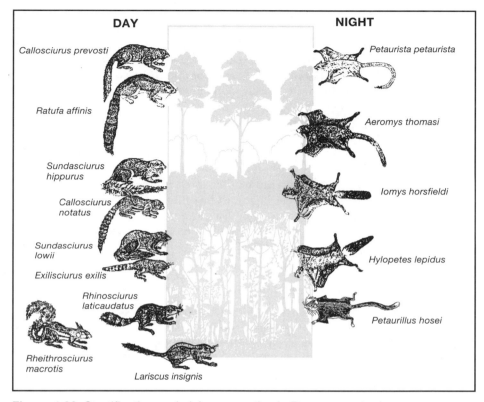

Figure 4.22. Stratification and niche separation in Bornean squirrels.

to diet (fig. 4.22). The most spectacular is the endemic tufted ground squirrel of Kalimantan, *Rheithrosciurus macrotis,* a large animal with a bushy tail and great tufts of dark hairs on its ears. Borneo also has the smallest squirrel in the world, the tiny pigmy squirrel *Exilisciurus exilis* which is a mere 11 cm long and is found in the lower forest levels, where it feeds on bark (Payne et al. 1985).

Rainforest birds also show clear vertical stratification (Wells 1971; Wells et al. 1979). In the top of the canopy, hornbills, barbets and pigeons feed on fruits and insects. The group inhabiting the middle of the canopy includes trogons, woodpeckers and bulbuls. Undergrowth birds include pittas, thrushes, babblers and pheasants. Niches are not always clearly distinct. Thus insectivorous understorey birds form mixed-species foraging flocks, except when they are breeding (Wells 1978).

Invertebrates occur at low population densities in tropical rainforest, an adaptation to avoid predation (Elton 1973), but also show strong vertical zonation (Sutton et al. 1983). At Bukit Timah, Singapore, the small, biting flies Ceratopogonidae and Homoptera and their predators are dominant in the top of the canopy. Canopy epiphytes are exploited by a large fauna including moss-feeding moths. In the understorey Psocoptera, ants, termites and most dipteran families are dominant. Insects fly through the canopy horizontally and vertically along favoured flight paths, where webspinning spiders lie in wait (Murphy 1973). In Brunei forest aphids Homoptera and Agolonidae were most abundant in the higher levels of the canopy (Sutton 1980). In lowland rainforests at Gunung Mulu, Sarawak, there are marked differences between butterfly families in the top of the canopy and at forest edges (most swallowtails Papilionidae, milkweeds Danaidae and Pieridae) and those of the forest interior (most Satyridae, Riodinidae and skippers Hesperiidae), while other families, such as Nymphalidae and Lycaenidae, show no such stratification (Holloway 1987). It may be that flowers adapted to different pollinators are commonest at different levels in the canopy (Appanah 1980).

Competition and Niche Separation

Studies of invertebrates and vertebrates give some insight into the complex inter-relationships between plants and animals. Plants have evolved various adaptations to encourage pollination or seed dispersal by different groups of animals. Among seed dispersers there may be little competition for food; for instance, birds and bats feed mainly on quite different sorts of fruits. However, squirrel numbers in West Malaysia may be limited by competition with primates for some foods (MacKinnon 1978; Payne 1979). Competition is likely to be most intense between closely related species (box 4.6).

Life at the various forest levels requires different lifestyles, different methods of travel, and choices of sleeping and breeding sites. The pioneer ecologist Charles Elton declared that every animal has its own "address" and "profession"; together, these describe its niche. No two closely related animals can exploit the same niche; this is illustrated elegantly by the study of frugivorous birds and primate communities.

Differences in time of activity, diet and foraging techniques have been demonstrated for closely related sympatric species of birds in New Guinea (Diamond 1972, 1973). Fruit-eating pigeons in New Guinea belonged to different guilds, taking different sizes of fruit (fig. 4. 23). Studies in Malayan and Bornean rainforests revealed similar differences in birds feeding on figs (box 4.5). Hornbill species at Kutai show different foraging strategies, thereby reducing competition between species (Leighton and Leighton 1983).

The monkeys and apes of the Sundaland rainforests are some of the most conspicuous and best-studied members of the rainforest fauna (MacKinnon 1974a, 1976; Chivers 1980; Rijksen 1978; MacKinnon and

Box 4.6. Niche separation in tropical squirrels.

For two or more species of mammals to live in the same habitat, their use of resources must be sufficiently different to avoid the competitive exclusion of one species by another. Such differences in lifestyle as ground-living or tree-dwelling, active by day or night, insect-eating or frugivorous, are obvious means of ecological separation between squirrel species. However, species occurring naturally in the same habitat often appear to be utilising the same food resources, but closer study reveals that each occupies a somewhat different niche.

The situation is well illustrated by the squirrels found in the Bornean rainforests. Of the 34 squirrels known from Borneo, 14 are nocturnal, and the rest active by day. The diurnal squirrels can be divided into terrestrial, arboreal and climbing categories, with different species showing different use of the various forest strata. The three-striped ground squirrel *Lariscus insignis* and shrew-faced ground squirrel *Rhinosciurus laticaudatus* feed on the ground or search for insects on fallen wood. This is also the habitat of the largest of the Kalimantan squirrels, the tufted ground squirrel *Rheithrosciurus macrotis*. Among the more arboreal squirrels, the slender squirrel *Sundasciurus tenuis* and Low's squirrel *Sundasciurus lowii* are most active on the tree trunks of the lower forest levels. Tiny pigmy and sculptor squirrels *Exilisciurus* forage on small and medium-sized trees, searching for insects, mosses and lichens. The plantain squirrel *Callosciurus notatus* and horse-tailed squirrel *Sundasciurus hippurus* feed mainly in the lower and middle forest levels but nest in the upper canopy. The two largest diurnal species, the giant squirrel *Ratufa affinis* and Prevost's squirrel *Callosciurus prevostii,* live and feed highest in the canopy. The nocturnal squirrel fauna is almost as diverse as that found by day, with the larger flying forms found in the upper levels of the forest and the smaller forms, including the pigmy flying squirrels, in the lower strata.

The rainforest squirrels show different food preferences when food is abundant but considerable overlap when it is scarce, when all species rely heavily on bark and sap. The largest canopy species are primarily fruit eaters. None of the smaller forest squirrels, apart from the horse-tailed squirrel, are seed specialists, unlike African or temperate forest species of comparable size. The squirrels of the middle layers are opportunistic feeders, eating both insects and fruit, whereas the ground squirrels are mainly insectivorous in diet; these species overlap somewhat with the tree shrews Tupaiidae, rather than with the arboreal squirrels. The Sunda squirrels (*Sundasciurus* species) feed mainly on bark and sap, while most of the beautiful squirrels (genus *Callosciurus*) are opportunistic feeders on a variety of plant material supplemented, by the smaller species, with insects. At night the larger flying squirrels eat fruit and leaves but not insects, while the smaller species are mainly insectivorous.

In Borneo only two large diurnal species forage and feed in the canopy layer, but in West Malaysia these two share this habitat with a third, the giant squirrel *Ratufa bicolor*. In Malaysia the three largest species of squirrels show less clear ecological divergence than the smaller species of the lower forest levels. All are fruit-eaters, but *Ratufa affinis* shows more use of the middle canopy levels and takes a significant proportion of leaves in its diet, whereas *R. bicolor* and Prevost's squirrels are often seen feeding at the same fruit trees. In Malaysia *R. bicolor* and Prevost's squirrel avoid competition by adopting different foraging patterns. While the larger giant squirrel is at an advantage in competitive situations, it has higher energy demands in terms of both basic metabolism and "travelling", whereas Prevost's squirrel can afford to spend less time feeding each day and can travel further to food trees. Perhaps the absence of *Ratufa bicolor* in Borneo can be attributed to the fact that the island's forests are less rich in "squirrel-type" fruits, so that there is only room for two large diurnal squirrels instead of three; the two which do occur are those showing least niche overlap elsewhere.

Sources: MacKinnon 1978; Payne 1979

218 THE LOWLAND RAINFOREST OF BORNEO

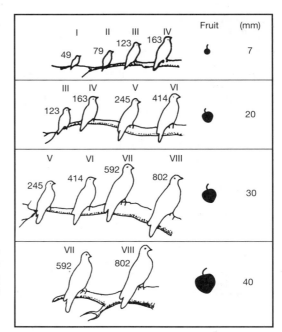

Figure 4.23. Ecological separation by feeding guilds of pigeons in New Guinea forests. Different-sized pigeons take different sizes of rainforest fruits.
Source: Diamond 1973

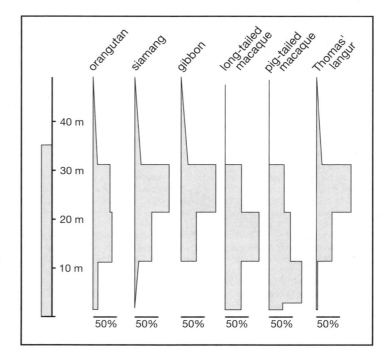

Figure 4.24. Vertical use of the canopy by six primates at Ketambe, North Sumatra. (After Rijksen 1978.)

MacKinnon 1980, 1984; Marsh and Wilson 1981; Galdikas 1978). Borneo has 13 species of primates found in lowland rainforest, including: three apes, the orangutan and two allopatric species of gibbons; five species of langurs, *Presbytis melalophos (femoralis)*, *P. rubicunda*, *P. frontata*, *P. hosei* and *P. cristata*; the proboscis monkey *Nasalis larvatus*; two macaques; and two nocturnal species, the tarsier and the slow loris. Recently a white langur has been discovered in the Danum Valley, Sabah, but this is probably not a new species but a hybrid between *P. rubicunda* and *P. hosei*, or a northern form of *P. frontata* (Johns pers. com.).

Fieldwork on the primates of Borneo and other Sundaland forests has now extended over almost two decades, giving insight into how these communities coexist (fig. 4.24). The way in which the six sympatric higher primates of Malayan and Sumatran rainforests exploit forest resources has been studied in some detail (Chivers 1980; MacKinnon and MacKinnon 1980; Rijksen 1978), as has the comparative ecology of Asian apes (box 4.7).

The quantity of food available limits the numbers of animals that a forest can support (table 4.4), with the amount at the times of greatest shortage being critical. The greater proportion of langurs (leaf monkeys) in Malayan rainforests, 82% by weight compared to 7% each for gibbons and macaques (Marsh and Wilson 1981), is probably because young leaves are never in as short supply as fruit (fig. 4.25). While all Asian diurnal primates eat a mixture of leaves and fruits, and can change their diets to less preferred foods at times of food shortage, langurs eat a larger proportion of leaves (Curtin 1980; MacKinnon and MacKinnon 1980; Bennett 1984).

To cope with the considerable amounts of foliage in their diet, langurs have evolved sacculated stomachs; these allow them to digest leaves more efficiently than any other primate. Bacteria in the stomach break down cellulose (a major component of all leaves) to release energy. They also deactivate toxins produced by the tree to protect its leaves from predation. Plant defence compounds are found in all trees but occur at higher concentrations in forests on nutrient-poor soils, where it is costly for trees to replace leaves eaten by herbivores (Waterman and Mole 1989). Since so many rainforest leaves are defended, many langurs must supplement their diets with other plant material, such as fruit and flowers.

Trees that fruit at times of general food shortage are especially important for primates, and no genus is more important than figs (Snow 1981), especially the large strangling figs. Figs made up 22% of the diet of gibbons at Kuala Lompat (Gittins and Raemaekers 1980) and 17% for *Hylobates agilis* in Malaysia (Gittins 1979), as well as being a main food item for orangutans (MacKinnon 1974a; Rijksen 1978). In Bornean forests some figs are fruiting at all times of year and some individuals fruit more than once a year. Animals congregate at big trees to feed on the crop. Primates show good knowledge of fig trees in their home range, and regularly check on the status of fruiting. For instance, the senior author has observed gibbons at

Box 4.7. Niche separation in Sundaland primates.

Primate communities on Borneo are typical of West Malesian rainforests. Usually such a rainforest community has two (in North Sumatra, three) apes of different sizes, two macaques and a variety of langurs or leaf monkeys. Studies in Kuala Lompat, West Malaysia, show how six diurnal primates can coexist in the same rainforest by different use of forest space and foods. The two gibbon species, the siamang *Hylobates syndactylus* and white-handed gibbon *Hylobates lar*, are both territorial with overlapping home ranges. Both take the same foods, especially fruit, but they show quite different foraging strategies. The siamang is twice the size of the smaller gibbon but travels only half as far each day and feeds at only half as many feeding sites. Overall, however, the siamang feeds for twice as long as the smaller ape. Since travel is costly in terms of energy, the siamang follows efficient routes between larger fruit sources, whereas *Hylobates lar* can afford to exploit smaller, dispersed food sources.

Both macaque species take a wide variety of foods, including insects, fruits and leaves, but the two species use different areas of forest and different strata within the forest. The long-tailed macaque *Macaca fascicularis* is a riverine species which ranges into lowland forest to feed but returns to sleep in tall, riverside trees at night; this may be a defence against predators. The larger pig-tailed macaque *Macaca nemestrina* is mainly found in hill forests. Although it feeds at higher canopy levels, it uses the lower forest layers and ground for travel and thus travels further and more "cheaply" in terms of energy expenditure than the more arboreal long-tailed macaque. *Macaca fascicularis* is an adaptable opportunist and will eat a great range of food items, though it may often concentrate on one abundant food source.

Two leaf monkeys, the dusky langur *Presbytis obscura* and the banded langur *Presbytis melalophos*, make up the Kuala Lompat primate community. Both take large proportions of young leaves in their diet, but this food source is common in the forest and not limiting. The two species, however, use canopy space quite differently, with the dusky langur spending more time in the upper canopy layer. The banded langur took the widest range of food types of all the primates studied and showed the greatest correlation between utilisation of foods and their relative frequency in the forest. Banded langurs also showed the most even use of forest space and occurred at higher densities, with groups having overlapping home ranges. The banded langur can probably be regarded as the most ecologically successful primate in this habitat.

Although all the Malayan primates eat fruit, the Kuala Lompat primate community can be explained in simple terms as consisting of two fruit specialists (gibbons), two opportunists (macaques) and two generalists (langurs), able to exploit a common food source, leaves. The gibbons' territoriality can be explained in terms of their reliance on fruit as the main item of their diet, but both also eat substantial amounts of leaves. The siamangs take a larger proportion of leaves than the smaller gibbons.

In Borneo the orangutan *Pongo pygmaeus* replaces the siamang as the largest primate in the community. The role of the smaller ape is taken by the agile gibbon *Hylobates agilis* in West Kalimantan and the endemic Bornean gibbon *Hylobates muelleri* over the rest of the island; both are fruit specialists. Both macaques occur on Borneo, as do several species of langurs which are all predominantly leaf-eaters. The orangutan and the two gibbons both rely heavily on fruit and would seem to be competing for this resource. The orangutan's foraging strategy, however, is quite different from that of the territorial gibbons. Orangutans are not restricted to small, defensible territories but wander over large ranges. They are usually solitary, though a few may gather briefly in a fruiting tree. Orangutans travelling together in large groups would simply not find enough food to meet their needs. Orangutans exhibit sexual dimorphism, with adult males about twice the weight of adult females. Although this dimorphism has probably arisen as a consequence of male rivalry, it has the effect of reducing feeding competition between the two sexes. A comparison of feeding ecology of male and female orangutans mirrors the situation described for siamangs and gibbons, with male orangutans generally travelling less and feeding longer than the females. Like the smaller gibbons, female orangutans are able to be more selective and to feed at more, smaller fruit sources.

Sources: MacKinnon 1976; MacKinnon and MacKinnon 1980; Rodman 1973

Kuala Lompat following a regular fig route which took in all major figs in their area. Orangutans show a similar knowledge of fruit trees, including durians within their range (MacKinnon 1971), and also congregate at figs; they may find these sites by the clamour of feeding hornbills.

Figs are important hornbill foods; Cranbrook (1982) observed large groups of hornbills, as many as 50 birds of Borneo's eight species, feeding on one fig at Gunung Mulu. In Kutai, East Kalimantan, large strangling figs

Table 4.4. Approximate densities of various large birds and mammals in Bornean lowland forest. (* Long-tailed macaques and proboscis monkeys are more or less restricted to riverine forest, so figures given here are per km of riverbank.)

Species	Individuals per km^2	Group size	Groups per km^2	Biomass (kg) per km^2
Great Argus pheasant *Argusianus argus*	3	–	–	?
Helmeted hornbill *Rhinoplax vigil*	0.4	2	0.2	?
Rhinoceros hornbill *Buceros rhinoceros*	1.1	2	0.4	?
Bushy-crested hornbill *Anorrhinus galeritus*	4	6	0.7	?
Giant squirrel *Ratufa affinis*	11	–	–	18
Banded langur *Presbytis melalophos*	20	13	1.5	120
Maroon langur *Presbytis rubicunda*	10-33	10	1-3	60-190
Langurs (combined)	78-110			440-570
Long-tailed macaque *Macaca fascicularis*	30*	15	2*	120*
Pig-tailed macaque *Macaca nemestrina*	6	30	0.2	27
Gibbon *Hylobates agilis/muelleri*	6	3.5	1.5	22
Orangutan *Pongo pygmaeus*	1.5	–	–	36.5
Proboscis monkey *Nasalis larvatus*	60*	25	2.5*	600*
Clouded leopard *Neofelis nebulosa*	0.02	–	–	4
Elephant † *Elephas maximus*	0.1	15	–	300
Rhinoceros *Dicerorhinus sumatrensis*	0.1	–	–	8

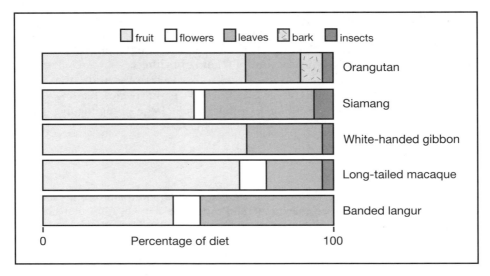

Figure 4.25. Diets of sympatric rainforest primates in North Sumatra, showing overlap and potential for competition.
Source: Rijksen 1978

are important food sources, especially as they fruit continuously at times of shortage. Since such figs commonly occur in large timber trees such as dipterocarps, they are rare in logged-over forests, and this can have serious consequences for forest mammals and birds (Leighton and Leighton 1983).

Despite the dominant role they play in Bornean forests, dipterocarps are not an important food source for vertebrates other than pigs *Sus barbatus* (Medway 1972). Mass migrations of bearded pigs following the crop of the illipe *Shorea* in Sarawak and East Kalimantan are well documented (Pfeffer and Caldecott 1986). Fruit-eating bats and parakeets have been known to eat the whole fruit crop of *Dryobalanops aromatica* in West Malaysia (Wyatt-Smith 1963). After mast fruiting years in Gunung Palung, pigs migrated into the area, and there was a change in foraging strategies of orangutans and other primates feeding on the other fruit trees that fruited at the same time (Leighton in prep.). At times of food shortage, monkeys and gibbons feed occasionally on the young leaves of a few dipterocarps (Chivers 1980), as do orangutans (MacKinnon 1974a), but most mammals and birds do not feed on dipterocarps (Medway 1972b).

Soil and Litter Communities

Green plants (autotrophs) provide a food substrate for animals, fungi and bacteria (heterotrophs). Those heterotrophs which break down dead organisms are called decomposers. They are especially active in forests decaying from the mature to the gap phase of the forest growth cycle, and in the forest floor layer. Decomposition produces plant nutrients and carbon

---

Box 4.8. Gliding — a rainforest adaptation.

Apart from birds, bats are the only other vertebrates to have developed true, powered flight. Gliding, however, has been developed independently among six different animal groups found in the tropical rainforests of Borneo. The ability to glide confers several advantages. Gliding animals are able to cover considerable distances between trees without having to descend to the ground and climb up again. This saves much work and energy so that the glider is able to cover a wider range each day in search of food. It also avoids the forest floor and any predators which may be roaming there.

The most commonly seen forest glider is the flying lizard *Draco*. There are several species of *Draco*, all searching for insects on tree trunks but living at different heights in the forest. The male *Draco volans* is territorial. He has a colourful flap of skin, yellow or orange, under his throat, and erects this "flag" to warn off intruders or to attract blue-throated females. The ribs of the flying lizard extend beyond its body wall to form six bony rays, which are covered in lateral flaps of skin. These flaps are spread when the animal glides but are folded away neatly when the lizard is running or resting.

The flying gecko *Ptychozoon kuhlii* also raises lateral extensions of its skin to form a gliding membrane. After a glide these are folded away under its belly. The gecko's flattened shape has two advantages. It lends the body extra buoyancy during a glide and helps to render the gecko almost invisible when flattened against a tree trunk.

An animal's ability to glide depends upon increasing the surface area in contact with the air. Gliding animals have different ways of achieving this. Flying frogs such as *Rhacophorus palmatus* have greatly extended webbing on their feet which allows them to parachute down gently from tree to tree. The flying snake *Chrysopelea paradisea* slithers quickly along a branch and, without checking its speed, launches into space. As it takes off the snake flattens its body and "swims" through the air.

The gliding performances of the flying lemur *Cynocephalus variegatus* and the many flying squirrels are even more spectacular. These arboreal mammals launch themselves from a high trunk, and a single glide may cover 100 metres. Both the flying lemur and flying squirrels have evolved a wide flap of skin, a flight membrane or patagium, which extends between their outstretched limbs and in the flying lemur also encompasses the tail. In flight the animals assume the shape of a kite. The flying squirrels can increase their flight surface area by means of a stiff cartilage rod which extends beyond the hand, drawing the patagium with it. This rod and the squirrel's tail may be used for steering. The flying lemur gains additional span because of the great length of its fingers and toes.

These gliding mechanisms, though valuable for travel, make the gliders clumsy climbers and hence very vulnerable to fast-flying birds of prey. It is probably because of this clumsiness that both flying lemurs and flying squirrels are active only at night, when they are less likely to be spotted by predators. Squirrels hide in tree holes during the day, but the flying lemur remains in the open, clinging to a tree trunk or branch. Its mottled, green-tinged coat blends with the bark background, providing perfect camouflage. Because it has no permanent den, the flying lemur must carry its young from birth; when the mother rests beneath a bough, her patagium provides the infant with a natural hammock.

dioxide which become available to green plants again. Rates of decomposition can be measured from soil respiration, from the ratio of litterfall to amount of litter on the forest floor, and by the rate of weight loss of litter enclosed in mesh bags. Anderson and Swift (1983) studied decomposition rates in different lowland forest formations in Sarawak and found total weight losses of 40% to 50% after 50 weeks, which is comparable to results from temperate, acid deciduous woodlands. The decomposition of leaf litter in the tropics is not as fast as is commonly supposed (Collins et al. 1984a).

Litter-feeding invertebrates are important detritivores. By reducing litter to tiny particles, they increase the surface area for bacterial and fungal decay and mix the litter with mineral soil. The main detritivores at Mulu were termites (62% of total litter individuals), earthworms (at 4%, fewer than in temperate soils) and beetles (3%). The major predators were ants (20%). Millipedes, woodlice and molluscs, which are common in temperate soil microfauna, were less important at Mulu, though land crabs were common in alluvial forest and may play a substantial role in litter removal (Collins 1979b, 1980b). Termites are probably the most important decomposer invertebrates in tropical rain forests and were common in all Mulu forests up to 1,860 m (Collins 1989). Termites feed on a wide range of foods (table 4.5).

Table 4.5. A summary of the number of genera and species of termites found in the forests of Mulu, with an analysis of their feeding habits.

Forest type	Alluvial	Kerangas	Dipterocarp	Montane
Total genera recorded	20	16	30	6
Total species recorded	31	25	59	10
Dry wood-feeders	8	3	10	0
% of total species	26	12	17	0
Rotten wood-feeders	21	17	36	8
% of total species	68	68	61	80
All wood-feeders	22	18	38	8
% of total species	71	72	64	80
Soil-feeders	12	9	24	2
% of total species	39	36	41	20
Leaf-feeders	4	1	4	0
% of total species	13	4	7	0
Lichen and moss-feeders	2	2	3	0
% of total species	6	8	5	0

Source: Collins 1984

Although termites may look superficially like ants, they belong to a quite different family, the Isoptera, and are most closely related to cockroaches. Keys to the termites of Sabah and Sarawak are provided by Thapa (1981) and Collins (1984). Like ants, termites are social insects; indeed, although highly social behaviour has evolved independently at least 11 times in the Hymenoptera (ants, bees, wasps), it occurs only once in the rest of the insects, in the termites. Termites form large colonies with a million or more members, all colony members being siblings except for the parents, the "royal pair" (Spragg and Paton 1980). The parents originate from other termite nests from which they fly with many thousand others to find a mate. Termite swarms commonly occur after rain and provide a food bonanza for predatory amphibians, reptiles and birds.

After landing a termite loses its wings, then male and female mate and build a "royal cell" in a crack in the ground or tree trunk. After copulation the female begins laying eggs from which mobile larvae develop. The royal pair must feed the first larvae, but as soon as these are large enough to forage for food and build the walls of the nest, the royal pair can devote themselves entirely to egg production. The queen produces thousands of eggs a day.

The termite colony includes blind and sterile workers as well as "soldiers" which protect the foraging columns of workers and guard the colony's nest. The soldiers have such large jaws that they cannot feed themselves and have to be fed by the workers. The latter also feed the larvae and the royal pair as well as collecting the queen's faeces. All colony members continually exchange food and saliva, which contain pheromones. This continual interchange of chemicals through the colony helps to coordinate colony activities. Thus all larvae, though potentially fertile termites of either sex, receive food impregnated with queen-derived pheromones which inhibit their development, producing sterile workers. The soldier termites produce similarly repressive pheromones, but if the number of soldiers falls, the level of pheromone is reduced so that new soldiers can be recruited. At times the queen also reduces the release of her repressive hormones, and eggs develop into fertile, winged termites which leave the nest in a swarm. A new cycle of colony formation then begins.

Most invertebrate detritivores of the lowland forest soil depend on free-living fungi and bacteria to break down indigestible plant material into a form they can use, but many termites carry in their guts symbiotic protozoa and bacteria that decompose cellulose. At Mulu the termites are represented by three families; Kalotermitidae, Rhinotermitidae and Termitidae. The dry-wood termites (Kalotermitidae) are usually confined to dead wood in the canopy and are of minor importance for forest decomposition processes. The damp-wood termites (Rhinotermitidae) may constitute 20% to 30% of the total termite density at Mulu and are important in the destruction of large logs and boles (Collins 1984).

The family Termitidae dominates many rainforest soils and can have a substantial effect on leaf and wood litter decomposition and nutrient cycling. All four subfamilies within the Termitidae have bacterial gut symbionts that assist in digestion of the components of dead plant material. The Termitinae and Nasutitermitinae feed on a wide range of dead wood materials and decomposing litter down to the finest remnants of soil organic matter and humus. The Apicotermitinae, which are rare in Southeast Asia, are exclusively soil feeders, feeding on soil organic matter and humus. The fourth subfamily, Macrotermitinae, feed mainly on fresh litter which, on excretion, is used to cultivate subterranean fungus combs or "gardens" of *Termitomyces* fungus, a fungus unknown outside termite nests. The fungus breaks down the cellulose and lignin, after which the combs are eaten by the termites. Macrotermitinae, like other animals, do not have the necessary enzymes to digest lignin or cellulose, but the fungus can do this. The termites thus benefit from this symbiosis by being able to utilise fresh litter as food before it has been decomposed by fungi or bacteria, in this way realising greater assimilation efficiency for important nutrients such as nitrogen (Collins 1983).

The termites' relationship with the fungus has made these insects so successful at litter decomposition that they are reported to be responsible for 32% of leaf litter removal in a forest in Peninsular Malaysia (Matsumoto and Abe 1979). At Mulu, however, termites remove only 1% to 4% of total annual litter production, probably because there are few fungus-farming Macrotermitinae present. In lowland dipterocarp forests at Mulu, termites occurred at densities of 1,500 animals (or 2.4 g) per square metre (ridge) and 390 animals (or 0.9 g) per square metre (alluvium) but were more common in heath forest (Collins 1983). In secondary forest and in the drier climate of logged forests, Macrotermitinae may become increasingly important (Collins 1980).

Forest Floor Community

All the debris from above comes to rest on the forest floor: fallen fruit and seeds, dead leaves, animal faeces and corpses, broken branches and even whole trees blown over by strong winds. All this material is broken down and decays to form a rich forest litter, the home of numerous invertebrates which work over the detritus.

The forest floor is shaded from the sun's hot rays so that temperature and humidity remain constant. This ever-damp habitat is home to many small, water-loving animals. Blood-sucking leeches *Haemadapsa* loop over the fallen leaves or hang from the vegetation, waving sinuous bodies in the air to sense the approach of any animal that will provide them with a meal. Attached by its suckered mouth-piece, the leech rides round on its unwitting host until its body is distended with blood. If left undisturbed,

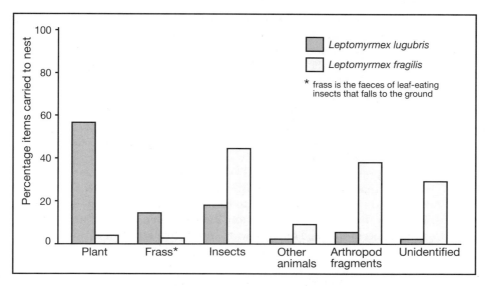

Figure 4.26. Ecological separation between two species of rainforest ants in Papua New Guinea, with respect to food items carried to the nest. The two species select different proportions of the available food. (After Plowman 1981.)

Haemadapsa leeches will feed for about 80 minutes, and after feeding may be six times heavier than before the meal. A single meal may provide sufficient sustenance for three to eight months (Fogden and Proctor 1985). Although unattractive and often unwelcome fellow-travellers, leeches leave very clean wounds, do not transmit disease and are increasingly appreciated for their value to medicine.

Sickle-headed terrestrial flatworms in many colours flow across the damp leaves and over slimy logs. They feed on organic material ingested through an eversible mouth on the underside of the body. These curious, primitive worms can multiply by splitting off buds from the tail.

Within the province of the forest floor are other smaller habitats. Niche separation between species is just as evident here as in the rainforest canopy (fig. 4.26). Corpses and dung are two habitats that provide special niches and support distinctive animal communities. Scavenging beetles, flies and ants all feed on the decomposing matter, which also provides food and shelter for dung beetles and developing maggots. Dung beetles (Scarabaeidae) can comprise half of the arthropod biomass on the forest floor and show enormous species diversity. They play a major role in the decomposition of organic matter by fragmentation, burial, partial assimilation, transportation of micropredators and parasites and aeration of the soil. Dung beetles may even help to disperse seeds that are rolled away in their balls of dung (Hanski 1983). Seeds are transported only a few metres but are provided with ideal growing conditions, a well-aerated, cultivated soil with rich manure incorporated.

Some animals, including nest-building termites, create their own microhabitat where climatic conditions are different from those prevailing in the surrounding environment. The mud-dauber wasp *Eumenes* patiently fashions and provisions a chamber of clay in which it seals a single egg. Clusters of these "pots" are attached to rocks or tree trunks. The whole process is a lengthy ritual. First the wasp must collect a drop of water from a pool or stream and mix it with dry earth to form a mud pellet. The pellets are carried singly to the site of construction and cemented together until the nest chamber is complete. Then the wasp hunts through the canopy, catching and paralysing spiders or caterpillars which will provide a store of fresh food for the developing larva. After the egg is laid, a final mud pellet is used to seal the entrance, and the larva is left to feed on its living larder.

Dead wood is a habitat in itself and a fascinating one to study. The animals and plants that live in and on the dead wood help to break down the woody material and recycle nutrients into the forest ecosystem. The organisms of this detritus food chain include armies of ants and termites, pill millipedes, fierce-jawed stag beetles, shiny rhinoceros beetles and beautiful, green longhorn beetles that are sometimes used for ornamental jewellery. Flat fiddle beetles, so named because of their shape, live among the plate-like bracket fungi that grow on rotten wood. They belong to the carnivorous family Carabidae and were once considered so rare that the Paris Museum paid 1,000 francs for a single specimen. In fact, these beetles are quite common but are rarely seen since they are active only at night, when the female bores a hole in the fungus to house her egg. The developing larva feeds on tiny insects that wander into its chamber.

Fungal hyphae penetrate the layers of rotting wood and help to break it down. Overnight luminous toadstools, fairy clubs and delicate cup fungi appear, and are eaten by ants and beetles. The extraordinary maiden's veil fungus is draped in a netlike tent and exudes a foul smell of decay that attracts blowflies, *Charaxes* butterflies and grasshoppers to feed on its slimy cap, where they pick up spores and carry them to new locations, thereby playing a vital role in fungus dispersal.

There are numerous predators on the leaf litter. Scorpions, spiders and armies of ants scavenge among the dead leaves. Gaudy sun spiders and large orb spiders weave delicate webs among the low vegetation to trap flying insects. Crab spiders blend with the flowers on which they rest, and prey on bees and butterflies attracted to the nectar. Legs outstretched, the dung spider sits motionless on a white silk mat. It is coloured to mimic the white excrement of a bird; insects attracted to the "dropping" are caught and eaten. With its camouflage, the dung spider is safe from the attentions of spider-hunting wasps (MacKinnon and MacKinnon 1974).

Not surprisingly, many invertebrates have developed defences against their predators: spines, hard shells, fierce jaws and acrid chemicals. Pill millipedes roll up when disturbed and form a perfect sphere of hard, impenetrable cuticle. Giant red and black millipedes curl up like fossil ammonites, leaving only a multitude of hard legs exposed; they also eject a foul-smelling acrid orange oil to deter predators.

During the day tree shrews Tupaiidae scour the litter of the forest floor, probing among the dead leaves and fallen tree trunks for beetle grubs, earthworms and insects. They also feed intensively on fruit when it is available, swallowing only the soft pulp and juice and spitting out the indigestible parts, just as bats do (Emmons and Biun 1991). Once regarded as the first rung on the primate evolutionary ladder that leads to humans, the tree shrews are now recognised as a distinct order Scandentia. Tree shrews have a peculiar "absentee" system of maternal behaviour whereby the mother spends only a few minutes of alternating days with her young for the first month of their lives. However, the mother spends much time with her young after they have left the nest. The absentee system probably serves to prevent predators from finding the young.

Fallen fruit provides a food supply for many creatures on the forest floor: pigs, deer, rats, tortoises and a host of forest insects and butterflies. Many of the forest trees also exhibit the strange phenomenon of cauliflory; they bear their flowers and fruits on the lower trunk so that the fruits are accessible to ground-living ungulates, which help to disperse the seeds.

Birds of the forest floor include the brightly coloured pittas, thrushes, babblers and pheasants, which are mainly insectivorous in diet but often congregate beneath a fruiting tree to feed on insects attracted by the rotting fruit. Among the many colourful pheasants and partridges, the Great Argus cock *Argusianus argus* is particularly handsome. He attracts his females by strutting and showing off his fanned and splendidly decorated wings. To give maximum benefit to his performance, he maintains a dancing ring about seven metres wide. He keeps this open space on the forest floor clear of all leaf litter. Argus pheasants are very shy, and while the "kawau" call is a common sound of the forest, it is rare to see a displaying bird (Davison 1981).

Argus feathers are sought for decorating helmets and other ceremonial garments of Iban warriors, and the pheasant motif is often used in Iban tattoo. Legend tells how the pheasant acquired his spectacular plumage. The pheasant and the coucal agreed to decorate each other as disguise against their enemies, and the coucal painstakingly painted beautiful "eyes" on the pheasant's tail. When it came to the coucal's turn to be painted, the lazy pheasant merely tipped the pots of pigment over the coucal, leaving him with a plain, black body and brown wings (Harrisson 1960).

Tiny mousedeer *Tragulus,* no bigger than cats, feast on the fallen fruit. Active by day and night, mousedeer roam the forest in pairs or alone (Davison 1980). They are not true deer but are believed to be similar to the deer's evolutionary ancestors. Instead of antlers, mousedeer have developed long, curved canine teeth for defence. The forest-dwelling muntjak (barking deer) are intermediate between the mousedeer and true deer; they have both long canines and short, pointed antlers.

Larger Herbivores

Apart from the tragulids, other forest herbivores include two species of muntjak *Muntiacus muntjak* and *M. atherodes,* sambar deer *Cervus timorensis* and the wild cattle known as banteng or tembadau *Bos javanicus.* Banteng move around in small herds consisting of several immature males and many cows and calves, all led by one bull. Like other Asian wild cattle, banteng prefer to feed on open forest fringes and in grassy clearings, and their spread through Southeast Asia is thought to have been assisted by people clearing and burning forest (Wharton 1968). Although banteng are increasingly rare in the wild today, their genes live on in their domesticated relatives, Bali cattle.

The largest forest herbivores, the elephant *Elephas maximus* and rhinoceros *Dicerorhinus sumatrensis,* are bulk feeders requiring huge amounts of greenery each day. They look for quantity rather than quality. Both exploit a food source found above the forest floor, leafy branches and twigs. Elephants have a special adaptation for reaching food above their heads, the long sensitive trunk derived from the nose and upper lip.

Elephants are unusual in having a matriarchal (female-dominated) society, with lone bulls living peripherally to the close-knit mother-family groups. In open feeding grounds elephants may congregate in herds, but in the forests they travel in small groups of one or more related females with their immature offspring, totalling about three to twelve individuals. Elephants are browsers, feeding mainly on monocotyledons such as palms, grasses, bamboos and wild bananas supplemented with herbs, shrubs and the leaves or bark of some trees. In Peninsular Malaysia palms and grasses constituted 75% of the elephant's diet (Olivier 1978). Primary dipterocarp forest is not prime habitat for elephants, and they prefer foraging on forest edges and in open clearings, even human-made ones, which sometimes leads to conflicts with farmers. Elephants travel great distances along traditional pathways between favourite feeding areas, mud wallows, drinking pools and mineral licks (Olivier 1978). Even when the forest has been cleared, they will still follow these routes and may cause considerable damage in cultivated fields. As more areas of natural habitat are cleared, elephant numbers decline. Today there may be about one thousand elephants left in Sabah (Davies and Payne 1982). In Kalimantan elephants are found only on the Sabah/East Kalimantan border in the Ulu Sembakung forests, most of which are designated as timber concessions.

The Sumatran rhinoceros *Dicerorhinus sumatrensis* was probably the creature Marco Polo saw on his return journey from China, and described as a unicorn. In fact, it has two horns and is the smallest living rhinoceros. It is also the hairiest, with a coarse, russet coat. Today rhinoceros are found in Borneo only in small, scattered populations living in rugged terrain such as the Gunung Lotung area in Sabah and Pulong Tau, Sarawak. It is doubtful if there are any rhinos left in Kalimantan. Sadly, they are now extinct in Kutai National Park, East Kalimantan, which was created to protect this species (Santiapillai and MacKinnon 1991). Even in Sabah the rhino population is estimated at no more than thirty individuals (Davies and Payne 1982); such a small and scattered population is unlikely to be viable in the long term.

The two major field studies on the Sumatran rhinoceros were remarkable for their lack of rhino sightings (Borner 1979; van Strien 1985). Much of what we know about this animal has been deduced from the tracks of its passing. Rhinoceros wander singly or in pairs over wide ranges, browsing on saplings. They visit favourite mud wallows regularly, and this habit has made them especially vulnerable to poachers seeking their valuable horn. Unfortunately, this terrible trade is unlikely to stop while rhino horn fetches its current inflated price of more than $12,000 per kilogram, for sale as an aphrodisiac and cure-all much sought by the Chinese. Other parts of the rhino are all alleged to have medicinal properties, and hoof, skin and even urine, fetch a good price, but none are as valuable as the horn (Martin and Martin 1989). Ironically, this protuberance is not real horn at all but a mass of keratin, like a giant fingernail, with no proven pharmaceutical properties.

Large herbivores throughout the world seem to experience a shortage of sodium in their diets. Carnivores do not have the same problem because flesh contains about 0.1% sodium. The availability of various minerals in Borneo soils is low in many areas. Elephants and other large mammals, including rhino and even apes, are known to use areas where higher concentrations of mineral salts are available and even to excavate caves in mineral-rich earth (MacKinnon 1974b; van Strien 1985). In Sabah the past and present distribution of rhinos is closely correlated with the distribution of salt sources such as saltwater springs, salt licks and mud volcanoes (Davies and Payne 1982). Payne (1992) also suggests that the patchy distribution of orangutans in Sabah, and elsewhere in Borneo, may be related to soil minerals, with these animals absent in nutrient-poor areas.

The Forest at Night

By night the forest is used by a community of animals quite different from those active by day. In West Malaysia more than 90% of the birds are diurnal, whereas about 80% of the mammals are crepuscular (active at dawn and dusk) or nocturnal (Harrison 1962).

The most successful group of night animals are the bats. These are the only mammals to have developed true powered flight, and this gives them the advantage of great mobility so that they can forage over great areas of forest (chapter 6). As night falls the large fruit bats (flying foxes) *Pteropus vampyrus* leave their roost trees and stream off in wide phalanxes in search of food. They are a spectacular sight, especially when the long ribbon of bats curves and winds, following the meandering river below. Fruit bats feed on a range of fruits and are important seed dispersers. However, not all bats feed on fruits. Among the Bornean bats there are insect eaters, bats feeding on nectar and pollen, carnivores feeding on other vertebrates and even bats that feed on fish, but there are no blood-sucking bats like those found in tropical America.

At night, especially after a heavy rainstorm, the forest reverberates to the croaking chorus of thousands of frogs as they come together to mate and spawn. Rainforests harbour enormous numbers of amphibians. Borneo has at least 183 species (Inger 1966) of which half live in the trees. Arboreal frogs are generally slender-bodied, with the tips of digits expanded into suction discs which help them to grip as they leap among the branches. The clinging force of these discs derives from friction and wet adhesion.

Tree frogs lay their eggs in a wet, frothy mass in a tree hole or glued to the surface of a large leaf, often overhanging a forest puddle. The foam is made from a small amount of fluid secreted by the female and is whipped up by her energetic kicking until it hardens like a meringue. The froth protects the eggs from evaporation, and the developing tadpoles remain within this protective sac until heavy rain washes them down into a tree hole or to a pool below. Here they complete their metamorphosis and grow into tiny tree frogs, fully equipped for arboreal life (Halliday and Adler 1986).

Night in the forest is the time to see two of the least-known primates, the tarsier *Tarsius bancanus* and the slow loris *Nycticebus coucang*. Unlike the diurnal primates, which feed mainly on fruit and leaves, the two lower primates are more insectivorous. Because of their nocturnal habits, lorises are difficult to observe. A two-year study in West Malaysia compared their feeding and foraging strategies with those of flying squirrels and palm civets, their possible competitors for food (Barrett 1984). Lorises feed mainly on insects and fruits, whereas the large flying squirrels take fruit, flowers and immature leaves. Palm civets forage opportunistically for a wide range of food including fruit and small animals.

The different nocturnal mammals also have very different means of locomotion. This influences their use of the canopy as well as their ranging behaviour. Flying squirrels, as a result of gliding, are able to range widely to exploit large food sources, and are active principally in the higher forest

Figure 4.27. Different foraging patterns and use of forest resources by three nocturnal mammals, thereby avoiding competition.
Source: Barrett 1984

layers. Palm civets range widely by descending to the ground; when arboreal, they are restricted mainly to larger branches. Slow lorises need a continuous substratum and are small and dexterous enough to move freely on small branches in the forest understorey, covering small home ranges of five to ten hectares (Barrett 1984). These differences in habitat use, and thereby foods encountered, reduce competition between the species and encourage niche separation (fig. 4.27). The slow loris forages mainly in the lower forest layers, where its slow locomotion has probably evolved for predator avoidance and as a means of catching prey. When a loris approaches an insect, it moves forward so slowly and deliberately that the prey does not seem to notice its advance. Suddenly, a hand shoots out and the loris grabs the insect.

The tarsier also feeds on insects but travels very differently, always in the lower layers of the forest. Here it leaps from one vertical trunk to the next. An adult tarsier is tiny, weighing no more than 150 g. For such a small creature its jumps are spectacular, sometimes as much as four metres in one bound. The tarsier is well adapted to its role as a vertical, clinging leaper, with long, spindly limbs and broad, flat pads to the fingertips. These pads resemble those of the tree frogs and serve the same purpose; they help the animal grip as it leaps from one support to the next. The tarsier needs good night vision and has enormous eyes, so large that it cannot revolve them in their sockets but must turn its whole head to look from side to side. The Ibans believe that the tarsier can swivel its head completely. It was not a good omen to encounter a tarsier on a head-hunting expedition since it was regarded as a clear warning that the hunter was likely to lose his own head!

Little is known about wild Bornean (western) tarsiers. The first preliminary study of *Tarsius bancanus* indicated that these small, nocturnal insectivores had a pattern of overlapping solitary home ranges, rather like those of bushbabies (Fogden 1974). More extensive studies under semi-natural conditions (Niemitz 1979, 1984) and in the wild (Crompton 1984) have shown that the tarsier is a pair-living territorial species using scent to mark territories. Male and female range solitarily, but both scent-mark at specific sites (Niemitz 1984). Bornean tarsiers do not use calls in a territorial context, unlike the pair-living Sulawesi tarsiers *Tarsius spectrum* who perform highly synchronised territorial duets, analogous to those of gibbons, in the early morning before they retire to their sleeping sites (MacKinnon and MacKinnon 1980b, 1984). Tarsiers are unusual among nocturnal prosimians because they eat exclusively animal food (in the case of *T. bancanus*, 90% arthropods, 10% vertebrates: birds, bats and snakes). Niemitz (1979) has suggested that the western tarsier occupies the ecological niche of a small owl in dense undergrowth where owls are unable to fly. Indeed, tarsiers show striking convergence with owls: large eyes and absence of a reflective tapetum; ability to rotate the head to look backward; and use of hearing rather than sight to locate prey. They also feed on a similar range of species, their locomotion is completely noiseless and they ambush their prey while moving fairly close to the ground.

The nocturnal arboreal mammal fauna of a habitat usually includes species of varying sizes which exploit different food resources. In general, small species consume mainly insect/animal prey, larger species supplement this diet with fruit and gums and the largest, nocturnal arboreal mammals are increasingly folivorous, obtaining protein from foliage and energy from fruit and/or foliage (Hladik 1979; Barrett 1984). Differences in body size are associated with differential habitat use, so reducing potential competition where there is dietary overlap. Many species may show physiological adaptations for their niche, such as an enlarged caecum and colon to aid the digestion of foliage as in the red giant flying squirrel *Petaurista petaurista*.

At night the niche occupied by the tree shrews during the day is exploited by other insectivores: the shrews, smallest of all mammals, and the moonrat *Echinosorex gymnurus*. The largest of the insectivores, the moonrat is so evil-smelling that predators avoid it; indeed, it advertises its presence with noisy shuffles and bold markings. In Sumatra the moonrat is grey, but in Borneo this mammal is a striking yellowish white and very conspicuous against the dark forest. The moonrat feeds on earthworms, insects and a few snails, occupying almost exactly the same niche by night that the common tree shrew uses by day (Langham 1983).

A specialised insect eater, the pangolin or scaly anteater *Manis javanica* is distinguished from all other Old World mammals by a unique and almost impenetrable coat of overlapping, horny body scales. When frightened, the pangolin rolls up into a tight ball, tail over head, so that a predator is faced with an armoured sphere that is impossible to prise open. Usually nocturnal, the pangolin sleeps during the day in an underground burrow. It feeds exclusively on termites and ants taken from nests in trees, at ground level or below the ground. The pangolin rips nests open with its sharp claws and probes the exposed galleries with its sticky tongue to extract the insect eggs and grubs. The pangolin is a versatile climber and can hang from a branch by its prehensile tail to explore and excavate nests in trees. The tail can serve another function; an infant pangolin rides on its mother's tail like a tiny jockey.

Many of the forest snakes hunt at night. Some, like the king cobras, yellow and black-banded kraits, cat snakes and brightly coloured coral snakes, feed on other snakes as well as on rats and frogs. Large king cobras up to five metres in length are quite common and completely unafraid of people. They rear up to face the challenger instead of slipping away as most other snakes do. The smaller cobras and pit vipers hunt rats, the pit vipers detecting warm-blooded prey by means of heat receptors on the front of the head. All of these snakes are poisonous, though this is not typical of snakes as a group. Borneo has at least 125 species of land snakes (Haile 1958; Stuebing 1991), with 50 poisonous species, but only the cobras, kraits and six pit vipers are really dangerous to people. Snake venom contains several active ingredients which serve not only to kill the victim by impeding respiration or destroying the blood vessels but also act as digestive juices.

At the head of the food chain are the cats and other predatory carnivores. Neither tiger nor leopard occur on Borneo; here the largest cat is the beautiful clouded leopard *Neofelis nebulosa*. Although often described as semi-arboreal, the clouded leopard probably only uses trees for resting and spends most of its waking hours moving on the ground, both by day and night. Such flexibility in activity allows a top predator to optimise access to food resources (Rabinowitz et al. 1987). Clouded leopards feed on monkeys, orangutans, wild pigs, deer, porcupines and even fish.

The four smaller cats feed on small birds and other animals, each taking a slightly different range of prey according to its own size and habitat. The small, nocturnal leopard cat *Felis bengalensis* takes small mammals and large insects, while the diet of the marbled cat *F. marmorata* includes rats (Payne et al. 1985). The flat-headed cat *F. planceps* probably feeds mainly on fish, whereas little is known of the habits of the rare bay cat *F. badia*. Most of the cats are camouflaged by their colouring. Stripes and spots break up their outlines and help them to merge with the streaky patches of light and shadow found on the forest floor. To be fully effective, such camouflage must be accompanied by appropriate behaviour; the cats lie in wait for victims or stalk prey cautiously.

Bears reached tropical Asia from the Palearctic, but only one species, the sunbear *Helarctos malayanus*, is found in Sundaland. Although it belongs to the family Carnivora, the sunbear is an omnivore, eating fruit, grubs, birds' eggs and fledglings, and honey. Sunbears are expert climbers and will scale great heights to reach the nests of bees and termites. With powerful claws and strong teeth, the sunbear rips open tree trunks to reach the hidden nests of stingless meliponid bees and to feast on the sweet contents.

Just as there are herbivorous mammals of all shapes and sizes, so there are different carnivores occupying many different niches. The mustelids (stoat family) are found throughout the world. In the Bornean rainforest they are represented by the golden red weasel *Mustela nudipes* and the larger, arboreal, yellow-throated marten *Martes flavigula*. The marten raids bees' nests for honey, nectar and grubs but also preys on crustaceans and larger prey, such as an occasional squirrel or mousedeer. The *teledu* or stink badger *Mydaus javanensis* has specialised on a diet of earthworms and cicada larvae (Payne et al. 1985). The badger's dramatic black-and-white markings may act as camouflage or serve as a warning to would-be predators. The badger can also squirt an obnoxious liquid from its anal gland, another deterrent to predators.

Civets are primitive carnivores; in fact, many live on fruit and insects rather than flesh. Members of the civet family have diversified to fill several niches. Both the tangalung *Viverra tangalunga* and the banded palm civet *Hemigalus derbyanus* are terrestrial and nocturnal. While the tangalung takes a wide variety of foods, including invertebrates and small vertebrates from the forest floor, the banded palm civet feeds on insects, earthworms and other small animals. Apart from the fish-eating otter civet *Cynogale bennettii*, a lithe and graceful swimmer, the other civets are active both in the trees and on the ground. The linsang *Prinodon linsang* has sharp, pointed teeth like a cat's, ideal for killing small birds, mammals and reptiles. The common palm civet *Paradoxurus hermaphroditus* has broader teeth to cope with a diet rich in fruit. The largest civet, the binturong *Arctictis binturong*, is also a fruit eater. Although mainly active at night, the binturong can sometimes be seen at midday feasting in a fig tree, using its thick, prehensile tail as an extra limb for support.

The ecology of the forest at night is just as complex as by day (fig. 4.28). Some of the largest mammals like elephants, banteng, rhinoceros, pigs and deer roam the forest both by day and night, but the arboreal fauna is more clearly divided into diurnal and nocturnal animals. A few nighttime creatures, like the pangolin, exploit niches that are vacant by day but most, like the bats and the moonrat, share resources that are exploited by other forest residents during the daytime.

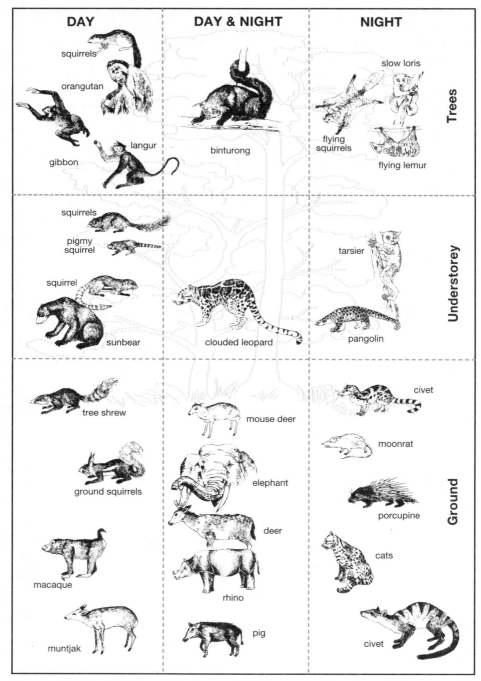

Figure 4.28. Stratification of the day and night communities of non-flying mammals in Bornean rainforest. (After MacKinnon 1972.)

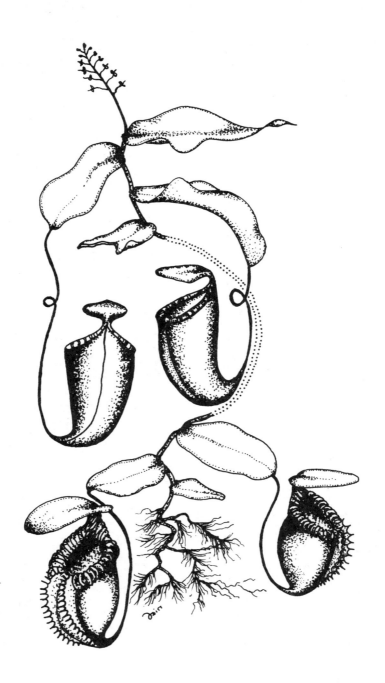

Chapter Five

Other Lowland Forest Formations

HEATH FOREST OR KERANGAS

Heath forest is the most distinctive and easily recognised of all lowland rainforest formations. The greatest extent of heath forest in the Far East is found in Borneo, where it is commonly known as *kerangas* after the Iban term for land that will not grow rice. The first scientific description of Bornean kerangas forests is that of Beccari (1904) in his book *Wanderings in the Great Forests of Borneo*, written after the scientist had visited an area in Sarawak with white crystalline sand and a peculiar vegetation of small, stunted trees.

Heath forests occur round much of the Bornean coastline on old raised terraces and inland, mainly on sandstone plateaux. In Sarawak, Sabah and Brunei heath forest is often found on dip slopes in hilly country, where sandstone beds lie close to and parallel to the surface (Whitmore 1984a). In Indonesian Borneo a broad swathe of heath forest runs diagonally through Central and East Kalimantan. The original area of Bornean heath forests is estimated at 66,882 km^2, with 48% remaining, (MacKinnon and MacKinnon 1986); 24,750 km^2 of kerangas forests remain in Kalimantan (MacKinnon and Artha 1981). The distribution of heath forests in Borneo is shown in figure 5.1.

Heath forests are found on soils derived from siliceous parent materials. These soils are inherently poor in bases, highly acidic, commonly coarsely textured and free-draining, and are often described as white sand soils. Such soils usually originate from ancient, eroded sandstone beaches that were stranded either by uplift of land or falling sea level, for example during the mid-Pleistocene (Burnham 1984). They are often covered in a superficial layer of peat or humus which is quickly lost once the natural vegetation is cleared. Although kerangas soils are freely draining, if they become waterlogged *kerapah* forest forms in the swampy conditions. In structure and physical character this is heath forest, though some tree species are particularly common at kerapah sites (Whitmore 1984a).

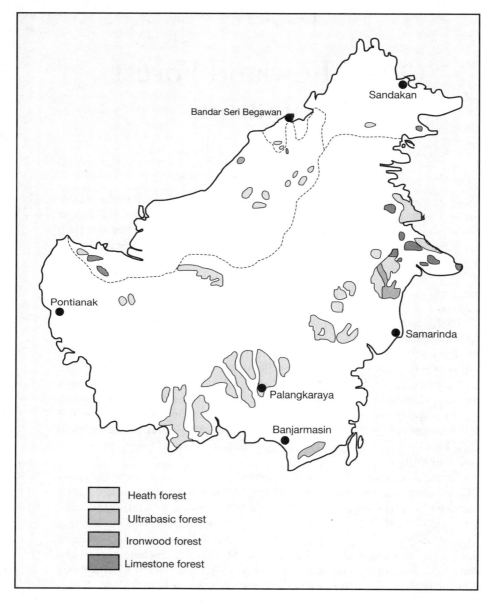

Figure 5.1. Sketch map of Borneo showing heath, ultrabasic, ironwood and limestone forest formations.

Forest Structure and Composition

Detailed studies have been made of heath forests in Kalimantan (Kartawinata 1978, 1980a) Brunei and Sarawak (Brunig 1965, 1974; Proctor et al. 1983). Heath forest has distinctive structural and vegetation characteristics, with trees generally shorter and smaller than those of mixed rainforest (Whitmore 1984a; Kartawinata 1978, 1980a). It is strikingly different from lowland dipterocarp forest in structure, texture and colour. Heath forest has a low, uniform, single-layered canopy formed by the crowns of large saplings and small poles. On aerial photographs the flat canopy is distinctive because of its pale tone and fine texture, reflecting tree-crown texture and small leaf size. Trees are often densely packed and difficult to penetrate (fig. 5.2). In an East Kalimantan study (Riswan 1982) heath forest had between 454 and 750 trees of 10 cm dbh per hectare (fig. 5.3).

Trees of large girth and big, woody climbers, including rattans, are rare. Buttresses are smaller, but stilt roots are commoner than in evergreen rainforest. Small, thin climbers are common, as are epiphytes. Many heath forest trees have thick leaves, and small-leaved trees are more common than in evergreen rainforest. In Sarawak *Casuarina nobilis* is often

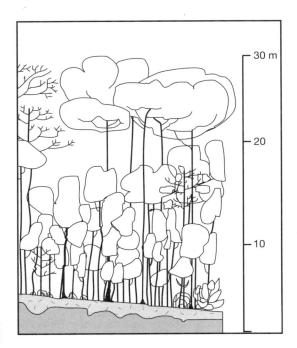

Figure 5.2. Profile diagram of dense kerangas forest in Brunei, Borneo. Trees reach up to 20 m height. Lowland dipterocarp forest is in the background. (After Whitmore 1984a.)

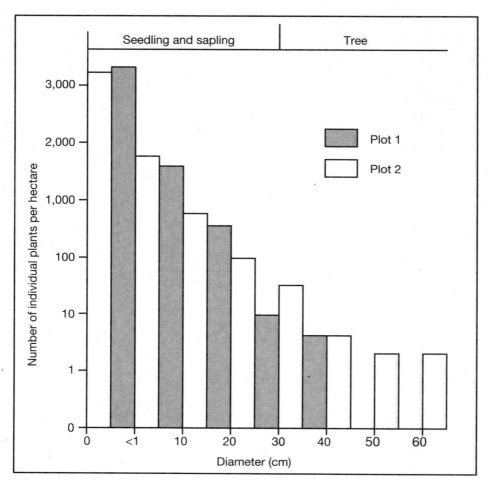

Figure 5.3. Diameter class distribution of trees in primary kerangas forest plots at Samboja, East Kalimantan. (After Riswan 1982.)

found in kerangas, as are *Dacrydium* and *Podocarpus*, both conifers with scalelike or needlelike leaves (Jacobs 1988).

In Bako National Park, Sarawak, Brunig (1965) recognised a series of heath forest types related to decreasing soil depth and increasing variability of water supply. Under the most favourable conditions heath forest was similar to evergreen rainforest, with dipterocarps prominent among the large trees, canopy 27-31 m high, and palms common. At the other extreme, the canopy was only 4.5-9 m high, there was only one dipterocarp, patches of single tree species were frequent, palms were rare and bryophytes were common.

In Kalimantan kerangas forest varies from tall, closed forest similar in structure to adjacent lowland mixed forest, to open scrubby vegetation or *padang* (Kartawinata 1978). Heath forests often form part of a mosaic with other forest formations. The open padang type occurs on the white sands at Sebulu, East Kalimantan, forming a mosaic within mixed dipterocarp forest (Kartawinata 1978). Trees are small and short, often with crooked stems and thin crowns, with the largest tree not more than 30 cm in diameter and less than 15 m high. Most trees are scattered and have diameters of 10-20 cm; the canopy is discontinuous, with crowns of sclerophyllous leaves that appear pale greyish green. Shrubs and saplings grow in clumps, and the forest floor either is covered with herbs and dwarf shrubs or is completely bare. Among the larger trees the dominant species are *Cratoxylum glaucum, Dactylocladus stenostachys* and *Tristaniopsis obovata*.

Tall, closed kerangas forest also occurs at Sebulu (Kartawinata 1978). On aerial photographs this forest formation appears more like lowland mixed dipterocarp forest. Trees grow to about 40 m tall, with diameters of up to 110 cm. Dominant tree species include *Eugenia palembanica, Ilex hypoglauca,* and *Cotylelobium malayanum*. In the shrub and sapling layer *Barringtonia sumatrana, Calophyllum soulattri* and *Shorea teysmanniana* are common. Many species of the open kerangas forest also occur here.

Padang forest, the most open type of kerangas, occurs near Buntok and Bukit Raya in Central Kalimantan, at Padang Lawai, East Kalimantan, and at Mandor, Pasir Panjang, Bengkawang and Sintang in West Kalimantan (Kartawinata 1978). Padang vegetation also occurs on parts of the Anamabas islands. This vegetation type is often regarded as heath forest degraded by fire or felling (Whitmore 1984a; Janzen 1974), but it may also be a natural vegetation type, growing on extremely impoverished soils. Characteristic species include the trees *Casuarina sumatrana, Cratoxylum glaucum, Dacrydium elatum, Baeckia frutescens* and *Tristaniopsis obovata*, as well as *Vaccinium baccanum* and *Nepenthes* species in the herb layer. Numerous species of orchids grow in the shade of shrubs, including species of *Dendrobium, Eria, Coelogyne, Bulbophyllum* and *Liparis* (Kartawinata 1978).

In the Sampit area, Central Kalimantan, tracts of dwarf kerangas forest occur over white sand in the centre of the peat swamp forests (Dilmy 1965). This dwarf forest is dominated by *Tristaniopsis obovata* and *Agathis borneensis*. The kerangas forests gradually merge into a transition *Agathis* forest and eventually to peat swamp forests dominated by *Shorea uliginosa* and *Palaquium*.

Heath forests are generally less species-rich (fig. 5.4) than other lowland rainforests. Thus 123 species of trees were recorded in a one-hectare plot of kerangas at Gunung Mulu, compared to 214 species in a plot of the same size in adjacent dipterocarp forest (Proctor et al. 1983a; see fig. 6.4). Com-

> **Box 5.1.** Mandor Nature Reserve.
>
> The nature reserve at Mandor, West Kalimantan, covering an area of 3,080 ha, is mainly heath forest (kerangas) with about 500 ha of peat swamp forest. The reserve was established during the Dutch colonial period and has remained relatively undisturbed to the present day. The State Forest Company (Inhutani) is extracting timber from adjacent forest, concentrating on meranti *Shorea*, rengas *Gluta renghas* and keruing *Dipterocarpus*. The provincial government operates a tourist park (Taman Wisata) adjacent to the reserve, and village lands abut the other sides. Local people exploit the adjacent forests for firewood, and land is being cleared for agricultural fields and small-scale gold-mining operations, but these are outside the reserve boundaries.
>
> Mandor is a fine example of heath forest. The forest canopy is low, with trees about 10-15 m high. Trees are densely packed with many thin poles and few trees more than 30 cm in girth. Species diversity is lower than in adjacent stands of lowland dipterocarp forest, although several of the same species are present.
>
> Where forest is cleared, white sands remain. Pools and streams are tea-coloured with runoff from the peat layer. Kerangas soils are nutrient poor, and many kerangas plants employ special strategies for increasing nutrient uptake. The creeper *Dischidia*, with enlarged leafsacs, is common at Mandor, as are ant plants. Ants shelter within these plants, which themselves benefit from the association by obtaining nutrients from the ants' faeces. Other plants employ different strategies to live on the sandy, nutrient-poor soils. The feathery-leaved *Casuarina* trees have nitrogen-fixing nodules on their roots.
>
> The Mandor forests are carpeted in pitcher plants, with at least five species recorded from the area. Clusters of *Nepenthes ampullaria* nestle in patches of *Sphagnum* moss. Tiny *N. gracilis* "pitchers" form around a rosette of leaves. The slighty bulbous flagons of *N. mirabilis* hang from tendrils that wind up shrubs. *N. rafflesiana* and *N. villosa* pitchers, vibrantly coloured with blotches of red, attract ants and other insects into their liquid-filled pitchers. Enzymes in the liquid digest the trapped victims and release nutrients to the plants.
>
> Although only a small reserve, Mandor supports some interesting plants including numerous species of gingers and orchids. White ground orchids and purple-leaved slipper orchids grow in clumps on the forest floor. The Mandor forests, especially the peat swamp, are home to gibbons, proboscis monkeys, long-tailed macaques and several species of birds.
>
> Mandor is less than the 5,000 ha usually regarded as the minimum size for a viable conservation area, but it is well protected and a fine example of Kalimantan's remaining heath forest. There are good trails through the reserve, and it is easily accessible to visitors from Pontianak.

pared with temperate forests, however, heath forests are still very rich in species. Brunig (1974) recorded 849 species of trees (from 428 genera), 133 shrubs, 96 herbs, 100 epiphytes and 55 lianas from the heath forests of Sarawak and Brunei. About a quarter of the trees (220 species) also occur in lowland dipterocarp forest; the dipterocarps *Shorea* and *Hopea* are espe-

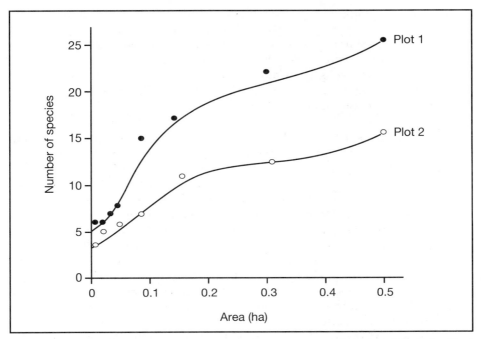

Figure 5.4. Species area curves in the primary kerangas forest at Samboja, East Kalimantan.
Source: Riswan 1987

cially common. The open padang formations are much less species-rich, with a different species composition; 48 of the 83 plant species recorded from padang in Sarawak and Brunei were not found in heath forest (Specht and Womersley 1979).

A noteworthy feature of heath forest is the abundance of genera of Australian affinity. Representatives of the family Myrtaceae are prominent in heath forest, especially *Tristaniopsis* and *Eugenia* (Whitmore 1984a). Shrubby plants such as *Vaccinium* species, *Rhodamnia linerea* and *Baeckia frutescens* and several species of the Rubiaceae and Melastomataceae are often common in wetter *kerapah* communities and on open, exposed hilltops. Also some species develop dwarf forms on sites with extreme nutrient and water deficiencies. Thus *Calophyllum nodosum*, normally a medium-sized tree, is only pole height in Bako National Park, Sarawak.

Lowland heath forest has many features in common with moss forest in upper montane zones (table 5.1) and with peat swamp forests, which also grow on poor, acid soils. At least 146 tree species are common to both heath forest and peat swamp forest (Brunig 1973), including the big timber trees *Shorea albida, S. pachyphylla, S. scabrida, Dryobalanops rappa* and *Cratoxylum amborescens*. Four types of peat swamp forest in Sarawak shared between 27% and 70% of their species with heath forest; the greatest similarities were with short peat swamp forest, which occurs on the most nutrient-poor soils (Whitmore 1984a). Heath forest also shares some common tree species, such as *Casuarina nobilis*, with upper montane forest and the summits of limestone hills.

Plants with supplementary means of obtaining minerals are common in heath forests. *Casuarina nobilis* has root nodules which contain nitrogen-fixing bacteria. The most conspicuous epiphytes in heath forest are the myrmecophytes (ant plants) *Hydnophytum* (fig. 5.5) and *Myrmecodia*, both in the Rubiaceae. They have thick, hollow tubers. Experiments using radioisotope tracers have shown that ants bring both organic and inorganic compounds into the chambered tubers and that these nutrients are absorbed by the host plants (Huxley 1978; Wallace 1989). Other epiphytes include the creeper *Dischidia*. Its convex leaves are pressed close to the host stem, forming hollows beneath, where ants live. In these symbiotic associations, the ants receive shelter, and the plant receives nutrients in the form of waste food, faeces and ant corpses. The endemic orchid *Bulbophyllum beccarii*, which grows on tree trunks in kerangas forests, adopts another strategy to maximise its access to nutrients. The large leaves are scoop-shaped and project obliquely from the trunk to catch plant litter; a special group of roots grows from the pseudobulb base into the trapped litter.

The ground flora in heath forest is sparse, with many mosses and liv-

Table 5.1. Forest formations on Mount Dulit, Sarawak.

Heath forest	Moss forest (upper montane)	Mixed forest
Lower levels usually well illuminated	Lower levels usually well illuminated	Lower levels deeply shaded
Undergrowth dense	Undergrowth dense	Undergrowth thin
Ground flora poor in species	Ground flora scanty	Ground flora moderately rich in species
Bryophytes rather abundant at lower levels	Bryophytes very abundant at lower levels	Bryophytes not abundant
Buttressed trees comparatively uncommon	No buttressed trees	Buttressed trees abundant
Conifers common	Conifers abundant	Conifers absent
Casuarina nobilis present	*Casuarina nobilis* present	*Casuarina nobilis* absent
Tristaniopsis spp. frequent	*Tristaniopsis* spp. frequent	*Tristaniopsis* spp. absent or rare

Source: Richards 1936

Figure 5.5. The ant plant *Hydnophytum*, with the enlarged stem cut open to show the chambers inside. Ants live in the chambers. The plant benefits by receiving nutrients from the ant faeces.

erworts as well as insectivorous plants such as pitcher plants *Nepenthes*, sundews *Drosera* and bladderworts *Utricularia* growing on the nutrient-poor soils. Like the ant plants, the insectivorous plants obtain nutrients from insects, though in this case by carnivory. The carnivorous habit is probably an evolutionary response to growing in habitats containing little or no available nitrogen. The sundews *Drosera* have leaves covered with long, red, motile, gland-tipped hairs which entrap insects and digest them by secreting proteolytic enzymes and ribonucleases (Heywood 1985). The bladderworts *Utricularia* bear bladders, each a hollow bag at the end of a stalk, with a small entrance near to, or opposite, the stalk. Projecting bristles around the entrance are so arranged that an insect on the plant stem will be guided towards the bladder mouth. The entrance is closed by a hermetically sealed valve which bears four hairs. If an insect touches these hairs, the valve is triggered, and a rush of water drags the animal inside where it is digested (Mabberley 1985). Six species of pitcher plants are also commonly found in heath forests (Smythies 1965), but the majority are montane species (see table 7.2).

Box 5.2. Myrmecotrophy: plants fed by ants.

The plant genera *Hydnophytum* and *Myrmecodia* (Rubiaceae) from Southeast Asia and Northeast Australia are well known for their swollen, tuber-like organs, which are inhabited by ants of the genera *Iridomyrmex* and *Crematogaster*. The colonies live in special chambers known as domatia and pack adjacent tunnels with their waste material. Two landmark experiments which offered radioactively labelled prey or honey water to foragers from tuber-inhabiting colonies showed that nutrient ions were absorbed from the colony waste, translocated and incorporated into the plant tissues.

At least 150 plant species in the Far East have specialised plant structures that appear to be adapted for the accommodation of ant colonies. There are two kinds of domatia: domatia in blades and petioles are more temporary and numerous, whereas those in twigs, stems and tubers are more permanent. Domatia-inhabiting ants may be aggressive towards herbivores or their eggs and thus protect the host plant from predation. In addition, ant colonies accumulate waste materials that are rich in plant nutrients, especially nitrogen and phosphorus. Ant-fed plants are often small epiphytes growing in nutrient-poor environments, but they can also be shrubs, small trees and even large trees, growing on a variety of soil types in the interior and fringes of forest.

There are also many plant structures that are routinely occupied by ant colonies, although they are not primarily domatia. These include the gigantic bowls of leaves made by the epiphytic ferns *Platycerium* and *Asplenium*. Many plant species have nests associated with their roots: these include many orchids and the rhizomatous masses of the fern genera *Lecanopteris* and *Solanopteris*.

The precise nature of the nutrients available to plants from ant colonies is unknown. Diverse compounds are derived from ant and prey corpses, excreta, and plant debris and animal faeces brought in by foragers. Ammonia and carbon dioxide generated by active colonies may be available to the plant. It has been shown experimentally that ambient ammonia is incorporated into amino acids and subsequently transported to various parts of plants. In *Dischidia* stomata are abundant on the inner surface of the flask-like leaves; respiration, particularly by the brood and the microorganisms associated with the nest material, may produce carbon dioxide useful to the plant tissues. The presence of fungi and bacteria within domatia may facilitate nutrient breakdown and transport.

Those ant species involved exhibit certain traits which may be of special significance in myrmecotrophy.

1. They nest in aerial parts of the plants, not in the ground.
2. The ants possess behavioural and morphological adaptations for excavating domatia with small entrance holes, which exclude other invertebrates that do not benefit the host plant.
3. Ant behaviour results in colony wastes being stored close to absorbent plant surfaces.
4. The ants have a diet high in animal tissue, or at least containing minimal amounts of plant material from which the release of nutrients is slow.
5. Ants forage away from the domatia-bearing plants so that nutrient gain through ant activity is derived independently of the root system. Ant species that provide protection to the host ("ant guards") may be especially useful by taking herbivorous prey and recycling nutrients to the host plant.
6. Ants prune encroaching vegetation and thus promote the vigour of the host plant.

(Continued on next page.)

> **Box 5.2.** Myrmecotrophy: plants fed by ants *(continued)*.
>
> 7. Ants colonize a plant in such a way as to promote full and harmonious occupation of domatia: either through polygyny, when a single colony with multiple queens occupies an entire plant; or through polydomy, in which a single colony occupies many nests.
> 8. Ants tend homopterans: many domatia are inhabited by homopterans, and the ants feed on the honeydew they secrete.
>
> The selective advantage for myrmecotrophs is a significant gain in nutrients that are in short supply. Ant services may be of most value when they include both myrmecotrophy and defence. Nutrient gains in carnivorous plants, although crucial, are marginal because of the costs of insect-trapping organs and secretions. Myrmecotrophy is widely assumed to be a mutualism, but little is known about potential selective advantages to the ants.
>
> *Source: Beattie 1989*

Two possible explanations have been offered for the curious structure and xeromorphic character of heath forest. They may be caused by nutrient deficiency on very acid soils (Richards 1952), or they could be adaptations to minimise water loss and reduce heat load during periods of drought (Brunig 1970, 1971). At first sight the first explanation seems the more likely. Several features of heath forest suggest inorganic nutrient deficiency, including its rapid degradation if cleared or burned; the presence of plants with supplementary means of obtaining mineral nutrition (ant plants and insectivorous plants); and the number of sclerophylls, which can be related to shortage of nitrogen and phosphorus (Beadle 1966).

Studies in Gunung Mulu National Park, however, showed that the soil nutrient pool in heath forest was similar to that in other adjacent forest formations, as was rate of litterfall (Proctor et al. 1983c; table 5.3). Research in East Kalimantan also showed that nutrients under virgin heath forest are comparable in amount to those under dipterocarp rainforest, although there is rapid degradation after clearing (Riswan 1982). Although the heath soils were not poorer in inorganic nutrients, the Mulu studies revealed several other constraints to plant growth and species diversity. Heath forest soil is extremely acid with a pH less than 4 at the surface, which is toxic to many plants. Moreover, the annual rate of return of all measured inorganic nutrients in fine litterfall was considerably less than in valley evergreen forest and, in the case of nitrogen, was only half (Proctor et al. 1983c). Phenols are abundant in heath forest leaves and litter; these may be toxic or may inhibit uptake when they leach into the soil (Whitmore 1990).

There is now increasing evidence that the very striking characteristics of heath forest are probably adaptations to survive drought. Although heath forests occur in humid tropical climates with no dry seasons or very short,

Box 5.3. Plants as predators: pitcher plants.

The insectivorous pitcher plants *Nepenthes* trap, drown and digest insects; the sundews merely trap and digest them. These and many myrmecophytes are found not only in heath forest but also in other forest formations poor in available nutrients, such as the centre of some peat swamps and ridge-crest forests on mountains.

Borneo is the centre of distribution for the genus *Nepenthes*, with 28 of the known species occurring on the island. Pitchers may be long and elegant, fat and bulbous or of various intermediate shapes, but in every case they resemble a drinking cup. This "pitcher" is not a flower but a modified leaf that holds water. A single plant may bear two or three kinds of pitchers of different shapes; the upper pitchers are longer and more funnel-shaped (fig. 5.6.). The lip of the pitcher is ribbed and overhanging, with nectar glands situated on the inner lip. The lid, which also has nectar glands, projects above the pitcher but does not prevent rainwater from entering. Insects lured into the pitcher by the nectar slip on the smooth waxy lip and fall into the water below. The inner wall of the pitcher is divided into two zones; the upper half is smooth and waxy. This slippery surface and the overhanging lip make it almost impossible for insects to climb out again. The lower zone has numerous glands secreting acidic enzymes which digest the trapped insects. Thus nitrogen and other nutrients are released for absorption by the plant. One remarkable pitcher on Mount Kinabalu is said to have contained two litres of liquid and the corpse of a rat (St. John 1862/1974).

The pitchers of *Nepenthes* are a world in miniature, supporting a community of insect larvae and other invertebrates resistant to the digestive enzymes. Most of these animals feed directly on drowned insects, or indirectly by preying on bacteria and other micro-organisms that have fed on the corpses. In Peninsular Malaysia 55 species of insects (mostly Diptera and mosquitoes) and three species of spiders have been recorded from *Nepenthes* pitchers, of which two-thirds normally live and breed there. The complex food web is well illustrated in *Nepenthes ampullaria*, where 27 species were recorded: three carrion feeders, six detritus feeders, 14 filter-feeding mosquitoes, two aquatic and one terrestrial predators and a parasitoid. Two predators, a spider and the larva of a mycetophilid fly, catch insects entering or leaving the pitcher (fig. 5.7.); the other predators have aquatic larvae. Since many of these species spend only the early part of their life cycles inside pitchers, they represent an export of nutrients which would have otherwise been available to the plant (Beaver 1979a, b), but they assist the digestive process by breaking down the accumulating detritus of dead animals (Kitching and Schofield 1986).

Some differences between the faunas of different *Nepenthes* can be related to the structure of the pitchers; others may be related to habitats. Most species (82%) have aquatic larvae, 79% of which are not known to breed in other habitats. A pitcher is colonised as it opens, and there is a rapid rise in numbers of individuals and species for the first few weeks corresponding to the period of maximum prey input. Food webs in the pitchers of *N. ampullaria* and *N. albomarginata* were similar (fig. 5.8.), but the species composition of the pitcher communities was different (table 5.2.).

Source: Beaver 1979a, 1979b, 1983

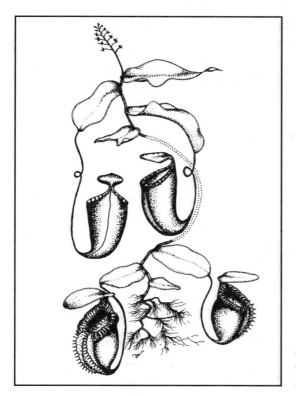

Figure 5.6. A pitcher plant showing the different shaped flasks on upper and lower leaves.

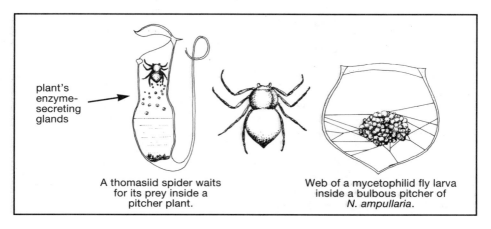

Figure 5.7. Predators in pitcher plants.

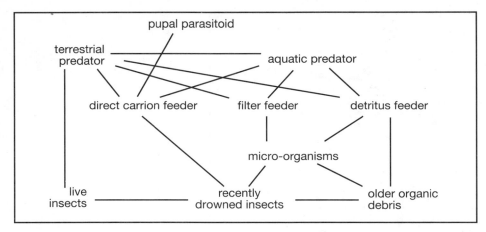

Figure 5.8. Food webs of *Nepenthes albomarginata* and *N. ampullaria* in Pulau Pinang, West Malaysia. (After Beaver 1983.)

Table 5.2. Functional and taxonomic components of the food webs in *Nepenthes ampullaria* and *N. albomarginata* in Pulau Pinang, West Malaysia.

Functional component	*Nepenthes ampullaria*	*Nepenthes albomarginata*
Direct carrion feeder	*Nepenthomanyia malayana* (B)	*Pierretia urceola* (B)
	Megaselia? bivesicata (X?)	*Endonepenthia schuitemakeri* (B)
	Xylota spp. (X)	*Megaselia? nepenthina* (B?)
		Nepenthosyrphus? capitata (B)
Filter feeder	*Tripteroides tenax* (B)	*T. tenax* (B)
	T. bambusa (P)	*T. bambusa* (P)
	T. nepenthis (B)	
	Culex acutipalus (B)	*C. curtipalpis*
	C. hewitti (B)	
	C. navalis (B)	
	Uranotaenia moultoni (B)	
Detritus feeder	*Dasyhelea ampullariae* (B)	
	D. nepenthicola (B)	*D. nepenthicola* (B)
	Anoetidae (2 spp.)	Anoetidae (2 spp.)
Aquatic predator	*Corethrella colathicola* (B)	*Taxorhynchites klossi* (B)
		Lestodiplosis syringopais (B)
Terrestrial predator	*Xenoplantyura beaveri* (B)	*Misumenops nepenthicola* (B)
Terrestrial parasitoid	*Trichopria* spp. (X)	*Tachinaephangus* (B?)

B - nepenthebiont; obligate residents of pitchers.
P - nepenthephil; species that live and breed in pitchers but are also found in other similar habitats.
X - nepenthexene; species that enter pitchers but normally feed outside.

Source: Beaver 1983

irregular ones, heath forest may suffer periodic water shortages (Brunig 1971). The sandy heath soils drain quickly, have a low capacity to retain water and are often shallow over an impervious iron or humus pan (Burnham 1984). Many heath forest characteristics (low uniform canopy, dense tree crowns, pale or shiny leaves, small leaves) may be adaptations to restrict water loss through transpirational cooling during drought periods. Transpiration of water helps leaves to stay cool and below the lethal temperature at which protoplasm denatures. Many of the physical features of heath forests minimise heat load or optimise cooling (Whitmore 1990). Experiments on water loss suggest that individual heath forest species are no better adapted than those of lowland evergreen rainforest to resist desiccation (Peace and MacDonald 1981). Instead, the structure of heath forest reduces water loss from the forest as a whole.

Table 5.3. Fine litter: fall, average amount and disappearance in various rain forests.

Forest formation and place	(a) Fine litterfall (t ha^{-1} year^{-1})	(b) Forest floor (t ha^{-1})	(c) Decay factor, k (a)/(b)
Lowland evergreen rainforest:			
Mulu, Sarawak: ridge	7.7	5.9	1.3
valley alluvium	9.4	5.5	1.7
Pasoh, W. Malaysia	10.6	3.2	3.3
Penang, W. Malaysia	7.5	4.9	1.5
Heath forest:			
Mulu, Sarawak	8.1	6.1	1.2
Forest over limestone:			
Mulu, Sarawak	10.4	7.1	1.5
Freshwater swamp forest:			
Tasek Bera, W. Malaysia	9.2	4.8	1.9
Montane rainforests:			
New Guinea, 4 close sites (2,500 m)	6.2 - 6.6	4.2 - 6.6	1.0 - 1.5

Source: Whitmore 1990

Heath Forest Regeneration

Heath forest is easily degraded, by felling and burning, to an open savanna of shrubs and scattered trees over a sparse grass and sedge ground layer, a formation often called *padang*. Padang consists of a low, shrubby vegetation in which the tallest trees usually reach only some 5 m. Padang vegetation grows very slowly (Riswan 1982) – only 1-2 m of plant growth was observed over 30 years in padang at Bako National Park, Sarawak – and it seems unable to revert to heath forest (Janzen 1974). However, charcoal has been reported 50 cm below the soil surface in heath forest in East Kalimantan (Kartawinata 1978), so it is possible that heath forest can regenerate from padang vegetation created by fire, but may take a very long time to do so.

Regeneration after disturbance in a heath forest in East Kalimantan was characterised by poor development of seedlings attributable to acid conditions and the low nutrient content of the soil (Riswan and Kartawinata 1988). Resprouting after coppicing was the main feature of vegetation recovery; such regeneration should give a floristic composition almost the same as that of the original habitat if destruction is not recurrent or severe (Riswan 1982). Seedling survival in degraded heath forest was very poor, except for species adapted to kerangas conditions, such as the epiphyte *Hoya multiflora*, the terrestrial fern *Schizaea dichotoma* and the climbing pitcher plants *Nepenthes*. Clear-cutting with burning is more destructive than clear-cutting alone; the addition of nutrients in the form of ash did not seem to accelerate the recolonisation by seedlings or the growth rate of resprouts (Riswan 1982). Heath forest sites are increasingly difficult to reforest, the greater the degree of disturbance of the original forests and soils. In Badas, Brunei, where *Agathis borneensis* seedlings were planted in degraded logged heath forest, seedlings only survived where there was a substantial amount of organic matter in the soil to supply the superficial feeding roots of the trees (Whitmore 1984a).

After logging, secondary growth takes the form of slow-growing xeromorphic shrubs and trees, and the possibility of re-establishing productive forest on such degraded sites is slight (Mitchell 1963). Nevertheless, if carefully managed, some richer heath forests on better soils can provide a continuing source of timber. Heath forests at Menchali in Malaysia have been logged sustainably since early this century; the removal of only a few trees per hectare has proved non-degrading (Beveridge 1953; Wyatt-Smith 1963). However, the logging of the valuable conifer *Agathis borneensis*, in heath forests in Central Kalimantan (Manaputty 1955) has not been sustainable, and little work has been done on re-establishment after felling.

Heath forests are a fragile ecosystem, easily and irreversibly degraded by human disturbance. Heath forest sites cannot sustain agriculture. At best they grow only one poor agricultural crop and quickly become degraded to open sandy savannas, as can be seen around Palangkaraya, Central Kalimantan. Once forest is felled and burned, the soil degenerates very quickly. The surface humus layer is either eroded, burned or oxidised. Heath

> **Box 5.4.** *Agathis*, a tropical conifer.
>
> Conifers are more usually associated with montane forest formations, but three conifers are locally abundant in heath forest in Kalimantan - *Agathis*, *Dacrydium*, and *Podocarpus*. *Agathis* has the most tropical distribution of all conifers. Like most other conifers in the Far East, it belongs to a southern hemisphere family, the Araucariaceae. Two species occur in Borneo: *Agathis borneensis* in the lowlands, and *Agathis dammara* on a few high mountains such as Kinabalu. In lowland forests in Kalimantan *A. borneensis* occurs as a solitary tree, or in large stands where it is the dominant species. The largest *Agathis* stands, now logged out, were found in heath forest around Sampit, Central Kalimantan. The total area of the Sampit *Agathis* forests was 30,000 ha, with almost pure pockets of this species as large as 5,000 ha. Elsewhere in Kalimantan *Agathis* is found on ultrabasic rocks, on limestone and occasionally in peat swamp forests. It is less successful in species-rich lowland forests but seems to thrive in more florally impoverished habitats where competition is less.
>
> *Agathis* produces a valuable timber, and the resin from the bark is also marketed as copal. It has good potential as a plantation species and as a tree for enrichment planting in logged-over or low-grade forest. This conifer is fast growing. In plantations in West and Central Java the species showed a high growth increment of 25 m^3/ha/year (Whitmore 1977). To maintain rapid growth, seedlings must be cleared to keep the canopy open until seedlings are well established.
>
> *Source: Whitmore 1984a*

forest podzols are of low clay content (Tie et al. 1979), and much of the cation-exchange capacity lies in the organic matter. The small amount of clay is rapidly washed downwards through the coarsely textured soil, leaving bleached sands of almost pure silica over a humic pan. The soil may become even more acid. Without vegetation cover or a surface humus layer, the white sand becomes extremely hot in the sun, creating conditions adverse to plant growth. The fragility and rapid degradation of the soil explains the infertility of kerangas sites and accounts for the failure of some poorly located transmigration sites in Central and East Kalimantan.

Fauna

Animals in heath forest and padang vegetation face similar problems to those in peat swamp forests. Poor soils cause low productivity, and plants tend to defend their leaves and young shoots with toxic or unpalatable compounds to make them less vulnerable to predators. High levels of phenols or tannins in the leaves not only deter animal consumers but may give one species a competitive advantage over other species which at some stage of their development may be more sensitive to high concentrations of tannins in the litter leachates (Brunig 1974). Leaves collected in litterfall at

Mulu contained high levels of polyphenols. This would seem to be a protection against predators, yet estimates of invertebrate leaf browsing were similar to those in other forest formations (Proctor et al. 1983; Anderson and Swift 1983; see fig. 4.18).

In Africa there was approximately twice the concentration of phenols in leaves of trees growing on white sands compared with trees on other lowland soils. This high phenol content appeared to be an effective deterrent to larger herbivores such as monkeys. Black colobus monkeys avoided almost all the leaves, except the most nutritious ones, where the benefits of eating a good food source outweighed the cost of ingesting defence compounds (McKey 1978; McKey et al. 1978). Primates such as orangutan and macaques have been recorded in heath forest habitats in Borneo (Davies and Payne 1982; Galdikas 1978) but in low numbers. Since heath forest often interdigitates with other forest formations, they may cross this habitat while feeding mainly in surrounding forests.

Because the forest is less species-rich and less food is available, animal communities are considerably reduced in heath forest and padang vegetation. In Malaysia mammal density was estimated at only one per 2 ha, and bird density at one per 0.4 ha (Harrison 1965). In Gunung Mulu heath forest supports only about half the density of the same avifauna as surrounding lowland evergreen forest, with the addition of one species, the ashy tailorbird *Orthotomus ruficeps* (Cranbrook 1982). This is a common bird found in a wide range of open, coastal and secondary habitats up to 1,500 m (MacKinnon 1988).

A comparative survey of amphibians and reptiles in Sarawak revealed interesting differences between forest formations (table 5.4). The heath

Table 5.4. Amphibians and reptiles in sample areas of heath and evergreen rainforest in Sarawak.

		Heath forest	Lowland evergreen rainforest	
		Nyabau	Pesu hill	Labang valley
Species numbers:	frogs	24	53	33
	lizards	13	31	33
	snakes	13	35	38
	turtles	0	1	2
	Total	**50**	**120**	**106**
Percentage of species shared with the other two habitats:		92	33	37
Percentage of species not shared with the heath forest:	frogs		60	48
	lizards		64	65
	snakes		77	74

Source: Whitmore 1984a

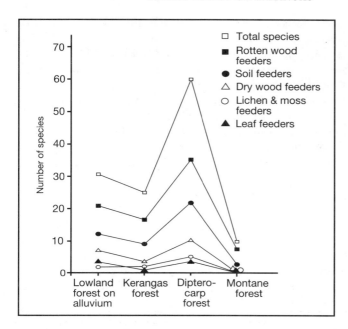

Figure 5.9. Species richness and feeding guilds of termites in different forest habitats.
Source: Collins 1984

forest had no turtles and less than half of the number of species of frogs, lizards and snakes found in other Bornean forests (Lloyd et al. 1968). Heath forests supportted only a third of the number of snake species found in dipterocarp forest. Moreover, there were fewer endemic species in heath forest. Most snakes feed mainly on small vertebrates, and their scarcity in the heath forest probably indicates a shortage of food, which becomes more marked along the food chain. Food scarcity least affects the frogs, which are secondary consumers feeding mainly on insects.

Among invertebrates there is a similar picture of reduced species richness in heath forest. This formation has the smallest number of dung beetles (Scarabaeidae) of any forest type at Mulu (Hanski 1983); most of those present are saprophages such as the Oxytelinae or predators such as the Staphylinidae (Hanski 1989). Cicadas, which are so characteristic of other lowland forests, are absent. The kerangas butterfly community is less diverse than in dipterocarp forest (Holloway 1984). Some groups of moths show maximum diversity in heath forest, but, in general, this habitat has a reduced assemblage of insects similar to that in upper montane rainforest in the same area (Holloway 1986).

The abundance and biomass of major detritivores, predatory soil invertebrates and litter invertebrates in three forest types at Mulu are shown in table 5.5. The organic soils of kerangas are dominated by termites, with beetles next in importance (Collins 1984). Detritivore biomass is greatest in kerangas forest, partly as the result of larger size of individual detritivores. Beetles, especially, are considerably larger in kerangas than in dipterocarp forest. Although the total numbers of soil termites are greatest in kerangas, they represent only a few species compared with very high species richness in dipterocarp forests (Collins 1979b; fig. 5.9).

For most animal groups studied, heath forests are less rich in species than neighbouring mixed lowland forest, a pattern one might expect in view of the less diverse flora and the fewer niches for exploitation. This

Table 5.5. The abundance and biomass of major detritivorous and predatory soil and litter microfauna in three lowland forest types, dipterocarp (DF), alluvial (AF) and kerangas forest (KF). Freshwater crabs have not been included.

	Abundance m^{-2}			Biomass mg m^{-2} (wet weight)		
	DF	AF	KF	DF	AF	KF
Detritivores/root feeders						
Megascolecidae	25	29	23	627	739	387
Moniligastridae	1	13	1	18	264	2
Isopoda	17	26	20	31	31	34
Diplopoda	3	17	21	11	110	73
Symphyla	27	9	8	10	10	5
Diplura	44	20	17	48	43	33
Blattodea	8	9	10	121	162	598
Isoptera	1,148	254	1,408	1,818	522	3,579
Coleoptera*	63	45	35	494	361	1,433
Diptera juvs.	10	3	10	15	31	38
Others	19	9	17	216	25	105
Total detritivores	1,365	434	1,580	3,409	2,298	6,284
% of overall total	70	64	63	76	86	89
Predators						
Gnathobdellida	(0.3)	(0.5)	0	1	12	0
Arachnida	57	22	25	171	128	109
Chilopoda	43	22	40	370	79	313
Dermaptera	2	1	1	13	8	5
Formicidae	471	202	861	527	134	353
Total predators	573	247	927	1,082	361	780
% of overall total	30	36	37	24	14	11
Overall totals	1,938	681	2,507	4,491	2,659	7,064

* The bulk of coleopteran records were for detritivorous or root-feeding species, although a few were predators.

Source: Collins 1984

> **Box 5.5.** White sands and blackwater rivers.
>
> The streams and rivers that drain kerangas and kerapah forests are tea-coloured or opaque black due to the presence of organic colloids. Such blackwater rivers are found throughout the tropics, draining podsolised white quartz sand soils or peat swamps. Tropical blackwaters are usually very acidic (pH between 3 and 4.5), contain many fewer inorganic ions than do clear, white or muddy waters in the same drainage basin, have low oxygen content, have low light penetration, and contain high concentrations of dark brown "humic acids". The low nutrient content of blackwater rivers is partially explained by the fact that they drain soils that are not formed by the weathering of base-rich rock. The high acidity of the water is due to the presence of humic acids. The polyphenolic humic acids are also well-known chelating agents for inorganic ions and may be preventing nutrient uptake by plants. The low oxygen concentration is probably due to the lack of aquatic plants.
>
> The productivity of blackwater rivers is very low. The fish and invertebrate biomass is greatly reduced in Amazonian blackwater rivers, compared with other lakes and rivers in the Amazon (Roberts 1973; Marlier 1973). In Malaysia blackwater rivers have a distinct and impoverished fauna, with only about 10% of the fish fauna found in other rivers (Johnson 1967). Algae are generally rare except for a few species that may be locally abundant, and macrophytes are often absent. Invertebrates such as water fleas, annelid worms, rotifers, nematodes and protozoans are rare (Johnson 1968). Of 15 fish species found in blackwater rivers, nine were air-breathers or lived near the surface (Johnson 1967), and almost all insects were air-breathers (Johnson 1968). This is perhaps not surprising in view of the low oxygen levels in blackwater rivers. The fish biomass is low, at 0.5 g/m^2, compared to an estimated fish biomass of 18 g/m^2 in the clear-water river Gombak (Bishop 1973). This low biomass can be attributed to the paucity of plants and invertebrates and therefore food availability for fishes.
>
> This impoverishment seems to be related to the high level of phenolic compounds in the ecosystem, derived from secondary plant products in the vegetation (Janzen 1974a). The low productivity of blackwater rivers is usually attributed to their low nutrient level or to lack of trace elements and low pH, but may also be due to the toxic effect of the humic acids (Janzen 1974a). Rainwater runoff into blackwater rivers is exceptionally rich in "humic acids" and probably other toxic compounds because the leachate from fresh vegetation and decomposing litter on kerangas soils is exceptionally rich in phenols and other plant defensive chemicals, which are not broken down by the resident plant and animal community. Humic acids are difficult to degrade and persistent to a greater degree than other chemical plant debris.

trend extends even to the freshwater fauna, which is similarly reduced in species diversity. The streams draining heath forests can be described as blackwater. Like peat swamp blackwater rivers (chapter 3), they are less species-rich than the streams that drain other lowland forests, and many fish found in them are facultative air breathers (Johnson 1968).

Ironwood Forests

Borneo is famous for its ironwood forests. Ironwood, *ulin* or *belian*, *Eusideroxylon zwageri*, is prized for its strength and durability and used for blowpipes and sirap tiles for house roofs. Ironwood belongs to the family Lauraceae and is found throughout southern Sumatra, Borneo and the southern Philippines. The warm, red wood can resist rotting for 40 years, or even for a century in dry conditions. It is used for piers, bridges, road foundations and flooring. The ironwood tree has large leaves and heavy fruits, like large nuts, which litter the forest floor.

Ironwood forest is a characteristic forest type within lowland dipterocarp forest, but this formation can vary in composition from site to site and from island to island. A nondipterocarp, the ironwood tree is widespread and common in the lowland dipterocarp forests of East Kalimantan. It is usually a tree of the main canopy and grows on sandy or alluvial soils, along river terraces or on undulating to hilly terrains. Scattered trees also occur on clay hillsides up to 360 m (Whitmore 1984a). Ironwood trees in Borneo do not form a distinctive, monodominant forest, unlike the ironwood forests in Sumatra, where researchers found 81 ironwood trees out of the 84 trees of dbh greater than 15 cm in a half-hectare patch (Whitten et al. 1987a). In Borneo ironwood trees usually occur as scattered individuals (fig. 5.10) and small stands but may become dominant to form a distinct forest type in association with dipterocarp species (Riswan 1982). Densities for ironwood were estimated at 33 trees of over 20 cm dbh per hectare at one site in Kalimantan (Laan 1925). Ironwood trees are a characteristic tree of the distinctive *empran* forests on river levees, which may carry commercial timber yielding up to 90 m^3/ha, a very heavy stocking level.

The original area of ironwood forests for the whole island of Borneo has been estimated at 1,440 km^2, with only about 40% remaining (MacKinnon and MacKinnon 1986). In Kalimantan only about 30% of the original area of this habitat type remains (MacKinnon and Artha 1981), and even most of this is very disturbed as ironwood trees have been logged selectively and removed.

Some of the best remaining areas of ironwood forest in East Kalimantan are found in Kutai National Park. Mixed ulin-meranti-kapur forests occur on poor to moderately well-drained soils in the western half of the park (Wirawan 1985). The forest has an open canopy layer at a height of 30-35 m and a closed subcanopy layer at 20-25 m. This thick subcanopy restricts the amount of light reaching the ground, and the shrub and ground layers are generally sparse and poor in species. Ironwood occurs throughout the area, especially on the lower slopes and along rivers and streams, where it forms pure stands. On the ridges kapur *Dryobalanops* and meranti *Shorea* often form a single dominant species stand. Other tree species scattered throughout this forest type include *Intsia palembanica*, *Eugenia* spp. and

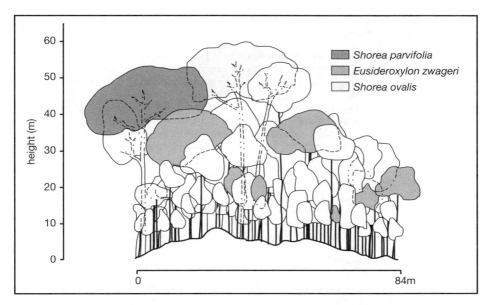

Figure 5.10. A profile diagram of a lowland dipterocarp forest with scattered ironwood trees at Lempake, East Kalimantan, Indonesia. (After Kartawinata et al. 1984.)

Palaquium spp.

In lowland evergreen rainforest, many trees, including dipterocarps, fruit gregariously to swamp their predators with a superabundance of fruit so that, even though many seeds are lost to predators, sufficient remain to germinate and to provide new stock for regeneration. A second strategy may be common in tree species occurring at relatively high densities, namely the production of large and toxic seeds almost continually throughout the year (Janzen 1974). Ironwood seems to adopt this latter strategy, but the model is untested, as no chemical analyses have yet been made of ironwood fruits. Ironwood has an irregular flowering season which peaks around the middle or end of the dry season, and fruits are found approximately three months after flowering (Koopman and Verhoef 1938). The seed, the largest of all dicotyledon seeds, is about 14 cm long, weighs about 230 g and is ovoid like a rugby ball. It often produces a sprout over 1 m tall before the leaves develop. To make such a start possible, the food in the seed must be in a form suitable for quick utilisation. This requires a high moisture content, rendering the seed vulnerable to fungal attack. Most large seeds need high humidity to remain viable, but they retain this viability only briefly, perhaps for a maximum of three to four weeks.

The larger its food store, the better equipped is the seed for germina-

> **Box 5.6.** Monospecific dominance in tropical rainforests.
>
> Tropical rainforests are characterised by high diversity; indeed, it has even been suggested that high species diversity may be necessary for ecological stability. The occurrence of monospecific dominance, as seen in ironwood stands, seems to contradict this thesis. In fact, examples of old-growth forest, where a single canopy tree species contributes more than 80% to the total basal area, occur throughout the tropics. The extensive areas of Malaysian and Sumatran forests dominated by the dipterocarp *Dryobalanops aromatica* and mangrove forests are other Asian examples.
>
> It was once believed that forest associations dominated by a single big tree species occurred on sites with nonoptimal soil conditions (Richards 1952). This model would explain the dominance of one or a few species in the mangrove association, where dominant species have to cope with salinity, tides and rapid colonisation. However, it does not hold true for ironwood and other evergreen rainforest associations, which occur on a variety of well-drained soils. It is possible that some monodominant climax formations may be the result of recolonisation after a dramatic disturbance such as a storm or widespread drought and fire. Major environmental changes affect the species pool by causing local extinctions of some rainforest trees, and may confer a competitive advantage on some survivors.
>
> On the other hand, if a forest remains undisturbed, the slow-growing, shade-tolerant species will come to dominate an increasingly large proportion of the canopy. Able to persist in shade, these individuals are potentially self-replacing. The best competitor, or the species most resistant to prevailing stress, will steadily increase in abundance. Accordingly, efficiency of dispersal is sacrificed in favour of large, heavy and poorly dispersed seeds that produce robust, shade-tolerant seedlings. It is interesting to note that ironwood and many other tree species that form monodominant stands in mature tropical rainforest do have heavier seeds and lack animal dispersal agents. If such a species is a self-replacing competitive dominant, then the composition of the old-growth forest where it occurs is determined by its unique presence in the species pool, and its dominance must reflect a long period without forest disturbance.
>
> *Source: Hart 1990*

tion in the shade but the heavier it is for animals to transport and disperse; as a result, ironwood trees often germinate in the shade of a parent tree. The large number of seeds and seedlings found in ironwood forests is evidence of very low seed predation, although porcupines are known to cause some damage in ironwood seedbeds (Whitten et al. 1987a). Prevost's squirrel *Callosciurus prevostii* may feed to some extent on ironwood fruits, and they are sometimes also damaged slightly by small beetles.

Ironwood trees are slow growing, with a k-strategy lifestyle; they produce relatively few offspring but make a large investment in each seed to ensure its successful germination and survival (Jacobs 1988). Young ironwood trees of 25 to 30 years may have trunks no thicker than 10 cm, so slow is their growth. Ironwood is never an emergent species, but it grows slowly, gradually expanding its crown and slowly renewing its leaves. The mature leaves are thick and coarse, an adaptation against predation. The tree may be 50 to 60 years old before it flowers and fruits. This has important implications for harvesting and conservation: harvesting mature ironwood trees before they have made a reproductive contribution will lead to eventual extinction of the species.

Because of its hardness, ironwood is an extremely valuable commercial timber, yet careless cutting regimes have greatly reduced this valuable resource (Peluso 1992). In parts of Kalimantan this species is already regarded as endangered. Major building projects that use durable ulin, such as the fine regatta stands for canoe races at Bangkirai, near Palangkaraya, are likely to become increasingly rare. For continued supplies of ironwood, urgent attention needs to be paid to the possibilities of establishing and maintaining plantations of this valuable species. When ironwood is felled, numerous coppice shoots grow from the remaining stump. Used carefully these could provide plantation stock, as cuttings grow more quickly than seedlings (Beekman 1949).

Forest over Ultrabasic Rocks

Ultrabasic (often called ultramafic) rocks, such as serpentines and the associated rock peridotite, are dense, igneous in origin, contain less than 45% silica and are rich in iron, magnesium and heavy metals. The soils that develop on ultrabasic rocks are often infertile due to a deficiency in phosphorus, potassium and calcium, and to potentially toxic concentrations of magnesium, chromium, cobalt and nickel (Proctor and Woodell 1975). Unlike in Sulawesi, which has the most extensive ultrabasic formations in the world, ultrabasics are rare in Borneo and account for only 1% of the land area. Several outcrops occur in Sabah, covering 3,500 km^2 (less than 5% of the state). In South Kalimantan ultrabasic soils occur in parts of the Meratus Mountains and on Pulau Laut.

In the Far East the vegetation of ultrabasic soils varies from stunted grasslands and shrublands to species-rich large-stature rainforests (Proctor 1992). The best-studied areas of this type of vegetation in the Far East are in New Caledonia, where forests over ultrabasics are distinctive in structure and sharply demarcated from adjacent lowland forests. However, this is mostly not the case in Borneo. Forests on the ultrabasic rocks of Ranau, Sabah, are high forest formations similar in structure to those in adjoining lowland dipterocarp forests (Whitmore 1984a). The Ranau forests support several dipterocarp species, including a tallest tree of 54 m (Fox and Tan 1971). In Sabah the dipterocarps *Dipterocarpus geniculatus, D. lowii, Shorea andulensis, S. laxa* and *S. venulosa* were recorded on ultrabasic soils but not on adjacent rocks. However, these species are not confined to ultrabasic soils; they also occur on other soils in Brunei and Sarawak (Ashton 1964; Meijer 1965).

Elsewhere in the lowlands of Sabah, especially over shallow soils, forests over ultrabasics tend to be low and similar in structure and physiognomy to heath forests (Whitmore 1984a). Similarly, the forests on ultrabasic formations at 2,400 m on the summit trail of Mount Kinabalu show a dramatic decrease in canopy height. Here the tree *Dacrydium gibbsiae* (which is codominant with *Leptospermum recurvum*) and the fern *Schizaea fistulosa* are restricted to this ultrabasic rock (Meijer 1971).

Proctor and Woodell (1975) reviewed the distinctive features of ultrabasic soils likely to affect plant growth. A major factor is their poverty in all major nutrients other than magnesium. This may determine the distinctive vegetation of the ultrabasic rocks in the Solomon Islands (Lee 1969) and the stunted grasslands and shrublands at 600 m altitude in the Talaud Islands, off Sulawesi (Lam 1927). The local dominance of *Casuarina* stands on ultrabasic formations on the islands of Darvel Bay, Malawali and Banggi Island, Sabah (Fox 1972), can probably be attributed to the tree's ability to fix nitrogen. On the other hand, the lowland forests of the ultrabasic mountain Gunung Silam in Sabah are species-rich and of high stature in spite of a high Mg/Ca ratio and substantial concentrations of nickel (Proctor et al. 1988).

Gunung Silam: an Ultrabasic Mountain

In Borneo detailed information about vegetation on ultrabasic soils is available only for Gunung Silam in Sabah. The environment, structure and flora of this small, ultrabasic mountain have been well studied (Proctor et al. 1988, 1989). The vegetation ranges from large-stature lowland evergreen forest to stunted, myrtaceous lower montane forest (Proctor et al. 1989; Proctor 1992). All forests were rich in tree species, ranging from 19 species in a 0.04 ha plot at 870 m, to 104 species in a same-size plot at 480 m. The Silam forests show a dramatic change in species composition between 610 m, where dipterocarps are common, and the stunted forest at 770 m, where dipterocarps do not occur (Proctor et al. 1989). This may be attributed to the "Massenerhebung effect", the compression of forest zones on a small mountain (chapter 7). Thus forest stature seems to be determined more by altitudinal factors, perhaps including hydrological factors, rather than soil toxicity.

Elsewhere in the Far East ultrabasic rocks support many rare or endemic plant species (Proctor and Woodell 1975). In New Caledonia, for instance, endemism is high on all substrates but highest on ultrabasic soils, with 79% of the 944 species that occur on ultrabasic soils restricted to them (Morat et al. 1984). In contrast, very few species appear to be endemic to ultrabasic soils in Borneo. Only two tree species found on Gunung Silam, *Borneodendron anaegmaticum* and *Buchaninia arborescens*, seem endemic to ultrabasic strata. Several pitcher plants and at least one species of orchid also appear to be endemic to ultrabasic soils in Borneo. Several palm species and at least one bamboo are confined to ultrabasic soils, and one species of *Salacca* palm is unique to Gunung Silam (Proctor et al. 1988).

Foliar analyses have shown that a high proportion of plant species on ultrabasic soils in New Caledonia have an unusual chemical composition, with more than 21% exhibiting high levels of manganese (more than 1 mg/g) and 5% showing high levels of nickel. However, on Gunung Silam only five species of plant showed high levels of manganese, and only one had high levels of nickel. Studies on Gunung Silam also show that the rates of forest processes, such as growth, litterfall and herbivory are not unusually low. This implies that the presumed widespread tolerance to the ultrabasic soils has been achieved without excessive metabolic cost (Proctor et al. 1989). Resistance mechanisms may involve: a capacity to maintain leaf functions at low foliar, major-nutrient concentrations; an ability to regulate foliar Mg/Ca quotients, either by exclusion of magnesium or preferential uptake of calcium, or both; and an ability to restrict foliar nickel concentrations, except in old leaves which may act as excretory organs for this element (Proctor et al. 1989).

Pleihari-Martapura Wildlife Reserve

In the Meratus Mountains forest over ultrabasic rocks is similar in structure and species composition to other areas of adjacent lowland and hill dipterocarp forest, although species richness is somewhat lower than for some other Kalimantan forests (fig. 5.11). A team from KPSL-UNLAM made a one-year study of an area of lowland and hill forest in the Pleihari-Martapura Reserve, in the southern Meratus Mountains. Permanent tree plots were established in hill forest over ultrabasic rocks at altitudes of 425 m, 525 m and 625 m. Changes in dominant species were observed between different altitudes (fig. 5.12). There was some fruiting or flowering of forest trees in most months of the year, but peak periods of flowering and fruiting were generally associated with drier months (figs. 5.13a and 5.13b).

The forest on the ultrabasic ridge and other forested areas within the reserve are rich in orchids, with several species endemic to the Meratus Mountains. One species found on the ultrabasic ridge, believed to be *Porphyrodesme hewitti*, is a new genus for Borneo. Other orchids observed on the ridge include *Flickingaria fimbriata, Dendrobium* spp. (common by the river), *Aeredes odorata, Pholidota imbricata* (which commonly indicates dry conditions), *Coelogyne rochusennii* and *Bulbophyllum* and *Eria* spp. (Lamb pers. com.).

Pleihari-Martapura Reserve protects the watershed of the Riam Kanan reservoir. Small streams tumble down the ultrabasic ridge in a series of

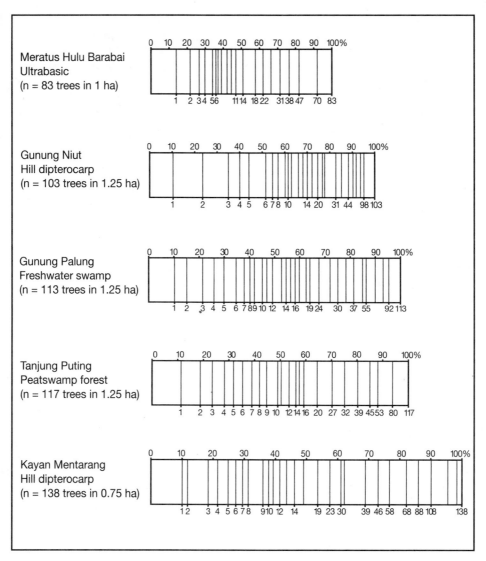

Figure 5.11. Relative abundance of tree species in different forest formations, showing species diversity. (N.B., at no site does one species contribute more than 14% of the trees. In the Meratus Mountains on ultrabasic soils, 61 species (73% of total) make up just 38% of the forest.)

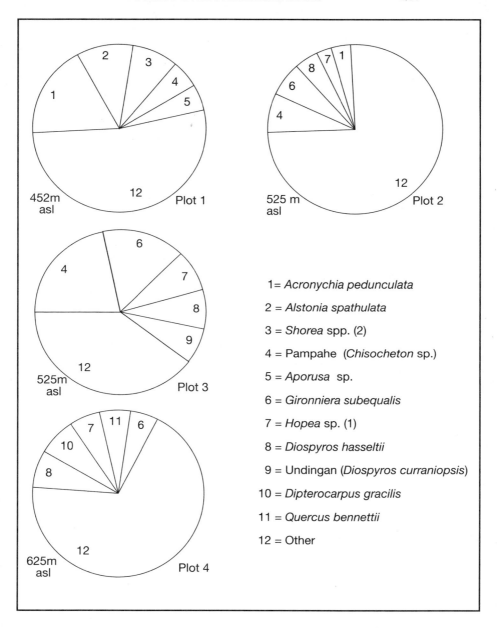

Figure 5.12. Dominant species at various altitudes above sea level (asl) in Pleihari-Martapura Reserve. (Plot size: 40 m x 100 m = 0.4 ha.)

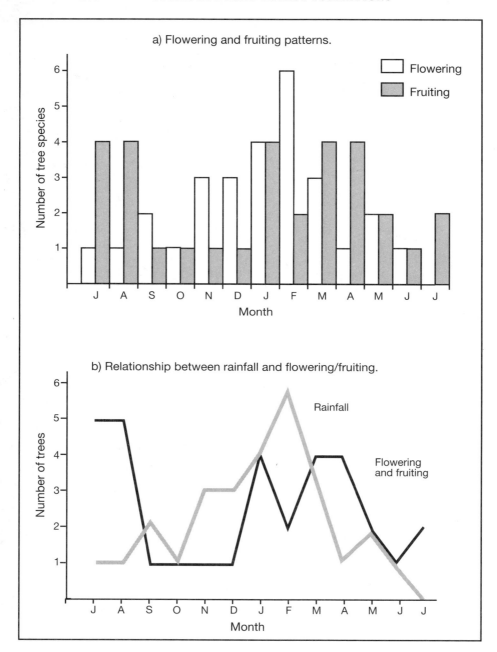

Figure 5.13. Flowering and fruiting of trees in Pleihari- Martapura Reserve from July 1988 to July 1989.

waterfalls to feed the Kala'an River, one of the several small rivers that feed the reservoir. At the edge of the reserve the forest is gradually being eroded by illegal agricultural activities as villagers clear fields to plant peanuts, a valuable cash crop. A narrow fringe of forest along the river is made colourful by the purple flowers of *Lagerstroemia speciosa*, an increasingly rare sight along Bornean rivers. In 1988 KPSL-UNLAM initiated a twelve-month study of the fish fauna of the Kala'an River to monitor changes in species composition and distribution along the river, correlated with deforestation and changing environmental conditions (see chapter 3, especially fig. 3.16).

Pleihari-Martapura has a varied fauna, though densities of primates and other mammals are generally low compared with those of other Kalimantan lowland and hill forests. This may be due to the relatively small number of large fruit trees encountered during surveys and probably also to overhunting by local villagers. The mammal fauna of the reserve include three Bornean endemics: the Bornean gibbon *Hylobates muelleri*, the maroon langur *Presbytis rubicunda* and the tufted ground squirrel *Rheithrosciurus macrotis*. Because the reserve's primates and other large mammals are very wary of human observers, it has been difficult to collect accurate data on range sizes or feeding strategies. Nevertheless, gibbon ranges in this forest seem to be larger than those recorded elsewhere for this territorial frugivore (Leighton and Whitten 1984). The maroon langur is primarily a leaf eater, though, like most langurs, it also takes large quantities of fruits when these are available (Supriatna et al. 1986).

Land Use on Ultrabasic Soils

The vegetation over ultrabasic soils in the Far East is very variable, but it is still not clear why ultrabasic soils should bear stunted vegetation in some areas and well-developed forest in others. For example, the large-stature forest on Gunung Silam is on soils with a high Mg/Ca quotient, while the low stature forest on Mount Kinabalu is on soils with low values for both magnesium and calcium (Proctor 1992). The infertility of ultrabasic soils is well known to local people, who usually avoid cultivating them, but increasing population pressure and demand for land mean that some ultrabasic areas are being opened for cultivation. Analyses of these soils have shown an increasingly unfavourable chemistry with depth, so that they may become very difficult to manage as surface layers are eroded. Conservation of vegetation over ultrabasic rocks would protect watersheds and the water chemistry of upland rivers. Ultrabasic vegetation may also be a potential source of important chemicals for medicinal and other uses (Proctor 1992).

Chapter Six

Limestone Habitats

Flying over the rainforests of Kalimantan and Sarawak, the traveller is sometimes surprised by the sight of jagged limestone piercing the sea of green canopy. These limestone formations were laid down long ago as calcium carbonate secreted by the corals, brachiopods and crinoid lilies on ancient coral reefs beneath the surface of prehistoric seas. As tectonic movements on the Sunda Shelf created the islands of the Indonesian archipelago, these ancient reefs and other sedimentary rocks were compressed and uplifted to create much of the island of Borneo. Over the aeons of time, climate and rain have sculpted and shaped the island, eroding away the softer rocks to mould the landscape into mountain chains, crested ridges and wide alluvial plains. The processes of weathering have created particularly dramatic landscapes in limestone areas.

Spectacular ranges of limestone hills, huge isolated boulders and limestone outcrops are scattered throughout the whole island of Borneo (Whitmore 1984b). Limestone outcrops emerge from the old marine sediments and ultrabasic rocks of the Meratus Mountains in South Kalimantan (Madiapura et al. 1977). Limestone crags rise above the rainforest at Bau, Mulu and Niah in Sarawak. The Niah Caves are important archaeological sites, as are the caves and rock shelters in the limestone cliffs at Madai, Gomantong and Tingkayu in Sabah. The most extensive limestone formations in Borneo (and indeed in the whole of Southeast Asia outside Irian Jaya) are the Mangkalihat karst ranges in the Sangkulirang Peninsula in East Kalimantan (Voss 1983; RePPProT 1987). Here squat limestone hills, dramatic escarpments, deep, river-eroded gorges, dissected pavements and sharp needles of weathered rock create a breathtaking landscape, a wilderness area still barely touched by development and human hands (fig. 6.1).

Limestone areas have a characteristic karst landscape which occurs in two forms in the tropics (Verstappen 1960). Tower karst consists of isolated hills, 100-150 m or higher, with steep, cliffed sides (slopes of 60° to 90°) and often riddled with caves. The appearance of the outlying hills of tower karst in Mulu National Park in Sarawak has been likened to "rotten and decaying teeth in receding gums" (Osmaston and Sweeting 1982). Cockpit karst, as seen in the Sangkulirang area, consists of a regular series of conical or hemispherical hills with less steep sides (slopes of 30° to 40°).

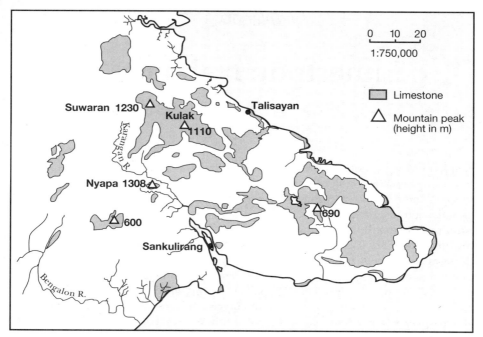

Figure 6.1. Distribution of limestone in the Sangkulirang Peninsula, East Kalimantan, the most extensive area of limestone in Borneo.
Source: Voss 1983

These limestone massifs were thrown up as a result of powerful geological activity, and the forces of nature are still shaping them today. Small amounts of atmospheric carbon dioxide dissolve in falling rain to produce a weak solution of carbonic acid which slowly reacts with the limestone (fig. 6.2). Gradually surface streams carve out gullies and eventually disappear down faultlines and sinkholes to flow underground through hidden caverns eaten out of the bedrock.

All exposed blocks of limestone of whatever size eventually are weathered into characteristic razor-sharp crests which are very difficult to traverse. The most spectacular examples in Borneo are a small group known as the Pinnacles on the northern end of Gunung Api, in Gunung Mulu National Park, Sarawak. At an altitude of 1,200 m they stand 45 m high, jagged splinters of eroded rock piercing the forest canopy (fig. 6.3).

As yet there have been no detailed studies of the flora and fauna of limestone formations and caves in Kalimantan. However, considerable information has been collected in the Gunung Mulu National Park in Sarawak (Anderson et al. 1982). From 1977 to 1978 a joint Royal Geographical

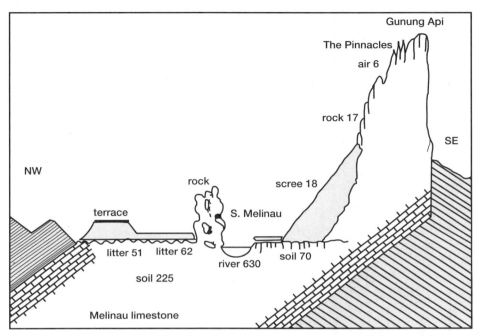

Figure 6.2. Cross section through the Melinau Valley, Gunung Mulu National Park, Sarawak. Figures show rates of solution of limestone tablets, expressed in mm/1,000 years. (After Osmaston and Sweeting 1982.)

Society and Government of Sarawak expedition conducted detailed ecological research in Gunung Mulu National Park, and a wealth of material has been published in a variety of scientific journals including two special issues of the Sarawak Museum Journal (Jermy and Kavanagh 1982, 1984). Subsequent caving expeditions in Mulu (Waltham and Brook 1979) and East and South Kalimantan (ESFIK 1983; KPSL-UNLAM 1989c; Mapala Sylva 1989) have yielded additional information on cave ecology and cave fauna. Surveys of the flora and fauna of the Sangkulirang limestone areas must be regarded as a research priority for the future.

Soils and Drainage

Limestone landscapes provide a diversity of soils and habitats (Anderson 1965; Chin 1977; Anderson et al. 1982; Anderson and Chai 1982). Soils on limestone are often richer in bases, especially calcium and magnesium, and with a higher cation-exchange capacity than other lowland forest soils

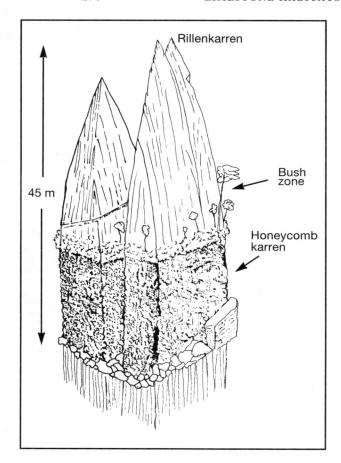

Figure 6.3. Schematic representation of two pinnacles in Gunung Mulu National Park. (After Osmaston 1980.)

(Proctor et al. 1983c). On gentler slopes, and in hollows, clay-rich, leached brown-red latosols are formed (Burnham 1984). The steeper parts of karst limestone are mostly bare limestone with a little soil in fissures. On wet mountains an acid, peatlike soil may form on crests and shelves, as on Gunung Api in Sarawak (Tie et al. 1979). The composition and depth of soil varies according to the purity of the parent material and the topography (Crowther 1982b, 1984). Although alluvial soils at the base of limestone hills are derived from other rocks, they are under the influence of runoff and erosion from the limestone, and may support characteristic vegetation types. In Sarawak the lowland evergreen forests at the base of limestone hills support several characteristic tree species, notably *Eusideroxylon melagangai, Gonystylus nervosus* and *Mammea calciphila* (Whitmore 1984a). In more populated areas natural forests have been cleared around the limestone outcrops, and the base-rich fertile soils are cultivated right to the foot of the limestone cliffs, as around Batu Hapu in South Kalimantan (KPSL-UNLAM 1989c).

Because of the porosity of the underlying rock and rapid drainage through it, most of the habitats of limestone hills are drier than those in surrounding country (though deep crevices and ravines may be highly humid). In general, vegetation on limestone must cope with drier and harsher conditions than on other substrates.

Soils over limestone are generally thin, and deficient in minerals other than calcium and magnesium. Studies of four lowland forests at Mulu revealed that soils of limestone forests contained considerably less nitrogen, phosphorus, potassium and sodium (Proctor et al. 1983c). The lack of phosphorus, in particular, may be a limiting factor for plant growth. Soils over limestone in the tropics are exceptionally poor in aluminium compared with those on surrounding substrates, and it may be that some calcicoles (chalk-loving plants) are aluminium-sensitive. This is true for some European species that grow on limestone (Whitmore 1984a). These factors, together with the limestone topography, influence the type and nature of the vegetation and account for the large number of endemic species found in limestone areas (Chin 1977).

LIMESTONE FLORA

No detailed studies have yet been made of Kalimantan limestone areas, but preliminary botanical surveys in Sangkulirang and elsewhere in Borneo (Anderson 1965) suggest that Kalimantan limestone habitats support a rich flora with many limestone-specific species. This accords with findings from elsewhere in Southeast Asia (Henderson 1939; Chin 1977, 1979, 1983a, 1983b). Although limestone habitats in Peninsular Malaysia occupy only 260 km^2, a mere 0.2% of the land area, they support 1,216 species of flowering plants, about 13% of the total Malayan flora. More than one-fifth of the plants recorded from these limestone areas are found only in Peninsular Malaysia, and half of these endemics are confined to limestone (Chin 1977). A full 355 species (more than a quarter of the total plant list) are characteristic of the Malayan limestone flora, with 254 species (21% of the total recorded) restricted to limestone. Of those found on limestone, but not confined to it, many are rock crevice plants rather than calcicolous (chalk-loving) species, and also occur at rocky sites on other substrates. Interestingly, 50 of the species confined to limestone in Malaysia are species of monsoon Asia that can compete successfully in the humid tropics only in the periodically dry limestone habitats (Whitmore 1984a).

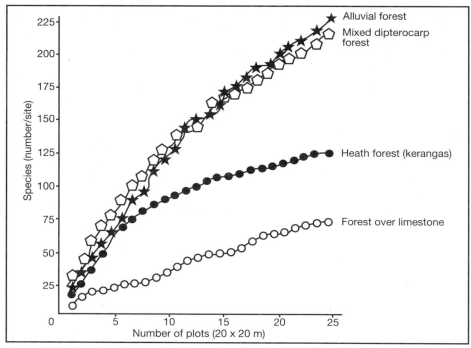

Figure 6.4. Species-area curves for trees >10 cm dbh in four forest types in Gunung Mulu National Park, Sarawak.
Source: Proctor et al. 1983a

Table 6.1. The numbers of identified tree species in common between each site.

Forest type	Alluvial	Kerangas	Mixed dipterocarp	Limestone
Alluvial	223	36	31	12
Kerangas		123	27	5
Mixed dipterocarp			214	12
Limestone				73

Source: Proctor et al. 1983a

Table 6.2. The estimated total forest above-ground biomasses of the alluvial, kerangas, mixed dipterocarp and limestone forest sites.

Forest type	Estimated total above-ground biomass (t ha^{-1})
Alluvial	250
Kerangas	470
Mixed dipterocarp	650
Limestone	380

Source: Proctor et al. 1982

The limestone flora of Sarawak is very rich in species (Anderson 1965). This richness is due to the distinctive herbaceous flora associated with limestone cliffs, and the open understorey of limestone forests on thin soils and steeper slopes. Many orchids, for instance, grow in limestone habitats, and several slipper orchids *Paphiopedilum* and jewel orchids are limestone endemics. In contrast, the tree flora of limestone forests is rather poor in species (fig. 6.4), with generally fewer trees and fewer species than forests on

Table 6.3. The contribution of the ten most abundant tree families expressed as percentage basal area (B.A.) and the actual numbers of individuals in each of these families in five forest types in Gunung Mulu National Park, Sarawak. Only individuals with dbh of 10 cm or above were counted.

	Alluvial forest		Gley-soil part of alluvial forest		Dipterocarp forest		Heath forest		Forest over limestone	
	B.A. %	No.	B.A. %	No.	B.A. %	No.	B.A. %	No.	B.A. %	No.
Abundant on all sites										
Dipterocarpaceae	13.0	33	11.5	8	43.2	114	42.9	92	47.4	140
Euphorbiaceae	8.9	97	10.8	42	2.8	70	3.0	65	5.8	89
Myrtaceae	6.4	44	4.5	15	4.0	65	11.0	74	5.1	47
Sapotaceae	4.8	30	0.9	4	4.8	52	3.3	38	4.1	39
Present on all sites										
Anacardiaceae	4.2	12	4.0	6	1.5	20	2.1	12	0.1	1
Annonaceae	2.0	18	1.3	5	0.5	4	2.6	32	0.7	9
Burseraceae	0.8	13	1.0	4	3.1	35	0.3	3	0.3	2
Ebenaceae	4.7	37	10.5	30	0.3	8	0.5	7	1.0	16
Guttiferae	5.2	42	5.4	19	2.2	35	17.0	170	0.7	7
Lauraceae	5.5	28	8.6	14	2.6	20	1.5	8	1.8	13
Leguminosae	15.1	43	23.6	25	5.1	15	1.7	18	0.2	4
Meliaceae	2.1	23	4.2	20	0.0	1	0.1	3	3.6	50
Myristicaceae	2.7	29	1.9	4	2.4	49	1.1	25	0.3	4
Sapindaceae	1.3	10	2.5	7	0.6	8	0.2	2	9.5	25
Sterculiaceae	2.7	10	6.4	9	0.4	4	0.1	2	3.3	14
Absent from forest over limestone										
Celastraceae	1.2	8	1.2	4	1.6	15	1.8	43	–	–
Fagaceae	1.7	11	2.0	3	2.6	15	0.9	21	–	–
Thymelaeaceae	0.8	3	0.2	1	0.3	6	2.9	16	–	–
In forest over limestone only										
Combretaceae	–	–	–	–	–	–	–	–	1.5	7
Absent from heath forest and forest over limestone										
Bombaceae	2.9	3	–	–	0.1	1	–	–	–	–
Absent from alluvial and dipterocarp forests										
Crypterionaceae	–	–	–	–	–	–	1.8	5	0.0	1
Absent from heath forest										
Tiliaceae	0.2	2	0.2	1	0.3	4	–	–	7.8	32

Source: Proctor et al. 1983a

deeper soils (Crowther 1982b; Proctor et al. 1983a, 1983b). Species composition is also different on limestone (tables 6.1 and 6.2). Of the 73 species recorded in Mulu limestone forests, 60% did not occur in other lowland forest formations (Proctor et al. 1982). The relative paucity of tree species (table 6.3) can be associated with thin soils, high calcium levels, rapid drainage and low humus content, which determine low soil moisture content, and the often rugged terrain. Most of the few dipterocarps associated with limestone in Sarawak occur in ravines along igneous intrusions and in deep soil pockets on ledges and in sinkholes (Whitmore 1984a). Though relatively poor in tree species, and therefore of limited value for logging, limestone areas are important for plant conservation because of their overall species richness and high levels of plant endemism.

LIMESTONE FORESTS

The ecology of the limestone habitats in Gunung Mulu National Park have been well studied and can be regarded as typical of limestone forests elsewhere in Borneo, including Kalimantan.

Soils on these limestone hills are very thin, comprising silty or clay loams mixed with limestone gravels in crevices and between boulders, and loams on the scree slopes (Baillie et al. 1982). Organic matter tends to increase with altitude. The vegetation is distinctive, with numerous endemic calcicolous species. Five limestone habitats with characteristic vegetation and flora can be distinguished on the mountains of Gunung Api and Gunung Benarat: lowland scree forest; lowland limestone cliff communities; lowland limestone forest; lower montane limestone forest; and upper montane forest (Anderson et al. 1982; Anderson and Chai 1982).

1. Lowland Scree Forest

The forest on the generally steep, boulder-strewn scree slopes is fairly open with few, widely spaced trees, dominated by massive emergents that occasionally exceed 5 m in girth and 50 m in height (Anderson and Chai 1982). Emergent tree species include *Azadirachta excelsa*, *Scorodocarpus borneensis* and *Sindora coriacea* with its prickly, cactus-like fruits. Ironwood and *Pometia pinnata* are common as dominants, though rarely very tall. Among the trees of the middle and lower storeys, the trunk-fruiting *Baccaurea lanceolata* and the black-barked *Diospyros cauliflora* are common, as well as *Paranephelium nitidum* and *Teijsmanniodendron pteropodum* (Anderson et al. 1982). The screes of eroded rocks and damp, calcareous soils provide a wide range of habitats for a rich shrub and herbaceous flora, including several species of Urticaceae (nettle family). The herb *Chloranthus officinalis* is typical of damper sites, while on the limestone rocks the fern *Antrophyum parvulum* and trumpet-flowered Gesneriaceae are common (Anderson and Chai 1982).

2. Lowland Limestone Cliff Communities

The sheer limestone cliffs, exposed to hot sun and generally lacking soil or moisture, are an extreme habitat for plant growth. Nevertheless, a few scattered shrubs, and even small trees, root in rock crevices and shelves. Many species, including herbs in the Gesneriaceae, are able to withstand dry periods by reversible desiccation of the tissues (Whitmore 1984a). The limestone cliffs are typical habitat for the tree *Fagraea auriculata* and also for the fig *Ficus tinctoria*, which usually grows as an epiphyte. Cascades of the fern *Phanerosus sarmentosa*, 5-6 m long, drape the cliffs while smaller ferns are found in the shadier spots. In sheltered ravines and on the lower slopes the cliffs are often clothed in a heavy herb layer. Clusters of dark-leaved Gesneriaceae with flower trumpets of scarlet, white or purple are common, especially the calcicolous genera *Monophyllaea*, *Paraboea* and *Epithema*.

The chalk-loving palm *Salacca rupicola* is found in crevices in cliffs and in the dwarf forest on more exposed limestone crags above 1,000 m. It is the only palm known to be confined to limestone in Borneo (Dransfield 1984). It is not known from any locality outside the Gunung Mulu National Park (Anderson and Chai 1982). The occurrence of other palms found on limestone is probably not directly related to the chemical characteristics of the rock. Thus *Caryota* is common on low limestone slopes because of the many openings in the canopy. These are caused by tree falls and vegetation loss due to landslips on the unstable slopes (Dransfield 1984).

3. Lowland Limestone Forest

On very steep limestone slopes, with gradients greater than 45°, a dense, irregular forest occurs. Small trees and shrubs cling precariously, their roots penetrating to great depths in crevices (Anderson et al. 1982). Below 800 m on moderate slopes the forest is higher and mainly dipterocarp, dominated by large emergent trees which may reach 40 m in height. Dipterocarps such as *Hopea andersonii*, *H. dasyrachis* and *Shorea multiflora* are common, also the nondipterocarps *Brownlowia glabrata* and *Palaquium sericeum*. Larger trees tend to be heavily buttressed. The shrub layer is sparse. Large woody climbers such as *Derris* and *Phanera* occur but are less common than in mixed dipterocarp forest; shade epiphytes are rare. Rattans are also rare; a hairy form of *Calamus javensis* is the only rattan recorded for the Mulu limestone (Anderson and Chai 1982).

4. Lower Montane Limestone Forest

The summits of limestone hills, even at relatively low altitudes, are covered in a deep mat of peatlike humus held together by tree roots which anchor them to the limestone beneath. This soil is acidic (pH 4.5) and supports low forest, sharing several species such as *Casuarina nobilis* with lowland heath forests. The similarity to heath forests suggests a response to the same factors, namely periodic water stress and lack of nutrients (Whitmore 1984a).

As on Gunung Mulu and other Bornean mountains, the vegetation on the limestone crag of Gunung Api (1,700 m) shows some altitudinal zonation (table 6.4). With increasing altitude both the height of the canopy and the average size of trees decrease; the number of dipterocarps also decreases (Anderson and Chai 1982). Lower montane forest begins at about 800 m, where terrain is generally steeper and more rugged with large limestone blocks. The dense forest is composed of small trees, few exceeding 150 cm in girth. The canopy height is less than 25 m. The only dipterocarp is *Hopea argentea*, and the forest consists mainly of noncalcicolous species growing on the deep mat of humus. Trees are often bent and leaning.

The light canopy allows luxuriant growth at ground level. Typical calcicolous herbs (mainly Gesneriaceae and Urticaceae) occur on exposed limestone, while begonias are found among the herbs and shrubs on deep litter. Large woody climbers are absent in the lower montane limestone forest, but epiphytes, mainly orchids, ferns and aroids, drape the lower stems of trees (Anderson et al. 1982).

5. Upper Montane Forest

On Gunung Api and the other limestone peak of Gunung Benarat, upper montane forest begins at 1,200 m at the start of the cloud layer. It is similar to the upper montane forests on the adjacent sandstone mountain, Gunung Mulu (Anderson and Chai 1982; see chapter 7). Small trees are interspersed with shrubs on the broken terrain. Abundant bryophytes cover the deep humus on the ground and drape the low forest. On the humus layer calcicolous plants give way to calcifuges (chalk-shunning species), the same species as are found on Gunung Mulu. These include shrub rhododendrons and the conifers *Dacrydium beccarii* and *Phyllocladus hypophyllus*.

Extensive areas of almost pure *Pandanus* on the limestone peaks may result from the destruction of former vegetation by fire. The large pitcher plant *Nepenthes stenophylla* occurs in open localities. Above 1,500 m the low, open forest on the ridge is dominated by *Leptospermum flavescens, Dacrydium* and *Phyllocladus. Casuarina* and rhododendrons also occur, and ant plants grow on the *Leptospermum* trunks, all species adapted to nitrogen-poor environments (Anderson and Chai 1982).

Table 6.4. Vegetation types and soils on Gunung Api.

Vegetation type	Altitudinal range; rock type; slope	Soil	Vegetation structure
Lowland forest over limestone	170-1,000 m; cemented scree, 30-40°	Immature, shallow organic soils of slightly acid to neutral pH	Canopy 25 m, emergents to 40 m (at 180 m). Mesophyllous rainforest, scant herb layer, few epiphytes. Low diversity. Dipterocarps dominant.
Lower montane forest	800-1,200 m; dissected blocks, < 50°	Some stable organic soils	Canopy 10 m, emergents to 25 m. Dense, stunted, mesophyllous forest with rich herb layer, many epiphytes (orchids, ferns, aroids). Dipterocarps absent (except *Hopea*), *Parishia* abundant.
Upper montane forest over limestone	1,200-1,710 m; pinnacle and tower karst	Organic soils of variable depth	Canopy 0.4-10 m with microphyllous species. *Nepenthes* and bryophytes abundant.
Dacrydium facies	1,200 m; tall pinnacles; wide fissures	Organic soils, stable or suspended by tree roots across crevices in karst	Canopy 10 m. Dense moss forest. Mosses and liverworts on shaded pinnacles. *Dacrydium beccarii* dominant.
Phyllocladus facies	1,500 m; N slopes of peaks; tower karst	Stable and suspended organic soils	Canopy 3-5 m. Sclerophyllous shrub association, more open, few epiphytes. *Phyllocladus hypophyllus* frequent.
Leptospermum facies	1,500 m; S slope of northern peak; tower karst	Stable and suspended organic soils	Canopy 4-8 m. Open, gnarled trees and dense shrubs amongst pinnacles. *Leptospermum flavescens* dominant, also conifers.
Pandanus facies	1,500 m; exposed, possibly burnt, ridge crests; reduced tower karst	Shallow organic soils	Canopy 1 m. Tangled prostrate pandan stems. *Pandanus* spp. dominant.
Summit facies	1,710 m; level limestone	Shallow organic soils	Canopy 0.4-2 m. Gradation of dense shrubs to low, dense carpet of shrubs, occasional orchids and ferns.

(Adapted from Anderson and Chai 1982.)

Because of the low density of large dipterocarps and the difficulty of the terrain in limestone ranges, forests on limestone hills have little commercial value, though some attractive herbs (balsams, begonias and Gesneriaceae) may have horticultural potential (Whitmore 1984a). Limestone hill forests do, however, serve an important protective function, maintaining the hydrological functions of the watershed. Moreover, because of their species richness and high levels of endemism, they are important areas for conserving floristic diversity (MacKinnon and Artha 1981). The more accessible lowland dipterocarp forests on the karstic plains in the Sangkulirang Peninsula are already heavily exploited for logging and rattan collection. Timber extraction has affected at least 30% of the area (RePP-ProT 1987), and future logging and planned transmigration schemes will open further areas of forest.

Forests on limestone appear to be particularly susceptible to fire damage, probably because of the generally drier conditions imposed by rapid drainage. In 1983 fires swept through large areas of forest on the limestone hills of the Sangkulirang Peninsula (Lennertz and Panzer 1983). The forest on the high limestone mountains of Api and Benarat in Sarawak is damaged occasionally by fire, apparently started by lightning strikes (Anderson and Chai 1982). After fire the ground may remain bare for several years before a slow succession of bryophytes and ferns re-establishes, and shrubs begin to grow in pockets of soil and litter (Anderson 1965).

THE FAUNA OF LIMESTONE HILLS

There is no special vertebrate fauna associated with limestone forests in Kalimantan. Surveys by KPSL-UNLAM in the Sangkulirang Peninsula recorded typical Bornean lowland forest fauna: banteng *Bos javanicus*, orangutan *Pongo pygmaeus*, Bornean gibbon *Hylobates muelleri*, sambar *Cervus unicolor*, muntjak *Muntiacus muntjak* and mousedeer *Tragulus* spp. This is not surprising since the limestone formations in the survey area tend to be blocks within a sea of lowland forests.

In Sumatra and mainland Asia such inaccessible and inhospitable outcrops are home to the serow *Capricornis sumatraensis*, a goat-antelope which is a relative of the European chamois. Only such a sure-footed animal could make its living among the precipitous limestone crags, picking a route along well-used tracks and sheltering in caves. Reports by roadbuilders of a large, goatlike animal with unbranched horns in the limestone hills about 100 km north of Samarinda have led to speculation that the serow may occur here too, a possible first record for Borneo (Payne et al. 1985).

Apart from cave fauna, the only animals known to be restricted to limestone habitats are certain molluscs (fig. 6.5). Studies of 28 isolated limestone hills in Peninsular Malaysia recorded 108 species of snails, of

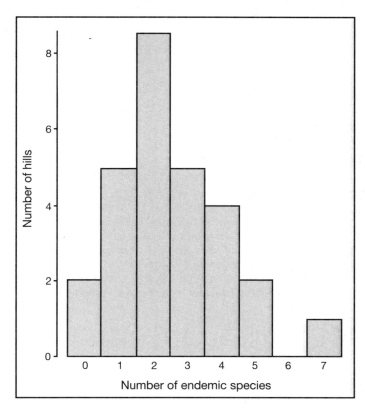

Figure 6.5.
Distribution of endemic snails on limestone hills in Peninsular Malaysia. (After Tweedie 1961.)

which 70 (66%) were restricted to single limestone hills (Tweedie 1961). Soil samples from limestone hills in Kalimantan and surveys in the Gunung Mulu National Park have revealed an interesting range of invertebrates from limestone habitats (table 6.5). In general, invertebrate fauna such as beetles tend to be less rich in limestone habitats than in other lowland forests (Stork 1986). However, there is a distinctive upper montane butterfly fauna in the park, more numerous on the summit of the limestone peak of Gunung Api than on the sandstone and shale of Gunung Mulu (Holloway 1984). Such pockets of invertebrate endemism probably also occur on Kalimantan limestone formations and make these habitats particularly important for conservation (fig. 6.6).

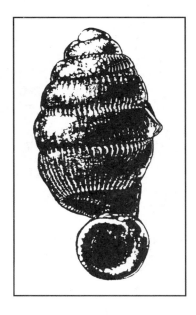

Figure 6.6. Sketch of *Diaphera porrecta* (Streptaxidae), collected by a team from KPSL-UNLAM from Batu Hapu, a limestone cave in the Meratus Mountains. Previously this snail was known only from a single type specimen from an unknown locality (Vermeulen in litt.). Limestone areas in Kalimantan are little explored but can be expected to show high species endemism.

Table 6.5. Snail shells found in soil samples from the Batu Hapu limestone caves near Binuang, Kalimantan Selatan.

Phaedusa dohertyi	*Liadertia indifferens*
Arinia inexpectans	*Liadertia* cf. *baritensis*
Palaina [*cylindropalaina*]*	*Liadertia* aff. *fraterna*
Diplommatina aff. *spinosa*	? *Microcystina* spp.
Ditropis sp.	*Mycrocystina vitreiformis*
Leptopoma bicolor	*Lamprocystis* spp.
Platyraphe linitus	? *Everettia consul*
Alycaeus jagori	*Videna planorbis*
Pupina hosei	*Videna* spp.
Cyclotus [*pterocyclos*] *trusanensis*	*Videna metcalfei*
Opisthostoma [*opisthostoma*]*	*Dyakia* spp. (juvenile)
*Georissa**	*Pupisoma* cf. *orcella*
? *Aphanoconia borneensis*	*Pupisoma orcula*
Hypselostoma spp.	*Charopa* cf. *vicina*
Diaphera porrecta	*Philalanka carinifera*
Lamellaxis clavulinus	*Pilsbrycharopa kobelti*
Lamellaxis gracilis	*Charopa* aff. *caloglypta**
Liadertia acutiuscula	

* *Possible new species.*

Soil samples collected by team from KPSL-UNLAM. Identifications provided by J. Vermeulen.

Figure 6.7. Major cave systems and rock shelters in limestone massifs on Borneo.

CAVES

Constant weathering processes have shaped both the external landscape and the interior of the limestone formations to carve out complex systems of underground caves and labyrinths. These caves are not only fascinating ecosystems but have played an important role in early human history and in the culture of many Bornean peoples to the present time. Figure 6.7 shows major cave systems in Kalimantan and elsewhere in Borneo.

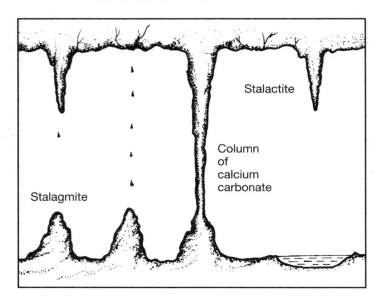

Figure 6.8. The formation of architectural features in limestone caves. Water dripping from the cave roof evaporates to form stalactites and stalagmites from the residual calcium carbonate. Eventually a column is formed.

Caves are formed by groundwater seeping through cracks in the rock and slowly dissolving the limestone. Falling rainwater contains carbon dioxide absorbed from the atmosphere, and, as the groundwater flows through soil it picks up more carbon dioxide and humic acids; this weak acid solution dissolves the rock. The simple process of limestone solution can create a diverse range of features. Surface streams create gullies and vanish down fault lines, eating away at the calcium carbonate of the bedrock. Large rivers carve out steep-sided gorges such as the spectacular Melinau Gorge in Sarawak and the cliff-lined stretches of the Karangan River in East Kalimantan. Smaller rivers and streams vanish down sinkholes to flow through underground caverns. Rainwater entering widely-spaced vertical fractures in the rock gradually enlarges small caverns into larger caves. As the undercutting continues, the cliff face collapses to create huge caves supported by natural arches. Underground rivers widen the caves and undercut the arches, and roof falls enlarge the caverns upwards. Experiments with limestone tablets show that limestone dissolution occurs fastest in rivers, fast in shallow soil and relatively slowly above ground (Walsh 1982; fig. 6.2). Cave erosion proceeds rapidly. The dissolved limestone is mostly carried away in runoff, but some calcium carbonate is redeposited by evaporation to create dripping stalactites and stalagmites in curious shapes and forms (fig. 6.8).

Borneo has some of the most famous and best-known caves in Southeast Asia. The Niah Caves of Sarawak have been studied in detail and are the source of much archaeological information (Medway 1977b; Harrisson 1958, 1972; Majid 1982). More recently the Royal Geographical Society Expedition to Sarawak explored more than a hundred kilometres of caves in Gunung Mulu National Park, including the longest cave system (Clearwater Cave) and the largest cave chamber in the world (table 6.6). The

Table 6.6. Lengths and depths of Mulu caves.

English name	Malay name	Surveyed length		Length	Depth
		RGS 1978	Mulu '80		
Southern Hills					
Deer Cave	Gua Payau	1,760 m	400 m	2,160 m	+220 m
Deer Water Cave	Lubang Sungai Payau	885 m		885 m	+3 m
Mayday Cave	Lubang Darurat	900 m		900 m	61 m
Snake Cave	Lubang Ular		780 m	780 m	31 m
Green Cave	Lubang Hijau	2,895 m	550 m	3,445 m	320 m
Gunong Api					
Cave of the Winds	Lubang Angin	2,300 m	4,250 m	6,550 m	+140 m
Clearwater Cave	Gua Air Jernih	26,330 m	11,255 m	37,585 m	+330 m
Leopard Cave	Gua Harimau Bintang	3,375 m	3,375 m		+78 m
Imperial Cave	Gua Imperial		705 m	705 m	-1 m
Solo	Sendirian	–			-259 m
Tiger Cave	Gua Harimau	1,000 m			104 m
Good Luck Cave	Lubang Nasib Bagus		2,900 m	2,900 m	+423 m
Prediction Cave	Lubang Ramalan	610 m		610 m	-46 m
Wonder Cave	Gua Ajaib	4,770 m		4,770 m	104 m
Gunong Benarat					
Benarat Caverns	Lubang Benarat	1,355 m	4,870 m	6,225 m	299 m
Sakai's Cave	Lubang Sakai		1,120 m	1,120 m	-140 m
Terikan Rising Cave	Gua Sungai Terikan	775 m		775 m	+12 m
Terikan River Cave West	Gua Sungai Terikan Barat	2,270 m		2,270 m	+30 m
Terikan River Cave East	Gua Sungai Terikan Timor	3,840 m		3,840 m	+61 m
Blue Moonlight Bay Cave	Gua Teluk Cahaya Bulan Yang Biru		9,400 m	9,400 m	119 m
Gunong Buda					
Beachcomber Cave	Lubang Penghuni Pantai		1,965 m	1,965 m	+30 m
Turtle Cave	Lubang Penyu		1,155 m	1,155 m	+38 m
Compendium Cave	Lubang Ular Dan Pangga		1,910 m	1,910 m	+34 m
Other minor caves		375 m	5,525 m	5,900 m	
TOTAL EXPLORED CAVES		50,065 m	50,160 m	100,225 m	

Source: Eavis 1981

Sarawak Chamber, discovered only in 1980 and three times as big as any cavern explored previously, is almost two kilometres wide, with a floor area equivalent to 17 football fields. In Kalimantan the limestone formations of the Meratus Mountains and the Sangkulirang region are also honeycombed with caves, but as yet these have been little explored (Madiadipura et al. 1977). The Sangkulirang ranges may hold cave systems as extensive as those in Mulu. A joint French-Indonesian caving expedition exploring the limestone hills in the P.T. Sangkulirang forestry concession at Pengadan found one cave, accessible only through a narrow slit in the cliff face, which extends for more than a kilometre deep into the hillside (ESFIK 1983). Who knows what further explorations may reveal?

Not all caves in Borneo occur in limestone outcrops, but by far the most extensive and spectacular galleries occur in such formations. A few caves are also found in sandstone, such as the important swiftlet caves of Berhala in Sabah. All caves share the same characteristic features: definable limits; low light levels; and relative stability of climatic factors such as temperature, humidity and air flow (Bullock 1966).

Cave Communities

Caves are dark, damp places that echo to the ceaseless dripping of falling water. Because of the lack of sunlight, few green plants grow there, only mosses and ferns in illuminated areas near the cave mouth. Fungi, however, pervade the whole cave region. Nevertheless, caves support abundant animal life and unique faunal communities.

Cave animals can be divided into three groups:
- **troglobites** are obligate cave species which can survive only in cave environments;
- **troglophiles** are species that live and reproduce in caves but are also found in similar dark, humid microhabitats outside;
- **trogloxenes** are species that regularly enter caves for shelter but normally feed outside. Cave swiftlets and bats belong in this category and are at the centre of the cave food web.

Explorations of the Mulu caves recorded the presence of at least 27 species of troglobites and about 70 species of troglophiles (Holthuis 1979; Roth 1980; Peck 1981; Chapman 1984). In addition, a few other species (e.g., rats and small shrews) may occur in caves but are more normally resident outside (Chapman 1985). Some small "cave-adapted" species may not be restricted to caves at all but occur elsewhere in underground habitats. The earless monitor lizard *Lanthonotus borneensis*, which moves like a snake, was first discovered at Niah in 1961 but has been found more recently in coastal floodplains. It is probably widespread in Borneo but frequently

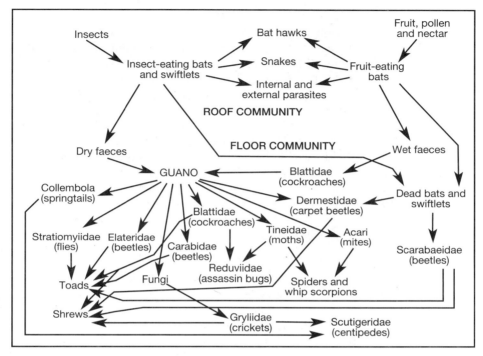

Figure 6.9. Simplified food web of a cave ecosystem.

undetected because it remains dormant underground during the daytime (Proud 1978).

Caves are ecological microcosms but derive all their energy inputs from outside. Some animals feed on plant roots attached to the cave roof or on plant material washed into the cave by underground streams. The major providers of food, however, are the swiftlets and bats that feed in the surrounding forests but roost and nest in the caves. The other animals of the cave community depend, directly or indirectly, on the bats and swiftlets that use the caves for shelter, living off them as predators and parasites or as detritivores supported by the piles of guano, rotting corpses and insect debris that they drop. Within the cave the animals can be divided into two communities: a roof community which includes the bats and swiftlets and animals feeding directly on them as parasites and predators; and a floor community living on the faeces and detritus below (fig. 6.9).

Bats

Many species of bats roost in limestone caves (fig. 6.10). Of the 77 species of bats recorded from Sabah, 27 are known to roost mainly in caves, and at least seven more species use caves occasionally (Francis 1987). At most sites bats outnumber swiftlets. Most of these bats are insectivorous, like the leafnose and horseshoe bats, or predatory, like the lesser vampire bat *Megaderma spasma*, feeding on smaller bats. At Niah the total population of 470,000 insect-eating bats includes the Bornean horseshoe bat *Rhinolophus borneensis*, the roundleaf bat *Hipposideros diadema*, the naked bat *Cheiromeles torquatus*, Horsfield's bat *Myotis horsfieldi*, the bearded tomb bat *Taphozous melanopogon*, Cantor's roundleaf bat *Hipposideros galeritus* and the lesser bent-winged bat *Miniopterus australis*. It has been calculated that these bats eat approximately seven tons of insects a day and are probably important in controlling insects, including agricultural pests such as the cocoa moth. The only nectar-eating or pollen-eating bat known to roost at Niah is the cave nectar bat *Eonycteris spelaea*, important for its vital role in pollinating durian flowers (Start and Marshall 1976).

For navigating the dark cave galleries and for tracking their prey in the forest at night, the insect-eating bats (Microchiroptera) have a highly developed system of sonar echolocation, one of the most amazing phenomena of the animal world. The bat issues a continuous series of high-pitched sounds, and these sound waves strike objects and rebound. The echo picked up by the bat's large, specially shaped ears indicates the position and distance of the objects, so that the bat can adjust the direction and speed of flight accordingly. Many insectivorous bats fly with the mouth open, emitting cautious squeaks, but the leafnose and horseshoe bats keep their mouths closed and emit sounds through their curiously shaped noses, rather as people do when humming.

The bats vary the rate of sound emission depending on their activity. At rest a bat may produce 5 to 10 sounds per second. During flight it produces 25 to 30 sounds per second until it meets an obstacle, when the rate increases to as many as 60 to 250 sounds per second (Stebbings 1984). This ability to alter wave frequency also enables bats to use longer wavelengths to locate distant objects. Each individual bat recognises its own sound and those of its own species, and is not confused by others, even when in a flock of several thousands.

Although all the Microchiroptera have well-developed echolocation, not all species roost in caves, and those which do so show significant differences in their use of the cave habitat. The sheath-tailed bats *Emballonura alecto* and *E. monticola* usually roost near cave entrances but may also roost outside, under fallen trees or buttresses. They have larger eyes than many bats and may use vision to detect potential predators. Some of the tomb bats, especially *Taphozous melanopogon*, sometimes roost in caves but prefer lighter areas near the entrance. Colonies of this species are found at Batu

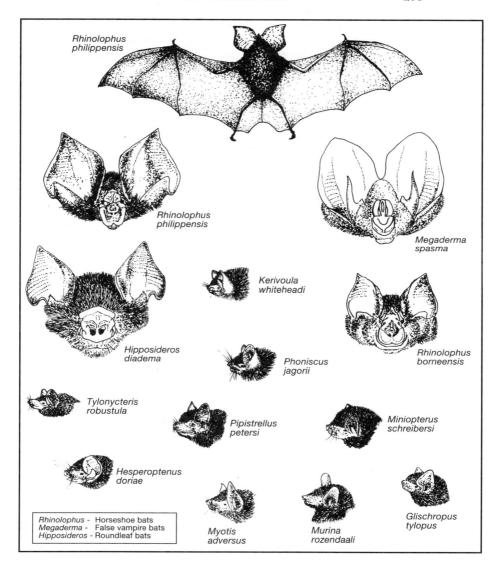

Figure 6.10. Some cave-dwelling bat species found in Kalimantan. (After Payne et al. 1985.)

Hapu in the Meratus Mountains in South Kalimantan. Another species of larger tomb bat, *Taphozous theobaldi*, was discovered recently by a team from KPSL-UNLAM in caves at the Batu Hapu massif, a first record for Borneo (Robinson pers. com.).

All of the bent-winged bats *Miniopterus* rely heavily on caves (Payne et al. 1985). Some species of *Miniopterus* form very large nursery roosts. Young bats are born blind and naked. Usually a single infant is born at a quarter of its mother's weight. For a day or two the infant may be carried by the mother, but after this it is left in the cave while she forages for food in the forest. When the females leave the caves to feed, the young bent-winged bats huddle together to keep warm.

Free-tailed bats also often roost in caves. The ugly naked bat *Cheiromeles torquatus* belongs to this family. Free-tailed bats feed mainly on insects, flying high over the canopy like swiftlets, but the two groups are unlikely to compete for food since the bats are feeding on nocturnal insects which are different species from insects active during the day when the swiftlets hunt. The wrinkle-lipped bat *Tadarida plicata* lives in huge colonies of up to a million or more individuals, and forms the large, spectacular flocks that can be seen emerging in ribbon formation from the Gomantong caves in Sabah and Deer Cave at Mulu (Francis 1987). Colonies of these bats produce massive piles of guano, often collected by local farmers for fertiliser.

The great flying foxes *Pteropus vampyrus* are not cave dwellers. Some of the smaller fruit bats (Megachiroptera) do roost in caves, although only a few can enter areas of total darkness. Surveys in Sabah uncovered two species of rousettes, *Rousettus amplexicaudatus* and *R. spinulatus*, as well as the cave nectar bat and the dusky fruit bat *Penthetor lucasii*, living deep within caves (Francis 1987). Rousettes are unusual among fruit bats in possessing a simple form of echolocation. They use their tongues to give broad-frequency clicks similar to those of swiftlets. These calls are quite different from the calls used by other bats, but enable the rousettes to detect and avoid large obstacles and to fly in total darkness. However, most fruit bats do not possess echolocation and rely on sight for navigation, and on sight and smell to find the nectar and fruit on which they feed. They are confined to more open caves or cave mouths where there is sufficient light for their manoeuvres.

Swiftlets

All four of the swiftlet species breeding in Borneo nest in caves and are central to the cave community: the black-nest swiftlet *Collocalia maxima*, the mossy-nest swiftlet *C. vanikorensis* (*salangana*), the white-nest swiftlet *C. fuciphaga* (*vestita*) and the white-bellied swiflet *C. esculenta*. The giant swiftlet *Hydrochous gigas* is also recorded from Borneo but is not known to breed there. It cannot echolocate and is a solitary nester on mountain cliffs in Java (MacKinnon 1988).

Swiftlets are aerial insectivores, catching all of their food (mainly insects) on the wing. They are efficient fliers and fly continuously when they are outside the caves, often foraging over considerable distances. For instance, most swiftlets which roost on Pulau Mantanini, Sabah, fly to the mainland to feed, a distance of 20 km each way (Francis 1987). Swiftlets are opportunistic feeders but seem to take mainly ants, termites and mayflies (Harrisson 1974; Langham 1980; Hails and Amirrudin 1981). Often they are seen circling around fig trees at the times when the young fig wasps emerge. They may prefer swarming insects because these can be captured in a relatively short period of time, thereby giving greater reward for effort. When feeding their young, swiftlets return to the cave with a large food ball in their mouth. Each food ball contains 100 to 1,200 insects, with a total weight of about half a gram (Langham 1980).

Like bats, most cave swiftlets are able to echolocate. The only other bird which can do this is the oilbird *Staetornis caripensis* from South America, which also lives in caves. The ability to echolocate allows swiftlets to nest in dark cave galleries where they are sheltered from outside weather and are relatively safe from predators. Since they can navigate the cave entrance and galleries in total darkness, they are able to forage over wide distances and often return to the caves several hours after sunset. So efficient is their echolocating system that cave swiftlets have been found more than four kilometres inside the Mulu caves. Of the Bornean cave-dwelling swiftlets, only the white-bellied swiftlet cannot echolocate. As a result it usually nests in cave entrances or in chambers that are relatively well-lit. Thus, in most caves, its breeding sites do not overlap with those of other species. This swiftlet also forages closer to home than the other species and always returns to roost before total darkness (Francis 1987).

Swiftlets call in the frequency range between 2 and 15 Hz, which is quite audible to people. They produce broad-band clicks, unlike the narrow frequency calls of bats. Since the strength of the echo is related to the frequency of sound and the size of the object being detected, the low frequency sounds used by swiftlets are not very good for detecting small objects. After numerous complex experiments, no swiftlet has been shown to detect obstacles less than 4 mm in diameter. This precision is considerably less than that of most echolocating bats (Fenton 1975).

Bornean swiftlets live in colonies throughout the year, even outside the breeding season (Medway 1960b). A million or more swiftlets may roost and nest in the same cave, building their cup-shaped nests on rocky ledges. Because of their very short legs, swiftlets are unable to take off from the ground and so, unlike other swifts and swallows, cannot collect mud for nest building. Instead, the birds make their nests out of gelatinous, sticky saliva, which solidifies when it dries (Medway 1962). This saliva is produced as a thin thread by sublingual salivary glands which enlarge during the nest-building season. All species use at least some sticky saliva in their nests.

Swiftlets may mate for life. Both sexes participate in nest building and raising young. At night swiftlets roost in pairs on the nest (Francis 1987). Different species select different sites for nests; this choice is influenced by the ability of swiftlets to echolocate, the nature of the nest saliva, and social factors (Francis 1987). Mossy-nest swiftlets have the softest nest saliva and require some sort of ledge to support their nests. Both black-nest and white-bellied swiftlets have very firm nest cement and often build their nests on cave ceilings or under an overhang, often with several nests attached to each other. All species avoid sites that are constantly moist or affected by running water after rain, as water softens the nest cement and causes nests to fall. Birds seem to prefer nesting with others of their own species, and new pairs build near existing nests; this may be an adaptation against predation. Nest collecting (a form of predation by humans) may also influence the distribution of nests; white-nest swiftlets seem to choose the higher, less accessible sites in the Gomantong Caves (Francis 1987).

Edible nests are produced by only two of the cave swiftlets, the black-nest swiftlet *Collocalia maxima* and the white-nest swiftlet *C. fuciphaga*. The white-bellied swiftlet and the mossy-nest swiftlet both incorporate so much plant material in their nests that they are of no commercial value. The hardened saliva is much prized for the famous birds' nest soup of Chinese cookery. Chinese traders have been visiting Borneo for more than a thousand years to buy the birds' nests (Medway 1957). Nests are collected only at certain times during the breeding season, once very early, since birds will rebuild, and again late in the season when many of the fledglings have already left the nest. Even so, there is usually a considerable loss of eggs and young.

Collecting methods vary according to the size of the cave. In some places nests can be reached with a pole from the ground. Elsewhere elaborate networks of bamboo ladders are constructed to enable the collectors to harvest the highest ledges. In the Great Cave at Niah in Sarawak, collectors scale vertical poles of jointed bamboo to terrifying heights of 50 m or more. The nests are dislodged with a knife attached to a long pole, and the collector also carries a flaming torch to illuminate the hazardous operation (Medway 1960b; Francis 1987).

The nests of black-nest swiftlets are those most commonly collected. These "black" nests have feathers embedded in the saliva and must be carefully cleaned before they are cooked. More expensive are the "white" nests of the less common white-nest swiftlet; these are almost pure saliva. One would imagine from the high price (up to $1,000 per kilogram) and hazardous collecting techniques that swiftlet nests must have a wonderful flavour. In fact, after cleaning and cooking, they are quite tasteless, merely giving texture to the other ingredients in a dish of fruit or meat. Nevertheless, these highly priced delicacies are prized by the Chinese as a tonic.

This valuable harvest should be sustainable, yet at closely monitored

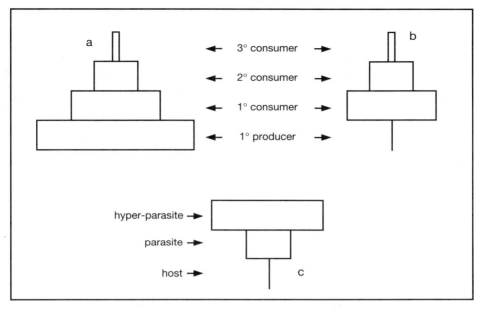

Figure 6.11. Pyramids of numbers: **a** - with a large number of primary producers, **b** - with a single primary producer, **c** - the case of parasites and hyperparasites. Bats and swiftlets are both predators and prey, as secondary consumers (a and b) and hosts to parasites (c). (After Phillipson 1966.)

caves such as Gomantong and Niah, swiftlet populations seem to be declining (Francis 1987). With increasing human populations, easier access to caves, and changing land use, pressure on this valuable resource will increase. Wise management strategies must be developed, taking into account the breeding biology and feeding requirements of the birds. Since 1987 all collecting of swiftlet nests at Niah has been banned to give the swiftlet populations a chance to recover. Collection of swiftlet nests in Kalimantan continues virtually unmonitored.

Roof Community

The roof community includes not only the cave swiftlets and bats but also the animals that prey on or parasitise them (fig. 6.11). Two species of snakes are regularly found in caves, *Oxycephala gonyosoma* and the marbled cave racer *Elaphe taeniura*. Both are harmless to humans and feed mainly on bats or other small mammals, and occasionally on swiftlets. *Oxycephala gonyosoma* is usually found near cave entrances where it catches bats flying out (Francis 1987), but *Elaphe taeniura* occurs further inside the caves. This species may reach two metres in length. It is a good climber and ascends the cave walls to the bat roosts and birds' nests to eat the young. Other bat predators include Brahminy kites *Haliastur indus*, peregrine falcons *Falco peregrinus* and bat hawks *Machaeramphus alcinus*, which hover outside the cave mouth in the early evening, ready to swoop on the bats as they leave. Bat hawks may also prey on swiftlets but probably catch mostly sick and injured birds and thus help to maintain the health of the population (Francis 1987).

High on the cave walls the bats and swiftlets are relatively safe from most predators, but even there a large, wingless cave cricket *Rhapidophora oophaga* feeds on young swiftlets and sucks the yolks from eggs. The caterpillars of the moth *Pyralis pictatis* devour and weaken the saliva nests so that they collapse, spilling their precious contents on to the cave floor below.

Bats and swiftlets are hosts to many external parasites, which feed from their host by sucking blood or eating flakes of dead skin. Some of these ectoparasites remain on the hosts when they leave the cave to feed, while others remain behind in the swiftlet nests. Some, such as the spiderlike, wingless nycterbiid flies, live almost their entire life on bats (Marshall 1971). Others, such as bat flies, bed bugs (Cimicidae) and chigger mites (Trombiculidae), spend only part of their life cycle on bats. Most species of parasitic insects are found on only one or two closely related host bat species (Marshall 1980). The naked bat *Cheiromeles torquatus*, one of the least attractive of all mammals, has its own special parasite, a large hairy earwig that feeds on the scurf of the bat's skin. This earwig is probably the largest ectoparasite ever found on a mammal. At about a quarter of the length of its host, it is a heavy burden, equivalent to a human permanently encumbered with a lobster. The extra energy costs of flying with this additional weight must be substantial.

Floor Community

On the cave floor a crowd of scavengers scours the fallen droppings, corpses, eggs and insect debris which fall from above. Cave snakes, civets, rats and monitor lizards all live within the caves and feed on young birds and bats that fall to the ground. Enormous numbers of small cave cockroaches *Pynoscelus striatus*, beetles, flies and springtails feed on the fallen faeces of bats and swiftlets. In places the guano seems to be writhing with cockroaches, as many as a hundred per square metre (Ko 1986). Tineid moths (close relatives of the clothes moths that ravage European wardrobes) hover over the fresh guano and lay eggs there. Tineid caterpillars, which carry cocoons around with them, feed on bird and bat guano as well as on dead bats, birds and invertebrates. At least 20 species of tineid moths are known to be cave dwelling, with different species within the same cave exploiting different food sources (Robinson 1980).

These dung eaters (**coprophages**) are caught and eaten by a multitude of predatory animals. Spiders lurk in crevices, trapping cockroaches in their webs. Whip scorpions *Sarax* run crabwise over the cave walls, seizing springtails and other prey with long, grasping pincers. Long-legged centipedes *Scutigera* feed on cockroaches as well as the cave cricket *Rhapidophora oophaga*, itself a scavenger and predator on swiftlet eggs and young nestlings. *Chaerilus* scorpions also prey on cave crickets, as does the rare huntsman spider *Heteropoda*, which leaps and pounces on prey. Assassin bugs (Reduviidae) use long forelegs to ambush small flies and moths. The bug stabs its prey with its long, daggerlike proboscis, injects a paralysing venom, then sucks out the victim's body fluids. The cave gecko *Cyrtodactylus cavernicolous* forages over the cave floor in the darkest passages, feeding on moths and flies. Small, shiny darkling beetles *Cercyon gebioni* scurry among the milling throng, collecting and consuming decomposing bodies.

In the dry, light cave mouths, predatory antlions (Myrmeleontidae) make their lairs. With slender abdomen and membranous wings, adult antlions resemble dragonflies, but with butterfly antennae. Their conical traps can also be seen in the dry soils under the overhang of house roofs or beneath the floors of stilted Dayak longhouses and Buginese homes. The larva makes a pit by backing downwards in a spiral into the light, sandy soil and flicking out the loosened sand with its head. There it sits, hidden at the bottom of the steep-sided pit, its enormous, curved jaws ready to catch any insect that falls. Trapped insects struggling to escape only slither further down the loose sand to the waiting antlion. After sucking out its victim's juices, the antlion tosses out the empty husk and repairs the trap. At night females hover above the dry soil, depositing eggs beneath the surface.

Streams within caves support their own biological communities, including many species such as decapod crustaceans, fish, snails, and insect larvae, introduced accidentally with floodwaters. Larvae of midges (chironomids), caddis flies and water moths (Trichoptera) are common, and provide an abundant food supply for a rich spider fauna. The spiders scavenge along the high-water mark (Chapman 1982).

Nine species of fish have been recorded in caves at Mulu, and at least one of these, the catfish *Silurus furnessi*, may be largely confined to caves (Chapman 1985). Cave fish often either show regression of the eyes or are totally blind. Many cave fishes are pale and colourless. There appears to be no selective advantage in sporting bright colours in a world of darkness.

The Mulu Caves

The caves in Gunung Mulu National Park have been explored and studied in greater detail than any other cave formations in Borneo or elsewhere in Southeast Asia (Brook and Waltham 1978, 1979; Waltham and Brook 1979;

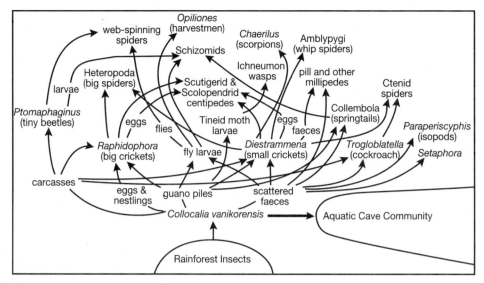

Figure 6.12. The mossy nest swiftlet *Collocalia vanikorensis* plays a crucial role in the life of the Mulu caves as the main importer of energy from the tropical rainforest outside. This simplified food web shows the relationships of the swiftlet-dependent community.

Eavis 1981; Chapman 1980, 1982, 1985). The Melinau limestone within the park has extensive cave habitats. The main gallery of Deer Cave is 1.8 km long and more than 100 m wide. Sarawak Chamber in Good Luck Cave has a floor area of 24 hectares; Clearwater Cave is 48 km long. So far 150 km of cave passages have been explored and mapped, and future explorations are likely to double this figure (Chapman 1985).

These important subterranean habitats support a rich collection of animals, including at least 31 vertebrate species (Chapman 1985). Of these, 13 species of bats, three of swiftlets, two of lizards and perhaps two fish species use the caves as shelter and for breeding, but feed outside the cave system. Among the more spectacular are the cave swiftlets *Collocalia*, which roost in tens of thousands in each of the major caves, and huge numbers of wrinkle-lipped bats *Tadarida plicata*, which stream out of Deer Cave at night and fly in sinuous ribbons over the forest.

The large-scale transfer of food energy from forest to caves by bats and swiftlets, mainly in the form of their droppings, supports a great variety of invertebrates within the caves (fig. 6.12). Two hundred cave-associated invertebrate species have been recorded at Mulu; twenty-seven of these, mainly arthropods, may be limited to caves (troglobites). They include seven crustaceans (four isopods, an amphipod, a prawn and a crab), seven

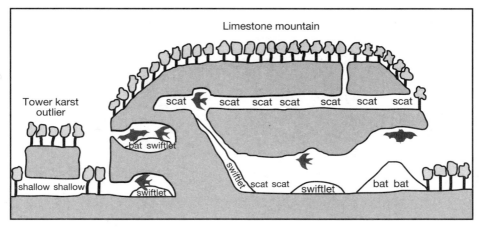

Figure 6.13. The four main habitat types within the Mulu caves. (After Chapman 1982.)

millipedes, four spiders, a beetle, and a cockroach (Chapman 1982). Many Mulu troglobites belong to very ancient groups which have largely disappeared from the modern land surface and are now represented by a few widely scattered species. Thus the nearest relatives of the millipedes are found in Indochina and Australia.

While some cave-dwelling invertebrates are found almost everywhere, others seems to be restricted to a few cave passages. The critical factors determining species distribution are food supply and abundance, and aspects of the cave environment such as microclimate (temperature, humidity, air movement and evaporation rate), both at ground level and in rock crevices where cave invertebrates occur (Chapman 1981).

There are four identifiable component habitats within the Mulu caves (fig. 6.13), and each supports a characteristic assemblage of invertebrate species, with very little species overlap (Chapman 1981, 1982, 1983). The four cave habitats can be labelled as:

- **scat habitat:** a damp, relatively low-energy habitat where the main food supply is scattered swiftlet faeces;
- **shallow habitat:** a shallow, tower-karst cave habitat where the main food supply is from flood sediment and tree roots;
- **swiftlet habitat:** in swiftlet guano;
- **bat habitat:** in bat guano.

The most widespread is the **scat association** consisting of up to 23 species of cave-evolved troglobites and troglophiles, including millipedes, woodlice, a beetle, two spiders and a cockroach (Chapman 1981, 1983). Members of the scat association occur in damp passages and in habitat patches where there are scattered swiftlet faeces, but not on large beds of

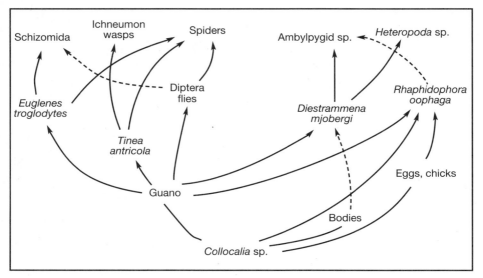

Figure 6.14. The swiftlet community food web.

swiftlet guano where they are excluded by guano specialists of the swiftlet association. Individual species of the scat association vary in their tolerance to drying air currents (the smaller animals tending to be less tolerant) so that cave topography and microclimate influence species occurrence (Chapman 1982, 1983).

In contrast, the guano specialists of the **swiftlet association** occur wherever swiftlet guano is plentiful, but are absent from the very damp situations favoured by scat members. The swiftlet association (fig. 6.14) includes one or other of two small crickets (but never both together), the large, egg-eating cave cricket, a moth, small beetles, flies, and predators on all these, including a powerful huntsman spider (Chapman 1981, 1983).

The **bat association** has been found on guano in a single cave beneath a large roost of free-tailed bats (fig. 6.15). It is dominated by two species of cockroaches, three species of moths (related to the familiar clothes moths) and a number of quite large beetles. Only two of the species identified also occur on swiftlet guano. It is possible that each bat species in Mulu supports a different guano-associated fauna (Chapman 1982).

The isolated tower karst outliers support a **shallow association** consisting of a few of the larger, more mobile troglophiles found in other Mulu caves and also several forest floor species. Such associations can be regarded as transitions between cave and forest communities (Chapman 1983). Most of the species are damp-loving and use the cave as a shelter and possibly as an egg-laying site, without being totally dependent on cave-based resources. Invertebrates found here include ants, harvestmen and

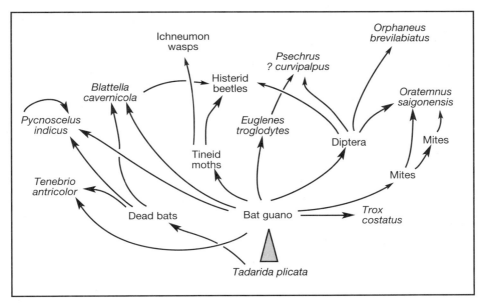

Figure 6.15. The bat community food web.

long-legged centipedes, plus the swiftlet-guano crickets and the huntsman spider. Predators in the shallow habitat include a *Scutigera* centipede, toads, a frog *Rana bladii* and a gecko *Cyrtodactylus pubisculus* (Chapman 1982).

Species represented in the four associations are shown in table 6.7. The distribution of the four groups within the Mulu caves is controlled by the presence of draughts, bats and swiftlets (Chapman 1985). Whereas bats tend to roost near cave entrances, swiftlets are found throughout the Clearwater Cave system, though they prefer to roost in drier areas. The rare deep-cave species are little threatened since these troglobites have a wide distribution, but species dependent on the faeces of locally distributed bats may be more difficult to conserve. For example, the hairy earwig *Arixenia esau* is found in Mulu only below a single roost of naked bats at Deer Cave. This is a popular tourist site, and the earwig and other local species could be disturbed by visitor activities.

Caves and Human History

Not only animals find safety and security in limestone caves. Stone Age humans chose to live in dry, sheltered cave mouths, where they were protected from the weather and secure from large predators like the sabre-

Table 6.7. The composition and species diversity of the faunal groupings in the Gunung Mulu caves.

Species	scat	Association swiftlet	shallow	bat	Sites (Deerwater)	(Skylight)
Pink leech					10	
Armadillo solumcolus	3					
Paraperiscyphis platyperaeon	3					
Setaphora parvicaputa	5					5
Glomeridesmid millipede	1					
White polydesmid millipede	8					
Pyrgodesmid millipede	9					
Cambalopsid millipede					200	
Hyleoglomeris spp.		6				
Scolopendrid centipede		1			1	
Red scutigeromorph	2	1				
Black/yellow scutigeromorph			8			
Oratemnus saigonensis				1,050		
Lychas spp.		2				
Amblypygid	1	37	15		20	
Schizomid	110	420			100	10
Spermophora spp.	3	122				
Pholcus spp.		1				
Ochyroceratid spider	30	370				
Nesticid spider		150				
Pecullid spider		10			10	
Ctenid spider	1					
Heteropoda spp.	22	100	40		60	6
Psechrus spp.				2		
"Acarine" opilionid	4					
Phalangodid opilionid	9					
Symploce cavernicola				7,000		
Pycnoscelus indicus				350		
Rhaphidophora oophaga	84	100	96		50	20
Diestrammena mjobergi	965	10,030				2
Diestrammena sarawakana			92		7,600	10
Diestrammena cf. *sarawakana*			10			
Nala ornata					10	
Mycetophilid larvae	200				20	
Guanobious dipteran		1,500		2,300	540	
Trichoptera					55	
Ptomaphaginus chapmani	1,449	50				
Hister spp.				1,400		
Gnathoncus spp.				70		
Trox costatus				1,030		
Tenebrio antricolor				560		
Anaulacus spp.					100	
Hyphaeron spp.					40	
Euglenes troglodytes	20	835	6	250	260	
Lithocharis vilis	7					
Anotylus spp.			20			
Rhaparochromine shieldbug		1,000				
Pheidologeton silenus			2,500			
Crematogaster spp.			1,000			
Monomorium spp.			100			
Pheidole longipes						400
Stenogaster spp.			20			
Ichneumonid wasps		28				
Tinea antricol	1	1,730		3,200		
Tinea porphyropa				26,300		
Crypsithyrodes concolorella				500		
Total Number	**3,108**	**16,493**	**3,907**	**45,012**	**9,096**	**463**

Source: Chapman 1985

Table 6.8. Mammal species identified among archaeological remains from the West Mouth, Subis Cave, Niah.

INSECTIVORA
Echinoserex gymnurus	moonrat
Hylomys suillus	lesser gymnure
Crocidura cf. *fuliginosa*	white-toothed shrew
Tupaia cf. *minor*	(lesser) treeshrew

CHIROPTERA
Pteropus cf. *vampyrus*	flying fox
Rousettus amplexicaudatus[1]	Geoffroy's rousette
Roussettus spp.	roussette
Eonycteris spelaea[1]	cave fruit bat
Rhinolophus trifoliatus	trefoil horseshoe bat
R. borneensis[1]	Borneo horseshoe bat
Hipposideros diadema[1]	diadem roundleaf horseshoe bat
H. galeritus[1]	Cantor's roundleaf horseshoe bat
Miniopterus cf. *tristis*[1]	bent-winged bat
M. cf. *blepotis*	bent-winged bat
M. australis[1]	bent-winged bat
Tadarida plicata	wrinkle-lipped bat
Cheiromeles torquatus	naked bat

DERMOPTERA
Cynocephalus variegatus	flying lemur

PRIMATES
Nycticebus coucang	slow loris
Presbytis spp. (*melalophos* gp.)	langurs
P. cristata	silvered langur
Macaca fascicularis	long-tailed macaque
M. nemestrina	pig-tailed macaque
Hylobates cf. *muelleri*	gibbon
Pongo pygameus	orangutan

PHOLIDOTA
Manis palaeojavanica[2]	giant pangolin
M. javanica	Malay pangolin

RODENTIA
Ratufa affinis	giant squirrel
Callosciurus prevostii	Prevost's squirrel
Sundasciurus lowii	Low's squirrel
Rheithrosciurus macrotis	tufted ground squirrel
Rattus muelleri	Mueller's rat
R. sabanus	long-tailed giant rat
Hystrix or *Thecurus* spp.	porcupine
Trichys lipura	long-tailed porcupine

table continues

1. Recorded in this cave in modern times.
2. Now extinct.
3. Tigers are not found on Bornea and these teeth may have been imported.

Table 6.8. Mammal species identified among archaeological remains from the West Mouth, Subis Cave, Niah. (*continued*)

CARNIVORA	
Canis familiaris	domestic dog
Helarctos malayanus	sun bear
Mustela nudipes	Malay weasel
Melogale orientalis	ferret badger
Lutra sumatrana	hairy-nosed otter
Amblonyx cinerea	Oriental small-clawed otter
Viverra tangalunga	Malay civet
Arctictis binturong	binturong
Hemigalus derbyanus	banded palm civet
Herpestes spp.	mongoose
Panthera tigris[3]	tiger
Neofelis nebulosa	clouded leopard
Felis bengalensis	leopard cat
PERISSODACTYLA	
Tapirus indicus[2]	Malayan tapir
Dicerorhinus sumatrensis	Sumatran rhinoceros
Elephas maximus	Asian elephant
ARTIODACTYLA	
Sus barbatus	bearded pig
S. scrofa	domestic pig
Tragulus napu	large mouse-deer
T. javanicus	small mouse-deer
Muntiacus muntjak	barking deer
Cervus unicolor	sambar
Bos javanicus (or B. bubalis)	banteng (or buffalo)
Capra hircus	goat

1. Recorded in this cave in modern times.
2. Now extinct.
3. Tigers are not found on Bornea and these teeth may have been imported.

Sources: Medway 1958, 1977b; Harisson 1966

tooth cat. The earliest known evidence for humans in Borneo is a human skull from Niah, believed to be more than 35,000 years old (Harrisson and Harrisson 1971; Bellwood 1978). Excavations in lower cave levels at Niah have revealed much about these Stone Age communities (chapter 1). Early *Homo sapiens* at Niah used stone axes, flaked and ground knives, and carved horn utensils. Their middens reveal that they ate molluscs and many different animals, including orangutans (Harrisson 1972; Medway 1977b; Majid 1982; table 6.8).

Since 1980 staff of the Sabah Museum have carried out excavations at cave shelters in the Madai and Baturong limestone massifs and at nearby open sites at Tingkayu on the shores of a former Pleistocene lake (Bellwood 1984, 1988). In the large rock overhang of Hagop Bilu, at Baturong, evidence of human habitation dates from about 17,000 years ago. Humans inhabited the nearby Madai caves from 11,000 to 7,000 years ago and again from 3,000 to 2,500 years ago. Indeed, the cave shelters here are still used seasonally by Idahans collecting birds' nests from the caves. Evidence from the Tingkayu sites shows that people in Borneo were already making stone tools as long as 28,000 years ago (Bellwood 1988).

Archaeological research in the Tingkayu, Baturong, and Madai regions shows that, between 18,000 and 7,000 years ago, people lived in caves and rock shelters, collected shellfish (table 6.9), made simple flake tools of

Table 6.9. Marine shellfish species represented in the Madai and Baturong caves; these species were eaten by early human inhabitants of Sabah.

Family	Species	Habitat
GASTROPODS		
Cerithiidae	1. *Cerithium aluco*	sandy intertidal
	2. *Cerithium* spp.	sandy intertidal
Cypraeidae	3. *Cypraea arabica*	intertidal reefs
Muricidae	4. *Chicoreus permaestus*	tidal mudflats, mangroves
Neritidae	5. *Neritina* spp.	estuarine, brackish water
	6. *Nerita lineata*	intertidal reefs
Potamididae	7. *Telescopium telescopium*	tidal mudflats, mangroves
Strombidae	8. *Strombus canarium*	tidal mudflats
	9. *Strombus luhuanus*	intertidal reefs
	10. *Lambis lambis*	intertidal reefs
Trochidae	11. *Tectus fenestratus*	intertidal reefs
Turbinidae	12. *Turbo necnivosus*	intertidal reefs
BIVALVES		
Arcidae	13. *Anadara antiquata*	shallow estuarine mud
	14. *Anadara granosa bisenensis*	shallow estuarine mud
Cardiidae	15. *Vasticardium alternatum*	sandy intertidal
Corbiculidae	16. *Batissa violacea**	tidal mudflats, mangroves
	17. *Atactodea striata*	tidal mudflats, mangroves
Ostreidae	18. *Crassostrea* spp.	tidal mudflats, mangroves
Placunidae	19. *Placuna pladenta*	shallow estuarine mud
Solenidae	20. *Siliqua radiata*	unknown
Spondylidae	21. *Spondylus nicobaricus*	intertidal reefs
	22. *Spondylus anacanthus*	intertidal reefs
Tridacnidae	23. *Tridacna crocea*	intertidal reefs
	24. *Hippopus hippopus*	intertidal reefs
Veneridae	25. *Gafrarium pectinatum*	intertidal reefs

* this species may also include specimens of *Polymesoda* (*Geloina*) spp.
Source: Bellwood 1988

chert and hunted a wide range of animals. These included the Javan rhinoceros *Rhinoceros sondaicus* and the wild dog *Cuon alpinus*, now both extinct in Borneo. The caves seem to have been abandoned between 7,000 and 3,000 B.P. This may reflect a shift towards coastal settlements, perhaps because the warmer and wetter climatic conditions of the early Holocene led to an expansion of lowland rainforest across areas which may have been slightly drier and more open during the late Pleistocene (Bellwood 1985). As the vegetation changed, early people may have preferred to move to more open coastal sites where fishing and food gathering were easier. After 3,000 B.P. the inhabitants of Madai and Baturong joined a network of agricultural societies with pottery, metallurgy (dating from 2,200 B.P. when the early Metal Period or Iron Age began here) and increasing involvement in trade (Bellwood 1988).

Since time immemorial, caves have been regarded as sacred, and valued as places of worship, the home of ancestral spirits and burial sites for the dead. The walls of the Painted Cave (Kain Hitam) at Niah are adorned with red haematite paintings, depicting "ships of the dead" which bear departed souls. This theme is common to many Dayak peoples, including the Ngaju of Central Kalimantan with their "ship of the dead" culture (Scharer 1963). Boat-shaped coffins also tell of these caves' long-term use as burial caves. Humans are believed to have used the Niah caves as long ago as 37,500 B.P., and some of the coffins in the Painted Cave date from the Mesolithic period 20,000 to 4,000 B.P. The Tapadong caves in Sabah were also prehistoric burial sites.

Cave burials continued in limestone caves in Sabah and in the Sangkulirang area in East Kalimantan until very recently. On the Segama River in Sabah, the local Dusuns buried their dead in the limestone cave of Batu Balus. After a period of mourning, the well-wrapped body was placed in a heavy coffin of *Intsia palembanica* wood within the cave and surrounded with pots and weapons for the use of the departed soul on his journey to the afterworld (MacKinnon 1974b). Batu Balus and the Tapadong caves were allegedly protected by dragons, a legend which arose from the fact that king cobras, up to five metres long and able to stand taller than a human, lived within the caves. Archaeological investigations of burial caves have provided valuable artifacts, Chinese pots and burial jars that reveal much information about the past cultures of Bornean tribes and their trading ventures (Harrisson and Harrisson 1971).

CAVES AS A RESOURCE

Caves are often regarded as damp, eerie and rather threatening places, but nevertheless, they are a valuable resource. They provide economic benefits to local communities, sites for archaeological and biological research, and opportunities for recreation, education and tourism, as well as an important focus for some traditional cultures.

The bats and swiftlets at the centre of the cave food web may be of considerable benefit to local farming communities. Some fruit bats roost in caves, including the cave nectar bat *Eonycteris spelaea* and the greater nectar bat *Eonycteris major*. These bats fly long distances daily in search of flowering trees. They feed on pollen and nectar and are important pollinators for many of the region's fruit trees, including durians, guava (jambu) and petai. The many insectivorous bats that roost in caves may benefit local people by preying on insects that are agricultural pests or vectors of disease. The bats from Niah are estimated to consume 7,500 kg of insects a day, thereby exercising some degree of insect control.

Swiftlet populations may also be important in controlling insect populations, including insect pests (Francis 1987). The 1.5 million swiftlets that nest at Niah are believed to consume in the order of 11,000 kg of insects a day, even more than the bats. Nests of the white-nest and black-nest swiftlets also provide a valuable harvest. Even the less valuable black nests fetch about $400 per kg at Batu Niah.

Edible birds' nests are a valuable export from Indonesia, with more than 62,000 kg exported in 1983 (de Beer and McDermott 1989). A detailed study of the collection and trade of swiftlet nests in Sabah estimates the value of exported nests to be more than $1 million each year (Francis 1987). However, harvests seem to be declining. The total harvest of black nests from Gomantong, Batu Timbang and Madai, the only Sabah caves to produce large numbers of black nests, is now only about 15,000 kg, whereas the total exports from Sabah of black nests in 1977 and 1978 reached almost 24,000 kg per year (Francis 1987). It is difficult to get accurate figures for collection of swiftlet nests in Kalimantan, but exports in the early 1970s were estimated at about 12,000 kg per year (Jessup and Peluso 1985). Export figures have declined considerably in recent years, which may reflect a decline in harvests due to overexploitation or simply an increase in unrecorded trade. Some black nests are exported from East Kalimantan to Sabah, and white nests sell in Sangkulirang for more than Rp 1 million ($500) per kilogram.

Swiftlet nests have been exported from Borneo for several centuries. Trade in birds' nests with the Chinese probably began in the sixteenth century (Medway 1963) and was at its peak from 1750 to 1850, during the height of the Sulu Sultanate (Piper 1988). Rights to harvest birds' nests from specific caves are usually controlled by families or groups of villagers, so that this resource is more privately controlled than other forest products (Jessup and Peluso 1985).

Guano deposits formed from the droppings of bats and swiftlets contain concentrated phosphates and are extracted as fertiliser in many areas. Although guano is widely believed to be a valuable fertiliser, it is very variable in quality and chemical composition. To grow crops a consistent fertiliser with the right balance of nitrates, potassium and phosphates is required. Guano contains considerable phosphate but is too poor in other minerals to be a complete fertiliser; moreover, only some of the phosphate is water soluble (Wilford 1951). Guano can be used on a small scale in home gardens, but much of its value is as an organic mulch rather than as a fertiliser.

At Niah Cave, where the Sarawak Museum operates a guano cooperative, huge quantities of cave soil have been removed, in many places down to the rock. Careless mining makes access difficult for visitors, prevents the harvest of many edible birds' nests by undermining collectors' climbing structures, has damaged the archaeological potential of the caves, and has affected the invertebrate fauna (Chan et al. 1985; Francis 1987). The guano from the Niah caves is used on the Sarawak black pepper fields but fetches very low prices. Guano collection seems to be on the decline both in Sarawak and in some areas of Kalimantan. In the past guano was collected from the Batu Hapu Cave in South Kalimantan, but this practice has now stopped, probably because of the greater ease of using more reliable commercial fertilisers.

Aside from their economic importance, caves are of great value for scientific research. Archaeological excavations at caves in Sarawak and Sabah have told us much about human prehistory in Borneo, about the early fauna and about climatic conditions (Harrisson 1966; Medway 1977b). Caves are valuable sites for biological research since their relatively simple ecosystems provide a model for studying ecological inter-relationships. Cave systems are often scenically very beautiful, with marvellous grottos, spectacular stalagmites and stalactites and other limestone formations. They have considerable potential for tourism and recreation.

CONSERVATION OF CAVES AND LIMESTONE FORMATIONS

Limestone formations and cave systems form a distinct ecosystem of high conservation value. The limestone flora is rich in species and shows high levels of endemism (Chin 1977). Isolated limestone hills may also show high levels of species endemism among invertebrates, such as in the molluscan fauna of the Batu caves, West Malaysia (Tweedie 1961). Many invertebrates are endemic to caves. The Mulu expedition discovered five new species of crabs endemic to the Mulu caves (Anderson et al. 1982; Chapman 1984), as well as numerous other invertebrates confined to cave systems, including several new species and endemics (Chapman 1982). Caves may also harbour local populations of rare species such as the blind, hairy earwig, which lives as a parasite on the naked bat. Such local populations are particularly vulnerable to extinction if their habitat is disturbed.

Box 6.1. The Sangkulirang limestone formations.

The Sangkulirang Peninsula, an area of thick Tertiary sediments, is the most extensive area of limestone in Borneo. The scenery is spectacular, with tower karst where sheer-sided limestone cliffs rise for several hundred metres above the surrounding plains. The limestone hills vary from isolated blocks to extensive mountain ranges rising to irregular summits of 600 m, with the highest peak at 1,100 m. The limestone scarp followed by the aircraft flight path from Samarinda to Tanjung Redeb is a dramatic spectacle of crested ridges, limestone pavements and jagged needles of eroded rock. This is one of the few areas in East Kalimantan where forests are relatively untouched and rivers run clear.

The forests of the limestone hills harbour an interesting flora, not represented elsewhere in the Kalimantan protected area network. The proposed Sangkulirang nature reserve (*cagar alam*) includes several scenically beautiful, and botanically unique, forested limestone hills within the Sangkulirang Peninsula. Other limestone formations to the west, around Muara Ma'au, lie within 100,000 ha of forest currently designated as protection forest. No ground surveys have been made to mark boundaries, and the entire 200,000 ha of the proposed Sangkulirang reserve system is merely a series of vague lines drawn on a map (MacKinnon and Artha 1981).

Parts of the limestone forests, especially on the peninsula itself, were damaged by the 1982-83 fires that raged through East Kalimantan. The area is further threatened by logging, timber stealing and the potential for quarrying lime to produce cement. In this latter context it is heartening that the builders of the PROJASAM road were not allowed to quarry limestone for road building. Transmigration plans also threaten the lowland forests of the peninsula, although these will probably not be developed within the next ten years. Other threats include the proposed extension of the PROJASAM road north to Tanjung Redeb, and the quarrying of limestone to mix with cargoes of coal exported from the Sengatta River mines.

No detailed floral surveys have been made in the area, but the limestone flora is expected to be of special interest. Impressive limestone caves and galleries occur within the limestone hills, and some of these in the PT Sangkulirang concession have already been explored by a French team (ESFIK 1983). In the Muara Wahau area at Gunung Kembang there are old burial caves, some with fifth-century Hindu statues (Boyce 1986). Many caves have important populations of swiftlets whose edible nests are harvested and fetch a high price for birds' nest soup (Jessup and Peluso 1985). Rattan and gaharu (aloe wood) are also collected extensively in the area. The lowland forests of the Sangkulirang Peninsula are rich in wildlife, with gibbons, banteng, proboscis monkeys, orangutans, pig-tailed macaques, bears and crocodiles. The peninsula supports an interesting and unusual semi-montane bird fauna, although most of the area is below 200 m, and the highest summits reach only 600 m (Holmes pers. com.).

Protection of the hill forests, limestone escarpments and mountain ranges will also protect the watershed and water quality, especially important if the planned transmigration and agricultural schemes go ahead. Floral and faunal surveys should be conducted as a matter of urgency, to identify the best remaining areas of forests and finest limestone areas so that these areas can be excluded from logging concessions. The limestone areas of Sangkulirang are some of the most valuable areas for conservation in Kalimantan (MacKinnon and MacKinnon 1986). The establishment of a well-protected reserve in this area would strengthen the whole Kalimantan reserve system and help to protect biodiversity by ensuring that all major habitats were included within the reserve network.

In addition to their value in conserving endemics, and thereby maintaining biological diversity, caves are important because of the shelter they afford swiftlet and bat communities, both of which bring benefits to adjacent rural communities.

The structure and communities of caves are complex but fragile, and particularly vulnerable to disturbance. Cave habitats are threatened by several factors: vandalism and excessive visitor use (KPSL-UNLAM 1989c); overharvesting of guano and birds' nests (Francis 1987); and direct destruction of the limestone to make lime, cement and marble chips. In the Sangkulirang area limestone formations are threatened by proposed transmigration schemes, proposals for cement factories, and the use of limestone for road surfaces.

Limestone areas have interesting and often very beautiful landscapes. Limestone hills and their associated formations and caves have considerable tourist potential, though great care needs to be taken to ensure that visitors do not destroy the very resource they come to enjoy. Many tourists visit Gunung Mulu National Park in Sarawak solely to visit the spectacular Deer and Clearwater caves, and many also make the arduous climb up Gunung Api to view the Pinnacles. Niah Cave has also been developed as an important tourist site. Visitor use brings benefits to the local economy, increases the understanding of the need for conservation, and engenders local and international support for the conservation areas. With proper management the Sangkulirang and other East Kalimantan limestone formations could also become attractive venues for tourists interested in visiting the more remote areas of Indonesia and seeing Dayak villages and old burial caves.

Borneo's most spectacular limestone scenery is found in the Middle Baram Valley (including Mulu National Park) and the Sangkulirang limestone formations of East Kalimantan. Although much of the island's limestone habitat is relatively untouched, only 4% is protected within national parks and reserves (MacKinnon and MacKinnon 1986). At present the only major conservation area in Borneo that protects extensive areas of limestone habitat is the Gunung Mulu National Park in Sarawak. Limestone hills and caves are also protected in the smaller Niah National Park in Sarawak. Proposals to create a major reserve in the Sangkulirang limestone (MacKinnon and Artha 1981) should be implemented as soon as possible.

Chapter Seven

Mountain Habitats

Unlike Java and Sumatra, where the landscape is dominated by volcanoes, Borneo has few very high mountain peaks. Much of the island, especially in Kalimantan, consists of vast expanses of flat, alluvial plains extending down to low-lying coastal swamplands. Mountain chains extend in an inverted Y-shape through the centre and north of the island. The Schwaner Mountains of Central Kalimantan join the western chain of the Kapuas Hulu Mountains along the Sarawak/Kalimantan border and then extend northwards via the Iran Range to the Crocker Range of Sabah. A southeastern fork extends into the Meratus Mountains of South Kalimantan, an area of floral distinctiveness.

Few of Borneo's mountains exceed 2,000 m, and only five are higher than 2,500 m. Three of these are in Sabah: Mount Kinabalu, at 4,101 m Borneo's highest mountain, Trus Madi (2,597 m) and Tambuyukon (2,576 m) in the Crocker Range. At 2,987 m Gunung Makita (Batu Ikeng), near Long Nawan in East Kalimantan (Voss 1983), and Gunung Siho (2,550 m), in the Kayan-Mentarang reserve, are the highest peaks in Kalimantan. Borneo's highest mountains are shown in figure 7.1.

Most of Borneo's mountains fall within one central biogeographical unit (see fig. 1.22) which contains the distinct montane flora and fauna of the island (MacKinnon and MacKinnon 1986). The Bornean mountains can be likened to islands in a sea of lowland rainforests. Because of this isolation, a number of unique montane species have evolved. Of Borneo's endemic bird species, 23 (73%) are montane, as are 21 of the island's 44 endemic mammal species (tables 1.6 and 1.7). With the exception of Mount Kinabalu and Gunung Mulu, many of Borneo's mountains are relatively unexplored. This is particularly true in Kalimantan, where few of the mountain ranges have been surveyed scientifically. Bukit Raya, 2,278 m, is probably the best studied of the Kalimantan high mountains and is noted as an area of rich biodiversity (Nooteboom 1987; K. MacKinnon 1988).

Mount Kinabalu is known to be particularly rich in plant and animal species, with many species once thought to be endemic to the mountain. At least some of this "uniqueness" may be attributed to the fact that Kinabalu is so much better explored than any of the other Bornean peaks. Certainly, as more of the Kalimantan mountains are explored, the known

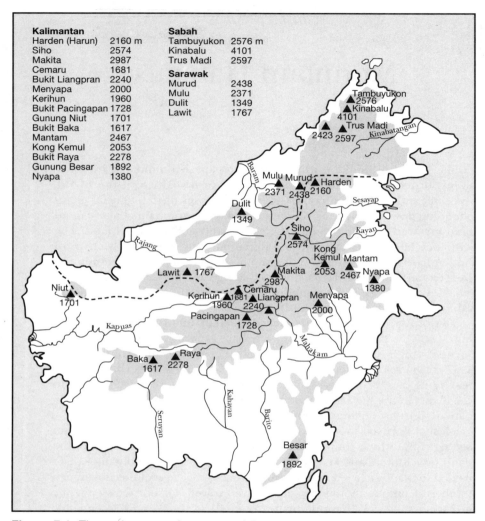

Figure 7.1. The main mountain ranges of Borneo, showing major peaks (height in metres above sea level).

ranges of some montane species are likely to extend. Recent expeditions to Bukit Raya, Gunung Niut and Barito Ulu have added to the Kalimantan list several montane birds, previously recorded only from Mount Kinabalu and other high mountains in Sabah and Sarawak (Prieme and Heegaard 1988; Rice 1989; Wilkinson et al. 1991).

Climate

The physical and biological differences between the hot, humid, lowland forests and the exposed mountain habitats determine which species can survive there. With increasing elevation the climate becomes both cooler and moister. For every 1,000 metres rise in altitude, the temperature drops about 5°C, which is equivalent to a ten-degree shift in latitude away from the equator. This is one reason why the mountain flora of tropical mountains includes plants more normally associated with temperate regions. Apart from an overall drop in temperature, montane species must cope with much more drastic daily changes in temperature — depending on cloud cover, time of day and amount of water vapour in the air — than lowland species.

At high altitudes the sun's rays lose less energy as they pass through the thinner atmosphere. Insolation at ground level is very intense, so that the temperature near the ground is much higher than the surrounding air temperature. It is this warmer microclimate that allows plants and smaller animals to survive in alpine zones and in such inhospitable habitats as the bare granite summit of Kinabalu. The ground heat is lost quickly by radiation at night, and daily temperature ranges may be as much as 15°C to 20°C at very high altitudes. Occasionally frost forms on the summit of Mount Kinabalu at night. Signs of glaciation on Mount Kinabalu indicate a cooler climate in the past (Myers 1978).

Other climatic factors that affect the distribution and form of plants on mountains are humidity, rainfall and the effects of wind. Rainfall is generally higher on the side of the mountain facing the prevailing wind, and higher on mountain slopes up to 1,500 m than in the surrounding lowlands. Rainfall totals are not the only indication of how much water the vegetation receives. Many plants at higher altitudes also derive moisture from the water droplets in clouds that adhere to their leaves and stems. The dew-point at which condensation occurs, causing clouds or dew to form, depends on both the temperature and the initial moisture content of the air. Since the percentage saturation of a mass of air increases as temperature falls, forests at high altitudes experience high relative humidity, especially at night, when water condenses on leaves.

During the wet season the peak and upper mountain slopes may be enveloped in clouds for days on end. In the drier months, when the air is not saturated with water vapour, a belt of cloud may form at about 2,000 m or lower, while the summit of higher mountains like Kinabalu remain clear. In the moisture-saturated environment of the upper mountain slopes mosses and lichens thrive, creating spectacular moss forests. The height at which cloud habitually settles on the mountain is important because cloud prevents bright sunlight from raising leaf temperature and also reduces the amount of radiation available for photosynthesis.

Winds on upper mountain slopes are another factor with which plants and animals have to cope, particularly in the exposed habitats above the treeline. Winds have both a cooling and a drying effect, reducing the ambient temperature and causing increased water loss from transpiring plants. Plants, and animals, able to survive at high altitudes have evolved physical, physiological and behavioural adaptations to cope with the extreme environment. Plants tend to decrease in size with altitude and on the highest slopes produce stunted, ground-hugging alpine forms that survive in sheltered hollows.

It has been suggested that the high levels of ultraviolet radiation received by subalpine and alpine zones, combined with the altitudinal shifts of vegetation during the Pleistocene, may have led to accelerated rates of mutation and speciation among montane plants (Lee and Lowry 1980). This is one possible explanation for the high levels of endemism found in the upper vegetation zones of tropical mountains. Of the subalpine plants on Mount Kinabalu 40% have been found only on that mountain (Smith 1970) but again this may reflect the more intensive survey efforts on Kinabalu rather than floral uniqueness.

Mountain Soils

The character of mountain soils changes with increasing altitude, generally becoming more acid and nutrient-poor, especially where acid peat is present. Decomposers such as termites occur at lower densities in high montane forests (Collins et al. 1984b), and the rate of decay and mixing of leaf litter is therefore much slower than at lower elevations. As a result, peat tends to accumulate in wetter localities in the cloud belt or upper montane zones, where waterlogged soils and anaerobic conditions further inhibit decay.

The soils on the summit, ridge tops and knolls, which receive water only from the atmosphere, are drier and more nutrient-poor than soil in dips and on the lower slopes, which receive input from groundwater percolating from rocks above (Burnham 1984). Differences in bedrock composition and climate are the major factors influencing soil formation at different heights up a mountain, with steepness of slope and openness of vegetation cover also important factors. Low temperatures slow down the process of soil formation through reduced evapotranspiration, slower chemical reactions and reduced density of soil organisms (Burnham 1984).

Productivity and Nutrient Cycling

With increasing altitude there is a decrease in biomass, plant growth rate and rate of production of leaf litter. In any type of forest there are two types of mineral turnover: rapid cycling in the small litter (leaves and twigs) and in throughfall (rain reaching the forest floor); and much slower cycling from breakdown of dead trunks and large branches.

A study of productivity and nutrient recycling in a montane rainforest in New Guinea revealed that rainfall provided the major input of minerals to the montane forest (Edwards and Grubb 1982). The amount of throughfall which reaches the ground varies with rain intensity. Throughfall contains minerals leached from the leaves and bark surface and from decomposition products. The other source of minerals is litterfall, but trees in mountain forests may re-absorb as much as half of the minerals from their leaves before shedding them. Thus lower quantities of minerals cycle through leaves in montane forests than in lowland forests. In the New Guinea study about 15% of nitrogen and 31% of phosphorus found in live leaves was re-absorbed before abscission (Edwards and Grubb 1982). The recycling of these two minerals may indicate they are limiting in at least some montane soils.

Whereas the lower limit of a particular vegetation type is determined by the ability of individual constituent species to compete with species of lower forest formations, the upper limit depends on a variety of factors including mean temperature and diurnal variation, soils, aspect and exposure to wind. Although low temperature, periodic drought and high radiation are the factors primarily responsible for limiting montane forest distribution and growth, the mineral supply is also limiting. In upper montane forests trees are stunted and gnarled, similar to those found on infertile soils in the lowlands. The slow rate of litter decay and subsequent humus mineralisation probably locks up plant nutrients (especially nitrogen and phosphorus) in forms unavailable for plant growth. Forest types are found at lower elevations, and forest growth is less on sites such as ridges, likely to be poor in nutrients (Corner 1978). The smallest trees occur on peaty soils of very low pH, which give poor conditions for plant growth; some of the species found on peat soils in upper montane soils also occur in similarly acid, infertile kerangas soils. Several plant species (e.g., ant plants, rhododendrons, sundews and pitcher plants) have adopted strategies to overcome the lack of nutrients in montane soils (Argent et al. 1988).

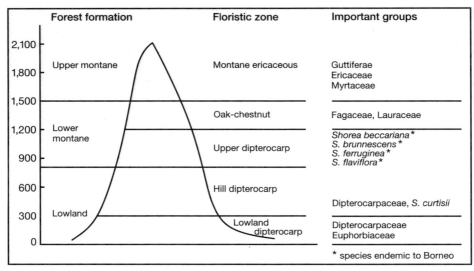

Figure 7.2. Vegetation zones on the main mountains of Borneo. (After Whitmore 1984a.)

Zonation of Mountain Forests

As one climbs a tropical mountain, one passes through successive, distinct vegetation zones with different species, different structure and different appearance (fig. 7.2). On the lower slopes the mountain is clothed in rainforest similar to that found in the surrounding lowlands, though, as one climbs higher, certain species are absent. Above this zone are forests dominated by chestnut *Castanopsis* and oak associations *Lithocarpus* and *Quercus*. Myrtaceae are also important in the lower montane forests. Even higher is the stunted upper montane forest, which gives way to the ericaceous belt and alpine meadows and, ultimately, on a high mountain like Kinabalu, to bare open slopes where plant life can survive only in sheltered hollows. Many plants cross the ecotones (transition zones) but may occur in different forms at different altitudes. Thus above Sayat-Sayat, at 3,700 m on Kinabalu, mountain gelam *Leptospermum* is a ground-hugging shrub only 15 cm high, but lower on the mountain it occurs as a small tree.

Such vegetation zones can be recognised on all tropical mountains, though they are not dictated by altitude alone (Bratawinata 1984). On an isolated low mountain all zones of flora and vegetation are narrower, while on an isolated high mountain, or in the central part of a mountain range,

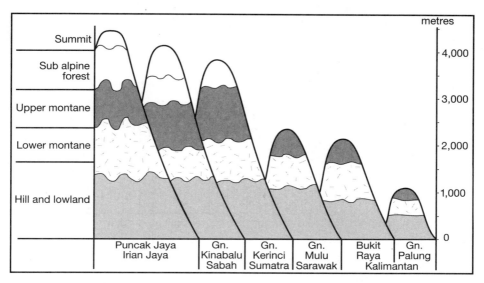

Figure 7.3. Vegetation zones on Bornean mountains, and other Indonesian peaks, to show the compression of vegetation zones on smaller mountains (the Massenerhebung effect).

they are broader (fig. 7.3). This is the **Massenerhebung effect**, first noted in the European Alps. Thus a small mountain close to the sea, such as Gunung Palung, has moss forest at about 800 m, whereas this upper montane forest begins at about 1,200 m on Gunung Mulu and even higher, around 1,800 m, on Mount Kinabalu.

Three major and parallel changes in forest structure and appearance occur with increasing altitude. There is a decrease in forest height and, more slowly, in biomass. The giant emergents of the lowland rainforest are absent from lower montane forest, and there is a further decrease in height in upper montane and subalpine forest (fig. 7.4). This progressive diminution has been described as a three-layered forest being replaced by a two-layered, and ultimately a one-layered forest (Robbins 1968, 1969). Secondly, leaf size decreases. Thirdly, the number of epiphytes increases, especially bryophytes and filmy ferns. There are also parallel changes in the flora (table 7.1).

The most dramatic change, which usually occurs over a relatively short distance, is from a forest of trees with larger or medium-sized leaves (mesophyll) and an uneven, billowing canopy surface, to forest with a lower, flattish crown surface. This upper montane rainforest is dominated by more slender trees with small leaves and often with gnarled limbs and very dense

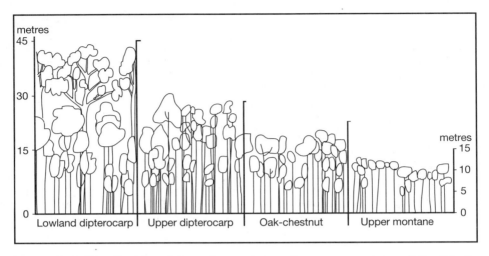

Figure 7.4. Altitudinal forest formation series on Bornean mountains. (After Wyatt-Smith in Whitmore 1984a.)

Table 7.1. Characteristics of four types of forest found on mountains.

	lowland forest	lower montane forest	upper montane forest	subalpine forest
Canopy height	25-45 m	15-33 m	1.5-18 m	1.5-9 m
Height of emergents	67 m	45 m	26 m	15 m
Leaf-size class‡	mesophyll	notophyll or mesophyll	microphyll	nanophyll
Tree buttresses	common and large	uncommon, small, or both	usually absent	absent
Trees with flowers on trunk or main branches	common	rare	absent	absent
Compound leaves	abundant	present	rare	absent
Leaf drip-tips	abundant	present or common	rare or absent	absent
Large climbers	abundant	usually absent	absent	absent
Creepers	usually abundant	common or abundant	very rare	absent
Epiphytes (orchids, etc.)	common	abundant	common	very rare
Epiphytes (moss, lichen, liverwort)	present	present or common	usually abundant	abundant

‡ - the leaf-size classes refer to a classification of leaves devised by Raunkier (1934) and modified by Webb (1959). The definitions are: mesophyll 4,500-18,225 mm^2; notophyll 2,025-4,500 mm^2; microphyll 225-2,025 mm^2; nanophyll less than 225 mm^2. An approximate measure of leaf area is 2/3 (width x length).

Source: Whitmore 1984a

subcrowns (Whitmore 1984a), and can be clearly distinguished on aerial photographs. Upper montane forest is frequently less than 10 m tall and sometimes is described as **elfin forest**. On outlying spurs and isolated peaks upper montane forest occurs at lower elevations and may abut directly onto lowland evergreen forest. On larger, higher massifs there is usually an intermediate formation, lower montane rainforest. This differs from lowland rainforest by having a lower canopy with fewer, smaller, emergent trees and fewer trees with buttresses. Big woody climbers are usually absent, though vascular epiphytes are abundant, and cauliflory is much less common, as are species with compound leaves (Whitmore 1984a).

The upper slopes of tropical mountains are often enveloped in cloud, and conditions are very moist. Trees, ground and rocks are covered in moss. The bog moss *Sphagnum* is sometimes found as one of the ground mosses, frequently associated with ground orchids *Corybas* and bladderworts *Utricularia*. Such moss forest is usually found in the upper montane rainforest but can occur in the lower montane rainforest zone on isolated mountains.

Bryophytes, mostly liverworts and filmy ferns, grow luxuriantly in the moss forest. At lower altitudes filmy ferns are confined to the most humid places in the forest, often near streams. On upper mountain slopes they occur everywhere that clouds gather at night and where conditions are perpetually wet. Filmy ferns are unique in that each frond, apart from its veins, consists of a single layer of cells. In all other ferns, and in all higher plants, leaf blades have a layer of cells (epidermis) on each surface with a system of cells and air spaces between, and openings (stomata) in the epidermis to allow exchange of gases, including water vapour, with the atmosphere. In the filmy ferns, however, the protective epidermis is lacking, and the green cells are in direct contact with the air, so the ferns can only survive in very humid conditions (Holttum 1978).

Changes in forest structure and form at different altitudes are paralleled by changes in species composition (Whitmore 1984a). Families found mainly or entirely in the tropics are restricted mostly to altitudes below 1,000 m. These include Anacardiaceae, Burseraceae, Capparidaceae, Combretaceae, Connaraceae, Dilleniaceae, Dipterocarpaceae, Flacourtiaceae and Myristicaceae (Whitmore 1984a). The mountain flora of Borneo has been derived from both Asia and Australasia (fig. 7.5). Several families of predominantly temperate distribution are found only above 1,000 m (van Steenis 1962, 1972b; Cockburn 1978). These include northern-hemisphere herb genera such as bittercress *Cardamine*, bedstraw *Galium*, gentian *Gentiana*, buttercup *Ranunculus* and violet *Viola*. Genera of southern origin include *Drimys, Gaultheria, Gunnera, Haloragis* and *Nertera*. Tree families commonly found at middle and higher elevations include Aceraceae, Araucariaceae, Clethraceae, Ericaceae, Fagaceae, Lauraceae, Myrtaceae, Podocarpaceae, Symplocaceae and Theaceae. Many of these families are

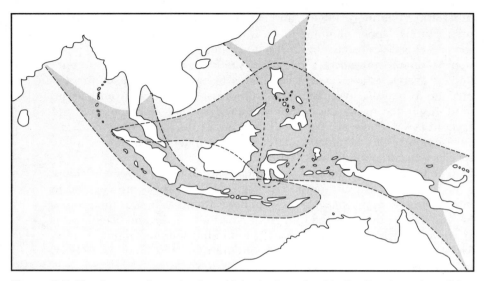

Figure 7.5. The three main routes by which plants arrived in the Sunda region. (After van Steenis 1936, 1972.)

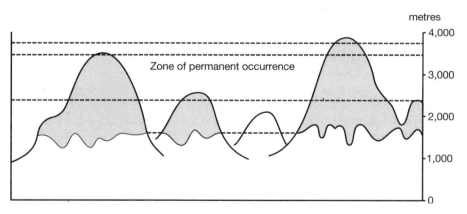

Figure 7.6. The zone of permanent occurrence on mountains and its significance for the distribution of species. Shaded areas represent present distribution of a hypothetical species. The population on the lower slopes depends on a source of seeds from the zone of permanent occurrence. Thus one mountain in a range may lack a species of plant even though that plant occurs at the same altitude on a neighbouring peak.

ZONATION OF MOUNTAIN FORESTS 325

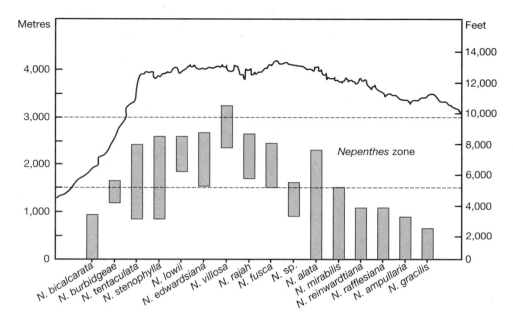

Figure 7.7. Altitudinal distribution of pitcher plants *Nepenthes* on Mount Kinabalu.
Source: Kurata 1976

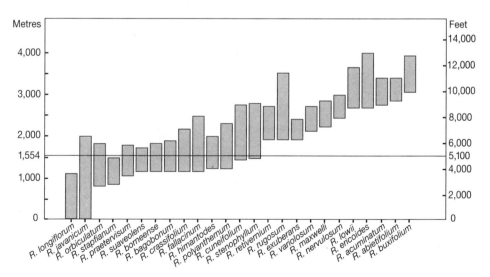

Figure 7.8. Altitudinal distribution of rhododendrons on Mount Kinabalu.
Source: Argent et al. 1988

Table 7.2. Distribution of *Nepenthes* species by forest types.

Species	Lowland	Montane Forest	Heath Forest	Peatswamp Forest	Occurrence
N. alata		X			Kinabalu only
N. albomarginata	X	X	X	X	widespread
N. ampullaria	X		X	X	widespread
N. bicalcarata			l	X	local in heath forest
N. burbidgeae		X			Kinabalu, Tambuyukon
N. decurrens	X				Baram only
N. edwardsiana		X			Kinabalu, Tambuyukon Trus Madi, 1,500-2,500 m
N. fusca		X			1,200-2,500 m
N. gracilis	X		X	X	widespread
N. hirsuta		X	X		600-1,000 m
N. leptochila		X			foothills 500-825 m
N. lowii		X			1,650-2,600 m
N. macrovulgaris		X			ultrabasic, 300-1,200 m
N. maxima		X			600-2,745 m
N. mirabilis	X		X		widespread
N. muluensis		X			Gunung Mulu
N. neglecta	X				Labuan only
N. northiana	X				Bau limestone only
N. pilosa		X			Batu Tiban, 1,200-1,800 m
N. rafflesiana	X		X	X	widespread
N. rajah		X			Kinabalu, Tambuyukon
N. reinwardtiana	X	X			epiphytic on rocks and trees in lowlands
N. stenophylla		X			600-2,660 m
N. tentaculata		X			240-2,550 m
N. veitchii	X	X			100-1,200 m
N. villosa		X			Kinabalu 2,340-3,240 m
Totals	**10**	**18**	**7**	**5**	**25 species**
Kalimantan only					
N. boschiana		X			Mount Sakumbang only
N. campanulata		X			E. Kalimantan limestone
N. clipeata		X			Mount Kelam only
N. ephippiata		X			Batu Lesong, Bukit Raya
N. mollis		X			Mount Kamoel only
Total		**5**			**5 species**

l = local

Sources: Smythies 1965; Phillipps and Lamb 1988

Box 7.1. Rhododendrons.

Rhododendrons are characteristic of the upper montane flora and illustrate some of the strategies that plants have evolved to survive in conditions that are less equable than those of the surrounding lowland forests. Worldwide there are about 900 species of rhododendrons, with about one-third, the *Vireya* rhododendrons, confined mainly to Southeast Asia. Of these, 50 species occur in Borneo. Twenty-five species of rhododendrons grow on the slopes of Mount Kinabalu, and five of these are not known from any other mountain (fig.7.9).

Vireya rhododendrons have scales, the function of which may be to help protect young, developing leaves from intense solar radiation, which is especially strong on tropical mountains. Other problems encountered by upper montane plants are how to ensure pollination and seed dispersal in a habitat where animals are less common than in tropical lowland forests. Rhododendrons have long-tailed seeds which are dispersed by wind. The long tails increase the buoyancy of the seeds and allow them to float on air currents. As wind dispersal is unreliable, rhododendrons must produce large numbers of seeds; this requires efficient mass transfer of pollen. The pollen grains are stuck together in groups of four; these tetrads are joined in strings held with sticky, viscous threads. A single pollinating animal brushing against the stamen may remove all the pollen from a pollen sac, and sometimes all the pollen from one flower. The pollen can then be transported as a mass to a receptive stigma. This method of mass pollen transfer increases the chances of successful pollination in a habitat where each flower may be visited infrequently by only a few pollinators.

Although various hybrid rhododendrons are known, the likelihood of cross-pollination occurring between species is reduced by: the physical separation of rhododendrons by geography or altitude; different flowering times; and different flower shapes and colours attracting different pollinators. For New Guinea rhododendrons, insects are important pollinators at lower altitudes, and vertebrates higher up. At cooler temperatures on higher mountain slopes, warm-blooded birds are less restricted in their movements than cold-blooded insects. The bright reds and yellows of the rhododendrons are characteristic of bird-pollinated flowers, and birds seem to be important rhododendron pollinators on higher mountains in Sabah. It is uncertain whether bats ever pollinate these flowers. Nor is it known how the birds cope with the rhododendron nectar, which is generally poisonous.

Rhododendrons occur mostly on acid soils and peats. Like other members of the Ericaceae family, rhododendrons form a mycorrhizal association between their roots and a fungus, which enables them to obtain nutrients in acid conditions. Nitrogen uptake is in the form of ammonium ions and not nitrate, the more usual form of nitrogen for plant growth. Alternative strategies employed by other plants in nutrient-poor situations, in lowland forests as well as in montane formations, include having bacterial nodules on the roots to provide nitrogen (as in *Casuarina*), symbiotic associations with ants to utilise ant waste products (in *Myrmecodia* and other ant plants), or trapping and digesting insects to obtain nitrogen (as in pitcher plants *Nepenthes*).

Source: Argent et al. 1988

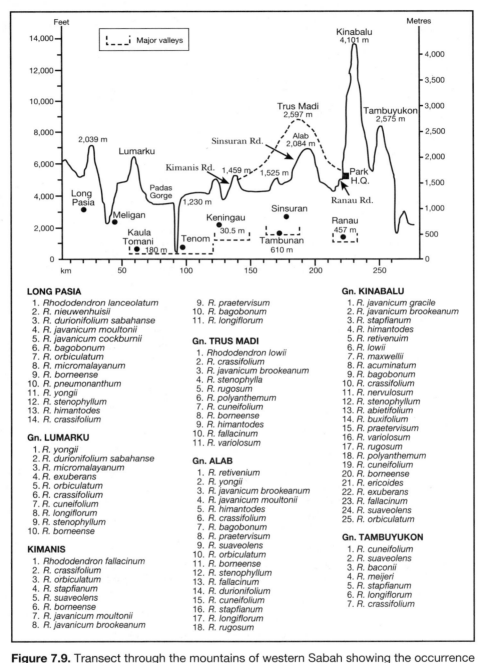

Figure 7.9. Transect through the mountains of western Sabah showing the occurrence of different species of *Rhododendron*.

Source: Argent et al. 1988

well represented in temperate latitudes (Meijer 1971; van Steenis 1972b; Whitmore 1984a). A plant species recorded on one mountain sometimes may not occur at a similar altitude on a neighbouring peak (fig. 7.6).

Upper montane forests share many common species and features of structure and appearance with heath forests (Richards 1936). Forests on Brunei mountaintops share several species with heath forests on rocky sandstone ridges at various elevations, including several species of *Eugenia, Rhododendron, Vaccinium, Horsfieldia polyspherula* and *Tristaniopsis pentandra* (Ashton 1964). Yet none of the heath forest dipterocarps extends above 1,500 m, and several species of mountain ridges are not found in lowland heath forest. It seems that some species occur only in lowland or montane habitats, whereas others respond to environmental features (e.g., a periodically dry environment, acid and nutrient-poor conditions, or both) that are common to both heath and montane forests (see chapter 5).

A few plants span the complete altitudinal range from lowlands to upper montane rainforest, for example, *Dacrydium beccari subelatum*, which is found on mountain hillsides, in lowland heath forest and in the Lawas peat swamp forest of Sarawak (Whitmore 1984a). More commonly, species are restricted to a single forest formation. Analysis of 1,759 species of plants from 43 families in Peninsular Malaysia showed that most species are restricted in occurrence to a single forest formation, with only 3% occurring in both the lowlands and lower montane forest and fewer than 1% extending from the lowlands to upper montane forest (Whitmore 1984a). Within plant groups many species show altitudinal zonation. This can be seen for lichens at Gunung Mulu (Sammy 1980), and pitcher plants (fig. 7.7) and rhododendrons (fig. 7.8) on Mount Kinabalu. Twenty-three of the 30 *Nepenthes* recorded for Borneo are montane species (table 7.2); almost every major mountain has its own form or species.

ZONATION OF ANIMALS ON TROPICAL MOUNTAINS

Just as the mountain plants can be zoned according to altitude, so too can the animals, although few are exclusive to any one zone. The main problems facing animals on upper mountain slopes are adverse climate, lack of shelter and shortage of food. Mammals can overcome the heat loss problem by growing thick, shaggy coats, but their distribution is still limited by food availability determined by the upper limit of their favoured food trees or animal prey. Because of the generally harsher conditions at higher altitudes, many fewer species of plants and animals are found on upper mountain slopes than in the surrounding lowlands. Indeed, the greatest richness of mammal species is concentrated in lowland forests below 350 m. According to Stevens (1968), 52% of the mammals of Peninsular Malaysia

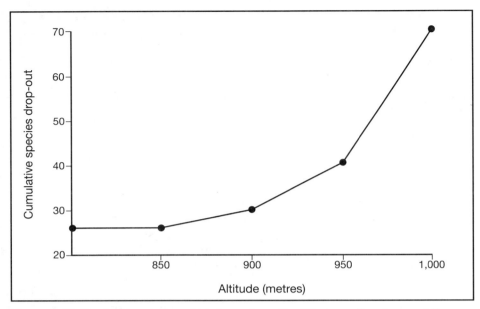

Figure 7.10. Decline in number of bird species with altitude on the slopes of Gunung Mulu, up to the lower edge of the hill buffer zone. (After Wells et al. 1979.)

do not occur above 350 m, and 81% are restricted to altitudes below 660 m.

On Gunung Mulu (2,371 m), Sarawak, there is a progressive reduction in bird species with increasing altitude, ranging from 171 species recorded in the lowland rainforest to only 12 species at the boundary of the upper montane forest at 1,300 m, where a different avifauna takes over (Wells et al. 1979); most species drop out above 900 m (fig. 7.10). Similarly, on Mount Benom in Peninsular Malaysia, only 3% of lowland birds reached 1,200 m (the lower part of the oak/chestnut forest), and the species of the upper parts of the lower montane forest were quite distinct from the lowland species (Medway 1972c). Nevertheless, these zones are not rigid; some "lowland" species occurred up to 850-900 m at Barito Ulu, where they overlapped with species characteristic of hill and submontane forests (Wilkinson et al. 1991). In New Guinea, Kikkawa and Williams (1971) found a distinct discontinuity in bird species at about 1,500-2,200 m corresponding to the change from lower to upper montane forest. Table 7.3 lists Bornean bird species dependent extensively or exclusively on montane forest.

Within the lowland forest at Mulu, there was a dramatic break at 130 m where 26 lowland bird species dropped out, just where the mountain begins to rise sharply from the undulating lowlands (Croxall 1979; Wells et al. 1979). Breeding success for most species also seems to decrease from this steepland boundary, with reduced fledgling success. Possibly the steep-

ZONATION OF ANIMALS ON TROPICAL MOUNTAINS   331

Table 7.3. Bird species dependent extensively or exclusively on montane forest.

Species	Southeast Asia	Borneo	Peninsular Malaysia	Sumatra	Java
Ictinaetus malayensis		X	X	X	X
Accipiter virgatus		X		X	X
Spilornis kinabaluensis	X	X			
Caloperdix oculea	X	X	X	X	
Arborophila hyperythra	X	X			
Haematortyx sanguiniceps	X	X			
Ducula badia		X	X	X	
Macropygia ruficeps		X	X	X	X
Cuculus sparverioides		X	X	X	
Cuculus saturatus		X	X	X	X
Otus spilocephalus		X	X	X	
Otus brookei	X	X		X	X
Glaucidium brodiei		X	X	X	
Batrachostomus poliocephalus	X	X		X	
Collocalia gigas	X	X	X	X	X
Harpactes whiteheadi	X	X			
Megalaima monticola	X	X			
Megalaima pulcherrima	X	X			
Picumnus innominatus		X	X	X	
Psarisomus dalhousiae		X	X	X	
Calyptomena whiteheadi	X	X			
Coracina larvata	X	X		X	X
Chlamydochaera jefferyi	X	X			
Pericrocotus solaris		X	X	X	
Chloropsis flavinucha	X	X			
Pycnonotus flavescens		X			
Dicrurus leucophaeus		X			
Oriolus cruentus	X	X	X	X	X
Oriolus hosei	X	X			
Cissa thalassina		X			X
Dendrocitta occipitalis	X	X		X	
Trichastoma pyrrhogenys	X	X			X
Napothera crassa	X	X			
Stachyris nigriceps		X	X	X	
Garrulax lugubris	X	X	X	X	
Garrulax palliatus	X	X		X	
Garrulax mitratus	X	X	X	X	
Pteruthius flaviscapis		X	X	X	X
Yuhina castaniceps		X			
Brachypteryx montana		X		X	X
Zoothera citrina		X			
Zoothera everetti	X	X			
Turdus poliocephalus		X		X	X
Myophoneus glaucinus	X	X		X	X
Seicercus montis	X	X	X	X	
Phylloscopus trivirgatus	X	X	X	X	X
Orthotomus cuculatus		X	X	X	X
Cettia whiteheadi	X	X			
Cettia vulcania	X	X		X	X
Bradypterus accentor	X	X			
Rhinomyias ruficauda		X			
Rhinomyias gularis	X	X			
Eumyias indigo	X	X		X	X
Ficedula hyperythra		X	X	X	X
Ficedula westermanni		X	X	X	X
Muscicapella hodgsoni		X	X	X	X
Rhipidura albicollis		X	X	X	
Pachycephala hypoxantha	X	X			
Arachnothera juliae	X	X			
Dicaeum monticola	X	X			
Zosterops atricapilla	X	X		X	
Chloracharis emiliae	X	X			
Erythrura hyperythra		X	X		X

Source: Wells 1985

land forest here has inadequate food resources, and the bird population is maintained by continuous immigration from the lowlands. This could have important implications for conservation and reserve design to protect montane ecosystems.

The decrease in species numbers up the mountain appears to result from loss of individual species as suitable resources and habitat diminish. On Mulu only six species pairs show congeneric, competitive replacement, far fewer than reported for New Guinea mountains (Kikkawa and Williams 1971). At Mulu pittas and *Rhiomyias* flycatchers are not replaced by others of the same genera, as in Peninsular Malaysia (Wells et al. 1979). Although the total number of species drops with increasing altitude, individuals of a single montane species may be common, up to five pairs per hectare on Mulu (Croxall 1979).

In a study of New Guinea forests, the niches occupied by forest birds changed up the mountain. With increasing altitude the proportion of tree-nesting insectivorous species increased, tree-nesting frugivores decreased and omnivores stayed about the same. The proportion of predatory birds decreased, while the proportion of ground-living birds increased (Kikkawa and Williams 1971). These changes can be explained by decreasing food resources, especially of fruit and small vertebrates, up the mountain.

Mammal surveys at Mulu revealed decreasing species richness with altitude. Primates were restricted to lowland forest and lower montane formations. In Sabah surveys showed primate biomass to be considerably less in upland areas above 500 m (Davies and Payne 1982). Some species, such as long-tailed macaques and orangutans, were not recorded in upland forest sites, and all other primates, except pig-tailed macaques, were much less common (table 7.4; fig. 7.11). Primate distribution on Mount Benom, West Malaysia (fig. 7.12), could be related to food supplies and the energy costs of travel in mountain habitats. The altitudinal limit for the white-handed gibbon *Hylobates lar* was lower than that for the larger siamang, and could be related to the smaller gibbon's greater reliance on fruit sources, which became increasingly scarce at higher altitudes. Similarly, the banded langur *Presbytis melalophos*, with its greater dietary versatility, was found at higher altitudes than the dusky langur *Presbytis obscura* (Caldecott 1980).

Invertebrates show similar altitudinal zonation. The numbers of species of butterflies decline with altitude on Mount Kinabalu (fig. 7.13) and Gunung Mulu, especially towards the lower/upper montane forest ecotone. Species composition also changes. There is a distinctive upper montane forest butterfly fauna (Holloway 1984). Changes in soil macrofauna also occur with increasing altitude on Gunung Mulu (Collins 1980a). There is a marked decrease in total numbers of soil invertebrates, due mainly to a gradual reduction in the abundance of ants and termites (figs. 7.14 and 7.15). The lowland forest is dominated by termites, beetles and earthworms, with ants the dominant predators. Earthworms and some beetles replace ter-

Table 7.4. Biomass density estimates of diurnal primates in Southeast Asian primary rainforest habitats (numbers represent kg/km²).

AREA	Lowland (below 150 m)				Upland (150-450 m)			Highland (450-900 m)	
	SABAH 4 sites (a)	KALIMANTAN Kutai (b)	W. MALAYSIA 5 sites (c)	W. MALAYSIA K. Lompat (d)	SABAH 6 sites (a)	SABAH Segama (e)	SUMATRA Ketambe (f)	SABAH 3 sites (a)	W. MALAYSIA 4 sites (c)
ORANGUTAN	38	135	–	–	9	150	140	+	–
GIBBONS	52	58	87++	106++	52	46	175++	14	56++
LANGURS	111	78	614	956	90	162	135	24	170
LONG-TAILED MACAQUE	+	19	49	232	37	21	144	–	–
PIG-TAILED MACAQUE	50	42	+	+	44	63	133	56	+
TOTAL (kg/km²)	251	332	750	1294	232	442	727	94	226

+ present, abundance unknown
– absent
++ includes siamang (*Hylobates syndactylus*)

Sources: (a) Davies and Payne 1982, (b) Rodman 1973, (c) Marsh and Wilson 1981, (d) Chivers and Davies 1979, (e) MacKinnon 1974, (f) Rijksen 1978

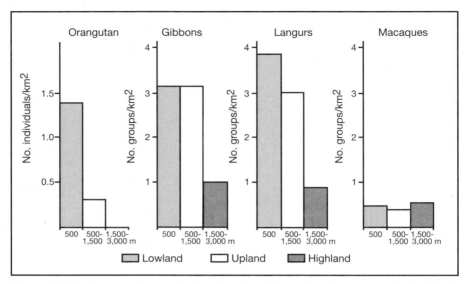

Figure 7.11. Changes in population density of the primate community in Sabah with altitude.
Source: Davies and Payne 1982

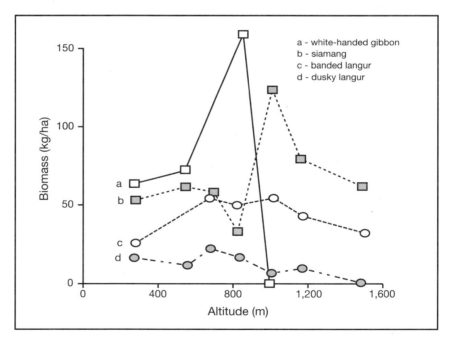

Figure 7.12. Changes in biomass of four primate species on Mount Benom, Peninsular Malaysia. (After Caldecott 1980.)

ZONATION OF ANIMALS ON TROPICAL MOUNTAINS

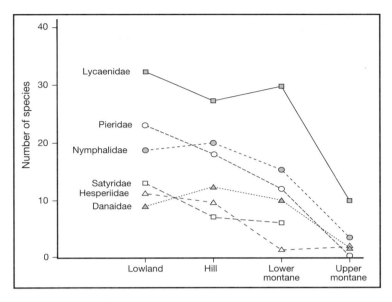

Figure 7.13. Decrease in number of butterfly species of different families with altitude on Mount Kinabalu. (After Holloway 1978.)

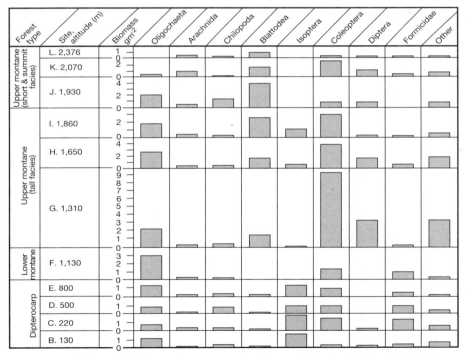

Figure 7.14. Biomass of the main groups of soil macrofauna at different altitudes on the west ridge of Gunung Mulu. (After Collins et al. 1984.)

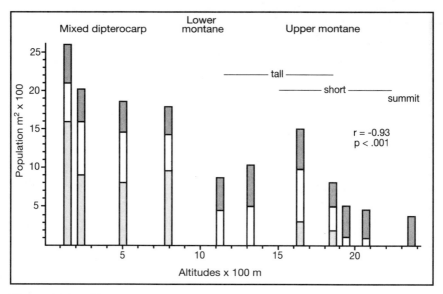

Figure 7.15. Changes in populations of soil macrofauna with increasing altitude along the west ridge of Gunung Mulu. The social insects are dominant: termites (light shading); ants (white); others (dark shading). (After Collins et al. 1984.)

mites as the dominant detritivores in the lower montane forest (at 1,130 m), with ants still the major predators. On the peat soils of the upper montane forest beetles become dominant at 1,300 m, where they are associated with earthworms, fly larvae and cockroaches. Above this altitude, in the more exposed, stunted vegetation zones, several invertebrate groups drop out (Collins et al. 1984; Collins 1989). On the ultrabasic mountain of Gunung Silam, in Sabah, the occurrence of soil invertebrates also shows an altitudinal effect. Earthworms are common on the lower mountain slopes but are markedly absent above 610 m (Leakey and Proctor 1987).

The ecology and distribution of 66 species of dung beetle were studied in different forest formations at Gunung Mulu (Hanski 1983, 1989). Species richness was highest in lowland forests, with the exception of *kerangas* forest, where dung beetles are scarce because of the scarcity of food resources such as dung and carrion (fig. 7.16). Species numbers began to decrease above 200 m, and no species were found at the summit of Gunung Mulu (2,376 m). The community in the lower montane rainforest was transitional between the lowland and upper montane communities, which both had distinct species compositions (Hanski 1983). Although the upper montane forests had few species of dung beetles, they

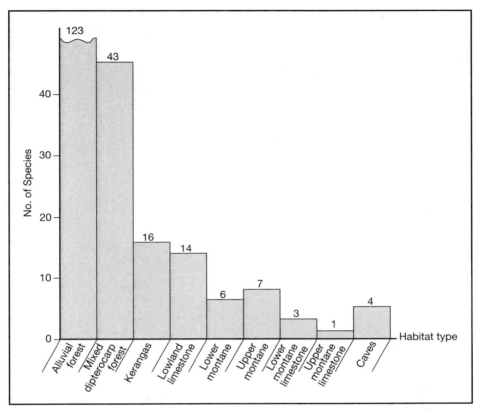

Figure 7.16. Species of carabid beetles collected in different habitat types on Gunung Mulu, Sarawak. (After Stork 1986.)

were very abundant.

The herpetofauna of mountains also shows altitudinal distribution, although the faunal zones are far less distinct than those of other animal groups (Smith 1931, Inger 1978, Dring 1980). On Mount Kinabalu 40 species of amphibians were found above 1,000 m, as were 27 species of lizards and 36 species of snakes, but no reptiles were found above 2,400 m (Smith 1931). Of Kinabalu's frog fauna, 11 species have not been caught above 1,100 m, six others reach their upper limit at 1,300 m, 10 at 1,800 m, five at 2,400 m and three more at 3,200 m (Inger 1978). On Mount Kerinci, in Sumatra, no reptiles or amphibians were found above 2,000 m, and no amphibians above 3,100 m (table 7.5). On Mount Benom, an isolated granite peak in West

Table 7.5. Altitudinal distribution of amphibians and reptiles on Mount Kinabalu.

	1,000	1,100	1,300	1,500	1,650	1,800	2,300	3,100 m
AMPHIBIANS								
Megophrys monticola					*	*		
Megophrys nasuta	*							
M. gracilis	*					*	*	
M. hasseltii						*		
Bufo leptopus	*	*				*	*	
Simomantis latopalmata	*							
Chaperina tusca	*	*						
Rana macrodon	*							
R. kuhli	*	*		*			*	
R. palavanensis	*							
R. luctuosa	*							
R. nicobariensis	*							
R. chalconota	*							
R. jerboa	*	*				*	*	
R. whiteheadi	*	*	*					
R. tuberilinguis	*	*						
Rachophorus leucomystax	*	*	*					
R. acutirostris	*							
Philautus petersi	*			*	*	*		
P. spiculatus	*							
P. mjobergi				*		*	*	
P. tuberilinghuis						*		
P. pintus					*			
P. amoenus						*		
Kalophrynus pleurostigma					*			
Nectophryne altitudinis						*	*	
Nectophryne misera						*	*	
Leptobrachella baluensis					*			
LIZARDS								
Gymnodactylus baluensis	*	*				*	*	
Hemidactylus frenatus	*							
H. garnoti	*							
Hemiphyllodactylus typus	*							
Gekko monarchus	*							
Draco cornutus	*							
D. obscurus	*							
D. fimbriatus	*							
D. formosus	*							
D. quinquefasciatus	*							
Japalura nigrilabris	*	*		*		*	*	
J. ornata	*							
Calotes cristatellus	*	*	*					
Tropidophorus beccarii	*							
Mabuya multicarinata	*		*					
M. multifasciata	*							
Lygosoma nieuwenhuisi	*							
L. variegatum	*	*						
Lygosoma kinabaluensis					*	*	*	
Trimeresurus sumatranus						*		
SNAKES								
Sibynophis geminatus	*							
Zaocys fuscus	*							
Natrix saravacensis	*		*			*		*
N. flavifrons	*		*					
N. chrysarga	*							
N. murudensis	*						*	
Elaphe flavolineata	*							
Gongylosoma baliodeira	*							
Calamaria vermiformis	*					*		
C. leucogaster	*				*			
C. pendleburyi	*							
Passerita prasina	*							
Psammodynastes puverulentus	*							
Amblycephalus laevis	*				*			
Bungarus flaviceps	*							
Naja naja miolepis	*							
Maticora intestinalis	*							
Trimeresurus gramineus	*	*						
T. chaseni	*							
Oreocalamus hanitschi					*			

Source: Smith 1931

Malaysia, no lizards were found above 1,000 m, but five snakes were recorded at this altitude, one of which occurred above 2,000 m. Similarly, seven species of frogs or toads were found above 1,000 m, but only one above 2,000 m (Grandison 1972).

The cooler, moister montane forests support smaller and less diverse fauna than the evergreen lowland rainforests. The carrying capacity of the montane forest is less, and the total animal biomass per hectare is lower. The steepland boundary seems to be an important limit to lowland species (Wells et al. 1979). At present mountain forests are less threatened than lowland forests, but this does not mean that the full range of mountain flora and fauna in Borneo is secure. Without adjacent reservoirs of lowland habitat as a source of immigrants, most mountain forests are unlikely to be adequate to sustain viable populations of lowland faunas, even for species that range over a wide altitudinal range.

Mount Kinabalu - Summit of Borneo

Mount Kinabalu rises dramatically above the coastal plains of northwest Sabah and well deserves the title "summit of Borneo". At 4,101 m, it is the highest peak in Southeast Asia, indeed the highest mountain between the Himalayas and the snow-capped peaks of Irian Jaya, New Guinea. The mountain is split by Low's Gully, a gaping gorge more than 1.5 km deep and 1 km wide, which divides the mountain into two blocks, so that it resembles an old volcano. In fact, it is the youngest nonvolcanic mountain in the world, a granite intrusion formed 15 million years ago by the hardening of a huge ball of molten rock that was forced beneath the sedimentary rocks of the Crocker Range. One million years ago this pluton was thrust upward through the Crocker Range by tectonic movements. The process continues today, and Mount Kinabalu grows half a centimetre each year (Jacobson 1986). The sandstone and shale that once covered the granite have been eroded to reveal the underlying rock.

During the Pleistocene glaciers covered the mountain's summit. They scoured and smoothed the granite plateau, leaving only the jagged peaks above the ice sheets. At Paka Cave (at 3,250 m) clusters of mixed-sized boulders mark the moraine pushed by the head of the advancing glacier. The ice sheet melted less than 10,000 years ago, but since then weather and rain have sculpted the mountain's crags further, to create carved pinnacles and deep abysses (Jacobson 1986). On cold, clear nights ice forms on the Wishing Pool at the foot of Low's Peak.

As there is no record of local people climbing the mountain, the honour of being first to climb the mountain goes to the Englishman, Sir Hugh Low (a government official in the service of Rajah Brooke), who reached the summit plateau in 1851. Although he never reached the

Box 7.2. Orchids.

Orchids are found at all levels of the forest and are common epiphytes in the upper montane forest. It has been estimated that in the tropical belts of the world with high mountain ranges more than 10% of the flowering plants are orchids. The whole island of Borneo has about two thousand species, about one-tenth of the world total. Mount Kinabalu is believed to have between 700 and 1,000 species, ranging from primitive forms like *Goodyera* to the terrestrial *Spathoglottis* and slipper orchids *Paphiopedilum*, and epiphytic orchids on trees, such as, *Dendrobium, Eria, Bulbophyllum* and the beautiful *Coelogyne*, with its hanging inflorescence. These epiphytic orchids have evolved several mechanisms to retain water, such as having large pseudobulbs and thick, fleshy or hard, leathery leaves, as well as insulating tissue on the exposed roots. Other orchids, such as *Arachnis* and *Trichoglottis*, have overcome the problem of water transport along a long stem by producing roots along the stems at the internodes. These adventitious roots are able to anchor the orchid to the tree, as well as absorbing moisture and nutrients from the bark and moss. *Vanda* orchids have shorter stems, very long root systems and straplike leaves to reduce water loss. The beautiful moon orchids *Phalaenopsis* have even shorter stems with long roots, and broad fleshy leaves to catch the little light that penetrates the shady forest where they live. Yet other orchids are saprophytes, depending completely on symbiosis with fungi that break down organic matter to supply their nutrients.

The different vegetative characters of orchids are related to the environment in which they live: soft, fleshy leaves to store water, broad leaves to catch light. These characters do not reveal taxonomic relationships. Orchids are identified by their flowers, whose structure is often complex, designed to attract pollinating insects. Unlike other flowering plants, orchids have the anthers and stamens fused to form a column. One petal is modified to form a labellum or lip. This acts as a landing pad for visiting insects, or sometimes as a trap so that the insect can only leave the flower by brushing against the stigmas and anthers, effecting pollination. Many orchids have nectaries, scent glands and other devices to attract insects, sometimes even mimicking insects themselves to attract specific pollinators.

Orchid seeds are minute and produced in large quantities. They are wind dispersed. For successful germination and growth, orchid seeds must land at a site where they can establish a symbiotic mycorrhizal relationship with certain fungi. Such fungi are found in dead bark on trees and also in mosses, where they break down dead tissues. Fungal hyphae invade the embryo and provide it with sugars and other nutrients for growth. This symbiosis continues during the establishment of the plant.

For growth orchids require light and plentiful moisture, as well as the mycorrhiza that provide nutrients. Several species of orchids may be found together at a suitable site. Competition from other plants, such as ferns on branches, may restrict colonisation by orchids, but some orchids have even been found growing in the birdnest fern *Asplenium nidens* and the staghorn fern *Platycerium*. In tall forest few species are found on the ground because of lack of light; most species occur as epiphytes high in the canopy. Orchids are especially common in the stunted elfin forest and moss forest, where the increased light and moisture produce ideal growing conditions.

Source: Lamb and Chan 1978

highest peak, it is named after him, as are a Kinabalu pitcher plant, a rhododendron and a squirrel. Low and his party camped for the night beneath the overhanging rock ledge of Paka Cave, but today's summiteers can spend the night in warmth and comfort at the park accommodation at Panar Laban, at 3,353 m.

In 1964 Kinabalu National Park was gazetted to protect Mount Kinabalu and its wealth of extraordinary plant and animal life. The park's 745 km^2 (291 square miles) include lowland rainforest at 450 m as well as the mountain itself, thus protecting a wide range of habitats at different altitudes. The visitor leaves the lowland rainforest at Poring for montane chestnut and oak forest at the park headquarters, then climbs through cloud forest and rhododendron thickets before reaching the granite summit, where a few hardy alpines struggle to survive on the windswept pavement. The park includes one of the richest and most remarkable assemblages of plants in the world, with representatives from more than half the families of all flowering plants (Jacobson 1986). The mountain boasts 1,000 or more species of orchids (Lamb and Chan 1978), 25 species of rhododendrons (Argent et al. 1988) and 610 species of ferns, including many spectacular tree ferns (Beaman et al. 1991). The ferns illustrate well the mountain's floral richness. One-fifth of the world's pteridophytes, 200 more species than are known for the whole of North America, have been recorded in Kinabalu National Park (Beaman et al. 1991). Sixteen species of insect-eating pitcher plants occur in the park, 10 on the mountain above 1,500 m (Kurata 1976). Pitcher plants commonly seen on the ascent route of Kinabalu include *Nepenthes lowii* from 1,500 m to 2,500 m, *N. edwardsiana* from 2,400 m to 2,850 m and *N. villosa* from 2,850 m to 3,300 m. Kinabalu also has the richest fig flora of any comparable area in the world, with 78 species. Most of these are not found above 1,800 m, but one species, *Ficus deltoidea* var. *kinabaluensis*, has been recorded on rock slabs at 3,100 m (Corner 1978).

A climb to the summit of Kinabalu is an enchanting experience for the visitor, moving up through the different vegetation zones. Below 1,300 m, lowland and hill dipterocarp forests dominate the landscape. Around the park headquarters (at 1,554 m) chestnut and oak forests, dominated by *Castanopsis, Quercus* and *Lithocarpus*, are home to numerous birds and squirrels, remarkable for their tameness. These forests are less species-rich than the dipterocarp forests but are still diverse, with more than 40 species of oaks recorded along the trails around the headquarters (Jacobson 1986). From here the route to the summit winds along the southern ridge of the mountain, a distance of 8.5 km through rich oak forests to elfin and rhododendron forest and finally to the bare summit. One of the park staff holds the remarkable record of having reached Low's Peak and raced back down to Timpohon Gate (at 1,830 m) in less than three hours, during the annual Climbathon race. Normally visitors take at least two days over the ascent, staying overnight at Laban Rata

(3,353 m) before leaving for the final ascent at 3 a.m. They climb the staircases of gnarled *Leptospermum* roots to Sayat Sayat (3,810 m), then travel on with the aid of guiding ropes over the granite plateau, to reach Low's Peak (4,101 m) in time to watch the sun rise. As well as enjoying the excitement and achievement of the climb, the visitor to Mount Kinabalu can see many of the phenomena characteristic of tropical mountains.

The vegetation on Mount Kinabalu can be divided into distinct altitudinal zones. Ridges, because of their exposure and thinner and poorer soils, carry plants of upper altitudes at lower levels. Conversely, sheltered gorges support forest more usually found at lower altitudes. Thus there is oak forest *Lithocarpus havillandii* at 3,000 m on sheltered parts of the east ridge, whereas the main oak forest lies between 1,300 m and 1,800 m (Corner 1978). The lowland and hill dipterocarp forests extend up to about 1,300 m. Above this lies the montane oak forest or lower montane forest, rich in oak and chestnut and the curious celery pine *Phyllocladus hypophyllus*, the most primitive living conifer. The celery pine is easily recognised by its wedge-shaped "leaflets". In fact, the stems of the celery pine are flattened to look like leaves, the true leaves being only small scales. At lower altitudes the celery pine is a tree 30 m high; at higher altitudes it grows in dwarf form as a bush (Corner 1978).

With increasing altitude fewer species of plants are able to survive the harsher climate and poor soils. The moss forest of the upper mountain zone begins at 1,800 m and extends up to 3,200 m. At this altitude the mountain is often enveloped in mist, and mosses and other epiphytes drape trunks and branches. The stunted trees allow more light to reach the forest floor, and this results in a profusion of plant growth: orchids, rhododendrons, rattans and bamboos. Above 2,600 m the gnarled, stunted trees

---

Box 7.3. The trig-oak, a missing link.

The trig-oak *Trigonobalanus verticillata* is a massive oak, recognised by its hollow base with many sucker shoots. It is found in the lower montane forest on steep slopes and broad ridges and is an indicator of this vegetation zone on the mountain. The trig-oak was only discovered in 1961 and is one of the great botanical discoveries of the century, providing a link between the beeches and oaks and also a link with the southern beeches (*Nothofagus*). It has the triangular nuts of north temperate beech trees (*Fagus*) but borne in acorn cups, and has the leaves and timber of tropical oaks (*Quercus*, *Lithocarpus*). The seedlings are like those of beech (Corner 1978). The trig-oak has an interesting growth habit. It grows a single, stout trunk up to 27 m high, with sucker shoots sprouting from the base. This trunk dies and rots away, to be replaced by the larger suckers that have grown into saplings. Instead of one trunk, there is a cluster of saplings round a hollow; gradually, some of these stems die and are replaced, and the hollow enlarges.

of mountain gelam *Leptospermum recurvum* and the conifer *Dacrydium gibbsiae* form curious elfin forest, brightened by flowering rhododendrons. Higher than 3,200 m, most of the summit is bare rock, scoured by ice during the Pleistocene and supporting very little plant life. Only a few plants can survive the harsh rains, intense sunlight and blasting winds of the summit. Dwarf shrubs such as *Schima* and *Leptospermum* and herbs such as Low's buttercup are found in sheltered pockets where soil has accumulated. Club-mosses trail the ground, and grasses and sedges grow in rock crevices, but the last 300 m to the summit is a bare pavement.

The mountain flora of Borneo seems to have been derived from both Malaysian and Australasian sources. From an analysis of about 900 cold-adapted mountain species, van Steenis (1972b) concluded that there were three tracks by which plants arrived in the Sunda region during the geological past (fig. 7.5). Tens of millions of years ago the spine of Borneo was much higher than it is today and supported a rich alpine flora. As the sedimentary rocks were eroded away, Mount Kinabalu was thrust up and became a refuge for these plants of cooler climes (Cockburn 1978). The mountain flora is rich in living fossils such as the celery pine and the trigoak, the missing link between the oaks and the beeches. At higher altitudes several plants appear that are more normally associated with temperate latitudes, species like Low's buttercup *Ranunculus lowii*, blackberries and raspberries *Rubus, Potentilla*, eyebrights *Euphrasia* and gentians *Gentiana*, which were isolated here on Borneo's highest mountain when the climate warmed at the end of the Ice Age (Smith 1980).

The Fauna of Mount Kinabalu

The animals of Kinabalu National Park are as varied as the flora, though less conspicuous. There are more than 100 mammals and 300 birds recorded for the park, including orangutan, gibbon, mousedeer, clouded leopard and ferret badger. Perhaps even the elusive Sumatran rhino occurs here, hunted out of most of Southeast Asia by poachers seeking its valuable horn. As with the plants, the greatest diversity of mammals and birds is found in the lowland forests around Poring, but Kinabalu's rarest species are found high on the mountain. Several species are believed to be endemic to Mount Kinabalu itself.

Whereas the lowland flora and fauna in Borneo are predominantly Asian in origin, the montane flora consists of a mixture of Himalayan and Australian elements, but the picture for animals is rather different. For birds and moths, at least, the upland species are almost entirely of Himalayan affinity. This disparity between floral and faunal affinities is surprising. If the montane community is an ancient relict community, now much reduced in extent because of the warmer climate (becoming a Pleistocene refuge), one would expect the plants and animals to have a common origin (Holloway 1970). It seems probable that at least some of the montane plants are of much more ancient Gondwanaland origin and have gradually spread here from other parts of the central mountain chain (Cockburn 1978), whereas the animals are more recent immigrants from Asia, having crossed to Borneo on dry land connections at times of lowered sea level during the Pleistocene.

Very few butterflies are recorded above 1,800 m on Kinabalu, and most of the species are found in similar lowland and mountainous localities throughout the Sunda region. Thirty-seven species of butterfly were collected at 1,800 m, but several of these are known to occur in various habitats from lowlands to 2,000 m. Only 16 species were characteristically upper montane, 10 of these being blues, hairstreaks and coppers of the Lycaenidae. The pierids, especially *Delias*, occur at high altitudes, whereas the skippers Hesperiidae occur mainly in lowland and hill forests (table 7.6). The butterfly fauna of Borneo is very similar to that of Peninsular Malaysia, Sumatra and Java (Holloway 1978, 1984).

Moths are more abundant at high altitude, and many species are recorded only from Kinabalu. Here moths show an altitudinal zonation that corresponds with that described for plants and birds. In the lowlands and foothills the pattern is similar to that for butterflies, with mainly widespread Oriental species, many characteristic of secondary growth forest. Species restricted to Sundaland are numerous in the foothills and the lower montane forest (1,200-1,800 m). One-third of the lower montane species are also found in the Himalayas, and another third are recorded only from Kinabalu (Holloway 1978). However, many of these supposed endemics may be found to occur on other Bornean mountains when these

Table 7.6. Number of butterfly species at different altitudes on Mount Kinabalu.

Family	Lowland	Foothill	Lower montane	Upper montane	Total species
Papilionidae	15	8	7	–	22
Pieridae	18	19	15	1	35
Danaidae	10	11	8	1	20
Satyridae	13	10	4	–	22
Nymphalidae	23	17	13	1	55
Lycaenidae	34	28	33	10	92
Hesperiidae	13	11	2	1	33
Others	2	4	3	–	11
Singletons, all families	45	37	33	2	
Total species	171	141	115	16	290

Source: Holloway 1978

are surveyed thoroughly.

Above 2,500 m endemism in moths increases to 50%, and most other species are of northern temperate or Himalayan origin, with only 5% of species shared with the mountains of eastern Indonesia. The high altitude moth fauna is similar to that of high mountains in Luzon, Seram and Sumatra. Only four species are known from the summit zone. The different vegetation zones are progressively poorer in species, decreasing from 220 species in lower montane forest to 73 in upper montane, 26 in the subalpine zone (at 2,500 m) and four on the summit. This reflects the increasing isolation of the altitude zones from areas of similar habitat (Holloway 1978).

On Kinabalu, as on other Bornean mountains, the resident birds can be classified into four zones of distribution (Smythies 1960):

- **higher montane birds** are found only on mountains exceeding 1,500 m, but on these mountains they may occur well below the 1,500 m contour;
- **montane birds** are found on mountains above 1,000 m and may sometimes be found on those below 1,000 m;
- **lowland birds** are found on plains and hills, but some occur up to 1,500 m;
- **submontane birds** link the lowland and montane zones.

At increasing altitudes on the mountain the variety of birds decreases. Above 2,500 m only a few specialised species are found, mainly insectivores or birds taking a mixed diet of insects and vegetable matter. Four of these species of the subalpine zone are not found below 2,500 m: the Kinabalu friendly warbler *Bradypterus accentor*, the mountain blackeye *Chlorocharis emiliae* (both Borneo endemics), the mountain blackbird *Turdus poliocephalus* and the mountain bush warbler *Cettia vulcania*. Several of the montane birds of Mount Kinabalu were long believed to be endemic to that mountain, but all are now known to occur on Gunung Trus Madi (Smythies 1978), and some species have recently been recorded on Kalimantan mountains. The black-breasted thrush *Clamydochaera jefferyi* has been recorded for Gunung Niut and Bukit Baka, and the mountain barbet *Megalaima monticola*, the mountain wren-babbler *Napothera crassa* and the mountain tailorbird *Orthotomus cuculatus* have been recorded on Gunung Niut (Prieme and Heegard 1988; Rice 1989).

The mammal fauna of Kinabalu shows distinct altitudinal zonation (figs. 7.17 and 7.18). More than one hundred species of small mammals have been recorded from the lowland and lower montane zones of the mountain. The diversity of species is about the same in both zones, probably because of similarity of habitats. Among these species, 13 are arboreal and non-flying, 17 are bats, 11 are gliders (flying lemur and flying squirrels) and 33 are terrestrial (shrews, rodents, mousedeer and tree shrews). About one-third of all species collected are terrestrial and two-thirds are arboreal (Lim and Muul 1978). Among the non-flying mammals, 60% are arboreal. Davis (1962)

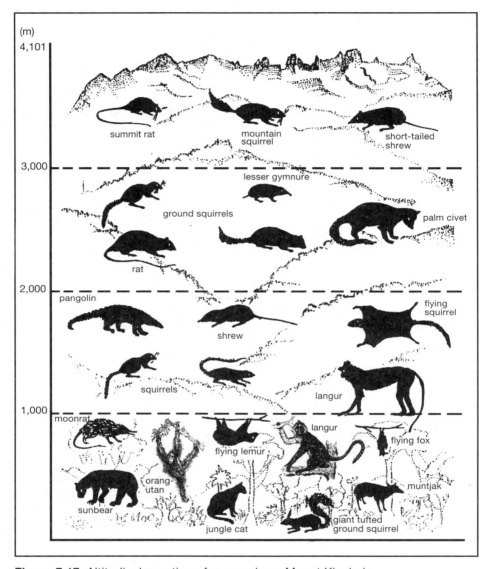

Figure 7.17. Altitudinal zonation of mammals on Mount Kinabalu.

reported that 45% of non-flying mammals collected in North Borneo were arboreal.

The greater diversity and multiple layers of lowland and hill forests provide more niches for animals than the less diverse, lower-canopied, upper montane forests. In the lower forest zones, food resources in the

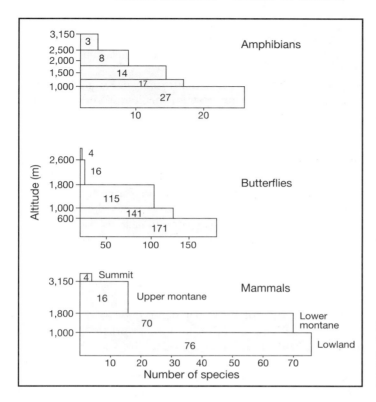

Figure 7.18. Reduction in number of animal species with increasing altitude on Mount Kinabalu.

canopy – fruit, flowers, new leaves and insects – are abundant and are exploited by many mammal species. Frugivorous, nectarivorous and insectivorous bats are common in the canopy zone. Langurs, gibbons, orangutan, flying lemur, and many of the tree squirrels and flying squirrels forage in the canopy layer and are adapted to an arboreal lifestyle. Most of these species are confined to the lower forest zones, although orangutan have been recorded as high as 2,000 m on Kinabalu (Lim and Muul 1978). The middle layers of lowland forests provide large quantities of plant material and insects for semi-arboreal and scansorial tree shrews, rats, civets, weasels, and some of the ground squirrels. Insectivores and ungulates forage on the forest floor. Carnivores prey on the smaller vertebrates.

The lower canopy height of montane forests provides fewer forest strata for animals to exploit than in lowland forests. Montane forests are also less species-rich and have a different floral composition. Whereas the lowland forests are rich in dipterocarps and "primate food trees", such as figs and durian, lower montane forests are dominated by oaks and chestnuts whose fruits are exploited by squirrels and other rodents. The composition of the mammal fauna of the mountain changes according to food availability, with a greater proportion of highland mammals being wholly or partly insectivorous.

Most of the species in the higher mountain zone on Kinabalu overlap with those in the lower montane zone. Sixteen species are recorded from the upper montane zone, and only four reach the summit (Lim and Muul 1978). The short-tailed shrew and Kinabalu shrew, found in the summit zone, are also recorded from other altitudes from the lowlands upwards, but the Bornean mountain ground squirrel and summit rat seem to be more exclusively montane species (Lim and Heyneman 1968). All the rodents and insectivores recorded in the two upper zones are terrestrial. No canopy forms of these groups nor large carnivores have been recorded from the highest elevations. This is probably due to lack of food resources. Interestingly, as on many other high mountains, small insectivorous shrews are found on the mountain's upper slopes. Since small mammals lose body heat very quickly because of their greater surface area, they must compensate with high energy intake and eat almost their own body weight of insects in a day. In cold mountain regions, with even greater heat loss, shrews must have particular problems in balancing their energy budgets.

Mount Kinabalu is both species-rich and an important centre for endemism. Fourteen (48%) of Borneo's endemic birds occur on Mount Kinabalu, as do 29 (65%) of the island's endemic mammals. The park includes a wide range of habitats and is a major centre for biodiversity on Borneo. Half of all Bornean birds, mammals and amphibians, including many rare and endangered species, have been recorded within park boundaries. Two-thirds of all Bornean reptiles are found here. Floristically, the park is just as rich, with at least half of the known Bornean species of figs, orchids, rhododendrons, mosses and ferns represented (see table 14.1). The Mount Kinabalu National Park is one of the richest and best-managed protected areas in Southeast Asia and truly can be described as a conservation area of global importance.

BUKIT RAYA

At 2,278 m Bukit Raya in Central Kalimantan is the highest mountain in the Schwaner Range and the third highest peak in Kalimantan. It is the only mountain over 2,000 m in Kalimantan that has been studied in any detail (Nooteboom 1987). The Bukit Raya area consists of younger, probably mid-Tertiary volcanic rocks, mainly of quartz diorite (Molengraaff 1900; van Bemmelen 1970). The mountain rises very steeply from the surrounding lowland forests and uplands. Environmental characteristics change with altitude and so do the soils. This is an area of high rainfall (more than 4,000 mm a year), and all soils are strongly leached; at high altitudes conditions are favourable for peat formation. For plant growth the most important ecological qualities of the soils are depth, availability of nutrients and availability of water. On Bukit Raya the nutrient content of all soils is low. Soils below 1,500 m are all over 1.5 m deep, but above that altitude soil depth is often less than 1 m, and the soils are poorly drained and waterlogged; this limits root growth (van Reuler 1987).

As a general rule, the more extreme the conditions, whether from increasing altitude or from changes in climate or soils, the less varied the species composition of the vegetation, and the more dominant one or two species become. This decrease in diversity and change in floral composition is well illustrated on Bukit Raya.

The vegetation on Bukit Raya can be zoned vertically according to altitude and soil types. Lowland dipterocarp forest occurs below 400 m, with dipterocarps making up about 30% of the canopy trees (40% of the trees over 30 m). Among the trees with a girth of more than 2.4 m in a 2.5 ha plot, more than 60% were dipterocarps (Nooteboom 1987). Above 400 m the number of dipterocarps decreases gradually. Moss forest occurs from about 1,600 m upwards. On ridges and the summit ridge, ericaceous vegetation is found, including many *Vaccinium* species, rhododendrons, *Leptospermum flavescens*, *Xanthomyrtus*, *Myrica javanica* and a few Rubiaceae, *Hedyotis*, *Randia* and *Urophyllum* (Nooteboom 1987).

The vegetation above 1,000 m shows greater similarity to that on Gunung Mulu than to the much higher Mount Kinabalu (Maloney 1987). As on Mulu, lower montane forest is dominated by Fagaceae (oaks and chestnuts). Upper montane moss forest occurs at 2,000 m, dominated by Myrtaceae, and with the primitive conifer *Phyllocladus hypophyllus*. The commonest palms on Bukit Raya are rattans, feather palms *Areca* and fan palms *Corypha*, which all show some altitudinal zonation. Of the rattans found in the Bukit Raya area, *Korthalsia rostrata* occurs only in the lowlands, whereas *Calamus caesius* and *Ceratolobus subangulatus* are still quite common above 600 m. Four other common species reach altitudes of more than 1,000 m (Mogea 1987).

Little is known about the animal life of Bukit Raya, though preliminary surveys of Bukit Raya and the adjacent Bukit Baka forests suggest a rich mammal and bird fauna, including several endangered species such as orangutan in the lowland forests (Rice 1989). Traditional Orang Ot (Penan) people are also known to live in the area (Schwaner 1853). These people, with their hunter-gatherer nomadic lifestyle, are an integral part of the forest ecosystem. Neighbouring Dayak societies also rely on these forests for harvests of minor forest products such as *gaharu* (aloe wood), rattan, damar and medicinal plants.

Bukit Raya and Bukit Baka are already designated as conservation areas, and in 1992 the area was declared a national park. A forest research project has started at Bukit Baka, sponsored by USAID. With good protection and management, these important sites will protect a variety of features including rich lowland dipterocarp forests, interesting mountain flora and many of Kalimantan's endemic montane species.

Effects of Disturbance on Montane Habitats

Clearing mountain forests can lead to serious erosion and environmental degradation: landslips, increased water runoff and increased sedimentation rates in rivers. The consequences of such clearance are well recognised by the Indonesian government, and all forests above 600 m or on steep slopes are officially protected as protection forest (*hutan lindung*). Unfortunately, illegal timber felling, sometimes sponsored, still occurs along with agriculture and other developments. Most of this clearance occurs on lower mountain slopes, destroying forests that not only protect watersheds but harbour the majority of plant and animal species found on mountain ranges.

Upper montane habitats are particularly vulnerable to disturbance from either natural events or human activities, since regeneration is extremely slow. In 1972-73 Sabah experienced a severe drought which lasted for four months. In some areas of stunted upper montane forest, above 2,700 m, all woody plants died (Lowry et al. 1973). Although five years later the vegetation was slowly recovering, some species coped better than others, and it seemed that some effects of the drought might persist for a century (Smith 1979). The regeneration of subalpine vegetation following a fire was studied in some detail in Costa Rica; the results are relevant. The fire killed the above-ground parts of shrubs such as *Vaccinium*, but suckers grew from the bases; this growth, however, was extremely slow. The soil surface cleared by the fire was not colonised by fast-growing pioneer species but instead remained mostly bare. Regeneration of Costa Rican oak forest was also slow (Janzen 1973).

At high altitudes decomposition processes are also slow, due to the absence or paucity of many lowland decomposers such as termites and earthworms (Collins et al. 1984). This means that dead vegetation, or abandoned human waste, decomposes only slowly. Plastic bags do not decompose at all. Visitors to montane habitats should be particularly careful not to destroy these fragile habitats by clearing vegetation, collecting orchids, pitcher plants and other plants, starting fires or abandoning litter.

Table 7.7. Mountain conservation areas in Borneo.

Site	Status	Area km²	Altitude m	Habitats	Special interest	Conflicts
Kalimantan						
Gn. Niut	C.A.	1,400	150-1,709	lowland-montane	flora	hunting, logging, agriculture
Gn. Palung	T.N.	700	0-1,160	lowland-montane, mangroves	flora + fauna research	logging, agriculture
Bentuang/Karimun	C.A.	600	300-1,960	hill + montane	transfrontier	hunting
Bukit Baka	T.N.	1,000	400-2,278	hill + montane	fauna + flora research	logging concessions
Bukit Raya	C.A.	1,100	100-2,278	lowland-montane	rich flora, tribal people	logging concessions
Batikap	H.L.	7,400	500-1,736	montane	rich flora, watershed	
Pleihari-Martapura	S.M.	300	200-1,170	ultrabasic	endemism, watershed	agricultural encroachment
Meratus Hulu-Barabai	pr	2,000	100-1,907	lowland-montane	floral endemism	agriculture, logging,
Kayan-Mentarang	C.A.	16,000	200-2,200	lowland-montane	biodiversity, culture	logging, agriculture
Ulu Kayan	pr	8,000	100-2,200	lowland-montane	flora, fauna	logging, cultivation
Sangkulirang	pr	2,000	100-1,385	limestone lowland	endemism	logging, hunting, quarrying
Gn. Beratus	pr	1,300	70-1,231	lowland + hill	flora	logging, cultivation
Ulu Sembakung	pr	5,000	130-2,143	lowland-montane	rich flora, elephant, rhino? trans-frontier	logging, agriculture
Apo Kayan	pr	1,000	500-954	montane + agroeco-systems	ethnobotany, biosphere reserve	hunting, changing agriculture
Sabah						
Crocker Range	N.P.	1,400	1,000-2,423	montane	flora, endemism	agriculture
Kinabalu	N.P.	750	900-4,101	hill-montane	biodiversity, alpine	agriculture, recreation
Trus Madi	pr	1,845	1,000-2,597	montane	flora, endemism	agriculture
Sarawak						
Pulong Tau	pr	1,645	500-2,000	hill-montane	transfrontier, rhino	logging
Lanjak Entimau	N.P.	1,700	500-1,500	hill-montane	watershed, transfrontier	
Gunung Mulu	N.P.	528	30-2,376	limestone montane	limestone, flora + caves, tribal people	

T.N. = N.P. = national park, S.M. = wildlife reserve, C.A. = nature reserve
H.L. = protection forest, pr = proposed

Sources: MacKinnon and Artha 1981; MacKinnon 1990

Mountains as Centres for Biological Diversity

Mountains are areas of great biological interest, representing Pleistocene refuges and centres of speciation and endemism. Endemism on mountains is particularly high for those groups, such as amphibians and invertebrates, that are poor dispersers. The relict highland forests of Tanzania, for instance, show high levels of endemism among some invertebrate, amphibian and reptilian groups, with endemism in the East Usambara mountains varying from 2% in mammals to 95% for millipedes (Rodgers and Homewood 1982). Thus clearing montane vegetation, particularly montane forest, may lead to extinction of unique species.

Borneo's mountains, with their surrounding lowland forests, constitute some of the most important sites for biodiversity on the island. More than 10% of the island's montane habitats are gazetted or proposed as conservation areas (MacKinnon and MacKinnon 1986). Important mountain reserves include Bukit Raya/Bukit Baka, Gunung Palung, Gunung Karimun and Bentuang, Gunung Niut and Kayan-Mentarang in Kalimantan, Gunung Mulu in Sarawak, and Mount Kinabalu and the Crocker Range in Sabah (table 7.7). Together these mountain reserves represent most of the major habitat types of Borneo, encompassing lowland, limestone and montane rainforests and subalpine ridges, and provide protection for more than half of the island's recorded plant and animal species.

Chapter Eight

Borneo Peoples - Migrations and Land Use

People have lived in Borneo for at least 40,000 years. The first Australoid races probably reached the island during interglacial periods of the Pleistocene, travelling over the dry land ridges that connected the Greater Sunda Islands to the Asian mainland. Niah in Sarawak was a site of human habitation from 40,000 to 20,000 B.P. (Harrisson 1959; Harrisson and Harrisson 1971; Majid 1982), the earliest evidence of human settlement in Borneo. At least 28,000 years ago people were using caves and rock shelters at Madai and Baturong in Sabah, and later established lakeside settlements there, harvesting animals and shellfish (Bellwood 1984, 1988). These early people probably also harvested the hill sago palm *Eugeissona utilis* and were already utilising forests and rivers, hunting wildlife, fishing streams and gathering forest food plants and other products. Their way of life was probably similar to that followed by the seminomadic Penan today (Sellato 1989b).

Much of what is known of these early people comes from evidence exposed by excavations of limestone caves and rock overhangs, where they sought shelter (Majid 1982; Bellwood 1988). They used simple stone tools and also may have had a well-developed bamboo and rattan technology, but such tools and hunting weapons have rotted away with no trace. Early in their history Bornean people seem to have been able to produce fire. A quartz pebble, believed to be a firestriker, was found in strata dated as between 20,000 and 40,000 years old at Niah (Majid 1982). Fire was a useful means of keeping warm and cooking food and, later, of clearing forest areas for cultivation. Although Paleolithic humans probably had little impact on the overall vegetation of the island, they were already hunting many of the mammals. Overhunting may have led to extinction of species such as the tapir *Tapirus indicus*, whose sub-fossil remains occur at Niah (Medway 1959).

These early human inhabitants probably began to tend patches of hill sago, much as the Penan do today. Also they may have practised some fruit tree and tuber cultivation. These were probably the earliest attempts at agriculture in Borneo. With the security of a regular harvest from sago and tubers, permanent settlements formed along coasts and rivers, and the people began to fish and to collect freshwater molluscs (Bellwood 1988).

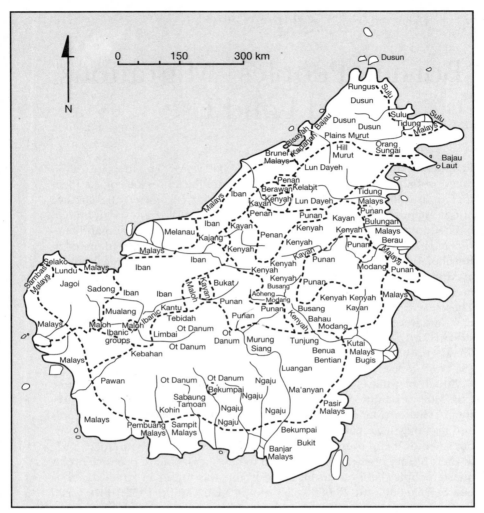

Figure 8.1. Ethnic groups of Borneo. (After Sellato 1989a.)

Later waves of migrants, the Mongoloids, brought rice cultivation from mainland Asia about 5,000 years ago (Glover 1979). Permanent settlements and rice harvesting necessitated the clearing of fields. A system of shifting cultivation evolved.

Borneo has been the focus for several waves of human migration, creating the patterns of peoples and settlements seen today (fig. 8.1). The various Dayak groups and the Penan are considered to be the original inhabitants of Borneo (Avé and King 1986). Over the centuries several tribes have moved gradually further inland into the mountainous interior, often retreating before the warlike advances of later arrivals such as the Iban Dayaks. Later waves of immigrants and traders settled on the coastal plains: Arabs and Malays, Chinese and Indians who came to trade and still dominate much of the island's commerce today. Borneo's rich potpourri of peoples is best illustrated by Sarawak, the most mixed society of all the island's states. In 1980 Dayaks accounted for 44% of the Sarawak population, with the other two indigenous groups, Malays 20% and Melanau 6%, together totalling slightly less than the Chinese 29% (Hong 1987).

Penan: Harvesters of the Forest

"Penan" is a collective term to describe several groups of indigenous hunter-gatherers, sometimes referred to as Punan, Ot, Ukit or Bukitan (Avé and King 1986). The Penan are a secretive, seminomadic people travelling the forested interiors of Sarawak and Kalimantan, gathering sago, hunting wild meat and collecting other forest produce (Needham 1953; Langub 1974; Kedit 1978; Hoffman 1986; Brosius 1986; Sellato 1989a).

The Penan hunter-gatherer lifestyle is probably similar to that of early humans. Traditionally Penan do not live in permanent settlements, but build temporary shelters in the forest, hunting and harvesting an area before moving on. Penan are knowledgeable forest dwellers who establish temporary camps close to patches of hill sago palm *Eugeissona utilis*, which they harvest as a staple food (Brosius 1988). Harvesting sago is one of the most time-efficient ways of gathering food. In Siberut, West Sumatra, where sago is a staple food, one day's work harvesting the palm provides a person with enough food for 17 days (Whitten and Whitten 1981). The Penan move around within the boundaries of a particular area and practise a system of forest management that ensures the sustainability of the resource base (Brosius 1986; Langub 1988). They rotate their extraction of sago palms between sago groves and only remove one to two trunks from a clump, leaving the palm to resprout.

Although relying mainly on a diet of sago, the Penan also hunt wild pigs and other forest animals with ironwood blowpipes, and fish with plant poisons. They collect forest products such as swiftlet nests, damar, rattan, rhinoceros horn, camphor, beeswax, gaharu (incense wood), hornbill "ivory", bezoar stones (gall stones of monkeys), bear claws and bile, and feathers, all of which are exchanged with their downriver Dayak neighbours for goods such as salt, metal items and textiles. These trading relationships are often of long standing and strengthened by intermarriage, and sometimes the two groups claim common historical and cultural roots (Avé and King 1986; Rousseau 1988). The economic relationships and evidence of cultural and linguistic exchange between the two groups has recently fuelled a controversy over whether the Penan are truly primitive hunter-gatherers, as most students believe (Brosius 1988), or whether they have abandoned a more settled lifestyle so that they can trade in forest products (Hoffman 1986, 1988). Whatever the case, the number of Penans following a nomadic lifestyle is on the decline.

In recent times the Indonesian, Sarawak and Sabah governments have actively encouraged the Penan to abandon their nomadic lifestyle and settle in permanent settlements. This has drawn them increasingly into the cultures and societies of the settled Dayak farmers. The process is being hastened as their forest homes are cleared for timber. Like other Borneo peoples, the Penan are affected by the logging concessions that clear their traditional lands (Hong 1987), and have recently started peacefully blockading the timber roads in Sarawak, a campaign that has earned them world-wide attention and sympathy (Popham 1987). Dennis Lau has produced a beautiful photographic essay of the Penans' vanishing lifestyle (Lau 1987).

Dayak Groups

"Dayak" literally means people of the interior, and is a collective name for a diverse group of tribal peoples who differ in language, art forms and many elements of culture and social organisation. They are mainly shifting cultivators of hill rice who dwell beside Borneo's rivers, many but not all in longhouse societies, and observe native customary laws or adat (Hong 1987; Jessup and Vayda 1988). There may be as many as three million Dayak people throughout Borneo (Avé and King 1986), about a third of the island's total population. Most of these people live in the lowland river basins and alluvial plains. According to Rousseau (1988), just over 130,000 Dayaks live in the upper river reaches of the mountains of central Borneo (40,000 in Sarawak, 90,000 in East Kalimantan, 2,000 in Sabah and fewer than 1,000 in Brunei), and these numbers are declining as more communities migrate to take advantage of the prosperity of the coastal regions.

The largest Dayak groups include the Iban and the Ngaju of Central Kalimantan. There are more than 368,000 Iban in Sarawak alone (Hong 1987), and they also live in West Kalimantan, Brunei and Sabah. The Iban had a reputation as fearsome fighters, headhunters and migratory people, whose aggressiveness led to migrations and resettlements of other native communities (Hose and McDougall 1912). They are related culturally, linguistically and historically to several groups in the Kapuas basin in West Kalimantan, including the Kantu, Seberuang, Bugau, Mualang and Desa. Other Dayak groups living in Central Kalimantan include the Ot Danum in the upper river valleys, and the Ma'anyan and the Luangan along the upper and middle reaches of the Barito (Avé and King 1986; Sellato 1989b).

In the interiors of East Kalimantan and Sarawak, along the upper reaches of the Kayan, Mahakam, Rajang and Baram rivers, are found the Kayan-Kenyah groups of people. The Kayan are found widely over the island; a small group of Kayan live along the Mendalam River, a tributary of the upper Kapuas in West Kalimantan. The Kenyah comprise several different subgroups with differences in language and culture. The Kayan and Kenyah have long-established social, economic and political interrelationships, and there has been considerable intermarriage. The Modang of the Mahakam basin are probably an offshoot of the Kayan-Kenyah complex (Avé and King 1986).

The Kajang live in the Seventh Division of Sarawak. The coastal Melanau of Sarawak, Malay sago cultivators, are related to the Kajang groups, as are the Berawan, the highland Kelabit and the Lun Bawang and Lun Dayeh. The last two extend into East Kalimantan, as do the Murut people of Sabah. The Bidayuh, or Land Dayaks, are found extensively in the lower Kapuas basin in West Kalimantan, especially upstream of Sanggau, along the Sekayam River and into Sarawak. The Maloh group (including the Embaloh, Taman and Kalis) live in the interior of West Kalimantan. Like other interior peoples, they show great artistic skill in beadwork, metalwork, painting and woodcarving (King 1985). In the lower reaches of the Kapuas are several Dayak groups, including the Selako of Sambas, and Kendayan inland of Pontianak, who speak languages related to Malay but regard themselves as culturally distinct from the coastal Malays (Avé and King 1986).

Most of the peoples of Sabah have closer linguistic links with the Philippines. One of the most important Sabah groups is the Dusun or Kadazan, who grow rice in *sawah* (wet fields) in the western coastal plains but are shifting cultivators in the hills. The Dusun of Brunei are related to the Kadazan of Sabah. Dusun also live in the upper waters of the Barito. Elsewhere in Borneo there are several distinct racial groups whose precise ancestral and linguistic links remain unclear, as in the Melawi River region in West Kalimantan (Avé and King 1986).

There is a wealth of literature on Bornean tribal peoples (Helbig 1955; Avé et al. 1983; Boutin and Boutin 1984; Rousseau 1988). Colonial administrators and explorers have written many fascinating descriptions of traditional Dayak lifestyles, culture and agriculture (Ling Roth 1896; Hose 1926; Hose and McDougall 1912; MacDonald 1985; Harrisson 1959; Bock 1985). More recent publications have dealt with: different Bornean societies in relation to their environments (King 1978; Appell 1976); the Maloh (King 1985) and Kantu (Dove 1985) in West Kalimantan; the Ma'anyan (Hudson 1972) and Ngaju (Scharer 1963) of Central Kalimantan; the Kenyah in Kalimantan and Sarawak (Whittier 1973, 1978; Vayda et al. 1980; Colfer

1981; Jessup 1981; Chin 1985); the Kayan (Conley 1974; Rousseau 1978); the Kelabits (Harrisson 1959c; Schneeburger, 1979) and Penihing (Nieuwenhuis 1904, 1907; Sellato 1989a); the Iban in Sarawak (Freeman 1970; Padoch 1982; Sutlive 1985); the Bukit people of the Meratus Mountains (Lowenhaupt-Tsing 1988); and the Murut and Dusun, or Kadazan, of Sabah (Rutter 1929; Williams 1965).

Dayak Migrations

There is little information on Borneo from precolonial times. The origins of the Dayaks are obscure. The first Dayaks, especially the Kenyah and Kayan, may have arrived in southeast Borneo as early migrants from the Asian mainland and then migrated up the Barito valley (Harrisson 1959c), with the Kayan reaching Sarawak in the fourteenth century (Hose and McDougall 1912). The history of Borneo was marked by large-scale migrations, warfare, headhunting and conquest. Vanquished people were driven off their land, or forged links with the invaders through intermarriage. Wars of conquest were waged by Kayan, Modang and Kenyah migrating from the headwaters of the Mahakam in the eighteenth and early nineteenth centuries (Jessup and Vayda 1988). This invasion was led by the great chief Bo Ledjo of Long Glat (Bouman 1924). Later they were forced to retreat from the Rajang valley after intermittent warfare with the Ibans.

The nineteenth century saw large-scale migrations of the Iban Dayaks, who moved into the Rajang valley from the south, probably from the Kapuas basin. By 1947 there were 114,000 Iban in the Rajang basin. One *bilek* family moved more than 300 km in the lifetime of one family member (Freeman 1970). As they moved the Iban cleared forest for rice cultivation, changing the landscape along the river valleys. Epidemics and warfare were probably key factors in these migrations. Having migrated up the Saribas and Rajang rivers, the Iban attacked the Kayan in the headwaters in 1863, and continued to advance north and east. War parties and headhunting raids drove other tribes from their lands. Longhouses were burned and iron implements cast into the river. By the early 1900s headhunting Dayaks had penetrated into the remote headwaters of the Rajang, Kayan, Mahakam and Kapuas.

The Iban, Kayan and Kenyah were especially notorious as headhunters (Hose and McDougall 1912), but headhunting and the capture of slaves were widespread throughout the island except in the very south. Faced with such Iban attacks, the Lugats retreated to the Anap, the Ukit to the Mahakam, and the Kenyah and Kayan into the Apo Kayan (Freeman 1970). Warfare and headhunting raids maintained populations at low levels in the highlands.

Heads were collected from victims of all ages, women, children and old people surprised during furtive ambushes, as well as fighting warriors slain in hand-to-hand combat (Hose and McDougall 1912; St John 1974). The decapitated corpses were left while the raiders fled with their trophies. Sometimes a threatened longhouse would buy off a headhunting party with slaves who were then beheaded to propitiate the souls of dead comrades. Heads were required for community rituals such as the Iban *kenyalang* and Kenyah *mamat* ceremonies for cleansing and strengthening the village (Jessup and Vayda 1988). Young men also headhunted to prove their bravery to a bride-to-be, but this practice was more common among the Kenyah than the Iban (Freeman 1970).

Headhunting formally ceased in the Kayan River basin in 1924 but continued in Sarawak until 1940. It was encouraged by the British during World War II so that many Japanese heads were taken (Harrisson 1959b). Omens were especially important to headhunting parties, and a Brahminy kite *Haliastur indus* seen flying from left to right across the route was a strong prohibition against further advance. To counteract this, Harrisson (1959b), who travelled the river with Kenyah in World War II, always kept one man in the boat facing backwards so that his view of the hawk was favourable.

Coastal Communities

Around Borneo's coasts many of the people are Malays, a heterogeneous group, united by a common language and religion, Islam. The Kutai of East Kalimantan and Banjarese of South Kalimantan belong to this Malay group. Many Malay peoples are derived from traditional peoples who converted to Islam. Others mixed and intermarried with Muslim migrants from outside Borneo, in particular from Sulawesi, Java and Sumatra. Malay communities live mainly on the coast and along major rivers, and subsist by small-scale trading, sea fishing, growing rice and other agriculture (Avé and King 1986). They also have a history of migrations. At the invitation of the Sultan of Kutai, and after poor harvests and heavy taxation by the Dutch, Banjarese migrants moved to the Mahakam lakes area in the 1870s to help open up the swamplands for wet rice agriculture (Potter in press). Seasonal migrations associated with rice harvests are also common in South Kalimantan (see box 10.9).

Gold was responsible for one of the major migrations to Borneo, an influx of Chinese gold miners to Sarawak and West Kalimantan in the late eighteenth century (Jackson 1970). Their descendants make up the large Chinese communities in these provinces today (see chapter 13). Gold output reached a peak between 1790 and 1820. In 1810 gold production from West Kalimantan exceeded 350,000 troy ounces and was valued at over 3.7 million Spanish dollars (Raffles 1817). Output began to fall in the 1820s and declined further over the next two decades (Jackson

1970). Faced with decreasing returns from gold mining, the Chinese turned increasingly to trade and to agriculture, farming, pigs and vegetables. From 1880 the Chinese began to establish plantations of pepper, gambier and, after 1910, rubber. Chinese immigration increased again from 1920 so that by 1930 Chinese made up one-third of the population of West Kalimantan, the largest Chinese agricultural community in Indonesia (Jackson 1970). Perhaps their most significant contribution to the agriculture of West Borneo was the creation of large, rice-growing colonies on coastal alluvium. Thus despite its long-standing contacts with Java, West Borneo received its sawah system direct from southern Asia (Jackson 1970). Today the Chinese are particularly active in the trading and commercial sectors in Sarawak and in the main Kalimantan towns, especially Banjarmasin and Pontianak.

Kalimantan has also received a regular influx of migrants from the other Indonesian islands. There have been movements of Bugis from South Sulawesi to South and East Kalimantan since at least the sixteenth century. These sea-faring Bugis engaged in trade and sea fishing. In the early 1950s a new wave of Bugis migrants arrived in East Kalimantan, spurred to migrate by hardships suffered during Kahar Muzakar's Islamic rebellion in South Sulawesi (Vayda and Sahur 1985). They were later joined by relatives and began to clear large areas of lowland forest along roads, for farming pepper *Piper nigrum* and other cash crops (see box 12.3).

Javanese colonists had already established themselves in large numbers in Kalimantan in the fourteenth and fifteenth centuries, during the height of the Majapahit empire in East Java. Their influence can still be discerned in the languages and cultural and court practices in the states of Banjarmasin, Sambas and Kotawaringin (Avé and King 1986). New migrants continued to arrive, and many successfully reclaimed rice fields in the tidal wetlands of South Kalimantan (Collier 1977, 1980).

Considerable numbers of poor farmers from Java and Bali have also been resettled in Kalimantan as part of the government-sponsored transmigration programme. Many manual workers in the oil and timber industries are also migrants from Java. Madurese are found in West, Central and South Kalimantan. The Dutch brought Madurese to Kalimantan at the end of the last century as part of their colonisation schemes. Madurese are found in a wide range of occupations including driving becaks, farming, cattle breeding, fishing and labouring (Sudagung 1984). They are efficient dryland farmers (see box 13.3).

On the northwest coast of Sabah and down the east coast, there are many Bajau, formerly known as sea gypsies, who roamed the Sulu Sea in search of fish and other marine products (Sopher 1965; Rutter 1930). Bajau also reside in the southern Philippines and in eastern Indonesia, especially coastal Sulawesi. Many of the Sabah Bajau have now turned to land-based economies, farming and raising water buffaloes (Avé and King 1986).

Although the Bajau, Buginese, Javanese and Madurese are all Muslims like the Malays, these groups have retained their separate identities. Other migrants to Borneo include Christian migrants from Flores and Timor, who have settled in Sabah, and a substantial community of Indians and Arabs (especially in East Malaysia) who have settled in coastal trading communities.

SHIFTING AGRICULTURE: SUBSISTENCE FARMING

The cultivation of hill (unirrigated) rice is the cultural and economic cornerstone of many Dayak peoples (Freeman 1970; Rousseau 1977; Chin 1984). Swidden agriculture, to cultivate hill *padi*, is the main subsistence activity in interior Borneo. Other food crops are grown interspersed with the hill rice to ensure a continuous food supply.

The cycle of cultivation (fig. 8.2) usually starts in May with slashing of the undergrowth, then felling of trees. Fields are burned in August, sown in September and harvested in February during a short dry season (King 1985). The Dayak choose the time for the various activities by reference to the stars, for example Orion's belt (Sellato 1989b), or to animal omens (Harrisson 1960). Men or women work in teams and use only simple tools: an axe, an adze, a chopping knife, a dibbling tool and a rice harvesting knife. Rice is the source of all life, and the harvesting cycle is accompanied by many taboos and rituals (Sellato 1989a). After the harvest a major feast is held.

Shifting agriculture has had a long history in Borneo. People have been clearing forest since prehistoric times (Flenley 1979), and the practice became more extensive after the arrival of rice cultivation in the archipelago. Major forest clearing was already evident by 2,500 B.P. (Maloney 1985). Early cultivators had access to fire and finely made stone tools (Glover 1979), but the arrival of metal tools and ironworking technology revolutionised forest exploitation. With iron tools, large trees could be felled and greater areas cleared for cultivation. Swidden agriculture is still the main economic activity of most of the inland indigenous people of Borneo. Some 36,000 households are involved in shifting agriculture in Sarawak (Hong 1987) and an estimated 500,000 households throughout Kalimantan. Shifting cultivation is especially important in Central Kalimantan, with around 30% of the population involved in such agriculture.

Swidden agriculture is a system of extensive, rather than intensive, land use. It is not the same as the generally unsustainable slash-and-burn agriculture that is practised increasingly by pioneer farmers and new immigrants in many parts of Borneo (fig. 8.3). Swidden agriculture is a form of shifting cultivation characterised by: rotation of fields rather than crops; clearing by fire; absence of draught animals and manuring; use of human

Month Year	Swidden type		
	Primary forest	**Secondary forest**	**Swampland**
February 1975	Selecting		
March	Slashing		
April			
May	Felling		
June		Selecting	
		Slashing	
July		Felling	
August	Burning	Burning	
September	Planting	Planting	Burning Planting
October			Transplanting and weeding
November		Weeding	
December			
January 1976			
February	Harvesting	Harvesting	Harvesting
March			
April	Carrying	Carrying	Carrying

Figure 8.2. The Kantu' calendar of work in the major swidden types during the 1975-1976 farm year.

Source: Dove 1985

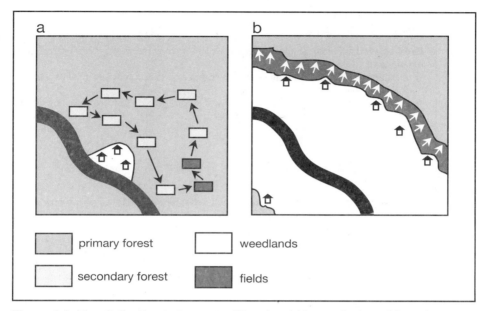

Figure 8.3. The distinction between traditional swidden agriculture (a), and the now prevalent shifting agriculture (b), which is endangering many forested areas. (After Rijksen 1978.)

labour alone; simple tools; and short periods of soil occupancy alternating with long periods of fallow (Chin 1987). Many swidden farmers are rational and sophisticated users of their natural environment (Padoch 1982; Dove 1985). Swidden agriculture can support 10-50 individuals/km^2, on average about 25 people/km^2 (Avé and King 1986). Sustainability depends on farmers cultivating for subsistence rather than for cash crops, and on cultural and adat restraints (Kartawinata et al. 1984).

The length of period of cultivation of a given plot and the duration of the fallow period are the key factors in maintaining soil fertility under swidden agriculture. The great luxuriance of tropical forests does not necessarily indicate fertile soils (Richards 1952; Whitmore 1984a). When forest is cut and burned, there is an initial input of nutrients into the soil from the ash. However, the soil is now exposed, and minerals are leached from the soil by rain. Nutrients are also removed by growing plants.

If fields are small and surrounded by primary or regenerating forest, most of the soil and nutrients washed from the cultivated field will be trapped by vegetation at the field boundary (Hamilton and King 1983). The residual effect of the forest on the organic content and aggregate stability of the soil generally lasts for the first year of farming so that soil loss

> **Box 8.1.** Shifting agriculture.
>
> **Extensive** (sustainable) shifting cultivation systems may be characterised by combinations of:
> - Gardening in forest clearings in primary forest or, more commonly, in secondary forest;
> - Short periods of cropping, alternating with longer periods of forest fallow;
> - Low human or animal energy input per unit of energy output (but use of fire adds large energy input);
> - Use of relatively simple tools;
> - Multiple cropping;
> - Low human population density and/or "traditional" people; and
> - Little disturbance of the soil surface.
>
> **Intensive** (unstable) systems that approach sedentary agriculture may be characterised by combinations of:
> - Use of man-created grasslands or degraded bush fallows for gardening or grazing;
> - Long periods of cropping, alternating with shorter or no periods of fallow;
> - High energy input per unit of energy output;
> - Use of technically more advanced tools (e.g., chainsaws);
> - Tendency toward monoculture cash crops;
> - High human population density and/or "new colonists"; and
> - Greater soil disturbance, which can include rough terracing, ditching, mounding and other modifications of terrain or use of livestock.
>
> The impact of shifting cultivation on the hydrology, erosion and nutrient status of forested watersheds depends on the period of cultivation, length of fallow, cultivation practices and crop type.
>
> *Source: Hamilton and King 1983*

remains low (Morgan 1985). The rate at which the soil loses its fertility depends on the degree of protection and cover provided by its new vegetation cover, first the crop and then the regenerating forest. Mixed crops and the cover afforded by weeds and secondary vegetation protect the soil from erosion, even on steep slopes with heavy rainfall. Furthermore, the shifting cultivators' practice of not turning over all the soil, but simply using a dibble stick to plant seeds, also helps conserve soil (Morgan 1970).

The greatest concentration of nutrients is in the top 7.5 cm of soil. Under normal shifting cultivation, loss by erosion removes at most one-tenth of this fertile topsoil and will have little effect on availability of nutrients. Trials in Sarawak on land with a seven- to ten-year fallow showed that hill rice can be cultivated even on 30° slopes without significant soil loss to erosion (Hatch 1980).

Fields are abandoned to fallow after one to three years of cultivation, when crop yields decline (fig. 8.4). This decline can be rapid, from a yield

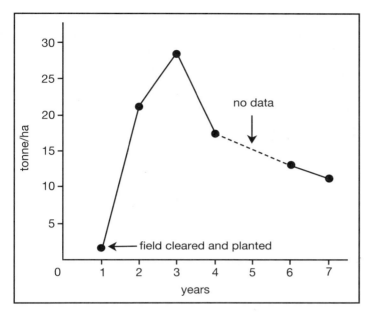

Figure 8.4. Dryland rice yields in successive years in swiddens cleared of rainforest, Central Kalimantan. (After Driessen et al. 1976.)

of 300-400 kg of rice per hectare on newly cleared lands in Malaysia to 250 kg and then 160 kg in succeeding years (Grist 1953). Although soil fertility declines with successive crops, it seems that weed growth is the main reason that shifting cultivators abandon their fields and open new areas (Seavoy 1973). To harvest an adequate rice crop, they need to spend a considerable amount of their time on weeding, almost a quarter of their labour for Kantu cultivating re-opened fields (table 8.1). The settled Penan of the Silat River in Sarawak do not weed their fields, and harvest enough rice for only two or three months (Chin 1987). Similarly, the Semai and Temiar Senoi of Peninsular Malaysia, who do only a little weeding, have low rice yields and abandon their fields after one year (Dentan 1968; Cole 1959). The amount of time that must be spent on weeding is the main factor limiting the extent of land a family can cultivate (Dove 1985; Freeman 1970). Women play an important economic role in opening and weeding the rice fields; this gives them status and a share in decision-making in the cooperative structure of the longhouse (Colfer 1981).

Problems with weeds are fewer in small forest clearings than in more open areas that are close to abandoned fields and a source of weed seed. Weed growth is greater in the second and third years of cultivation; it also increases with a shorter fallow period. After a single rice crop, the minimum fallow period is estimated to be 12 years; if a second rice crop is taken, involving further clearing and burning, the fallow must be well over 15 years (Freeman 1970). The fallow period allows the growth of woody perennials and pioneer trees (building-phase forest), which shade out the weeds and return organic matter to the soil in the form of leaf litter (Uhl et al. 1982). If the fallow is too short, there is still a reservoir of weed species and their seeds in the secondary growth, and weeds can become a serious problem even during the first year of cultivation.

Declining yields have been attributed to: an increase in weeds; increase in pests and diseases; erosion of topsoil and loss of humus and nutrients; and a change in composition of soil flora and fauna (Chin 1987). Short cropping periods interspersed with long fallows mitigate these factors. Shifting cultivation with adequate fallows can be an efficient and ecologically sustainable form of agriculture on poor soils (table 8.2). Even during the fallow period the land may produce food and income for the farmer. For example, rubber or fruit trees produce crops during the interim, and rattans planted before the field is abandoned can be harvested at clearing time (Mayer 1988).

Table 8.1. Overall labour inputs per hectare throughout the swidden cycle.

Swidden stage	Swidden type		
	Primary forest	**Secondary forest**	**Swampland**
1. Selecting	0.6	0.8	0
2. Slashing	8.0	9.2	11.5
3. Felling	12.0	4.0	0
4. Burning	1.4	3.6	10.5
5. Planting	12.9	14.7	91.9
6. Weeding	0	48.8	
7. Guarding	1.0	4.0	11.0
8. Harvesting	29.9	49.2	86.5
9. Carrying	3.8	4.9	9.6
10. Harvesting of relishes	(not included)	(not included)	(not included)
11. House making	3.2	4.0	13.0
12. Toolmaking	15.5	24.6	40.9
Totals	**88.3**	**167.8**	**274.9**

Figures in person days per hectare.
Source: Dove 1985

Table 8.2. Comparative production data from Kalimantan farmers and other swidden groups.

	Kantu'[+]	Iban[+]	Hanunoo	Lamet	Raiapu Engga	Lua' Karen
Labor inputs/ha						
Primary forest	88 days	138-175 days	545 days	NA	NA	NA
Secondary forest/ grassland	168-275 days	123-163 days	501 days	NA	1,108 days	NA
Area cultivated						
per worker per year	0.97 ha	0.53 ha	1.28 ha	0.46 ha	NA	NA
per consumer per year	0.59 ha	0.32 ha	0.61 ha	0.31 ha	NA	NA
Rice yields						
per hectare	457-1,322 kg	810 kg	2,410 kg	1,335 kg	NA	NA
per worker	4.5-5.2kg	5.4kg	4.5kg	NA	NA	NA
As a % of requirements	30-40%	32%	NA	NA	NA	29% 59%

Note: All workdays have been recalculated based on the 350-minute workday of the Kantu'. All rice is measured in a threshed and winnowed but unhusked state.
NA = Data not available.
[+] = Swidden cultivators in Borneo.

Source: Dove 1985

Shifting Agriculture: Ecologically Sound or Environmentally Damaging?

Shifting agriculture has been described variously as a major cause of forest destruction (a view widely held by government planners and foresters) or an ecologically sound way of utilising forest soils. Evidence can be found in Borneo to support both views. Part of the problem arises from the fact that the term "shifting cultivation" is used to describe a wide range of agricultural practices, from the ecologically sustainable swidden agriculture, described above, to slash-and-burn systems where large areas of land are cleared (often for cash crops), cultivated until the soil is exhausted and then abandoned.

Shifting cultivation in Borneo, and elsewhere in Indonesia, has been blamed for: a heavy loss of timber due to felling of primary or old secondary forest (Lau 1979); soil erosion, resulting from the clearance of steep hillsides; and flooding caused by rapid water run-off from areas cleared of natural vegetation.

It has been claimed that 400,000–750,000 ha of Indonesian forests are cleared every year by shifting cultivators, compared with 100,000 ha cleared by logging, and that this agricultural clearance represents a considerable loss in lost timber revenue (Hasan in litt.). In Sarawak the annual value of timber lost to shifting cultivation through clearance of 60,000 ha of forest was estimated at $100 million in 1976 (Lau 1979; Chan et al. 1985). A survey by the Sarawak Soils Division calculated that a typical six-person family of shifting cultivators obtained an annual income of M$550 per hectare of prime forest destroyed, whereas its value in commercial timber would be M$3,600, with extra employment opportunities of M$1,800 per year in permanent production forest (Lau and Chung 1978). Since 1976 logging followed by agricultural conversion has accounted for 40% to 45% of Sabah's deforestation, and shifting cultivation is blamed for more than half the state's annual forest loss (Reppeto 1988). From these figures the "costs" of shifting cultivation would seem to be high in both economic and environmental terms.

But are these figures correct? How much primary forest do swidden farmers destroy? The majority of shifting cultivators farm in secondary forests, less than twenty years old and with little or no commercial timber, since this land is easier to clear (Jessup 1981; Padoch 1982; Chin 1985; Cramb 1989; Kartawinata et al. 1989). Most Dayak communities clear both primary and secondary forests, and there are good ecological and cultural reasons for having swiddens in both (Dove 1983). Secondary forests usually burn better, but there are more problems with weeds, and therefore their cultivation requires more labour. In the past, when warriors had to guard the female weeders against attacks by headhunters, the cultivation of secondary areas tied up even more of the work force (Dove 1983).

Only a small fraction of the total area of Borneo under shifting cultivation is newly cleared from primary forest. Hong (1987) estimated that less than 18,000 ha of primary forest was cleared for swidden agriculture each year in Sarawak, one-fifteenth of the 270,000 ha logged by the timber industry in 1985. Moreover, when shifting cultivators clear new land, they remove useful timber for their own use (Freeman 1970) so that not all felled timber "goes up in smoke". The Baram Kenyah protect desirable species when fields are cleared, and even plant some useful trees in the forest clearings that are left to regenerate (Chin 1985).

The widespread effects of soil erosion and flooding attributed to shifting cultivation rarely relate to the activities of the "traditional" swidden cultivators associated with this form of agriculture. They do not clear sites on ridge tops and steep slopes, which are usually less fertile and more difficult to cultivate. Old-growth patches are often left in areas where shifting cultivation is practised (Jessup 1981; Vayda et al. 1980; Kartawinata et al. 1989). This mosaic with secondary forest has good soil-holding characteristics (Nye and Greenland 1960), and patches of secondary forests serve as a seed source for the recolonisation of fallow fields. Since fields follow land contours, many fields are irregular in shape. Thus the field boundary is often as close to the centre of the field as in smaller cultivated patches, allowing rapid recolonisation from seed sources in adjoining forest (Vayda

> **Box 8.2.** People and forests in East Kalimantan.
>
> Rapid economic growth in East Kalimantan over the last twenty years has been accompanied by an increase in the rate of forest exploitation, and large areas of forest have been cleared. The MAB (Man and the Biosphere) programme compared forest use by two Dayak communities and examined how traditional practices are changing with the timber boom.
>
> Several thousand Dayak people live in longhouse communities in the Apo Kayan plateau. They practise long fallow or forest-fallow shifting cultivation, clearing forest to grow a main crop of hill rice. After cropping for one or two years, old fields are left to revert to secondary forest, usually for at least 15 years and at some sites for more than 30 years. The Apo Kayan has been inhabited for centuries, with most cultivated sites being allowed to return to fallow, then recleared many times. Nowadays almost all the forest cleared is secondary forest. Such traditional shifting cultivation is not usually destructive in terms of soil fertility and forest cover. Fields are abandoned when weeding becomes a problem. Fallow periods must be long enough to suppress weeds and to prevent short-term degradation of the forest into scrub, but also sites should be left occasionally for longer periods, of 40 to 50 years, to prevent a decline in fertility and an increase in weed species (Whittier 1971).
>
> Migrants from the Apo Kayan established settlements in the lowlands of the Telen River, and mainly practise shifting cultivation of dry rice. Migrations from the Apo Kayan to the lowlands have occurred often during the last 200 years, encouraged by economic factors and the difficulties of obtaining trade goods such as salt, cloth and steel tools in the highlands. Opportunities to work and trade have attracted settlers, particularly to the navigable rivers of the lowlands (Colfer 1982; Jessup 1981; Whittier 1973). Migration had reduced the population of the Apo Kayan to 9,000 by 1980, compared to 16,000 in 1928 and 12,500 in 1970 (Jessup 1983). This has led to reduced pressure on forest and wildlife.
>
> Whereas almost all forest cleared in the Apo Kayan is secondary, most fields at Long Segar in the lowlands are cleared from primary forest and cultivated for one or two years. In the Apo Kayan about 0.3 ha is cultivated per person, compared to about 0.4 ha per person at Long Segar. The villagers of Long Segar grow rice for sale as well as for home consumption, and the availability of chain-saws and fuel allow them to make much bigger fields (up to 300 ha, compared to 30 ha in the Apo Kayan). Such large fields recover more slowly and are further from the forest seed source. Large fields also burn better and more thoroughly than small ones, and this may destroy the seed bank and delay regeneration. Agricultural practices in the Apo Kayan (where rice is grown mainly for subsistence) are less damaging to forest, especially primary forest, than in Long Segar, where shifting cultivators are degrading fairly extensive areas of forest.
>
> *Source: Kartawinata et al. 1984*

et al. 1980).

In conservation terms traditional swidden agriculture, involving the clearance of small areas for short periods with long fallows, can be regarded as a relatively benign use of marginal land. Chemical treatments are not used, so it is nonpolluting. The soil surface is little disturbed. Logs and the forest edge trap soil, so losses of nutrients and minerals by leaching and erosion are small, especially if the land is returned to fallow after one or two years. Limited and localised land clearance has little impact on native plants and animals, and forest regenerates quickly in the small clearings. However, shifting agriculture may lead to a change in the species composition of secondary forests, favouring trees adapted to fire, cutting and other disturbances associated with shifting cultivation (MacKie et al. 1987). A shifting agricultural system allows for maximum impact of natural predators on pests of cultivated crops, and traditional intercropping diversifies the yield. If abandoned permanently, old clearings will regenerate, though succession to natural forest may take many decades (Riswan and Kartawinata 1987).

Traditional shifting agriculture at low population densities can be ecologically sound, as in the Apo Kayan (Kartawinata et al. 1984; Jessup and Vayda 1988). The system is sustainable, provided that the associated conditions of low crop yields and low population density remain socially and economically desirable. The critical factor is the length of the fallow period. In a subsistence system 90% to 95% of a clan's range is usually lying fallow at any one time (Rappaport 1968, 1971). Even so, not all swidden cultivation in Borneo has been sustainable. The extensive grasslands of the Kapuas foothills are testimony to poor land use in the past. The need to open new lands may have been one of the main reasons for the large-scale migrations in Borneo's past. The Baleh Iban of Sarawak cultivated land intensively for two to six years, with no fallows or very short ones, then moved on to clear new fields (Freeman 1970). However, not all Iban groups are so destructive. Elsewhere, Iban communities have adapted their land practices to local conditions, and in some areas they clear only secondary forests (Padoch 1982).

Populations in the interior of Borneo have always been low, and in the past shifting cultivation was usually a sustainable form of land use. Today the pattern is changing. Increasing population pressure, new markets for cash crops, and the introduction of new technology, such as chainsaws and mechanised transport, have opened up greater areas to exploitation (Kartawinata et al. 1984; Jessup 1981; Jessup and Vayda 1988). The timber boom of the last two decades and the value of Borneo's commercial hardwoods have precipitated a conflict between foresters and agriculturalists. The large-scale deforestation apparent in Kalimantan today is often the work of new settlers or cultivators, enlarging their holdings for cash crops rather than subsistence farming. Settlers move in to clear forest lands round new townships and mining camps, and fields and pepper plantations follow the new logging roads (Petocz et al. 1990). In areas where there is an increased demand for land, the fallow period is shortened, and the land does not have time to recover; this leads to degradation of the environment, often resulting in alang-alang grasslands (fig. 8.5). However, even these

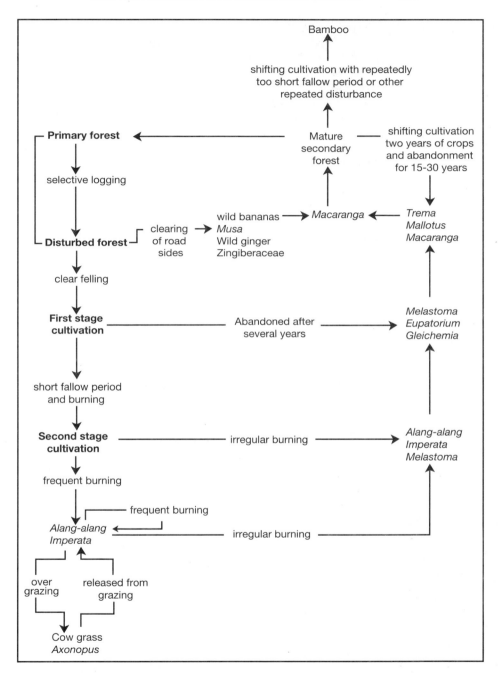

Figure 8.5. Land use and forest succession.
Source: Whitten et al. 1987a

unproductive lands can be useful to local communities. Several Dayak groups regularly fire *Imperata* grasslands to encourage new grazing for deer and other animals, which are then hunted (Seavoy 1973; Dove 1983).

Different agricultural practices require different inputs and have different ecological impacts (table 8.3). One way to take pressure off forests is to intensify cultivation, to grow higher-yielding staples or cash crops. Rice, especially sawah rice, is capable of high yields but is a highly labour-intensive crop. The Kelabits of Sarawak and West Kalimantan and the Lun Dayeh of East Kalimantan have successfully shifted to irrigated rice cultivation, as have the lowland Dusun in Sabah (see box 12.2). However, in areas of poorer soils, it is more appropriate to adopt a system of shifting agriculture, growing hill padi with a mixture of vegetable crops, both because of scarcity of labour and because it frees members of the community for other activities. Cash cropping offers ways to intensify shifting cultivation or to replace it with permanent cultivation, and has been tried with some success in Sarawak (Chin 1987). Rubber and pepper are already

Table 8.3. Comparison of some characteristics of different types of food procurement systems. (Numerical estimates are rough approximations.)

Characteristics	Foraging[1]	Shifting Cultivation	Intensive Agriculture
Human population			
Density (km^2)	less than 10	25-50*	200-500
Mobility			
settlements	temporary	temporary or permanent	permanent
food collecting area	temporary	temporary	permanent
Probable food limitation	energy	protein	protein, energy or none
Production system			
Food production (kJ/m^{-2}/year^{-1})	inapplicable	250 - 500*	100-5,000
Energy balance (output : input)	+ve, less than 5:1	10:1-30:1	+ve 5:1 to -ve 1:5
Fossil energy subsidy	none	negligible	great
Nutrient cycles	undisturbed	medium-term balance	highly disrupted
Effect on ecosystems			
Local species diversity	unaffected	maintained	greatly reduced
Successional changes	unaffected or utilised	utilised	opposed
Environmental manipulation	little to none	great, temporary	great, permanent

*Allows for fallow areas.
1. e.g., Penan

Source: Deshmukh 1986

major cash crops in Sarawak and Kalimantan (chapter 12). Ironically, crops such as pepper create even more problems of land use (Vayda and Sahur, 1985; see box 12.3). Recent developments in mixed crop/tree farming offer better solutions for permanent cropping on poor soils; these are discussed in chapter 12.

Harvesting the Forest

Patchworks of shifting cultivation tend to follow the river valleys, leaving areas of undisturbed forest in between. Many Bornean peoples exploit the forests for food by hunting and trapping, fishing and gathering. In some communities, for example, the Kenyah of Sarawak, these activities may be as important as agricultural activities, or even more important for providing food other than starch staples (Chin 1985).

There is no comprehensive survey of the wide range of forest plants used by Bornean communities. In the Malay peninsula about 2,400 plant species have economic uses (Burkill 1935; Jacobs 1982), with one-third being edible. About 800 species of the plants and ferns of Sarawak have edible parts; many are used by local peoples (Chin 1985). The Kenyah of East Kalimantan harvest at least 62 species of forest food plants, some of which are cultivated in home gardens (TAD 1981). A large number of forest food plants are found in secondary forest or abandoned ladang; indeed, the Kenyah exploit secondary forest more than primary forest for food (table 8.4).

Forests provide both food staples and emergency foods used when rice harvests are poor. The hill sago *Eugeissona utilis* may have played a key role as a food for early immigrants and is a staple for the nomadic Penan today. Sugar is extracted from the sap of the sugar palm *Arenga pinnata* and the young seeds of *Borassodendron borneensis*. Bornean forests are rich in wild fruit trees including mangoes *Mangifera*, mangosteen *Garcinia*, rambutans *Nephelium*, durian *Durio*, langsats *Lansium* and rambai *Baccaurea* (Meijer 1969). Fourteen of the 17 indigenous mangoes are sold in local markets in Kalimantan; five occur only in the wild (Bompard 1988). Several species of durians are harvested, including the wild red durians *Durio dulcis* and *D. graveolens*, and the yellow-fleshed *D. kutejensis*. Durians are so valuable that Dayaks lay claim to individual forest trees, and only the owner can harvest the fruit from the tree. Many are sold on local markets, including the prized *casturi*, endemic to South Kalimantan. Wild breadfruits and jackfruits *Artocarpus elasticus, A. integer* and the delicious cempedak *A. cempeden* occur in the forest and are cultivated by Kenyah Dayaks, as are rambutans *Nephelium eriopetalum* and *N. mutabile* (TAD 1981). As well as collecting wild fruits for home consumption and sale, many Dayak communities tend wild fruit trees and palms in the forest (Padoch and Vayda 1983) or culti-

Table 8.4. Some forest plants used as food by Kenyah Dayaks in Long Sungei Barang (Apo Kayan) and Long Segar (Telen River) - East Kalimantan.

Scientific name: FAMILY, Genus, species	English name	Part	Habitat
AMARILLYDACEAE			
Curculigo orchioides	–	fruit	2°
ANACARDIACEAE			
Mangifera caesia	mango	fruit	1°, G
Mangifera torquenda	mango	fruit	1°, G
APOCYNACEAE			
Dyera costulata	jelutong	sap (exudate)	1°
ARECACEAE			
Calamus javensis	rattan	sap	1°
C. ornatus	rattan	fruit	1°
Daemonorops mirabilis	rattan	tip	1°
Korthalsia echinometra	rattan	tip	1°
Eugeissona utilis	sago (bertam)	pith	1°, 2°, G
Pinanga kuhlii	wild betel	fruit (nut)	1°, 2° (?)
BAMBUSACEAE			
Bambusa spp.	bamboo	tip/shoot	1°, 2°
Dendrocalamus asper	bamboo	tip/shoot	1°, 2°
Schizostachyum blumei	bamboo	tip/shoot	1°, 2°
BOMBACEAE			
Durio dulcis	red durian	fruit	1°, G
D. graveolens	red durian	fruit	1°, G
D. zibethinus	durian	fruit	1°, G
COMPOSITAE			
Erechtites hierarchifolia	–	shoots	2°
CYPERACEAE			
Cyperus bancanus	reed	shoot	2°
ELAEOCARPACEAE			
Elaeocarpus spp.	–	fruit	1°, 2° (?)
EUPHORBIACEAE			
Antidesma montanum	–	fruit	1°, 2° (?)
Baccaurea spp.	rambai	fruit	1°, 2°, G
Elateriospermum tapos	–	seed	1°
FAGACEAE			
Castanopsis tungurut	chestnut	fruit (nut)	1°
Lithocarpus sundaicus	tan-oak	fruit (nut)	1°
GRAMINAE			
Setaria palmifolia	foxtail millet	seed	2°
LEGUMINOSAE			
Parkia speciosa	locust bean (petai)	seed	1°, G
Pithecellobium bubalinum	–	seed	1°
MARANTHACEAE			
Stachyprynium jagorianum	–	shoot	2°, G

Table 8.4. Some forest plants used as food by Kenyah Dayaks in Long Sungai Barrang (Apo Kayan) and Long Segar (Telen River) - East Kalimantan (continued).

Scientific name: FAMILY, *Genus, species*	English name	Part	Habitat
MELIACEAE			
Aglaia gangga	–	fruit	1°, 2° (?)
Lansium spp.	langsat	fruit	1°, 2°, G
MENISPERMATACEAE			
Albertisia papuana	–	leaf	1°
MORACEAE			
Artocarpus dadah	breadfruits	fruit	1° (G?)
A. elasticus	and	fruit	1° (G?)
Artocarpus spp.	jackfruits	fruit	1°, G
Ficus variegata	fig	bark, fruit, shoot	1°, 2°
MUSACEAE			
Musa acuminata	wild banana	fruit, shoot	1°, 2° (G?)
MYRTACEAE			
Eugenia spp.	guava	fruit	1°, 2°
PANDANACEAE			
Pandanus spp.	pandan	fruit	1°, 2°
PASSIFLORACEAE			
Passiflora foetida	passion fruit	fruit	2°
POLYGALACEAE			
Xanthophyllum spp.	–	fruit	2°
POLYPODIACEAE			
Diplazium esculentum	fern	young leaf and stem	2°
Nephrolepsis biserratus	fern	young leaf and stem	2°
Stenochlaena palustris	fern	young leaf and stem	2°
PROTEACEAE			
Helica serrata	–	leaf	1°, 2°(?)
ROSACEAE			
Rubus spp.	wild raspberry	fruit	2°
SABIACEAE			
Meliosma nitida	–	fruit	1°, 2° (?)
SAPINDACEAE			
Dimocarpus cinerea	–	fruit	1°, 2°(?), G
Lepisanthes alata	–	fruit	1°, 2°(?)
Nephelium eriopetalum	–	fruit	1°, 2°, G
N. lappaceum	rambutan	fruit	1°, 2°, G
N. mutabile	–	fruit	1°, G
Nephelium spp.	–	fruit	1°, G
ZINGIBERACEAE			
Nicolaea speciosa	ginger	shoot	(1°?), 2°, G

1° = primary forest
2° = secondary forest, including old fields
G = home gardens

Source: TAD 1981

vate these fruit trees in home gardens (Seibert 1988). Penan in East Kalimantan actively disperse fruit seeds and plant fruit trees along their migration routes; later, they return to harvest the fruit or to hunt wild pigs attracted by the fruit (Sellato 1986a).

Yams (*Dioscorea* species) grow wild in the forest and may be important "famine" foods when rice harvests fail. The locust beans (*petai*) of the

---

Box 8.3. Forest management by local villagers in West Kalimantan.

Smallholder agriculture in West Kalimantan ranges from wet-field (irrigated paddy rice) and dry-field production of annual crops to various agroforestry regimes such as trees and crops planted in home gardens, fruit and rubber gardens, forest enrichment planting and other forest management practices. The Dayak villages of Sungai Bongkang and Bagak in West Kalimantan illustrate different strategies of forest management.

Most people in Sungai Bongkang are directly dependent on old-growth forest, either as a reserve for swidden agriculture or as a source of forest products. Swidden cultivation as practised in Sungai Bongkang combines field crop production with the production of tree crops and old growth forest management. Such 'swidden-fallow agroforestry' (Deneven and Padoch 1988) combines intensive and extensive forest management techniques. After clearing a swidden from old growth, the household cultivates the land for one or two years, then makes a decision on its future use. The swidden plot can be left in unaltered fallow for future cultivation; the owner, and other villagers, will use wild or encouraged successional species. Alternatively, the fallow plot can be planted with fruit trees, including durian and/or illipe nut trees, or the swidden fallow can be planted with a cash crop such as rubber, cocoa, illipe nut or pepper.

Villagers in Sungai Bongkang manage three categories of forest: forest gardens planted in former residential sites which are dominated by fruit trees; enterprise plots: old swidden fallows planted with predominantly economic trees which generally have no local use; and forest reserves: old growth forest that has never been cleared, used primarily for construction materials and as a reserve area for swidden. The distinction between these categories may be blurred over time. In all these managed forests trees may be deliberately planted or wild trees may be tended. Planted trees include durian, other fruits, and various *Shorea* species for illipe nuts. Found/managed trees include *Agathis* for damar and other resin, latex-producing species, ironwood and honey trees.

The villagers of Bagak have only limited access to forest but own irrigated rice lands and have good roads to markets; privately held lands in Bagak are planted predominantly with economic trees (fruit and rubber). For most Bagak villagers, forest dependence is on intensively managed mixed-forest gardens. In both villages a variety of rights and claims to particular trees, forest products and landuse types are recognised, including limited-access village common property, descent-group common property and private property. Whether or not a resource has become privatised depends on a combination of three factors: the economic value of the resource, the amount and intensity of labour required for its management, and the sociocultural value of the product. Land-use and forest-use histories in the two villages have led to different management practices and rights of access, with less common access to marketable products in Bagak.

Source: Peluso 1992

forest emergent *Parkia speciosa* have protein-rich seeds which are eaten locally and sold on local markets; the young leaves and part of the flower stalk are also edible. A wide range of leaves, shoots and roots of wild plants are eaten as vegetables. The young fronds and stems of tree ferns *Cyathea contaminans* and ferns *Diplazium, Nephrolepis* and *Stenochlaena* are boiled or fried. *Diplazium esculentum* is rich in vitamin A (TAD 1981). Ferns, bamboo shoots, wild gingers and the hearts of wild banana are important relishes (Dove 1985; de Beer and McDermott 1989). Many species of macrofungi are also important foods; 43 species are regarded as edible by the Iban of Sarawak (Sather 1978). Several species are collected for sale on local markets, and the "black mushroom" *Auricularia auricula-judae* is especially prized (Chin 1981).

Illipe nuts (*tengkawang*) of the genus *Shorea* are harvested from the forest (see chapter 9). The nuts are eaten, and oil is extracted for cooking and commercial use. Candlenuts (*kemiri*) *Aleurites moluccana* provide seeds which are used for spices in cooking and are pressed to give oil for candles, lamps, soap and hair care. Many other flavourings and spices are derived from plant parts (de Beer and McDermott 1989). A substitute for salt is provided by plant ash from burning ferns, grasses, bamboo shoots and palm midribs (Ruddle 1978). The bark of wild cinnamon *Cinnamomum buhrmanii*, source of "cassia vera", is collected in the Meratus Mountains and elsewhere for home use and for sale to traders. This spice and many others are used as ingredients in medicinal preparations (de Beer and McDermott 1989).

Many rural communities in Borneo depend on wildlife for a large proportion of their diet. Wild game is hunted extensively in Sarawak and the inland areas of East Kalimantan. Charred remains of mammals from Niah show that wild meat has always been an important source of food to Bornean peoples (Medway 1969). The Kenyah of Sarawak regard practically all mammals and birds as potential sources of food (Chin 1985). Wild meat is of considerable economic significance in Sarawak. Hunters take an estimated annual harvest of one million wild pigs, 23,000 deer *Cervus unicolor* and 31,000 muntjak *Muntiacus muntjac* (Caldecott and Nyaoi 1985; Caldecott 1988). Other hunted animals include mousedeer, monkeys, squirrels, civets, pangolin, porcupines, hornbills, binturong, sunbears and flying fox. Most of these are killed with shotguns, although traditional blowpipes are more practical for small animals such as squirrels. The adult Kenyah of the Kayan River in Kalimantan have also abandoned the blowpipe for hunting, though male children use the blowpipe and poisoned darts to hunt small animals (Whittier 1973). The harvest of wild meat is an important food source for rural communities but is threatened by logging and forest clearance (box 8.4).

Riverine fish represent another important subsistence resource which is largely dependent on intact forest, since fish stocks are strongly influenced

Box 8.4. Wild game meat in Sarawak.

The indigenous people of the interior of Sarawak, as elsewhere in Borneo, are largely dependent on shifting cultivation, growing mainly hill rice. Wild game meat is an important supplement to this diet, and animals are trapped or hunted with dogs and spears, blowpipes and shotguns. All medium to large animals, including squirrels and monkeys, are hunted in Sarawak, but deer, muntjak, wild pig and mousedeer make up 60%-90% of the harvest. The single most important prey species is the bearded pig *Sus barbatus*, a forest species.

A total of about 18,000 metric tonnes of wild meat is harvested in Sarawak every year. This is equivalent to about 12 kg of meat for every person in the state, though the amount varies greatly between areas. The cost of replacing this meat with other supplies would be at least M$100 million per year and probably much more. In rural communities wild animal meat contributes almost 150 g per person per day.

The clearance of forest for logging has led to a decline in many populations of wildlife, including wild pigs and deer. Wild pigs are affected mainly by loss of feeding and breeding grounds and by damage to food trees, such as oaks and dipterocarps. Deer are vulnerable to increased hunting pressure along logging roads and around salt licks. Since bearded pigs range over large areas and make long distance seasonal migrations to areas of mast fruiting, a decline in numbers due to logging may affect several rural communities. In logged-over areas meat harvests decline, falling from about 54 kg per person per year to about 18 kg within ten years. The impact may be made worse by reduction in stocks of river fish due to increased sediment loads from soil erosion and oil pollution as a result of logging activities. As numbers of pigs and deer decline, other species may suffer from increased hunting pressure.

Source: Caldecott 1988

bearded pig *Sus barbatus*

by water quality. Other sources of animal protein are several species of frogs *Rana*, monitor lizards *Varanus* and pythons *Python reticulatus*. Only a few species of invertebrates are taken as food items: river crabs (Potamonidae); river prawns *Macrobrachium*; and one species of mollusc *Cipangopaludina* found in streams. None of the terrestrial molluscs are regarded as edible. A few insects are eaten, including short-horned grasshoppers and locusts (Acridiidae), long-horned grasshoppers (Tettigoniidae), beetles and especially the larva of the large palm weevil *Rhyncophorus ferrugineus*, which is considered a delicacy (Chin 1985). Honey is collected from the nests of wild bees *Apis dorsata* (MacKinnon 1975; Chin 1985).

Shifting cultivation is a form of subsistence farming, and many children suffer from nutritional problems (Anderson 1977, 1982). Most Dayak communities are heavily reliant on hill rice which provides 75%-80% of their protein but few vitamins or essential minerals. Protein energy malnutrition and anaemia are widespread in East Kalimantan (TAD 1980). Vitamin A deficiency is also likely. Endemic goitre is common in isolated upriver communities such as the Kayan Hulu (TAD 1981). Forest foods make up for many of these dietary deficiencies (table 8.5). Dayak children in Long Segar (Muara Wahau) were less healthy than those in Long Ampung (Kayan Hulu) who ate larger quantities of non-rice foods, many of them wild plants collected from the forest (Colfer 1981). The importance of forest foods as a dietary supplement cannot be overstressed (Chin 1985),

Table 8.5. Nutritional content of some forest plants and related home-garden species.

English	Indonesian	Latin (botanical) Scientific	Energy (kcal)	Protein (g)	Vitamin A (I.U.)
Source of energy					
Sago palm	sagu	*Metroxylon sagu*	353	0.7	0
Sugar palm	aren	*Arenga pinnata*	368	0	0
Breadfruit	sukun	*Artocarpus communis*	111	1.5	20
Jackfruit	nangka	*A. heterophyllus*	106	1.2	330
	(seed)		165	4.2	330
	cempedak	*A. integer*	116	3.0	200
Yam	ubi manis	*Dioscorea alata*	101	2.0	0
Taro	tales	*Colocasia esculenta*	64	0.6	0
Source of protein					
Locust bean	petai	*Parkia speciosa*	142	10.4	200
Lead tree	lamtoro	*Leucaena glauca*	148	10.6	4.6
Terap tree	benda	*Artocarpus elastica*	N.I.	N.I.	N.I.
Source of vitamin A					
Fern	paku daun	*Diplazium esculatum*	35	40	2,881
Illipe nut	tengkawang	*Shorea* spp.	N.I.	N.I.	N.I.
Yellow durian	lai	*Durio kutejensis*	N.I.	N.I.	N.I.

N.I. = not investigated
Source: TAD 1980

yet their continued availability depends on maintaining forests.

Forest plants are also harvested for use in the longhouse: canes or rattans for mats, baskets and numerous other items; ulin *Eusideroxylon zwageri* for construction and roofing; plants for handicrafts; fibres for weaving and wrappings; poisons for killing fish and game (e.g., blowpipe poisons such as *Antiaris toxicaria* and *Paratocarpus venenosus*); and plants for rituals and for medicines (TAD 1981; Pearce et al. 1987; Leaman et al. 1991). Animal skins, antlers, claws, boar tusks and bird feathers, especially those of the hornbills, are used for decorative or ritual purposes.

With increasing exposure to modern lifestyles and availability of alternatives, communities are utilising wild plant products less frequently, so it is important to document their knowledge before it is lost. Treatments for a wide range of illnesses and diseases are still obtained from plant extracts. These plants are easily available, familiar and free in regions where modern medicines are often difficult to obtain. Moreover many of the treatments are very effective (Leaman et al. 1991). Some plants may repay further study and pharmaceutical investigation, for example, *Pityogramma*, *Blechum*, *Nephrolepis*, *Urena* and *Celrodendrum* (Pearce et al. 1987). Other plant products with known physiological effects include the root of *Derris elliptica*, an insecticide and fish poison, *Goniothalamus*, an insect repellent with antimicrobial action, *Diospyros*, a fish poison, and the poison present in *Parartocarpus* latex. Many of these plants are still collected exclusively from the forest and will disappear as forests are cleared.

Forest products are collected for trade as well as home use (chapter 9). Surplus fish, wild meats and other forest products are traded for cash in many areas, and gathered mainly for this purpose in some. Some minor forest products are obtained as a byproduct of hunting and other forest activities. They include antlers of deer *Cervus unicolor* and muntjak; bezoar stones, especially from the grey langur *Presbytis hosei*; hornbill "ivory" from the casque of the helmeted hornbill *Rhinoplax vigil*, the only hornbill with a solid casque; tail feathers from the helmeted hornbill, the rhinoceros hornbill *Buceros rhinoceros* and Bulwer's pheasant *Lophura bulweri*; claws, canines and gall bladders of the sunbear *Helarctos malayanus*; scales of the pangolin *Manis javanica*. These items have been traded for many centuries; hornbill "ivory", for instance, was one of the most valuable imports into China before modern times.

Many rural communities depend on forests for food and other harvestable products. The way these communities utilise forests provides valuable lessons for agriculture, agroforestry and conservation.

Box 8.5. Traditional medicines of the Kenyah of East Kalimantan.

The Kenyah of the remote Apo Kayan plateau of East Kalimantan gather many of their own medicines from local plant species. Informants identified 213 plants they regard as medicinal species, used to treat at least 66 distinct health problems. A large number of these plant preparations are commonly used to treat wounds and illness, to purge worms and other parasites, and for many other medical and ritual purposes.

The Kenyah are traditional hunters, gatherers and farmers, using a long-fallow system of swidden cultivation. This system creates mixed patches of currently cultivated fields, fallow fields and regrowth (secondary) forest within the extensive upland forests of the plateau and high river valleys. Some medicinal plants can be found only in the undisturbed forests. These are normally collected by the men while hunting wild pig and deer, or on expeditions to cut wood for construction and to gather rattan. Other medicinal plants are cultivated by the women in home gardens or encouraged to invade recently cleared and burned fields. Most medicinal species known to the Kenyah are found in the secondary forests created by the traditional farming system.

The people of Long Sungei Barang, one of the Kenyah villages on the Apo Kayan plateau, know more than 200 local medicinal plants. A few are collected for their commercial value. Two important examples are gaharu and *Eurycoma longifolia*, a tree known throughout Indonesia as kayu pasak bumi, famous for its bitter taste and valued for its various medicinal and tonic properties, especially as an aphrodisiac. Most Kenyah traditional medicines, however, are used locally since imported medicines are rarely available and are very costly.

Many medicinal plants used by the Kenyah appear to be locally indigenous species. *Cissus simplex*, a liana in the grape family, is abundant in the disturbed secondary growth forests. This vine is known to the Kenyah as aka keleput laso' - the vine (aka) like a blowpipe (keleput) that treats burns (laso'). The abundant sap from the heartwood is blown from a cut stem directly onto burned flesh to relieve the pain and aid healing. Many members of the ginger family Zingiberaceae are also used medicinally by the Kenyah, to treat complaints as varied as malaria and stomachache. Many of these species have not yet been identified or studied by taxonomists.

Some of the medicinal plants used in Long Sungei Barang occur throughout Southeast Asia. Others were introduced from other parts of the world and are cultivated for their value as foods as well as medicines. *Litsea cubeba*, a tree of the laurel family, grows wild in the secondary forests of the Apo Kayan plateau; its fruit and bark are used to treat fever and chills. In Java parts of this tree are used to prepare balms and salves. The bark contains the aromatic compound citronella as well as various alkaloids.

The Kenyah cultivate pineapple *Ananas comosus*, an introduced species native to South America, for its fruit. The young stalks are used as a vegetable when other local food is scarce. The women make strong thread from the leaf fibres. The juice from an unripe fruit mixed with honey is a popular treatment for intestinal worms, especially for children, and pregnant women understand its power to induce abortion. These and similar medicinal uses are reported throughout the world.

Toxicity data suggests a strong relationship between plant chemistry and use, but also suggests that selection of plants by the Kenyah often includes more subtle criteria, such as the traditional belief in spirits and curses as sources of disease and misfortune. Of the species used by the Apo Kayan Kenyah, 40% are new to science.

The study of plants used as traditional medicines may lead to the discovery of "new" drugs and more effective treatments for disease. The great variety of local species used by the Kenyah of the Apo Kayan plateau indicates the great potential of Kalimantan forests as sources of new medicines, and confirms the importance of traditional knowledge of forest resources.

Source: Leaman et al. 1991

Omens and Augury

Living close to the forest, the Dayak peoples have a special reverence for animals, using them as important symbols in many rituals and omens in daily life (Freeman 1960; Harrisson 1960). For the Iban, for example, there are many mammals, birds, reptiles and insects whose behaviour may be meaningful in certain circumstances (Freeman 1970).

Bird augury is especially important in Dayak culture. Robert Burns, a grandson of the poet, visited the interior of Borneo in 1847 and wrote of the importance of the hornbill *Buceros rhinoceros* as an omen bird to the Kayan (Harrisson 1970). In Ngaju culture the rhinoceros hornbill sits at the head of the tree of life (Scharer 1946). It is not difficult to see why this bird should have captured the imagination of the Bornean natives, with its large size and spectacular upcurved casque. Possibly the male bird's habit of imprisoning his mate in a dark hole at breeding time also appealed to the warriors' ideas of male dominance (MacKinnon 1975). Ibans made stylised wooden carvings, exaggerating the horn into a sweeping spiral festooned with decorative figures and embellishments. These carvings were used as the centrepiece of headhunting rituals and harvest ceremonies. Although hornbill carvings are still used in religious festivals, the birds themselves no longer have any special significance.

Bird augury is not static, and the hornbill has been supplanted as an omen in Iban life by the handsome Brahminy kite *Haliastur indus*, believed to be the reincarnation of Singalong Burong, the god of war. In his avian guise, Singalong Burong imparts few warnings himself but acts through the intermediary of his sons-in-law, each believed to be the reincarnation of a god in bird form. The seven main Iban omenbirds are the rufous piculet *Sassia abnormis*, the banded kingfisher *Lacedo pulchella*, the scarlet-rumped trogon *Harpactes diardi*, the red-rumped trogon *Harpactes duvauceli*, the crested jay *Platylophus galericulatus* and the maroon woodpecker *Blythipicus rubiginosus* (Freeman 1970). These birds and the white-rumped shama *Copsychus malabaricus* are important omen birds for several other Dayak groups (Maloh, Kayan, Kenyah, Punan Gang and Bidayuh) as well as some Asian societies such as the Nagas of Assam, a possible indication of a common ancestry (Banks 1988). The Kenyah also revere the Brahminy kite and erect wooden replicas of this bird on poles in their villages to show good will and respect.

Apart from their role in augury, birds play an important part in prescribing the times for agricultural activities. The Kelabits of the central highlands time the planting of their crops according to the arrival of winter migrants from Asia. Different months in the Kelabit calendar are named after the various migrants. When the yellow wagtail *Motacilla flava* is seen on the open uplands, it is time to prepare the rice fields for planting. In October, when the brown shrike *Lanius cristatus* arrives, the

Box 8.6. Uses of forest plants in Sarawak.

Use	
Abortificants and contraceptives	*Cyrtandra angulerii*
Aphrodisiac	*Euricoma longifolia*
Artifacts (shields, masks)	*Alstonia spatulata*
Bark cloth	*Xanthophyllum scortichinii*
Basketwork	Bamboo, rattan, pandans, palms
Beverages	*Nypa fruticans, Uncaria* sp.
Bird lime	*Artocarpus elasticus*
Boats	*Dehaasia brachybotrys*
Building	*Eusideroxylon melangangai, E. zwageri*
Carvings	*Myristica eliptica*
Caulking	*Melaleuca leucadendron*
Charms	*Horsfieldia polyspherula*
Cordage	*Ficus benjamina, Goniothalamus malayanus*
Containers	Bamboos
Cooking and eating utensils	Bamboo, *Nepenthes* pitcher, *Pandanus*
Decoration and ornaments	*Lepisanthes divacirata, Dicranopteris linearis, Lycopodium cernuum*
Drugs	*Aetoxylon sympetalum*
Dyes and tannins	*Oroxylum ridicum, Castanopsis borneensis*
Exudates, damar	*Hopea, Shorea*
Fibres	*Arenga pinnata*
Fish intoxicants/poison	*Derris eliptica, Fordia coriacea*
Fishing rods	*Salacca vermicularis*
Fruits	*Durio, Baccaurea, Nephelium*
Furniture	*Calamus scipionum*
Gums	*Dyera costulata, Artocarpus elasticus, Palaquium gutta*
Incense	*Begonia holttumii, Aquilaria microcarpa, Eleiodoxa conferta*
Insecticides	*Derris eliptica*
Juices and saps	*Ventilago malaccensis*
Medicines	*Pityrogramma calomelanos, Aralidium pinnatifidum*
Musical instruments	*Shorea atrinervosa, Eusideroxylon zwageri, Intsia palembanica*
Oils	*Shorea beccariana, S. fallax, S. hemsleyana, Mesua ferrea*
Poisons	*Parartocarpus venenosus, Antiaris toxicaria*
Resins	*Hopea beccariana, Shorea albida, S. argentifolia*
Ritual	*Cordyline fruticosa, Goniothalamus macrophyllus*
Rope and string	*Tetrastigma pedunculare, Polyalthia glauca*
Salt and sugar making	*Nypa fruticans*
Spices	*Cinnamon pathenoxylon*, gingers
Starch	*Metroxylon sagu, Eugeissona utilis*
Stimulant	*Piper*, betel
Thatching	*Shorea flaviflora, Cratoxylon arborescens, Nypa fruticans*
Tool making	*Eusideroxylon zwageri, Melanorrhoea beccarii*
Vegetables	*Stenochlaena palustris, Gnetum gnemon, Thespesia populnea*
Vinegar	*Nypa fruticans*
Wearing apparel	Rattans, *Pandanus*
Weapons	*Koompassia excelsa, Prainea limpato, Eusideroxylon zwageri*
Weaving	*Gossypium barbadense*

Source: Pearce et al. 1987

farmers must hurry to finish their sowing (MacKinnon 1975).

Terrestrial mammals play a large role in the spiritual, verbal or artistic life of most Bornean peoples. The barks of deer and mousedeer are significant omens which may even lead to the separation of a newly married couple (Ling Roth 1896). The rhinoceros is such an awesome beast that Dayak men often go through the painful process of inserting a *palang* or cross piece through the penis to emphasise their manhood and to emulate the male rhinoceros (Harrisson 1964). The clouded leopard, largest of the forest cats, is also an animal of ritual significance. Clouded leopard skins are worn for ceremonies. A carving of a clouded leopard may adorn the top of a *belawang* pole and, traditionally, this required a freshly taken head, or the sacrifice of a slave, during the ritual Iban ceremony for cleansing and strengthening the village (Freeman 1970).

The gibbon is more highly revered than any other diurnal arboreal mammal. All the Land Dayak peoples of southwest Borneo attach some significance to the gibbon as an omen, though the message it imparts does not depend on direction as with most other omens. It is a bad sign if gibbons are silent, but loud, persistent cries during a drought herald the approach of rain. If a man is lost, gibbon calls will lead him home (Harrisson 1966). The orangutan holds a special place in Dayak life, featuring in many legends (MacKinnon 1974b). When headhunting was officially banned, many Dayak tribes turned to collecting orangutan heads instead of human ones. Some beautifully carved skulls can still be found, hanging in Dayak longhouses where they are sometimes used in healing rituals (Medway 1976; Allan and Muller 1988). Sadly, there seems to have been a recent upsurge in hunting of orangutans to obtain skulls for sale as tourist souvenirs.

Traditional Dayak beliefs are disappearing fast along with the cultural and technological restraints that protected Borneo's forests and wildlife. Many Bornean mammals are threatened by forest loss. Orangutans and gibbons are kept as household pets, and many other forest mammals and birds are regarded as pests that damage and destroy crops. Fortunately, large areas of relatively untouched habitat still exist, especially in the mountainous interior. Many of these areas are proposed reserves, and it would be fitting if the Dayak peoples, with their special knowledge of the forests and their inhabitants, could be engaged to guard these areas and to guide visitors. Wilderness tourism to areas such as Kayan-Mentarang may be one way of bringing new income to the inland peoples, while conserving the forests that have long provided their livelihoods.

Transmigration and Resettlement

Since colonial times, in the early 1900s, large numbers of people have been moved under government schemes from the densely settled islands of Java, Bali, Madura and Lombok to the sparsely populated Outer Islands, including Kalimantan, to relieve pressure on overstressed watersheds and to open new land for agriculture. By 1985 at least 2.5 million people had been relocated to new transmigration settlements. The number of unassisted migrants who moved to the Outer Islands during the same period may be two to three times as great (Ross 1985).

Resettlement to Kalimantan began in 1921 when 250 families from Central Java were moved to a site at Barabai, South Kalimantan, but the land was unsuitable, and malaria claimed many victims (Avé and King 1986). Undeterred, the Dutch continued with their colonisation project, settling Madurese in the uplands of South Kalimantan and Javanese in the tidal swamps. The first transmigration to East Kalimantan dates from 1938 when 26 families were moved from East Java to Jambayan in the Kutai area (Avé and King 1986). Since 1950, which was the beginning of large-scale organised transmigration, more than 120,000 families were moved to over a hundred sites in Kalimantan (Whitten et al. 1987c; MOF/FAO 1991). Many were Javanese, and most were sent to East Kalimantan (50,000 people or 12,000 families). In West Kalimantan more than 37,000 settlers, mainly Javanese, arrived between 1971 and 1982.

Transmigration has not been wholly successful, and some settlements have failed, often because of poor site selection. Part of the large Rasau Jaya transmigration colony, in the Kapuas delta south of Pontianak, has been abandoned because the land there was unsuitable for farming (Avé and King 1986). Similarly, Banjarese farmers abandoned the land at a government resettlement scheme on white sand soils at Keluang Pantat in the Pasir district of East Kalimantan (Vayda et al. 1980). The natural vegetation of white sand soils is kerangas forest; these areas are usually totally unsuitable for agriculture (Kartawinata et al. 1984).

Elsewhere settlements have suffered from the consequences of poor site preparation. Because of time constraints, sites are often cleared with heavy machinery. If land-clearing operations are not done carefully, they may result in removal of topsoil, severe soil compaction and the blocking of natural waterways with debris, all of which can cause environmental problems (Ross and Donovan 1986). Manual or semimechanical clearance, though labour-intensive, is often less damaging. Even so, much soil fertility may be lost, since felled trees are rarely burned *in situ* (the most ecologically sound method) but are often pushed with topsoil into hollows and water courses to level the land (Whitten et al. 1987c). The early establishment of cover crops helps to minimise erosion and compaction of the soil.

Transmigration is costly. Official estimates place the cost of relocating

Box 8.7. The cows of Long Segar: a lesson in development aid.

Long Segar village in East Kalimantan is a resettlement project, part of Proyek Resetelman Penduduk Long Segar/Long Noran (RESPEN). Long Segar has generally prospered under the government-sponsored RESPEN scheme, but much of the aid could have been directed better if the settlers, Christian Kenyah Dayak, had been consulted in the planning stages of the resettlement.

It is government policy to encourage sawah cultivation in this area because wet rice gives higher yields with reduced labour (per crop) from permanent fields. The people of Long Segar have abandoned sawah cultivation and returned to cultivation of hill rice as their staple food because the area is unsuitable for sawah cultivation. The yellow podzol soil is shallow and infertile, the uplands are subject to drought and erosion and the lowlands are subject to flooding. After the third year of planting, the rice crop is inadequate, and sometimes alang-alang replaces the last rice crop. These problems possibly could be overcome by importing fertilisers, building terraces, and using water pumps, but these solutions are expensive.

Accommodation for the resettlement villagers has been provided according to a standard village plan, with each household given a separate house. Traditionally Kenyah people live in longhouses, sharing food and gathering on the open front of the longhouse for community discussions in which all longhouse members participate. Moving into separate households has led to a breakdown of traditional Kenyah society. Households now only share food with immediate neighbours, and since no one has a porch large enough for all members of the community to come together, women and children are now excluded from the community decision-making process. Sharing of childcare responsibilities is more difficult with nuclear households, and women have less time for productive activity outside. Women's decreased involvement in economic activities and in decision-making undermines their traditionally high status in the community, since political and economic activity go hand-in-hand with high status.

The disparity between what the government agency RESPEN and the Kenyah themselves believe to be appropriate technology for the resettled villagers is best exemplified by the story of the cows of Long Segar. In 1975 RESPEN brought 16 cows to Long Segar as draft animals to help in the cultivation of the sawah. These 16 cows were to belong to the community as a whole and act as seed stock. Sixteen families were given a cow. Contracts stipulated that the first calf was to be given to another member of the community (who signed another contract), the second calf belonged to the owner of the cow, the third to the owner of the bull and the fourth began the cycle again. If no calf was born to the cow, she belonged to the contract owner after seven years. In this way it was hoped that each family in the community would soon own a cow. The seed stock could not be sold until all members of the community had a cow. Similar government-sponsored schemes to introduce cattle to village communities are being implemented throughout Kalimantan (Boring pers. com.).

Unfortunately, the cow introduction scheme at Long Segar has been a failure. After five years, and five calves, only 10 cows remain in the village where they run wild, causing damage to gardens; the others have died accidentally or have been slaughtered after damaging gardens of an adjacent Kernyanyan community. The people of Long Segar earn their livelihood from swidden agriculture, a way of life that involves periods of intense activity when the villagers live in their ladangs, and

> **Box 8.7.** The cows of Long Segar: a lesson in development aid *(continued).*
>
> periods of little agricultural activity when the people remain in the village. Tethered or corralled cows need daily care to provide food and water, and access to the bull. Since it is not practicable to transport the cows to the *ladangs*, a trip often undertaken in small canoes, the cows are allowed to run wild to fend for themselves but in so doing cause considerable damage to village gardens. Since the cows are community-owned and the contracts state that any damage they cause will be paid for at the time of sale (a long time in the future with only 10 cows and 229 families!), no individual can be held responsible for the garden damage.
>
> The cows were never requested nor wanted and have become a nuisance to Long Segar and adjacent communities. The people are afraid to dispose of them outright, as they were provided by the government, but, because of their agricultural lifestyle, are unable to devote the time and care that the cattle require. They represent a good example of development assistance, planned from outside and above but inappropriate to local conditions. The government agency's intentions were well-motivated. RESPEN has spent approximately Rp 500,000 per household in Long Segar but failed to consult people on its plans for the community. Rather than promote wet rice cultivation, in association with cattle for draft and beef, aid may have been better spent on promoting cash crops, such as coconuts, sirsak, oranges (all of which grow well) and possibly coffee, as well as on agroforestry schemes to ensure a steady supply of bamboo, rattan and other necessary weaving materials. Such projects would allow the community to continue traditional swidden agriculture to meet their daily rice requirements. The message is clear. Experimental schemes should be prepared in consultation and collaboration with the recipients, to determine what is appropriate to their needs and situation.
>
> *Source: Colfer and Soedjito 1988*

a family at nearly $10,000 (Reppeto and Gillis 1988). In 1984 the transmigration programme accounted for 2.9% of federal development expenditures, more than the share of all public health and family planning programmes (Ross 1982, 1985). In view of these high costs it is imperative that transmigration settlements succeed, and indeed many do.

Poor site preparation, poor soils, inappropriate crop selection or development options can all lead to low yields and hardship for the settlers. The government emphasis on growing sawah rice in transmigration settlements has not always been appropriate on Kalimantan's poor soils, which are often more suited to tree crops. Some of the more successful old transmigrant villages in marginal lands have succeeded because they adopted local systems of agriculture that were already proven. Thus Javanese settlers at Purwosari used the Banjarese system of cultivating tidal wetlands (Collier 1977; see chapter 10). It is encouraging that attention is now turning more to tree crops (RePPProT 1990). The government plans to place 80% of future transmigrants in rubber projects and other commercial tree-cropping projects, to settle areas unsuitable for annual crops (Repetto 1988).

Many transmigration settlements, especially in East Kalimantan, are remote from coastal settlements and urban centres. Farmers are able to grow crops but may have little access to good transport or markets to sell their produce. Transmigration sites on the brink of failure because of constraints such as poor soils, low crop yields and lack of access to markets are now the focus of rehabilitation projects under the Second Phase Development programme funded by the World Bank; Sebamban in South Kalimantan is one such project.

Because of the failures cited above, transmigration has attracted some severe criticism, both within Indonesia and internationally (Secrett 1986). Those opposing transmigration on environmental grounds argue that it is a major cause of forest loss. In almost all provinces the land allocated to sponsored transmigrants amounts to less than 10% of the conversion forest and less than 1% of the total forest area. For all Kalimantan provinces, except South Kalimantan, the figures are much lower, amounting to less than 1% of conversion forest; in South Kalimantan as much as 6.6% of conversion forest (which includes degraded alang-alang) has been allocated for transmigration (Whitten et al. 1987c). It should be noted, however, that certain types of forest, especially those on shallow peats, are more likely to be cleared for settlement.

It has been suggested that the total land cleared as a result of a transmigration programme may be five times as great as originally planned (Ross 1984). Certainly unassisted migrants, encouraged to follow the government-sponsored migrants, clear large areas of adjacent forest lands. The settlers themselves, mostly Javanese, may perceive forest as a source of ghosts and animal pests and are quite happy to see it felled, even though scrub may harbour more pigs and rats than natural forest (Whitten et al. 1987c). The problem may be further exacerbated by transmigrants moving off unsuitable sites and by local people who join in the general free-for-all of forest clearance to extend their holdings. The environmental consequences of such shifting agriculture can be serious, leading to degradation of hillsides and forested lands. Sometimes transmigrants on poor land turn to other activities, such as the gold mining that threatens parts of Tanjung Puting National Park in Central Kalimantan. It is essential therefore that transmigration planners consider the likely long-term impacts on habitats and wildlife of locating settlements near primary or little-disturbed forests (see chapter 9). Transmigration settlements and wildlife reserves do not make good neighbours.

As well as immigrants from other islands, the transmigration programme accepts a significant proportion of local farmers. In both Kalimantan and

Sarawak isolated inland groups are encouraged to move downriver to more accessible settlements, for ease of administration and better access to schools and medical care. They are encouraged to abandon shifting agriculture to farm permanent fields and grow sawah rice.

In East Kalimantan about 10,000 Dayak families, mainly Kenyah, have been relocated since the beginning of the 1970s (Avé and King 1986). Finance for these schemes has come mainly from the taxation of timber companies. Dayak villages have also been resettled in South Kalimantan and, at a slower pace, in West and Central Kalimantan, sometimes to move them out of timber concessions. By 1980 1,500 Dayak families had been resettled in South Kalimantan, including the Dayak of the Jorong uplands who formerly controlled the production of *sirap* (ironwood shingles) there (Avé and King 1986). In West Kalimantan a number of Kendayan Dayak have been moved from hilly regions to the lowlands. Although well intentioned, these schemes have not always been successful in their avowed aim of halting forest destruction. Often the resettled people, with greater access to modern technology and markets for cash crops, cause more environmental damage than the people of the parent village (see box 8.2).

Governments have sponsored tree-crop schemes under the PIR project in Kalimantan (Perkebunan Inti dan Rakyat: Nucleus Estate and Smallholder Scheme) and SALCRA projects in Sarawak. Not all of these schemes require forest clearance; many are on *Imperata* grasslands. Each participant obtains two hectares of rubber or other cash crops, one-quarter of a hectare for a house and house-lot, and one hectare for his own use. For the first eight years he must maintain his rubber until it has come into full production, and during this time he receives a large loan (about $3,500 in West Kalimantan) which he must pay back with interest. The Javanese appear to value the project, but the indigenous Dayaks are less enthusiastic, because they lose their independent status as shifting cultivators, and there is uncertainty over continuing rights to the land (Avé and King 1986). In 1992 the Ministry of Forestry and Ministry of Transmigration signed an agreement to involve transmigrants, and resettled local communities, in the Industrial Timber Estate (HTI) programme, which aims to increase timber production from plantations on Kalimantan and other Outer Islands.

Kalimantan has long been a destination for various groups of migrants who have come without government assistance. An especially interesting group in West Kalimantan is the Madurese, from the small island of Madura off the northeast coast of Java. Madurese have been migrating to the west coast of Kalimantan since the beginning of this century and now comprise 2.5% of the population of the province (Sudagung 1984). Some of them are not permanent migrants but move back and forth periodically between Madura and Kalimantan. Most Madurese are farmers, rather better than the Javanese at cultivating the dryland hill soils of Kalimantan, as they are used to poor quality land in their home island, Madura. There they customarily establish dry fields (*tegalan*). This form of agriculture is more appropriate to much of Kalimantan than irrigated cultivation. Many Madurese in Riam Kiwa, South Kalimantan, farm alang-alang lands on the hills but cultivate sawah in valley bottoms (Potter 1987). Madurese also keep livestock, rearing cattle which graze on the alang-alang grasslands.

THE FUTURE

Borneo is a rich potpourri of peoples. Though few in number, they have altered the island landscape, creating rolling grasslands where there was once natural forest. Forest clearance for logging and intensive agriculture is destroying rich natural ecosystems and replacing them with unsustainable agricultural systems. It is no accident that fertile Java can support human densities of about 800 people/km^2 whereas Kalimantan supports human populations 50 times lower. Kalimantan's generally poor soils are not suited to intensive agriculture, yet increasing populations demand increasing productivity. The main challenges for Borneo's people in the future will be to harvest forest resources on a sustainable basis and to develop appropriate agroforestry systems combining cash and staple crops.

Chapter Nine

Forest Resources

Over 60% of the land surface of Borneo is still under natural forest (MacKinnon and MacKinnon 1986). With such extensive forest resources it is not surprising that timber is a major source of revenue for Kalimantan and the East Malaysian states of Sabah and Sarawak. Oil-rich Brunei has less need to harvest its forest for timber and still retains more than 70% forest cover. Although Borneo's forests are now mainly valued for their timber, they also provide many other goods and services of benefit both to local communities and to the national economy.

In 1968 Kalimantan was estimated to have 77% forest cover, with 41,470,000 ha of forest, about 34% of the total forest area of Indonesia (Avé and King 1986). In 1984 the national forest map (TGHK) was produced, mapping areas of production forest and of other forest categories. Using satellite imagery and aerial photographs from 1982 to 1988, the National Landuse Project (RePPProT) mapped forest cover in TGHK areas throughout the Outer Islands (table 9.1). At this time forest cover in Kalimantan was still estimated at 73%, although it was obvious that large areas of production and protection forests had been cleared of forest. Forest loss was not evenly spread throughout Kalimantan. While South Kalimantan retained only 48% forest cover, 91% of East Kalimantan remained under forest (RePPProT 1987). By 1990, with better baseline data, forests were estimated to cover only 34,730,000 ha, or 63% of Kalimantan (MoF/FAO 1991; fig. 9.1). This represents a forest loss of 7 million hectares over twenty years.

Forests for Timber

Timber is a major export earner for Indonesia, second only to oil, with much of the exported timber coming from Kalimantan. Most wood is harvested from natural forests rather than plantations. Timber exploitation has a long history in Kalimantan and was already important during Dutch colonial times. Starting in 1904 a number of timber concessions were granted in the upper Barito and the self-governing lands of the east coast, especially Kutai (Potter 1988).

By 1914 80% of the timber floating down the Barito was from dipterocarps, while wood coming from the east coast was mainly ironwood (van Braam 1914). The large eastern tracts of dipterocarps were much more inaccessible and difficult to exploit, and several early attempts failed in spite of heavy investment (Potter 1988). In 1924 Dutch colonial officers produced a forestry map for the Residency of South and East Borneo (covering the provinces of Central, South and East Kalimantan), which showed 94% of the residency as forest, or 75% when swamp, strand and secondary growth were excluded. Figures for the extent of forested lands published in 1929 were still the basis for giving timber concessions in 1975 (Hamzah 1978; Potter 1988). Even during colonial times forest conservation was already a matter of concern. Four forest complexes were established as hydrological reserves in southeast Borneo: the mountains of Pulau Laut and three reserves covering the Meratus Mountains from north to south (van Suchtelen 1933).

Table 9.1. Summary of forest cover within existing TGHK forest categories.

Region	Forest Categories						
	Nature reserves	Protection forest	Limited production	Normal production	Conversion forest	Unclassified areas	Totals*
Kalimantan							
Total area (km^2)	36,355	65,469	118,292	132,246	115,325	68,147	535,834
Forest cover (km^2)	33,147	60,082	101,695	110,311	74,027	16,335	395,597
%	91.2	91.8	86.0	83.4	64.2	24.0	73.8
Sumatra							
Total area (km^2)	41,114	64,121	66,576	69,319	86,760	147,419	475,309
Forest cover (km^2)	34,670	42,753	46,919	48,995	36,290	23,608	233,235
%	84.3	66.7	70.5	70.7	41.8	16.0	49.1
Sulawesi							
Total area (km^2)	14,426	43,202	45,877	14,222	16,241	52,177	186,145
Forest cover (km^2)	12,008	33,368	35,472	10,027	10,193	11,626	112,694
%	83.2	77.2	77.3	70.5	62.8	22.3	60.5
Maluku/Nusa Tenggara							
Total area (km^2)	8,033	29,681	24,339	14,151	51,782	30,773	158,759
Forest cover (km^2)	5,498	20,626	18,658	8,968	29,170	5,294	88,174
%	67.9	69.5	76.7	63.4	56.3	17.2	55.5
Irian Jaya							
Total area (km^2)	73,027	109,169	45,893	77,242	96,401	13,068	414,800
Forest cover (km^2)	61,721	95,083	41,759	70,441	77,940	2,639	349,583
%	84.5	87.1	91.0	91.2	80.8	20.2	84.3
Totals for Outer Islands							
Total area (km^2)	173,173	311,995	301,345	307,637	366,959	309,710	1,770,819
Forest cover (km^2)	147,004	251,912	244,503	248,742	227,620	59,502	1,179,283
%	84.9	80.7	81.1	80.9	62.0	19.2	66.6

* These forest totals include some secondary forest as well as natural forests. Note that large areas of production and protection forests have been cleared of forest.

Source: RePPProT 1990

FORESTS FOR TIMBER 397

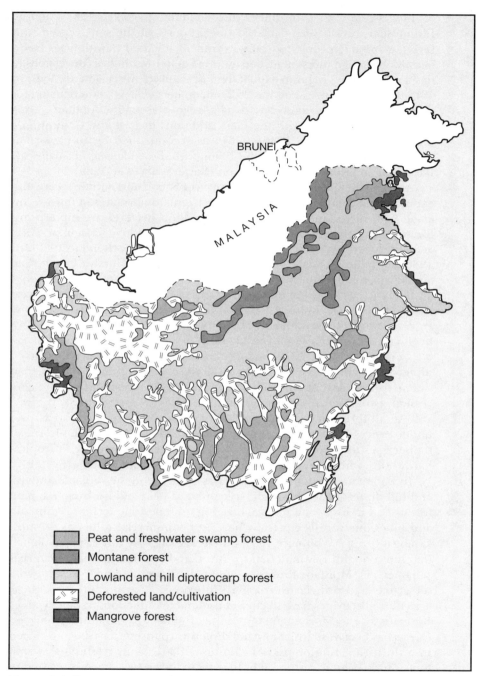

Figure 9.1. Forest resources in Kalimantan.
Source: REPPProT 1990

Large-scale cutting of timber in Kalimantan began in 1967 when all Indonesian forests were declared the property of the state. Faced with severe economic problems, the government initiated new policies based heavily on foreign investment, and awarded generous timber concessions to foreign companies eager to exploit the vast stands of tropical hardwoods. By 1972 the area of concessions was 26.2 million hectares. As world demand for cheap tropical hardwoods grew, this figure increased to more than 31 million hectares by 1982, with the main additions in East and Central Kalimantan (Avé and King 1986). This was accompanied by an important change in logging methods in the 1970s, with the outlawing of small-scale harvesting in favour of mechanisation (Kartawinata et al. 1989).

At the same time the Malaysian states of Sabah and Sarawak were also experiencing a timber boom. Sabah has 4.7 million hectares of forests, covering 64% of its land area. About two million hectares are dipterocarp forests, with potential commercial yields as high as 140 m^3/ha; in practice, actual recoveries have been about half this yield or less (Repetto 1986). By 1980 3.2 million hectares, virtually all of Sabah's productive forests, had already been logged or placed under logging concession (FAO 1981).

Sarawak has some 9.4 million hectares of forest, 75% of the land area. Swamp forests account for 1.5 million hectares, 16% of the forested estate. By now, most of the valuable stands of ramin *Gonystylus bancanus* have been cut, although a significant amount of commercially valuable *Shorea albida* remains. After 1970 logging shifted toward harvesting the mixed dipterocarp hill forests, which by 1978 were yielding more timber than swamp forests. At that time only about 41% of hill forests were under concession, compared to almost all swamp forests (Repetto 1988). By 1986 86% of all forest in Sarawak (excluding parks and protected areas) was under concession; today the figure is even higher. Much of the Bornean timber goes to Japan, which absorbs 29% of the world's trade in tropical hardwoods, with 75% of this coming from Sarawak and Sabah (fig. 9.2).

In Kalimantan extensive logging has occurred in the valuable lowland and hill dipterocarp forests, in large areas of peat and freshwater swamp forests and in ironwood forests. Logging on limestone terrain is difficult, and limestone forests generally have less commercial value, as do most kerangas forests (Whitmore 1984a). Coastal mangrove forests are exploited for chipwood, for raw materials for rayon and for local building materials (chapter 11). Montane forests are designated as protection forests to conserve soils and protect hydrological functions. Logging and subsequent agricultural activities have led to a considerable reduction in some types of Bornean forests. More than 60% of the original area of natural ironwood forests has been lost, while lowland dipterocarp forests have been reduced by a third (MacKinnon and MacKinnon 1986). Many freshwater swamp forests have been so disturbed by logging that they have been converted to open swamp habitats, as in the Sungai Negara wetlands (South Kalimantan) and Mahakam basin (East Kalimantan).

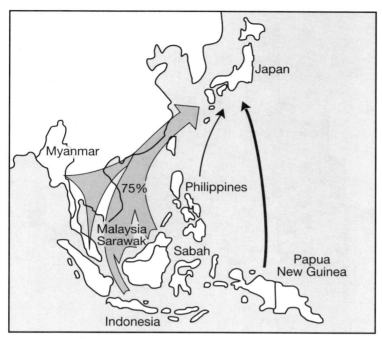

Figure 9.2. Main supplies of tropical timber to Japan in 1987. Three-quarters of all tropical timber products imported into Japan come from the island of Borneo. Logs are exported from Sabah and Sarawak, and a large proportion of Indonesian plywood, veneers and sawnwood exports to Japan come from Kalimantan.
Source: Nectoux and Kuroda 1989

Although Borneo has more than three thousand tree species, only a few are harvested for commercial timber. Loggers focus on fewer than a hundred species in Borneo (Burgess 1966), with exports predominantly of 12 species (Myers 1984). In Kalimantan the most valuable timber trees are ironwood *Eusideroxylon zwageri* and dipterocarps, including meranti *Shorea* spp., merawan *Hopea* spp., kapur *Dryobalanops* spp. and keruing *Dipterocarpus* spp. The most valuable timber trees extracted from swamp forests are ramin *Gonystylus bancanus* and the dipterocarp *Shorea albida*, which occurs in western Borneo. Among the legumes, *Intsia bijuga*, *Intsia palembanica*, *Pericopsis mooniana* and *Pterocarpus indicus* provide valuable timber. Over the last 16 years meranti has accounted for more than 70% of the timber harvest in Kalimantan.

In 1971 the whole of Kalimantan produced 1.16 million cubic metres of wood, which increased to 14 million cubic metres in 1976 and to 18 million cubic metres by 1978. After 1979 the export of wood declined substantially

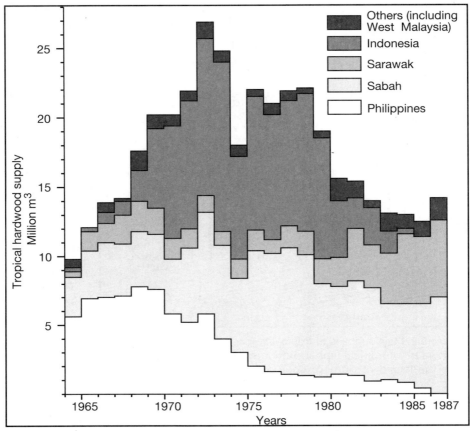

Figure 9.3. The rise and fall of Indonesian log exports. The Malaysian states of Sabah and Sarawak have been the main suppliers of logs to Japan since 1978. Between 1971 and 1978 Indonesia was the main supplier. In 1974 Indonesia exported 11.5 million cubic metres to Japan (47% of tropical log imports) with eight million cubic metres from Kalimantan.

Source: Nectoux and Kuroda 1989

(fig. 9.3) because of a fall in demand on the world market and a change in policy in Indonesia, which reduced or eliminated exports of logs and raw timber in favour of plywood and processed wood products. Nevertheless, logging has continued apace (table 9.2). The effects of logging can be seen clearly as one flies over Kalimantan. Where there was once almost continuous forest, now there are grassy hills, patchworks of cleared agricultural fields and yellow logging roads extending like fungal hyphae into the remaining forest. The 80-minute flight from Banjarmasin to Balikpapan crosses forest for only about one-fifth of the journey. Flights over the lowlands of East Kalimantan reveal similar landscapes: forest dissected by logging roads, large gaps created by logging, agriculturalists and the 1982-83 fire, and muddy rivers, the consequence of upstream logging activities and land clearance.

By 1990 remaining areas of production forest in Kalimantan covered 16.74 million hectares, of which 10.79 million hectares remained unlogged; another 5.97 million hectares of conversion forest was available for clear felling prior to conversion to other land uses (table 9.3). The extent of remaining forest cover in Kalimantan was estimated to be only 63%, compared with almost 77% in 1968. Even if the earlier figures are not totally reliable, this decrease gives some impression of the scale of forest destruction. It would mean that over twenty years some 70,000 km^2 (7 million hectares) of forest were cleared as a result of logging and subsequent slash-and-burn agriculture; this does not include damage to adjacent areas of forest. This is equivalent to losing an average 3,500 km^2 every year, or 1,000 ha of forest every day.

Forest loss is occurring especially fast in East Kalimantan. In 1978 this province, with less than 1% of Indonesia's population, produced nearly 25% of the country's total export earnings, mainly from timber and petroleum products. Many migrant workers have come to East Kalimantan to find work on the logging concessions and in the sawmills. As the forests are opened by logging roads, new immigrants and settlers encroach along them, clearing *ladangs* (small agricultural fields) as they go. Their slash-and-burn agriculture is far removed from the traditional swidden practices of

Table 9.2. Log production in Indonesia from 1981 to 1985/86 (cubic metres).

Province	1981	1982	1983/84	1984/85	1985/86
Sumatra	4,553,821	3,093,252	3,871,277	4,142,437	3,704,911
Java	833,220	869,946	1,326,069	1,481,910	475,182
West Kalimantan	1,323,460	1,083,512	1,904,858	1,866,996	1,691,041
Central Kalimantan	2,962,555	3,117,765	3,060,727	2,688,181	3,419,456
East Kalimantan	2,856,560	3,282,371	3,097,924	4,124,001	3,557,383
South Kalimantan	810,404	664,585	750,057	596,829	514,192
Total Kalimantan	7,952,979	8,148,233	8,813,566	9,276,007	9,820,072
Bali & Nusa Tenggara	21,892	–	26,873	18,860	25,802
Sulawesi	533,261	194,188	206,118	123,231	210,169
Maluku	1,216,435	742,531	790,022	731,121	780,025
Irian Jaya	763,818	328,363	174,643	189,143	173,789
East Timor	–	–	–	–	–
INDONESIA	15,875,426	13,376,513	15,208,568	15,962,709	14,551,950

N.B. The figures in the table represent only logs from HPH concessions on which royalties were collected. Total log production, including logs from conversion forest, is about 10 million cubic metres more each year.

Source: *MoF/FAO 1991*

Table 9.3. Estimated condition of mixed hardwood natural forest with management potential in 1990, by regions of the Outer Islands. (Area in million hectares).

Region	Regular and limited production forest			Conversion forest	Tidal forest with management potential	Forest within nature reserves	Other forest	Total forest	Forest as percentage of land area
	Unlogged	Logged	Heavily logged						
Sumatra	5.66	1.81	0.90	3.60	0.39	3.47	4.55	20.38	43
Kalimantan	10.79	4.03	1.92	5.97	0.89	3.04	8.09	34.73	63
Sulawesi	2.63	0.19	0.80	1.33	0.14	1.20	4.04	10.33	52
Maluku	1.73	0.22	0.11	1.96	0.20	0.41	1.40	6.03	70
Irian Jaya	9.55	0.10	0.03	6.09	0.53	6.17	11.18	33.65	82
Nusa Tenggara	0.08	0.01	0.01	0.03	0.01	0.13	1.70	2.36	29
Total	30.07	6.36	3.77	18.98	2.16	14.42	30.96	107.48	60

Source: *MoF/FAO 1991*

established Dayak villages. These new settlers include Bugis pepper planters and spontaneous transmigrants from South Kalimantan and South Sulawesi (Kartawinata and Vayda 1984). Indigenous Dayaks living adjacent to logged areas and transmigration schemes have also taken advantage of the new access given by roads, and have cleared and burned extensive ladangs (for example, along the new roads along the east coast and inland from Sangkulirang to Muara Wahau). Where nearby urban centres, and oil and timber camps, provide markets for fruit and vegetable crops, cultivators may clear new ladangs to grow cash crops. Thus both logging, and also the settlers that follow behind, are changing the landscape of much of Kalimantan.

The pattern is similar in northern Borneo. Since 1976 logging followed by agricultural conversion has accounted for between 40% and 45% of Sabah's deforestation. According to government sources, shifting cultivation has been responsible for more than half of the state's annual forest loss, and the total area affected by shifting cultivation amounted to three times the area of forest logged through 1980 (Repetto 1988). This figure can be disputed since shifting cultivation often involves clearance of secondary forests (see chapter 8). Nevertheless, logging also contributes to this category of forest loss by opening up new lands to pioneer farmers.

Effects of Logging on Forest Structure and Dynamics

Even under modern intensive methods of logging, only a small proportion of rainforest trees are commercially valuable and will be removed for timber. Thus, in an average dipterocarp forest, about 14 trees are felled per hectare; the ITCI concession in East Kalimantan removes between 8 and 12 trees per hectare. The figure may be higher in richer dipterocarp associations containing *Dryobalanops aromatica* and *Shorea curtisii*.

Although only a few trees are removed, far more are damaged by the logging process (fig. 9.4). The chosen timber trees are all large emergents, with crowns often 15 m across. When they fall they destroy or damage a considerable part of the surrounding forest, including smaller fruiting trees that are of value to wildlife. The number of adjacent trees pulled over by the falling timber is particularly high in areas where lianas link neighbouring trees. Although any one species of commercially valuable tree may be rare, timber trees are not scattered uniformly across the forest. Logistically it is easier to log in pockets where several suitable trees occur relatively close together. Logging may thus leave an irregular pattern of islands of undisturbed forest separated by open, logged and disturbed areas, and by logging tracks. Studies in Malayan forests showed that the felling of even 10% of the trees in an area of lowland forest can result in destruction and damage to at least 55% of the other trees (Burgess 1971; figs. 9.5 and 9.6). Only 35% of the forest was left undisturbed after logging

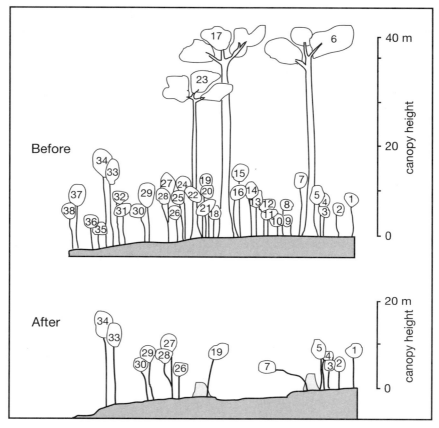

Figure 9.4. Forest profile before and after logging.
Source: Johns 1983a

activities. In East Kalimantan figures for forest damage from logging vary from 15% to 50% (Tinal and Palanewen 1978; Kartawinata 1980; Abdulhadi et al. 1981).

Besides damaging the remaining forest trees, the logging operation totally removes the vegetation from a substantial part of the logged-over area. At a study site in Malayan hill forest, logging roads and landings occupied 6% of the logged area, and a further 6% was covered by soil removed during logging operations (Burgess 1973). Studies in East Kalimantan (Kartawinata 1980) and Sabah (Meijer 1970) estimated that 20% to 30% of a logged area was bare soil, composed of roads and log yards. High lead logging by cables on hillsides in Sabah causes substantial damage

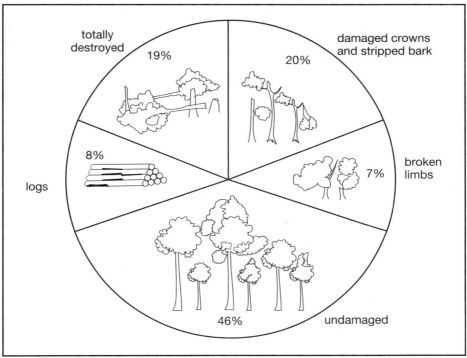

Figure 9.5. The effect of extracting 8% of trees by selective logging on an area of forest on South Pagai, Mentawai, West Sumatra. (After Alrasjid and Effendi 1979.)

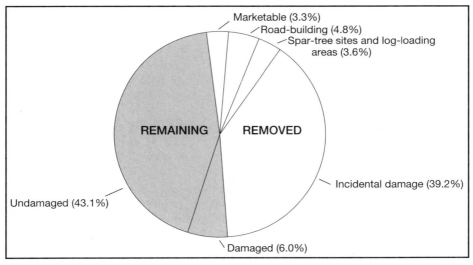

Figure 9.6. Tree loss and damage in a logged forest, West Malaysia.
Source: Johns 1983a

to older advanced growth but less to seedlings.

Heavy machinery churns up the ground and compacts and bares the soil on logging roads. The infiltration rate of water is much reduced on the compacted soils of logging roads (Abdulhadi et al. 1981; Hamzah 1978), leading to increased runoff and soil erosion (Liew 1974). The silt load of streams increases dramatically; in an East Kalimantan study it was more than thirty times greater in logged areas than in unlogged parts of the same watershed (Hamzah 1978). This may lead to increased sedimentation rates in reservoirs, irrigation canals and water filtration plants downriver.

Different stages of selective commercial logging in a small catchment in the Ulu Segama, Sabah, caused considerable changes in output of sediment and water (Douglas et al. 1992). Comparisons of suspended sediment yield between the logged catchment and a nearby unlogged area showed that the amount of sediment from the logged area was four times greater after a logging road was built, five times greater after logging within 37 m of the road and 18 times greater in the five months immediately following logging of the remainder of the catchment. However, a year after logging ceased the largest monthly sediment yields were only 8.6 times those of the undisturbed catchment, indicating some degree of recovery.

The impact of logging roads may be more detrimental to the regeneration of the forest than the actual felling damage (Burgess 1971; Meijer 1970). Between 30% and 40% of established seedlings may be destroyed by tractors during log haulage (Johns 1988). Seeds cannot germinate in compacted soils, and soil erosion may also occur, delaying colonisation. Shade-tolerant seedlings cannot survive along open roadsides. Instead, the bare ground is colonised by woody climbers and pioneer shrubs and trees. Colonisation may take several years in hilly country. Perhaps the most far-reaching deleterious effect of logging roads, however, is that they provide easy access to logged forests, which are then not allowed to regenerate but are cleared for agriculture by immigrant farmers.

Tropical forest recovery after logging depends on a variety of factors, the most important being overall damage to seedling stocks, presence of mature seed-producing trees and the presence of natural pollination and dispersal agents. Dipterocarps are wind-dispersed, but many other timber trees rely on animals for both pollination and seed dispersal. It is therefore not only ecologically preferable, but often to the foresters' advantage to let forests regenerate naturally and to maintain as far as possible their natural complement of animal life (Johns 1988). The economic expediency of using fast-growing alien tree species to re-establish ground cover may disturb the natural regeneration process.

If selectively logged forest is not further disturbed, natural regeneration will take place slowly. Light-demanding seedlings will grow up in areas where crown cover has been removed. However, the forest that regenerates after logging will have a different species composition from the original

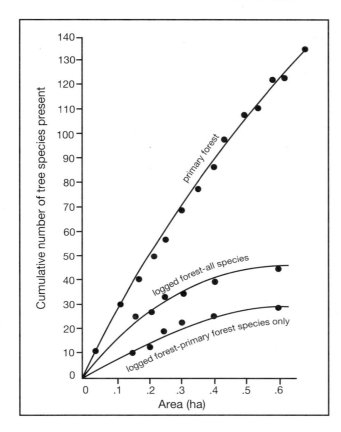

Figure 9.7.
Tree species diversity in primary and logged forest in Silabukan Forest Reserve, 20 years after logging. (After Davies and Payne 1982.)

stand in many places, and may be less rich in the emergent species that are valued for timber (Riswan 1982; fig. 9.7). In theory, selective logging on a rotation system should be sustainable. In practice, the degree of damage to remaining trees and the environmental disruption caused by poor management practices means that commercial logging, as currently practised, is not sustainable.

The Sustainability of Timber Harvests

Government regulations for the Indonesian timber industry have been formulated to encourage a sustainable harvesting system. Officially, no more than 10% of timber should be taken out of a concession area; trees of

> **Box 9.1.** Commercial forest management systems.
>
> Several systems have been developed to manage tropical forests for a sustained yield. These systems combine rules concerning felling (type, size and age of trees to be felled; environmental protection; logging techniques) and silvicultural practices (cutting climbers; poisoning undesirable species; enrichment planting). Unfortunately, many of the systems are not strictly applied, and much current forest management is unsustainable.
>
> The choice of system depends on the type of forest, national forestry policy and the demands of the timber market. Two types of forest management system have been employed in Borneo.
>
> A monocyclic system, such as the Malayan Uniform System used until recently in Sabah and Peninsular Malaysia, tends to organise the felling of all saleable trees into one logging operation. The cleared area is then left to regenerate naturally until mature trees are available once more for harvest (in 60-80 years). Such logging creates large gaps in the canopy and causes extensive damage. Surviving trees of no commercial value, which could compete with valuable seedlings, may be eliminated (by poisoning for instance) to encourage natural regeneration. In practice, cleared areas rarely do regenerate; more often, they are converted to other forms of land use.
>
> The polycyclic or selective systems are based on repeated logging on a rotation cycle of 25 to 40 years. In principle these systems are applied in Indonesia (Selective Cutting System or Tebang Pilih Indonesia TPI) and Sarawak (Liberation Thinning System). Only fully mature trees of commercial value are felled and extracted. Smaller trees (less than 50 cm dbh in normal production forest (HPH) and less than 60 cm in limited production forest (HPT)) are left to mature to produce the next harvest. In addition to the diameter limits, the TPI stipulates a minimum number of commercial species that must remain after logging: 25 trees/ha of dbh 20-49 cm for HPH/HPT. This system relies on saplings for regeneration and sustainability. It requires a full inventory prior to logging, careful felling and silvicultural treatment to encourage sapling growth and for enrichment planting. This system creates smaller canopy gaps and, in theory, increases productivity.
>
> In practice the concessionaires tend to obey the minimum diameter limits since the plywood mills will often not accept logs of less than 60 cm. However, the loggers are under no obligation to take a range of species, so they are inclined to harvest only those species which fetch a high price at the time of operations; these practices change species composition and forest structure. Moreover, the system is rarely properly regulated, and there is always short-term financial pressure to carry out repeat logging before the end of the rotation cycle needed for regeneration. The Ministry of Forestry recognises the fact that only a small proportion of the forest is being managed properly.

less than 50 cm diameter should be left standing. This "selective" logging system is designed to ensure that sufficient numbers of good-sized trees remain so that natural regeneration can take place. Concessions are granted for a period of 20 years; the concession-holder can undertake one cutting, but the logged-over area should then be left for a minimum period of 35 years before being logged again. (In Sabah and the Philippines the cutting cycle has been established at 40 years.)

In fact, it is doubtful if the Indonesian rotation cycle is long enough to allow for regeneration of commercially valuable trees. It may take 100 years or more for some canopy trees to mature (Riswan et al. 1985; Soedjito 1988). Moreover, although trees over 50 cm dbh are large enough to cut, some may be relatively young and may have never fruited; this is especially true for dipterocarp species, which flower and fruit irregularly, often with several years between fruiting seasons. Thus many dipterocarps of sufficient girth to be cut may not have made any reproductive contribution to future forest regeneration.

High grading (removal of all large trees) could result in removal of young, big trees, and depletion of fast-growing and genetically superior stock. The extensive damage caused by logging, and removal of trees that have not yet fruited, is likely to rule out the prospects for commercial forestry beyond the second rotation. In any case, the value of timber from the second rotation depends on minimising damage to medium-sized trees during the first cut. The second cycle, with cutting after only a short interval, removes even more of the parent stock needed for regeneration. Current logging practices will certainly not lead to sustainable harvests.

The short concession period, lack of regulation and the existing levy system all encourage poor logging practices. Since harvests are limited to large stems of commercially valuable species, large forest areas must be logged to obtain the desired output. Royalties are based on removals rather than on the volume of merchantable trees in a stand, with no distinction for species, grade or size, so concessionaires harvest only the most valuable species. There is little incentive to protect immature, inferior or less marketable varieties (Gillis 1988), while heavy damage is inflicted on new saplings and seedlings, decreasing the probability of regeneration (Abdulhadi et al. 1981). Commonly loggers re-enter stands for a second logging during the duration of the concession. The situation is made worse in East and South Kalimantan by the operations of many unlicensed loggers, who enter areas that have already been selectively logged by licensed timber companies (Potter 1989). These problems are not confined to Kalimantan. High grading in Sabah has led to 72% of uncut trees being damaged by logging; in Sarawak damage is estimated at 42% (Repetto 1988).

Since the 1970s the Indonesian government has promoted domestic forest-based industries, in an attempt to conserve forest resources and maximise revenues from the timber harvest. In 1978 the export tax on logs was doubled from 10% to 20%, while most sawn wood and all plywood were exempted. In 1980 log exports were further restricted, and finally they were banned totally in 1985. These policies stimulated rapid growth in the number of operating and planned plymills in Indonesia, from 16 in 1977 to 182 in 1983 (Repetto 1988). Plymills have been established at Samarinda, Banjarmasin and Pontianak.

Although recorded timber harvests for Indonesia declined from a peak of 25 million cubic metres in 1979 to almost half that amount in 1982, they are increasing again now. By 1988 plymills required 20 million cubic metres of log feedstock, and sawmills required 18 million cubic metres more, a total of roughly 38 million cubic metres of logs. Of this it was estimated that only 1.3 million cubic metres would be available from plantations, and natural forests would have to provide the remaining 36.7 million cubic metres, almost 1.5 times more than the previous peak in 1979 (Repetto 1988). Government projections predicted a harvest level in 1989 of three times the 1979 peak. The continuing demand for wood will put increasing pressure on Kalimantan's remaining forests and the goods and services that they provide. The expected production from plantation forests is optimistic if not unrealistic.

OTHER FOREST PRODUCTS

As with any other change in land use, timber exploitation has hidden costs; destruction of forests for timber reduces their value for other activities. As forests disappear, so do such valuable natural products as rattan, resin, fish, game meat, honey, wild fruits, and pharmaceutical and cosmetic compounds. Many of these resources are harvested sustainably from Bornean forests and provide food, income and employment for local communities. The value of such minor forest products is often under-rated. In 1982 exports from Indonesia of forest products other than wood earned $120 million, more than the export values of copper, aluminium, tea, pepper or tobacco (Repetto 1988). Table 9.4 shows export figures for nontimber forest products.

Rattans

The spiny climbing rattan palms (subfamily Lepidocaryoidae) are the most important of the minor forest products harvested in Indonesia. One hundred and thirty-seven species of rattans are known from Borneo, occurring at various altitudes from sea level to 2,900 m on Mount Kinabalu. A few species occur over a wide altitudinal range, but most can be distinguished as either lowland or montane species. Limestone hills normally have a poor rattan flora, though a species related to *Calamus marginatus* seems to be confined to limestone (Dransfield 1974). Kerangas forest is poor in rattans, and peat swamps in Borneo support only a few species, notably *Korthalsia flagellaris*. Most species are found in lowland forest where rattan seedlings are abundant, awaiting a gap in the canopy which would allow them to surge up towards the light. Light-demanding species such as

Korthalsia macrocarpa grow along riverbanks. In Bornean secondary forest the larger species of *Korthalsia* and *Daemonorops* are often abundant, along with *Calamus scipionum*. These species are very tolerant of disturbance and may even benefit from some opening of forest. In poor secondary forest, and alang-alang fields, rattans are very scarce or absent (Dransfield 1974).

Like other rainforest plants, rattans have complex inter-relationships with forest animals. Many rattans harbour ants in their leaflets, spines and stems; the ants help to protect the plant from predation. In most of these ant/rattan relationships, the ants farm mealy bugs as a source of honeydew. Rattan flowers are scented and depend on insects, including ants, beetles, thrips, bees, wasps and flies for pollination. Birds, monkeys, apes, and civets are probably all important dispersers of rattan seeds.

Table 9.4a. Exports of non-timber forest products from Indonesia (in tonne).

Commodity (species)	1983	1984	1985	1986	1987	1988
Rattan, all forms	82,500.0	91,400.0	90,000.0	108,900.0	143,900.0	199,144.0
Illipe nuts	16,905.6	542.0	225.0	573.0	14,323.3	15,977.0
Shorea spp.						
Illipe oil	66.2	–	69.1	331.2	1,977.6	–
Shorea spp.						
Kemiri seeds, shelled (1)	391.9	415.7	415.5	482.9	393.5	–
Aleurites moluccana						
Kemiri seeds, in shell (1)	4.4	34.4	25.7	11.6	15.1	–
Aleurites moluccana						
Cassia vera (1)	18,843.5	18,963.0	15,845.0	2,029.0	no data	no data
Cinnamomum burmannii						
Jelutong latex	1,980.7	2,235.0	2,659.0	1,148.0	2,498.8	–
Dyera costulata						
Gutta percha latex	–	5.0	–	–	3.4	–
Palaquium spp.						
Damar oleoresin	2,473.2	2,426.9	2,773.7	1,604.2	1,829.1	3,179.0
Dipterocarpus spp.						
Copal resin	1,719.4	1,876.0	2,012.0	2,029.0	2,810.5	–
Agathis dammara						
Dragon's blood - rattan resin	2.2	6.5	7.9	3.7	12.9	–
Daemonorops spp.						
Other resins	2,871.8	33,532.1	6,478.0	2,633.7	6,062.7	–
Gum damar	405.1	427.0	1,150.0	3,889.0	1,211.2	–
Dipterocarpaceae spp.						
Other gums and balsams	545.7	0.1	15.7	67.2	16.6	–
Gaharu wood	39.0	252.6	53.8	17.9	151.2	–
Aquilaria spp.						
Charcoal	32,993.9	39,863.0	38,571.0	36,788.0	34,453.8	23,632.0
Aren fibres - palm leaves	564.9	700.8	639.1	641.7	749.2	–
Arenga spp.						

(1) A significant fraction of these products is derived from cultivated sources.
Source: de Beer and McDermott 1989

Rattan species may have solitary stems or clusters of stems. This characteristic is important economically since it distinguishes between one-harvest rattans, such as *Calamus manan*, and reharvestable species, such as *Calamus caesius* and the *Korthalsia* species. Rattans can be cultivated; the oldest rattan plantations in the world are found in Central Kalimantan (box 12.4). In East and Central Kalimantan, Dayak farmers plant rattans in their ladangs before letting them go fallow for seven to ten years. By the time the fields are brought back into use, the rattan can be harvested for the first time. If the demand for rattan drops, the plants can be left growing till the market improves again. A rattan plant cut for the first time 15 years after planting is said to produce three times as much rattan as a plant first cut after seven years (Weinstock 1983). As wild supplies diminish, these small-scale plantations could play an increasingly important role in

Table 9.4b. Exports of non-timber forest products from Indonesia (in U.S.$1,000).

Commodity (Species)	1983	1984	1985	1986	1987	1988
Rattan all forms	87,100.0	94,700.0	98,400.0	109,100.0	211,560.0	–
Illipe nuts	8,026.6	290.0	186.0	464.0	5,131.0	–
Shorea spp.						
Illipe oil	126.7	–	163.4	824.7	2,946.1	–
Shorea spp.						
Kemiri seeds shelled (1)	156.8	190.6	173.1	215.9	173.1	–
Aleurites moluccana						
Kemiri seeds in shell (1)	9.8	15.5	20.0	6.0	4.8	–
Aleurites moluccana						
Cassia vera (1)	23,086.9	22,390.0	18,254.0	1,263.0	no data	no data
Cinnamomum burmannii						
Jelutong latex	3,314.6	4,289.0	5,514.0	1,388.0	2,183.4	–
Dyera costulata						
Gutta Percha latex	–	44.6	–	–	119.1	–
Palaquium spp.						
Damar oleoresin	1,382.9	1,324.0	1,212.2	523.5	651.9	–
Dipterocarpus spp.						
Copal resin	869.0	944.0	1,563.0	1,263.0	1,541.7	–
Agathis dammara						
Dragon's blood, rattan resin	0.7	4.6	14.4	11.2	15.7	–
Daemonorops spp.						
Other resins	1,413.6	2,573.4	3,647.8	2,115.2	2,361.5	–
Gum damar	162.6	196.0	486.0	1,644.0	506.7	–
Dipterocarpaceae spp.						
Other gums and balsams	227.2	0.4	13.6	31.6	9.4	–
Gaharu wood	121.4	805.3	441.9	109.7	179.2	–
Aquilaria spp.						
Charcoal	1,942.0	2,666.0	2,600.0	2,751.0	2,682.5	–
Aren fibres palm leaves	486.2	732.8	853.2	971.8	851.4	–
Arenga spp.						

(1) A significant fraction of these products is derived from cultivated sources.
Source: de Beer and McDermott 1989

providing rattan.

Indonesia produces more than 75% of the world's rattan supply. Exports doubled between 1968 and 1977, and again in the decade to 1988, with rattan earning more foreign exchange than any other forest product except logs (Peluso 1983b, 1986). From 1981 to 1987 the value of exports of rattan cane rose from $72.9 million to $212 million, almost a doubling of value in real (inflation-adjusted) terms (de Beer and McDermott 1989). Nearly half of these exports came from Kalimantan, with the largest production from East Kalimantan (Peluso 1986). By far the greatest stocks of rattan are collected from the wild, and there is increasing concern whether this harvest is sustainable. Collectors have to travel farther to find suitable supplies, and are cutting younger stems, so that immature canes make up an increasing proportion of the rattan sold to dealers.

Indonesian rattan was already important on the world market by the mid-1800s. Buginese traders from Sulawesi visited Kalimantan to exchange salt, cloth and tobacco for rattan collected by upriver Dayaks and some Kutai people. The Sultan of Kutai exacted a tax on the trade of all forest products leaving by the Mahakam River. Local village headmen were paid fees by collectors and traders, and supervised the extent of the harvest. When East Kalimantan officially became a province in 1957, these taxes were transferred to the provincial authorities, so there was no local incentive to check that harvesting was sustainable (Peluso 1983a, 1983b). Better access to upriver areas by motorised boats, the opening up of remote areas by commercial logging operations, an influx of new immigrants lured by the timber boom and the allocation of timber-felling rights exclusively to large companies, all led to an increase in rattan collecting. In the late 1970s, when world demand for rattan increased and the Philippine government restricted harvesting, Indonesia became the major supplier. Increasing numbers of traders began to buy rattan, leading to overharvesting, with serious depletion of wild stocks (Peluso 1986).

From January 1989 the Government of Indonesia has banned the export of unprocessed rattan in an attempt to conserve stocks, and to retain more added value in the country. Initially this led to falling prices for Indonesian collectors and cultivators and put less pressure on wild stocks, but also gave less incentive for farmers to cultivate rattan. In the long term the sustainability of rattan harvests will depend on improved regulation of the trade and greater investment in rattan as a plantation and buffer-zone crop.

Resins and Incense Wood

Jelutong is a latex tapped from the tree *Dyera costulata*; rubber is extracted from the milky sap. Jelutong trees are scattered through the swamps of Central Kalimantan and were particularly important as a source of rubber

> **Box 9.2.** The economic values of some forest species.
>
> Apart from commercially valuable dipterocarps, Bornean rainforests harbour many other species of plants and animals of value to local and national economies.
>
> - Rural communities in Sarawak eat considerable quantities of wild meat (mostly bearded pig) conservatively estimated to be worth $50 million every year, at current prices. The loss of this cheap source of protein would have a crippling effect on many village economies. If the wild meat had to be replaced with imported, tinned food, or domesticated pigs and cattle, there would be an adverse effect on Sarawak's balance of payments.
> - In the U.S.A. alone as many as half of all prescriptions written every year contain a drug of natural origin; these have been valued at well over $3 billion. The two richest areas in the world for natural production of the complex chemicals that may be used in pharmaceuticals are tropical rainforests and coral communities. Many local communities in Kalimantan and elsewhere in Borneo rely on forest plants for medicines and drugs.
> - In Sabah, studies suggest that high densities of wild birds in commercial *Albizia* plantations limit the abundance of caterpillars that would otherwise defoliate the trees. The birds require natural forest for nesting.
> - Long-tailed macaques *Macaca fascicularis* are used for testing live vaccines, an essential part of immunisation programmes which have been instrumental in saving the lives of hundreds of thousands of children worldwide every year. These monkeys, often regarded as pests, have a considerable economic value for medical research. The silvered langur *Presbytis cristata* is also used for biomedical research on human diseases.
> - Edible nests of black-nest and white-nest swiftlets *Collocalia (Aerodramus) maximus* and *Collocalia fuciphaga* are harvested extensively in Borneo and have been traded for centuries. In Sabah alone the export trade in edible nests is estimated to be worth between $600,000 and $1.2 million per year.
> - Rattans are the most valuable forest export after timber, earning Indonesia $212 million in 1987. About 90% of the world's rattan comes from Indonesia and 75% of this harvest from Kalimantan. Most rattans are collected from the wild, so loss of forest leads to loss of this valuable resource.
>
> Whereas profits from timber exploitation tend to accrue to large business interests outside the province, harvesting minor forest products brings income and employment to local communities. Continued harvests depend on forests remaining relatively undisturbed.

before plantations of Brazilian rubber *Hevea brasiliensis* were established throughout Southeast Asia. Jelutong was an important trade item in South Kalimantan from the early 1900s, so important that its exploitation was discussed heatedly in the Dutch parliament and by the top levels of the colonial administration (Potter 1988). At one time concessions were given for the tapping of jelutong within prescribed areas.

Other locally and commercially important plant exudates include gutta percha from *Palaquium*, damar resins from various dipterocarps (Chin 1985), and resins from *Agathis borneensis* and *A. dammara* (Whitmore 1979). Damar is used locally to seal boats. Gaharu, a resinous product of the tree

> **Box 9.3.** Gaharu (aloe wood or incense wood).
>
> Gaharu, a fragrant wood, is caused by a pathological condition in living *Aquilaria* trees when the heartwood is infected by fungus. In Borneo gaharu is produced by at least three species, *Aquilaria beccariana, A. malaccensis* and *A. microcarpa*. Trees are felled to obtain the fragrant heartwood. The grade of gaharu is determined by its colour (the darker it is, the more valuable) and intensity of its fragrance.
>
> The trees that yield gaharu are scattered in lowland and hill rainforest, up to 800 m in the Apo Kayan. They are not abundant anywhere, so collectors must cover a wide area. Collecting involves felling many trees to obtain a small quantity of good quality heartwood. Field collectors believe in the relationship between occurrence of gaharu and outward signs of decay, but gaharu can also be found in apparently healthy boles. Thus many healthy trees are also felled that contain no gaharu. Collection is wasteful and invariably leads to depletion of the resource.
>
> Gaharu has been traded for centuries throughout Southeast Asia and with China and the Middle East. It is used in Chinese and Malay pharmacoepias to treat a variety of ailments, especially those associated with pregnancy and childbirth. The Kenyah Dayaks believe that shavings of gaharu swallowed with water will cure indigestion and stomachaches. It is used in incense and cosmetics and is regarded as a stimulant, tonic and carminative (cure for flatulence). It is also used in the manufacture of perfume in India and Japan, which is probably why it commands its present high price: in 1988 Rp 800,000 to Rp 1 million ($500) per kilo, according to collectors in Sangkulirang and the upper Barito.
>
> It may be possible to produce gaharu commercially. The fungus *Cytosphaera mangiferae* was isolated from the diseased tissues of standing *Aquilaria agallocha* trees in India and reinoculated into wooden blocks from the same species under laboratory conditions. These then exuded an aromatic, dark, resinous product (Jalaluddin 1977).

Aquilaria malaccensis and other species in the family Thymelaeaceae, is also collected in Bornean forests for its medicinal properties and value as incense (box 9.3).

Also commercially valuable is *kayu putih* oil extracted from the bark of *Melaleuca leucodendron*, of the Myrtaceae family. *Melaleuca* occurs naturally in swamp forests and disturbed areas and forms dense stands in areas subject to periodic burning, as in South Kalimantan (chapter 10). Kayu putih oil is widely used to treat a range of ailments in Indonesia and was an important ingredient of some European medicines (Burkill 1935).

Illipe Nuts

Illipe nuts (*tengkawang*) are another valuable forest resource. They are collected mainly from the forest, though a few Dayak communities also cultivate *Shorea* trees. In Sarawak 11 species of *Shorea* are important producers of illipe nuts (Anderson 1975), with most exports from *Shorea macrophylla*.

In West Kalimantan illipe nuts are harvested mainly from *Shorea macrophylla*, *S. beccariana* and *S. amplexicaulis*. In South Kalimantan the main producers are *S. pinanga* and *S. palembanica*.

Illipe nuts contain an edible oil which is sometimes used locally to flavour rice, but most nuts are exported, for use in confectionery as an alternative to cocoa butter, for soap making and for use in cosmetics, medicines and animal feeds. Like other dipterocarps, illipe trees flower and fruit irregularly at intervals of two to seven years. The single-seeded fruit germinates two to three days after falling (Sim 1978). On germination the nut's oil content decreases rapidly, so ripe illipe nuts must be collected as soon as possible. Fruits are collected from the forest by Dayaks, who shell and dry them in the sun, then sell them to Chinese traders. Illipe nuts are an important source of income for some communities. Collecting, processing and sale all take place over a period of about six weeks. In a good year one Kenyah longhouse in Sarawak collected more than

---

Box 9.4. Valuation of a rainforest.

A study of an area of Amazonian rainforest in Peru showed that the most economically productive form of land use was to retain the area as forest for long-term harvesting of products such as fruits, oils, medicinal plants and native rubber. A systematic inventory of one hectare of the Mishana forest revealed the presence of 275 trees of which 72 species (26.2%) yield products, such as edible fruits and rubber, with a market value in the nearby town of Iquitos. The forest also contained medicinal plants, lianas and several understorey plants of commercial importance in the region, but these species were too small to be included in the sample.

Given that both fruit and latex can be harvested every year, the total financial value of these resources is considerably greater than the current market value of one year's harvest. Under a low-intensity harvesting scheme, it was calculated that one hectare is worth $6,820; $490 of this total comes from timber harvested at a sustainable yield of 30 m^3/ha every 20 years. These amounts reflect the cumulative values of harvests over 50 years, calculated as current value (net present value). If all the merchantable timber were extracted in one operation, the value of one hectare would be only $1,000.

Using identical investment criteria, the net present value of another Amazonian site, a one-hectare plantation of the tree *Gmelina arborea* was estimated at $3,184, or less than half the value of the natural forest. The value of cattle ranching was even lower, $2,960 before deduction of the costs of weeding, fencing and animal care. These results clearly indicate the importance of non-wood forest products. These resources not only yield higher net revenues per hectare than timber, but they can be harvested with far less damage to the forest. Both ecologically and economically, the wisest use of these tropical forests is conservation, with sustainable utilisation of non-timber forest products. These results provide a valuable comparison for other tropical rainforests.

Source: Peters et al. 1989

10,000 kg of nuts, worth M$17,000. The income to each household was more than that from gaharu collection or from rubber on village plantations (Chin 1985). Such harvests are only sustainable while forests remain relatively undisturbed.

Forests for Water and Soil Conservation

Natural vegetation cover plays an important role in regulating the behaviour of water drainage systems. Particularly important is the "sponge effect" by which rainfall is trapped and held by catchment forests and natural grasslands, and thereby released more slowly into river systems, reducing the tendency for floods in periods of heavy rainfall and continuing to release water during periods of dry weather. These functions, which depend on ground cover, are reduced or lost when the vegetation of upland catchments is destroyed. After felling the interceptive role of the

---

Box 9.5. The effects of logging.

The impact of logging on watersheds depends on several factors:
- the amount of canopy removed;
- the amount of biomass removed (including how much slashed vegetation remains);
- logging methods;
- timing: wet or dry season;
- soil conditions and topography;
- extent, nature and use of roads, skid trails and landings;
- methods of slash disposal and site preparation;
- how quickly regeneration occurs or reforestation is carried out;
- whether buffer strips of forest are left along rivers;
- climatic events, such as heavy rains, after logging.

The initial direct impacts of harvesting logs (deforestation) are:
- removal of tree canopy, understorey growth and litter reduces soil protection, and leads to greater raindrop impact and bare soil;
- logging activities may change soil properties due to soil compaction, loss of organic material, and so on. This results in reduced infiltration and increased likelihood of soil erosion;
- removal of trees reduces transpiration, increases air movement and changes temperature. This changes evapotranspiration, usually reducing it;
- removing trees reduces root mass and soil shear strength so that landslips are more likely to occur on slopes;
- in cloud forest logging reduces the water capture functions of forest, thereby reducing precipitation.

Source: Hamilton and King 1983

canopy is lost, and the water-retaining capacity of the litter layer and root zone is reduced, so there is increased runoff from cleared lands. This leads to more water entering streams and rivers, greater variation in high and low flow rates and increased likelihood of floods immediately downstream.

Although the short-term consequences of logging may result in local flooding, as forest cover regenerates the prelogging hydrological regime will be restored. Much of the peak-flow effect may be caused by poorly located and designed roads which speed water flow off-site; good siting and conservative logging can reduce the effects on downstream flooding. If forest is not allowed to recover after logging (for instance, if it is cleared for cultivation or other land use), the effects on the hydrology of the watershed may be more severe, especially if the land becomes degraded. Unfortunately, logged-over forests in Kalimantan and elsewhere in Indonesia are often converted to other forms of unsustainable land use, as can be seen in the spread of alang-alang grasslands in South Kalimantan; these now cover more than one-quarter of the province (Donner 1987). Overgrazing and indiscriminate burning over long periods are bad for both soil and water conservation.

One of the more serious consequences of forest clearance, especially in hilly terrain, is the increase in soil erosion and the likelihood of landslips. Tree-root shear strength is important in maintaining slope stability in hilly areas (O'Loughlin 1974). Just how important can be seen from an incident in North Sumatra where logging of a small area had tragic consequences. After 10 hectares of forest were cleared on a steep slope in the Alas valley in 1981, a landslip occurred, blocking the river; when this dam burst, the sudden flood killed 13 people and washed away rice fields, part of a village, several new bridges and roads (Robertson and Soetrisno 1982).

There is conflicting opinion as to whether cutting of rainforest may also affect local climate (Hamilton and King 1983). Studies in the Amazon suggest that very large areas of lowland forest may regenerate some of their own rain (Salati et al. 1979) and that large-scale and permanent deforestation may reduce or alter rainfall patterns (Salati et al. 1981). If these findings are applicable elsewhere in the tropics, then the large-scale clearance of lowland forests (rather than selective logging) and conversion to open areas over much of South, West and East Kalimantan may lead to local reductions in rainfall. Certain forests, such as mountain forests in permanent cloud cover, do "capture" and condense atmospheric moisture and thereby increase precipitation. Cutting down such cloud forests results in loss of this extra precipitation, though it is restored as the forest regrows. If cloud forests were converted to another land use, this moisture would be removed from the water budget of the watershed, so protecting cloud forests makes good ecological and economic sense.

It is vital, therefore, that if forest lands are logged or converted to

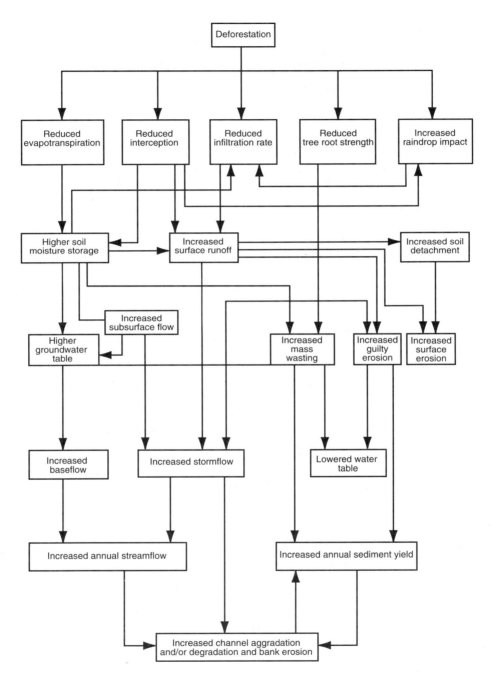

Figure 9.8. Some likely hydrologic changes of deforestation. (After Cassells et al. 1983.)

Box 9.6. Forest functions.

For the natural system	For the social system (people)

Protective

- soil protection by absorption and deflection of radiation and precipitation
- conservation of humidity and carbon dioxide by decreasing wind velocity
- sheltering, and providing required conditions for, plant and animal species

- sheltering agricultural crops against drought, wind, cold, radiation
- conserving soil and water
- shielding people against nuisances (noise, sights, smells, fumes)

Regulative

- absorption, storage and release of CO_2, O_2 and mineral elements
- absorption of aerosols and sound
- absorption, storage and release of water
- absorption and transformation of radiant and thermal energy

- improvement of atmospheric conditions in residential and recreational areas
- improvement of temperature regimes in residential areas (through roadside trees, parks)
- improvement of the biotope value and amenity of landscapes

Productive

- efficient storage of energy in utilisable form in biomass (plants and animals)
- self-regulating and regenerative processes of wood, bark, fruit and leaf production
- production of a wide array of chemical compounds, such as resins, alkaloids, essential oils, latex, pharmaceuticals

- supply of a wide array of raw materials to meet growing human demands
- source of employment
- creation of wealth

In the case of tropical forests the first two functions, protective and regulative, are extremely important and not very well known while the third, productive – is largely underestimated and underused.

other land uses, such activities should employ a sound soil and water conservation regime. Both soil erosion and water losses are reduced by increasing vegetation cover. Selective logging on a rotation system occurs relatively infrequently, and any hydrologic or soil impacts are diminished as natural vegetation regrows (Hamilton and King 1983). Problems occur when this natural regeneration is not allowed to occur, through poor logging practices, or by settlers clearing logged-over lands for often inappropriate forms of agriculture (fig. 9.8). The trend to establish forest plantations of rapidly growing species which are harvested totally with large equipment on very short rotations may also have adverse effects on the water flow and soil erosion rates in watersheds (Hamilton and King 1983). Traditional shifting agriculture, with its system of mixed crops and long fallow periods maintaining vegetation cover, is generally more ecologically sound.

FOREST DISTURBANCE AND WILDLIFE

Disturbance and clearance of rainforests affect not only the soils and hydrology of the area but also the animal communities that depend on those forests. The majority of mammal and bird species found in Borneo are dependent on mature forest. Many insects also are dependent on tropical forests and sometimes restricted to one or a few rainforest tree species; in Panamanian forests 945 species of beetle were recorded from just one species of tree (Erwin and Scott 1980).

Alterations of natural habitats result in changes in plant and animal communities, with corresponding changes in densities of resident species. Many species are eliminated, while a much smaller number, mostly weed or pest species such as grasses, rats, and squirrels, benefit by increasing their populations or inhabiting new habitats (Johns 1983a).

Some 81% of rainforest mammals in Southeast Asia are confined to forests below 660 m (Stevens 1968). Yet it is these lowland forests that are most threatened by logging operations and the ensuing land clearance. When forest is cleared completely by clear-felling or for large-scale agricultural projects, the majority of animals that live there will eventually die. Studies in Peninsular Malaysia have estimated that current levels of clear-felling of primary forests may exterminate as many as 31,000 gibbons, 45,000 macaques and 346,000 langurs every year (Marsh and Wilson 1981). The species most affected by forest clearance are those with very localised distributions, those with specialised diets, and territorial species. Some species may be able to move out into adjacent undisturbed habitats, but if these habitats are already sustaining populations at carrying capacity, they will be able to absorb very few surplus animals. Overcrowding of animals may lead to stress and reproductive failure. In Sabah migratory movements of orangutans caused by logging activities led to a rise in local population density, increased social interactions and a drop in the birth rate (MacKinnon 1974a).

The rainforests of Borneo are extremely complex ecosystems. The forest depends for its maintenance on interactions between plants and animals, particularly for pollination and seed dispersal of trees. Clearing or disturbance caused by logging will disrupt or even destroy these natural inter-relationships. Of 760 species of trees in 40 ha in Brunei, only one was wind pollinated (Ashton 1969); all the rest relied on animal pollinators, especially insects. Insects which occur at low densities and have very localised distributions are in particular danger of becoming locally extinct, perhaps with serious consequences for the pollination cycles of some rainforest trees. The effects of logging or conversion of forests to plantations varies with insect group (Holloway et al. 1992). Thus plant-feeding moths show significant loss of diversity in simplified habitats, whereas beetles exploiting a more uniform resource base, such as dung or carrion, show much less change in diversity and faunistic composition.

Some forest-dependent mammals and birds play a crucial role in pollination cycles of species of commercial importance. Durians in Southeast Asia are seasonal in their fruiting and depend on nectar-feeding bats (especially *Eonycteris spelea* and *Macroglossus minimus*) for pollination. When durians are not in flower, these bats depend on other wild tree species for nectar. Loss of forest habitat can lead to local loss of bats and failure of the valuable durian crop (Start and Marshall 1976).

Where forests are selectively logged and not further disturbed, they retain much of their biodiversity. Which species survive, and in what numbers, will depend on the extent and structure of remaining forest, the length of time since logging and the specific niche requirements of the species concerned.

Undisturbed forest is typically dark, humid, cool and wind-free, with scattered patches of early successional vegetation in tree-fall gaps and along waterways. Recently logged forests are well-lit, drier and hotter at midday, and experience greater wind turbulence. In unlogged forest less than 2% of the incident sunlight reaches the forest floor; this may increase to more than 90% in recently logged forests. Since seeds of many rainforest trees germinate only in shady conditions, light levels affect natural regeneration processes. Increased light is associated with increased soil and air temperatures, and humidity may drop from above 80% in primary forests to less than 50% in opened forests. These factors also affect the distribution and viability of many invertebrates and damp-loving species such as amphibians and reptiles.

The commencement of logging operations may be marked by a decline in animal population densities, as the more mobile species move out of the logging area to avoid disturbance. At least 48% of the original mammal

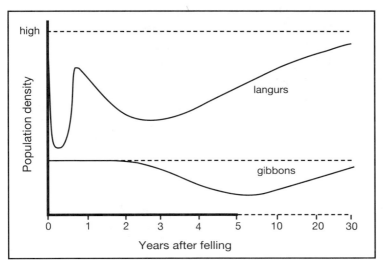

Figure 9.9. Changes in population density of langurs (non-territorial) and gibbons (territorial) after selective logging.

Source: Marsh and Wilson 1981

species disappeared from logged forest in Peninsular Malaysia (Stevens 1968), and disturbed forest contained only 40% of the species of primary forest (Harrison l969). Similar findings have been reported for birds (Harrison 1965; Wells 1971; Diamond 1973, 1975). Bornean species particularly disturbed by logging operations include orangutan (MacKinnon 1974a; Rijksen 1978), slow loris (Barrett 1984), sun bear (Wilson and Wilson 1975), flying squirrels *Petaurista elegans* and *Aeromys tephromelas* (Muul and Lim 1978; Barrett 1984) and stream-inhabiting birds such as forktails and kingfishers (Johns 1986).

Most rainforest mammal species can survive in logged forests, especially if logging operations are light and the forests are left to regenerate naturally. Thus all Bornean primates, and most other species of nonflying mammals, have been recorded in logged forests, although population numbers often decline immediately after logging.

In Peninsular Malaysia primate densities were lower in logged forest than undisturbed forest, but the effects differed for different primates and with the age of the secondary stand (Marsh and Wilson 1981). In particular, group size and population density of langurs seem to decline after logging (Marsh and Wilson 1981; Davies and Payne 1982). There seems to be an initial decline in total primate densities until one to two years after logging when most species, except gibbons, show some recovery, with animals recruited from adjacent areas of undisturbed forest (fig. 9.9). How-

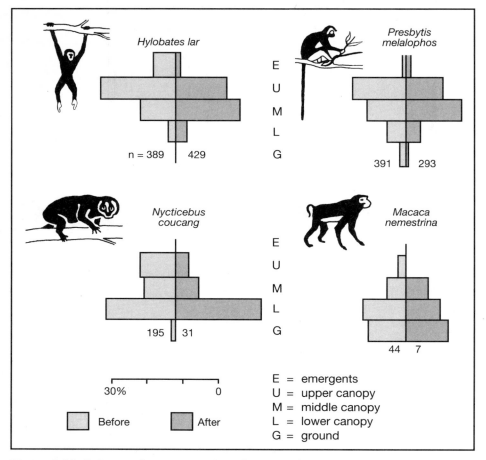

Figure 9.10. Patterns of canopy use by primates before and after logging at a study site in Peninsular Malaysia. (After Johns 1983a.)

ever, total primate densities are significantly lower in logged forest distant from unlogged areas (Johns 1988).

Animals' response to logging varies with the intensity and method of logging and with the diet and behaviour of the species concerned (Wilson and Wilson 1975). The incidental damage caused by logging operations may cause a critical reduction in food resources for some animals, so that they cannot survive in the logged forest. Figs are a key species for many frugivores, providing food when other fruits are scarce. Many fruiting figs are epiphytic, so removal of host timber trees reduces the abundance of figs and can lead to a drop in the numbers of highly frugivorous species such as gibbons and hornbills (Leighton and Leighton 1983). Other animals show dietary shifts to cope with reduced availability of certain food types.

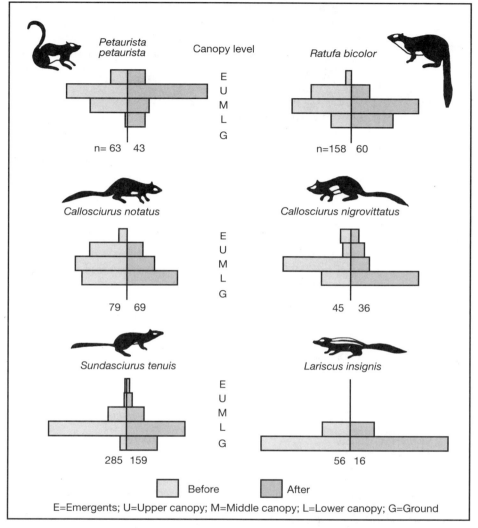

Figure 9.11. Patterns of canopy use by squirrels before and after logging at a study site in Peninsular Malaysia. (After Johns 1983a.)

Thus primates shift from frugivory to folivory, eating increasing amounts of leaves and other plant parts (MacKinnon 1974a; MacKinnon and MacKinnon 1980); some squirrels shift from frugivory to feeding mostly on bark and sap (Payne 1980). Particularly among primates, this results in changes in social organisation to adopt the most efficient foraging strategy (Johns 1981). Both primates and squirrels show changes in foraging and ranging patterns in the different forest strata to compensate for changes in forest structure and disruption of arboreal pathways (figs. 9.10 and 9.11).

Bat species decline as a result of logging. The larger bats persist better than smaller species, but only the Rhinolophidae and Hipposideroidae appear to maintain numbers successfully. The smaller species, which glean insects from foliage, are particularly susceptible to forest disturbance. Species that travel long distances nightly from cave roosts or other colonies (e.g., flying foxes *Pteropus vampyrus* and diadem roundleaf bats *Hipposideros diadema*) probably occur throughout logged forest, subject to the availability of food sources such as flowering trees and shrubs. A few species are able to utilise habitats created by logging. Sheath-tailed bats *Emballonura* spp. roost under road bridges and culverts, and the hollow-faced bat *Nycteris javanica* roosts in hollow, fallen logs (Johns 1988). The general decline in bat species associated with logging could have important implications for forest regeneration, since bats are important as pollinators and seed dispersers.

For some species logging may provide increased food resources. Long-tailed macaques and silvered langurs are often more numerous in secondary habitats (Marsh and Wilson 1981; MacKinnon 1983), where the large amount of new leaf growth and insects provides an abundant food source (fig. 9.12). Predators, like the clouded leopard, may be more abundant in logged-over forests, where they feed largely on the rats which are abundant in disturbed forest. Sunbears also seem common in some logged

Table 9.5. "Extinct" and colonising bird species in logged forest.

a.	"extinct" species	b.	colonising species
	Otus rufescens		*Falco severus*
	Hirundapus caudacutus		*Clamator coromandus*
	H. giganteus		*Phodilus badius*
	Harpactes kasumba		*Caprimulgus indicus*
	Ceyx erithacus		*Eurystomus orientalis*
	Lacedo pulchella		*Mulleripicus pulverulentus*
	Halcyon concreta		*Dryocopus javensis*
	Buceros bicornis		*Cymbirhynchus macrorhynchus*
	Sasia abnormis		*Corydon sumatranus*
	Hemipus hirundinaceus		*Pycnonotus goiavier*
	Pericrocotus igneus		*Corvus enca*
	Malacopteron affine		*Prinia rufescens*
	Stachyris poliocephala		*Orthotomus sepium*
	S. leucotis		*Lanius cristatus*
	Macronous ptilosus		*Arachnothera crassirostris*
	Copsychus pyrropygus		*Zosterops palpebrosa*
	Enicurus leschenaulti		*Lonchura leucogastra*
	Muscicapa mugimaki		
	Culicicapa ceylonensis		
	Rhipidura perlata		
	Prionochilus percussus		
	Dicaeum concolor		

Source: Johns 1983a

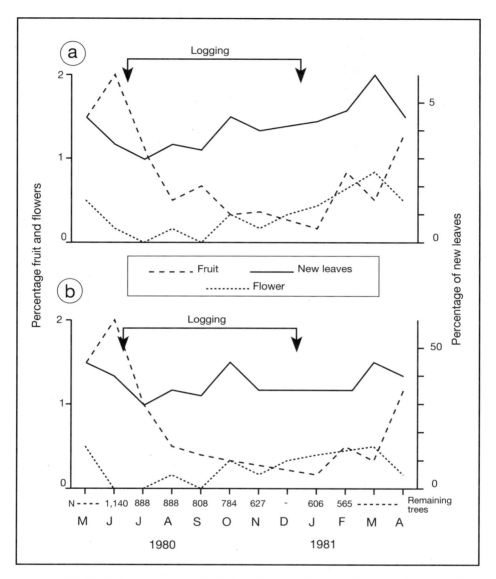

Figure 9.12. Fruit, flower and new leaf abundance at Sungai Tekam as a) a proportion of trees sampled, b) proportion of original sample. New leaf production increases among trees remaining after logging, due to increased light availability from opening up the canopy. Young leaves provide an abundant food supply for many partly folivorous animals and may help to buffer the impact of logging disturbance on some mammal populations. (After Johns 1983a.)

forests (Johns 1983a).

A few rainforest animals, mainly browsers and grazers, prefer more open disturbed and secondary forest and may increase in numbers after logging activities. The lush growth of grasses, sedges, small pioneer trees and giant herbs after logging increases the food supply for species such as elephant, banteng and deer. In spite of the abundant food supply, however, rhinos are often absent from logged forests, probably because they are avoiding the presence of people.

Ecosystem disturbance typically causes reduction in number of some

Table 9.6. A comparison of feeding guild membership among the avifauna of unlogged and selectively logged forests in West Malaysia.

Feeding guild	Number of species			
	observed before logging	considered "extinct"	colonisers	observed after logging*
Terrestrial frugivore	8	0	0	2
Arboreal frugivore	26	1	0	19
Arboreal insectivore/frugivore	43	3	6	28
Insectivore/nectarivore	10	0	1	8
Bark insectivore	12	0	2	8
Terrestrial insectivore	11	1	0	6
Foliage insectivore	37	9	4	17
Sallying insectivore	24	3	1	18
Sweeping insectivore	7	2	1	6
Raptor	14	1	2	9
Piscivore	3	2	0	1

* Likely to be underestimated due to the preponderance of rare species.
Source: Johns 1983a

Table 9.7. Changes in the relative abundance of bird feeding guilds after logging in a West Malaysian rainforest.

Feeding guild	Before logging		After logging	
	n	% total	n	% total
Terrestrial frugivore	15	1.1	13	1.0
Undergrowth frugivore	32	2.4	4	0.3
Canopy frugivore	331	25.3	517	38.6
Undergrowth insectivore/frugivore	93	7.1	63	4.7
Canopy insectivore/frugivore	171	13.1	207	15.3
Insectivore/nectarivore	36	2.8	28	2.1
Bark-gleaning insectivore	80	6.1	43	3.2
Terrestrial insectivore	49	3.7	25	1.9
Foliage-gleaning insectivore	366	28.0	274	20.4
Flycatching insectivore	105	8.0	111	8.3
Piscivore	5	0.4	0	0
Raptor	26	2.0	56	4.2
Total numbers	1,309		1,341	

Source: Johns 1983a

> **Box 9.7.** Logging and wildlife in the Sungai Tekam Forestry Concession, Pahang, West Malaysia.
>
> The Sungai Tekam Forestry Concession is a 315 km² commercial, sustained-yield logging concession, and has been the site of various studies. The concession is a typical hill dipterocarp forest, of which 3% to 5% is unloggable. These patches, located on steep slopes, plateaux or waterlogged areas, act as wildlife refuges. They contain a number of vertebrates, such as babblers Timaliidae, that are largely lost from recently logged forest. Populations of such intolerant species are able to move outwards from the small refuges to recolonise older logged forest. Evidence from another site, Pasoh Forest Reserve, Negeri Sembilan, suggests that all species have recolonised within 25 years of logging.
>
> Undisturbed patches at Sungai Tekam are used as a concentrated food source by various frugivorous birds such as hornbills Bucerotidae and green pigeons *Treron*, but these birds also range widely over logged areas. They feed on a few kinds of fruit only produced by trees of the early succession, but appear able to travel between the fruiting canopy trees that remain. Most mammals have sufficiently variable diets that they are able either to adjust their foraging to take account of the differing availability of food types, as with monkeys and squirrels, or, as with elephants, tapirs and gaurs, to make use of the regenerating undergrowth itself. Gaurs are wide-ranging and able to travel between both primary and logged patches; they may retreat to the primary refuges to avoid human disturbance, but are more commonly seen in logged than in primary forest.
>
> Results so far from the continuing studies at this site indicate that few vertebrate species are lost by the logging of 18 trees per hectare, which causes a total damage level of 31% of trees destroyed. Almost all the vertebrates that avoid recently logged areas are retained in the primary forest patches left because of the difficult terrain. Whether the small percentage of the area left unlogged is sufficient to maintain populations in the long term is unclear. Nevertheless, the fact that such a large proportion of the vertebrate fauna currently persists in recently logged forest with these very small primary forest refuges is significant.
>
> *Source:* Johns 1983a

species (and increases in others) rather than species deletions, but sometimes populations are so reduced that this leads to local extinctions (table 9.5). The endemic large tufted ground squirrel *Rheithrosciurus macrotis* was not observed in logged forests in Sabah, where it may be replaced by the smaller four-striped ground squirrel *Lariscus hosei* which has a similar insect-eating diet (Johns 1988). Among forest birds declines in abundance are most marked among terrestrial and flycatching species, and those that glean understorey foliage. Logging may also seriously affect those species that feed along fast-flowing streams, which may become silt-laden or dammed by logging activities (Wells 1971). Generalist frugivores/insectivores and species using the upper levels of the canopy (canopy frugivores) seem little affected by logging (tables 9.6 and 9.7). This is surprising since frugivores may face a reduction greater than 50% in numbers of food trees in logged forest. Nevertheless, hornbill densities in Sabah remained similar in logged and unlogged habitats, even for a fig specialist such as the helmeted hornbill (Johns 1988). Sometimes a logged forest is even able to support a greater species richness than an undisturbed forest because of the wide variety of habitats created within the logged area. Thus twice-logged forest at Tabin in Sabah contained both pittas, deep forest species, and buttonquail, a characteristic scrub and grassland species (Johns 1988).

Although many rainforest species are able to adapt to the changed conditions of logged forest, the effects of logging on rainforest communities depends on the degree of disturbance and damage to forest structure. This effect may be logarithmic, so that higher disturbance levels, and loss of a few more trees due to carelessness or bad management, may result in a substantial increase in loss of animal species. Thus the number of species persisting under the 80% damage levels of cable logging may be only half the number of those surviving the 62% damage levels of conventional logging (Johns 1988). Clearly the value of logged forests for conservation depends on the ways in which they are managed to minimise forest disturbance. Moreover, while many vertebrate species may be able to survive in patches of less damaged forest if forest is logged only once, repeat logging for small-sized timber stocks, which is now happening in Kalimantan concessions, has severe and permanent effects on wildlife.

Forests that have been selectively logged once exhibit a similar species diversity to primary forest but a different species composition. Species that are intolerant of disturbance are replaced by species adapted to the conditions of regenerating forest (Johns 1981). Nevertheless, logged forests support a high proportion of mature forest species, including many that could not survive in small, isolated forest reserves. Only large areas of undisturbed primary forest can be expected to retain the whole array of rainforest species, but well-managed forest, selectively logged on a sustained-yield basis, could play an important role in wildlife conservation, especially if special protection is given to fruit trees (table 9.8), mineral springs, oak forests and other feeding grounds. However, its conservation value diminishes drastically with poor logging practices, further logging and changing land use.

Logged forests make useful habitat corridors, linking reserves and other areas of undisturbed natural habitats (chapter 14). "Islands" of remaining forest within cleared land will lose species over time, the rate depending on the size of the remaining block of forest and its distance from other blocks of forest. However, the rate of local extinctions can be mitigated by immigrations from remaining forest habitats, including areas that have been logged selectively but allowed to regenerate.

Logged forests are also used by animals that migrate seasonally, sometimes moving over long distances far outside reserve boundaries. Elephants, for instance, probably migrate from Sabah south into the Ulu Sembakung area in East Kalimantan. Elephants, rhinos and banteng range over large

Table 9.8. Important food trees for wildlife and people.

TREE FAMILY	Genus and species	Ungulates	Of special value to: Primates	Birds	Human
ANACARDIACEAE:	Bouea		+		+
	Buchanania	+	+		
	Dracontomelon	+	+		
	Koordesiodendron		+		
	Mangifera	+	+		+
ANNONACEAE:	Cananga			+	
	Mezzettia	+			
	Polyalthia		+	+	
	Xylopia		+		
BOMBACACEAE:	Durio	+	+	+	+
BURSERACEAE:	Canarium			+	+
	Dacryodes				+
DILLENIACEAE:	Dillenia	+	+		
DIPTEROCARPACEAE:	Dipterocarpus	+			
	Dryobalanops aromatica	+			+
	Shorea beccariana	+			+
	S. fallax	+			+
	S. hemsleyana	+			+
	S. macrantha	+			+
	S. macrophylla	+			+
	S. mecistopteryx	+			+
	S. palembanica	+			+
	S. pinanga	+			+
	S. richetia	+			+
	S. seminis	+			+
	S. splendida	+			+
	S. stenoptera	+			+
EBENACEAE:	Diospyros	+			+
EUPHORBIACEAE:	Baccaurea		+	+	+
	Sapium		+		
FAGACEAE:	Castanopsis	+			+
	Lithocarpus	+			
	Quercus	+			
GUTTIFERAE:	Calophyllum		+		
	Garcinia	+	+		+
	Mesua ferrea	+			+
LEGUMINOSAE:	Dialium		+		+
	Intsia palembanica		+		
	Koompassia		+		
	Parkia		+		+
	Pithecellobium		+		+
MELIACEAE:	Aglaia		+	+	+
	Chisocheton			+	
	Dysoxylon			+	
	Lansium	+	+		+
	Prunus		+		
	Sandoricum	+			+
	Xylocarpus	+			+
MORACEAE:	Artocarpus	+	+	+	+
	Ficus	+	+	+	
	Parartocarpus		+		+
MYRISTICACEAE:	Horsfieldia			+	
	Myristica			+	
MYRTACEAE:	Eugenia		+		
OXALIDACEAE	Sarcotheca		+		
PALMAE	Calamus	+	+	+	+
POLYGALACEAE:	Xanthophyllum		+		+
RUBIACEAE	Nauclea		+		
	Randia		+		
SAPINDACEAE:	Dimocarpus		+		+
	Nephelium	+	+	+	+
	Xerospermum		+		+
SAPOTACEAE:	Ganua	+			+
SIMAROUBACEAE:	Irvingia	+			
TILIACEAE:	Grewia		+		
	Microcos		+		
URTICACEAE:	Sloetia		+		
VERBENACEAE:	Vitex		+		

Source: Caldecott 1988

areas to visit mineral salt licks. Bearded pigs in Borneo show large-scale seasonal migrations; huge herds of thousands of pigs ford rivers and cover great expanses of forest, following the fruiting seasons of favoured food trees (MacKinnon 1974b; Caldecott and Pfeffer 1986). Similar migratory movements are seen among fruit bats and have been reported for some populations of orangutans (MacKinnon 1974a).

Production forests can play an important role in maintaining wildlife populations, but their long-term value depends on long rotation cycles (of perhaps fifty years or more) to allow forests to regenerate naturally. Shorter logging cycles result in progressive changes in forest regeneration and forest composition, with associated loss of floral and faunal diversity and abundance. Replanting of logged areas with fast-growing alien species may disturb regenerative processes and lead to gradual breakdown of the food web. It is vital, therefore, to conserve some primary forests in an unlogged state.

Since most of the lowland forests of Borneo are being logged, or will be logged in the near future, it is timely to consider the best course of management to ensure the minimum disturbance and long-term survival of wildlife resources. Production forests can never fulfill all the conservation objectives of protected areas, but they can be an important supplement to primary forest reserves. Production forests surrounding fully protected areas will effectively increase the conservation estate to help preserve much of Kalimantan's lowland wildlife.

Reforestation

Large-scale programmes of reforestation are being planned in Kalimantan and elsewhere in Indonesia to meet needs for fuel, timber and other wood products. In most respects replanting trees on opened lands (reforestation), or planting them (afforestation) on areas that have been without forest for a long time (e.g., Riam Kiwai in South Kalimantan), should produce the opposite effects to those of removing trees. Such reforestation schemes should take pressure off natural forests, make unproductive lands more productive, minimise soil erosion, decrease runoff after rainfall, reduce the risk of flash floods and restore nutrient budgets to impoverished soils. However, there is little evidence that planting trees increases rainfall and water supplies, except where fog or cloud capture of moisture may result (e.g., in mountain cloud forests). Indeed, in most places reforestation of open land results in lower water tables and reduced stream flow, especially during the dry season (Hamilton and King 1983). Although storm flow will be reduced, even large-scale reforestation programmes will not prevent flooding, though they may reduce it in watersheds where a major part of the catchment is severely degraded with gullies, compacted

> **Box 9.8.** Watershed management: Riam Kanan, South Kalimantan.
>
> Riam Kanan has been identified as one of the eleven critical watersheds in Indonesia. It protects the water for the Riam Kanan dam, completed in 1972. The project was designed for power generation, flood control and irrigation. A reforestation programme was begun in 1973.
>
> The Riam Kanan River is dammed where it flows from the Meratus Mountains, creating a 9,000 ha reservoir. A 30 megawatt (MW) hydroelectric plant provides power to Banjarmasin and much of the southern part of South Kalimantan. The 100,000 ha watershed above the dam ranges from low hills to peaks at 1,200 m (Gunung Aurbunak). Rainforest still covers the highest hills, but most of the lowlands have been converted by unsustainable agricultural practices to *alang-alang* grasslands of *Imperata cylindrica*. Many of the soils are highly weathered and acidic.
>
> The size of the human population has more than doubled since 1962, and by 1985 8,000 people lived in 14 villages in the hilly lowlands around the reservoir. Most families grow rice, peanuts and vegetables in small plots at the forest edge, usually converting the forest to grass by repeated cropping and burning. Villagers do not practise traditional long fallow shifting cultivation, and their fields now extend well within the boundaries of the protection forest and Pleihari-Martapura game reserve. The rich alluvial soils, traditionally used for wet rice culture and groves of fruit trees, were drowned by the reservoir. Cattle grazing now occupies much more land than cropping in Riam Kanan. The forest provides rattan and minor forest products; commercial logging began in the upper watersheds in 1981. Local people and seasonal immigrants mine gold and diamonds in alluvial deposits along forest streams in the dry season.
>
> The government began the reforestation programme on the premise that tree plantations would reduce erosion and slow the sedimentation rate while increasing dry season water levels. The goal of the programme is to convert all grassland (some 35,000 ha) to man-made forest, but few seedlings from the hundreds of hectares planted annually have survived the recurrent grass fires. The project gained an added sense of urgency after the severe drought of 1982-83, when electricity production fell from 30 MW to 5 MW because lowered water levels limited production.
>
> There is little surface erosion associated with any of the many agricultural practices in Riam Kanan, although some very steep fields immediately adjacent to the shore probably contribute sediment to the reservoir. Mining appears to contribute large amounts of sediment to heavily mined tributaries. Overgrazed hills adjacent to the reservoir are subject to rill erosion in heavy rains. Buffer strips around the reservoir, and control of mining activities, could overcome these problems. The extensive, and so far unsuccessful, reforestation programme, which has been directed at the most level land in the valley to permit mechanical planting, cannot be justified on the basis of erosion control.
>
> *Source: Schweithelm 1987*

soil, and almost all rainfall is quickly channelled into streams and rivers. In such cases reforestation will certainly reduce erosion rates and the accumulation of river sediments.

One way to replenish forests after logging is by enrichment planting and reforestation of cleared areas such as log ponds. Forestry companies in Indonesia are legally obliged to reforest their concession areas; they must pay a levy which is refundable if restoration is completed satisfactorily. Unfortunately, few firms comply with this regulation, and those that do reforest areas considerably smaller than the areas they log. For instance, in 1979 the large ITCI-Weyerhauser concession in East Kalimantan was replanting 1,000 ha per year with fast-growing trees while logging an estimated 30,000 ha per year (Avé and King 1986).

Provincial reforestation programmes have usually concentrated on relatively small areas and have often failed to achieve their targets of regreening critical lands. In South Kalimantan, where in 1974 almost 1.2 million hectares of forest were under exploitation, yielding a production of 542,000 m^3 of wood, a figure of 45,000 ha was set for that year for reforestation; this was less than 4% of the area of exploited forest. In practice only 192 ha were replanted in 1974. Between 1976 and 1980 a target figure of 24,200 ha was set, but by 1980 only 13% of the proposed hectarage had been replanted. In West Kalimantan for 1979-80, only 9,960 ha of a planned 50,000 ha of replanting was actually achieved (Avé and King 1986). This trend has continued so that by the end of 1988 regreening and reforestation programmes had planted only 250,000 ha in Kalimantan, whereas almost three million hectares of critical lands remained unplanted (table 9.9). Moreover, reforestation schemes have often failed when wild fires from agricultural fields have swept through reforested areas in alang-alang lands, destroying the young trees.

Both timber firms and provincial forestry projects usually choose to plant with fast-growing exotic species rather than slow-growing native hardwoods. Eucalyptus, acacias and pines have all been tried with some success.

Table 9.9. Regreening and reforestation in critical lands in Indonesia, 1988.

Region	Reforestation programme in critical areas (hectares)			
	Sub-programme "Regreening"		Sub-programme "Reforestation/Rehabilitation"	
	Area planted by end of 1988	Remaining critical area to be planted in 1989	Area planted by end of 1988	Remaining critical area to be planted in 1989
Sumatra	1,323,003	2,298,600	493,580	1,405,900
Java	3,045,126	1,188,500	–	–
Nusa Tenggara	468,811	1,225,900	124,654	1,034,500
Kalimantan	137,693	1,165,300	205,772	1,798,300
Sulawesi	835,016	965,200	395,893	1,099,300
Maluku	4,896	330,400	1,915	305,400
Irian Jaya	–	95,800	–	186,800
Total	5,814,545	7,269,700	1,221,814	5,830,200

Source: MoF/FAO 1991

These fast-growing monocultures may protect the soil and restore vegetation cover, but the resulting forest is less species-rich than the original forest, is susceptible to wild fires and provides only an impoverished habitat for wildlife. Because dipterocarps fruit rarely, produce seeds that are difficult to store and have low seedling success, raising dipterocarp seedlings for replanting logged forests has always been difficult. However, recent tissue-culture experiments at ITCI and Wanariset in East Kalimantan are producing much larger numbers of healthy dipterocarp seedlings, suitable for replanting deforested lands and recreating natural forests (Smits et al. 1992).

Forest Fires on Borneo

From late 1982 to 1983 drought and fires affected an estimated 3.6 million hectares of forest (an area the size of Belgium) in East Kalimantan (fig. 9.13). Estimates by a West German team indicated that the damaged area included 800,000 ha of primary lowland rainforest, 550,000 ha of peat swamp forest, 1.2 million hectares of selectively felled forest and 750,000 ha under shifting cultivation, including secondary forest. The loss of standing timber and growing stock was estimated at more than $5 billion (Lennertz and Panzer 1983). Satellite imagery from 1983 indicates that during the same drought Sabah lost almost a million hectares of tropical vegetation (Malingreau et al. 1985; Beaman et al. 1985). The Borneo fires were major ecological events with profound and lasting impacts on the environment, vegetation and wildlife.

The immediate cause of the fire was a combination of severe drought, destructive logging practices and slash-and-burn agriculture. From July to November 1982 rainfall in Kalimantan was only about 40% of the normal level, and there was almost no rain from February to mid-May 1983. Such a severe drought probably occurs only once every 50 to 100 years (Leighton and Wirawan 1986) and was undoubtedly related to the El Niño climatic events at the time. It is interesting that in 1877 the Swedish explorer Bock, travelling through East Kalimantan, reported a similar severe drought which caused famine in the Kutai district.

During the protracted dry season, drought stress caused evergreen trees to shed their leaves, which led to an accumulation of dry litter on the forest floor. This, together with dead wood left after felling in logged areas, provided abundant fuel. The fires were started by local people and new settlers, who were clearing land by burning prior to planting crops. Fire is a common tool in traditional slash-and-burn agriculture. In the exceptional drought conditions the fires spread into large areas of rain-

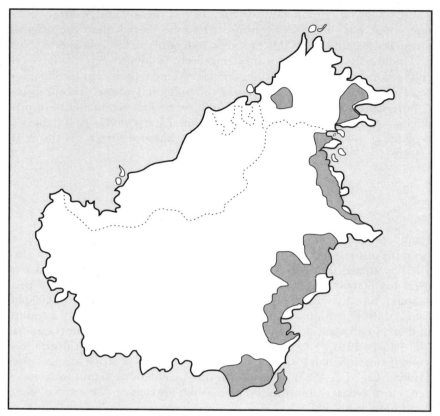

Figure 9.13. Areas of forest affected by drought and fire in 1982-1983.

forest. The most extensive fires occurred from August to October 1982 and from March to May 1983. Smoke covered Kalimantan for four weeks in early 1983 and caused airport closures as far afield as Surabaya, Java, 400 km to the south, and Singapore and West Malaysia, more than 1,500 km to the west.

The extent of damage caused by the fire varied with forest type. Peat swamp forest suffered the most damage. With the drought, the water table fell in the large inland freshwater swamps, and the top peat layer was exposed, and dried out. The shallow-rooted vegetation dried out, and many trees fell. Dry peat and dead wood fed ground and surface fires over large areas of the swamp forest. Highly inflammable resins caused some trees to burn like spectacular 60 m torches. Fire spread through both the peat and underground coal seams to penetrate new areas of forest.

In dryland forest, trees were more susceptible to drought on the drier

Box 9.9. Some possible effects of burning on five ecological processes.

Ecological process	Effect
Natural succession	• Curtailment of natural succession and ecosystem evolution • Vegetation pattern reflects pattern of burns, the mosaic containing different successional stages • Creation of bare areas, facilitating invasions by weeds and exotic species • Local breakdown of ecological balance between species • Progressive reduction of species diversity, regulated by fire tolerance • Progressive increase in uniformity, with fewer ecosystems and specialised niches • Migration and concentration of herbivores in areas with a flush of new nutritious plant growth
Organic production and decomposition	• Loss of biomass • Reduced energy capture and primary production due to leaf loss • Reduced secondary production, at least until new flush of plant growth occurs • Reduction of organic turnover by decomposition
Nutrient circulation	• Loss of elements in smoke, windblown ash, etc. • Diminution and simplification of nutrient cycle • Reduced retention of nutrient capital in organic matter • Reduced significance of litter layers in decomposition • Increased loss of elements by surface runoff and leaching • Changed rate of nitrogen fixation
Water circulation	• Reduction in rainfall interception • Increase in throughfall • Reduction in transpiration • Reduction in moisture in upper soil layers due to greater evaporation • Increase in water discharge
Soil development	• Increase in soil erosion as vegetation cover is lost • Loss of organic matter • Formation of base-rich soil surface layer • Increase in pH of soil surface layer, affecting micro-organisms (e.g., nitrifying bacteria) • Darkening of soil with charcoal and loss of vegetation, resulting in higher soil temperature • Death and decomposition of plant roots • Increase in nutrient loss by leaching • Possible progressive long-term decline in soil nutrient capital • Increased salinity as trees are lost

Source: Ovington 1984

ridges and slopes than on wetter alluvial soils. Fire killed more trees in areas of secondary forest than in primary stands (Riswan and Yusuf 1986). Climate, site, vegetation and human factors combined to create a mosaic of areas affected to varying degrees by drought and single or repeated fires. Such vegetation mosaics can still be clearly distinguished in aerial surveys of Kutai National Park, an area that suffered severe fire damage. Although the primary forest in the park was damaged by drought and fire, selectively logged forest suffered comparatively more damage. Logging had opened the canopy and created a drier microclimate; the debris from logging provided fuel for the fires.

Soil quality at the burned sites has deteriorated due to death of soil organisms and increased nutrient leaching. Soil erosion was 10 to 20 times greater in burned than in unburned areas in Kutai National Park (Shimokawa 1988). When the rains returned in 1983, soil erosion from exposed areas increased the sediment load of the Mahakam River and threatened the important inland fisheries. The almost complete destruction of the peat swamp forests, which had acted as a giant sponge regulating water flow, led to increased flooding in downstream riverside villages (Mackie 1984).

Fire and Forest Ecosystems

Studies conducted in Kutai since 1983 give some information on the effects of the drought and subsequent fires on rainforest vegetation and fauna (Wirawan 1984; Leighton and Wirawan 1986; Tagawa and Wirawan 1988). The fires caused more damage in areas of logged forest than in undisturbed forests, where fire swept mainly through the undergrowth and only burned the crowns of those trees heavily laden with lianas (Tagawa et al. 1988). In primary forest about 25% of canopy trees were killed, mainly from drought, while over 90% of understorey treelets and lianas were killed, mostly by fire (Leighton and Wirawan 1986). Mortality was 40%-60% among the canopy trees that produce fruits and seeds important to vertebrates, and crown dieback affected the majority of surviving fruit trees; this had important consequences for frugivorous birds and mammals.

Fires caused extensive damage to about half the lowland forests within Kutai National Park and led to local extinctions of species of some animals and plants (Petocz et al. 1990). Large numbers of animals died during the fires, especially less mobile creatures such as nestlings and forest floor reptiles and amphibians. Surviving animals have had to contend with decreased food resources due to the dieback of fruit trees during the drought and extensive forest destruction by the fire. Some species populations, particularly of hornbills and squirrels, have declined (Leighton and Wirawan 1986). A few species such as deer and banteng ultimately may benefit from the fire, as regenerating vegetation provides them with an increased food supply of herbs and grasses. Woodpeckers are also common in the burned forests, feasting on wood-boring beetles in the dead, standing wood left by the passage of the fire.

Most populations of frugivorous vertebrates, including hornbills, fruit pigeons and barbets, declined in numbers after the fires. This is not surprising given the high fruit-tree mortality (44% of canopy fruit trees) and loss of productivity among surviving fruit trees. Two of the five territorial hornbill species were no longer represented, and the other three species had declined in numbers. Since hornbills are territorial, it is unlikely that the missing groups could have moved and survived elsewhere, since adjacent appropriate habitat was already occupied by other hornbills. There was also a decline in numbers of three species of diurnal seed-eating squirrels *Ratufa affinis, Callosciurus notatus* and *C. prevostii*, which had been common in the forest prior to burning (Leighton and Wirawan 1986).

Population densities of the large-bodied primates were the least affected of the arboreal vertebrates. Nine species of monkeys and apes and two prosimians are recorded from Kutai (Wirawan 1985). Several species such as red, grey and white-fronted langurs were found immediately after the fire but were much less common three years later in fire-affected areas. Proboscis monkeys also vanished from isolated pockets of mangrove forest where they were recorded immediately after the fire (Doi 1988). Orangutans were still common and breeding after the fire, but, since few fruiting trees remained in the Mentoko area, they had changed their feeding behaviour and were eating large quantities of bark and young leaves immediately after the fire, and the stems and young leaves of secondary trees (Suzuki 1992); these are normally less favoured foods (MacKinnon 1974a; Rijksen 1978). The orangutan's omnivorous diet and dietary flexibility have allowed it to survive in these less-than-optimal conditions. For this large, arboreal ape the changes in forest structure following the fire have also been important. Orangutans are now often seen walking on the ground between patches of forest in more open areas.

Invertebrate populations in Kutai were also affected by the fire. In forest soils earthworms were dominant, whereas in open, previously burnt areas many kinds of beetles were found. Generally, density and biomass of soil and litter invertebrates decreased gradually from the unburnt forest soils to the open *Imperata* grassland areas (Yajima 1988). Similar results were observed in northern Thailand, with a drastic decline in density and biomass of soil macrofauna following burning (Watanabe et al. 1983). Termites play an important role in the decomposition cycle of tropical forest soils (see chapter 4), but in Kutai termites and their mounds were rare, even several years after the fire (Yajima 1988). This is surprising in view of the large amounts of dead and rotting wood on the forest floor. Although the soil and litter fauna seemed to recover quickly after the fire, species composition changed, with species such as snails occurring only in areas of unburnt forest (Yajima 1985). Butterfly collections in burned and unburned areas show that certain species are gradually making a comeback, presumably as the vegetation recovers and food plants become available (table 9.10).

It is difficult to predict the recovery of the Kutai forests. After normal rains recommenced, seeds of a limited number of fast-growing species germinated, and three years after the fire the area was colonised with secondary pioneer species such as *Macaranga trichocarpa* and *Anthocephalus chinensis* (Tagawa et al. 1988). Although regeneration helped to stabilise forest soils, widespread erosion occurred, and steep slopes and formerly stable stream channels still showed continued slippage five years after the fire.

Regenerated forest often has a different species composition from that

Table 9.10. Butterflies as indicators of forest recovery in Kutai National Park after the 1983 fires.

a. Species abundant in June 1983 immediately after the fires.

   Salectara liberia (Pieridae)
   Euploea sylvester (Danaidae)
   E. mulciber (Danaidae)
   E. leucostictos (Danaidae)
   E. diocletianus (Danaidae)

b. Species very rare in 1983, and decreasing after 1983.

   Ixias undatus (Pieridae)
   Hebomoia glaucippe (Pieridae)
   Appias nero (Pieridae)
   Hypolimnas bolina (Nymphalidae)
   Polyura delphis (Nymphalidae)
   Charaxes solon (Nymphalidae)
   C. durnfordi (Nymphalidae)

c. Species found only in unburned forests, and increasing there.

   Prothoe franck (Nymphalidae)
   Agatasa calydonia (Nymphalidae)
   Kalima paralecta (Nymphalidae)
   Thaumantis klugius (Amathusinae)
   Atrophaneura neptunus (Papilionidae)

d. Species very rare or absent after the fires, but apparently recovering in 1986 and 1988.

   Trogonoptera brookiana (Papilionidae)
   Troides helena and *T. amphrysus* (Papilionidae)
   Papilio demolion (Papilionidae)
   P. palinurus (Papilionidae)
   Paranticopsis delessertii and *P. megarus* (Papilionidae)

Source: Suzuki (in prep.)

of the original forest stand. In Kutai many of the mature canopy trees were destroyed by the fire, so are not contributing seed for forest regeneration. Seed banks in the soil have also been destroyed in intense pockets of fire. It is possible that many of the less common canopy hardwood trees may become extinct locally. Species such as dipterocarps and Fagaceae, which require mycorrhizal associations, may not regenerate because fungal spores were eradicated by the fires.

The fast-growing plants that now dominate the regenerating vegetation produce fruit eaten mostly by small birds and only rarely by primates, hornbills or other larger birds and mammals. Many primates have increased their dependence on leaves, but other mammals and frugivorous birds that cannot switch to folivory are likely to remain rare. The situation is particularly serious for fruit and nectar feeders. Many frugivores also rely heavily on figs. Germination sites on tree limbs for the most important *Urostigma* figs have become rare as a consequence of the fires (Leighton and Wirawan 1986). The lack of fruit resources may lead to further declines in numbers of some frugivorous animals.

Although canopy trees were lost to drought in unburned primary forests in Kutai, most areas maintain healthy populations of plants and animals, which can act as sources for recolonisation of regenerating areas of burned forest. However, there is a heightened risk of further forest fires in the previously burned areas because of the large amount of dead wood on the floor and the domination of understoreys by softwood pioneer trees susceptible to burning (Kartawinata 1980b). Indeed, in 1987, during another very dry spell, fires again swept Kalimantan in a "sea of fire" (Tempo 1987).

As well as their effects on vegetation and wildlife, the forest fires also had a high cost in economic and human terms. A large tract of land was degraded, and timber worth billions of dollars was lost. There was very little attempt at timber salvage, and only poor quality timber was left. The livelihoods of many thousands of local people were threatened, especially those dependent on collection of forest products (Mackie 1984). Paradoxically, the fires also helped some local cultivators by opening up large areas of forest without further clearing.

Cash crop growers also suffered from the fire. The peppercorn loss was estimated at almost $2 million (Mackie 1984). Trade suffered in minor forest products such as rattan, since wild stocks throughout East Kalimantan were destroyed, and new rattan vines take about nine years to reach harvestable size. Most sectors of East Kalimantan's economy are linked in some way to tropical forest resources, or depend on the forests' environmental functions, such as watershed protection (Mackie 1985).

For East Kalimantan the 1982-83 fire was an ecological and economic disaster. The most visible culprits were the shifting cultivators who lit the fires. However, the conditions that made the forest vulnerable to fire were two decades of deforestation, encouraged by government land-use policies to open a frontier region to large-scale commercial exploitation.

CONSERVATION OF FORESTS

The arguments for conservation of forests and wise management for their

---

Box 9.10. Goods and services provided by tropical forests.

- **Ecological diversity.** Natural forest areas maintain examples of the different types of natural community, landscape and land form, and protect the full range of animal and plant species and their genetic variability.
- **Timber.** Forests provide a sustainable yield of wood products for domestic use and export earnings.
- **Other forest products.** Forests provide a sustainable harvest of minor forest products including rattan, damar resin, fruits, nuts, honey, edible plants, and pharmaceutical and cosmetic compounds. In 1987 such non-wood products in Indonesia were valued at more than $238 million.
- **Water and soil conservation.** Forests protect watersheds and ensure an adequate quality and flow of fresh water. They control erosion and sedimentation, and are especially important where these affect downstream investments that depend on water for transportation, irrigation, agriculture and fisheries, and recreation.
- **Climate regulation.** Forests moderate the climate both locally and globally: they influence the composition and heat-retaining capacity of the atmosphere and the heat exchange characteristics of the earth's surface.
- **Wildlife.** Forests support fish and wildlife, which are vital food sources for local communities as well as generating income and employment. In Sarawak alone pig and deer meat is estimated to be worth M$100 million annually.
- **Integrated rural development.** Forests support the integrated development of rural lands; trees can be used to rehabilitate degraded lands and to diversify production systems through agroforestry.
- **Recreation and tourism.** Forests provide opportunities for outdoor recreation and tourism for local residents and foreign visitors.
- **Education and research.** Forests provide opportunities for formal and informal education, research, and the study and monitoring of the environment in natural areas.
- **Cultural heritage.** Forests are part of the national heritage. They contribute to the folklore and traditions of local peoples, protect historical features and enhance the quality of the environment.
- **Options for the future.** Lands kept under forest retain natural processes and ensure open options for future changes in land use.

Source: Poore and Sayer 1988

sustainable use are irrefutable. Forests play many roles: they provide soil stability and protection; they enhance soil fertility; they regulate the quantity and quality of water flow; they affect local and global climates; and they act as purifiers, removing pollutants, including carbon dioxide, from the atmosphere. Primary rainforest also provides a livelihood for local communities, and is a source of food, timber, building materials, medicines and other utilities. In addition, forests improve the quality of life and provide opportunities for recreation, research and tourism.

These benefits can be seen and appreciated in the present. Forests can also provide future benefits as reservoirs of genetic resources. The value for domestication of some forest species, such as fruit trees and banteng, has already been proved. Past experience suggests that many forest species not yet utilised, and perhaps still undiscovered, may prove of future value for purposes as yet unrecognised. For instance, when peat swamps were first logged in Sarawak, ramin *Gonystylus bancanus* was considered unmarketable and was left unfelled to be poisoned; now it is the most valuable species in these forests (Whitmore 1984a).

Moreover, as long as forests are left undisturbed or are harvested sustainably in an ecologically sound manner, the options for future land uses are left open. Once forests are damaged and cleared their ecological diversity declines, species are lost and irreversible ecological and environmental changes occur. Borneo's rich biodiversity is protected within the island's forests. Only 3% of Borneo's forests are included within gazetted conservation areas and wildlife reserves (MacKinnon and MacKinnon 1986). It is imperative that the extent of protected forests within Kalimantan should be increased, and that protection and management of existing national parks and reserves should be improved, to conserve part of Indonesia's rich national heritage.

Chapter Ten

Wetland Resources

Wetlands are a major habitat in Kalimantan, covering more than 10 million hectares, about 20% of Kalimantan's land mass (MacKinnon and Artha 1981). The major wetland habitats in Kalimantan are freshwater and peat swamps and coastal mangroves. The Kapuas, Mahakam and Barito (Indonesia's longest rivers) have extensive floodplains with associated swamps and lake systems. Areas of wetland habitat are shown in fig. 10.1. This chapter focuses on the ecological and economic importance of freshwater habitats. The value of coastal mangroves is discussed in chapter 11.

THE ECOLOGICAL IMPORTANCE OF WETLAND HABITATS

As Indonesia's population increases and economic development proceeds, wetland habitats are becoming a focus for regional development strategies. Kalimantan wetlands, and especially the tidal swamplands, have long attracted the attention of developers because of their potential for growing rice. Already 1.2 million hectares of tidal swamplands in Kalimantan have been developed for agriculture, and a further 1.4 million hectares are believed suitable for agricultural development (Euroconsult 1986). This latter area is 9% of all the tidal swamps in Indonesia. Before these wetland areas are converted, it is timely to ask whether agriculture is the best and most appropriate land use for them and to consider what benefits are derived from natural wetlands.

Natural wetlands provide people, directly and indirectly, with an enormous range of goods and services: staple food plants, commercial timbers, fertile grazing land, support for inland and coastal fisheries, flood control, breeding grounds for waterfowl and fuel from peat. These "hidden" values are rarely quantified and are often overlooked in regional development plans.

Wetland habitats are extraordinarily productive. Primary production in open marshes may be twice as great as in tropical rainforests (Odum 1971). Moreover, wetland plants are highly productive in waterlogged conditions where other plants cannot grow. Many wetland trees have specialised tissues or organs to obtain and transmit oxygen to submerged roots. Many open-

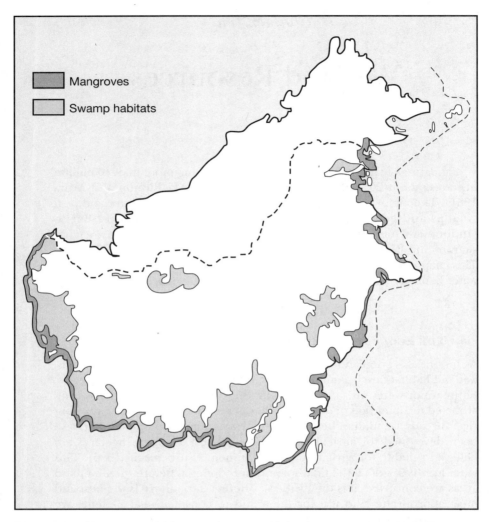

Figure 10.1. Mangroves, tidal swamplands, and inland swamp habitats in Kalimantan.

water plants have large leaf areas and little wood or thickened tissue, so that more of the plant is devoted to photosynthesis, providing energy for growth, than in many land plants. Some wetland plants have unusual physiological adaptations, such as the ability to stimulate alcoholic fermentation in roots to provide energy, at the same time avoiding alcohol poisoning (Maltby 1986).

Many wetland soils are alternately wetted and dried. This increases the release of nutrients and speeds the turnover of organic matter (Maltby 1986). Water moving past the plants, especially in tidal wetlands, provides a steady or pulsed supply of nutrients, even though the nutrient concentration in the water may be low. This water movement removes dead matter, keeping plant communities healthy and vigorous. Finally, and most obviously, wetland plants in healthy wetlands rarely suffer from lack of water.

Wetlands produce a high sustainable yield of plants that are useful to people. They also support a wide variety of animals through grazing and food chains, both within the wetland itself and beyond its boundaries through the action of currents and tides. Flood plains and lake systems in Kalimantan support productive freshwater fisheries (Zehrfeld et al. 1985; Giesen 1987; Chaeruddin 1990). Flood plains are also important spawning, nursery and feeding grounds for fishes. Bornean river systems are rich in species; 290 species of freshwater fish are recorded from the Kapuas and its tributaries (Roberts 1989), and Sarawak rivers support at least 100 fish species (Chan et al. 1985). The Kalimantan inland fisheries can be regarded as a national resource, providing food for local consumption and export to Java (Zehrfeld et al. 1985).

Deep swamp areas once may have been important for the estuarine crocodile *Crocodylus porosus* and the false gavial *Tomistoma schlegeli*, though these species have now been hunted almost to extinction in many areas. Shallower and seasonally dry areas provide grazing for cattle and wild ungulates, including banteng *Bos javanicus* and sambar deer *Cervus unicolor* (Giesen 1987). Swamp forests support rare species such as the orangutan *Pongo pygmaeus*, the endemic maroon langur *Presbytis rubicunda*, the clouded leopard *Neofelis nebulosa* and the bay cat *Felis badia*, while riverine habitats harbour the proboscis monkey *Nasalis larvatus*, the otter civet *Cynogale bennetti*, the pesut or freshwater dolphin *Orcaella brevirostris*, and a host of birds. Many wetland habitats are species rich, and isolated water bodies, such as oxbow lakes, may harbour locally endemic fish species.

Changes in land use can disrupt the ecology of wetland habitats. Wetlands in Kalimantan are being drained and converted to agriculture and aquaculture. Swamp forests are felled and mined for peat. Water movements are altered by the building of dams, and drainage and irrigation canals. All of these developments alter the wetland habitats and their ecological processes, and may have long-term and far-reaching effects on productivity both in the wetlands themselves and in adjacent lands. They also reduce and fragment the area of wetland habitats, leading to a decline in species diversity and loss of some species of known or potential economic value.

Water Flow and Water Quality

Wetland habitats play an important role in hydrological functions, regulating water flow, purifying water supplies and acting as a sponge to release water gradually to adjacent lands. Freshwater forests function as water reservoirs and sedimentation basins for the large rivers. Extensive areas of peat swamp play a major role in the hydrology of the lowland river basins, storing excess water, reducing flooding and regulating water flow (Klepper 1990). Deforestation, drainage, and removal of the peat can seriously disturb the water regime in these habitats and adjoining lands.

Wetlands slow the velocity of water and increase its "residence time" in the ecosystem, thereby enabling biological, physical and chemical changes to occur. They remove suspended sediments which might otherwise silt up rivers and irrigation canals or, because of high turbidity, might reduce the production of phytoplankton in lakes and estuaries. Moreover, a reduction in water flow rate leads to greater sedimentation, so swamps help to build up subsiding coastlines in delta regions.

Peat swamps and other wetlands improve water quality by filtering out surplus nutrients and toxins. Excessive amounts of nutrients, such as nitrogen or phosphorus from fertiliser runoff, can cause rapid algal and plant growth and can lead to eutrophication and subsequent deoxygenation as dead plants decompose. Wetland plants can absorb these nutrients, and wet, low-oxygen soils favour denitrification by certain bacteria that transform nitrates into gas, which diffuses into the atmosphere. In a similar manner, large amounts of phosphorus are inactivated by bonding to inorganic ions, especially aluminium and iron. In agricultural areas with acid sulphate soils this can be a major disadvantage, tying up fertilisers so that nutrients are not available to plants (Dent 1986). Studies in Pulau Petak by a team from KPSL-UNLAM reveal that at least one hectare of primary forest is required per hectare of cultivated land in order to provide sufficient fresh water to leach acids from the soil and to maintain soil fertility (Klepper et al. 1990).

Some wetland plants can remove heavy metals, pesticides and other toxins from water, and can transform and fix contaminants so that they do not enter the food chain or groundwater (Maltby 1986). Natural wetlands can treat, and render harmless, coliform bacteria (from faeces) and suspended solids created from human waste (National Research Council 1976). Aquatic plants can be utilised in the treatment of sewage and waste water. Experiments have shown that duckweed *Lemna* spp. can remove half the nitrogen, 67% of the phosphorus and almost all the heavy metals from polluted water (Maltby 1986). Water hyacinth, usually regarded as a nuisance species, can be used to treat raw sewage and as a pollution filter (National Research Council 1976). In just 24 hours it can remove more than 75% of the lead in contaminated water, and also absorbs cadmium, nickel, chromium, zinc, copper and iron as well as pesticides and other

Box 10.1. Making aquatic weeds useful.

Water is an important resource. Aquatic weeds can cause problems by clogging rivers and causing floods, blocking canals and pumps in irrigation projects, hindering hydroelectricity production, wasting water in evapotranspiration, increasing waterborne diseases and interfering with fish movements and fish culture. There is no simple way to reduce infestations since herbicides may adversely affect the environment and fish production. Yet aquatic weeds can also have considerable value as highly productive crops for animal feed, human food, fertiliser, fuel and energy production, paper and fibres.

Animal feed
Duckweeds are eaten by fish, ducks, geese, wildfowl, cattle and pigs. Fish are attracted to the fringes of water hyacinth beds, and the grass carp, a quick-growing fish with highly prized meat, feeds on submerged weeds. Water hyacinth can also be used as feed for water buffalo. Fresh aquatic weeds usually contain too much moisture to be efficient foodstuffs. When water hyacinth is used as a feed, 95 kg of water must be excreted by the animal for every 100 kg of fresh weed eaten. Pressing out some of the water and ensiling the residue is a promising technique for developing this cheap and readily available resource as an animal feed.

Fertiliser
Many aquatic weeds contain appreciable quantities of nitrogen, phosphorus, potassium and other plant minerals. The ash of water hyacinth, for instance, contains 30% potash, 7% phosphoric acid and 13% lime, and makes an excellent fertiliser. Aquatic weeds can be used as green manure on farmland to benefit crops and improve soil texture. In China, Indochina and India *Azolla* is an important biofertiliser. Floating mats of aquatic vegetation are planted with vegetables at Danau Bangkau in South Kalimantan. Rice is grown on vegetation mats in swamps in Vietnam. Rotting vegetation on the mat provides nutrients for the crop.

Pulp, paper and fibre
These can be produced from fibrous, reed-like aquatic weeds. Paper can also be made from water hyacinth. Aquatic weeds may be used for thatch, furniture, mats and baskets.

Energy
Water hyacinth can be fermented to produce methane gas. The water hyacinth harvested from one hectare will produce more than 70,000 m^3 of biogas. This fuel burns readily and can be used for cooking, heating and as a source of power. Experimental biogas plants have been established in some villages in South Kalimantan.

Wastewater treatment
Some aquatic weeds can extract compounds containing nitrogen and phosphorus, which are common pollutants in waterways. Plants such as water hyacinth can be used to treat sewage, and dissolved nutrients are recovered for reuse. At experimental sites in Florida, U.S.A., 20-40 tonnes of wet water hyacinth can be harvested per hectare per day, removing the nitrogenous waste of over 2,000 people and the phosphorus wastes of more than 800. Duckweeds such as *Wolffia*, *Lemna* and *Spirodela* also show promise for recovering nutrients from wastewater. In India water hyacinth has been used to clean tannery effluents. In Malaysia the tiny floating fern *Azolla* is used to treat wastewater from sugar refineries and rubber processing plants.

Aquatic plants as food
Only one aquatic plant, rice, is widely used as a food crop. Water spinach *Ipomoea aquatica*, *Limnocharis* eaten in Java, lotus *Nelumbo nucifera*, Chinese water chestnut *Eleocharis dulcis*, taro *Colocasia esculenta*, swamp taro *Cyrtosperma chamissonis* and arrowhead *Sagittaria trifolia* are all Asian aquatic food plants with potential for increased exploitation. Duckweeds, too, can be used as foods. *Wolffia arrhiza*, the smallest flowering plant on earth, is used as a vegetable in Burma, Laos and northern Thailand.

Sources: National Research Council 1976; Maltby 1986

toxins (Maltby 1986). The pollutants are then removed simply by lifting the water hyacinth from the water. Moreover, since water hyacinth readily absorbs nitrogen, phosphorus, potassium and other nutrients from water, it makes an excellent fertiliser or green manure (National Research Council 1976; see box 10.1).

The role of wetlands in regulating water flow, improving water quality and treating waste water may far exceed their economic value from any other form of land use. In India Calcutta's sewage undergoes natural purification in the marshlands east of the city where lakes and pools act as oxidation ponds. Water hyacinth augments the water treatment, and coliform bacteria from faeces are reduced by 99%. In addition to cleaning waste water, these ponds are stocked with carp and tilapia and are highly productive fisheries (Ghosh *in* Maltby 1986). The swamps around Banjarmasin in South Kalimantan play a similar role, treating raw sewage and other pollutants and rendering them harmless while providing a cheap source of protein for local communities.

Plant Harvests from Wetlands

Many wetland plants are economically important; some are food staples. Rice, most of which is grown on flooded soils, is the primary food resource of over half the world's people and occupies 11% of all arable land worldwide (Maltby 1986). Rice was introduced to Kalimantan from China and mainland Asia, brought by early human migrants (Bellwood 1985). While cultivation of hill *padi* on drylands is widespread, wet rice farming was restricted until recently to certain areas and racial groups (see chapter 12). Only a few Dayak groups farm wet rice or *sawah* rice (Dove 1985; Padoch 1988), but the Banjarese are accomplished sawah cultivators. In West Kalimantan irrigated wet rice farming was restricted to Chinese districts (Jackson 1970). The ecology of sawah rice fields is discussed later in this chapter.

The sago palm *Metroxylon sagu* probably originates from New Guinea and the Moluccas but is cultivated and grows semi-wild in freshwater swamps along Borneo's coasts. Natural sago stands occur in freshwater swamps with soils of more than 70% clay and up to 30% organic matter (Flach 1983). The starch stored in the palm trunk can be used as a staple food when made into sago flour. The palm is harvested just before flowering, when the stem contains a maximum of food reserves; the starch between the fibres is beaten out and collected (Stanton and Flach 1980). Sago is used widely in eastern Indonesia; it is a food staple for a quarter of the population of Irian Jaya and over 100,000 Papuans (Tan 1980). It is an important secondary food for some Bornean communities, especially the Melanau of

Sarawak (Morris 1953) and the Bisaya of Sabah (Ruddle et al. 1978).

The sago palm may have been introduced to Borneo in prehistoric times and was probably a main staple for early human communities (Avé and King 1986). Sago starch contains about 80% carbohydrate and serves as an adequate substitute for rice. In Indonesia and Malaysia the starch is used commercially to manufacture cakes, noodles and *kerupuk*. It can also be used as a paste for papers and textiles or as a raw material for the production of ethanol, high-fructose syrup and adhesives (Westphal and Jansen 1989). Pith and young trunks provide forage for animals; the "bark" can be used as wood or fuel, the leaves for thatch, and the leaf stalks and midribs to weave walls and fences and to provide fibre for cordage. Young leaves are woven into baskets, and the growing point of the palm can be eaten as a vegetable, palm cabbage (Ruddle et al. 1978).

Sago palm is one of the most underexploited resources of Southeast Asia. Sago production is a highly efficient way of provisioning a community. Sago productivity per unit of labour is high, with outputs similar to those of shifting cultivation and higher than in sedentary agriculture or hunter/gatherer economies (Townsend 1974; Ruddle et al. 1978). Sago starch is mainly harvested from semi-wild stands, as in South Kalimantan, but many swamps in Kalimantan and elsewhere in Indonesia are suitable for sago cultivation (box 10.2). At present sago plantations cover only a very small area, 130,000 ha in Indonesia (20,000 ha in Kalimantan), about 35,000 ha in Sarawak and Sabah, and about 1,000 ha in Brunei (Flach 1983; Westphal and Jansen 1989). The high yield of starch (generally 7-9 tonnes/ha/year but possibly as much as 30 tonnes/ha/year) opens up exciting prospects for producing large amounts of food with little environmental impact (Maltby 1986). Through bioconversion, using microorganisms as chemical converters, sago can be used to produce either alcohol or protein. Recent trials suggest that it may be possible to convert 14.5 tonnes of starch, a reasonable yield for one hectare, into 3.5 tonnes of protein (Maltby 1986). Thus there may be advantages in cultivating high-yield starch plants like sago in little-disturbed wetlands, rather than draining swamps to plant crops richer in protein but with much lower yields.

Several other commercially important crops are wetland species. Oil palm *Elaeis guineensis*, a tree originating in West African wetlands, is one of the world's most important sources of edible and soap-making oil. It yields more oil per hectare per year than any other vegetable (Masefield et al. 1969). Breeding has extended the range of the plant into increasingly drier environments. Oil palm plantations have been established in Sarawak and Sabah and at transmigration sites in West and Central Kalimantan (see chapter 12).

Several wetland plants are used by local communities for food and medicines. Kangkong *Ipomoea aquatica* probably originated in tropical Asia but is now grown widely as a leaf vegetable throughout southern and

> **Box 10.2. Sago orchards of the Melanau.**
>
> The Melanau people of the Oya, Mukah and other rivers along the coast of Sarawak have traditionally cultivated the sago palm for subsistence and commerce. They have cultivated the palm successfully on acidic peat swamp soils, soils generally regarded as unsuitable for agriculture.
>
> Melanau sago orchards are established by felling the swamp trees and burning the dried vegetation to clear fields. Suckers of the sago palm *Metroxylon sagu* are planted in the swampy fields. The young sago palms are often destroyed by agricultural pests such as ants, monkeys, pigs and sun bears, but once the palms are established they require very little work. Sago orchards are weeded only once a year for the first two to three years, and occasionally thereafter.
>
> In the early years of the plantation vegetables are often planted between the young palms. After 10-15 years, depending on soils and other variables, the palms are mature. The Melanau cut the mature palms, either before or during flowering, or in the early stages of fruit development, when starch yields are highest. The trees are felled, cut into logs and rasped to remove the sago pith, which is then washed to remove the starch. Processing is a tedious task, but some of the work is now mechanised. A palm of 30 m length provides a family with sufficient food for 5-6 weeks.
>
> The practice of interplanting perennial and annual crops is a form of agroforestry. While waiting for the sago palms to mature, the Melanau produce a fast-growing crop in the partial shade afforded by the growing palms. The palms protect the soil and vegetable crop from the full force of the rain, while the mixed cropping seems to protect young suckers from attacks by pests. Reportedly the Melanau have produced sago from the same lands for centuries with no evidence of environmental deterioration.
>
> *Sources: Morris 1953; Padoch 1988*

Southeast Asia (Westphal and Jansen 1989). This water spinach is often grown in wet-rice fields but also grows wild in wetland habitats such as the Sungai Negara wetlands in South Kalimantan (Giesen 1990), where it is eaten by lakeside communities and harvested for sale in local markets. Kangkong is rich in iron and used medicinally as a sleeping draught. In the Danau Bangkau area stems of the waterlily *Nymphaea nouchali* are eaten as a vegetable, and flour is made from its seeds (Giesen 1990).

Wetlands are also a major source of nonfood plants. Reeds (mainly *Phragmites karka*) and sedges are used in thatching, weaving, handicrafts, paper production and even bedding (AWB/KPSL-UNLAM 1989). Uses for various aquatic weeds are described in box 10.1. Swamp forest trees provide timber, building materials and firewood for local communities (box 10.3).

The bark of the swamp forest tree *Alseodaphne coriacea* is collected in South Kalimantan to manufacture mosquito repellent (AWB/KPSL-

UNLAM 1989). Swamp forests can be exploited for rattans, gums, resins, tannins, oils and medicines as well as for edible fruits such as mangosteen *Garcinia* and durian. Commercially valuable latex is derived from the swamp jelutong *Dyera polyphylla*, an endemic species restricted to peat swamp forests in Borneo (Prescott-Allen and Prescott-Allen 1982). A survey of the Kapuas lakes area in West Kalimantan revealed that 81% of more than 200 plant species investigated in the Danau Sentarum wildlife reserve were useful to local communities, with 30% being used as food items, 6% for medicinal purposes and 27% for building materials (Giesen 1987).

Box 10.3. *Melaleuca* - the paper bark tree (gelam).

Gelam *Melaleuca cajuputi* (formerly *leucodendron*) is often found in pure stands in degraded freshwater swamp forests throughout South and Central Kalimantan. It is an interesting tree because of its tolerance to adverse soil conditions such as acidity, salinity, and waterlogging. It is extremely fire-resistant, and fire encourages its regeneration. The thick bark protects the cambium from excessive heat, and, even if damaged by fire the tree rapidly grows new shoots. Fires trigger the release of seeds from fruits which remain on the parent tree after ripening. Burning also creates a favourable environment for germination, since competing vegetation has been cleared, the soil is exposed and nutrients are more readily available (Wade 1981).

Melaleuca has a wide range of uses. Cajeput oil, extracted from the leaves, has germicidal and other medicinal properties. It is used as an insect repellent and in the manufacture of soap, throat pastilles and tiger balm ointment (for rheumatism). Flowers are a source of nectar for honey bees, and the dried fruits can be used as a substitute for black pepper. The bark may be used for caulking boats, for torches, and in poultices on infected wounds. It can also be used as packing and potting material and for cork granules in linoleum. Gelam wood is hard and fairly heavy. It may be used to provide posts and poles for temporary buildings or for foundation piles under groundwater level as it is not durable when exposed to air. Gelam can also be used for oars and shipbuilding. It is widely used for firewood and charcoal.

The gelam forests of Kalimantan are exploited for firewood and provide a substantial part (about two-thirds) of the income for many villagers in South Kalimantan (Koesmawadi 1984). Collection and transport of firewood is labour intensive, and returns per hectare of land are low, about 1 m^3 of wood per year compared to 10-20 m^3/ha/year from commercial plantations on fertile soils elsewhere. Nevertheless, gelam wood is probably one of the major products of Pulau Petak, at least for the north and central parts of the island.

Melaleuca has considerable potential as a plantation tree in acid sulphate soils and other poor soil conditions, which are unsuitable for rice-growing and commercial timbers. It is being used with some success in Vietnam, to reforest areas of coastal forests defoliated with Agent Orange.

Sources: Brinkman and Vo Tong Xuan 1986; Klepper et al. 1990

Timber from Swamp Forests

Peat swamp forests are an extremely valuable forest resource in Kalimantan, especially West and Central Kalimantan, and in Sarawak (Chan et al. 1985). Between 20 and 30 peat swamp species yield commercial timbers. Table 10.1 lists some species of commercially important trees from Kalimantan wetlands. The most valuable swamp tree in Kalimantan is the ramin *Gonystylus bancanus*, the second major source of timber in Indonesia after the dipterocarps meranti *Shorea* spp. and keruing *Dipterocarpus* spp. Other important timber species are *Copaifera palustris*, and the dipterocarps *Dryobalanops rappa*, *Shorea platycarpa*, *S. rugosa*, *S. scabrida* and *S. uliginosa*.

Table 10.1. Some commercially valuable trees of the wetlands of Borneo.

Family	Species	Indonesian name
Anacardiaceae	– *Campnosperma auriculata*	terentang
	– *Gluta renghas*	rengas
Araucariaceae	– *Agathis borneensis*	
Bombacaceae	– *Durio carinatus*	durian
Combretaceae	– *Terminalia phellocarpa*	
	– *Lumnitzera littorea*+	
	– *Lumnitzera racemosa*+	
Crypteroniaceae	– *Dactylocladus stenostachys*	
Dilleniaceae	– *Dillenia grandiflora*	simpur jangkang
Dipterocarpaceae	– *Anisoptera marginata*	
	– *Dipterocarpus coriaceus*	keruing
	– *Dryobalanops oblongifolia*	petaling
	– *Dryobalanops rappa*	kapur
	– *Hopea mengerawan*	merawan
	– *Shorea albida*	
	– *Shorea palembanica*	meranti merah
	– *Shorea platycarpa*	
	– *Shorea rugosa*	
	– *Shorea scabrida*	
	– *Shorea uliginosa*	
Ebenaceae	– *Diospyros* spp.	
Leguminoseae	– *Koompassia malaccensis*	kempas
	– *Copaifera palustris*	sepetir
Loganiaceae	– *Fagraea fragrans*	tembesu
Malvaceae	– *Hibiscus tiliaceus*+	
Meliaceae	– *Xylocarpus granatum*+	
	– *Xylocarpus moluccensis*+	
Myrtaceae	– *Melaleuca cajuputi*	gelam
Rhizophoraceae	– *Bruguiera gymnorhiza*+	bakau
	– *Combretocarpus rotundatus*+	perepat darat
	– *Rhizophora apiculata*+	
	– *Rhizophora mucronata*+	
Sapindaceae	– *Pometia pinnata*	matoa
Sonneratiaceae	– *Sonneratia alba*+	perepat laut
Sterculiaceae	– *Heritiera littoralis*+	
Theaceae	– *Tetramerista glabra*	punak
Thymelaeaceae	– *Gonystylus bancanus*	ramin
Verbenaceae	– *Avicennia* spp.+	

+ mangrove species

Source: Silvius et al. 1987

These species make up more than 70% of all trees over 1.8 m girth in peat swamp forest (Whitmore 1984a). In peat swamp forests in Sarawak and West Kalimantan *Shorea albida* is an important timber tree. Freshwater swamp forest is generally less commercially valuable than dry lowland forest because of lower yields of timber and problems of access.

Working methods in peat swamp are much less damaging to the soil surface than operations in hill dipterocarp forests because tracks of wooden sleepers are laid to remove the logs, and working can continue throughout the year. With careful management peat swamp forest can be harvested on a sustained yield basis. Peat swamp forests will become even more important in the future as more of the dry lowland forests are cleared and brought into agricultural use. Mixed swamp forest can be managed on a uniform silvicultural system with a single clear-felling operation planned on a 45-year rotation, as practised in Sarawak (Lee and Lai 1977; Lee 1979).

Forest regeneration depends on an adequate source of seedlings remaining after felling. In 1977 in Sarawak peat swamp forest was being logged at 8,000 ha per year and yielding 11 m^3/ha, with removal of 17% of stems and damage to another 23% (Lee 1979). After logging and clearing of remaining trees, the outer zones of peat swamp forests regenerate well, especially fast-growing species such as *Shorea, Dryobalanops* and *Cratoxylum*. Ramin, however, does not regenerate naturally as well as the other species and seems to be shaded out by fast-growing pioneer species (Lee 1979).

Silviculture is more difficult on inner swamp forest zones. Mixed peat swamp forests regenerate better than pure stands of *Shorea albida*. Many *S. albida* seedlings survive logging but are soon killed by competition. In a typical area, 17 years after logging *S. albida* had declined from 28% to 2% of potential crop trees. After logging natural regeneration occurs, but the forest is dominated by fast-growing species such as *Cratoxylum, Litsea* and *Xylopia coriifolia* so that the new timber crop has a different species composition (Whitmore 1984a).

Extensive areas of swamp forest in Kalimantan have already been cleared for logging, agriculture and transmigration. Although only 1% of Kalimantan's land area has been allocated for transmigration, for peat swamps the figure is much higher, about 10% (Silvius et al. 1987). Any plans for forest conversion and peat drainage should take into account the high water-holding capacity of peat soil and its important role in hydrological regimes. This function may be protected best by retaining forest cover and following sound silvicultural practices (Andriesse 1974; Klepper et al. 1990).

Logging and burning to clear agricultural fields have already taken their toll of Kalimantan swamp forests. Peat swamps are very vulnerable to drought. As water levels drop, the peat layer dries out and burns readily if fires spread from agricultural fields. Large expanses of peat swamp forests were destroyed by burning during the severe drought of 1914 (van der Laan 1925) and again in 1982-83 (Mackie 1984). No areas of undisturbed peat swamp forest remain in South Kalimantan (Giesen 1990). In the Sungai Negara basin uncontrolled logging by local operators has resulted in a devastated forest with low canopy cover (25%-30%) and little chance of regeneration. Uncontrolled burning of agricultural fields has spread into logged peat swamp forests. After burning, these forests have regenerated into less species-rich, stunted secondary forest with inner swamp areas dominated by trees of *Combretocarpus rotundatus* or *Shorea balangeran* (Giesen 1990). Felling and burning has created a mosaic of *Melaleuca* swamp forest and sedge and grass swamp in place of freshwater swamp forest. The *Melaleuca* stands in the Sungai Negara wetlands and near Banjarmasin, South Kalimantan, are repeatedly cut for firewood and charcoal (Whitmore 1984a; AWB/KPSL-UNLAM 1989).

Food Chains and Animal Resources

The dry matter and nutrients in wetland plants enter food chains when plants are grazed by animals or when plant material is broken down by decomposers or eaten by detritivores. Micro-organisms and invertebrates feed on the dead plant material and are themselves eaten by larger animals. Many of the species at the top of the complex food web of the wetland ecosystem are economically important. Wetland food chains support the valuable fisheries of the Mahakam lakes, Kapuas lakes and Sungai Negara basin. Fish eat plants and algae from the lake waters. Seasonal flooding brings organic matter into the lakes from the vegetation on the flooded plain to make these the most productive freshwater fisheries in Indonesia. Floodplains are also important spawning, nursery and feeding grounds for fishes (Zehrfeld et al. 1985; Giesen 1987; Chaeruddin 1990).

Fish is the main source of protein in Indonesia, and the Kalimantan inland fisheries can be regarded as a major national resource. The Middle Mahakam lakes not only provide freshwater fish for local communities and other settlements in East Kalimantan but also produce more than a third of all dried freshwater fish imported into Java (Zehrfeld et al. 1985). Two-thirds of all freshwater fish caught in West Kalimantan come from the Kapuas lakes (Giesen 1987). The Sungai Negara lakes provide most of the 35,000-45,000 tonnes of freshwater fish caught annually in South Kalimantan, a yield of more than 3 tonne/ha of lake or 100 kg/ha if the whole swamp area is included in the calculation (AWB/KPSL-UNLAM 1989).

In the Sungai Negara wetlands of South Kalimantan fishermen burn the mats of sedge and water hyacinth to clear a passage for boats. Nutrients from the burned vegetation return to the lake ecosystem. With the input of new nutrients from the ash, there is a burst of algal growth on which many swamp fishes feed, and this maintains high levels of fish populations.

Detritus-feeding insects also increase in numbers with the flush of algal growth and provide a food resource for carnivorous fish and other secondary feeders. Among these animals at the head of the aquatic food chain are river turtles and water birds, which are caught and eaten by lakeside communities (AWB/KPSL-UNLAM 1989).

Other wetland animals of economic value include arowana fish *Scleropages formosus*, valued as aquarium fish (box 10.4). Crocodiles, monitor lizards *Varanus* and snakes are all collected for their valuable skins, much in demand for making shoes, handbags and other leather items. Crocodile farms have been opened in Banjarmasin and Pontianak, and attempts are being made to

---

Box 10.4. *Ikan siluk*, the dragonfish of West Kalimantan.

The Asian arowana or dragonfish belongs to the primitive family of bonytongue fish Osteoglossidae. The bonytongues are large, freshwater fish found throughout the tropics. Only one species, *Scleropages formosus*, occurs in Kalimantan. Because of their primitive appearance, bonytongues have become popular as ornamental aquarium fish. According to Chinese tradition, the arowana brings luck to its owners.

There are three colour varieties of the arowana in Kalimantan: green, golden yellow, and red. The red form is the rarest and most valuable, fetching prices ten times greater than those given for green dragonfish. A fully grown red arowana, 60 cm long and weighing 8 kg, is worth $2,000 or more in Jakarta, and even young, thumb-sized specimens sell for $1,000. In Indonesia the red arowana is found only in West Kalimantan, in and around the Kapuas Lakes (now part of the Danau Sentarum reserve).

Arowana were once caught and eaten by local fishermen, but over the last 10 years they have become far more valuable as aquarium fish. There is a substantial trade in arowana out of West Kalimantan, with 30,000 fish being exported annually in the early 1980s, although the quota set by the Conservation Department (PHPA) was only 20,000 fish. The extensive trade has led to a decline in fish stocks, and arowana are now caught infrequently by local fishermen. Nevertheless, the high price encourages continued harvesting.

The arowana eat frogs, small fish and insects, coming to the surface at night to feed but spending daylight hours on the river bottom. During the drier months arowana inhabit small streams that spill into the Kapuas lakes. As lake waters rise the arowana move into the larger lakes to breed, and return again to the smaller streams when water levels fall. These migrations make them vulnerable to fishermen using long nets stretched across the streams.

The Asian arowana is listed on Appendix 1 of CITES and is a protected species in Indonesia, but trade still continues, with most fish derived from wild stocks. Attempts are being made to rear arowana in captivity, and one firm in Pontianak is having some success.* Breeding arowanas is complicated since the fish is a mouth brooder. It is still not known which sex cares for the eggs. Young arowana may also take refuge in an adult's mouth when danger threatens. Adults have been caught with 30-40 young in the mouth. The young grow rapidly and are already 45 cm after two years.

At present all breeding stock are adults captured from the wild. Most of the fish traded also come from the wild. Wild-caught young are an important source of income to fishermen in the Kapuas lakes. So valuable are the arowana, that poaching has become a serious problem in Tanjung Puting National Park. The trade in arowana fish is estimated to be worth at least $2 million annually. As natural stocks become exhausted due to overharvesting, supplies will only be maintained if the species can be bred commercially.

* By late 1995 all red arowana for export were being captive bred.

Source: Giesen 1986

breed these reptiles in captivity. Crocodile skins may fetch $50 per metre in overseas markets, and the meat is also sold to local restaurants. Previously, clutches of eggs and young animals were collected from the wild, and this harvest and hunting has taken a serious toll of the wild crocodile population. It is now extremely rare to see crocodiles in Kalimantan rivers, although populations still occur in the Tanjung Puting and Kutai reserves and in the more remote reaches of some rivers. Pythons *Python reticulatus* are never found very far from water, and these and other snakes are hunted ruthlessly for their skins (Chaeruddin et al. 1990). In fact, snakes play a valuable role in preying on rats and other rodents which damage the rice harvests, but they are rarely credited with being beneficial to people.

Three species of monitor lizards occur in Kalimantan, but only the water monitor *Varanus salvator* is trapped for its skin, used for leatherwork and fancy goods. Surveys conducted by KPSL-UNLAM revealed that more than 100,000 skins were exported annually through South Kalimantan, with about 65,000 of these trapped within the province (Djasmani and Rifani 1988). The total quota for monitor exports is 114,000 monitors, but this is often exceeded. Monitor lizards are common throughout Kalimantan and may thrive near human habitations, where they scavenge on household waste. Field surveys conducted by teams from KPSL-UNLAM and PSL-UNTAN suggest that many wild populations are declining due to overharvesting, and trappers must now travel further afield to find their catch (Djasmani and Rifani 1988).

In Kalimantan Dayaks eat monitor meat and eggs, but there is little commercial trade in these products. Monitors are regularly offered for sale at the food markets in Kuching, Sarawak. There is some trade in monitor gall bladders, which are used in medicinal preparations; however, snake gall bladders are preferred and command higher prices (Luxmoore and Groombridge 1989). A few lizards are stuffed for the curio trade, and hands, feet and heads are used for making fancy goods such as key rings, for decorating belts and for other handicrafts. Large toads are also collected for the fancy goods trade. More than 56,000 toads were exported from South Kalimantan to Java for processing in 1986 (Djasmani and Rifani 1988).

Domestic livestock and wild deer and pigs graze in wetland areas in the shallow swamps, and on new grazing on dried-out lake beds. Water buffalo are kept by villagers in the lakeland communities. Other domestic animals, such as ducks, feed in the lakeland areas on water weeds, snails and frogs. Piles of discarded shells indicate that large quantities of *Ampullaria* snails are collected as duck food. The Alabio ducks bred in the Amuntai region of South Kalimantan are remarkable for their egg-laying abilities (box 10.6).

Box 10.5. Wetland resources of economic value.
- The Mahakam Lakes and other freshwater ecosystems in Kalimantan are an important source of fish for local communities and the main suppliers of dried freshwater fish exports to Java. In 1985, of 19,000 tons of dried freshwater fish imported to Java, 7,000 tons (37%) came from the Middle Mahakam Lakes of East Kalimantan, with a further 2,000 tons from South Kalimantan and 3,000 tons from Central Kalimantan, the two other main exporters. Overfishing may threaten these fisheries (Zehrfeld et al. 1985).
- The worth of products from a single female estuarine crocodile *Crocodylus porosus* and her lifetime's offspring has been estimated at $280,000 (Whitaker 1984).
- 75% of the resident population in the Upper Kapuas basin, West Kalimantan, earn their livelihood primarily from fisheries. In 1985 these fisheries were valued at $4.3 million, excluding sales of ornamental arowana fish. This catch is two-thirds of the total catch for the whole province (Giesen 1987).
- Among Kejaman households in Sarawak the most important animal food is fish, providing more than 40% of the side dishes eaten with rice. An estimated annual fish catch of 54-76 tons per year is taken from the lower Baram (Watson 1982), equivalent to 14.2-19.6 kg/ha/yr. The market value of this harvest is conservatively estimated at $150,000-$190,000.
- The value of exports of reptile skins (snakes, monitor lizards and crocodiles) from Kalimantan is considerable. In 1988 one Banjarmasin trader alone exported at least 54,000 skins of monitor lizards *Varanus salvator*, collected in South and Central Kalimantan. Purchased at a price of $5 per skin from local trappers, this harvest was worth at least $270,000 to the local economy (Djasmani and Rifani 1988).
- The river terrapin *Callagur borneoensis* and soft-shelled river turtles Trionychidae are collected and sold for food throughout Borneo. The eggs of river turtles are also collected and eaten.
- The peat swamp forest tree ramin *Gonystylus bancanus* is the most important source of commercial timber in Indonesia after the dipterocarps meranti and keruing.
- Indonesian exports of jelutong, a latex derived from the swamp forest tree *Dyera costulata*, were valued at $2 million in 1987 (de Beer and McDermott 1989).

Box 10.6. Alabio ducks.
In the Alabio swamps, north of Banjarmasin, Banjarese villagers have been breeding a local breed of duck *Anas platyrhynchos borneo* for egg production. The swamps have little free land for rice production, and commercial production of duck eggs has supplemented rural incomes since at least the beginning of this century. Egg producers can earn as much as 20 times the $200 per capita income of farmers with similar amounts of land in Java and Bali.

In the past the flocks of ducks were allowed to scavenge by day in the swamps but were caged at night to lay their eggs. In the dry season the ducks were kept caged to prevent them from damaging local rice fields. When human population densities increased and rice cultivation expanded, free-ranging ducks became a nuisance. The herders responded by caging their ducks throughout the year and providing them with sago and fresh fish. Dried fish, snails, rice bran and swamp vegetation are all used as duck feed. By using locally available foodstuffs rather than imported feed, and adapting their farming system, producers have been able to increase their flocks, increase egg production and maintain egg production throughout the year.

The Alabio ducks are prolific egg-layers. They have been so highly bred for laying that each female now produces at least 250 eggs a year, but the ducks are no longer good brooders. Instead, the people have evolved their own ingenious hatchery techniques. Eggs are spread out on mats to warm in the sun, then buried in sawdust and packed in large containers. Eggs are turned at regular intervals and kept warm until the ducklings hatch.

Source: Vondal 1987

Fisheries in the Middle Mahakam Lakes

The Kalimantan lake systems are some of the most productive freshwater fisheries in Southeast Asia (table 10.2). The Middle Mahakam Lakes area in East Kalimantan extends over 1.8 million hectares, including 500,000 ha of seasonally inundated swamp forest underlain by thick layers of peat, 50,000 ha permanent swamplands, and scattered lakes, many of them oxbow lakes (old bends of the river which have been cut off from the main river). The three largest lakes are the mixed-water lakes of Lake Jempang (14,600 ha), Lake Semayang (10,300 ha) and the blackwater Lake Melintang (8,900 ha). About one-third of the total area is permanently or seasonally flooded; this floodplain is very important for fisheries. Many of the lakes are seasonal, with a maximum water depth of 7-8 m at full flood, and they become increasingly of a blackwater type as the flood season progresses. Water levels typically fluctuate from 4 m to 6 m, with a double high water period occurring between October and May, after which water recedes in the dry season (Zehrfeld et al. 1985). Such seasonal inundation is typical of tropical floodplains. The fisheries ecology of such systems has been described well by Welcomme (1979).

Table 10.2. Productivity of lakes and rivers in Borneo, compared with other freshwater systems in Southeast Asia.

Area	Productivity kg wet weight/ha	Source
Middle Mahakam, East Kalimantan:		
open waters only	139.0	Dunn and Otte
total floodplain	20.0	1983
Upper Kapuas, West Kalimantan:		
floodplain	20.0-60	Giesen 1987
Danau Sentarum reserve	37.5	
Sungai Negara wetlands, South Kalimantan:		
lakes and swamps	100.0	AWB/KPSL-
open waters	3,000.0	UNLAM 1989
Baram River, Sarawak	14.2-19.6	Watson 1982
Lake Ciburug, Java	500.0-600	Hickling 1961
Lake Tempe, Sulawesi	40.0-1,000	Hickling 1961
Average inland waters, Indonesia	20.4	D-G Fisheries 1981
Grand Lac (Tonlé Sap) Cambodia	40.7	Welcomme 1979
Lower Mekong, Cambodia	15.5-27.8	Welcomme 1979

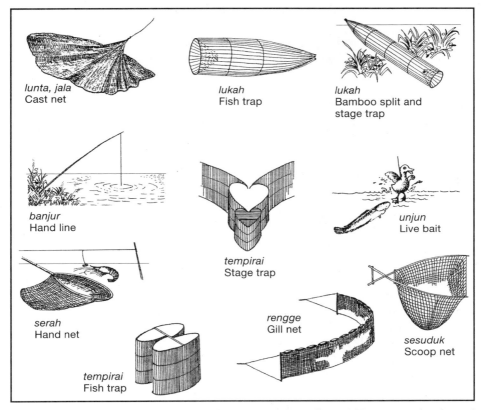

Figure 10.2. Fishing methods in the Mahakam lakes, Sungai Negara wetlands and other Kalimantan wetlands.

The Mahakam lakes and rivers are important fisheries, with between 20,000 and 35,000 tonnes of fish being landed yearly since the early 1970s (Zehrfeld et al. 1985). The oxbow lakes are generally the most productive wetlands, with many juvenile and larval fish. The slow-flowing large rivers, such as the Mahakam and the lower reaches of the Belayan, are also fished, but harvests are greater in the blackwater rivers that drain the swamps. These rivers are fairly productive, yielding especially whitefish (cyprinids) such as *belida* (*Notopterus*), *jelawat* carp (*Leptobarbus* sp.), *lempam* carp (*Puntius schwanefeldi*), *kendia* (*Thynnichthys*) and *patin* (*Pangasius*). In the swamps juvenile fish shelter beneath floating mats of vegetation; juvenile fish and small fish species are more abundant in this habitat (Zehrfeld et al. 1985).

East Kalimantan is presently the largest single supplier of dried freshwater fish for Java, with 6,000-9,000 tonnes being exported annually. Most of this harvest comes from the Mahakam lakes, caught using a variety of fishing methods (fig. 10.2). However, there is good evidence that the high

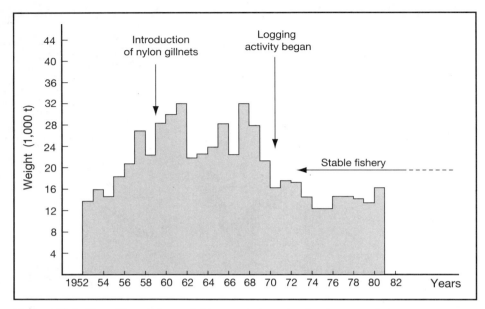

Figure 10.3. The effects of overfishing and changing land use on fish harvests in the Mahakam lakes area. Changing land use, through conversion of forests for logging or agriculture, impacts on fish catches, as does overfishing through improved technology.
Source: Dunn and Otte 1983

level of fishing (which has increased in intensity due to the introduction of commercial nylon gillnets) is affecting the numbers and species composition of the catch, and that many species are being overharvested (fig. 10.3; Zehrfeld et al. 1985). Constant harvesting, especially at high levels, impacts on recruitment into the fish population (fig. 10.4).

A marketing survey in 1985 revealed that the market share of nine of the 15 important fish species had declined dramatically. Species such as the belida and featherback *Notopterus* spp. and patin *Pangasius* spp. had almost disappeared from the market (Zehrfeld et al. 1985). Featherback females lay their eggs on submerged timber in slow-moving water, and they are guarded by the male. During this period he is very easy to catch, but without his protective presence the eggs would be eaten by smaller fish and other animals.

Heavy fishing in the Mahakam lakes led to declining harvests of whitefish (cyprinids), whereas the catch of blackfish (catfish, channids, anabantids) remained steady, or even rose. The differences in response to heavy levels of fishing arise because of the different behaviour and feeding patterns of the two fish populations. Residual populations of blackfish remain in isolated, inaccessible parts of the swamp forest that do not dry

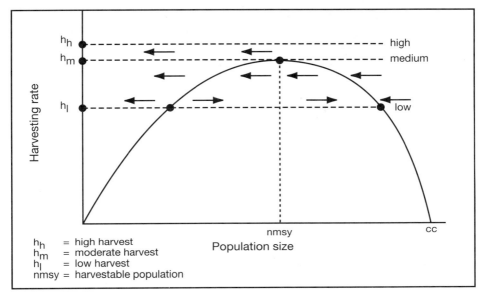

Figure 10.4. The effects of constant harvesting rates on the rate of recruitment to a population. Arrows indicate the expected changes in population under the different harvesting regimes. "cc" indicates the maximum population within the carrying capacity of the habitat. "nmsy" indicates the harvestable population for a maximum yield on a sustainable basis.

Source: Whitten et al. 1987a

out during the dry season. These populations cannot be fished out totally and provide the spawners for the next flood period. Whitefish species, however, return to the rivers during the dry season. Here they are very vulnerable, and populations are severely reduced when fishing pressure is high. Even the harvest of most blackfish species fell in 1983, when the long drought and subsequent fires affected even the most isolated swamp forest refuges (Zehrfeld et al. 1985). Differences in price for different species may also lead to changes in fishing patterns and to increased harvesting levels of the more valuable whitefish species.

Indications of high fishing levels and probable overharvesting are: the disappearance of certain species, such as belida and patin, from the market; several fish species breeding at smaller size; and decreased catches for unit effort, with catches in some areas reduced by as much as 80% (Zehrfeld et al. 1985). If the important Mahakam fisheries are to be maintained to yield high and sustainable harvests, some effective alternatives need to be introduced to reduce pressure on fish stocks. Measures which could be taken to conserve fish stocks include: limiting the use of small-meshed nets (less than 2 cm) to avoid catching small fry and thereby enhance recruitment; restricting total numbers of fishermen and amount of fishing gear per family; and establishing catch quota for certain fish species. Aquaculture in floating cages and ponds, better fish processing, and support for private fish hatcheries would also help to reduce pressure on the populations of whitefish, which are particularly threatened by overfishing. In the Sungai Negara wetlands *ikan gabus* (*Channa striata*) populations are showing similar signs of overfishing (AWB/KPSL-UNLAM 1989).

Aquaculture

Fisheries are some of Indonesia's most valuable natural resources, providing cheap and available protein for much of the country's population. As open-water catches have begun to decline and demand for fish has escalated, aquaculture is playing an increasingly important role in fish production. There are two types of aquaculture: brackish water culture (*tambak*), found in mangrove areas (chapter 11); and freshwater aquaculture including fish ponds, cage culture and mixed rice-fish culture. Although tambak covers much greater areas, both in Kalimantan and elsewhere in Indonesia, many more households are involved in some form of freshwater aquaculture to provide fish for domestic consumption.

In Kalimantan commercial freshwater fish ponds cover only a very small area, 2309 ha in 1989 (Buku Statistik Indonesia 1991). The area of ponds is being extended, even though most local fish consumption is still met from wild-caught stocks. The productivity of fish ponds depends on input of fresh water and nutrients. Water needs to be changed regularly to wash out wastes and sweep in new sources of food. The most important species reared in home fish ponds are common carp and gourami. Fry and broodstock are imported from Java. *Tilapia mossambica* is also reared in household ponds.

Elsewhere in Indonesia mixed rice-fish culture is very important, overall contributing 28% of the area of freshwater culture, mainly on Java (Silvius et al. 1987). Such multiple-cropping systems have been established in other parts of the world; for instance, in Louisiana, U.S.A., crayfish farming and rice cultivation are combined (Maltby 1986). After the rice is harvested the pond is flooded, and recently hatched crayfish feed on the rice stubble. After six months the crayfish reach marketable size and can be harvested; the new rice crop can then be sown. As the pond slowly dries out, the remaining adult crayfish, supplemented with additional stock for breeding, burrow deep into the soil until the pond is flooded again. Thus the rice crop can be harvested without damaging the breeding population (Maltby 1986). Multiple-cropping systems such as this have considerable potential for Indonesia. Milkfish are already reared in rice fields in Java. Combining fish farming and timber production in natural wetlands would preserve the basic wetland ecological conditions and landscape while providing a food crop and source of income.

Reclaiming Tidal Swamplands for Agriculture

Tidal influence may extend as far as 100 km upriver in the coastal swamplands of Kalimantan. These tidal swamplands include freshwater swamp, peat swamps and mangroves, and occupy the broad, flat and poorly drained zones between the mountains and the sea. Many of these lands already have been cleared of forest and converted to agriculture.

Reclamation, settlement, and cultivation of swamp areas requires a good understanding of the swamp ecosystems and soil properties. Schemes will only be successful with careful site selection, good management of water regimes, appropriate crops and adequate funding. Tidal swamps are fragile habitats requiring careful management. Freshwater swamps are nutrient-deficient and frequently anoxic. Seawater is toxic to most plants and can destroy soil structure, therefore it cannot be used for irrigation. However, at low concentrations brackish water may be beneficial in restoring soil nutrients and flushing acid soils. Inappropriate developments can lead to soil toxification and declining fertility, low water quality, salt intrusion and environmental degradation (Knox and Miyabara 1984). Moreover, developments in swamp areas may affect critically the ecological conditions in land outside the development areas.

The tidal swamplands of Kalimantan and other Outer Islands have long attracted developers because of their potential for the cultivation of wetland rice. The 30 km canal built by the Dutch between 1880 and 1890 to connect the Kapuas and Barito rivers, for instance, opened up the area for agricultural development, and local Banjarese made drainage canals from the swamp areas to the canals. Since then extensive areas of tidal swamplands in South, Central and West Kalimantan (Vayda 1980) have been converted to agriculture, both by spontaneous migrants, especially the Banjarese, and as part of government-sponsored transmigration schemes.

Reclamation of tidal swamplands usually starts with the digging of a main drainage canal, linked to secondary and tertiary canals, about one metre deep, with the bottom of the canals slightly above the low tide level. Tidal action forces water into the canals at high tide, while at low tide water drains out of the land into the canals. This tidal system, as used in South Kalimantan, provides water and nutrients for irrigation and flushes salts and acids out of the soils (Burbridge and Maragos 1985). Ponds at the ends of the secondary canals store leached products and maintain canal flow during low tide, thus accelerating the removal of leached salts.

The development of drainage systems, irrigation canals, rice-field plots and human settlements all have an impact on the environment. Local tide variations interact with river flow to zone the land and water into a series of distinctive environments. Gradients in salinity patterns impose a strong

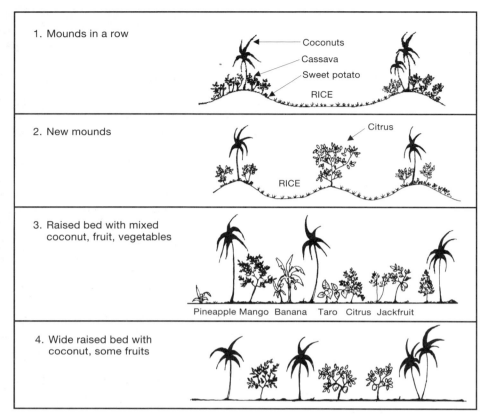

Figure 10.5. The Banjarese coconut/rice agricultural system in South Kalimantan tidal swamps, showing crop succession.
Source: KEPAS 1985

order on natural communities of plants and animals. Many different community types are found in a small area, and they can be stressed or destroyed easily if existing gradients are altered (Knox and Miyabara 1984).

Some species of aquatic macrophytes can serve as indicators of water quality in tidal swamps (Sastrosoedardjo et al. 1986). The aquatic weed *Eleocharis dulcis* is an indicator of permanent stagnant water with a low pH; *E. retroplexa* also indicates acid water. The attractive, pink-flowered *Melastoma malabatharicum,* relative of the "Singapore rhododendron", thrives on land of the poorest quality. Other plants such as *Commelina* and *Emilia* indicate low pH. Fallow land returning to secondary forest is dominated by the paperbark tree *Melaleuca cajuputi,* which is very tolerant of acid conditions and occasional fires.

Tidal swamplands with a properly managed water regime can be converted into productive agricultural lands. The Banjarese have been particularly successful with their system of combining crops of rice and coconuts, with coconuts becoming the main crop as rice yields decline (fig. 10.5). Other crops which grow well on raised land in tidal areas include kapok *Ceiba petandra*, kecapi (*Sandoricum koetjape*), and fruit trees such as rambutan *Nephelium lappaceum*, jackfruit *Artocarpus heterophylla* and mango *Mangifera indica*. Successful agriculture, however, depends on local soil conditions; soils vary in quality throughout the swamplands.

Acid Sulphate Soils

Many of the tidal swamplands pose problems for agriculture because of their potentially acid sulphate soils beneath the surface peat or alluvium layer. Many of these areas are freshwater or mangrove swamps where pyrite (FeS_2) is formed during soil formation. Pyrite accumulates in waterlogged soils that are both rich in organic matter and flushed by dissolved sulphate, usually from sea water (Dent 1986). Under naturally flooded conditions, oxidation cannot occur in these anaerobic soils. When the soils are drained, they are exposed to air and the pyrite oxidises to produce ferric ions and sulphuric acid in drainage water. Acid sulphate soils develop where the production of acid exceeds the neutralising capacity of the parent material and the pH value falls to less than 4, rendering the soils unfit for most crops.

Under these acid conditions, ferric and aluminium ions are strongly bound to organic matter and displace other cations, such as potassium, magnesium and calcium, which are important plant nutrients; these go into solution and are leached out of the soils, so are lost to crops. Phosphates, added in fertilisers, are also strongly bound to the iron and aluminium, so become unavailable to plants.

Acidity can be corrected by liming, but acid soils may require large quantities of lime and continued application, which is expensive. Lime neutralises the acid but frees aluminium and iron, which wash into surface water and may increase its toxicity. Because it neutralises soil acidity, the application of lime increases productivity of crops more than the application of fertilisers (Konsten pers. com.).

Acidity is also alleviated by flooding, which reintroduces anaerobic conditions. In the presence of easily decomposed organic matter pH rises, ferric oxides are reduced to ferrous salts and sulphates are reduced to hydrogen sulphide. Reduction in acidity following flooding alleviates aluminium toxicity, but the crop may still suffer from the presence of soluble iron and hydrogen sulphide. Floodwater standing on acid sulphate soils may become severely acid and can affect crops growing on neighbouring nonacid soils (Dent 1986).

Potential acid sulphate soils should be kept wet and not allowed to dry out. Reclamation of tidal swamplands therefore requires great care, to

> **Box 10.7. Acid sulphate soils.**
>
> Acid sulphate soils pose chemical, biologica, and physical problems for crops. Chemical problems for dryland crops include:
> - direct effects of severe acidity, primarily the increased solubility and toxicity of aluminium and possibly iron (III), manganese and hydrogen ions;
> - decreased availability of phosphate, caused by iron and aluminium-phosphate interactions;
> - nutrient deficiencies;
> - salinity.
>
> Under flooded conditions, for example under rice cultivation or fish ponds, acidity is reduced, but new problems include:
> - iron (II) toxicity;
> - hydrogen sulphide toxicity;
> - CO_2 and organic acid toxicity.
>
> Physical problems arise mainly through the inhibition of root development in acid sulphate horizons:
> - crops suffer water stress;
> - soil ripening is arrested. Clay and organic soils remain soft, unable to bear heavy loads, are poorly structured and therefore poorly drained;
> - field drains may be blocked by iron oxide (ochre) deposits.
>
> Unsuitable conditions for most micro-organisms impede the release of nutrients from soil organic matter. Crops under stress are especially susceptible to disease.
>
> *Source: Dent 1986*

provide only shallow drainage, and irrigation, to keep soils wet and to flush out the acid salts. Wetland rice is an appropriate crop for such areas since soils are kept water-logged, and acidification cannot occur.

A major constraint to development of swamps for agriculture is the lack of fresh water during the dry season; groundwater levels drop, and oxidation occurs. Furthermore, large quantities of water are required to flush the soils. In the early stages of reclamation conditions may be toxic from semi-decomposed materials and swamp gases. Later aluminium and iron toxicity factors increase. A delicate balance exists between the various destructive and regenerative processes at work in swampland soils; if this is not maintained, it is almost inevitable that land will have to be abandoned (Knox and Miyabara 1984).

Development of acid sulphate soils leads to poor crops, and the leaching of acid and other toxic elements into drainage waters, affecting vegetation, fish and other aquatic wildlife. The high acidity also affects the health and welfare of local communities; many people in acid areas have lost their teeth by the age of thirty, and suffer from hair loss (Konsten pers. com.).

Success and Failure in the Tidal Swamps

By the end of 1984 213,200 ha of tidal swamplands had been reclaimed under the transmigration scheme in Kalimantan (Euroconsult 1986). The Repelita IV target figures for Kalimantan are 150,000 ha. Another 150,000 ha proposed for transmigration settlements under Repelita V is now likely to be delayed while attention is paid to improving conditions at existing sites.

The first government-sponsored colonisation project for cultivation of rice in a swampy area in Kalimantan was Purwosari near Banjarmasin in 1939 (Hardjono 1977). The Purwosari settlement succeeded because the colonists used the traditional Banjarese system of reclaiming tidal swamplands, a system that had evolved in response to local conditions (Collier 1979). Instead of burning agricultural waste and spading soil deeply, farmers using this system cut up straw and weeds with hoes and leave them to rot by submergence. The field is then ready for planting. This maintains peat thickness and protects rice plants against acidity and oxidation of pyrite (box 10.8).

Other areas of land drained for transmigration settlements in South Kalimantan have not always been as successful as anticipated, and rice yields frequently fall after two or three years. In the transmigration settlement of Tamban Luar in South Kalimantan, yields of unhusked rice declined between 1970 and 1978 from 3 tons/ha to 1.9 tons/ha (Knox and Miyabara 1984), one-fifth of the yield obtained from non-swampland irrigated rice fields in Java (Koesoebiono 1983). Over the same period the peat layer had shrunk from 1 m to only 0.1 m depth. Decline in rice production was due to nutrient deficiency and competition from weeds. A poor watershed, long fallow period and declining soil fertility which favoured invading weeds combined to produce an environment where farmers could not produce a rice crop profitably. The same holds true for burned-over areas which become acidic. In some tidal swamp transmigration schemes rice yields have been so low that land has been abandoned (KEPAS 1985).

Apart from the marginal suitability of tidal swamplands for agriculture, development planners have consistently failed to evaluate the impact of the conversion of swamplands to agriculture upon extremely valuable freshwater fisheries and estuarine and coastal fish stocks. It can be expected that changes in land use will alter nutrient cycling, increase acid drainage, and reduce spawning and nursery areas. Moreover, the commercial value of the swamp forest is substantial and equates with that of many upland forest areas (Burbridge et al. 1981). In the long term the wisest and most profitable form of land use for tidal swamps may be to leave them under forest, thereby supporting swamp and coastal fisheries and ensuring a sustained supply of commercial timber and other forest products.

Box 10.8. Tidal swamplands and the Banjarese system of agriculture.

The Banjarese have been reclaiming and cultivating the tidal swamplands of southern Kalimantan for generations. They have evolved a successful multicropping system of sustainable agriculture in these marginal lands.

Samuda Kecil village, Central Kalimantan, is a small village, where local Banjarese and migrant Javanese have successfully created good agricultural land from tidally influenced swamp with a peat layer about 50 cm deep. Main agricultural activities are smallholder coconut plantations and rice fields. Intercropped with the coconuts are coffee, citrus fruits and other tree crops (fig. 10.5).

The system of land clearance and agriculture is well adapted to the local ecology and is productive and self-sustaining over a long period of time. Farmers select a small stream flowing into the river and widen and deepen it; they extend the stream with a canal following the natural drainage system. In 50 years this canal has been extended 10 km from the river. About one hectare of forest is opened for rice growing at a time. The drainage canal is extended into the area, and trees are felled and burned in the dry season; adjoining fields will be burned at the same time. After this rice seedlings are planted.

Only one crop of rice is grown per year since the canals are not adequate to use tidal influence in the dry season. Local varieties of rice are grown since the people feel that high yielding varieties are less suitable for these swampy, poorly drained fields. For a one-hectare field a seedbed 85 m x 85 m is planted, and after one month the seedlings are transplanted to a second seedbed. Planting is labour-intensive, taking two women seven hours a day for seven days. After another month the rice seedlings are transplanted, in December or January. It takes two women one month to transplant a one-hectare field, weeding as they go. Each year the weeds are more severe. Fields are not ploughed. Weeds are cut, sometimes under water, and the seedlings planted in a hole poked in the soil. After the first and second year the fields must be weeded first. By the fifth or sixth years it takes two people 15 days to weed a one-hectare field.

In May the harvest begins and lasts one month. Harvesting is labour-intensive since the rice matures unevenly and must be cut with a hand-held rice knife. Families with only one hectare harvest the rice themselves, families with more use hired labourers, who receive one third of the crop.

After two years and two rice crops a ditch or small canal is dug in the centre of the field, which until then is covered with water. In the third or fourth year farmers plant coconuts in the rice field on small mounds of soil, with shallow canals between each row of coconuts. The mounds are built up and canals deepened to drain the fields, protecting coconut roots from inundation. After three to five years the mounds connect and form a continuous ridge for the trees. Other crops are set between the coconuts: coffee, bananas and vegetables. Rice is planted in the shallow canals for three more seasons, then the coconut trees are too tall and shade out the rice.

At this time the farmers open up more forest, either next to their first field or at a new location. The optimum size of farm per family is 4 ha of coconuts and 2 ha of rice. Clearing and cultivating this land takes 20 years or more, probably the productive life of a farmer and his wife. Coconuts begin fruiting in the eighth year and are harvested five times a year, each tree averaging 20 coconuts per harvest with 135 trees on one hectare. Coconuts are a good cash crop, yielding copra and fresh nuts for sale. Farmers switch from rice to coconuts because of increasing costs (weeding) and declining yields; returns are higher for coconuts than for rice.

Although forest lies close to the village, the villagers do not fell timber nor harvest other forest products for sale. They do not have time for this, and already have an adequate income from agriculture. When a new family forms, they open a new area of forest and

> **Box 10.8.** Tidal swamplands and the Banjarese system of agriculture *(continued)*.
>
> begin the cropping sequence. Compared with government-sponsored transmigration schemes, the Banjarese rice/coconut system provides higher returns than the rice-only system of the transmigrants. The Banjarese system has fewer pest and disease problems since new ricefields are continually being opened. Moreover, apart from opening forest for fields there is little other exploitation or damage of surrounding forests, and this protects the hydrology and nutrient flow in the tidal swamplands. The government-sponsored Purwosari transmigration project is a success because the transmigrants have planted coconuts and rice using the Banjarese system.
>
> Source: Collier 1979

Peat Swamps for Agriculture

There are particular problems associated with drainage of peat swamps for agriculture. Deep peat is expensive and difficult to convert to agriculture and rapidly degrades (Anon. 1976). However, where the peat is shallow, as in coastal areas, swamp forest may be converted to sawah, coconut plantations and sago plantations. Water management is the key to success in these agricultural developments.

Apart from the planned but very limited implementation of a gigantic Kalimantan polder system between 1948 and 1953 in the Alabio swamps, empoldering has played a minor role in swamp reclamation. To reclaim peat areas, shallow drainage ditches are dug, and initially the vegetation is left to increase the rate of water extraction (Silvius et al. 1987). Valuable timber is extracted, and then the vegetation is felled, allowed to dry, and burned. Drainage canals are then dug.

As an area is drained its hydrology changes, affecting not just the reclaimed site but adjacent lands. Peat soils may shrink by as much as 25% to 30% in the drained zone (Driessen 1978), thereby reducing the water-holding capacity of the peat and increasing the risk of floods downstream. During the reclamation process, soil temperature and salinity increase, and organic compounds released into streams lead to reductions in levels of soluble oxygen and increased acidity (Haeruman 1986). As drainage waters become more acid, they can support less animal life, and the fish populations show a greater incidence of skin diseases (Haeruman 1986). Plankton samples taken in irrigation canals show different levels of species diversity, with highest levels at sample sites close to the river and fewer species found in the secondary canals where water conditions are more acid (University of Indonesia 1986).

Cutting and burning the vegetation interrupts the cycling of nutrients, destroys the peat layer and leads to reduced nutrients in the soil (Driessen

1978). Peat soils often require fertilising with phosphorus, potassium, magnesium and nitrogen, and farmers may employ controlled burning of the peat to release nutrients. Eventually peat soils may become so unproductive that they are abandoned. Regeneration of natural vegetation is generally poor in such areas and in drier peat areas may consist of almost pure stands of *Macaranga* and *Melaleuca cajuputi*.

It has been estimated that more than 50% of all freshwater swamp forests and shallow peat areas in Kalimantan eventually will be converted to agriculture (Silvius et al. 1987), leading to loss of habitat, fragmentation of habitat and increased risk of fire. When the natural forests are cleared and replaced by agriculture, the ecosystem becomes simpler, less species-rich and more susceptible to pests and diseases. Species that benefit and show a population increase are usually those regarded as pest and weed species, especially fast-breeding rats and other rodents, rice-eating birds, and weeds such as grasses and sedges. This development of weeds and animal pests is one of the main problems in newly reclaimed peat areas, requiring active use of chemical herbicides and pesticides (Haeruman 1986). Opening up wetland sites may also lead to a local increase in the incidence of malaria and skin diseases among settlers. Forest clearance brings the settlers into contact with animal-borne diseases, and opened wetland habitats provide increased breeding sites for vectors such as mosquitoes.

Peat for Fuel

Indonesia ranks fourth in the world for peat reserves, but peat is still little utilised as an energy source. To be commercially viable, peat layers must have an organic content of more than 65% and be more than 1 m thick.

Table 10.3. Peat resources surveyed for potential as fuel.

	Area (ha)	Thickness (m)
West Kalimantan		
Rasau Jaya	45,000	1 – 10
Sambas	6,000	1 – 6
Kubu	16,200	2
Jalan Galang	3,700	1 – 10
Terentang	8,600	2
Sintang	150	1 – 2
Central Kalimantan		
Sampit	7,100	1 – 4
Palangkaraya	4,000	1 – 6
Kanamit	7,000	1 – 3
South Kalimantan		
Kecamatan Gambut	400	1 – 3
TOTAL	92,650	

Source: JP Energy Oy 1987b

Peat has potential as a fuel for peat-fired electricity generators for small-scale power stations in Kalimantan (JP Energy Oy 1987a). The possibilities of exploiting peat as fuel for power plants at Pontianak (output 22 MW) and Palangkaraya (4 MW) are being investigated by the Ministry of Mines and Energy in collaboration with FINNIDA (Finnish Aid Agency). Kalimantan has extensive peat areas (table 10.3). The decision on whether or not to mine them will depend on energy demands in the provinces with utilisable peat (West and Central Kalimantan), and cost competitiveness in relation to other forms of fuel. There are, however, hidden ecological and environmental costs of peat mining.

One site at Kalampangan, 17 km southeast of Palangkaraya, has already been opened as a pilot peat-production area. The peat layer here is about 3.5 m deep, and it has been suggested that a 1 m layer of usable peat could produce as much as 6,700 MW/ha from fuel peat with a 40% moisture content. The peat harvested from a 200 ha plot would be sufficient to provide fuel for a small (4 MW) power plant for twenty years (JP Energy Oy 1987b).

Draining the peat prior to mining will lower groundwater levels both in the mined area and adjacent peatlands. It will also lead to increased water runoff and an increase in peak flow, though the areas of peat production will be so small that this will probably have little impact, except close to the production site. The peat layers also act as filters, and concentrate metals and other elements from the groundwater. Drainage and mining will lead to increases in suspended sediments in river waters, more soluble organic compounds and higher acidity; all of these will affect aquatic life. Increased sediment load decreases light penetration and reduces algal production, to the detriment of invertebrates feeding on algae, and the fish which prey on them. Peat particles may clog the filtering devices of filter-feeding invertebrates and the nets of web-spinning invertebrates. On the other hand, increased sedimentation may favour some detritus feeders, and increased water flow in the dry season will benefit the aquatic fauna. Generally, however, decreased species diversity and decreased abundance of benthic invertebrates can be expected downstream from peat exploitation areas (JP Energy Oy 1987b).

Fish populations may be affected by hydrological changes, increases in toxic substances and reduced oxygen concentration. Changes in spawning grounds will adversely affect reproduction, and reduced food availability because of water turbidity will also lead to a decline in fish populations. The quality of fish as a human food may also decrease.

Peat mining will affect water flow, lead to a decline in water quality and decrease fish populations, but will provide an alternative form of fuel and employment opportunities and perhaps will open more land to agriculture. As with all changes in land use, regional planners need to assess the costs and benefits of exploiting peat.

CREATION OF NEW WETLAND HABITATS

While logging, drainage and agricultural developments damage and change many wetland ecosystems and threaten some wetland species, new wetland sites are also being created in the form of reservoirs, lakes in dammed river valleys, sawah rice fields, and irrigation and drainage canals. These new sites can often be of great value to wildlife. Indonesia has some of the largest areas of artificial wetlands in the world, particularly on Java, where irrigated rice fields and fish ponds make valuable additions to wetland habitats. Irrigated rice fields have replaced natural vegetation in large areas of the tidal swamplands of Kalimantan and around inland lake systems in the Mahakam and Barito basins. Some highland Dayaks also cultivate wet rice (Dove 1985; Padoch 1988).

Rice Fields

Sawah rice fields are essentially modified swamps, created and maintained by farmers who manipulate the yearly or twice-yearly cycle of flooding, planting, harvesting (removal of organic material), and enrichment with fertilisers (nutrient input). If rice fields are abandoned, natural regeneration will begin as part of the forest growth cycle (see chapter 4).

Rice fields support a surprisingly diverse group of plants and animals, typical of freshwater swamp ecosystems (fig. 10.6). The ecology of rice fields has been little studied, but these habitats are known to support a rich aquatic fauna (Fernando 1977; Fernando et al. 1980). A study in Thailand found that a single field supported 589 species of plants and animals, 209 of which were considered rare (Heckman 1979). The species list included 38 flowering plants, 173 algae, 120 protozoans, 52 rotifers, 12 molluscs, 33 crustaceans, 21 dragonflies, 39 beetles and 18 fish. (The list would be even longer if it included species preying on the rice as insect pests and seed predators). No true plankton was present, probably because the water was too shallow. Mosquitoes are rare in rice fields with water more than a few centimetres deep because they are eaten by predatory fish (Heckman 1979). The high species richness of the Thailand rice field may be due to the fact that rice had been grown in that area for more than five thousand years in areas where rice has been grown for periods of only about a hundred years, the number of plants and animals is lower (Whitten et al. 1987b). Species numbers in rice fields in the newly opened tidal swamplands of Kalimantan could be expected to be lower because of the newness of the habitat and the often harsh and acid conditions.

Apart from the rice crop, many other species commonly found in rice fields are harvested by man, including water spinach (kangkong) *Ipomoea aquatica*, frogs, prawns, crabs, snails and fish. Leaves of the minute, floating fern *Azolla pinnata* can be used as animal feed. Nitrogen-fixing blue-green

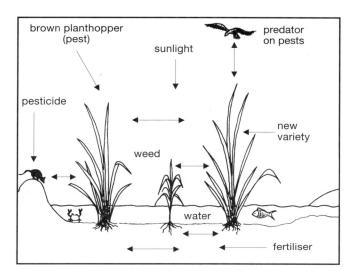

Figure 10.6. Tidal swampland rice field system.

algae in the leaves provide nutrient inputs to the rice-field ecosystem. Some fish are deliberately introduced into the ecosystem, and many have been cultured for food for at least two thousand years. Irrigation canals between the rice fields provide suitable habitat for economically important fish such as the snakehead *Channa striata*, betok *Anabas testudineus* and sepat siam *Trichogaster pectoralis* (introduced from Java). The different ecological requirements of some rice-field fish are listed in table 10.4. Snails,

Table 10.4. Ecological characteristics of ricefield fish.

	Ecological characteristics
Trichogaster pectoralis	Vegetarian, grazes on periphyton (algae and rotifers attached to plants, stones, etc.)
Aplocheilus panchax	Feeds in large schools on microscopic particles
Betta splendens	In small pools or shallow water, sit-and-wait predator on invertebrates
Trichogaster trichopterus	Occurs in deeper water in groups of three or four; searches for invertebrate prey
Monopterus alba	Nocturnal, lives in burrows, eats quite large prey
Clarias batrachus	Large predator on insects and other relatively large animals; also scavenges
Channa striata	Large predator on fishes and frogs
Anabas testudineus	Omnivore, feeds on phytoplankton and zooplankton, preys on other smaller fish and own fry

Source: Whitten et al. 1987b

frogs and other aquatic organisms are a source of food for ducks and wild water birds such as egrets and herons. Heavy applications of fertilisers, herbicides, and insecticides may improve the quality and quantity of the rice

---

Box 10.9. Rice harvesting at Gambut and Kertak Hanyar, South Kalimantan.

The extensive rice-growing areas lying immediately east of Banjarmasin were laid out during the 1920s and 1930s by Banjarese labourers from the Hulu Sungai, contracted to work on the new road from Banjarmasin to Martapura. The rice fields were created by clearing swamp forest and building shallow drainage canals (*handil*) to connect with existing streams. The area is an example of the success of Banjarese collective endeavour in opening up new lands, especially swamp lands. Such techniques were also used by groups of Banjarese migrants in the Mahakam lakes area of East Kalimantan, the Krian district of Perak in Malaysia (settled in the 1890s), and in parts of Riau and Jambi in Sumatra.

Today the rice harvests of Gambut and Kertak Hanyar, whose soils are mainly alluvial and more productive than those of the lower-lying tidal swamps to the west, form an important source of staple food for the city of Banjarmasin. While the permanent population of the district now stands at 50,000, the resident labour supply has always been inadequate to cope with the harvest in August and September. Each year thousands of Banjarese make their way to the district from the Hulu Sungai, 200 km to the north. They head for the homes of family, or the descendants of fellow villagers, to earn much-needed cash and to assist their kin in bringing in the crops. The harvest months at Gambut conveniently fall at different times from those of the main Hulu Sungai rice crops.

The composition of the harvesting group at Gambut may be seen as a barometer of economic conditions in the Hulu Sungai and elsewhere. In the uplands of Hulu Sungai rubber is an important source of cash income, but in 1989 rubber prices were low. Large numbers of labourers from Barabai and Kelua therefore trekked to Gambut. In the Simpur district of South Hulu Sungai rice crops failed in April-May, while at Amuntai, in the north, the rattan carpet industry experienced marketing problems. These setbacks, critical for people whose incomes are marginal, led many to Gambut for the first time. Approximately half of the 470 Banjarese studied had no family connections in the area; those who did stay with family were usually regular annual visitors.

It is possible to envisage a time when such annual migrations will no longer occur, when improved water control in the Hulu Sungai will allow more double-cropping of rice and when off-farm work is more lucrative, especially for the landless, or for those who must share-crop an area too small to feed a family. The group whose share of the resources in the Hulu Sungai is least adequate, including widows and newly established, young families, are likely to continue to look to Gambut as an important source of food and work, at least for one or two months of the year.

As promised irrigation schemes are extended into the Gambut district itself, the current concentration on local rice varieties, which are harvested slowly with the finger knife (the *ani-ani* or Banjarese *ranggaman*), may give way to high-yielding types more amenable to mechanisation. Already the Javanese employ the sickle, and can earn more in less time. At present, however, the people of Banjarmasin prefer to eat the local *siyam unus* and similar rice varieties now grown at Gambut. Harvesting these varieties is labour-intensive and requires an annual influx of extra labour. The annual migration is important socially as it reinforces and strengthens kin ties between the population of Gambut and Kertak Hanyar and their families in the Hulu Sungai.

Source: Potter in prep.

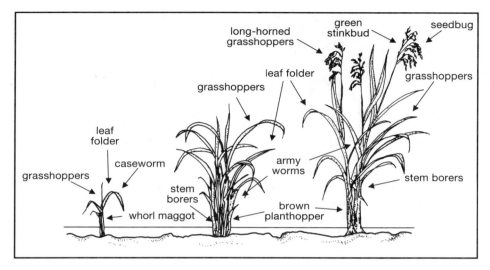

Figure 10.7. Insect pests on different parts of rice plants at different stages of growth. (After van Halteren 1979.)

crop but lead to reduction in other forms of aquatic life, thereby reducing the amounts of other food sources which can be harvested from this agroecosystem (Yunus and Lim 1971; Fernando 1977).

Most of the larger aquatic animals are adapted to living in water with little or no dissolved oxygen. Many rice-field fish, such as the climbing perch *Anabas testudineus*, have accessory breathing organs. Snails have lungs and gills, and many live close to the oxygen-rich surface layer; insects and spiders carry a bubble of air with them. Rice fields are allowed to dry out prior to harvesting. This means that newly flooded fields have to be recolonised with plants and animals or that species must have resistant dormant stages, able to withstand the "drought". Seeds, spores and insect and snail eggs lie dormant in the mud until flooding occurs. Snails, crabs and the swamp eel *Monopterus alba* can bury themselves in the mud and remain inactive during the dry period. Other species may migrate to more permanent water bodies. With reflooding, recolonisation can occur from nearby swamps and lakes.

Rice fields, like all new habitats, attract a host of species, not all of which are desirable additions to the ecosystem. Insect pests attack all parts of the rice plants, except roots under the water, throughout the plant's life cycle (fig. 10.7). Grasshoppers feed on rice as well as many other plants, while certain leafhoppers are known only from rice. Stemborers, the caterpillars of pyralid moths, bore into the rice stems, killing or damaging the plants and thereby reducing yields. The brown planthopper *Nilaparvata lugens* transmits a virus that devastates the rice crop and is considered to be Indonesia's major insect pest. All of these pests can be controlled by applications of insecticides, but insect populations gradually build up resistance to these chemicals. However, there is great potential for using more natural control methods for insect pests by changing environmental conditions or by encouraging natural predators. Such biological control requires a thorough knowledge of the ecology of the target species. Some species thrive on plants given large amounts of nitrogen, others prefer the microhabitat among short varieties of rice, while others are encouraged by a second crop of rice in a year (van Halteren 1979).

Rats cause at least as much damage to rice crops as do insect pests. It has been estimated that as much as 30% of a rice crop may be lost before harvest to pests, including insects, rats, other mammals and birds. Rats eat rice at all stages of its development, including ripening rice grains, growing shoots and even the stored crop after harvesting. Rats are probably responsible for 5%-6% loss of the annual rice yield (Whitten et al. 1987b), but individual fields may suffer much more damage than this. The most serious rat pest in Kalimantan is the rice-field rat *Rattus argentiventer* which damages crops throughout Southeast Asia. The Polynesian rat *Rattus exulans* and house rat *Rattus rattus* also prey on rice. Rat control is difficult, but crop damage may be reduced by careful land use (see chapter 12).

Rice fields are favoured habitat for a variety of birds which, together with rats, are probably the greatest seed predators on rice crops. Among the more serious bird pests are the Javan sparrow *Padda oryzivora*, whose Latin name tells of its rice-eating habits, and the chestnut munia *Lonchura malacca*. The Javan sparrow is not native to Kalimantan but was introduced from its native Java; wild populations have become established from escaped cage birds.

Several species of waterbirds frequent rice fields and may assist in keeping insect pests under control. Insect-eating rails, crakes and swamphens are secretive birds and have been little studied in Kalimantan. More conspicuous large rice-field birds are egrets and herons, but these birds are rare or absent from rice fields in many parts of Kalimantan as a result of shooting, poisoning with biocides and destruction of roosts. Although egrets and herons feed in the same areas and have some overlap of diet, they do not compete for most food resources (Whitten et al. 1987b). Cattle egrets and Javan pond herons avoid competition by feeding in the fields at different stages of the crop cycle, and show different feeding behaviours and animal prey preferences. These useful predators on the rice-field pests can be encouraged by leaving patches of forest for roosting sites.

Introducing Exotic Species

Although new wetlands can support a diverse flora and fauna, the colonising species found there are not necessarily those indigenous to the area, but are often more widespread opportunist species, with broad niche requirements, and well able to exploit a new environment. Some of the most successful species have been deliberate introductions. The blue-flowered water hyacinth *Eichhornia crassipes*, a native of South America, was introduced to the Bogor Botanic Gardens in 1894 and has since spread throughout Indonesia, clogging waterways, lakes and irrigation canals. It chokes propellers and may impede the passage of even large, powerful boats, as on the Alabio River between Danau Panggang and Babirik in South Kalimantan. Techniques for controlling and utilising this weed are now being developed (Soerjani 1980; National Research Council 1976).

Other exotics that have been deliberately introduced to new lakes and other water bodies include several kinds of fish. Reservoirs such as Riam Kanan in South Kalimantan and Batang Ai in Sarawak have been stocked with *Tilapia mossambica*. Other *Tilapia* species originally introduced to fish ponds are now found wild in streams elsewhere in Sarawak (Chan et al. 1985). They do not seem well adapted to Bornean rivers, but can be difficult to eradicate from fish ponds because they are prolific breeders, and their eggs remain viable in dry mud. The gourami *Helastoma temmincki* has invaded many of the lakes of the lower Baram and Tinjar, and the soft-shelled turtle *Trionyx sinensis* and American bull frog *Rana catesbiana* are found in Kuching ditches (Chan et al. 1985). So far there is little information on the impact of these introductions on the native fauna. Introducing any exotic species, however, carries the risk of adverse ecological effects because of competition with native species, the absence of natural population controls, and the possibility of introducing disease.

THE SUNGAI NEGARA SWAMPS - A CONVERTED WETLAND HABITAT

The Negara River basin in South Kalimantan lies between the Barito River and the Meratus Mountains, covering an area of about 6,000 km^2 of very flat land, only 3-4 m above sea level. Most of this (90%) is flooded annually or is permanently wet, and can therefore be classified as wetland. Vegetation types include riparian forest, remaining along parts of the Barito, Negara, Tabalong, Balangan, Tapin, Tapirai and Kajang rivers; gelam swamp forest dominated by *Melaleuca cayuputi*, mixed peat swamp forest, kerangas forest and kerapah; sedge and grass swamp and lake vegetation (Giesen 1990). The Sungai Negara basin has been identified as a major wetland site in Kalimantan (Silvius et al. 1987) and was surveyed in 1989 by a joint team from Asian Wetland Bureau, the Conservation Department (PHPA), KPSL-UNLAM, PSL-UNTAN and Kompas Borneo (PHPA/AWB/KPSL-UNLAM 1990).

In the central part of the basin lie three lakes, Danau Panggang, Danau Sambujur and Danau Bangkau. Danau Bangkau, the largest lake, lies in a depression between the Negara River and the alluvial terrain near Kandangan. The lake may have formed as river waters were cut off between the raised river levees and the alluvial terraces, or it may have been formed by a downward tectonic movement in the area (Klepper 1990). The material forming the bottom of the lake and surrounding swamps is a heavy, grey, marine clay, containing pyrite from about 50 cm depth, usually overlain by a shallow (40 cm) organic layer. To the southeast this marine sediment is covered by peat of approximately 2 m depth (Klepper and Asfihani 1990).

The lake is fed by runoff from the alluvial terrain in the east and by the Batangalai River from the northeast. It is connected to the surrounding swamps and the Negara River by the Batangalai River and several channels. Water fluctuations between the wet and dry seasons are considerable. In the wet season water depth is 2 m in most of the lake, and up to 3.5 m at the centre; in high floods the water level may be 1 m higher. In the dry season most of the lake bed dries out.

The original vegetation of the lake area was freshwater swamp forest, except probably in the deepest parts, with peat forest in the peat area to the southwest. At present only the peat area is still forested, but the flora is greatly impoverished because of repeated fires, with stands of stunted belangiran *Shorea belangeran*, a commercial timber tree.

In the swamps the vegetation consists of grasses, sedges, weeds and small shrubs. There is a gradual transition to the deeper parts of the lake, where the vegetation is floating or submerged. The floating mats of vegetation are known as *kumpai*. Kumpai may be associated with true aquatic, free-floating species such as *Eichhornia crassipes*, the water hyacinth. More mature kumpai, with thick sods up to 45 cm in thickness, includes more shrub and climber species. A number of sedge species, such as *Lepironia articulata* and *Thoracostachyum sumatrana*, are used by local people, mainly for mat-making. *Nelumbo nucifera* seeds are eaten, and the vegetation provides a habitat for numerous waterfowl and other animals, notably the water monitor (trapped for skins) and several turtles, also harvested for food. Large *Ampullaria* snails are collected and fed to domestic ducks. The prolific growth of aquatic vegetation blocks waterways and leads to transport problems in most years. In the dry season the vegetation is burned to clear the waterways and to return nutrients to the lake ecosystem.

The great productivity and prolific growth of the aquatic vegetation of the lake is utilised by local communities for a curious form of agriculture called *hampung*. Small rectangular islands (3 m x 12 m) cut from the

Figure 10.8. Relative density of blackfish and whitefish species in different habitats of the Sungai Negara wetlands.
Source: Chaeruddin 1990

floating mats of vegetation are towed to a convenient site close to the village and smothered with a thick layer of bottom mud and water hyacinth. When this decomposes, beans, pumpkins and cucumbers are grown on the rotting vegetation and nutrient-rich sediments.

The lakes are important inland fisheries, and a large proportion of the 35,000-45,000 tonnes of freshwater fish caught annually in South Kalimantan originates here. Expressing total catch per unit of lake area (approximately 12,700 ha), the yield is more than 3 tonnes/ha. Many fish also feed in the swamps (fig. 10.8). The catch per unit area of lake and

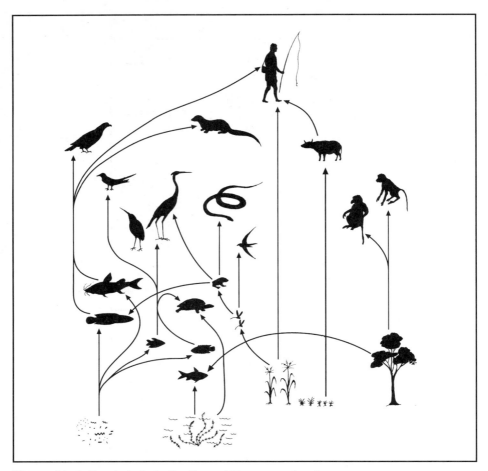

Figure 10. 9. Food chain in the Sungai Negara wetlands.

swamp together is 100 kg/ha (0.1 tonne/ha) (AWB/KPSL-UNLAM 1989).

Figure 10.9 illustrates the main food chains in the Sungai Negara wetlands. The major threats to the lake ecosystem are water pollution, hunting and overexploitation of fisheries. Water pollution from human waste and from fertiliser runoff from surrounding sawah may be responsible for the high phosphate concentrations and the abundant growth of the water hyacinth. Danau Bangkau is especially vulnerable to pollution as the water turnover is slow, with water remaining in the lake for one to two years.

> **Box 10.10. Threats to wetlands.**
>
> **HUMAN THREATS**
>
> *Direct*
>
> Drainage for crop production, timber production and mosquito control.
>
> Dredging and stream channelisation for navigation channels, flood protection, and reservoir maintenance.
>
> Filling for dredged spoil and other solid waste disposal, roads and highways, and commercial, residential, and industrial development.
>
> Construction of dykes, dams, and levées for flood control, water supply, irrigation, and storm protection.
>
> Discharges of materials (e.g., pesticides, herbicides, other pollutants, nutrient loading from domestic sewage, and agricultural and other land development) into waters and wetlands.
>
> Mining of wetland soils for peat, coal, sand, gravel and other materials.
>
> *Indirect*
>
> Sediment diversion by dams, deep channels and other structures.
>
> Hydrological alterations by canals, spoil banks, roads and other structures.
>
> Subsidence due to extraction of groundwater, oil, gas and other minerals.
>
> **NATURAL THREATS**
>
> Subsidence (including natural rise of sea level), droughts, hurricanes and other storms, erosion and biotic effects.
>
> Drought conditions in 1982/83 led to lower water levels, and the top layers of peat swamps dried out. Fires spreading from farmers' fields smouldered and spread through the peat layer, destroying large areas of peat swamp forests in Central and East Kalimantan.
>
> (Adapted from Maltby 1986.)

Hunting has already led to the extinction of crocodiles in the lake system. Some of the bird species and turtles are rapidly decreasing in numbers, probably in response to hunting, pollution and habitat disturbance (PHPA/AWB/KPSL-UNLAM 1990).

WETLANDS FOR CONSERVATION

Wetlands are generally very vulnerable ecosystems, depending upon the delicate balance of water levels and water flow, sedimentation, microclimate, and so on, all of which can be easily changed by on-site or neighbouring developments. Many wetland habitats are dependent upon factors outside their immediate boundaries. Changes in siltation levels and runoff patterns resulting from developments far upstream can affect the ecology of a river or lake. Conversely, changing land use and hydrological patterns within a wetland site can influence areas far beyond the development site. Mining peat swamps and converting them to agriculture can reduce water quality and disrupt water flow patterns in lower river basins. Draining tidal swamps can affect the nutrient input and balance of coastal mangrove forests and estuarine ecosystems. Pollution from a factory can spread through waterways to reach delicate wetland ecosystems. Introduced species are free to pass along the same interconnecting waterways to compete with indigenous species or change local conditions.

A new drainage or irrigation canal can affect water-table levels over large areas. Boreholes and wells result in lowered local water tables; grazing by domestic animals or fires set by man can transform wetland vegetation; a dam or barrage can cut off adult fish from their spawning grounds; water-soluble toxins in herbicides or pesticides drain into wetland sites. Excessive nutrient inputs from fertilisers in runoff from agricultural fields can pollute inland lakes, lead to eutrophication and deoxygenation, and destroy valuable freshwater fisheries. Pollution from human and industrial waste can destroy aquatic life in waterways far downstream from the developed site.

In Kalimantan waterways are important "highways" into the interior. The large amount of river traffic creates special problems. Powerful boats, travelling at high speed, create erosion of river and canal banks, increasing sediment loads. River traffic disturbs river wildlife and pollution, from oil and fuel spills, poisons aquatic food chains.

Water is one of the most valuable of all natural resources, essential for all forms of life. Lakes and rivers provide drinking water and water for domestic and industrial use. Surface waters are also important for agriculture; aquaculture of fish and other aquatic animals for food; transport; power generation; irrigation; recreation; and scientific research. Wetland habitats provide other services and resources: regulating water flow; preventing flooding; purifying water; and providing commercial timber and a host of useful plants and animals. Ironically, the very productivity of wetland sites has encouraged developers to convert them to other forms of land use. Too often, fragile ecosystems are being drained and converted to marginal agricultural lands. Yet the most productive future for many of these wetlands, especially areas where acid sulphate soils predominate, may be to retain them as natural forested wetlands, exploiting only their fisheries and forest products at sustainable levels.

As Kalimantan's wetlands come under increasing pressure from development, it is important that some areas are protected for conservation of

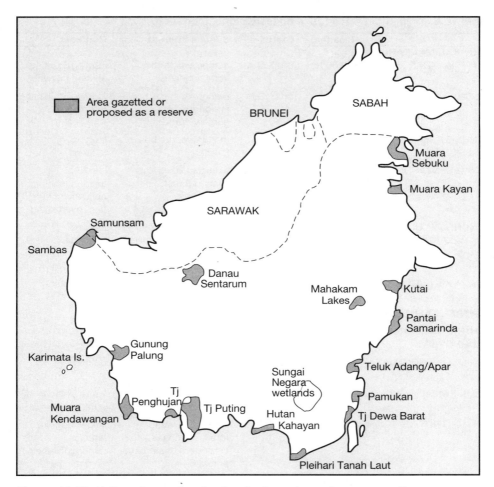

Figure 10.10. Kalimantan wetlands of major importance for conservation.

wetland resources. The viability of protected wetlands is, more than for most other habitats, dependent upon appropriate land-use practices outside the wetlands, for these affect the quality and quantity of water flow. Establishment of wetland protected areas is, therefore, only one element in a more comprehensive approach to water resource and land-use planning.

Wetlands are important to both human communities and wildlife. Many of the major wetland sites in Kalimantan are already included in gazetted reserves and conservation areas; others are proposed for protection (fig. 10.10). Table 10.5 lists the more important Kalimantan wetlands and their associated flora and fauna. Other important wetland sites in

Table 10.5. Kalimantan wetlands of conservation importance.

	Area (ha)	Status	Special interest	Conflicts
WEST KALIMANTAN				
HUTAN SAMBAS	120,000	prop. C.A.	mangroves, turtles, floral richness	logging, hunting, shifting cultivation
PALOH	176,548	prop. S.M.	mangrove important fish nursery, turtles	logging, collecting turtle eggs
GUNUNG ASUANSANG FOREST	28,000	H.L.	mangrove, dry beach forest	logging, hunting, ladangs
DANAU SENTARUM	80,000	S.M.	seasonal lakes and swamp forest, waterbirds, fisheries	logging, shifting cultivation, hunting, overfishing
GUNUNG PALUNG AND SWAMPLANDS	130,000	C.A. & prop. T.N.	freshwater and peat-swamp forest, mangrove, beach forest, rich flora and fauna	logging, ladangs
MUARA KENDAWANGAN	150,000	prop. C.A.	mangrove, freshwater and peat swamp forest	not threatened
CENTRAL KALIMANTAN				
CENTRAL KALIMANTAN MANGROVES	20,000	H.K. prop. H.L.	small blocks of mangrove/nipa	cutting
TANJUNG PENGHUJAN	40,000	prop. C.A.	mangrove, freshwater forest	cutting
TELUK KUMAI	3,900	prop. T.L.	mangrove/nipa, crocodiles, turtles	water traffic cutting
TANJUNG PUTING	300,040	T.N.	peat and swamp forest, kerangas, crocodiles, waterbirds, primates	illegal logging, ladangs, hunting/fishing gold mining
TANJUNG PUTING (extension)	70,000	prop. T.N.	mangrove, freshwater dolphin (?), dugong, turtles	logging, settlements
DANAU SEMBULUH	7,500		lake seasonal swamps	–
SOUTH KALIMANTAN				
HUTAN KAHAYAN	150,000	prop. C.A.	mangrove, freshwater/ peat swamp forest, fish spawning area	logging, clearing for agriculture
ALABIO POLDER	6,000	–	seasonal wetlands, waterbirds	hunting
DANAU BANGKAU	480,000	–	Barito basin, freshwater & peat swamps, waterbirds	changing land use, drainage, fishing, reed cutting

Table 10.5. Kalimantan wetlands of conservation importance *(continued).*

	Area (ha)	Status	Special interest	Conflicts
SOUTH KALIMANTAN *(continued)*				
PULAU KAGET	85	C.A.	mangrove, proboscis monkeys	cutting wood
PULAU KEMBANG	60	T.W	mangrove, proboscis monkeys	cutting mud for bricks
SOUTH KALIMANTAN MANGROVES	90,000	T.W.	mangrove	logging, hunting
PLEIHARI TANAH LAUT*	35,000	S.M.	mangrove	wood cutting, shifting cultivation, badly damaged, plantation
PLEIHARI MARTAPURA	36,400	S.M.	swamp forest, reservoir, floral endemism	ladang encroachment, logging
PULAU SUWANGI	500	prop. C.A.	mangrove, orchids	wood cutting
PULAU SEBUKU	14,400	prop. C.A.	mangrove	wood cutting concession
HUTAN BAKAU PANTAI TIMUR	66,650	prop. C.A.	mangrove/nipa, fish nursery	wood collecting
TANJUNG DEWA BARAT	16,250	prop. C.A.	mangrove	reclamation, cutting
TANJUNG KELUMPANG	13,750	prop. C.A.	mangrove	cutting
PAMUKAN	10,000	prop. C.A.	mangrove	cutting
EAST KALIMANTAN				
APAR BESAR	90,000	–	mangrove, freshwater swamp forest	logging
TELUK APAR & TELUK ADANG	130,000	prop. C.A.	good mangrove, freshwater swamp	logging
PANTAI SAMARINDA	95,000	prop. H.L.	Mahakam delta, mangrove, freshwater swamp forest	clearance, damaged, cutting
SUNGAI BERAMBAI	110,000	prop. C.A.	peatswamp forest	logging
PERAIRAN SUNGAI MAHAKAM	200,000	prop. C.A.	Mahakam lakes, freshwater swamp forest, freshwater dolphin,	river traffic, pollution, overfishing,
MUARA KAMAN	62,500	C.A.	peatswamp forest, ramin	settlement, logging
MUARA ALONG	small area		ox-bow lake, fruit bats	
KUTAI	320,000	S.M. T.N.	mangrove, freshwater swamp, rich bird and primate fauna	logging, oil exploitation, settlements, fire
PULAU BIRAH-BIRAHAN	small coral islands	prop. C.A.	coral reefs, seabirds, nesting turtles	collection of turtle eggs

Table 10.5. Kalimantan wetlands of conservation importance *(continued)*.

	Area (ha)	Status	Special interest	Conflicts
EAST KALIMANTAN *(continued)*				
KEPULAUAN SANGALAKI	small coral islands	prop. C.A.	mangrove, turtle nesting beaches tourism	collection of turtle eggs
MUARA KAYAN	80,000	–	mangrove, freshwater dolphin	logging for chipboard, damaged
MUARA SEBUKU	110,000	prop. C.A.	best mangroves in Kaltim, peat swamp, proboscis monkeys	logging

prop. - proposed
C.A. - Cagar Alam/Nature Reserve
S.M. - Suaka Margasatwa/Wildlife Reserve
H.L. - Hutan Lindung/Protection Forest
T.N. - Taman Nasional/National Park
T.L. - Taman Laut/Marine Park
H.K. - Hutan Konversi/Conversion Forest
T.W. - Taman Wisata/Recreation Park
* In 1992 S.M. Pleihari Tanah Laut was degazetted for conversion to plantation (HTI).

Sources: MacKinnon and Artha 1981; Silvius et al. 1987

Borneo include Samunsam Wildlife Reserve, Loagan Bunut and peat swamps in Sarawak. Many of these sites are of regional importance and are listed in the IUCN Directory of Asian Wetlands (Scott 1989), the Indonesian Wetland Inventory (Silvius et al. 1987) and the ICBP Inventory of Wetlands in East Asia (Karpowicz 1985).

Chapter Eleven

Coastal Resources

In Kalimantan 60% of the population lives in the coastal zone (fig. 11.1). Settlements and large towns are concentrated in the low-lying coastal plains and along the shoreline. Villages are often located near estuaries to take advantage of the availability of fresh water, sheltered moorings and rich coastal fisheries (Polunin 1983). Because of their accessibility, coastal areas are also increasingly the focus of industrial development activities.

The coastal zone is often seen as an unproductive resource or wasteland that needs to be reclaimed, yet coastal ecosystems provide a wide range of goods and services to human communities (Soegiarto and Polunin 1980; Polunin 1983; Saenger et al. 1983; Salm and Halim 1984; Salm and Clark 1984). Natural productivity in coastal ecosystems is high (chapter 2). Harvests and services such as coastal protection are provided free by these self-sustaining, self-renewing ecosystems.

Development processes can easily disrupt these natural processes. Coastal ecosystems are influenced both by offshore activities and by those far inland, so are particularly vulnerable to disturbance (Knox and Miyabara 1984). Because of this ecological fragility, regional developers must consider environmental factors and ecological inter-relationships when planning aquatic resource development and coastal resource management (Burbridge and Maragos 1985). Coastal developments, whether logging of swamp forest, drainage of tidal swamplands, infilling of estuaries, reclamation of land or exploitation of coastal oil fields, all affect natural ecosystems and have environmental costs that may outweigh some of the increased benefits they provide. An understanding of the ecology of coastal ecosystems and their interdependence on other offshore and inland ecosystems is essential for wise use of coastal resources (see chapter 2).

Estuarine and Coastal Fisheries

The natural productivity of estuaries and the shallow coastal seas around Borneo support rich coastal fisheries. Kalimantan's great rivers continually discharge sediments, minerals and nutrients into estuaries, replenishing materials to sustain high productivity. Estuarine productivity is more than

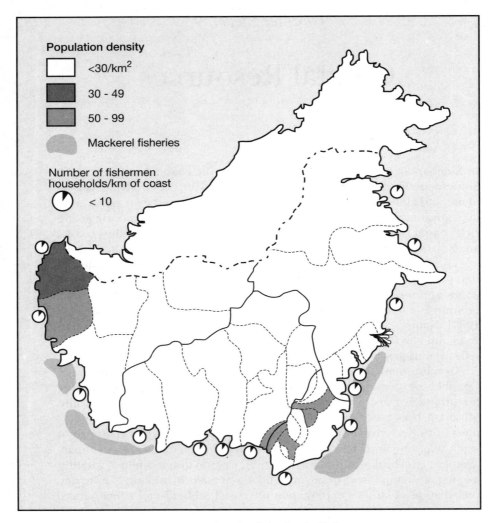

Figure 11.1. Population density and major fisheries in Kalimantan.
Source: Salm and Halim 1984

twice that found in most land ecosystems, and twenty times greater than in the open oceans (Knox and Miyabara 1984). Because of this high productivity, the coastal zone is ultimately the source of nutrients for Kalimantan's rich commercial fisheries, both inshore and far out to sea. Estuaries are some of the most valuable fishing areas in Indonesia, as well as being important nursery sites for larvae and juveniles of finfish and many shellfish. Two species of clupeid fishes are particularly important in estuarine fisheries: *Clupea macrura* (*ikan terubuk*) and *C. toli*, much prized for their flesh and eggs. They are caught in the open sea and also gill-netted in estuaries on their way upriver to spawn. Terubuk catches have declined in recent years, but large-scale terubuk fisheries remain in the Lupar estuary, Sarawak (Chan et al. 1985).

Fish is the main source of animal protein for the peoples of Indonesia (Knox and Miyabara 1984). Fish consumption per capita is higher in Kalimantan than in any other part of the archipelago, three times greater than in Java (Birowo 1979). Fishing is an important subsistence and income-generating activity for many households around Kalimantan's 8,000 km of coast. The greatest number of households relying on coastal fisheries are found in East Kalimantan (table 11.1).

The shallow waters of the Sunda Shelf support rich populations of pelagic (surface-water) fishes and demersal (bottom-feeding) creatures such as prawns and finfish. A high diversity of fish species is typical of tropical waters. The waters off the east coast of Malaysia, for instance, support at least 43 families of fish (Tan 1974) and 22 species of prawns (Lee 1972). More than 40 species of marine fish are economically important in Indonesia. The seas also provide a rich harvest of marine and brackish-water prawns, a variety of other crustaceans, bivalves (especially the blood cockle), cuttlefish, squid, turtles, dugongs, sea cucumbers, jellyfish and other invertebrates (Soegiarto and Polunin 1980; Burbridge and Maragos 1985).

Important pelagic fisheries are found off southwest Kalimantan and the east coast of South Kalimantan, around Pulau Laut and extending northwards (Salm and Halim 1984). The main species harvested are mackerels *Rastrelliger* spp. The South Kalimantan fisheries yield as much as 0.2 tons of fish per fisherman annually (Salm and Halim 1984). Sea fish are harvested both by small-scale artisans, putting out to sea in small boats and attracting the fish to the light of paraffin lamps, and by more commercial trawling

Table 11.1. Fishing households in Kalimantan, 1988.

Province	Coast length (km)	Fishing households (all)	Marine fisheries	Open water	Tambak and freshwater pond
W. Kalimantan	1,940	15,267	5,047	5,953	4,267
C. Kalimantan	737	29,099	3,327	25,772	–
S. Kalimantan (including Pulau Laut)	1,135	30,651	7,102	19,629	3,920
E. Kalimantan	4,242	22,551	8,502	9,376	4,673
Total	8,054	97,568	23,978	60,730	12,860

Source: Buku Statistik 1991

operations. In 1989, 214,534 tons of fish valued at $100 million were harvested off Kalimantan's coasts (Buku Statistik 1991). Apart from the east coast, most of Kalimantan's coastal waters are considered to be underexploited for many fish species (Collier et al. 1979).

Demersal fisheries yield large numbers of shrimp and associated finfish for local consumption and export. Prawn-fishing grounds are found all around Kalimantan's coasts, with the whole of the east coast being heavily fished, in places probably beyond the maximum sustainable yield (Salm and Halim 1984). In Kalimantan coastal prawn fisheries are dominated by commercial operations. Spiny lobsters (Palinuridae) are also harvested around Pulau Laut and along the north coast of East Kalimantan. In 1981 145 tons of fresh and frozen lobsters and crayfish, worth $913,000, were exported from Indonesia, with more than 10% of the harvest coming from Tarakan; this harvest increases annually (Salm and Halim 1984). The catch is collected from local fishermen by ships with cooling and freezing facilities. Mangrove crabs *Scylla serrata* and other crabs *Portunus* are also collected for local sales and export.

In addition to these commercial sea fisheries, fry of milkfish *Chanos chanos* are harvested from the coastal waters of eastern Borneo for *tambak* (brackish-water fish pond) aquaculture. Milkfish fry are collected seasonally from April to May and again from November to December along the coasts of East Kalimantan, and at Pulau Sebuku and Pulau Laut in South Kalimantan (Soesanto 1979).

Harvests from the Sea

Many species of molluscs are harvested along Kalimantan's coasts. Clams, cockles Lucinidae and scallops Pectinidae are collected for food; waste shells provide lime and animal feed. Giant clams *Tridacna* are collected for their meat and shells. Pearl oysters *Pinctada* and commercial trochus *Trochus niloticus* are harvested off the Karimata, Anambas and Natuna islands and on the east coast close to the Mahakam delta. The green snail *Turbo marmoratus* is also harvested on the east coast. These molluscs yield mother-of-pearl for jewellery, buttons, inlay work, paints and cosmetics. Ornamental helmet shells *Cassis cornuta* from the Natuna Islands are sold domestically. Other gastropods, especially the larger species, are collected as subsistence foods (Salm and Halim 1984).

Many species of marine algae flourish on shallow rocky substrates and reef flats, and are harvested for food or for commercial production of agar-agar (table 11.2). Yet others provide indirect but definite benefits to coastal communities. The latter include the nitrogen-fixing blue-green algae, which grow in seemingly dead areas of reefs, where they convert nitrogen into a usable form for plants and animals and thereby support high productivity on the coral reefs. Certain green *Halimeda* and red

Box 11.1. Management implications of harvest, hunting and mining activities on estuaries.

Fisheries
- Fishing gear and weirs that block the access channels to coastal lagoons or connecting channels between lagoons interfere with the spawning cycles and movements of fishes and shrimps.
- Active fishing gear dragged along the estuary floor causes damage to rooted vegetation and natural oyster beds.
- Fishing must be regulated at sustainable levels and should be controlled by seasonal quotas and size limits to preserve breeding stock.

Clamming
- This should be regulated by permits and quotas, and should be carefully monitored for impact on substrates and stocks.

Crocodile hunting
- This can be a sustainable industry only in areas that still retain abundant stocks. It is being replaced by crocodile farming.

Mining
- should be preceded by impact assessment studies;
- must control sedimentation resulting from disturbance of the substrate;
- should be prohibited in the critical habitats of valuable species;
- must avoid disruption of the hydrological regime in adjacent lands;
- should be discouraged or strictly controlled upstream of critical habitats;
- must be carefully regulated;
- should be accompanied by rehabilitation (mitigation) in damaged areas;
- should have plants for the disposal of spoil in an approved site **outside** wetlands and other critical habitats;
- must avoid interference with the flow of water through the system by pipes and service roads.

Forestry operations
- need careful evaluation to determine truly sustainable levels for both selective logging and clear-felling, especially in coastal mangroves;
- should employ methods that cause least damage to the substrate;
- must be accompanied by regeneration of mangroves, if necessary through replanting with propagules, weeding of undesirable species and disposal or other treatment of trimmings;
- should be confined to areas of least value to fisheries;
- should be accompanied by research and monitoring studies to determine effects on fisheries;
- need strict protection of adjacent sites as a possible supply of propagules.

Source: Salm and Clark 1984

Porolithon and *Lithothamnion* algae cement the reefs and contribute calcareous sediment to build and maintain the sandy beaches of coral islands, where turtles nest. Algae are also the main food for surgeonfish Acanthuridae, wrasses and parrotfish Scaridae and rabbitfish Sigamidae, which are harvested by local fishermen.

Algae collected for food include *Gracilaria, Caulerpa* and *Codium*. Important seaweeds for domestic and export trade include *Hypnea, Gracilaria, Gelidiopsis* and *Euchema*. These yield the gelling agent agar-agar. Dried *Euchema* is processed to extract carrageenan (Salm and Halim 1984). The main algae-collecting grounds off Kalimantan are the southern coast of Pulau Laut and the Berau coast in East Kalimantan (Salm and Halim 1984).

Table 11.2. Useful marine algae of the Indonesian region.

Use	Class	Species
Food	Chlorophyceae	*Caulerpa laetevirens, C. peltata, C. racemosa, C. sertularoides, Chaetomorpha crassa, C. javanica, Codium tenue, C. tomentosum, Enteromorpha compressa, E. intestinalis, E. prolifera, Ulva lactuca.*
	Phaeophyceae	*Chnoospora pacifica, Hydroclathrus clathratus, Dictyota apiculata, Padina australis, Sargassum aquifolium, S. granuliferum, S. polycystum, S. siliquosum, Turbinaria ornata, T. conoides.*
	Rhodophyceae	*Porphyra atropurpurea, Acanthophora spicifera, Bostrychia radicans, Caloglossa adnata, C. leprieurii, Catenella impudica, C. nipae, Corallopsis salicorna, Eucheuma edule, E. cottonii, E. gelatinae, E. horridum, E. serra, E. spinosum, E. muricatum, Laurencia obtusa, Gracilaria confervoides, G. crassa, G. taenoides, G. javanicus, G. blodgettii, G. arcuata, G. lichenoides, Grateloupia filicina, Halymenia durvilliae, Hypnea cenomyce, H. cervicornis, H. divaricata, Sarcodia montagneana, Gelidiopsis rigida, Gelidium* spp.
Agar-agar	Rhodophyceae	*Eucheuma edule, E. gelatinae, E. horridum, E. muricatum, E. serra, Gelidium amansii, G. latifolium, Gelidiopsis rigida, Gracilaria confervoides, G. eucheumoides, G. taenioides, G. lichenoides, Hypnea musciformis.*
Medicine	Cyanophyta	*Nostoc commune.*
	Chlorophyceae	*Acetabularia major.*
	Rhodophyceae	*Eucheuma serra, Hypnea musciformis, Rhodymenia palmata.*
Agriculture	Phaeophyceae	*Hydroclathrus clathratus, Padina australis.*
	Rhodophyceae	*Gracilaria confervoides.*

Source: Soegiarto and Polunin 1980

Sea cucumbers (*trepang*) of the family Holothuroideae are collected off the Anambas, Natuna and Karimata islands, and along the coast of Berau, Bontang and other parts of East Kalimantan. These holothuroids are dried in the sun, and the preserved trepang is sold to local and overseas markets. Trepang has been traded for centuries and was one of the commodities purchased by the early Chinese traders when they first visited Borneo.

Dugongs are occasionally captured off southern Kalimantan for their meat, oil and incisors. A few specimens have been shipped to the Ancol aquarium in Jakarta for exhibition. Although sperm whales *Physeter catodon* and sei whales *Balaenoptera borealis* have been recorded off northeast Kalimantan, these aquatic mammals travel through deep sea channels and are rarely recorded close to shore (Salm and Halim 1984).

Turtles and their eggs are harvested in the Berau Islands, East Kalimantan (box 11.2). Turtle eggs are also collected on the nesting beaches of Pleihari Tanah Laut, South Kalimantan, and at Paloh, West Kalimantan (table 11.3). At Paloh green turtles *Chelonia mydas* and hawksbills *Eretmochelys imbricata* are the most common species, but olive ridleys *Lepidochelys olivacea* (*karahan*) also use the nesting beaches. During the main laying season, from March to September, egg collectors travel the beaches every night, locating the nests by following the tell-tale tracks left by the female as she hauls herself up the beach to the dry sand above the tide line. A single female hawksbill lays 100-150 eggs in one night. Eggs are sold for Rp 100 (5 cents) each in local markets, so a single nest yields a profitable harvest to villagers on low incomes. Unfortunately, there are no accurate data on annual harvests of turtle eggs at Paloh. Estimates suggest that more than one million eggs are collected from Selimpai and the other four Paloh turtle beaches, a profitable venture for the collectors who, in 1989, paid an annual fee of Rp 15 million ($7,500) for collecting rights.

The number of turtles nesting at Paloh is said to be declining, probably due to overharvesting of eggs. To counteract this decline, the local Conservation Department (PHPA) has begun a programme of collecting 4,500 eggs each year of all three species and transplanting them to a secure hatching compound. New hatchlings are reared on a diet of fish and seaweed until six months old, when they are released into the sea. At this stage they are thought to be less vulnerable than new hatchlings to predation. Not all transplanted eggs hatch, and hatchling mortality prior to release is said to be between 10% and 15%, so probably no more than one young turtle is released for every 400-500 eggs harvested. Turtles suffer high predation rates naturally, and only a few individuals need to survive to replace the long-lived adults, thus maintaining population levels.

Turtles do not breed until they are about twenty years old. Most species nest annually, but individual females may not reproduce every year; on average a female green turtle nests every third year. The green turtle is the most prolific reptile known; in Sarawak green turtles have been recorded laying up to 11 clutches in a year, at ten-day intervals (Moll *in* Halliday and Adler 1987). Only time will tell whether the Paloh programme will help to maintain turtles nesting on Kalimantan's beaches. There is no tagging programme yet at Paloh, so it is impossible to know which females return to nest, or where they travel from there. However, turtles tagged on Malaysian beaches have been recorded nesting at Paloh.

Apart from their value as a harvestable resource, turtles could be exploited as a tourist attraction. Large numbers of tourists visit the nesting beaches on the east coast of Trengganu in Peninsular Malaysia, bringing valuable income to local communities. The turtle islands off Sabah, pro-

Table 11.3. Feeding and nesting grounds for green turtle (*Chelonia mydas*) and hawksbill turtle (*Eretmochelys imbricata*) in Kalimantan.

	Chelonia mydas				*Eretmochelys imbricata*			
	nesting site	eggs collected	feeding ground	turtle hunted	nesting site	eggs collected	feeding ground	turtle hunted
1. Sambas-Paloh	X	X	–	X	X	X	–	–
2. Pulau Lemukutan	X	–	–	–	–	–	–	–
3. Kep. Karimata (P. Buan, P. Serutu)	X	–	–	X	–	–	–	–
4. Kumai	X	–	–	–	–	–	–	–
5. Tj. Puting	X	–	–	–	–	–	–	–
6. Pleihari-Tanah Laut (Tj. Malacur)	X	–	–	X	–	–	–	–
7. Tanjung Selatan	X	–	–	–	X	–	–	–
8. Kep. Marabatua (P. Denaweng, P. Payung-payungan)	X	X	–	X	X	X	X	X
9. Kep. Laut Kecil (P. Pamalikan, P. Kunyit, P. Matasiri, P. Kalambau)	X	–	–	–	X	–	X	–
10. P. Birah-birahan (S)	X	–	–	–	–	–	–	–
11. Tanjung Layar	X	–	–	X	–	–	–	–
12. Kep. Sambergelap	X	–	–	–	–	–	–	–
13. P. Lari-larian	X	–	–	–	–	–	–	–
14. P. Ambo	X	X	–	X	X	X	–	X
15. Kep. Balangan, Kep. Mamuju	–	–	–	X	X	–	X	–
16. Pasir	X	X	–	–	–	–	–	–
17. Balikpapan	X	X	–	–	–	–	–	–
18. P. Birah-birahan (E)	X	X	–	–	–	–	–	–
19. Sangkulirang	X	–	–	–	–	–	–	–
20. P. Mataha, P. Bilangbilangan	X	X	–	–	–	–	–	–
21. P. Semama, P. Sangalaki	X	X	–	–	X	X	–	–
22. P. Maratua, P. Balikukup	X	X	–	X	X	X	X	–
23. P. Sambit, P. Palembangan	X	X	X	X	–	–	–	–

Kep. = Kepulauan   P. = Pulau   Tj. = Tanjung
For locations see figure 2.6. The island groups 20-23 are the Berau turtle islands.

Source: Salm and Halim 1984

Box 11.2. Harvesting turtle eggs in the Berau turtle islands, East Kalimantan.

The Berau turtle islands are coral islands lying off the coast of East Kalimantan, north of the Sangkulirang peninsula. The northern group of small islands includes Panjang, Derawan, Semama and Sangalaki. To the south lie the Karang Muara islands and Palembangan, Sambit, Bilang-bilangan, Mataha and Balik Kukup, an eastward extension of the coral reef of Karang Besar. Further east on the continental shelf lie the two oceanic atolls of Pulau Kakaban and Pulau Maratua. Kakaban is surrounded by a beautiful coral reef where hawksbill turtles *Eretmochelys imbricata* forage.

The Berau islands are important rookeries for green turtles *Chelonia mydas*. Turtles nest on several islands, but Pulau Sangalaki is the most important rookery, with large numbers of green turtles and occasional hawksbills nesting there. Apart from its green turtle rookery, Pulau Semama also has a nesting colony of lesser frigate birds *Fregata ariel*.

Turtles nest all year, with a peak in egg-laying from August to October. This corresponds to the nesting season on the Sabah turtle islands, some 600 km to the north. About 12,000 female green turtles are believed to nest on the Berau islands, and the total population of green turtles feeding around the islands is estimated at 25,000. Hawksbill turtles are also abundant grazing around the coral reefs. About two to three million turtle eggs are collected on the Berau islands each year. The annual rent paid for 1984 for egg-collecting concessions on the various islands gives a good indication of the main nesting sites and shows the importance of the rent as an income for the local government of Kabupaten Berau.

Annual rent 1984	
P. Sangalaki	Rp42 million
P. Bilang-bilangan	Rp28 million
P. Belambangan	Rp20 million
P. Sambit	Rp19 million
P. Mataha	Rp16 million
P. Derwan	Rp1.75 million
P. Semama	Rp1.25 million
Total rent, 1984	**Rp128 million ($125,000 in 1984)**

Approximately 60 tons (two million eggs or more) are collected each year, with 600,000 - 750,000 eggs from Pulau Sangalaki alone.

There has been a reported decline in the number of nests over the last few decades. Turtle numbers are said to have declined due to the disappearance of seagrass beds in the 1940s, perhaps due to release of oil from dumps at Tarakan and Balikpapan prior to the Japanese advance during World War II. Other factors are the predations of Balinese turtle hunters, and Filipino fishermen dynamiting the reefs. Since 1907, under regulations established by the Dutch resident, it has been illegal to catch marine turtles or harvest their eggs without a licence. Green turtles, and also hawksbills, are caught and slaughtered for their meat and shells. A survey in 1984 found 200 kg of hawksbill scutes at Tanjung Batu, ready for shipment to Ujung Pandang; this represented two months buying with a yield of 0.5 kg per turtle at a price of Rp 50,000/kg, a harvest of 400 turtles. More than 2,000 hawksbills are killed each year in the Berau islands alone. Turtles are wide-ranging species. Some of these turtles may breed off Sabah and northern Australia but are captured when they visit feeding grounds in Indonesia.

Since 1969 egg collectors have been required to restock the nesting population, and between 1969-1976 contractors released more than 6,000 juveniles, between six and nine months old, into the sea. In 1982 Pulau Sangalaki and Pulau Semama were gazetted as a wildlife reserve and marine park respectively, but the egg-collecting rights are still leased out. The Karang Muara Islands and Pulau Maratua are proposed nature reserves (*cagar alam*).

The Berau turtle islands and adjacent reef complexes, including the algae and seagrass beds, offer a unique opportunity for research on island-nesting turtles in Indonesia, especially green turtles and perhaps hawksbills. To date, little is known of the extent and pattern of turtle movements or the levels of harvesting that populations can sustain. Research should concentrate on surveys of the main beaches to monitor nesting, predation and hatching success, and a tagging programme to provide information on the ranging of turtles and the frequency of their return to the Berau islands to feed and nest.

Source: Schulz 1984

tected as a national park, are a popular tourist venue. The golden beaches and turtle nesting sites of West Kalimantan also have considerable potential for sensitive development for tourism.

The Value of Coral Reefs

Around most of Kalimantan's coasts, large rivers flow into the shallow seas of the Sunda Shelf, and the waters are too muddy or too freshwater for growth of coral, so reefs are absent or poorly developed. However, there are a few noteworthy coral reefs off Kalimantan, particularly the Karimata islands off West Kalimantan, the beautiful fringing reefs of the Berau islands, including Sangalaki, and the Semporna reefs, lying just within Sabah waters. Coral reefs are most valuable for their fisheries and other marine harvests, and for their potential for tourism.

The many fine subdivisions of food and space that support high diversity on the coral reef (chapter 2) make these reefs some of the world's richest fisheries, supporting a "standing crop" of fish five to fifteen times as great as the crops of productive North Atlantic fisheries (Stevenson and Marshall 1974). They also support other species harvested for food (crabs, lobsters, snails, clams, octopus, sea urchins, trepang and turtles), ornamental products (pearls, mother-of-pearl, coral, echinoderms, molluscs, turtles and ornamental fish) and industrial products (bulk coral, turtle shell and giant clams).

High levels of competition with other organisms in the crowded reef conditions has led to the evolution of some interesting interactions. One such interaction, well developed on coral reefs, is antibiosis, the production by one organism of substances that are harmful or repulsive to others (Burkholder 1973). Some of these substances are highly active biocompounds with medical applications. Thus certain reef-dwelling sea fans and anemones have been found to possess compounds with antimicrobial, anticoagulant, antileukaemic and cardioactive properties (Ruggieri 1976). Such species may produce important anticancer drugs or serve as models for the synthesis of effective new drugs.

Perhaps the most important function of fringing reefs along waveswept shores is the role they play in preventing coastal damage and erosion. This is a "free" service, with fringing reefs acting as "self-repairing breakwaters" (Johannes 1975). The ability of coral organisms to colonise new areas can be exploited by the establishment of artificial reefs. Old cars dumped in the sea off Brunei provide a skeleton for new corals, thus converting unsightly waste into useful and productive ecosystems. Coral reefs are important to coastal and inland communities for subsistence, income generation, research and recreation. The need for better protection and management of coral reefs in Indonesia is well argued by Soegiarto and

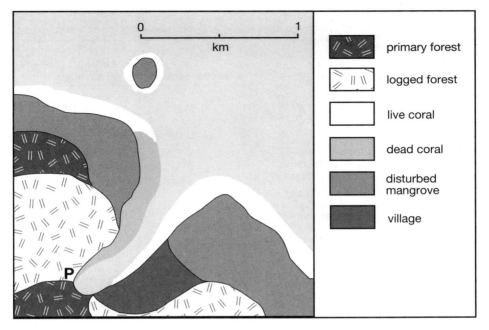

Figure 11.2. The distribution of living and dead coral around a logging company pier, 'P', on Siberut Island, West Sumatra. (After WWF 1980.)

Polunin (1980), Salm and Halim (1984) and Salm and Clark (1984).

Unfortunately, like most delicately balanced ecosystems, coral reefs are easily damaged. Coral mining for lime, fishing with dynamite, or even dredging and landfall schemes on the nearby coast can all cause irreparable harm. Inland erosion and siltation can also kill or damage offshore reefs (fig. 11.2). There is good evidence that increased soil erosion and sedimentation as a result of logging have smothered nearby coral reefs off Siberut, Sumatra (WWF 1980), and Pulau Satang Besar, Sarawak (Chan et al. 1985). Industrial development in East Kalimantan has also led to destruction of coral reefs. The fringing reefs near Balikpapan were mined to provide construction materials for the Balikpapan-Samarinda road, and coral reefs in Bontang Bay have been damaged by dredging and by the hot water outflow from the natural gas (LNG) plant (Dahuri 1990).

Tourism development can help to protect reefs by bringing income to local communities and stopping destructive exploitation. The Semporna reserve in Sabah is a popular tourist destination, and Sangalaki in the Berau islands has recently been developed for diving. The Karimata islands, off the coast of West Kalimantan, probably also have potential for tourism devel-

Box 11.3. Connections between coral reefs and neighbouring linked habitats.

Habitat neighbouring	Benefits	Potential damage	Linking mechanism
Reef flat	Introduces fixed nitrogen, dissolved and particulate organic compounds to reef food web As feeding ground and nursery for reef fishes, increases diversity and abundance of reef species		Transport by waves, currents, fishes and sea urchins Nocturnal/diurnal migration
Seagrass beds	Introduce dissolved and particulate organic compounds to reef food web. As feeding grounds and nurseries for reef organisms, increase diversity and abundance of reef species		Transport by currents, fishes and sea urchins. Nocturnal/diurnal migration
	Consolidate sediments, protecting reef from smothering	Destruction of sea grass beds by repeated anchoring releases sediments into the water column and increases turbidity; reef organisms can be smothered, resulting in decrease of diversity and abundance of reef species	Transport by waves and currents
Sand or mud flats	Feeding grounds for reef fishes; increase diversity and abundance of reef species		Nocturnal/diurnal migration
Mangroves, lagoons, estuaries	Introduce dissolved and particulate organic compounds to reef food web. As feeding grounds and nurseries for reef fishes, increase diversity and abundance of reef species		Transport by tidal flushing and currents. Nocturnal/diurnal migration
	Trap pollutants and silt, protecting reef from poisoning or smothering	Disturbance of substrate, and release of trapped silt and pollutants can result in smothering or poisoning of reef organisms	Transport by tidal flushing, stream flow, and currents

> **Box 11.3.** Connections between coral reefs and neighbouring linked habitats *(continued)*.
>
> | Habitat linked | Benefits | Potential damage | Linking mechanism |
> |---|---|---|---|
> | Beaches and dunes | | Sand released by erosion or destruction of binding vegetation may smother organisms | Transport by wind, waves and currents |
> | Watersheds | Regulate stream flow | Silt, floods and dilution of seawater caused by deforestation and erosion can stress organisms | Transport by streams and currents |
> | Urban or industrial developments | | Litter; domestic, chemical and thermal pollution; increased freshwater run-off can poison or physically damage organisms or cause eutrophication | Transport by streams and currents |
> | Agricultural development | | Silts, floods and dilution of seawater; pesticides, herbicides and fertiliser pollution can smother organisms or cause eutrophication | Transport by streams and currents |
>
> Source: Salm and Clark 1984

opment. Nevertheless, unless it is carefully controlled, tourism itself can be detrimental to the reefs. Careless anchoring of boats (Davis 1977), snorkellers and divers walking on the coral surface, and collection of coral and shell souvenirs can all lead to destruction of the reef and may cause long-term changes in reef communities (Woodland and Hopper 1977).

MANGROVES AS A RESOURCE

Mangrove swamps are often regarded as wastelands of little or no value until they are "developed", that is converted to some other land use. This approach fails to recognise the natural values of mangrove ecosystems. Indeed, mangroves and estuaries are probably the two most valuable coastal ecosystems in Kalimantan in terms of their benefits to human communities. The importance of mangrove as a resource derives both from its harvestable products, both traditional and commercial, and from the free services it provides, such as coastal protection and erosion control. There have been several comprehensive reviews of the value of Indonesian mangroves as a natural resource (Burbridge and Koesoebiono 1980; Soegiarto and Polunin 1980; Polunin 1983; Burbridge 1983).

Mangrove Fisheries

Because of their high productivity and physical structure (chapter 2), mangroves are an important habitat for many marine organisms. Mangrove swamps provide food, shelter and substrate for a number of species of commercial and subsistence importance, including: mangrove crabs *Scylla serrata*, prawns *Penaeus* species, the sergestid shrimps *Acetes*, used in *belachan* paste, and other crustaceans; many fish, including mullet, milkfish and barramundi; scallops, cockles, clams, mussels, oysters, and other bivalve and gastropod molluscs; and estuarine and marine reptiles such as monitor lizards and crocodiles. In addition, many coastal and offshore fisheries are dependent on mangroves serving as nursery and spawning grounds for many aquatic species, including milkfish and prawns, which are important in tambak culture. Organic material produced in mangroves is transported to open coastal waters by tidal currents, locally enriching pelagic and demersal fisheries (Saenger et al. 1983; Mann 1982). It is probable that most of the micro- and macro-fauna in the mangroves and surrounding coastal areas are dependent on the productivity of litter from mangrove forests (Ong et al. 1980).

At present the most valuable mangrove-associated species in Indonesia are penaeid prawns, which supported an export market worth more than $482 million in 1988 (Buku Statistik 1988). Prawn fisheries extend around the coast of Kalimantan, and in some places are already being overharvested (fig. 11.3). The juvenile stages for several prawn species live and feed among mangrove roots and adjacent vegetation while the adults breed offshore (fig. 11.4). The offshore production of penaeid prawns is closely correlated with the extent of mangrove forest, and any reduction in mangrove area leads to reduction in prawn production (Martosubroto and Naamin 1977). Indonesian mangroves are estimated to yield about 100 kg of prawns per year per hectare of mangrove forest (Turner 1977).

The influence of mangroves also extends far beyond the mangrove boundaries and subsidises offshore fisheries. An environmental impact assessment of a proposed development in the mangroves of Segara Anakan, the last major area of mangrove in south Java, suggested that clearing the mangroves would lead to loss of employment for 2,400 fishermen and an annual loss in income of $5.6 million at current prices, due to loss of mangrove input to offshore fisheries (Turner 1975). In 1978 550,000 tonnes of

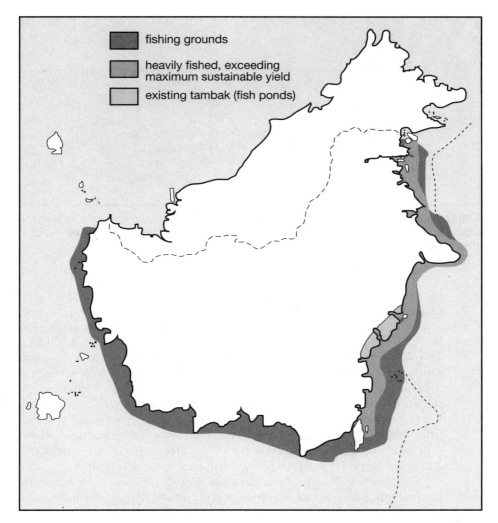

Figure 11.3. Prawn fisheries: penaeid shrimps.
Source: Salm and Halim 1984

harvested fish, approximately one-third of the marine catch and valued at $194 million, were of species directly linked to mangroves and estuaries during some stage of their life cycle (Salm 1981).

Mangroves for Commercial Timber and Chipwood

Mangrove habitats are increasingly threatened by conversion to other land

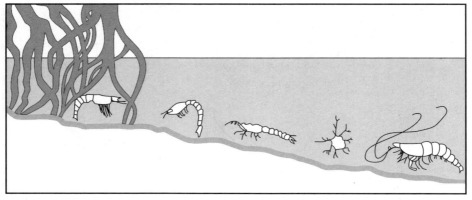

Figure 11.4. Adult prawns breed offshore but the larval stages move inshore with the tide. Young prawns feed and shelter for several weeks in the mangrove habitat until they are mature.

uses (table 11.4). All around Borneo's coasts mangrove forests are being exploited commercially for production of timber and chipwood. In Kalimantan 90% of all mangrove areas are allocated as forest concessions for felling (table 11.5). Note the inconsistencies between tables 11.4 and 11.5 for estimates of area of mangrove remaining but both sets of data show that much of the remaining mangrove is allocated as logging concessions.

Mangrove forests provide a variety of timbers of various qualities. As timber, mangrove-associated woods are often poor, but *Heritiera littoralis*, *Lumnitzera* and *Xylocarpus* give timber of good quality (Soegiarto and Polunin 1980). *Ceriops* and *Avicennia* yield highly durable wood, ideal for poles and pilings, and are used as building materials in coastal settlements. *Rhizophora* wood is preferred by local fishermen for boat-building.

Since the 1960s mangrove forests in Indonesia have been harvested increasingly as a source of chipwood and pulpwood. Most Kalimantan mangrove forests are now logged commercially for chipwood, but in some areas trees are harvested for pulp. *Excoecaria agallocha* yields good quality pulp, and *Camptostemon* and *Sonneratia* may also have potential as a source of pulpwood (Soegiarto and Polunin 1980). Elsewhere in southern Asia, species of *Heritiera* and *Rhizophora* are used as a source of viscose-rayon fibre.

Large-scale logging of Kalimantan mangroves began in 1972 when Chipdeco (Chip Development and Producer Company) obtained a concession of 85,000 ha on Tarakan Island, East Kalimantan. By 1980 Indonesia was exporting 300,000 m^3 of chipwood to Japan annually (Soegiarto and Polunin 1980), with one-third of this total provided by the Chipdeco mill in East Kalimantan (Burbridge and Koesoebiono 1980). Chipdeco has expanded its activities in East Kalimantan and holds extensive mangrove concessions around Tarakan and in the Sebuku delta to the north. On Tarakan Chipdeco cuts about 150 ha of mangrove per month, or 2,000 ha per year. Each cubic metre of wood provides 1.2 tons of chips, so for each 7,000-ton shipment, some 233,000 trees are needed (Jhamtani 1989). *Rhizophora* trees are the main species harvested for chips. Trees are felled in strips 50 m wide, and narrower belts of mature trees are left undisturbed between the harvested areas to provide seed for natural regeneration.

In theory such strip-cutting should be sustainable since *Rhizophora* begins fruiting at four years of age and generally regenerates well. The management of mangrove forest for sustained-yield harvesting of *Rhizophora* is possible on a 30-year rotation (Chan et al. 1982; Ong et al. 1985; Soegiarto 1985). The trees are usually harvested while they are still growing fast and before net primary productivity declines too much (fig. 11.5). Unfortunately, concession times in Kalimantan are usually shorter (20 years for Chipdeco), and the companies often log areas for a second time, a process more destructive than the first cut. Problems also arise when larger areas are cleared, and the sus-

Table 11.4. Mangrove distribution and land use.

Province	original area[1]	remaining area[1]	logging area	conservation area*	fisheries	allocated for expansion	*tambak* intensification
East Kalimantan	950,000	266,800	143,000	90,000	34,320	7,000	1,290
South Kalimantan	165,000	66,650	65,000	11,000	2,760	3,000	400
Central Kalimantan	20,000	10,000	10,000	7,000	–	–	–
West Kalimantan	425,000	40,000	39,500	17,000	10,000	2,000	–
Total	1,560,000	383,450	257,500	125,000*	47,080	12,000	1,690

[1] = mangrove and nipa
* areas of remaining mangrove within gazetted or approved reserve areas
– note that logging concessions often overlap areas proposed as reserves

Source: Jhamtani 1989

Table 11.5. Extent of mangrove habitat, logging concessions and maximum allowable cut in Kalimantan.

Province	Total area (ha)	Concession area (ha)	% habitat leased	max. annual allowable cut (m^3/ha)
East Kalimantan	150,000	143,000	95	21,000 - 126,000
South Kalimantan	75,000	67,500	90	77,000
Central Kalimantan	10,000			
West Kalimantan	40,000	39,500	97.5	19,000 - 22,000

Source: Burbridge and Koesoebiono 1980

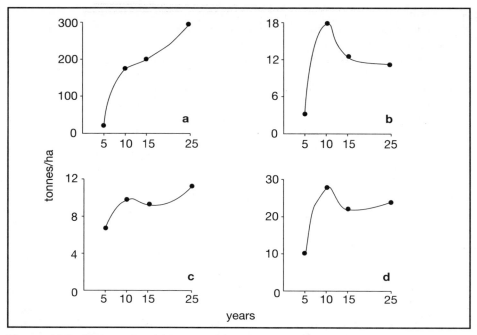

Figure 11.5. Changes in a) biomass, b) mean annual increment, c) litter and d) net productivity in four stands of *Rhizophora apiculata* in Peninsular Malaysia, harvested on a 30-year rotation. (After Ong et al. 1985.)

tainability of the woodchip industry must be questioned.

A survey of exploited mangrove areas in Sarawak in 1974 found less than 10% successful regeneration in the Rajang delta (Chai and Lai 1980). Similarly, in the Matang area of Peninsular Malaysia, 50% of the coupe must be replanted (Hamilton and Snedaker 1984). Inadequate natural regeneration has also been reported from Sabah and other parts of Indonesia. In areas where commercial species are clear felled, succession is sometimes dominated by less valuable species such as *Avicennia officinalis* (Soegiarto and Polunin 1980). Table 11.6 shows the average number of mangrove seedlings per hectare associated with different cutting times and levels of degradation. Successful natural regeneration depends on leaving an adequate number of seed trees within the harvested areas (Kent 1986).

In the Sebuku delta, where logged areas are interspersed with large stands of undisturbed mangroves, there is good natural regeneration of *Rhizophora*. Similarly, at Pulau Sinualan, regeneration was successful in a series of trial plots using a strip clear-cutting system, when a minimum of 70 seed trees were left per hectare; here regeneration was dominated by *Rhi-*

zophora apiculata, followed by *Bruguiera gymnorrhiza* (Riyanto and Tobing 1979). Regeneration probably depends on soil quality as well as extent of clearance. In areas of poor soils secondary growth consists mainly of the sea holly *Acanthus ilicifolius* and mangrove fern *Acrostichum aureum*, species which choke out young *Rhizophora* seedlings. This thick *Acrostichum* undergrowth must be cleared before the area can be recolonised with mangrove species.

Regeneration of mangrove will also be affected by land management outside the mangrove. Changes in seasonal water flow in rivers, and rates

Table 11.6. Average number of seedlings per hectare associated with different cutting times.

Site or crop condition	*Rhizophora* (seedlings/ha)	*Bruguiera parviflora* (seedlings/ha)	Locality
Six months before final felling	18,563	8,215	Matang Mangrove Reserve, Malaysia
Six months after final felling	8,579	1,205	Matang Mangrove Reserve, Malaysia
Twelve months after final felling	9,145	3,164	Matang Mangrove Reserve, Malaysia
Twenty-four months after final felling	8,342	150	Matang Mangrove Reserve, Malaysia
After second thinning	26,510	4,226	Matang Mangrove Reserve, Malaysia
After third thinning	18,977	1,110	Matang Mangrove Reserve, Malaysia
After second and third thinning in *Acrostichum* infested area	6,066	2,630	Matang Mangrove Reserve, Malaysia
Clear-felled *Acrostichum* infested area	1,770	222	Matang Mangrove Reserve, Malaysia
Clear-felled *Acrostichum* infested area	444	0	Matang Mangrove Reserve, Malaysia
One year after final felling	249	667	Rajang Mangrove Reserve, Sarawak
Three years after final felling	333	447	Rajang Mangrove Reserve, Sarawak
Six years after final felling	2,201	442	Rajang Mangrove Reserve, Sarawak
Two months after logging in *Bruguiera* forest		6,148	Sabah
Three months after logging in *Rhizophora* forest	8,699		Sabah
Mature *R. apiculata*	*R. apiculata* seedlings only 25,327		Rajang Mangrove Reserve, Sarawak

Source: Hamilton and Snedaker 1984

of accretion in coastal areas, will affect mangrove habitats. There appears to be poor regeneration of mangrove stands near major rivers whose watersheds are being developed. Conversely, natural regeneration is good, even under heavy harvesting, in coastal bays with small rivers or undeveloped watersheds (Burbridge and Koesoebiono 1980). Given the importance of maintaining sufficient mature trees as a seed source, it would be sound forestry practice for concessionaires to retain part of the concession as a conservation area. The undisturbed mangrove stand would provide a seed source for natural regeneration as well as a research site for monitoring regeneration and growth cycles. Such conservation areas should be established in the rich mangrove stands allocated as production forests in the Sebuku and Sebakis deltas of East Kalimantan.

Other Mangrove Products

Many coastal communities in Kalimantan harvest mangrove products for domestic use and local sale. Commercial and traditional products from mangroves range from construction materials, firewood, materials for fish traps, nipa thatch, charcoal and medicines to fish, sweetmeats and honey. Boxes 11.4 and 11.5 list a wide range of harvestable mangrove products and their uses.

Mangrove forests are an important source of firewood. More than half of all domestic energy needs in Indonesia are satisfied by fuelwood. In coastal areas this fuelwood comes mainly from mangrove forests. The quality of fuelwood varies according to tree species. *Rhizophora, Bruguiera* and *Ceriops* provide good quality firewoods, whereas *Avicennia* and *Sonneratia* are generally poor as fuel (Soegiarto and Polunin 1980). Mangroves are also a major source of charcoal. Unlike in Sumatra, where large quantities of charcoal are produced for export, charcoal production in Kalimantan is mainly for local use (Soegiarto and Polunin 1980). *Rhizophora, Bruguiera* and *Xylocarpus* are felled for charcoal production. Traditional forms of charcoal production, using a harvesting cycle of 35 years, allow regeneration and seem to be sustainable. However, increasing demand for pulpwood and chipwood are placing greater demands on areas traditionally harvested for charcoal.

Another mangrove product of considerable value is tannin. Mangrove bark contains 20%-30% tannin. Its extracts are used extensively in the leather industry, in medical treatments and in the wine and beer industry. Annually 10,000 tonnes of mangrove bark are used in tannin production in Indonesia (Soegiarto and Polunin 1980). Tannins are also used locally as a preservative for fishing nets. In the Philippines the sap of *Excoecaria agallocha*, a species also found in Kalimantan, is used to deaden nerves in dental cavities (Maltby 1986).

Nipa swamp *Nypa fruticans* often grows inland of the coastal mangrove

Box 11.4. Products of mangrove ecosystems.

Fuel
firewood for cooking, heating
charcoal
alcohol

Construction materials
timber, scaffolds
heavy construction timbers
railroad ties
mining pit props
boatbuilding materials
dock pilings
beams and poles for buildings
flooring, panels, clapboard
thatch and matting
fence posts, water pipes,
  chipboards, glues

Fishing equipment
poles for fish traps
fishing floats
fuel for smoking fish
fish poison
tannins for preserving nets and lines
wood for fish-drying or fish-smoking
  racks

Textiles and leather
synthetic fibres (e.g., rayon)
dyes for cloth
tannins for leather

Agriculture
fodder, green manure

Other natural products
fish   birds
crustaceans   mammals
shellfish   reptiles and reptile skins
honey   other fauna (amphibians, insects)
wax

Food, drugs and beverages
sugar
alcohol
cooking oil
vinegar
tea substitute
fermented drinks
dessert topping
condiments from bark
sweetmeats from propagules
vegetables from propagules
fruit or leaves
cigar substitute

Household items
furniture
glue
hairdressing oil
tool handles
rice mortars and pestles
toys
matchsticks
incense, mosquito repellent

Paper products
paper of various kinds

Other products
packing boxes
wood for smoking sheet rubber
wood for firing bricks
medicines from bark, leaves and
  fruits
cigarette wrappers from leaves

Source: Saenger et al. 1983

Box 11.5. Commercial and traditional uses of some Bornean mangrove species.

Species	Uses
Acanthus ebracteatus and *A. ilicifolius*	Fruit pulp used as blood purifier and dressing for boils and snake bites; leaf preparation for relief of rheumatism; leaf juice as hair preserver.
Acrostichum aureum	Litter for cattle; roof thatch; young fronds edible.
Aegiceras corniculatum	Poor firewood, paper, bark as fish-poison.
Avicennia spp.	Inferior firewood, rough walling, fuel for brickmaking, charcoal; cure for thrush; resin and ointment from seed applied to ulcers and tumours; bark used for treatment of skin parasites and gangrenous wounds.
A. alba	Bark and seed contain fish-poison, resinous exudate used in birth-control; seed ointment relieves smallpox ulcers.
A. marina	Young leaves used as vegetable; honey; soap.
A. officinalis	Seed and seedlings eaten; honey; charcoal; salt from wood ash.
Bruguiera spp.	Bark for tannin, dyes; poles, charcoal, firewood, building materials, chipboard, fishtraps.
B. cylindrica	Firewood and timber.
B. gymnorrhiza and *B. sexangula*	Fishing stakes, firewood, charcoal, telegraph poles; radicles as food; eye medicine from fruit; incense from knee-roots, condiment from bark; adhesive from bark; fruit chewed as betel substitute.
B. parviflora	Firewood, charcoal, tannins, paper.
Ceriops tagal	Timber, firewood, fuel for brick-making, charcoal, tannins, good dyes, paper; bark decoction said to stop haemorrhage; bark yields adhesive and net preservatives; dyes for batik and mat-making.
Cerbera manghas	Rubbing with fruit eases rheumatism; seeds contain medicinal oil; bark and sap contain a purgative.
Derris heterophylla	Weak fish-poison.
Excoecaria agallocha	Timber, flooring, panels, paper, pulp; sap and wood contain purgative; sap yields fish-poison; match sticks, boxes, firewood; incense, toys, honey.
Heritiera littoralis	Boat-building, timber, dock pilings, beams, chipboard, firewood; ground seeds cure diarrhoea; sap poisonous; glues, tannins.
Lumnitzera spp.	Timber, poles, heavy construction wood, chipboards, firewood; medicines; decoction of leaves for thrush.
Nypa fruticans	Leaves for thatch, mats; young leaves for cigarette wrappers; sap for sugar, alcohol and vinegar.
Oncosperma tigillaria	Piles, house posts, flooring, fish stakes; flowers added to rice as seasoning; fleshy fruit preserved; terminal buds for vegetable.

> **Box 11.5.** Commercial and traditional uses of some Bornean mangrove species *(continued).*
>
> | | |
> |---|---|
> | *Rhizophora* spp. | Timber, fishing stakes, piles, pit props, firewood, charcoal, tannin; decoction of *R. mucronata* bark used for haematuria, diarrhoea, dysentery, leprosy; prop-root bark and fruit sap as mosquito repellent; wine from fruit; honey from nectar. |
> | *Scyphyphora hydrophyllacea* | Fence posts, tool handles, firewood, charcoal, glues. |
> | *Sonneratia* spp. | Planking, boxes, wall boards, heavy construction material, firewood, hats. |
> | *S. alba* | Timber and firewood; pneumatophores as fishing floats; leaves for cattle food. |
> | *S. caseolaris* | Fruit eaten; sap as skin cosmetic; leaves as fodder for goats; pulp. |
> | *S. ovata* | Fruit edible, also used for sprain poultices; fermented juice said to check haemorrhage. |
> | *Xylocarpus* spp. | Timber (*X. moluccensis* highly prized), poor firewood; furniture, tannin, glues, paper; oil from seed for illuminant and hair; bark decoction for cholera; pencils, carvings, tool handles. |
>
> Sources: Soegiarto and Polunin 1980; Hamilton and Snedaker 1984

fringe (chapter 2). Nipa leaves are used for roof thatch, wrappings and cigarette papers. The nipa palm also provides fodder, fuel, alcohol, vinegar and sugar. Factories in Sabah and Sarawak produce sugar and alcohol from nipa palm sap (Hamilton and Snedaker 1984; Fong 1988). Wild nipa swamp can yield sugar at three tonnes/ha (Morton 1976).

Mangrove Services

While the value of marketable products of mangrove forests can be expressed in terms of rupiahs or dollars, the value of the "free" services provided by mangroves is more difficult to quantify and is often overlooked. Detritus from the mangrove ecosystem is the base of the food chain on which many coastal and inshore fisheries depend. The tangled mangrove roots trap soil particles and silt, the products of inland erosion, which would clog and suffocate offshore coral reefs and reef flats. New mangrove trees colonise the trapped silt and build outwards, creating new land on the shallow coastal shelf. This coastal mangrove/nipa fringe acts as a buffer against storm damage and wave action, and protects stabilised muds and river banks from the effects of tidal currents. It is noteworthy that damage suffered to life and property in the cyclone-ravaged Bay of Bengal in India is consistently less in the forested mangrove areas inland from the Ganges-Brahmaputra delta than in areas where mangroves have been cleared (Hamilton and Snedaker 1984). Moreover, as a living ecosystem, mangroves are self-maintaining and renewable. After sustaining storm damage, mangroves will regenerate and repair themselves at no cost, whereas a damaged sea wall requires expensive repair work. Once the mangrove is destroyed, replacing such free services as coastal protection and waste filtration requires considerable energy, technology and finance.

Mangrove resources are renewable only if the ecological processes governing the system are maintained. The health and natural regeneration of mangrove forests depend on three main factors: an adequate amount of water and the correct balance between fresh and salt water; an adequate supply of nutrients; and a stable substrate. Modification of one or more of these critical factors may severely impair or eliminate the renewability of the resource (Hamilton and Snedaker 1984). Yet all of these factors can be affected by events outside the mangrove habitat itself. The freshwater supply and its nutrients, for instance, may be altered by deforestation of the watershed, construction of major upstream water-storage areas, or river diversions.

Mangroves are further threatened when the harvest of direct products exceeds the sustainable-yield limit. In Kalimantan mangroves are particularly threatened by large-scale commercial wood chip operations, which far exceed other uses. Conversion activities such as agriculture, fish ponds, and residential and industrial developments are also increasing. These activities irreversibly alter the condition of the intertidal zone, and the three factors critical for regeneration are so modified that the renewability of the resource may be lost (Hamilton and Snedaker 1984).

Mangrove Management

Mangrove areas may be managed for their natural productivity, or they may be converted to other forms of land use such as aquaculture, agriculture, salt production, industrial and housing schemes, and port facilities. Throughout Kalimantan many mangrove areas have been cleared to construct brackish-water fish farms (tambak) or reclaimed for growing rice and other crops. Figure 11.6 shows the impact of local settlement on the mangrove swamps of the Mahakam delta, East Kalimantan.

From an ecological viewpoint, all plans to alter the mangrove resource should take into account the long-term value of the various management options. Even if mangrove forests are carefully managed for timber extraction, with artificial propagation and rotation, there will still be heavy costs

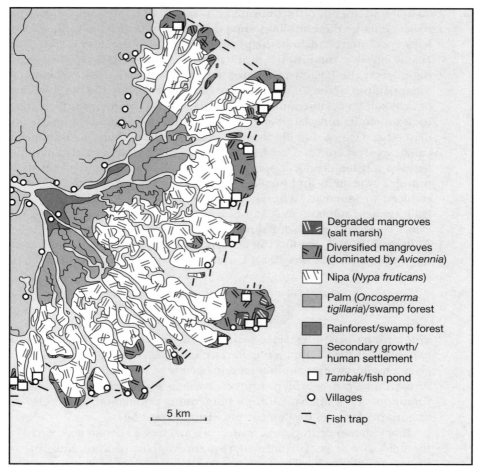

Figure 11.6. Sketch map of the Mahakam delta, East Kalimantan, showing the impact of local settlements on the natural vegetation. (After Eve and Guigue 1989.)

in non-timber benefits. Conversion to tambak may also have hidden costs. It has been estimated that the average coastal fish pond in Sumatra produces 287 kg of fish/ha/year but that the loss of one hectare of mangrove to aquaculture leads to a net loss of offshore shrimp and fish of approximately 480 kg/ha/year (Whitten et al. 1987a). Moreover, it must be remembered that single-purpose exploitation, whether for timber or tambak, effectively forecloses all other utilisation options.

Mangrove areas are generally not suitable for agriculture. Nevertheless, some coastal mangroves have been cleared for crop growing around coastal settlements, or by transmigrants settled in tidal swamplands, who seek to enlarge their holdings when the rice crop begins to decline. Problems encountered include the persistent activity of the mud lobster *Thalassina anomala*, and invasion by the mangrove fern *Acrostichum aureum*. In Kalimantan the biggest problem associated with mangrove reclamation for agriculture is the high failure rate of schemes due to the development of acid sulphate soil conditions (see chapter 10). Many mangrove areas have soil containing large amounts of sulphur as pyrite. When soils are drained and exposed to the air, oxidation releases sulphuric acid. Drainage canals exacerbate the problem. The soils become extremely acid and develop very high concentrations of soluble salts; this leads to problems in nutrient availability and uptake of fertilisers, resulting in crop failure or reduced productivity. The joint Dutch-Indonesian LAWOO project is studying acid sulphate soils and how to improve crop production in the Pulau Petak area in South Kalimantan (Lembaga Penelitian Tanah 1973; Andriesse 1974). Coconuts can be grown with some success in these areas on raised bunds (see box 10.8). Liming helps to improve crop yields but is expensive.

Conversion to Tambak

Fish farming may offer more benefits than agriculture. It has been estimated, for instance, that a one-hectare milkfish pond in the Philippines gives a higher profit margin than one hectare of rice field (Blanco 1972). Aquaculture can provide a source of cheap protein for domestic consumption, and employment for a burgeoning population, and is also an important earner of foreign currency from export sales.

Brackish-water fish ponds, *tambak*, are manmade ecosystems created in the tidal zone, especially in mangrove forests. Here prawns and fish are reared for sale for local and overseas markets; most of Indonesian prawn exports, for instance, are harvested from tambak. Indonesia has the potential to establish more than 800,000 ha of tambak for prawn and fish breeding (Maltby 1986). Most tambak production is concentrated along the coasts of north Java, north Sumatra and south Sulawesi. By 1989 Kalimantan had only 11,259 ha of tambak, concentrated in East and South Kalimantan (fig. 11.3). East Kalimantan has the most extensive areas of tambak, (9,251 ha) to the south of the Mahakam delta. Fish and prawn production from the Kalimantan tambak was valued at Rp 3,776 million ($220,000) in 1986 (Buku Statistik 1988). Although the area of tambak production is very small at present, a further 74,000 ha of Kalimantan mangrove is believed to have tambak potential (Directorate General of Fisheries 1987). Conversion of mangrove to tambak for prawn production is continuing in East Kalimantan and may prove to be a threat to the rich east coast mangroves.

The historical origins of tambak are uncertain, but such ponds probably existed in Java prior to 1400 (Schuster 1950). Unlike agriculture on land, tambak production is almost entirely dependent on wild populations, with ponds receiving larvae on the tide or being stocked with wild-caught larvae (Polunin 1983). Milkfish fry for tambak are collected along the east coast of Kalimantan in April-May and November-December (Salm and Halim 1984). Traditionally milkfish have been the major harvest from tambak, providing 55% of the total Indonesian production. Increasingly, however, attention is turning to rearing penaeid prawns, especially the tiger prawn *Penaeus monodon*, and common tilapia *Tilapia mossambica*. Other cultivated species include banana prawns, mullets, barramundi and mangrove crabs. Tambak production accounts for 6% of total Indonesian fish production by weight and 12% by value (Knox and Miyabara 1984), but Kalimantan's contribution to this harvest is very minor since the area of tambak is small, and mostly managed with traditional technology.

Tambak are built up from low coastal mud flats by excavation and the construction of protective dykes and channels. The ponds are created in swamplands from just below the mangrove fringe to the inland limit of tidal influence. Mangrove trees are sometimes planted along the dykes to provide shade, some protection from erosion, and detrital material for the pond ecosystem (Schuster 1950).

The control of water into and out of the tambak is the key to successful production. The water must be changed regularly to remove wastes and to sweep in new food sources and fish and prawn fry. Most Kalimantan tambak rely on natural stocking with larvae swept in by the high tide. Water flow is controlled by sluice gates in the pond walls. The bottom of the pond is about 0.4 m below the mean level of the spring high tides, and just above the mean low tide level, so that the pond can be drained to destroy predators, parasites and disease carriers. The pond bottom is then fertilised, and the pond is filled. Inputs of fresh water, sea water and rain create a wide range of salinity regimes in ponds. River sediments and rain introduce nutrients.

When the sluice gates are opened at high tide, a host of marine organisms are swept into the ponds, including the fry of many harvestable species such as milkfish, mullet *Mugil*, snapper *Lutjanus*, eels *Monopterus albus*, penaeid prawns and mangrove crabs *Scylla serrata*. More than forty species of fish have been found in Javanese tambak and their environs (Schuster and Djajadiredja 1952). Apart from these naturally occurring fry, common tilapia are sometimes introduced into tambak, though they are less profitable and more difficult to manage than milkfish. After three to five months, depending on the food supply and growth of fish, the pond is harvested. Milkfish yields range from about 150 kg/ha (0.15 tonnes/ha) to 2 tonnes/ha (Soegiarto and Polunin 1980). High yields require good quality stock, good water quality and high nutrient levels, achieved by adding a top-dressing of nitrogen and phosphate fertiliser. Prawns require supplementary feeding.

The tambak habitat and the surrounding area provide an environment for many species of birds, including migratory shorebirds and mangrove and coastal species. Storks and white egrets return to the mangroves to roost at night. During the day they feed on inland rice fields, where they play a useful role, preying on insect pests. Some of the large herons and storks also prey on the tambak fish, but other shorebirds feed on mud-dwelling invertebrates of no economic significance.

Tambak production can be increased by opening up additional land or water areas for ponds, or by intensification to increase the yield from existing tambak. Intensification is thought to have the lesser potential for ecological damage. As larger areas of mangrove are cleared, there is growing evidence that tambak production falls, a self-perpetuating process that eventually leads to lower harvests than in the original mangrove situation (Whitten et al. 1987a). Tambak production is also sensitive to water pollution, sedimentation, oil pollution and contamination from pesticides. The present harvest levels of 0.1 to 2.0 tonnes/ha are low when compared to the 10 to 12 tonnes/ha attained in Taiwan (Maltby 1986). Measures to increase tambak production include: improving the supply of fish fry and postlarval prawns; better control of fertiliser and pesticide use; proper pond construction and maintenance; proper canal design; flood control; pollution control; better marketing; and the avoidance of acid sulphate soil problems (Knox and Miyabara 1984). The use of floating net cages as fish-rearing pens is also becoming more common in Indonesian coastal wetlands but may create hazards for other wildlife.

As with conversion of mangroves to agriculture, the construction of fish ponds in mangrove swamps can result in problems from acid sulphate soils. When the pond is drained and allowed to dry out, these soils are exposed to air, and the pyrite they contain is oxidised to create highly acid conditions. Once the pond is refilled this can lead to: poor fertiliser response because phosphate is bound to the ferric and aluminium ions and therefore is unavailable to the nutrient cycle; low natural food production; and slow fish growth or fish deaths from acidity (Dent 1986). Further losses of stock may be caused by heavy rains which wash acids and soluble aluminium into the ponds from the banks (Dent 1986). The low pH and the toxicity of iron, aluminium and manganese inhibit the reproduction and growth of fish and other pond life, and may be the cause of the generally low productivity achieved in many brackish-water ponds (Singh 1980).

Acid sulphate soil conditions can be improved by liming, leaching and appropriate fertilisation. In the Pulau Petak delta, South Kalimantan, farmers excavate a pond and use the topsoil to build a dyke. Lime is then applied liberally to the pond, which is flushed with water several times to clear the acid (Konsten pers. com.). Nevertheless, fish crops may remain poor for up to ten years after the construction or deepening of ponds in acid sulphate soils (Dent 1986). Moreover, the acid waters created by such conversions may seep into adjacent nonacid soils and restrict crop growth, and may also affect animal life in offshore waters.

Logging for chipwood, and the conversion of mangroves to other uses, especially to tambak, are the two greatest threats facing Indonesian mangroves (Burbridge 1983). Often tambak projects are proposed and implemented without regard for suitability of soils or for the ecological consequences of the loss of the mangroves. It is paradoxical that in some cases mangroves, which provide stocks of milkfish fry and postlarval shrimp for traditionally operated tambak, are being converted to new tambak, thus reducing natural stocking to all tambak and requiring greater levels of intensive management to maintain high tambak fishery production levels (Polunin 1983; Collier et al. 1979). In an attempt to simulate a more natural system, the Forest Research Institute has developed a tambak-forest system, the *tambak tumpang sari* system. This multiple land use involves planting *Rhizophora* on a rectangular mound, surrounded by a ditch 4-5 m wide where fish can be cultured. This system is not very appropriate for milkfish, which feed on algae rather than detrital material and prefer unshaded waters (Achmad 1986). Nevertheless, it could be highly productive for other tambak fish and prawns, and has the added advantage of encouraging mangrove regeneration, thus preserving some of the natural benefits of mangroves. A similar system developed in Vietnam may be appropriate for Kalimantan. *Melaleuca* trees are planted on mounds and dykes in the tidal swamplands to provide firewood and honey from their flowers (IUCN 1985).

Before advocating clearance of mangrove swamps and other tidal wetlands for tambak or agriculture, it is worth considering the following facts. The average yearly animal protein production in swamps and marshes is 9 g/m^2, which is over three times the average for natural terrestrial ecosystems (Maltby 1986). Estuaries, which often depend on mangrove inputs, are twice as productive again as swamps and marshes (Turner 1980). The protein yield from fish dependent on mangroves is greater than that derived from the cultivation of crops, including rice, in the converted wetland. Even the high yields of various types of aquaculture may not be as high as those of the area of wetland that the fish ponds replace. The development of aquaculture exacts both environmental and ecological costs. Natural, highly productive fisheries, as well as the other support systems of wetlands, are provided without artificial energy inputs and at no environmental cost. Usually the most rational and valuable way to utilise mangroves is to leave them as natural wetlands and to harvest their products at a sustainable level.

Oil Spills and Coastal Ecosystems

Oil deposits are abundant in the coastal plains and shallow seas of Indonesia. Borneo has several major oil and gas basins; some offshore fields are already being exploited (chapter 13). The main oilfields in Kalimantan lie off the coast of East Kalimantan, with exploitation centred on Balikpapan and Tarakan for oil, and Bontang and Muara Badak for natural gas. Oilfields are also being exploited in Sarawak, at Miri and in Brunei.

By June 1989 daily production of crude oil in Indonesia was 1.3 million barrels (*Jakarta Post*, 16 June 1989). The Kalimantan fields provide about one-tenth of this production. Oil exploration and exploitation are complex and highly technological procedures. Mistakes, accidents and system failures have high environmental costs. Intertidal habitats such as mangroves, mudflats and reef flats, and species that span the land/sea interface (oysters, crabs, turtles, dugongs, sea turtles and seabirds) are especially vulnerable to fouling, poisoning and death from oil spills (Baker 1983). In exploited fields there is always the hazard of accidental blowouts of oil wells. The shipping lanes through the Macassar straits, through which the oil is transported, are studded with reefs and islands, with the consequent danger of shipwreck and oil spills. Oil drifting into mangrove areas clogs the lenticels on exposed roots, thus reducing the trees' intake of oxygen; in trees that survive, chronic stress is expressed in reduced productivity and leaf loss (Saenger et al. 1983). Large areas of coastal forests in Brunei are already badly damaged as a result of oil exploration and development activities.

Oil and natural gas deposits are valuable assets which will certainly be developed in Kalimantan. It is essential, however, that exploitation should incorporate all possible environmental safeguards to minimise damage to other coastal sectors. Not only do accidental oil spills threaten adjacent coastal environments, but also oil is carried on the currents to areas far removed from the original spill. These affected areas may include wetland nursery areas for commercial fisheries, turtle nesting beaches and even tourist beaches. Cleanup operations are neither cheap nor easy, as shown by the oil spill caused by the shipwreck of the tanker Exxon Valdez off Alaska (*New Scientist*, June 1989). The environmental consequences are long lasting; indeed, natural ecosystems are often damaged beyond repair.

Development and Coastal Resources

Conflicting demands for land use will continue to put increasing pressure on coastal habitats and resources (table 11.7). The development of the coastal zone is complex and can have effects far beyond the narrow coastal strip and beach and the delta/sea interface. Changing land use inland, or developing any part of the various coastal ecological zones such as mangrove or the beach-lagoon-estuary complex, can affect all other zones (Knox and Miyabara 1984; Burbridge and Maragos 1985). In Kalimantan the main issue for concern is the conversion of tidal swamp forests, including mangroves, which provide water purification systems, coastal

Table 11.7. Matrix relating important resources and environmental decision categories to key problems in coastal zones.

Decision category	a	b	c	d	e	f	g	h	i	j	k	l	m	n
Forest use allocation							*	*				*	*	
Method of forest exploitation	*	*	*	*	*		*	*	*	*			*	*
Land clearance timing/method	*	*	*	*			*	*	*	*			*	*
Crop varieties/patterns/methods	*	*	*	*				*	*	*			*	
Irrigation methods	*	*	*	*				*	*	*				*
Agricultural intensification	*		*		*	*			*					*
Animal husbandry				*				*						*
Fishing rights allocation					*							*	*	
Choice of fishing gear					*									*
Fish ponds siting/construction	*	*		*	*								*	*
Aquaculture intensification	*		*	*	*									*
Canal location		*	*								*	*		
Canal size	*	*	*	*									*	*
Field-level water control				*		*		*	*	*		*		
River alteration	*	*	*	*	*					*			*	*
Water pollution					*	*				*				*
Coral reef mining					*								*	*
Coral reef dynamiting					*								*	*
Population density	*			*	*	*	*	*			*			*
Settlement site	*	*	*	*							*	*		*
Drinking water supply				*	*									
Waste disposal					*	*								*
Vector control									*	*				*
Industrial sites	*	*								*	*	*		*
Toxic wastes	*				*	*								*
Shipping canals/channels		*	*		*	*								*
Port location	*	*			*						*			*
Road location	*	*									*	*	*	*
Tourist development		*										*	*	*
Mineral exploitation	*	*			*	*							*	*
Petroleum exploitation			*		*	*								*
Reserve designation	*	*	*	*	*	*	*	*						
Soil conservation	*	*						*	*					

a - Soil toxification/fertility decline
b - Land instability/river shallowing
c - Low water quality/salt intrusion
d - Catastrophic flood/drought
e - Fisheries decline
f - Brackish-water pond production decline
g - Forest product shortage
h - Vegetation succession/competition
i - Agricultural pest outbreaks
j - Disease and vector build-up
k - Stable infrastructure and environmental impacts
l - Multiple allocation of same resource
m - Overutilisation of resource base
n - Insidious environmental degradation

Source: Knox and Miyabara 1984

Box 11.6. Pollutants and their effects on coastal ecosystems.

Herbicides
- may interfere with basic food chain processes by destroying or damaging free-living phytoplankton or algal or sea grass plant communities and zooxanthellae in coral;
- can have serious effects, even at very low concentrations.

Pesticides
- may selectively destroy or damage elements of zooplankton or aquatic invertebrates; planktonic larvae are particularly vulnerable;
- may accumulate in animal tissues and affect physiological processes.

Antifouling paints and agents
- may selectively destroy or damage elements of zooplankton or poison aquatic organisms;
- only a major factor near major harbours, shipping lanes and industrial plants cooled by seawater.

Sediments
- smother substrate;
- smother and exceed the clearing capacity of some filter-feeding animals;
- reduce light penetration, which may alter vertical distribution of plants and animals in coastal waters;
- may absorb and transport other pollutants.

Sewage and detergents
- may interfere with physiological processes.

Sewage, nutrients and fertilisers
- may stimulate phytoplankton and other plant productivity beyond the capacity of control by grazing aquatic animals and thus modify the community structure;
- may cause eutrophication and consequent death of estuarine and reef organisms.

Petroleum hydrocarbons
- have been demonstrated to have a wide range of potential damaging effects at different concentrations.

Heated water from power stations and industrial plant cooling
- changes local ecological conditions; water temperature is a key factor in distribution and physiological performance of many organisms.

Hypersaline waste water from desalinisation plants
- changes local ecological conditions; salinity is a key factor in distribution and physiological performance of many aquatic organisms.

Heavy metals (e.g., mercury)
- may be accumulated by, and have severe physiological effects on, filter-feeding animals and fish in estuaries, coastal waters and reefs;
- may accumulate in higher predators.

Radioactive wastes
- may have long-term and largely unpredictable effects on the genetic nature of the biological community.

(Adapted from Salm and Clark 1984.)

Box 11.7. Coastal ecosystems in Kalimantan.

Category	Positive values, functions and uses	Detrimental uses and practices	Adverse environmental consequences
Agroecosystems	food production, livestock production, timber products, fuel, fish production.	short cycle slash-and-burn (swidden) agriculture, overgrazing by livestock, overexploitation/clear cutting of forests, inappropriate application of fertilisers, pesticides and herbicides, deforestation.	soil erosion, loss of soil fertility, sedimentation downstream, pollution and contamination within ecosystem and downstream, reduced production within ecosystem and downstream, infilling of wetlands affecting estuaries, seagrass, reefs.
Tambak	increased fishery production, increased income, increased living standards, increased per capita protein consumption.	unnecessary conversion of mangroves and other valuable coastal systems, water pollution, poor siting, excessive pesticide use.	loss of fishery resources, reduced prawn and milk fish fry (and natural stocking), reduced marketability.
Freshwater ecosystems (including lakes, marshes, rivers, wetlands, streams)	natural flood control and storage, water supply and recharge, nutrient and sediment sinks, waterbird habitat, food production, building and energy materials.	conversion to dry land, conversion to wet agriculture, sedimentation from soil loss, irrigation withdrawal, flood control channels.	increased flooding, degradation of habitat, infilling and water pollution downstream, infestation by aquatic weeds, reduced fishery yields, reduced water quality.
Beaches (including associated beach forests)	nesting habitat for birds and sea turtles, fishing habitat (recreational and subsistence), timber and fuel, shoreline protection, recreation, tourism.	coastal structures aggravating beach erosion, exploitation of eggs and adult turtles, excessive beach and reef mining, conversion of beach forest and dunes to urban use, oil pollution.	decline in sea turtles, loss of beaches, damage to coastal structures and buildings, reduced fishery values, water pollution from eroded beaches, loss of habitat, timber and fuel, increased coastal erosion, degraded tourism value.

Box 11.7. Coastal ecosystems in Kalimantan *(continued)*.

Category	Positive values, functions and uses	Detrimental uses and practices	Adverse environmental consequences
Estuaries (including associated mud flats and embayments)	nutrient influx to coastal waters, fisheries production, nursery and spawning areas for many coastal fish, links to mangroves, seagrasses, pelagic and demersal fisheries.	urban pollution (sewage, thermal), industrial pollution, hydrology modifications (upland irrigation and water withdrawal), conversion to tambak and dry land, overexploitation of resources.	reduced fishery production, reduced habitat for adults and fry of fishery species, infilling and sedimentation, reduced estuarine habitat, degradation of water quality.
Tidal swamp forests	habitat for fish, wildlife and plants, flood storage, links to mangroves, timber and fuel, links to rice culture, fisheries production.	excessive logging, conversion to tambak, conversion to dry land, reclamation and irrigation, destruction of mangrove buffer, transmigration sites.	reduced fishery yields, reduced timber and fuel, degradation of habitat, loss of wildlife and plants, reduced rice yields, disruption of hydrology, acid soils, loss of peat.
Mangroves	sediment filter, nutrient filter, fishery resources (fin and shellfish), net transfer of production to coastal fisheries, breeding and spawning grounds for many coastal species, nursery ground for coastal and estuarine species, links to seagrass and coral reefs, shoreline protection, buffer for tidal swamps, timber, fuel, tannin and other chemicals.	conversion to dry land, excessive upland soil erosion, overexploitation of wood, overexploitation of fishery resources, upland irrigation and water withdrawal, transmigration sites, conversion to tambak, oil pollution.	degraded coastal water quality, loss of most values, functions and uses, loss and degradation of habitat due to sediment infilling, reduced fishery production, reduced fry production, reduced nursery habitat, secondary impacts to reefs, seagrasses, swamps.
Seagrass beds	nutrient filter, net transfer of production to coastal fisheries, feeding habitat for green turtles and dugongs, nursery grounds for coastal fisheries, links to mangroves and coral reefs, fishery production, especially finfish.	coastal urban pollution --thermal and domestic sewage, industrial pollution, coral mining, excessive upland soil erosion, overexploitation of fisheries, inappropriate coastal development, construction and dredging, oil pollution.	degradation of habitat, loss of habitat due to infilling, loss of habitat due to changed hydrology, displacement of seagrasses, reduced fishery production, loss of fry and breeding habitat.

> **Box 11.7.** Coastal ecosystems in Kalimantan *(continued).*
>
> | Category | Positive values, functions and uses | Detrimental uses and practices | Adverse environmental consequences |
> |---|---|---|---|
> | Coral reefs | links to seagrass, mangroves, beaches, and coral islands; shoreline protection, beach sand replenishment, high productivity shellfish production, finfish production, spawning grounds for fish, nursery grounds for fish, mariculture, tourism and recreation, ornamental species (shells, corals, fish, etc.) seaweed harvesting. | excessive coral mining, sedimentation dredging and filling, overexploitation, industrial discharges, fishing with poisons and explosives, urban pollution, oil pollution, water pollution. | coastal erosion loss of nursery and spawning habitat for fish, destruction from sedimentation, filling and dredging, degradation from water pollution, destruction of habitat, reduced tourism value. |
> | Demersal ecosystems | high productivity in upwelling areas and coastal areas, high prawn and finfish production. | overexploitation, excessive coastal pollution, oil pollution, inappropriate fishing techniques. | loss of fishery habitat, reduced fishery production, degraded and tainted catch. |
> | Pelagic ecosystems | high productivity in upwelling areas, high yield migratory species. | overexploitation, oil pollution. | reduced fishery production, degraded and tainted catch. |
>
> Source: Burbridge and Maragos 1985

protection, and important inputs to estuarine and offshore food chains and fisheries production.

Practices detrimental to mangroves and other coastal forests include: conversion to dry land and tambak areas, especially as components of fisheries and transmigration programmes; excessive harvesting of mangrove wood for fuel, poles, timber, and wood chip production; water diversion projects that can reduce freshwater and nutrient input; blockage or reduction of tidal circulation; sedimentation from upland sources, especially deforested watersheds; and pollution from domestic, petrochemical and other industrial spills and discharges.

Since mangrove and other coastal ecosystems depend on input of nutrients and fresh water from outside, developments inland may affect these coastal habitats, a fact that should be taken into consideration when such developments are planned. For example, changes in hydrology caused by the building of a dam upstream could affect estuarine and mangrove-related fisheries. Erosion of deforested watersheds and increased sedimentation rates build up alluvial flats on the shallow coastal shelf and create silt-laden waters unsuitable for corals and seagrass beds, which need light to photosynthesise. Effluent from pulp and paper mills and oil refineries can be expected to lead to declines in offshore fisheries. Oil spillages and thermal pollution also affect the ecology of estuarine and marine waters. Natural gas plants and electricity-generating facilities discharge substantial amounts of hot water from their cooling towers, and this affects the temperature, and therefore marine life, in the immediate vicinity of the outlet (Nontji 1988). In 1989 natural gas from damaged pipes was found to be leaking into the aquifers and contaminating local water supplies in the Bontang region (*Jakarta Post*, 2 June 1989). Urban landfilling and discharge of domestic and industrial waste and wastewater, which affect and destroy coastal habitats, are becoming a growing problem.

The natural values of many coastal ecosystems are considerable. Mangroves, in particular, can be considered so valuable for their role in coastal protection, water purification, and essential inputs to offshore fisheries that they should not be converted to any other land use "except when overriding national priorities are involved and no other alternative is economically or environmentally feasible" (Hamilton and Snedaker 1984). Recognising the valuable services that mangroves provide, the Government of Indonesia has developed regulations to protect a mangrove "green belt" 50-200 m wide along the seashore (Achmad 1986; Anwar et al. 1986). More recently, and after much interdepartmental debate, it has been decided that the width of the green belt should be the difference in vertical height between low and high tide levels multiplied by 130; if the tide rises 1.5 m, the undisturbed mangrove belt should be 195 m wide (Haeruman pers. com.).

The need to conserve coastal habitats to protect natural resources becomes increasingly urgent as industrialisation and development activities expand in the coastal zone. Several coastal reserves have been gazetted or proposed (fig. 11.7). The Berau coral reefs, East Kalimantan, and the Karimata islands off West Kalimantan are designated as conservation areas but have little effective protection. To date, less than 2% of remaining mangrove habitats lie within gazetted and approved reserves within Kalimantan (MacKinnon and Artha 1981), and many of these areas are already disturbed. There are no sizeable tracts of protected mangroves in East Kalimantan, although this province has the richest and least-disturbed mangrove areas on the island. For the whole island of Borneo, only 0.9% of the

DEVELOPMENT AND COASTAL RESOURCES 527

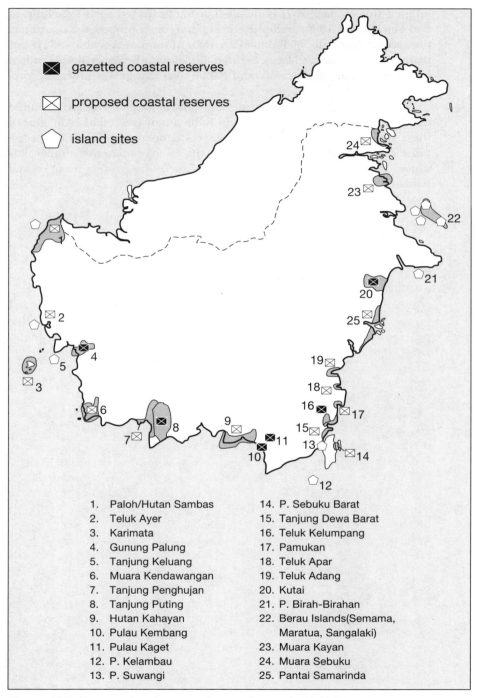

Figure 11.7. Proposed and gazetted coastal reserves in Kalimantan.

original area of mangroves is included within protected areas (MacKinnon and MacKinnon 1986). Extensive areas have been proposed as conservation areas, especially in Kalimantan (MacKinnon and Artha 1981), but many of these, such as Muara Sebuku, are already being exploited as logging concessions. In Sarawak and Sabah also mangroves are underprotected (Chan et al. 1985).

Even where areas of mangrove and tidal swamp forests are included within designated protected areas in Kalimantan (see table 10.5), few of these reserves have any effective protection or management. Since the value of mangrove fisheries far exceeds the value of mangrove timber, it is urgent that conservation measures are enforced to protect these coastal habitats.

Chapter Twelve

Agriculture and Plantations

Kalimantan covers 28% of Indonesia's land area but supports only 5.3% of the republic's population so it is not surprising that development planners regard the island as "underpopulated" and ripe for development and resettlement programmes. Average population density overall is 16 people/km^2 (1990). Most of the population is concentrated around major towns. The *kabupaten* of Hulu Sungai, South Kalimantan are some of the few areas of high population density outside the towns (Donner 1987; see table 1.11).

Kalimantan's population is growing steadily, mainly due to immigration from the other Indonesian islands, especially Java and Sulawesi. Annual population increase from 1971 to 1980 was 2.98% overall, varying from 2.11% in South Kalimantan to 5.73% in East Kalimantan. The latter, Indonesia's second largest province, had only 500,000 inhabitants in 1960, but with transmigration, spontaneous migration and population increase this rose by 140% to 1.2 million over the next twenty years (Kuswata and Vayda 1984). If present trends continue, the province's population will double in the next thirteen years (RePPProT 1990). The number of people that Kalimantan can support depends on the island's agricultural potential and its industrial base (chapter 13).

Agricultural Potential

People cannot live without food, and most food is derived from agriculture. As the human population on Kalimantan increases, more and more land is cleared for agriculture. Already 36% of West Kalimantan, for instance, has been converted to agricultural use (RePPProT 1990).

More than half the island of Borneo lies below 150 m in altitude, and these vast lowland plains would seem to have enormous potential for agriculture. However, the success of agricultural crops depends on three main factors: climate, especially the amount and distribution of rainfall; topography; and soil fertility. Kalimantan is one of the wettest regions in Indonesia, with more than 70% of the island experiencing seven to nine consecutive wet months and fewer than two dry months each year

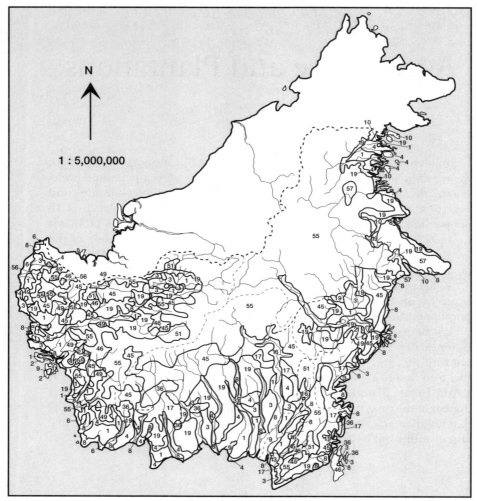

Figure 12.1. Land suitability for agriculture in Kalimantan. (Numbers relate to table 12.1.)

Source: Soil Research Institute 1973b

(Oldeman et al. 1980; see fig. 1.12). Yet many crops such as rice require a dry season for ripening and harvesting. It is noteworthy that those areas in Kalimantan with the most marked dry seasons, the southeastern and eastern lowlands, are also some of the most important areas for wetland rice.

Topography, including relief, slope and elevation, is one of the major physical factors affecting the use and management of land. Elevation influences temperature and rainfall; relief and slope influence erosion, soil drainage and runoff, ease of cultivation, and type of crops. Knowledge of soils and their associated characteristics is also basic to the development of agricultural programmes; the REPPProT project (REPPProT 1990) evaluated land suitability for agriculture and potential transmigration sites in Kalimantan.

The predominant parent materials of Borneo are the sedimentary rocks, shale and sandstone, which produce soils of low fertility. In the humid tropics high temperatures and moisture result in intense weathering and leaching, thus depleting basic cations and silica while iron and aluminium compounds remain. As a result many of Kalimantan's soils are acidic.

Although Bornean soils support luxuriant rainforests this is no index of soil fertility. When cleared, the soils often give poor agricultural yields. It is

Table 12.1. Land suitability for agriculture in Kalimantan.

Unit No.	Soil Characteristics	Cultivable Land (%)
Low-level plains < 100 m elevation, non- to slightly dissected, slopes predominantly 0 - 8%		
1	Wet strongly acid peat	40
2	Wet neutral peat	40
3	Wet acid humid clays	50
4	Wet extremely acid sandy alluvial soils	20
6	Wet acid alluvial soils	95
8	Slightly acid wet alluvial soils	95
9	Extremely acid wet alluvial soils	30
10	Neutral alluvial soils, poorly drained	95
13	Slightly acid clays	95
Low-level plains < 100 m elevation, slightly to moderately dissected, slopes predominantly 0-15%		
16	Strongly acid sandy clays, poorly drained	90
17	Strongly acid sandy clays	90
19	Strongly acid sands	75
21	Slightly acid clays	90
Hill < 500 m elevation, moderately dissected with slopes > 15%		
36	Strongly acid clays	40
Hills < 500 m elevation, moderately dissected with slopes > 30%		
45	Strongly acid humic sandy clays	15
46	Strongly acid sandy clays	15
49	Strongly acid sands	15
Low and high mountains above 500 m, moderately incised, with slopes > 30%		
51	Strongly acid sandy clays	25
Low and high mountains above 500 m, strongly incised, with slopes > 30%		
55	Strongly acid humic clays	5
57	Slightly acid clays	5

Source: Soil Research Institute 1973b

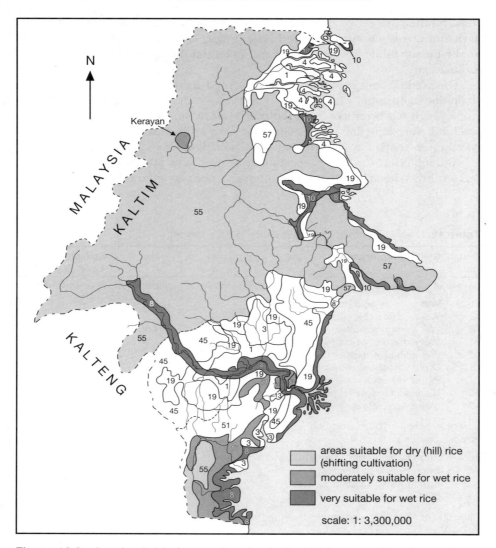

Figure 12.2a. Land suitable for growing rice in East Kalimantan. (Numbers refer to land units in table 12.1.)

Source: FAO 1974

commonly, and wrongly, suggested that this is because most of the inorganic nutrient capital is held in the forests rather than the soil. In fact a considerable fraction of the inorganic nutrient capital is in the forest floor and soil, held in the humus layer (Proctor et al. 1983; Proctor 1989; Whitmore 1984a). Repeated burning after forest clearance destroys this fertile humus layer, and nutrients are leached out of the exposed soil by heavy rain.

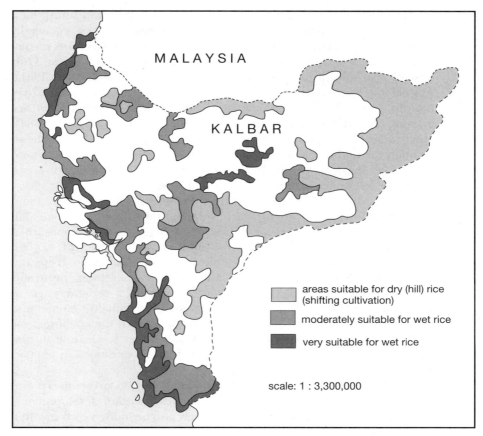

Figure 12.2b. Land suitable for growing rice in West Kalimantan.
Source: FAO 1974

The central highlands of Kalimantan have shallow acidic clay soils, often strongly leached, and steep slopes which are generally unsuitable for sustained agriculture (fig. 12.1). Lateritic soils and acid clays are characteristic of much of the nonvolcanic lowland of Borneo (table 12.1). These soils are fairly infertile, and unsuitable for most agricultural systems other than traditional shifting cultivation. It is difficult, because of land form and heavy rainfall, to bring them under irrigated and terraced rice production. The natural forest is always wet, and difficult to clear and burn with simple equipment. There is rapid regrowth of weeds and secondary vegetation. Furthermore there is a high chance of erosion when vegetation cover is removed, especially on steep slopes. It has been estimated that 30% of Kalimantan is susceptible to landslides; the most critical zones are hill and mountain areas where forests have been cleared or partly felled (RePPProT 1990).

Many coastal areas of Borneo, especially in the west and south, have extensive areas of lowland peat soils. Peat thicker than 1 m deep is generally regarded as unsuitable for agriculture, and the RePPProT reports recommend no agricultural developments on peats thicker than 75 cm. Only 40% of peat soils in Kalimantan are suitable for farming (Hamer 1981). Alluvial soils found along the main rivers, in inland basins and at some river mouths, provide some of Kalimantan's most fertile soils and best wet rice growing areas (fig. 12.2a and b). Podzols, low in minerals (Burnham 1984), cover wide areas, including a broad swathe running southwest to northeast through the middle of Central Kalimantan. These include the white sand soils of kerangas areas, which are totally unsuitable for agriculture and should be left under natural forest (Riswan 1988).

Many of the tidal swamplands and river deltas of South, Central and West Kalimantan have potential acid sulphate soils (Lembaga Penelitian Tanah 1973). The development of these swamp areas has to be approached with caution. If soils are allowed to dry out they will become acid, so good water management is crucial. Careful site selection, appropriate crops and adequate funding are also necessary for agricultural developments to succeed in these swamplands (KEPAS 1985). Converting a wetland system to agricultural use can upset the natural biological balance of animal and plant populations so that pests and weeds may reduce the economic viability of the new land use. Changes in groundwater levels, due to new drainage channels, may critically affect ecological conditions in habitats adjacent to the development areas.

Where natural systems such as Kalimantan forests are converted to agriculture the ability of those agroecosystems to sustain development is often low (Burbridge et al. 1981). There may also be hidden costs involved in such conversions:

- rapid loss of soil fertility after forest clearance which can only be compensated for by increasing investment in fertilisers;
- loss of sustained yields of forest products;
- increased soil erosion;
- alterations to hydrological regimes of watersheds.

All of these costs need to be assessed before further land is cleared for agricultural development.

The Indonesian Soil Research Institute estimated the extent of the most important soil types in parts of Kalimantan and evaluated their suitability for agriculture. Of the 54,000 ha assessed, 37,863 ha were regarded as suitable for certain crops, and 16,173 ha were unsuitable for agriculture (Donner 1987). Information taken from satellite surveys in East Kalimantan, a destination for many settlers, showed the land to be susceptible to serious erosive processes because of its geological structure, soils and rainfall patterns. Detailed investigations of more than twenty soil types in the province showed most of them to be unsuitable for cultivation (Voss

1979). Moreover, in spite of Kalimantan's apparently vast open spaces, there may be few areas available to accommodate additional transmigrant families. In South Kalimantan and the Sambas district of West Kalimantan much of the suitable land is already under permanent cultivation (Donner 1987; RePPProT 1987a, 1987b), and there is little room for expansion and new settlement.

Much of the conditionally suitable land can be cultivated only with

Table 12.2. Land utilisation factors.

Cropping/Land utilisation	Soil loss rating (L)
Bare cultivated soil	1.0
Irrigated sawah	0.01
Rainfed sawah	0.05
Upland crops (*tegalan*)	0.7
cassava	0.8
maize	0.7
beans	0.6
potato	0.4
groundnuts	0.2
rice	0.5
sugarcane	0.2
bananas (rarely as monoculture)	0.6
coffee (with ground cover)	0.2
yams	0.85
spices (chilli, ginger)	0.9
Mixed garden (multistorey, variable ground cover)	
high density	0.1
cassava/soybean	0.2
medium density	0.3
low density (*Cajanus* spp./peanuts)	0.5
Shifting cultivation	0.4
Estate production (poor ground cover)	
rubber	0.8
tea	0.5
oil palm	0.5
coconut	0.8
Natural forest (primary and well regenerated)	
high litter	0.001
low litter	0.005
Production forest	
clear felling	0.5
selective logging	0.2
Fish ponds	0.001
Shrubs/grassland	0.3
Nonvegetated badlands	0.95

"L" represents soil loss from given crop situation compared with loss from bare cultivated soil (L=1).

* Note that bare cultivated soil is 1,000 times more prone to soil erosion than natural forest with a good litter layer. A rubber plantation is 800 times as prone to soil erosion.

Source: Hamer 1981

care, for specific crops and using particular techniques (table 12.2). The vast area of podzolic soils could be unproductive if poor cultivation methods were used (Donner 1987). The areas considered appropriate for sawah rice cultivation are very limited (figs. 12.2a, b), yet this has long been promoted as the main crop at new transmigration sites. Tree crops are far more appropriate for many of Kalimantan's soils and are attracting increasing interest (Lahjie and Seibert 1988; RePPProT 1990). Transmigrant settlers at Muara Wahau in East Kalimantan are now supplied with coconut trees. The ecology and environmental impacts of some of Kalimantan's major crops are considered below.

AGROECOSYSTEMS

Agroecosystems are the most manipulated of all non-urban ecosystems, with farmers putting in high levels of energy (labour) and matter (fertilisers, biocides) to yield high levels of preferred crops. Humans are at the head of the food chain in this ecosystem, and other competitors for resources (herbivores, predators, weeds, pests and disease) are eliminated when possible by the use of trapping, weeding, biocides, and other control mechanisms. Not only is the ecosystem man-made and man-manipulated but also the main plant and animal components are man-selected and may have been genetically altered during domestication (Simmons 1981). Plant production is harvested either directly, through harvesting of crops, or indirectly by culling domesticated livestock that graze on wild or cultivated vegetation. In agroecosystems, as in their exploitation of natural forest resources (chapter 9), people are either herbivores or third-trophic-level carnivores (fig. 12.3). Farming ecosystems are determined by the needs of the cultivator and the uses of crops produced. Crop agriculture is far more important than livestock farming in most areas of Kalimantan.

Agroecosystems can be divided into shifting (seminatural) and sedentary types. During shifting agriculture, the natural system is manipulated over a limited area for a relatively short period of time (one to five years) and then allowed to revert to natural vegetation during the fallow period (chapter 8). Sedentary agriculture, such as sawah cultivation or plantation development, aims at permanent replacement of the natural ecosystems with man-made habitats. In general these man-made habitats are simpler than natural ecosystems and therefore more likely to be thrown out of balance by the predations of pests or by disease. Grazing systems represent a modification of wild vegetation rather than its replacement, though heavy grazing, including the use of fire, may lead to eradication of the wild vegetation.

In sedentary dryland agriculture the soil assumes an importance that it

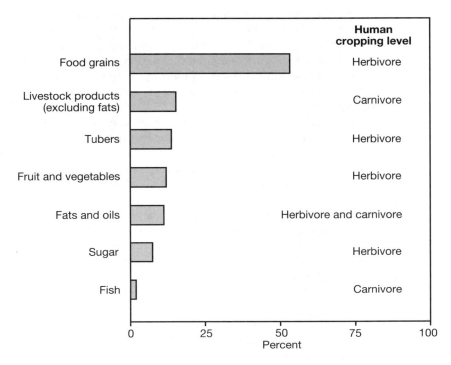

Figure 12.3. Human cropping levels for food. The harvest is dominated by grains, with wheat and rice both supplying one-fifth of world food energy. People operate as herbivores, relying on food grains, but offtake at other trophic levels is common, including high carnivore levels for some fish and wild pig and deer meat in Kalimantan. (After Simmons 1981.)

does not have in shifting systems; it is now the long-term reservoir for all nutrients, which are constantly depleted as crops are harvested and removed (Simmons 1981). To maintain fertility, nutrients must be returned to the soil, either in the form of organic waste or chemical fertilisers. In wetland rice systems, however, the soil is mainly a rooting medium for the plants, and the water supplies most of the necessary nutrients, which are often provided by blue-green algae which fix nitrogen. On nonacid soils, sawah rice cultivation can be sustainable over the long term because of this constant replenishment of nutrients from outside the system, through the waters used to flood the rice fields (Simmons 1981).

All ecosystems can be described in terms of productivity (yield), stability, and sustainability (the ability to continue production at the same level) (Whitten et al. 1987b). Agroecosystems can also be assessed for equity, that is, how equally their products, crops or income, are shared among the members of the community (Conway 1985). Agricultural developments are usually aimed at increasing productivity, but this often may be at the expense of stability, sustainability and equity. Traditional shifting agriculture, with its short periods of cultivation, long fallows, mixed cropping, and benefits shared among those who work the fields, scores highly for all points.

Intensive cultivation of a single crop may increase yield, but a monoculture may be less stable and more prone to disease and pest depredations. The extensive citrus orchards of West and South Kalimantan are vulnerable to the citrus phloem virus, which has necessitated the destruction of infected citrus bushes in Java and Sabah (Lamb pers. com.). Pepper yields a higher income than shifting cultivation does, but is not sustainable in one place over the long term (Kartawinata and Vayda 1984). A rubber or oil plantation is highly productive, stable and sustainable, since old trees are gradually replaced with new plantings, but the profits are not shared equally among all the participants. Different agricultural systems have different ecological and economic values.

Rice

Rice *Oryza sativa* is the main staple of 40% of the world's population. This grass evolved along the foothills of the Himalayas and probably was first cultivated in ancient India at least 9,000 years ago (Vergara and De Datta 1989). Rice cultivation probably began in Indonesia about 1500 B.C., brought by immigrants from the Asian mainland. It is most important as a food staple, but broken rice is used to make starch for laundry, and in food, cosmetics and textile manufacture. Beers, wines and spirits are made from rice, including the potent *tapei* drunk at Dayak festivals. The rice husk can be used as fuel, bedding and absorbent material, for building boards, in cement, and as a carrier for vitamins, drugs and toxicants. Charred rice hulls and ash are used to filter impurities out of water. Rice bran is a valuable poultry food (Vergara and De Datta 1989).

Rice is the most adaptable crop in the world, growing in a wide range of latitudes and environmental conditions. Different systems of growing rice have evolved to suit specific environments and socio-economic conditions. Rice grows best on fertile, heavy soils, but when grown in lowland submerged conditions (sawah) the physical properties of the soil are relatively unimportant so long as sufficient water is present. Soil pH before and after flooding of lowland fields is an important determinant of soil fertility, especially in the tidal wetlands of southern Kalimantan (see chapter 10). Sawah rice is also grown in swamp swiddens by the Iban and Kantu' (Padoch 1982; Dove 1985) and in irrigated rice fields by the Lun Dayeh in the Kerayan of East Kalimantan (Padoch 1988).

Box 12.1. Good agricultural practices.

Soil and water practices
- excessively steep areas retained under forest
- all sloping dryland crop areas terraced or strip cropped
- all sloping wetland rice areas bench terraced
- all terraces carefully designed and constructed
- provision for safe water disposal from terraced areas
- protection ditches around hill sides, where necessary, to protect lowlands
- soil conservation works planned on watershed basis, not by individual field or farm
- excessive weeding of upland crops avoided
- close contour planting and silt-trap drains where terraces are not feasible
- shape/treat all gullies to reduce erosive runoff flow velocities
- adequate but not excessive water application to irrigated areas
- integrated planning of water disposal (runoff) and water storage ponds where desirable for domestic use, livestock supply, fish production, etc.
- careful soil moisture management to exploit residual soil moisture after a wetland rice crop

Crop practices
- pest and disease control (annual and perennial crops)
- subject to crop hygiene, return all crop residues to fields or homestead gardens (directly or through animal manures)
- return all animal manure to improve soil structure and nutrients
- use best available seed varieties, avoid continued use of homegrown seed
- use balanced artificial fertilisers as recommended; avoid dependence on single input (e.g., urea) and nutrient imbalance
- precede nutrient-depleting crops (e.g., tobacco) with green manure crop
- plan balanced crop rotation or intercropping

Homestead gardens
- make maximum use of livestock rearing or subsistence tree cropping
- avoid excessive number of trees and overshading of smaller species
- cut out unthrifty or uneconomic individual trees and species

Livestock practices
- carry optimal number of draught animals
- replace, if necessary, with selected cattle and feed for meat production
- on small farms make maximum use of controlled poultry production to supplement income
- cage poultry over farm pond if fish are kept, or cage for manure recovery for house garden
- closely control all grazing by sheep and goats

General
- aim for balanced, uniform production schedule, with labour requirements spread throughout year
- on small farms, aim for full employment of farm labour, through:
  1) - livestock (especially poultry, fish, small animals)
  2) - production of materials for cottage industries (weaving, mats, etc.)
  3) - multiple cropping

Source: FAO 1984

> **Box 12.2.** Irrigated rice fields in the Kerayan.
>
> Highland Dayak farmers often plant patches of rice in natural areas of swamp, but the Lun Dayeh of the Kerayan in East Kalimantan grow sawah rice in a more intensive manner. Subsistence farmers, the Lun Dayeh practise both shifting cultivation and irrigated rice production.
>
> Although the Kerayan is a hilly area, the valley bottoms are flat, broad and well watered. By controlling water levels with canals, dykes and watergates, the Lun Dayeh have transformed these valleys into highly productive irrigated ricefields. They regularly produce good harvests of rice even though much of the Kerayan is covered in soils of low fertility.
>
> The Lun Dayeh spend little time preparing the land. They slash the rice straw and weeds and trample this organic matter into the soil. Much of the land clearing is done by water buffalo, which roam free through the fields before planting time, feeding on rice straw. The buffalo trample straw into the soil, break up clods of earth and fertilise the fields with their faeces. Young rice plants are prepared in seed beds, then transplanted into the fields.
>
> By keeping water levels high in the fields during the growing season, Lun Dayeh farmers virtually eliminate weeding. Good water control means that harvests rarely fail and that yields are high (2 tons/ha), enough to meet household needs. The Lun Dayeh system is highly productive in the Kerayan and is now being tried in other areas in East Kalimantan, notably the Apo Kayan, where the technology was introduced to the Kenyah by Lun Dayeh teachers. By practising both shifting cultivation and more intensive wet rice cultivation, the Lun Dayeh are able to maximise agricultural production on the relatively poor soils of the Kerayan.
>
> *Source: Padoch 1988*

Table 12.3. Effects of various conservation practices on the detachment (D) and transport (T) phases of erosion.

Practise	Rainsplash D	Rainsplash T	Control over Runoff D	Control over Runoff T	Wind D	Wind T
Agronomic measures						
covering soil surface	*	*	*	*	*	*
increasing surface roughness	−	−	*	*	*	*
increasing surface depression storage	+	+	*	*	−	−
increasing infiltration	−	−	+	*	−	−
Soil management						
fertilisers, manures	+	+	+	*	+	*
subsoiling, drainage	−	−	+	*	−	−
Mechanical measures						
contouring, ridging	−	+	+	*	+	*
terraces	−	+	+	*	+	*
shelterbelts	−	+	+	*	−	−
waterways	−	−	−	*	−	−

* good   + improves   − no effect

Source: Morgan 1986

In the upland areas of Kalimantan and northern Borneo, hill paddy (*padi*) is a more important crop; its cultivation is often taken to be synonymous with shifting cultivation (Chin 1984). There is a rich diversity of padi varieties associated with different Dayak groups, so the phasing out of shifting cultivation may lead to loss of genetic diversity (Chan et al. 1985). Because hill paddy can be grown on agriculturally poor land it utilises soil-terrain conditions that often cannot support any cash crop, making land productive when it would be otherwise closed to agriculture (Maas et al. 1979). Almost all shifting cultivators, including the Iban and Kantu, practice multiple cropping (Freeman 1970; Dove 1985). Hill paddy is typically interspersed at various stages of the agricultural cycle with a variety of other crops such as taro, maize, cucurbits, mustard, peppers, cassava, sugar cane, bananas and leafy vegetables (Freeman 1970; Padoch 1982; Chin 1984; Dove 1985; Hong 1987).

Seeds and cuttings of other crops are set at the same time as the rice crop, and benefit from the initial input of nutrients from the ash. Such an agricultural system is flexible, and well adapted to the Kalimantan environment (Dove 1985). (Compare this swidden system to sawah rice which is grown in extensive monocultures and requires massive inputs of fertilisers to make it independent of the natural mineral cycling processes, and inputs of chemicals to suppress competition from other plants and animals). Multiple crops save labour, since lands need to be cleared and tended only once, and may make better use of soil nitrogen and water than single crops (Chin 1987). Intercropping systems also seem more resistant to disease (Holdridge 1959; Greenland 1975).

Multiple cropping provides good soil cover and prevents soil loss (table 12.3). The scattered distribution of plants limits the spread of pests and diseases. With multiple cropping the cultivator obtains a continuous supply of food throughout the year, and labour requirements are also spread. Sometimes farmers set fruit trees in their ladangs before abandoning them; these trees can be harvested even during the fallow period and are protected when the field is reopened (Chin 1985). Traditional agroforestry systems such as the home gardens established round Dayak and other Indonesian villages are an extension of this multiple-cropping system.

Cassava

Cassava *Manihot esculenta (utilissima)* is indigenous to tropical America and was already cultivated in the first millenium B.C. The Portuguese brought it to Asia and it was probably introduced to Indonesia around 1810 (Bruijn and Veltkamp 1989). It is now a major crop in Indonesia, yielding 13 million tonnes per year. About 60% of production is used for immediate consumption, either fresh or dried, 25% is used for starch production to make *kerupuk*, cookies and other snacks, and 15% is exported. The storage roots consist almost exclusively of carbohydrates, mainly starch. Cassava roots can contain dangerous amounts of cyanogenic glucosides, a defence against plant predators. Glucoside content varies with ecological conditions and mineral supply. The first rains after a dry season cause a large increase in glucosides, as do high nitrogen content and low potassium in soils. Rasping and slow drying of the starch paste allows the lethal hydrogen cyanide to escape so that the roots can be eaten; boiling alone is not always safe (Bruijn and Veltkamp 1989).

Cassava can be grown on soils of very different physical and chemical characteristics, but growth and yield are poor on heavy clay and other poorly drained soils. Some cultivars are salt-tolerant. Cassava produces a reasonable crop on highly depleted or even eroded soils where other crops, like maize, are unproductive. Therefore it is often cultivated as one of the last crops before the fallow in the shifting cultivation cycle. Moreover, cassava has no distinct period of harvesting, so tubers can be collected as needed. They provide a valuable source of carbohydrate and some vitamin C when other staples are scarce. The leaves contain vitamins A and C and are a useful vegetable.

Fruit Trees

Many native Bornean fruit trees are harvested from the wild or managed in old-growth forests (Peluso 1992b), and several have been semidomesticated and cultivated in Dayak home gardens or orchards (Kartawinata and Vayda 1984; Seibert 1988). Twenty-four species of wild mango are known to occur in Borneo (Bompard and Kostermans 1992; fig. 12.4); several are cultivated (table 12.4). The endemic and much-prized kasturi *Mangifera casturi* of South Kalimantan is known only from cultivated specimens in the Martapura area. The mangosteen *Garcinia mangostana*, a favourite fruit of both Queen Victoria and wild orangutans (MacKinnon 1974a), is also extensively cultivated. Not only does the mangosteen provide delicious fruit, its rind and bark are used in traditional medicines, and its wood is prized for carpentry and the manufacture of rice pounders.

Borneo is also a centre of diversity for rambutans *Nephelium* with 16 of the 22 known species occurring on the island, including eight endemics (Seibert 1988). Rambutans *Nephelium lappaceum* are cultivated for their fruits, which can be eaten fresh or canned, or made into jam. The seed can be used to produce rambutan tallow, an edible fat also used for making soap and candles. The seed iself is edible but bitter and narcotic. The wood is used for timber. Rambutans have several medicinal uses. The fruit is effective in killing intestinal parasites; the roots are used for decoctions to treat fever; the bark is used as an astringent for disease of the tongue; and leaves are used for poultices for headaches. Dyes can be produced from young shoots and fruit skins.

Rambutans occur naturally in lowland rain forest and late secondary

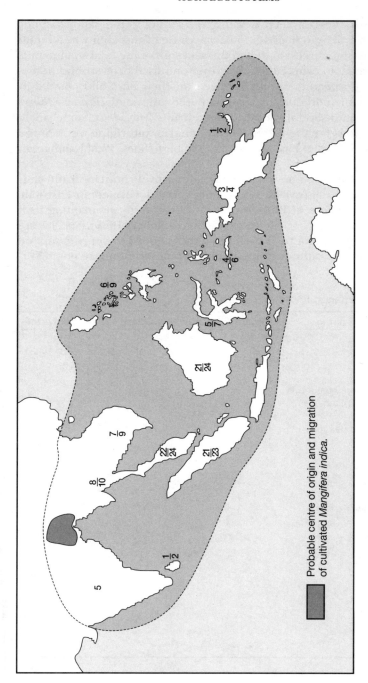

Figure 12.4. Distribution of wild *Mangifera* species and origin of cultivated mango *M. indica*. Number of wild species in country/island group above the line. Total number of species, including introduced species, below the line.

Source: Bompard 1988

forest. They grow best on deep, well-drained soils of fertile, sandy loam or clay loam with a pH range of 4.5 to 6.5. Because of this ability to tolerate acid conditions these trees do well in the acid soils of the tidal swamplands in South Kalimantan, where they are grown on raised beds around homesteads. Other Bornean fruits much prized for their flesh and planted in home gardens include durians, especially the cultivated *Durio zibethinus* (Meijer 1969), and several species of breadfruits *Artocarpus*. Some smallholder plantations for the production of tengkawang (illipe nuts) *Shorea* have been established in East Kalimantan (Seibert 1988), West Kalimantan and Sarawak (Chin 1985).

Fruit trees require minimum attention and provide both food and cash crops; many species also yield useful timber. Dayak villagers in East Kalimantan and elsewhere in Borneo practice agroforestry by planting fruit trees in home gardens or abandoned ladangs (Seibert 1988). Species are selected from the wild for their usefulness, then tended to improve survival and crop yields; this is the first step in plant domestication (see box 8.3).

Table 12.4. Species of mango found in Borneo.

Latin name	Local name	Kalimantan	Sarawak	Sabah	wild	cultivated
M. applanata		*	?	*	*	*
?M. aquaea			?	*		*
M. blommesteinii		E	*	*	*	
M. caesia		*	*	*	*	*
M. casturi	kasturi, kastuba, pelipisan	S				*
M. decandra		E	*	*	*	*
M. foetida	asam depeh	*	*	*	*	*
M. gedebe		E,S			*	
M. griffithii		*	*	*		*
M. indica						i
M. lagenifera			*		*	
M. langong		W	?	*	*	*
M. laurina	mpelem timun	*	*	*	*	*
M. macrocarpa		*	*	*	*	*
M. odorata		*	*	*		i
M. pajang			*	*	*	*
M. parvifolia		W	*	*	*	
M. pentandra				*		i?
M. quadrifida		*	*	*	*	*
M. rigida		*	*	*	*	*
M. rufocostata		*	?	*	*	S
M. subsessilifolia				*	*	
M. swintonioides		*	*	*	*	*
M. torquenda		*			*	E

i = introduced
E, S, W = provinces of East, South and West Kalimantan

Source: Bompard 1988

Pepper

Pepper *Piper nigrum* is native to the Western Ghats in Kerala, India, and is a good example of a rainforest species that has proved to be a valuable commodity for the human species. The pepper plant arrived in Indonesia two thousand years ago, brought by early Hindu colonists. Since then it has become an important smallholder crop. In Indonesia the main areas of production are East and West Kalimantan, Lampung (Sumatra) and the island of Bangka, together accounting for 95% of the crop (de Waard 1989). Some 60,000 ha of pepper in Indonesia yield 30,000-38,000 tonnes per year, with a value of $30 million to $120 million depending on price fluctuations. Early in the nineteenth century the crop was introduced to Sarawak, which produces 95% of the Malaysian crop. Pepper is now third among Sarawak's exports (Chan et al. 1985). Fruits are harvested from May to September during the drier months. To obtain black pepper, entire fruit spikes are picked when the fruit is fully grown but still green; for white pepper, fruit spikes are collected when some fruits have turned red or yellow. The aroma, flavour and pungency of dried peppercorns vary with region (de Waard 1989).

Pepper is a perennial woody climber that requires heavy rainfall (2,000-4,000 mm distributed throughout the year), warm temperatures and high relative humidity. The crop grows best below 500 m in Borneo on well-drained soils, ranging from heavy clay to light sandy clays. In West and East Kalimantan, and Sarawak, pepper is grown on poor soils (de Waard 1989). Cultivation is labour, and land, intensive and involves a substantial amount of initial investment for ironwood posts. The introduction of pepper as a cash crop in some swidden communities has led to a reduction of hill paddy farm size, as farmers have less time to spend on other crops and cultivate smaller plots more intensively (Keuning 1984). As a result there may be less demand for land, allowing longer fallow periods.

Pepper plants can produce abundantly for fifteen to twenty years, but crops decline after about ten years in East Kalimantan if no fertilisers are added to replace nutrients removed by the plants (Vayda and Sahur 1985). The practice of clearing all undergrowth and maintaining bare strips between pepper plants can lead to heavy soil erosion, which may have adverse effects downstream and also reduces the amount of topsoil available to the vine root systems, ultimately affecting yields (Maas et al. 1979). In recent years many pepper gardens in Sarawak and Kalimantan have been abandoned because of fungal attack. The soil-borne fungus *Phytophthora palmivora* attacks vines through the roots and underground stem. Abandoned pepper farms in Kalimantan would eventually return to natural forest (Riswan and Hadrijanto 1979; Soedjito 1988) but more often are burned over and become unproductive *alang-alang* grasslands.

> **Box 12.3.** Buginese pepper farmers.
>
> In the early 1950s a wave of Bugis migrants arrived in East Kalimantan, spurred to migrate by hardships suffered during Kahar Muzakar's Islamic rebellion in South Sulawesi. They began to clear lowland forest to plant pepper. By 1980 these migrants had converted about 1,170 hectares of forest to plantations along the Balikpapan-Samarinda road and planted 3.5 million pepper vines. The pepper farmers choose sites close to the main road, with water and with ironwood trees for support stakes for the pepper vines. Ideally sites have a slight slope for drainage since waterlogged roots develop a fungal disease.
>
> The Bugis farmers clear forest and burn the land during the drier months of August and September, then plant with rice or other annual food crops from which they subsist until the perennial pepper crop begins to produce. By the second year only a small patch of land will be set to rice and the rest converted to pepper. Plants do not give substantial yields until three years old. About 3,000 pepper plants are set to a hectare and separated by weeded strips of bare soil. This clean weeding and lack of interplanting have a detrimental effect on soil conservation. Soil erosion can be serious in areas of heavy rainfall (Hatch 1983). Other crops are not grown since they would compete for nutrients with the closely packed pepper.
>
> The Buginese farmers do not use fertilisers, and yields of pepper plantations on the poor red-yellow podzolic soils (ultisols) decline after about ten years. In Sarawak the use of fertiliser allows old pepper areas to be reused (Padoch 1982) but, in East Kalimantan, Bugis pepper farmers have always cleared new areas of primary or logged-over forest when old plantations were abandoned. Ash from the burning provides the only nutrients added to the soil. Such farming practices are neither environmentally sound nor ecologically sustainable.
>
> As yields decline, the Buginese abandon their pepper fields, often selling the land to Banjarese settlers whose agricultural efforts often fail on the exhausted soils. Once the land is abandoned, secondary growth develops but the land is rarely allowed to return to forest. Farmers clear and burn new fields, and the Bukit Soeharto area has suffered a series of wild fires that have destroyed secondary vegetation and reforestation schemes. Meanwhile the Buginese farmers are clearing new areas of primary forest for their pepper plantations.
>
> This environmental degradation is not being caused by farmers desperate to subsist but by people attempting to improve their economic and social status. Many of the so-called shifting cultivators clearing logged areas are not subsistence farmers but town businessmen, and even government officials, more interested in clearing the forests for timber sales than in cultivating crops. This is a far cry from the swidden agriculture practised in more remote areas.
>
> *Source: Vayda and Sahur 1985; Kartawinata and Vayda 1984*

Other Crops of Ladangs and Home Gardens

A variety of food and cash crops, fruit trees, spices and other useful plants are grown in agricultural fields (ladangs) and in permanent home gardens around village houses.

Taro *Colocasia esculenta* originated in Southeast Asia and was probably cultivated before rice. Most taro is used for human consumption, but it is also used in religious festivals and traditional medicines and is fed to livestock, particularly pigs. Taro, especially the tubers, is an important staple when other crops are in short supply. Taro tolerates a wide range of environments as diverse as upland fields, wet paddy fields with a continuous supply of moving water, and raised beds in poorly drained swamps.

Like taro, maize *Zea mays* is often grown in home gardens and ladangs. Maize can be grown on a wide variety of soils but requires high levels of nutrients. Because of its great physical and chemical demands on the soil, maize is often grown as a pioneer crop. Since a young crop leaves much of the ground uncovered, soil erosion and water loss can be severe (table 12.2). Maize therefore is not a suitable crop for steep slopes.

Soya bean *Glycine max* can be grown alone or intercropped with other crops such as cassava, maize and oil palm. It can be grown on the bunds between paddy fields or without tillage in rice stubble after the harvest. Soya beans are sensitive to low pH and require good weed control. Irrigation is essential during flowering and seed filling to ensure a good yield. Soya bean is being promoted as a smallholder crop for transmigration sites in Kalimantan. Like several other food staples, soya bean originates from northeast China where it was already being cultivated 13,000 years B.P. (Shanmugasundarum and Sumarno 1989).

The cash crop coffee (*Coffea arabica* and *C. robusta*) is a native of the understorey of tropical forests in Africa. In Kalimantan coffee is grown mainly in home gardens beneath fruit trees and other shading crops, which simulate the conditions of the coffee's natural habitat. The shade afforded by the fruit trees has a tempering effect on shoot growth and improves leaf retention but reduces flowering. Without shade, bushes bear more fruits. Crop husbandry therefore aims to provide only sufficient shade to maintain enough foliage to sustain the crop as well as new shoot growth throughout the season (van der Vossen and Soenaryo 1989). Over nine hundred insect pests are known to infest coffee, but these can be controlled by integrated pest management, using both chemical and biological controls.

Groundnuts *Arachis hypogaea* are a common crop in hillside fields. Most of the ladangs encroaching into Pleihari-Martapura reserve in South Kalimantan are cultivated with this cash crop. Groundnuts are not native to Borneo but originated in Bolivia, and were probably brought to Indonesia by the Dutch in the seventeenth century. In Indonesia this is a smallholder crop, with most of the production being used for human consumption. About 20% of the Indonesian crop is crushed for oil.

Groundnuts will grow on a variety of soils, though it is easier to harvest the underground pods from friable, well-drained soils. Groundnuts are legumes so do not require fertilisers as they fix nitrogen symbiotically. Even with their nitrogen-fixing properties, however, groundnuts are not a beneficial crop for soil fertility when grown alone. They provide little ground cover as protection against soil erosion. To achieve maximum yields competitive weeds must be eliminated. The bare soil between plants, and constant turning of soil as weeds are removed, increase soil susceptibility to erosion, especially on steep slopes. Fields are often harvested for only a few years and then abandoned. A more ecologically beneficial way of cultivating groundnuts is to grow them as a nitrogen-fixing understorey cover beneath other crops.

Before 1940 the kapok tree *Ceiba pentandra* was an important commercial crop, and Indonesia was the main producer of kapok fibre. Today this fibre has been replaced by synthetics, and the tree is grown only for local use, with its striking form a characteristic sight in village gardens. Kapok fibres from the fruit can be used for stuffing for mattresses, upholstery and life belts, and for thermal and acoustic insulation. The seeds contain an edible oil used for cooking, for making soap and as a lubricant. After oil extraction, the pressed seeds can be used as animal feed. Preparations of the leaves are used as medicines for fevers, coughs and venereal diseases. Kapok bark is a diuretic and is used against fever, asthma and diarrhoea (Zeven and Koopmans 1989).

The layered growth of plants in home gardens allows greater diversity of species as well as greater productivity, with shade-tolerant species such as coffee growing beneath fruit trees, and plants tapping different kinds and quantities of nutrient resources. These gardens mimic natural forest in their structure, and many of the plants cultivated are native forest species (TAD 1981).

PLANTATIONS

Most of Borneo's agriculture has traditionally been orientated towards food production for home consumption and a few smallholder cash crops, such as pepper, cloves, coffee and rubber. Increasing attention is now being paid to plantation crops and tree crops. The three most important plantation crops in Kalimantan are oil palm, rubber and coconuts. Native species such as rattans also have considerable potential as plantation crops (box 12.4).

Coconuts

The origin of the coconut *Cocos nucifera* is unknown. Fossil coconuts have been found from India to New Zealand, but the palm probably originated in Melanesia or eastern Indonesia (Westphal and Jansen 1989). Some dispersal may have occurred naturally when the buoyant fruits were carried by tides and currents to new beaches, but most of the coconut's spread is probably through the activities of people. The coconut is a valuable source of food and drink, and fruits were carried by many early seafarers. Few

Box 12.4. Cultivation of rattans.

Two species of rattans, *Calamus caesius* and *Calamus trachycoleus*, are cultivated in South Borneo. *Calamus trachycoleus* is cultivated along thousands of hectares of riverside alluvium in the Barito Selatan area of Central Kalimantan. As wild supplies dwindle (for example, of the valuable *Calamus manan* in Sumatra) attempts should be made to cultivate more species of rattan and to extend rattan plantations as buffer zones to remaining forests.

Calamus caesius and *Calamus trachycoleus* are planted in old ladangs or *belukar*, in cleared areas with some trees for support, such as in old rubber gardens. They are grown in areas with alluvial soils, at all times moist and usually flooded from three to five months a year. Ripe fruit is collected during the fruiting season (October-November), crushed and allowed to rot for one or two days, and washed to remove the scaly pericarp and flesh, a process performed in the wild by birds. Germination occurs after 10 days. At one month, when seedlings have well-developed roots and the first leaf is about 8 cm long, seedlings are transferred to a nursery on silty clay in light shade on the riverbank. The seedling grows rapidly and by 14 months may have a cane 1 m tall with seven to eight leaves and one sucker. At this stage it is transplanted to a garden cleared of undergrowth and with fairly open canopy. *Calamus trachycoleus* grows rapidly and needs little further care other than occasional clearing of the canopy and undergrowth, but *Calamus caesius* requires constant clearing of debris around the cane buds for maximum harvest. The first harvest can be cut seven to ten years after planting and every two years thereafter, normally during the driest months. Mature canes may be 15-20 m long in ten years and the colony may consist of ten canes. One hectare can produce 10.5 tonnes of wet rattan in a year (about 6 tonnes of dry rattan). These yields could be improved by more intensive planting and care. Attempts should be made to cultivate *Calamus scipionum* and *Calamus manan* on dry land; both species are harvested from the wild in Borneo.

Rattan has many uses, including the manufacture of furniture and rattan mats (*lampit*) for export. In the past the long internodes of *Calamus scipionum* were exported as malacca canes for walking sticks. Locally rattan stems are used for twine, baskets, mats and other handicrafts, carpet beaters, footballs, toothbrushes, mouthpieces for flutes, furniture and fish traps. Rattan leaves are woven into thatch for temporary forest shelters. Rattan thorns tied to house rafters will prevent birds and bats roosting and are tied to fruit tees to prevent the fruit being stolen. Thorns can also be used as coconut graters.

The "cabbage" of most rattan species is edible. The apex of *Daemonoropus melanochaetes* cooked with fish and coconut milk was a special dish reserved for guests at the Kraton in Yogyakarta, and the related *D. hallieriana* is sold in Kalimantan markets as a vegetable. Resin from the fruits of *Daemonoporus* species yields "dragons blood", a deep maroon resin formerly used for medicine and as a dye for Bornean handicrafts.

Source: Dransfield 1974

plants produce as many useful products as the coconut palm. It is a source of oil, medicine, fibre, timber, leaves for thatch and mats, wood for utensils and charcoal for fuel (Purseglove 1975).

Coconuts can grow on a wide range of soils, from coarse sand to clay, provided the soils have adequate drainage and aeration (Ohler 1984). They require regular rainfall; although the leaves are designed to minimise water loss and can withstand drought, long dry spells impede palm growth and nut production (Ohler 1989). Today coconuts are grown throughout Kalimantan, everywhere from sandy beaches to tidal swamplands and village gardens. The fine pinnate leaves and flexible stems allow trees to withstand strong winds, making them suitable for exposed coastlines. Until 1960 coconuts were the major source of vegetable oil in Indonesia, but soya bean and oil palm are now more important (Purseglove 1975). Coconut oil is derived from copra, the sun-dried endosperm or kernel of the nut, which contains 65% oil. Best quality oil is used for margarine and cooking oil; lower grades are used for soap. Secondary uses include detergents, cosmetics, resins, wax, candles and confectionery.

Indonesia has about 2.9 million hectares planted to coconuts but is no longer a major exporter of coconut oil, as most is used in the domestic market. With increasing competition from other vegetable oils, the coconut's importance as a plantation crop may decline. Nevertheless, it will always be important in rural areas because it grows under conditions where other food crops often do not perform well, and it produces many products used in local households (Ohler 1989).

Coconut plantations vary in copra production, from 0.5-1 tonne/ha in smallholder plantations to 3.5-4.5 tonne/ha in well-managed plantations of local tall palms in Indonesia (Ohler 1989). A stand of mature coconut palms allows considerable light to reach the ground, and multicropping can be practised, with plants of different heights and different nutrient and shade requirements all grown together. Coffee, cocoa, cloves, bananas, pineapple, ginger, beans, maize and even rice can be grown beneath the coconut crop. Intercropping is valuable to the smallholder because it increases his crop and his income. Multicropping is environmentally beneficial since it increases ground cover and thereby increases soil protection. It also may be an important factor for obtaining maximum returns from coconut plantations. Coconut production may increase because the palms benefit from tilling, weeding and fertiliser applied for the other crops. Coconuts are an important secondary crop in the rice-growing tidal wetlands of South Kalimantan (Collier 1979), Hulu Sungai Selatan and Tapin (see chapter 10).

More than 750 insect pests attack coconuts; a fifth of these are not known to attack other plants (Whitten et al. 1987b). One of the main pests is the rhinoceros beetle *Oryctes rhinoceros* (fig. 12.5). Its larvae tunnel through the unopened leaves, creating triangular gaps once the leaves

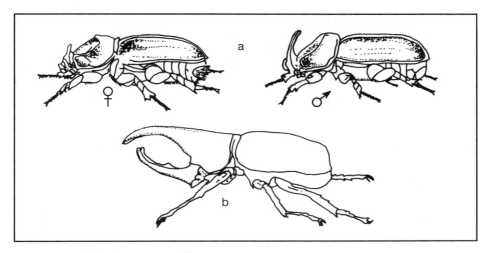

Figure 12.5. Rhinoceros beetles a) *Oryctes rhinoceros*, and b) *Xylotrupes gideon*, pests of coconut palms.

unfold. This pest can be controlled biologically by applying *Bacillovirus oryctes* to breeding places (Ohler 1989).

Weevils *Rhyncophorus ferrugineus* bore into the coconut stem to lay their eggs. Many caterpillars feed on coconut leaves. The bright colours of the nettle caterpillars Limacodidae warn predators of the unpleasant consequences of attack, and indeed the caterpillars bear barbed spines which release a painful toxin. Leaf-mining hispid beetles *Brontispa longissima* attack the coconut fronds but can be controlled biologically by introducing a parasitic eulophid wasp *Tetrastichus brontispae* which deposits its eggs in pupating beetle larvae. The emerging wasp grubs use the beetle larvae as a living food supply. The wasp can attack 60% to 90% of the pupae and 10% of the beetle larvae; about twenty wasps emerge from a single beetle pupa (Davis et al. 1985).

Rats live in the crowns of coconuts and cause considerable damage to young leaves and nuts; they also attack stored copra and can cause losses of 20% or more of the crop (Davis et al. 1985). One way to control rats is to encourage birds of prey, such as owls, in plantations. The plantain squirrel *Callosciurus notatus* also feeds on young nuts (Duckett 1982). Wild pig *Sus barbatus* can be destructive in coconut plantations, as can elephants, rooting up the young seedlings in order to eat the heart ("cabbage") of the palm. Even orangutans *Pongo pygmaeus* have been known to cause damage to young palms in the Muara Wahau transmigration site in East Kalimantan. The orangutans, displaced as their natural forest habitat was destroyed, began feeding on young palm hearts as an alternative food when their preferred fruit trees were no longer available. This led to the inevitable conflict between poor farmers and wildlife, a conflict made more complex by the fact that in this case the offending pest was a rare and endangered species. Unfortunately some orangutans were killed before the authorities intervened. This highlights the increasing problem of how to conserve Indonesia's rare wildlife as pressure on forests increases for logging and agriculture, and people and wildlife compete for limited resources.

Rubber

Rubber *Hevea brasiliensis* originated from the Amazon basin of South America and was introduced to Southeast Asia in 1876. Since then extensive areas of rubber have been planted. Today 92% of the world's rubber comes from Asia, with Malaysia, Indonesia, Thailand and Sri Lanka together accounting for about 80% (Ghani et al. 1989). In Indonesia 80% of all rubber is grown on smallholdings, but it is increasingly being planted as a crop in forestry plantations (HTI) and transmigration settlements. Rubber is an important smallholder crop and a source of rural income in West Kalimantan (Avé and King 1986) and Sarawak, where 1.6% of the state is under rubber (Chan et al. 1985). Rubber estates have been established at Danau Salak in South Kalimantan and in the Sampit area of Central Kalimantan as part of resettlement programmes (PIR). The tapping is done by smallholders. When tapped the rubber tree produces a milky white latex which is processed into sheet rubber for sale. The bulk of natural rubber is exported to industrialised countries and used for tyre production. Rubber is also used in car manufacture (e.g., fan belts), engineering components (e.g., building mounts), flooring, sports goods, toys, surgical gloves and condoms. Rubber wood, which used to be cleared and burned prior to replanting, is being used increasingly as a timber suitable for furniture, flooring, stairs and doors. The large areas of rubber plantations in Kalimantan, Sabah and Sarawak represent a large sustainable source of wood which could help take the pressure off natural forests.

Rubber is less demanding in terms of soil fertility and topography than other tree crops such as oil palm and cocoa, and can be planted on land which is not suitable for these crops. Rubber suffers from attacks by several major pests and diseases. Various fungal diseases attack the tree roots but can be controlled effectively by early establishment of cover crops. Several fungi also attack the leaves and branches. The roots are attacked by underground insect pests such as termites *Captotermes curvignathus* and grubs of *Melolonthis* bees. Mites and thrips cause defoliation of young leaflets, and sap-sucking insects, such as scale insects and mealy bugs, may also cause significant damage (Ghani et al. 1989). Slugs and snails (including the introduced giant snail *Achatina fulica*) and various mammals from rats to elephants can all cause damage to young plantings. Fire is the main hazard to newly planted rubber trees established in alang-alang grasslands as part of the PIR schemes.

Oil Palm

Oil palm *Elaeis guineensis* is a native wetland species, originating from the palm groves of West Africa. Oils are extracted from the palm kernels and from the fleshy fruit. The oil palm was introduced into Southeast Asia in 1848 through the Bogor Botanic Gardens, where one of the four original "mother" oil palms can still be seen (Hardon 1989). The natural environments of oil palms are swamps, river banks and other wet places; they thrive in open areas as they cannot compete with faster-growing species. Highest yields are achieved where rainfall is well distributed throughout the year. Oil palms can grow on a wide variety of soils, from sandy soils to lateritic red and yellow podzols, young volcanic soils, alluvial clays and peat soils. When the palms are first planted a cover of leguminous plants is established to protect the soil, provide humus, add nitrogen and suppress weeds.

Because the fruits are perishable and need to be processed quickly after harvesting, oil palm plantations are only commercially viable on a relatively large scale (as a single commercial holding or a cooperative of smallholdings) with a good transport system between the growing areas and processing plant. This requires heavy initial investment and a sufficient labour force. Oil palm is a relatively important crop in Sarawak and Sabah and is grown increasingly in Kalimantan, especially West Kalimantan (Dove 1986c). Plantations have also been established at transmigration settlements at Sampit, and at Kota Baru, South Kalimantan. Apart from its valuable fruits, oil palm may have potential for the manufacture of reconstituted wood products because of its low density fibres (Hardon 1989).

Yields of oil palm in Southeast Asia rose dramatically after the introduction of the palm's natural pollinator, a small beetle from the Cameroons. Initially overproduction caused prices to fall, but the market has stabilised again. Oil palms in Southeast Asia are remarkably free from pests and diseases. Occasional outbreaks of bagworms *Psychidea* and nettle and slug caterpillars *Limacodidae*, notably in Sabah, can be controlled with minimum applications of insecticide. The rhinoceros beetle *Oryctes rhinoceros* has readily adapted to oil palms, but destruction of its breeding places and provision of good ground cover keep it under control (Hardon 1989). The plantain squirrel can also cause significant damage to young palms.

Cocoa

Cocoa *Theobroma cacao* is native to the forests of tropical America, especially the Amazon, Guyanas and Orinoco, and was cultivated by the Maya people prior to the Spanish Conquest. The Spaniards brought it to Manila in the seventeenth century, and from there it spread via Sulawesi to Java and Peninsular Malaysia. The main cocoa-growing areas in Borneo are in Sabah, but experiments with the crop in East Kalimantan, and at Barabai on the slopes of the Meratus mountains, are proving successful. The main products of cocoa beans are chocolate, cocoa powder and butterfat, which is used for chocolate manufacture, cosmetics and pharmaceuticals.

Cocoa has potential both as a smallholder and plantation crop. It thrives in a variety of soils, provided that they are deep, well-drained and rich in organic matter (Wessel and Toxopeus 1989). It requires better soil than rubber or oil palm. The basaltic soils around Tawau, Sabah, provide especially good growing conditions for cocoa. Cocoa intercrops well with other tree crops including rubber and coconuts. Young cocoa trees require shade, but more sunlight is necessary to trigger fruiting. *Gliricidia* is often grown as a shade tree. Lightly shaded and unshaded cocoa trees require more fertilisers, especially nitrogen, than shaded cocoa. Problems associated with cocoa include its pest the cocoa pod-borer moth *Acrocercops cramerella*, against which pesticides must be used. The larva of this small moth bores into the cocoa pod to feed on the placental tissues, thus reducing or preventing normal bean development. During most of its life cycle the insect is protected within the pod, and therefore it is difficult to control. At the beginning of the twentieth century the cocoa pod-borer destroyed the cocoa industry in Central and East Java; this pest is now present in Sabah and East Sarawak. Research continues on methods of control, and also on the shade and nutritional requirements of cocoa in areas with low soil fertility (Wessel and Toxopeus 1989).

Timber as a Plantation Crop

The traditional source of supply for the wood-based industry (sawnwood, plywood, veneer and mouldings) has always been natural forest, with certain species being harvested selectively. As natural forests are depleted and the world demand for raw timber remains high, attention is turning to the viability of producing timber from plantations. Fast-growing species such as *Acacia mangium* can be thinned in the eighth year and felled as logs in the tenth year, giving as good a return as other plantation crops. In Malaysia plans are already under way to create large-scale tree plantations of fast-growing species such as *Acacia mangium, Gmelina arborea* and *Paraserianthes falcataria* to produce sawlogs. Sabah has a programme to establish 250,000 ha of forest plantations by the year 2000 (Tho and Salleh 1987).

The Ministry of Forestry in Indonesia has instigated a large-scale Industrial Timber Estate programme (HTI) for timber production, to relieve pressure on natural forests. Although the target for Repelita IV was 1.5 million hectares, by the end of 1988 only 69,000 ha had been planted (MOF/FAO 1991). Prospects for large-scale investments now seem good, and several HTI projects are planned for Kalimantan. Already Inhutani, the

State Forestry Company, has established commercial timber plantations of mahogany, meranti and eucalyptus on Pulau Laut. These trees are now 30 years old and are being harvested.

Greater reliance on commercial plantations, and utilisation of woods which are a secondary byproduct of other plantation crops, will help to reduce exploitation of natural forests and allow them longer times for regeneration. On good soils plantation crops can be harvested sustainably over long periods. (On poor soils yields are probably not sustainable due to loss of organic matter and nutrients). Some plantation species like *Acacia* and *Albizia* can be coppiced. Tree plantations have a shorter harvesting cycle than natural forests and may provide good soil cover, but care needs to be taken with harvesting. The trend to establish plantations of rapidly growing species, which are harvested totally with large equipment on very short rotations, may have adverse effects on the water flow rate and soil erosion rate in watersheds (Hamilton and King 1983). Plantations of fast-growing exotics are not as biologically rich as natural forests, but field studies begun at Riam Kiwa, South Kalimantan, suggest that areas of alang-alang replanted with *Acacia mangium* and *Gmelina* support a greater number of bird and insect species than do the open grasslands.

Several tropical hardwoods have potential as plantation crops. *Pterocarpus indicus*, a widespread leguminous tree, provides timber of good grain and grows rapidly from large woody cuttings (Whitmore 1972). At Wanariset in East Kalimantan the Tropenbos programme, in cooperation with Inhutani, has developed a method of propagating dipterocarp species vegetatively for replanting large forest areas (Smits et al. 1987, 1992). The establishment of native species in plantations may have considerable commercial potential and is more desirable ecologically than introducing exotic species. Among the most valuable timber woods of Kalimantan are the red merantis, including *Shorea johorensis*. This dipterocarp occurs naturally on fertile soils on well-drained alluvial soils and hillsides up to 600 m. Until recently the only way to propagate *Shorea* was from seeds; feasibility depended upon availability of seeds. Like all dipterocarps *Shorea* fruits infrequently at intervals of several years, and the seeds are viable for only a short time (Whitmore 1984a). Moreover seedlings must be infected with ectomycorrhizal fungi (chapter 4), which are killed by the high soil temperatures in open clearings.

Recent experiments with seedlings naturally infected with mycorrhizae, planted under shade, and with artificially infected cuttings, have shown good results (Smits 1989; Smits et al. 1987, 1992). In East Kalimantan forests *Shorea johorensis* yields an average of three to eight cubic metres of wood per hectare per year. One tree yields about 1.5 cubic metres of wood (Smits 1989). Because of its good growth and high quality *Shorea johorensis* is promising as a plantation tree (Seibert 1988). Nevertheless, plantations of native species such as *Shorea* will not yield as quickly as fast-growing exotics. Some wood can be removed by thinning after thirty years, for production of veneer and as construction wood, but the main crop can only be harvested about forty years after planting when trees have a diameter of 40 cm (Smits 1989). Use of selected clones of faster-growing dipterocarps with suitable mycorrhizal fungi will increase production. Plantations can be established in secondary or shrub forest by underplanting of faster-growing species or through line planting (Smits et al. 1992). Both methods simulate natural forest regeneration.

Plantation Ecology

Plantations are often promoted as an acceptable form of land use because they are said to fulfil various natural forest functions, such as protecting soil and water and providing habitat for native wildlife. While it is certainly true that some plantation crops can give good soil protection, especially in mixed cropping systems, this is by no means a general rule. Comparisons of soils under different crop regimes show that land under rubber is almost as prone to soil erosion as bare cultivated soil or *Imperata* (Hamer 1981) unless there is good ground cover (see table 12.3). It is the immediate ground cover of herbs and leaf litter which protect soils. Smallholder rubber gardens, where trees are planted close together and other vegetation is allowed to grow, approximate to mixed forest conditions more closely than do industrial estates with clear weeding. In plantations there may be considerable soil losses when the land is first cleared or when old trees are removed prior to replanting.

Successive plantation cycles are likely to alter the nature of the soil and reduce the amounts of inorganic nutrients (Bruijnzeel 1982). Comparison of nutrient cycles in a rainforest and a conifer plantation of *Araucaria* in Queensland, Australia, showed that, while litterfall as dry matter was about the same in both habitats, the quantity of nutrients reaching the ground in litter and throughfall was much less in the plantation, as was the rate of litter decomposition and nutrient release (Brassel and Sinclair 1983). Certain crops, such as pines, may also be less environmentally desirable because of their susceptibility to fire. As many of Kalimantan's natural forests disappear, demands for timber, pulpwood and cash crops will be met increasingly from plantations, so it is essential to understand the ecological costs and benefits of these ecosystems.

While a monoculture of rubber or a tree crop such as *Acacia mangium* or *Albizia* is ecologically and aesthetically preferable to an alang-alang grassland, plantation systems do not compare with natural forests in terms of species richness. In general, plantations support very few plant species, especially in a well-maintained estate where most plants that are competing for resources with the crop will be removed. An oil palm plantation may support no more than 15 species, including ferns growing on the palm trunks (Whitten et al. 1987a). Plantations are much simpler ecosystems, both in structure and species composition, than natural forests. With fewer food and shelter resources, plantations provide fewer niches for other plant life and animals. Moreover, since the plantation is in a building phase (Whitmore 1984a) and all dead material is usually cleared, there is a much reduced decomposer community. Insecticides and herbicides also help to maintain low species diversity. As a result there are fewer niches available for pollinators, leaf eaters, fruit predators, parasites and so on. The better managed and maintained the estate, the more resources are reserved for the crop and the fewer niches are available to other organisms.

Different plantations have different ecological characteristics, depending on the crop, whether there is a monoculture or mixed cropping system, and the level of management. Surveys of diversity and abundance of easily recorded mammals and birds within natural forests, selectively logged forests and plantations in East Kalimantan showed that both species diversity and density declined drastically in forestry plantations (Wilson and Johns 1982). Six large mammal species were recorded in plantations (table

Table 12.5. Mammal and bird species recorded in primary forest and disturbed habitats.

Habitats	Plantation	Scrub	Forest fringe	3- to 5-year-old logged forest	Recently logged	Primary forest
Bornean gibbon	–	*	*	*	*	*
Hylobates muelleri						
Maroon langur	*	*	*	*	–	*
Presbytis rubicunda						
Sambar deer	*	*	*	*	*	*
Cervus unicolor						
Barking deer	*	*	*	*	–	*
Muntiacus muntjac						
Mouse deer	*	*	*	*	–	*
Tragulus spp.						
Bearded pig	*	*	*	*	–	*
Sus barbatus						
Civets (Viverridae)	*	*	*	*	*	*
Sun bear	–	–	–	–	–	*
Helarctos malayanus						
Pied hornbill	–	*	*	*	–	*
Anthracoceros convexus						
Bushy-crested hornbill	–	–	–	*	–	–
Anorrhinus galeritus						
Rhinoceros hornbill	–	–	–	*	*	*
Buceros rhinoceros						
Helmeted hornbill	–	–	*	*	–	*
Rhinoplax vigil						
Argus pheasant	–	–	–	*	–	*
Argusianus argus						

Source: Wilson and Johns 1982

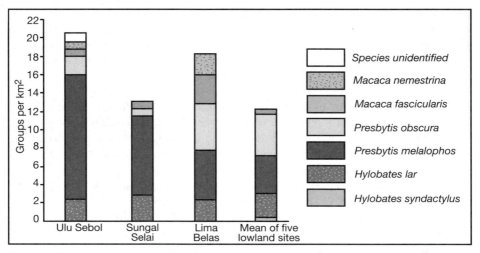

Figure 12.6. Primate group densities at three sites close to cultivation. These are compared with the mean density at five other lowland sites that were not close to new cultivation.
Source: Marsh and Wilson 1981

12.5) but none were resident there; all entered from areas of adjacent forest to feed or to visit waterholes. Even as occasional visitors to forestry plantations, some species can cause significant damage. Sambar deer knock over young trees in *Albizia* plantations, and maroon langurs *Presbytis rubicunda* peel the leaves off young seedlings. Clearance of forested areas for plantations leads to displacement of animals into adjacent remaining forest. Primate populations in such zones are often higher than in undisturbed lowland forests (fig. 12.6). These displaced animals may then become crop raiders.

Plantations with the greatest variety of habitat components can be expected to provide the most food resources and to support the greatest variety of wildlife; mixed plantations should support more mammal and bird species than do monocultures (table 12.6). Reduction of species diversity not only reduces the range of foods available to native wildlife, especially as they may not be able to eat exotics, but also the number of plant species of potential value to local human communities.

In general, monocultures such as rubber and oil palm are poor for wildlife. Rubber is a long-term crop, and the tree branches provide a substrate for epiphytic ferns and sites for nesting birds and squirrels. Occasionally some patches of natural or secondary forest are left in rubber plantations in poorly drained areas; these can act as a refuge for native birds and other wildlife. Tapping activities cause little disturbance to wildlife. On the other hand there is little food available for animals; rubber fruits once a year, producing hard, unpalatable fruits. Oil palm, however, produces oil-rich, colourful fruits which are attractive to wildlife, but the regular cutting of the fruits disturbs the only part of the plant where animals could shelter.

Few mammals other than rats and squirrels are actually resident within the plantation. Whereas other species are lost when forest lands are converted to plantations, these rodents may benefit as a result of increased habitat and food resources with reduced competition. The plantain squirrel *Callosciurus notatus*, normally an inhabitant of secondary forest and scrub, has adapted particularly well to oil palm plantations and is a major pest in newly planted plantations in Malaysia (Duckett 1982). In its natural habitat its diet consists principally of insects, supplemented with other small animals and fruit (Harrison 1962a). In oil palm plantations the squirrel benefits from the continuous fruiting of the palm, and fruits are the major food source. The squirrels nest in older palm trees and have few natural predators other than a few birds of prey such as the serpent eagle *Spilornis cheela* and some snakes, including the cobra *Naja naja*. Night-flying owls, which are effective predators on rats (Duckett 1976), are inef-

Table 12.6. Crops and their usefulness for wildlife, soil protection and other values.

Crop	Soil Protection	Wildlife	Cash	Food	Fuel	Other useful products (e.g., building materials)	Sustainability	Susceptibility to fire
Natural forest	***	***	*	*	**	**	***	–
Disturbed forest and secondary forest	***	**	*	*	**	**	***	*
Fast-growing firewood	**	*	*		**	*	**	**
Timber plantations	***	*	**		*	*	**	*
Mixed tree plantations	***	**	**	*	*	*	***	*
Fruit plantations	**	*	**	**	*	*	**	*
Coconuts	*	–	**	**	*	*	***	**
Rubber	*	–	**		*	*	**	*
Oil palm	**	–	**	*			**	*
Eucalyptus/Melaleuca	–	–	*		*	*	**	*
Maize/cassava	–	–	*	***	–	–	–	**
Cloves	*	–	***		–	–	**	*
Groundnuts	–	–	*	**	–	–	**	*
Cinnamon	**	–	**	–	*	–	**	*
Pepper	–	–	***	–	–	–	*	***
Coffee	*	–	**	–	–	–	*	
Hill rice	–	–	**	***	–	–	•	**
Tea	*	–	**	–	–	–	*	
Animal fodder (e.g., *Leucaena*)	*	–	*	*	*	–	**	**
Bamboo	*	–	*	*		*		***
Rattan	*	*	**	*		*	**	*
Sago	**	**	**	**		*	**	
Mangrove	**	**	**	*	*	*	**	**
Herbs and grasses (undercropping)	**	*	*	*		*	**	***

• = except on shifting system with long fallows
* = Some   ** = Moderate   *** = Good/high

fective against the diurnal squirrels.

Predators may also benefit from the changed habitat because of an increase in prey populations. The barn owl *Tyto alba* has been able to increase its range in Sumatra (Whitten et al. 1987a), and has increased in numbers in Peninsular Malaysia, where it helps to control rats in oil palm plantations (Duckett 1976). To date, however, this species has not been recorded in Borneo (Smythies 1960; Holmes and Burton 1987). The barn owl is a commensal of man, adapting to artificial man-created habitats. Its absence from Borneo can probably be explained by the fact that humans had little impact on the natural environment until long after the island was cut off from the Asian mainland. Because of its predations on plantation pests, this bird might be a valuable addition to the island's avifauna. As with any introduction of an exotic species, however, consideration would have to be given to the possible effects on other native species that occupy a similar niche.

Owls and other birds of prey hunt in plantations, but they also require areas of natural forest to breed and maintain recruitment. Surveys in Sabah have shown that many insect-eating birds that help to control pests in *Albizia* plantations are local species, recruited from neighbouring natural forests (Sheldon et al. 1992). Natural forests are also vital as alternative food sources for many of the bird and bat species that are important pollinators of prized fruit trees such as the durian. Yet again, we see the need for conservation of substantial areas of natural forests to help maintain the productivity of agricultural lands. Buffer zones established between forest areas and plantation crops palatable to forest mammals can reduce crop raiding and minimise man/wildlife conflicts (MacKinnon et al. 1986; Oldfield 1988b; see chapter 14).

Pest Ecology and Control

It has long been suggested that the rareness of individual tree species in rainforests is a protection against predation: it is more difficult for a leaf-eating predator to find many scattered trees of the same species than a clump (Janzen 1971). In species-poor forests pests, especially insect pests, can spread rapidly, attacking most of the plants. This holds true for both natural forests (Anderson 1961b) and man-made ecosystems such as plantations. Non-native species in plantations may be particularly vulnerable to predation since they have not had time to build up defences against predators in their new environment.

Most species that become agricultural pests occur in natural ecosystems at relatively low densities, and some may have specialist or ephemeral niches. With a change in land use and a superabundant food supply, their niche expands. Natural predators may not multiply as fast as the prey species. Free from natural checks such as predation and limited food availability, the pest species multiplies rapidly and achieves pest proportions.

Species reproductive strategies fall within a spectrum between two extremes. Species with an **r-selected strategy** produce large numbers of offspring with a short reproductive period and minimal parental investment in each offspring. Species with a **K-selected strategy** have a long reproductive period and produce few offspring, each receiving considerable parental investment and care. Thus an r-selected species favours quantity while a K-selected species aims for quality. Not surprisingly many pest species, especially insect pests, are r-strategists, since they are the species best able to take advantage quickly of a change in habitat and food supplies. Mammal pest species such as rats and pigs also fall into this category.

Rats are the most important group of pests hindering agricultural production in Southeast Asia (Soerjani 1980). Rats feed on almost all agricultural crops and also damage stored products. Rats are probably responsible for a 5%-6% overall loss of the annual Indonesian rice yield (Whitten et al. 1987b), and rats and birds together may cause more damage to rice crops than all other types of pest (Anon. 1976). Rice-field rats *Rattus argentiventer* are clearly r-strategists. A female rat can reproduce at two months old, has a short gestation period of twenty-one days and a large litter size (Lam 1983). Rats have short life spans of about four to seven months (Harrisson 1956) but one female is capable of producing twenty or more offspring in this time.

In response to a "plague" of rats in the Kelabit highlands, the late Tom Harrisson organised an airlift of stray cats from Kuching to be parachuted into the stricken area, an imaginative attempt at biological control. More usually poison is used, but rodenticides such as warfarin may enter the food chain and destroy non-target species such as birds of prey. Usually control methods such as poisoning or trapping have only limited, short-term success. Those rats which survive benefit from reduced competition for food and reproduce even faster, with larger litters, so the population increases rapidly. New immigrants also move into the cleared area from surrounding rice fields. Rats are best controlled by careful land management, in which rice is planted and harvested at the same time over large areas, and there are few scrub areas to provide alternative food supplies when rice is not available.

Wild pigs are a serious pest in Kalimantan, since even a lone animal can cause considerable damage. Monkeys, especially long-tailed macaques *Macaca fascicularis* and pig-tailed macaques *Macaca nemestrina*, can also become agricultural pests, raiding fields of rice and corn and fruit trees. Although monkeys take only a very small percentage of the overall crop, the consequences for individual farmers can be disastrous. Macaques are not r-strategists, but they are opportunistic feeders and take a much wider range of foods in their diet than most monkey species (MacKinnon and MacKinnon 1980); they occupy a broad niche. This opportunism allows them to take advantage of new habitats such as agricultural fields. Their travel strategy also allows them to exploit this new environment, since they naturally spend a large part of their time travelling on the ground, unlike more arboreal primates such as langurs and gibbons.

Agricultural crops are plagued by large numbers of insect pests. Most insects are r-strategists, able to multiply quickly to benefit from an abundant food supply. Spraying with insecticides is the usual method employed to control insect pests. Continued use of a biocide, however, sometimes leads to the target species developing resistance. As insects reproduce and multiply quickly in a fairly short time, this resistance can spread rapidly throughout the population. The resistance of mosquitoes to DDT is a well-known example. Many species of insects, mites and ticks are already resistant to one or more biocides. Asia has been identified as an area where pest resistance to chemical controls is likely to rise steeply, especially if biocide use reaches levels as high as those in many developed countries and Japan (Conway 1982).

If the ecology of pests, predators and crops (or their wild relatives) were better understood, better control measures could be developed (Salick 1983). Biological control is the most ecologically sound method of pest control. It involves controlling pest population levels by predators, parasites or pathogens, just as in natural populations. It can also be achieved by introducing behaviour-changing hormones, by sterilising males, or by using resistant crop varieties. Most biological controls are highly selective and do not cause environmental damage. There have been a few remarkable success stories, such as the introduction of the disease myxomatosis to control rabbits in Australia, and the introduction of a natural predator, the moth *Cactoblastis cactorum*, to attack the prickly pear *Opuntia* and other introduced cactus species in Australia and southern Africa. Biological controls have also been used successfully for some plantation pests such as rhinoceros beetles, but they cannot be the answer to all pest problems. Control is usually slow, pests are almost never wholly eliminated, and patterns of control are often unpredictable. Moreover if the "control agent" is being introduced into a new environment, it is also freed from its natural restraints and may be detrimental to local species. As with any introduction of an exotic, it is necessary to understand the species' ecology. Integrated pest control, combining biological controls with use of chemicals in a manner least disruptive to natural processes, is often the best way to control agricultural pests (Stern et al. 1959). At the same time further research is required to produce pest-resistant varieties of crops.

Careful management and manipulation of the ecosystem can often help to reduce pest levels. Predatory spiders, for instance, may play a useful role in pest control since they are generalist rather than specialist predators (Whitten et al. 1987b). This means that they may be effective at controlling populations of several pest species at the same time. Spiders seem to reduce pest levels significantly in rice and cotton crops, but it is the whole community of spiders that must be encouraged and not just a single species. Here a knowledge of spider ecology helps. Crops should be sprayed with biocides at midday when spiders are usually inactive and sheltering under vegetation.

LIVESTOCK

Animal domestication and husbandry are probably rather recent activities in Borneo. There is some evidence that captive orangutans were kept for meat in prehistoric times (Harrisson 1968) but there seem to have been no attempts to domesticate the island's native pigs (Caldecott 1988a) and wild cattle (banteng), which have been husbanded and bred elsewhere in Indonesia. Instead Bornean peoples have exploited wild meat by hunting pigs, and by deliberately burning and maintaining some areas of grasslands close to their fields to encourage deer and other ungulates (Seavoy 1975). The wild jungle fowl *Gallus gallus* does not occur in Borneo, so although many villagers keep *kampong* chickens these probably arrived rather late in human history and were already in a domesticated form. Similarly the domesticated pig *Sus scrofa* is not native to the island. Chickens are kept for ritual and for cock fighting, pigs for food and dogs for hunting, but all of these animals have arrived as domesticated species. The water buffalo *Bubalus bubalis*, raised in the swamplands of central Kalimantan, in the Kerayan highlands and in many coastal areas, is not native to the island but was introduced from mainland Asia.

The rearing of livestock has never been a major agricultural activity in Kalimantan, but recently the government has attempted to encourage villagers to keep cattle and goats. Cattle have multiple uses. They produce meat, fat, milk, hides and other minor products. They can also be used as draught animals to pull oxcarts and to plough rice fields. By trampling on wet rice fields after the harvest, cattle improve the soil; they also eat weeds and fertilise fields by defaecation. Traditionally the rice-growing Banjarese did not use cattle in their rice fields (see box 10.8) since the soils are thin and unsuitable for ploughing. Nor are cattle useful to shifting cultivators who must travel long distances to tend their fields and often spend several weeks in field shelters tending their crops (see box 8.7). Animal husbandry involves some degree of constant care, and this may be why the banteng *Bos javanicus* has never been domesticated in Borneo, although the related Bali cow is a useful draught animal on Bali, a rice-growing island. Shifting cultivation and animal husbandry, at least of large animals, have not been compatible.

Today as more of Kalimantan is converted to open swamps and grassland, there may be greater potential for cattle rearing. What does animal

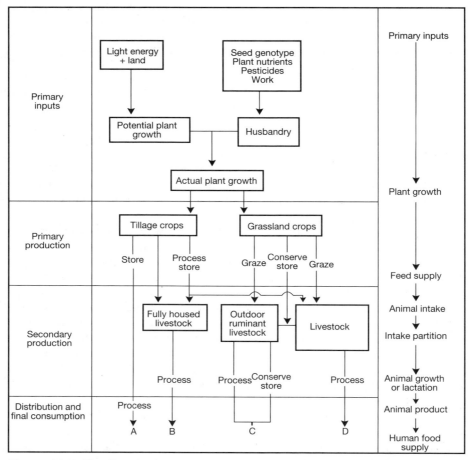

Figure 12.7. The four food chains which characterise agriculture. In chain A the absence of a box at the secondary production stage means that humans crop directly at the herbivore level. (After Duckham and Mansfield 1970.)

husbandry mean in ecological terms? Domesticated or semidomesticated animals forage on wild vegetation and are then culled for human use (fig. 12.7). The animals themselves may return a few nutrients to the system in the form of excreta, but most nutrients are removed from the site of intake. Where livestock are allowed to wander over large areas, as in nomadic pastoralism, they may exert relatively little pressure on natural vegetation even in marginal habitats. Sedentary animal husbandry, however, increases the environmental impact of grazing, sometimes dramatically as in the Sahel (Sinclair and Fryxell 1985).

Box 12.5. Some effects of grazing by domestic stock on five ecological processes.

Ecological process	Effects of grazing
Natural succession	• Modification of natural succession by treading and selective grazing leading to dominance of unpalatable species • Invasion of unpalatable tree, shrub and perennial species and expansion of grassland • Increased competition with native herbivores • Disturbance of native animal species by domestic stock
Organic production and decomposition	• Primary production diverted to ground level with loss of trees and shrubs • Reduction in total biomass and possibly in energy capture • Decrease in biomass of native animals • Natural decomposition process circumvented by grazing animal cycle • More of primary production diverted to large herbivores • Increased herbage intake leads to less litter and lower rates of decomposition
Nutrient circulation	• Reduction in nutrient pool with fewer nutrients in vegetation • Local and uneven reallocation of nutrients according to distribution of faeces and urine • Increased rate of nutrient circulation • Replacement of slow cycling through soil organisms by more rapid, plant-animal cycling pools • Initial stages of decomposition in rumen and gut of grazing animals • Loss of nutrient capital with removal of animal products (meat, milk, hides)
Water circulation	• Increased surface runoff • Reduction in interception and transpiration • Drying of soil surface layers • Increase in evaporation from soil surfaces with loss of vegetation cover
Soil development	• Localised overgrazing resulting in soil erosion • Increased exposure of soil, especially where animals congregate • Increased salinity with loss of trees and shrubs • Increased soil compaction due to treading

Source: Ovington 1984

Overgrazing and indiscriminate burning of vegetation (to encourage new growth) over long periods result in adverse hydrologic and soil movement effects. The impact of grazing on some ecological processes is detailed in box 12.5. The degraded alang-alang grasslands around the Riam Kanan Lake in South Kalimantan are burned regularly to produce new grazing for domestic cattle. Wild fires nibble away at forest edges and destroy regenerating shrubs and reforestation schemes, and the bare earth created by the burning is particularly prone to soil erosion. Regular burning prevents regeneration and the restoration of hydrological or soil protection functions. Devegetation caused by continuous grazing may have even more deleterious long-term effects than logging on a clear-cutting regime (Hamilton and King 1983).

Genetic Resources

A quick review of the main agricultural and commercial crops reveals a mixture of native species that have been domesticated or semicultivated (for example, many fruit trees and spices) with cultivars introduced from other parts of the tropics. Many of the latter are also important in the economies of countries other than Indonesia. Rice, for instance, provides food for 40% of the world's population. Of the several thousand species of the world's plants known to be edible, only 150 have become commercially important (Plotkin 1988), and fewer than twenty species produce most of the world's food (Vietmeyer 1986). Few of these staples can be grown readily or sustainably on Borneo's poor soils. On the other hand, native plants, including many fruit trees and palms, are already important sources of food, handicraft material, medicinal treatments and income. Many have considerable potential for domestication and extensive cultivation.

Natural forests and other natural habitats are a genetic storehouse, both for potential future crops and for providing genetic material to improve species that are already cultivated. Local communities often prefer to cultivate local strains, which may be less productive than higher-yielding varieties but are more resistant to pests and better adapted to local conditions. In the tidal swamplands of South Kalimantan farmers prefer local varieties of rice, believing them to be better tasting and more suited to the area's relatively poor soils. Agricultural improvement requires a constant search for material to improve yields and resistance to disease. This requires access to a gene pool of wild native species, species which will be lost if their habitat is destroyed.

Patterns of diet and use of forest products are largely a matter of tradition. Sago palm, cultivated by the Melanau of Sarawak, is probably one of the most underexploited resources of Southeast Asia (Flach 1983), with tremendous potential for more extensive cultivation. As demands for food and other natural products increase, many more species presently utilised only at the local level may achieve more importance. Already tropical fruits such as durians and rambutans are being planted in northern Australia where there is a ready market for the fruit. At present, research is concentrated on only a few important agricultural crops. Since many more plants have great potential as food and pharmaceuticals, and these are just as likely to be small herbs as large trees (Lowry 1971), it is vital to keep options open by protecting and maintaining species diversity. Biodiversity can be conserved both *in situ* in natural habitats and *ex situ* in botanical and zoological collections, or by retaining germplasm in gene banks. Germplasm collections have already been established for most major commercial crops in Southeast Asia. In collaboration with UNESCO and the PROSEA programme, Indonesia is establishing at the Bogor Herbarium a database of all useful native species, especially those used for food, medicines and other traditional uses.

Future Trends in Agriculture in Kalimantan

The enormous differences in population density between Borneo and more densely populated Indonesian islands such as Java can be related to the islands' different landforms and soils, and their capacity for sustaining agriculture. High population densities in Java depend on a combination of basic volcanic soils and regular and heavy rainfall, with a good dry season for ripening cereals and harvesting them. The population depends on rice cultivation and good harvests in irrigated fields (Mohr 1945). Borneo, in contrast, has few areas of recent vulcanism and generally has poor soils. Traditionally the staple crop has been hill padi rather than sawah rice. The unequal population densities of Java and Kalimantan strongly reflect the unequal agricultural potential of the two islands. It is likely that these variations were equally important in historic times (Bellwood 1985).

As Kalimantan's population increases and more land is converted to agriculture, it is important to choose those crops most appropriate to the island's soils and topography. Less than 10% of the island's terrain is suitable for wet-rice culture, and most of these lands are already under production (RePPProT 1990). Greater agricultural production in the future will depend on tree crops and innovative agroforestry projects that combine natural forest species and plantation crops (RePPProT 1990; Smits et al. 1992). Agroforestry practices play an important role in traditional land use systems in Kalimantan (Seibert 1988; Peluso 1992) and provide a good model for new settlement areas. Sustainable agriculture in many parts of Kalimantan may depend on development of agroforestry projects that closely simulate natural forests in structure and composition, with multiple crops, good ground cover and a multi-layered canopy. Combinations of introduced and native species can provide a wide range of products (timber, foods, fruits, fibres and medicinal herbs) for local communities, while serving as a genetic reservoir for many native species.

Chapter Thirteen

Development and the Environment

Indonesia is now the third most populous nation in Asia, after China and India, but the population of 180 million is not spread evenly throughout the republic. Kalimantan, with 9 million people (in 1990), is much less densely populated than Java, Sumatra or even Sulawesi. Overall Kalimantan has a population density of only 17 people/km^2, compared to more than 800/km^2 on Java (Buku Statistik Indonesia 1991). These differences in population density reflect the islands' different agricultural potentials. The coastal swamps, forest-covered interior and poor soils of Kalimantan, all unfavourable to settlement, until recently have prevented economic development of the extent found in Sumatra, Sulawesi and Java. In the last twenty years this picture has changed, with increasing development and investment in Kalimantan in accordance with the Repelita programmes.

The face of Kalimantan is changing fast. The island is rich in natural resources – timber, oil and other minerals. Logging and the export of timber, oil and other natural resources have made Kalimantan an important contributor to the Indonesian economy (fig. 13.1). East Kalimantan accounts for 21% of the country's export timber revenues. (Repetto 1988). Kalimantan's open spaces have attracted the attention of developers and regional planners for industry, transmigration, agriculture and plantations. Expansion in the industrial sector includes establishment of oil refineries, natural gas (LNG) industrial plants, and fertiliser factories in East Kalimantan, pulp and plywood mills in Pontianak, Samarinda and Banjarmasin, and the development of coal mining and other mining activities throughout Kalimantan. All of these economic developments have environmental costs and consequences.

Mineral Resources

Kalimantan is rich in mineral resources. Indeed, the name Kalimantan may be a reference to the diamonds and other precious stones for which the island has been renowned for centuries. Probably the first important exploitation of minerals was the mining and working of iron ore which occurs in local deposits throughout Borneo. With the introduction of iron-

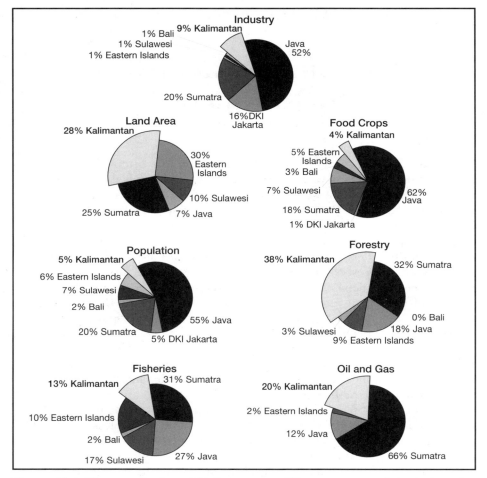

Figure 13.1. The contribution of Kalimantan to different sectors of the Indonesian economy.
Source: RePPProT 1990

working skills from mainland Asia between the fifth and tenth centuries A.D. (Bellwood 1985), local peoples developed sophisticated smelting and iron-working skills which spread rapidly inland from the sixth century onwards (Sellato 1989a). The Apo Kayan, the Mantalat River in the upper Barito basin, the Mantikai tributary of the Sambas and the Tayan tributary of the Kapuas in West Kalimantan all had iron ore deposits and became well-known sites for smelting and iron-working (Avé and King 1986).

Iron brought fundamental changes to the peoples of Kalimantan. With iron tools the forest could be cleared more easily; this was a key factor in the spread of shifting cultivation. *Mandau* or fighting swords, made from local iron by Dayak smiths, were much-prized trade goods between Dayak tribes (Wilken 1893). Trading towns grew up in the Sarawak River delta and Brunei, probably connected with the massive diffusion of iron technology along Borneo's coasts. By 1000 A.D. Santubong, Sarawak, was a major trade centre with an important iron-smelting industry (Avé and King 1986); by the fourteenth century Santubong was one of the richest towns in Southeast Asia (Sellato 1989a).

Gold and diamonds have also been collected since time immemorial and traded to the sultans' courts and to Hindu and Chinese traders. Traditionally Dayaks hardly ever manufactured or used gold ornaments themselves (Sellato 1989a), but trade in gold has influenced the island's culture. Gold was being exported from western Borneo at least as early as the thirteenth century; by the late seventeenth century Chinese traders were collecting cargoes of gold at Sambas (Hamilton 1930). Indian contact with the goldfields may have been even earlier (Sellato 1989a). The trade goods received in exchange, jars, beads and brassware, became an important part of Dayak ritual. The quest for gold also brought new immigrants. In the late eighteenth and early nineteenth centuries, thousands of Chinese miners flocked to work in the goldfields of West Kalimantan and Sarawak (Jackson 1970). When the gold workings were spent these Chinese, mainly Hakka, turned to farming and plantation work. Today West Kalimantan has one of the largest Chinese communities in Indonesia, and Chinese make up more than one-fifth of the population in Sarawak and Sabah.

The sultans' courts on the coasts were centres of craftsmanship, with expert gold, silver and brass workers. This technology was transferred to the inland tribes, and several Dayak groups became expert metal workers and craftsmen in silver and copper (Chin 1980). The famous *mandau* swords manufactured by the Kayan, Kenyah and peoples of the upper Mahakam are decorated with silver, copper or bronze decorations. The inland Maloh also learned silverwork and produced fine brass and bronze jewellery, much prized by other Dayak groups (King 1985). The lack of copper and zinc sources locally meant that these metals had to be imported, so bronze/brass industries were mostly limited to coastal areas (Sellato 1989a).

Today Kalimantan is still noted for its gold, diamonds and precious and semiprecious stones. Indeed, high gold prices and a renewed flurry of exploration for this valuable metal may push gold to new prominence as an export earner in the next decade. For the moment "black gold", oil, is Kalimantan's most important mineral export, but the island has considerable mineral wealth besides. East Kalimantan has useful deposits of gold, nickel, coal, lead and zinc, kaolin, antimony, phosphate, quartz sand, sulphur (in the Sangkulirang area), pyrites, gypsum, crystalline quartz, zircon, muscovite, clay, diamonds, iron sand, ruthenium, calcite and limestone (fig. 13.2). Central Kalimantan has gold, diamonds, coal, iron ore, limestone, kaolin, quartz sand and crystalline quartz (UNIDO 1984). South Kalimantan has rich deposits of nickel on Pulau Sebuku, iron at Gunung Ulin and Riam Pinang, and manganese in the Hulu Sungai. Coal seams occur under most of East and South Kalimantan, as a broad swathe through Central Kalimantan and in the upper Kapuas basin. Many of these mineral deposits are commercially viable.

Gold

"Borneo is not only one of the most fertile countries in the world but one of the most productive in gold and diamonds," wrote Raffles more than a century and a half ago (Raffles 1817). Like most of his contemporaries, and many developers since, he overestimated grossly the agricultural potential of the island, but the gold mines, at least during his time, were among the most productive in the world (Jackson 1970). The most significant deposits lay in the west between Sambas and Pontianak in West Kalimantan and around Bau in Sarawak. By the early nineteenth century West Borneo was the major single source of gold in Asia, yielding nearly one-seventh of total world output (Crawfurd 1820).

The deposition of solutions bearing gold and other minerals, especially copper and silver, occurred chiefly in association with the Tertiary intrusion of acid igneous rocks (Jackson 1970). Gold is extremely heavy, and the main concentrations are often found quite close to the primary veins with only the finest dust carried downstream. Some gold is to be found in almost all West Kalimantan rivers, and in many other Kalimantan rivers (Jackson 1970). For centuries gold was retrieved on a small scale by Dayak miners, panning for gold dust in the rivers. Gold is still collected in this fashion by local people in all four provinces of Kalimantan. *Dulang*, shallow wooden trays for gold panning, are on sale in local markets such as Martapura.

The first major commercial gold-mining operations were Chinese-run. In the gold rush of the late eighteenth and early nineteenth centuries, the richest and most accessible goldfields were worked first, with the largest mines between Sambas and Pontianak around Mandor. Then the Chinese migrated westwards into Landak territory in the Kapuas basin and, as these deposits were worked out, began to open very small mines in the interior. By the second half of the nineteenth century the Kalimantan gold-mining industry had faded rapidly, but it left a lasting environmental and cultural impact. The landscape was marred by large areas cleared of vegetation and honeycombs of abandoned, water-filled diggings, reservoirs, sluices and canals. Chinese immigration also changed the agricultural landscape of West Kalimantan and brought the development of irrigated

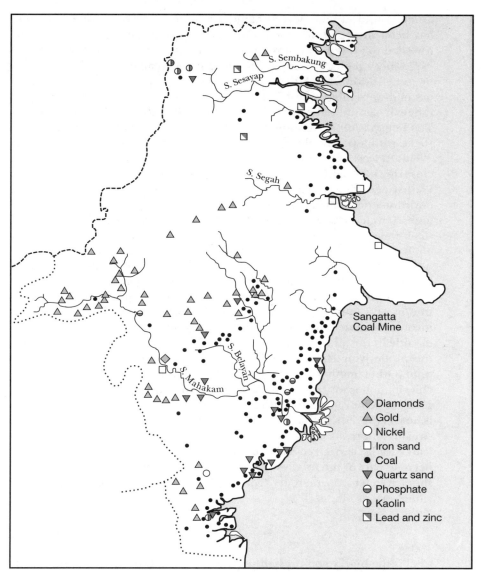

Figure 13.2. Mineral resources of East Kalimantan. (After Voss 1983.)

rice fields, vegetable gardens and pig rearing (Jackson 1970).

Kalimantan is now in the throes of a new gold rush. Since 1984, falling oil revenues and rising gold prices on the world market have encouraged a new round of commercial exploration, and the whole of Kalimantan has been divided into gold-mining concessions. Geological teams have travelled to even the most remote inland valleys to prospect for gold, and exploratory geological surveys are being made in all four provinces, mainly by joint Indonesian-Australian ventures. Commercial operations have opened at Sambas, West Kalimantan, in the foothills of the Schwaner range in Central Kalimantan, and on the Kelian River in East Kalimantan. The Kelian River site is estimated to have 55 million tonnes of gold-bearing rocks, with a potential yield of 1.9 g of gold per tonne (*Jakarta Post*, June 1989). At least one commercial operation in Central Kalimantan has had to close down because of the activities of illegal gold miners. Elsewhere, as at Pleihari, South Kalimantan, illegal gold miners are causing considerable environmental damage, excavating whole hillsides in the search for mother lodes. Careless disposal of cyanide and mercury, used in gold extraction by small-scale gold-mining units, is allegedly poisoning river water and causing ailments in some upper Barito communities.

The cyanide and mercury in gold-mining effluents are potentially harmful to aquatic life; fish in particular are extremely sensitive to minute concentrations of cyanide. Environmental factors can reduce the concentration of a number of pollutants, including cyanide. Natural processes involved include photodecomposition by sunlight, acidification by carbon dioxide in the air, oxidation, dilution, adsorption onto solids and seepage into underlying strata, as well as biological action over long periods of time. Other methods are increasingly expensive and complex. Alkaline chlorination is widely used at Canadian gold mines (Taylor and Sukarsono 1991) to remove toxic metals, but reagent costs are high, and the cyanide is not recovered. Simple disposal of cyanide or mercury-bearing waste into streams and rivers is particularly hazardous since these water-borne chemicals may be carried to sites far from the original source of pollution or become part of the food web. For instance, the vegetable *kangkong* (water spinach) takes up mercury from the water. Contaminated spinach can cause serious illness.

Coal

Coal is the highly compressed remains of partially decayed swamp vegetation, sandwiched between layers of marine sedimentary deposits. Primeval swamp forests covered much of the earth many millions of years before the first gymnosperm (conifer) or angiosperm (flowering) trees evolved. The earliest "trees" were in fact primitive ferns. Some, such as the giant club moss *Lepidodendron*, were as tall as 30 m. Fossil remains of these ancient forests have been found in the coal seams of Kalimantan and Sumatra, which formed about 300 million years ago during the Carboniferous period.

Coal deposits are widespread throughout South and East Kalimantan. Opencast coal mining under the control of the sultanates was already in operation in Kalimantan by the early nineteenth century, producing small quantities of low-quality coal for local use (Lindblad 1988). The small state mine at Palaran near Tenggarong, in the sultanate of Kutai, was a typical example. The first modern coal mine in Kalimantan was the Oranje Nassau, opened by the Dutch at Pengaron in South Kalimantan in 1849, more to establish their rights to the island's mineral wealth than because of its commercial potential (Lindblad 1988). Similarly the British set up the British North Borneo Company to run Sabah because they were interested in the coal mines of Labuan. These colonial rights were only established with some inconvenience. To attend the opening of the Pengaron mine, the governor-general J.J. Rochussen had to travel for 13 uncomfortable days in a small canoe (Fasseur 1979).

In 1888 the Dutch East Borneo Coal Company (Oost-Borneo Maatschappij) established a large, modern coal mine at Batu Panggal on the banks of the Mahakam. There were also small-scale indigenous operations at Martapura, along the Barito, along the upper Mahakam and at Berau. In 1903, with Dutch investment, the large Pulau Laut coal mine came into production and by 1910 was producing about 25% of all Indonesian output (Lindblad 1988). The production of the large Dutch mines was exported, whereas the smaller operations catered for the local markets. Poor quality coal and the availability of cheaper coal from Europe, especially Britain, eventually led to the decline of the big Dutch mines on Kalimantan. Recent discoveries of new coalfields have led to renewed interest in Kalimantan coal.

In 1982 total coal production in Indonesia amounted to only 482,000 tonnes, but the government has ambitious plans to produce 17 million tonnes annually during the next decade. Much of this production is expected to come from new mines in East and South Kalimantan, where six major concessions total four million hectares (Bachruddin et al. 1992). Coal is already mined in South Kalimantan at Sungai Pinang, Kintap and Pulau Laut, and a new opencast mine is opening at Batu Licin. East Kalimantan has the largest coal reserves, at Kutai, Pasir and Berau; in 1988 the province exported 428,000 tonnes valued at $13 million. Poor quality coal is mined on a 200,000 ha concession at Tenggarong, but some of the newly opened reserves are of very high quality. In 1989 Kaltim Primacoal opened a new coal mine on the Sangatta River in East Kalimantan, where the coal is of the best quality in the world, being very low in sulphur and yielding only 2% ash. Exploitation of the Sangatta coal seams will involve building a town and communications for three thousand families, although the mine is expected to be viable for only fifteen to thirty years. The environmental damage caused by the opencast mining operations will last much longer than the lifetime of the mine, the legacy of the production cycle of a nonrenewable resource (fig. 13.3).

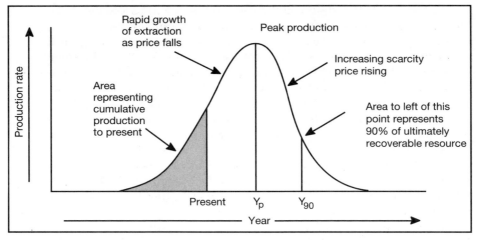

Figure 13.3. The production cycle of a non-renewable resource: a mineral such as gold ore or a fuel such as coal, oil or natural gas. (After Holdren 1975.)

Limestone

There are scattered limestone outcrops throughout the coastal mountains of Kalimantan and in the Sangkulirang peninsula (see chapter 6). In South Kalimantan, limestone deposits with potential for commercial exploitation occur at Batu Licin, Rantau, Tabalong and in the Meratus Mountains. Some of the Sangkulirang deposits are already being quarried for hardcore for roads, and plans to set up a cement factory are under consideration. Construction of the PROJASAM road from Muara Wahau to Sangkulirang used limestone for infilling the road, but this was dug from pits rather than from the limestone massifs. Kaltim Primacoal is quarrying some limestone for roads and jetties at the Sangatta coal mine.

Extensive commercial exploitation of limestone outcrops could lead to serious environmental disturbance and destruction of unique flora and cave habitats (see chapter 6). Local communities might also suffer economic hardship from loss of income from natural resources such as swiftlet nests, and from a failure of the durian harvest through lack of cave-roosting bats to pollinate the durian trees.

Diamonds

Diamonds are formed when pure carbon, deep within the earth's crust, is subjected to very hot temperatures and high pressures. Kalimantan's diamonds are thus a legacy from long-distant geological activity. These precious stones have long been collected and traded by local people. The nineteenth century explorer Schwaner writes of the Ngaju Dayaks of Central Kalimantan digging pits to search for diamonds (Schwaner 1853). Today most diamond-mining operations are still on a small scale, with local people digging pits, excavating the earth and washing the soil to recover diamonds and other precious and semiprecious stones. Such diggings can be seen at Cempaka in South Kalimantan, an area of cratered, water-filled earthworks. Diamonds are also mined at Loktabat, Karamunting, Karang Intan and Riam Kanan in the Meratus foothills, and at Kota Baru on Pulau Laut, but Martapura is the renowned centre for diamond mining. Kalimantan diamonds are of good quality, and the diamonds auctioned at the Friday market at Martapura are sent to diamond merchants all over the world, including the diamond-cutting centres of Antwerp and Amsterdam. The Acorn diamond mine, a joint Indonesian-Australian commercial venture, has opened near Martapura.

Other precious and semiprecious stones are mined in the Martapura area by local people panning the muddy sediments. Small diamonds and other stones are polished in Martapura and sold on local markets. Quartz, rutile, jasper, agates and amethysts are all found in this area (*Voice of Nature*, September 1989). Every local farmer and fisherman hopes to "strike it rich" one day. In June 1989 a fisherman, seeking a rock to weight his net, found a pretty stone and took it home. The rock turned out to be a 61.5 carat diamond worth Rp 200 million ($100,000) (*Jakarta Post*, 2 September 1989).

MINING AND THE ENVIRONMENT

Mining is one of the most environmentally damaging of all economic activities, scarring the earth's surface, destroying habitats and producing a large quantity of waste in the form of slag heaps, waste dumps and tailings. Mines and processing units for basic raw materials are a source of water contamination from ore tailings and chemicals. Open pits are a hazard to wildlife and people; abandoned workings filled with stagnant water can become breeding grounds for mosquitoes and other vectors of disease. Roads, quays, airstrips and waste dumps may impede natural drainage patterns.

Every mining operation is faced with the safe disposal of the waste products from its processing plants. Spoil heaps and tailings are sometimes allowed to contaminate streams. Hazardous materials such as chemical wastes and polychlorinated biphenyls (PCBs) should not be abandoned on site. After mining there should be an attempt to restore the land to protect watershed values, to control surface erosion and to encourage natural regeneration of vegetation and return of wildlife, by providing cover, food plants and water holes with good quality water.

Some of the most serious environmental consequences from mining arise from water contamination, either of surface water or of groundwater (Kelly 1988). The sources of water-borne contaminants from mining operations can be classified into two groups: acids generated from the exposure of iron sulphide minerals to the atmosphere; and contaminants which result from mining and processing of the ore and disposal of the waste. Processing of ores to recover valuable minerals involves a range of chemical processes and waste products, many highly toxic. Acidic effluents emanating from coal mines are a recognised environmental hazard. Acid drainage is caused by underground and surface water contacting metal sulphides, becoming acid and acquiring dissolved heavy metals. In sufficient concentrations these metals are toxic to fish and other aquatic life.

Chemicals and heavy metals in surface waters can become concentrated as they pass along the food chain. Chemicals in groundwater are an even more serious problem. Migration of contaminants from mine tailings into groundwater can result in serious problems long after mining has ceased. Once groundwater is contaminated it is very difficult, if not impossible, to decontaminate it, and it will remain polluted for decades. Unlike surface waters, groundwater is not cleansed by exposure to air or by dilution.

The treatment of chemical waste is beyond the scope of this book. The Environmental Management Development in Indonesia programme (EMDI) worked with the Ministry of Population and Environment (KLH) to prepare guidelines for monitoring and controlling toxic and hazardous waste (Nagendran 1991) and pollution levels, and to develop water quality standards (Taylor and Sukarsono 1991).

Petroleum

Petroleum deposits are the remains of microscopic plants and animals that were buried in the mud and sand of shallow prehistoric seas. These organisms were slowly decomposed by bacteria to leave a residue of hydrocarbon compounds which, under conditions of high temperature and pressure, were converted into oil and gas. This process is well illustrated in the 80 km wide Mahakam delta. Seismic studies show that since the middle Miocene (20 Ma ago) three major deltaic complexes have accumulated. Compaction of vegetation has led to its degeneration into hydrocarbons which accumulate in the underlying sandstone reservoirs (Combaz and de Matharel 1978). There are several oil and gas fields in the Mahakam delta. These include Attaka field (opened in 1970), with a potential production of 100,000 billion barrels per day (bbl/day), Bekapai (1972) 50,000 bbl/day and Handil field (1974) 120,000 bbl/day (Combaz and de Matharel 1978). The oilfields are also associated with gas. Oil and liquefied natural gas (LNG) are Indonesia's most important export commodities (Nontji 1986).

Indonesia is the second largest petroleum producer east of the Persian Gulf, surpassed only by the People's Republic of China. Oil production in Kalimantan dates back to 1897, when a former colonial administrator Menten (who also established the Batu Panggal coal mine on the Mahakam) started experimental drilling in Kutai, with remarkable results (Lindblad 1988). By 1902 both Shell, with concessions acquired from Menten, and Royal Dutch were producing oil from the Kutai fields in the Mahakam delta, and in 1905 oil was also discovered off Tarakan. In 1907 Shell and Royal Dutch combined their operations (Lindblad 1988). The Kutai and Tarakan basins are still the major oil-producing fields in Kalimantan today.

In 1980, with Indonesian exports of oil and gas at a peak, Indonesia produced 77.5 million tonnes of petroleum, less than 3% of total world production, and ranked thirteenth on the world producer's list. Production and processing of natural gas had increased to 12,000 million m^3 in 1980, less than 1% of total world production, but Indonesia has now become one of the most important producers and exporters of liquefied natural gas (Donner 1987). Oil and gas provide two-thirds of Indonesia's tax revenues and are a major source of foreign exchange. Most of this oil comes from the east coast fields of Sumatra, which in the early 1970s contributed 92% of Indonesia's total production (Burbridge et al. 1988), but East Kalimantan is becoming increasingly important as a supplier of oil and natural gas. In 1987 the total production from the coastal zone of East Kalimantan amounted to 90,369,381 bbl of oil and 15,240,399 million m^3 of natural gas (Dahuri 1990).

The most important oilfields in Indonesia are located in the Tertiary basins. Commercial oil production is generally confined to upper Tertiary beds (Miocene and Pliocene). The oilfields off southeast and northeast Kalimantan lie in the Tertiary geosynclines on the edge of the Sunda shelf. The main oilfields in Kalimantan (table 13.1) are in the Balikpapan-Samarinda region and around the island of Tarakan, both onshore and offshore. In the Barito basin in southern Kalimantan commercial amounts of oil were found in beds of Eocene age (Husin and Suradi 1973). Kalimantan's main oilfields are shown in figure 13.4.

Oil refineries have been constructed at Balikpapan (opened 1950) and Sungai Pangkalan (1971) in East Kalimantan. Oil from the South Kalimantan field at Tanjung is also piped to Balikpapan for processing. Balikpapan produces kerosene (18%), diesel oil (13%), gasoline (11%) and fuel oil (41%). Sungai Pangkalan's main products are kerosene (21.9%), fuel oil (49.5%), gas oil (9.7%) and gasoline (15.6%) (Burbridge et al. 1988). Liquefied natural gas facilities have also been built in South

Bontang Bay, the largest LNG plant in the country. Natural gas is used to produce urea and ammonia fertiliser at Bontang; a third fertiliser plant was opened by President Soeharto in 1989. The effect of these developments has been to create focal points of economic activity and associated urban development; Bontang is the fastest-growing town in Indonesia. The exploitation of oil and natural gas and associated products can have major environmental impacts, especially on local ecosystems.

Oil and gas developments cover large tracts of coastal lands and waters in the form of leases, but actual permanent production facilities, pipelines and refineries occupy a very small area. Nevertheless, their local impact on natural ecosystems, population concentrations and economic activity has been significant. Any exploitation of oil stocks runs the inevitable hazard of oil spills and leaks and the associated pollution. Construction work and the discharges of oil refineries, processing plants and factories can change the ecology of nearby habitats. Local fish stocks may decline because of increased water temperatures from discharge of cooling water from processing plants. Many species in tropical waters live at or near the upper limits of their temperature range, and a rise in water temperature of only a few degrees can prove lethal (Johannes and Betzer 1975). For some fish species, sex is determined by water temperature, so a change in temperature can change the demography of a population and lead to reduced breeding activity. Hot water discharged from the LNG plant at Bontang has killed local coral reefs and reduced other marine life (fish, crustaceans, echinoderms) near the outfall (Nontji 1988). Another problem is the disposal of mercury extracted as a solid waste during the gas liquefication process (Burbridge et al. 1988). Other potential sources of pollution associated with LNG plants are: carcinogenic smoke from gas flares at the LNG plant; urea dust from the fertiliser plants; and methylene and mercury from the olefin plant (Burbridge et al. 1988).

As industries expand there is increasing pressure on nearby land for development and on water resources for industry. The oil towns of Sangatta and Bontang are encroaching into the lowland forests of Kutai National Park (Wirawan 1985). The eastern fringes of the park are badly damaged

Table 13.1. Potential recoverable oil and gas reserves of Kalimantan.

Basin	Oil and gas reserves (billions of barrels of oil equivalent)
Ketungau/Melawi basin	0.0510
Barito basin	0.3199
Asam-Asam basin	3.1450
Kutai basin	22.9310
Tarakan basin	2.0650
Total	28.5119

Source: Nayoan 1981

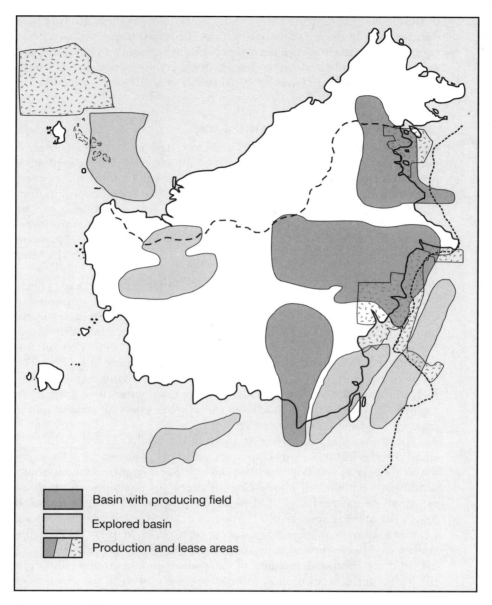

Figure 13.4. Kalimantan oil basins and oil fields.
Source: Salm and Halim 1984

by the activities of agriculturalists and settlers, many of them growing produce to supply these industrial centres. In an attempt to protect the aquifers and water supplies necessary for Bontang's industrial plants, the watershed behind the town has been declared protection forest where agriculture is banned (Wirawan 1985; Petocz et al. 1990).

Oil Pollution

Exploitation of oil, and other forms of energy, have environmental costs as well as economic benefits (table 13.2). In any area where oil is exploited, processed and shipped, it is inevitable that there will be some spillage and oil pollution. This was recognised early in the history of oil exploration in Indonesia. Legislation to protect the environment in Indonesia can be traced back to 1927 when the Storage of Oil Ordinance came into force. This was followed in 1967 by the Basic Mine Law; Pertamina has also set up an agency to coordinate all activities regarding oil pollution (Donner 1987).

Damage to living organisms from oil pollution may include: toxic effects leading to changes in metabolic functions such as respiration or photosynthesis; physical smothering; and tainting. The amount of damage to, and impact of oil pollution on, living resources depends on a number of factors including: volume and type of oil (fig. 13.5); climate and weather conditions; and interactions affecting oil toxicity (Baker 1983; Salm and Clark 1984). The frequency of oiling also affects the amount of damage. Thus plants and animals in coastal waters near ports and piping installations may suffer from many small spillages rather than from the dramatic effects of a major spill as experienced when a tanker runs aground.

The Showa Maru oil tanker accident in the Straits of Malacca in 1975 showed some of the serious consequences of oil pollution in Indonesian waters (Soegiarto and Polunin 1980; Baker 1982). Mangroves are especially sensitive to oil pollution. The oil clogs the trees' pneumatophores which act as organs of gas exchange. Oil pollution raises water temperature and lowers dissolved oxygen levels, leading to death of some mangrove trees. Leaves may also be damaged by oil (Lugo et al. 1978; Mathias 1977). Acute short-term effects of oil in mangrove swamps are likely to be: high mortalities of invertebrates; defoliation of mangroves; and death of seedlings. In the longer term, trees in most affected areas are likely to die. Such trees may be old and large, so that recovery by regeneration to pre-spill forest may take a long time (Baker 1983). Mangroves and other coastal and riverine ecosystems are also susceptible to other forms of industrial pollution (Saenger et al. 1983).

Oil pollution of sandy beaches is not usually a serious problem, but where it occurs it affects animal life. Organic pollution, by oil or domestic waste, can reduce the depth of the aerated layer of sand and so affect the zonation of beach biota. Oil on the shore between the sea and laying areas can coat adult turtles, contaminate eggs, and trap young hatchlings on their way to the sea (Baker 1983). On rocky shores, algae and seaweeds in rock pools and high up on the intertidal zone are particularly susceptible to oil pollution (O'Brien and Dixon 1976). Shore molluscs include: grazers (e.g., limpets, chitons and winkles); carnivores (e.g., whelks); filter feeders (e.g., mussels, cockles, oysters, clams and scallops); and deposit feeders. In

Table 13.2. Some environmental impacts of energy production.

	Health	Property	Social	Quality of life	Environmental services
Exploration					
oil/gas	–	–	–	invasion of wilderness	–
Harvesting					
coal mining	accidents, black lung	loss of farmland, subsidence	use of public lands	defaced landscape	acid drainage
offshore oil	accidents	–	–	oil on beaches	oil as a biocide
hydroelectric dam	dam collapse	loss of farmland	displacement of residents	loss of wild rivers	fish passage, wildlife breeding grounds
Processing					
oil refining	air/disease	air/crops	–	smells, visibility	pollution of estuaries
shale processing	air/disease	water consumption	–	waste piles	water pollution
Conversion					
coal power plant	air/disease	air/crops, buildings	–	noise, visibility	acid rain, CO_2/particles/climate
fission reactor	reactor accident that breached containment would produce all classes of impact				
Transportation					
oil tanker	fire	fire, collision	–	oil on beaches	oil as biocide
electrical transmission	electrocution	restriction on land use	–	unsightly towers	–
plutonium	leak/cancer	land contamination/quarantine	terrorism, nuclear bombs	–	–
Consumption					
automobile	air/disease	air/crops	suburbanisation	noise, visibility	paved environment, heat/climate
Waste management					
radioactive wastes	leak/mutations	land use	terrorism, sabotage	–	groundwater contamination

Source: Holdren 1975

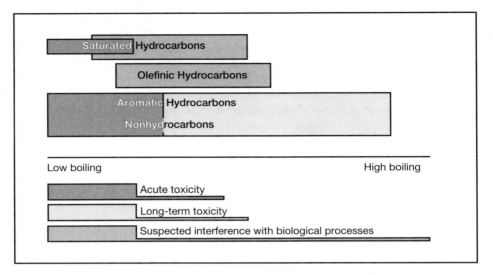

Figure 13.5. Diagram of the toxicity to marine organisms of different fractions of crude oil. The low boiling-point hydrocarbons are the most poisonous but are the first to evaporate. (After Blumer 1969.)

the intertidal zone and in shallower water any of these may be killed by smothering or through toxic effects of the oil. Filter feeders and deposit feeders are also particularly likely to ingest and accumulate dispersed or sedimented oil. Such tainting may affect many species of commercial importance. Once pollution ceases, however, concentrations of hydrocarbons in the tissues are usually reduced to low levels within thirty days (Baker 1983). Usually intertidal areas are denuded of algae and animals following severe oil pollution but are readily repopulated by algae once most of the pollutants have been removed. Algal cover then remains high for a long time due to the slow recovery of grazing animals. Many algae are of economic importance (Salm and Halim 1984); oiled or tainted algae lose their commercial value.

Oil slicks at sea also seriously affect marine animal life, by killing the phytoplankton and zooplankton at the base of the food chain as well as by the more immediate poisoning of birds, mammals and fish. Mammals and birds using intertidal areas or the sea-air interface are particularly at risk in oil spill areas. Oil on the coats of marine mammals and birds affects the animals' insulation and buoyancy. As the animals make a desperate attempt to clean their fur or feathers they are poisoned by the oil; they may also suffer from eating oil-contaminated fish and other marine foods. Wading birds and some land mammals may be poisoned by eating fish and carcasses contaminated by oil. Severe damage is likely to upper-reef corals in shallow water, especially if oil slicks coat the corals during extreme low tide. Coral reefs and seagrass beds off the coast of East Kalimantan were badly affected by oil spills during the 1940s (Schulz 1984). The environmental costs of oil spillages are high, and borne mainly by local communities in terms of health hazards, reduced fish catches and losses in prawn fisheries. Coastal communities are also at greater risk of storm damage, since destruction of mangroves and coral reefs reduces their effectiveness for coastal protection.

In fresh waters, oil spills affect wildlife as well as human health and wellbeing. Some spills, particularly of light products, may cause substantial fish kills in rivers, lakes and shallow sea areas (Green and Trett 1989). Oil pollution also kills plants and inhibits growth: by acting as a physical barrier and preventing gas exchange; through direct toxicity; by affecting soil properties; and by affecting growth and the ability to compete (Baker 1983). The plants most commonly affected by oil pollution are coastal communities like mangroves and seagrass species, but some terrestrial and freshwater species may also be at risk. Low concentrations of oil may inhibit photosynthesis and depress growth of phytoplankton in fresh waters. Attempts to clean up a spill may actually cause further damage to some life forms. Dispersants and detergents used in cleanup operations may have more damaging effects on phytoplankton than the oil itself (Baker 1983). Blue-green algae may increase in numbers after pollution.

Spills of light oils can cause large mortalities of organisms in streams and rivers. Spills (especially of heavier oils) in ponds and lakes with no through-flow, or with restricted flow, present the extra problem of the persistence of the pollutant and slow recovery rates. Spilt oil may penetrate to the water table and pollute drinking water. Petroleum spills also affect soil properties and microbial action. Fertilisers and ploughing can help the recovery of agricultural land after a limited oil spill (Baker 1983).

Industrial Developments

Industrial developments take up only a very small area of Kalimantan, but their local impact on natural ecosystems has been significant. Most industrial sites in Kalimantan are located in the coastal zone and near rivers. They produce wastes, including toxic chemicals, which are discharged into rivers and coastal waters (table 13.3).

In 1980 Indonesia introduced a new forest policy, banning the export of raw logs, to help to build up a national wood-processing industry. This led to the construction of many plywood mills in the major towns of Balikpapan, Samarinda, Banjarmasin and Pontianak. National pulp and paper production has also increased. Plywood mills, pulp mills, rattan factories, palm oil and rubber factories and sugar refineries all discharge their effluents into rivers near the coastal zone. There has been little detailed study of these practices in Kalimantan and other parts of Indonesia, but it is likely that industrial activities affect water quality. A study in Peninsular Malaysia monitored the effects of industrial effluents from a rubber factory outfall on water quality and macrofauna composition at various points along the river and in adjacent mangroves. Conditions were most unfavourable for aquatic life near the outfall, where effluent concentrations were highest. The effluent had caused several species of fish to disappear from the estuary, while others had moved downstream from the outfall (Seow 1976).

The lower Mahakam in East Kalimantan, which is badly polluted, is one

Table 13.3. Major waste products and their receiving environments.

	Environments into which wastes are discharged: X Environments into which wastes get transferred: O				
	Air	Fresh water	Oceans	Land	Clinical effects of residues on humans?
Gases and associated particulate matter (e.g., SO_2, CO_2, CO, smoke, soot)	X	O	O	O	yes
Photochemical compounds of exhaust gases	X	O	?	O	yes
Urban/industrial solid wastes			X	X	no
Persistent inorganic residues;					
lead (Pb)	X	O	XO	X	Pb-disputed
mercury (Hg)					Hg-definitely
cadmium (Cd)					Cd-definitely
Persistent organic compounds:					
oil			X	O	no
organochlorine residues	O	XO	X	X	disputed
pharmaceutical wastes		X	O		unknown
Organic wastes:					
sewage		X	X		possible bacteria carrier
Fertiliser residues:					
with nitrogen (N) and phosphorus (P)		O	O	X	yes, especially N
detergent with P		X	O		no
Radioactivity	X	O	X	O	yes
Land dereliction				X	no
Heat	X	X	XO		no
Noise	X				yes
Deliberate wasting (e.g., defoliation in CBW)		O	O	X	yes

CBW = chemical or biological warfare

Source: Simmons 1981

Box. 13.1. AMDAL: A guide to environmental assessment in Indonesia.

AMDAL Analisis Mengenai Dampak Lingkungan, (the analysis of environmental impacts) is an integrated review process to coordinate the planning and review of proposed development activities, particularly their ecological, socio-economic and cultural components. A parallel process, SEMDAL, applies to projects which were already underway in 1987, but had not yet assessed their environmental impacts.

The AMDAL process is regulated by Government Regulation No. 29 1986, the first environmental protection legislation promulgated under the key Indonesian environmental law, Act Number 4, 1982, which establishes the principle of sustainable development.

Overall coordination of the AMDAL is the responsibility of a new environmental agency (BAPEDAL - Badan Pengendalian Dampak Lingkungan). The authorities responsible for AMDAL implementation are 14 sectoral government departments and other institutions at the central level and 27 provincial governments at the regional level.

The goal of AMDAL is to facilitate and expedite economically sound, environmentally and socially acceptable development ventures. A series of AMDAL documents are prepared. The PIL or PEL determines if the project will have significant environmental impacts and therefore requires an ANDAL or SEL. A major project with known impacts may go straight to ANDAL/SEL without a PIL.

The ANDAL and SEL include:
- An elaborate and in-depth study of the potential environmental impacts of a proposed activity
- A plan of action to manage and mitigate the predicted impacts (proposed RKL and RPL design), for example through changes in project design or location
- Identification and evaluation of residual impacts, i.e., those which cannot be mitigated or managed.

The RKL sets out the design and operating requirements for mitigating environmental effects caused by potential or existing projects. The RPL assesses whether mitigative measures suggested in PIL/PEL or ANDAL/SEL and RKL documents are effective, and proposes appropriate modifications.

General relationship between AMDAL and project planning

AMDAL Stage	Project Cycle Phase
Initial screening of project	Planning and programme development
PIL	Pre-feasibility design
ANDAL and conceptual outline	Feasibility design
Detailed RKL/RPL	Detailed design and permits
Implementation of RKL/RPL; modification if necessary	Pre-construction: construction; operation; post-project evaluation

Source: BAPEDAL and EMDI 1992

of the rivers receiving attention under the Prokasih (Clean Rivers) programme. Only 20% of East Kalimantan coastal waters are fished, yet already industrial pollution may be affecting marine fisheries. Contamination of coastal waters has already resulted in a reduction in fish populations in the polluted waters off North Java and East Sumatra (Donner 1987).

Industrial Effluents and Pollution

Pollutants of water can be divided into four categories: pathogens, toxins, deoxygenators and nutrient enrichers (Prowse 1968). **Pathogens** are associated with untreated sewage and include a wide range of bacteria, protozoa and parasitic worms harmful to man and other organisms. **Toxins** are derived from industrial waste and from agricultural chemicals (table 13.4). Their effects may be immediately obvious and dramatic, such as the death of large numbers of fish, or cumulative over the long term. Moreover, because of the interconnections between different aquatic systems, the effects of such pollutants can be manifest in ecosystems far from the original source of contamination.

Deoxygenation arises as the result of natural processes of decomposition of organic matter by bacteria and fungi. When large quantities of organic wastes are released into water, deoxygenation and the subsequent death of animals by suffocation can be expected. **Nutrient enrichment** is known as eutrophication (chapter 3). Eutrophic habitats make productive fisheries. Accordingly fish ponds are often fertilised artificially to increase fish production. Naturally eutrophic systems are generally well-balanced, but the addition of extra nutrients (e.g., fertilisers in runoff from agricultural fields) can upset this balance and lead to spectacular algal blooms, as a result of high nutrient levels and favourable temperature and light conditions. The blooms are a natural response to environmental change, but when the water can no longer support high levels of algal populations, the

Table 13.4. The major types of pollution in Indonesian rivers and coastal waters.

Category	Example
Synthetic organics	Phenols in wood treatment
Chlorinated products	Pesticides in agriculture
Sedimentation	Siltation, coastal changes
Litter	Plastic bags, polystyrene
Micro-organisms	*Vibrio* and coliform bacteria
Trace metals	Mercury, lead, tin and copper
Biostimulants	Fertilisers, urban sewage
Fossil fuel compounds	Oil, tar balls, wastes

Source: Soegiarto and Polunin 1980

Box 13.2. Protection and wise utilisation of valuable ecosystems during development.

In planning developments, the following points should be considered:
(1) Areas of naturally high primary productivity such as wetlands, coral reefs, seagrass beds and lowland rainforests should receive high priority for preservation or management. Wherever destruction or serious alteration is contemplated, benefits and costs should be carefully weighed.
(2) Will the proposed development or alteration result in the loss of productive farmland on site or downstream?
(3) Will the sequence of ecological succession be seriously disrupted or retarded? What will be the consequences? Will it be possible for natural processes to repair the damage?
(4) Will important wildlife habitat be damaged or lost? Critical areas include: (a) feeding areas; (b) protective cover; (c) spawning and nursery areas, breeding grounds, nest sites.
(5) What changes in the animal community might be expected to accompany development? Will valuable species be lost? Will pests or disease vectors (e.g., mosquitoes) become more common?
(6) Are there endangered plant or animal species dependent upon the area to be developed? Can they be protected by preserving features necessary for their survival or by leaving pockets of habitat as refuges?
(7) Will animals or plants which form key links in foodwebs be properly protected?
(8) Does the area support significant or potential commercial fisheries? Is fish production important in the local economy?
(9) Can the area be utilised for aquaculture operations? Will the proposed development limit or preclude this option?
(10) Can a healthy ratio of subsidised to unsubsidised ecosystems be maintained? Will sufficiently large natural areas that are not dependent upon direct human management be retained? Have the political and economic consequences of failure to do this been anticipated?
(11) Can moderate diversity of animal and plant species be maintained? Avoid extensive monocultures in farming and forestry to limit large-scale disease and pest outbreaks.
(12) Chemical herbicides and pesticides are valuable when utilised properly. Avoid compounds which are known to be extremely toxic or persistent in the environment; use only in situations where there is no other suitable alternative, and apply at recommended levels.
(13) The accidental introduction of new plant or animal species must be avoided. Some introductions (e.g., many food plants) are valuable additions to the local biota, but any introduction should be planned with care.
(14) If the carrying capacity of either humans or animals is to be increased significantly, are basic needs such as water, food and shelter sufficient to meet the demand? Will the quality of life or the health of individuals be lowered?
(15) Does the area to be developed contain sites or ruins of historical or religious significance? Can these be preserved?
(16) Are there existing human inhabitants of the region? Will their traditional rights and needs be accommodated? To what extent are they dependent upon natural populations of plants and animals? Will the proposed development destroy or restrict access to these resources; if so, how will the people be compensated?
(17) Will areas of natural beauty be degraded or destroyed? Are these important to an existing or potential tourist industry?

(Adapted from Odum 1976.)

algae die and decay. The rapid decomposition of organic debris by bacterial action robs the water of its oxygen, sometimes to the extent that fish and other aquatic organisms die. Phosphorus is usually the limiting factor in eutrophication. Once the artificial inputs cease, phosphates become locked into the bottom sediments of the pool or lake, and natural balance is restored.

Industrial effluents range from organic substances, with a high biochemical oxygen demand (BOD), to those which are highly toxic or thermal. Palm oil, rubber, sugar-refining and other food-processing industries, all found in Kalimantan, have carbohydrate-rich wastes and can create severe oxygen depletion and destroy aquatic communities. Toxic substances are released by industries such as tanning, mining, chemical manufacturing, pulp, paper and plywood production, and pharmaceutical and pesticide manufacturing (Taylor and Sukarsono 1991).

Pesticides are a major pollutant, derived from land runoff during rain, and are particularly serious if they are not degraded within a short time of application. Modern pesticides have been designed to lose their toxicity only a few days or weeks after application, but older pesticides may remain in the environment for years. Nitrogen fertilisers have a certain toxic potential, as nitrates are converted to nitrites, which may affect respiratory functions. Although pollutants, especially organic pollutants, are broken down over time, ecosystems are complex but fragile. An increasing pollution load may damage some parts of the decomposer chain and thereby reduce the ecosystem's capacity for self-purification. Maintaining a buffer zone of riverine forest between crop lands and waterways reduces the likelihood of water pollution from agrochemicals (fig. 13.6).

Worldwide, more than 63,000 chemicals are in use, yet for many of them we have no knowledge of the ecological effects of their unrestricted use (Taylor and Sukarsono 1991). The fate of pollutants in water is controlled by many factors. Chemical and biological reactions can detoxify a pollutant or make it more toxic. Such reactions depend on factors such as alkalinity, acidity, and the amount of calcium in the water, and also the solubility of the pollutant. An insoluble precipitate is less available to aquatic plants and animals although the chemicals may still reach people through the detritus food chain. Aquatic organisms may be exposed to chemicals present in water, food or sediment. Water-soluble chemicals may enter the organism through the body surface, gills or mouth; chemicals in food may be ingested through the gastrointestinal tract.

The solubility and availability to living organisms of metal ions has led to concern. In high concentrations (in meat, fish and food plants) metals can be a hazard to human health. Some have severe carcinogenic effects, such as minamata disease named after a Japanese fishing community where many people developed tumours from eating fish with a high mercury content. Because metals do not degrade, they are transferred or stored in

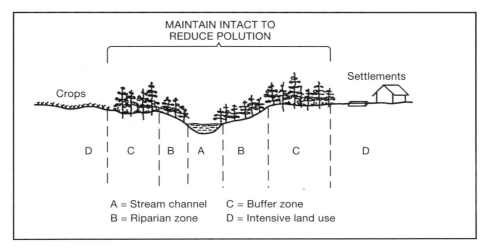

Figure 13.6. Maintaining a buffer zone of natural vegetation between waterways and human settlements and crop lands reduces the likelihood of water pollution.

the water and can become "available" under appropriate conditions. Lead is strongly attracted to particles, and concentrates in sediments; cadmium is likely to remain in water (Taylor and Sukarsono 1991). Biodegradation of pollutants by microorganisms (bacteria, fungi, protozoa and algae) is an important removal process in water and sediments. Microbial degradation is completed by digestion of materials by a number of benthic (bottom-living) invertebrates which are detritus feeders (roundworms, worms, snails and clams). Bioaccumulation and transfer of metals through the food chain may occur through these invertebrates. Bottom sediments tend to be a significant reservoir of organic matter with high bacterial populations. Microbial activity may facilitate the release of metals from sediments, by converting inorganic metal compounds to organometallic molecules which are even more toxic, such as the methyl compounds of arsenic, lead, mercury, selenium and tin.

A toxicant may be present only at very low levels in the water but can exert a large biological effect if it is taken up selectively and accumulated in an organism's tissues (table 13.5). This biomagnification results from retention and concentration of a pollutant that is resistant to degradation through the trophic levels of the food chain (fig. 13.7). Many organisms can activate or deactivate a pollutant (Connell and Miller 1984). Organisms in natural aquatic habitats are constantly exposed to changes in their environment. Species diversity and abundance may be affected by direct and indirect exposure to a toxic substance. Organisms are not equally susceptible to pollution and may be more susceptible at different stages in their life cycles. Thus the eggs, larval and juvenile stages of fish are the most sensitive to chronic exposures. Sublethal effects include increased incidence of disease and behavioural changes. Under some conditions, over long periods of time, local populations may adapt to the polluted conditions. However, tolerant organisms in contaminated areas may accumulate high concentrations of pollutants, and transmit these through the food chain to other less well adapted species, including humans.

The use of water for one purpose affects other uses. Degradation of water quality by municipal and industrial wastes, mining and oil extraction is a serious problem for water management. Industrial wastes can seriously impair water quality for drinking and for other domestic and industrial uses. At present most wastes are discharged directly into rivers and coastal waters, causing considerable pollution and affecting aquatic life and crops. There is urgent need to improve water quality standards in Indonesia (Taylor and Sukarsono 1991).

Table 13.5. Examples of biological magnification.

(a) Enrichment factors for the trace element compositions of shellfish, compared with the marine environment.

Element	Enrichment factor		
	Scallops	Oysters	Mussels
Ag	2,300	18,700	330
Cd	260,000	318,000	100,000
Cr	200,000	60,000	320,000
Cu	3,000	13,700	3,000
Fe	291,500	68,200	196,000
Mn	55,000	4,000	13,500
Mo	90	30	60
Ni	12,000	4,000	14,000
Pb	5,300	3,300	4,000
V	4,500	1,500	2,500
Zn	28,000	110,300	9,100

(b) The concentration of DDT in the food chain.	
Source	Parts per million
River water	0.000003
Estuary water	0.00005
Zooplankton	0.04
Shrimps	0.16
Insects (Diptera)	0.30
Minnows	0.50
Fundulus	1.24
Needlefish	2.00
Tern	2.8-5.17
Cormorant	26.40
Immature gull	75.50

Source: King 1975

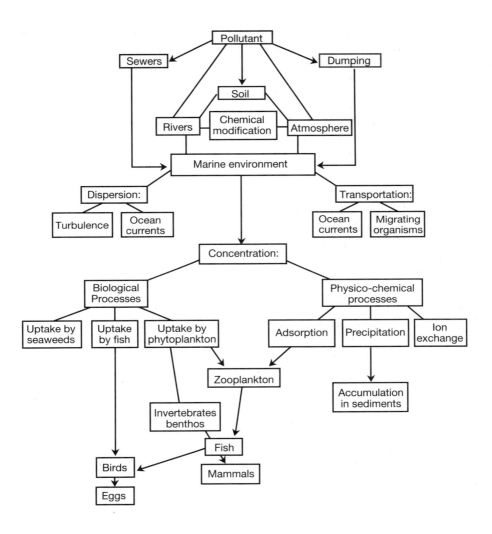

Figure 13.7. A flow diagram of the pathways by which a contaminant can find its way into the oceans and the ways in which it can be concentrated, with various lethal and sublethal effects.
Source: Macintyre and Holmes 1971

Forest Conversion

Mineral resources, such as fossil fuels, are a finite reserve and nonrenewable. Forests, however, are a renewable resource which, theoretically, can be harvested many times on a sustainable yield basis. In practice there are few genuine examples anywhere of sustainable management of tropical moist forests (Wyatt-Smith 1987). The Malayan Uniform System of management in Malaysia, and silvicultural methods in Surinam (de Graaf 1986), demonstrate that if logged-over forests are protected from agricultural settlers and premature repeat cutting, tropical moist forest can be managed as a renewable resource (Wyatt-Smith 1987). Unfortunately less than 1% of all tropical moist forests are managed on a sustainable yield basis (Poore 1989; Gillis 1988). Probably no Kalimantan forests are being managed for sustainable yield production, while Sabah and Sarawak are liquidating their forest resources or overcutting their permanent forests (Poore and Sayer 1988; Burgess 1988; Nectoux and Kuroda 1989). Bornean forests are thus being "mined" rather than managed sustainably.

Although industrial development is increasing in Kalimantan, most industrial activities affect relatively small areas of land and local ecosystems. Forest loss is much more extensive and the greatest environmental change in Kalimantan. Many of the ecological and economic consequences of forest loss have already been discussed in chapter 9. Forest conversion is also contributing to two major environmental problems facing Indonesia; changes in local and global climates, and the expanding area of critical lands.

Forest Loss and Climatic Change

Tropical forests play a significant role in atmospheric regulation mechanisms at the local, regional and global level (Prance 1986; Reynolds and Thompson 1988; Myers 1988c). Forests exchange moisture and energy with the atmosphere. Moisture is transferred from the soil to vegetation, where it is converted by energy into water vapour which returns to the atmosphere. Most of the energy comes from incident sunlight reflected off the vegetation (Myers 1988c). With forest loss there is less scope for evapotranspiration, and less moisture is returned to the atmosphere for recycling as rainfall.

Whether forest loss leads to a decline in local rainfall is a matter of some dispute (Myers 1988c). There is, however, some evidence that declines in forest cover are associated with changes in rainfall patterns in various areas in the tropics, including Western Ghats, India (Meher-Homji 1988), Peninsular Malaysia (Chan 1986) and parts of the Philippines

(Myers 1988c). Influence of forest cover on regional climate has been demonstrated for Amazonia (Salati and Vose 1984; Salati et al. 1981). At a global level, loss of tropical forests is contributing to global warming via the greenhouse effect (Woodwell et al. 1983; Myers 1988c).

The Greenhouse Effect

Global climate is presently getting warmer as a result of increases in carbon dioxide and other "greenhouse gases" in the atmosphere (Bolin et al. 1986). Levels of carbon dioxide have been increasing over the last hundred years and, if the present trend persists, the concentration will double from 0.3% in the middle of the last century to 0.6% by the middle of the next. Although these amounts are still very tiny, they have major consequences for global climate. Most of the sun's energy passes through the atmosphere to warm the earth. The warm ground reflects some of this radiation, some as invisible wavelengths in the infrared spectrum. Carbon dioxide and other greenhouse gases limit the earth's radiation of heat into outer space, causing heat to be trapped and reflected back to the earth's surface. The more carbon dioxide is in the atmosphere, the more radiated energy is absorbed, and the warmer the atmosphere becomes. This is known as global warming or the **greenhouse effect**.

As a result of the greenhouse effect, rainfall patterns, temperature and sea levels throughout the world are changing. Indeed, they have been changing over the last hundred years. As temperatures rise the great polar icefields are melting and shrinking, and sea levels rise. Sea level is currently rising by about 0.1-0.25 cm/yr and may rise even more over the next few years. If this trend continues, low-lying lands will be inundated by seawater. Just a few metres rise in sea level would submerge much of southern Kalimantan, where many of the tidal swamplands actually lie below sea level.

The buildup of carbon dioxide in the global atmosphere stems primarily from combustion of fossil fuels, contributing 5.2 gigatonnes per year (Myers 1988c). The destruction of forests and changing land use contribute to the rise of carbon dioxide levels in the atmosphere in two ways: directly, by burning of forests; and indirectly, since forests act as "carbon sinks". The amount of carbon dioxide released into the atmosphere from burning biomass, especially forests, is believed to be substantial, between 0.9 and 2.7 gigatonnes each year (Myers 1988c).

At the same time as carbon dioxide is being released into the atmosphere from burning of fossil fuels and forests, large amounts are being absorbed by natural reservoirs or "sinks". The tropical rainforests are major carbon sinks. Plants absorb carbon dioxide from the atmosphere and use it to make sugars during photosynthesis; plants also release some carbon dioxide during respiration. Forests are important carbon sinks because,

Box 13.3. The ozone layer – a global issue.

The ozone layer in the stratosphere acts as a screen protecting the Earth from harmful levels of ultraviolet radiation. Ozone (O_3) is created in a photochemical process which begins with the breakdown of molecular oxygen O_2 into two separate atoms as a consequence of radiation. Another molecule of oxygen combines with one of the atoms to produce ozone.

$$O + O_2 = O_3$$

Ozone is destroyed by several complex chemical reactions involving compounds of oxygen, hydrogen, chlorine and nitrogen, with the last three acting as catalysts at very low concentrations. It is also removed from the stratosphere by large-scale transport processes.

Chlorofluorocarbons (CFCs), used as propellants in aerosols, and as blowing agents, solvents and refrigerants, are considered to be the most significant ozone-depleting substances. Other industrially produced chemicals such as halons, methylchloroform and carbontetrachlorides also cause ozone depletion. Gases produced by natural decomposition processes also contribute.

One such gas is nitrous oxide, much of which is produced by bacteria in the soil. If the soil is overtreated with fertiliser, the release of nitrous oxide can increase. Other contributors are major ocean upwellings and the burning of fossil fuels. Methane (CH_4), also the most significant contributor to the greenhouse effect, comes from a variety of sources, including rice sawah, swamps, marshes and cow dung.

Most of the source gases are inert in the lower atmosphere but break down naturally when they have risen above the ozone layer and are exposed to ultraviolet light from the sun. Thus free chlorine atoms are released from CFCs, sink into the ozone layer and there react to remove ozone without consuming chlorine.

$$Cl + O_3 \longrightarrow ClO + O_2$$
$$ClO + O_3 \longrightarrow Cl + 2O_2$$

In general terms the main reactions causing the destruction of ozone are:

$$O_3 + O \longrightarrow 2O_2$$
$$O_3 + X \longrightarrow XO + O_2$$
$$O + XO \longrightarrow X + O_2$$

Net result $O_3 + O \longrightarrow 2O_2$

where the radical X may be H or HO or NO or NO_2 or Cl.

The radicals are produced in the atmosphere only during the daytime. At night they combine to form yet more chemicals, to create atmospheric "reservoirs" of long-lived substances such as O_3, NO_3, N_2O_5, $ClONO_2$, HNO_4. Because of their long lives (months to years), these chemicals can move down into the troposphere and be removed by rain.

As the protective ozone layer is removed – and a hole has already been detected over the Antarctic – more ultraviolet radiation will reach the earth. This can be expected to lead to a rise in the incidence of skin cancers. Increased ultraviolet radiation also has marked deleterious effects on marine phytoplankton, the base of the ocean foodchains.

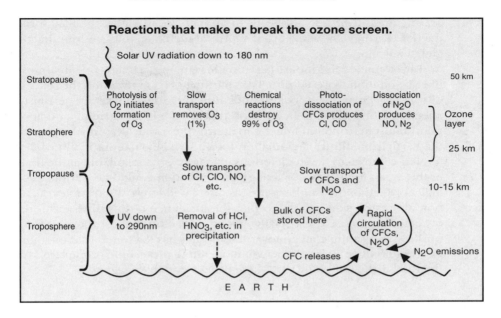

Reactions that make or break the ozone screen.

during the building phase, they fix a great deal more carbon dioxide than other ecosystems, such as open grasslands. In this respect young plantations in the building phase can also be important sinks for carbon dioxide. Reforestation programmes can therefore help global climate on two counts. The young forests absorb carbon dioxide, and the shade conditions produced by the forests help to cool the earth's surface and reduce the amount of heat that is reflected back into the atmosphere.

Tropical forest conversion is also a potential source for other greenhouse gases such as nitrous oxide and methane, which also trap infrared radiation in the atmosphere, thereby increasing global warming (Mooney et al. 1987; Myers 1988c). Of course, there are other sources of these gases, including hydrocarbon fuels and farmlands, but destruction of tropical forests may contribute more to atmospheric pollution than was previously recognised. Forest conversion may increase the levels of methane produced as a byproduct by termites, which are important decomposers of dead and dying wood. Termites are estimated to be eight to ten times more common in land disturbed by human activities than they are in primary forests (Collins 1980c). It has been suggested that termites produce about 150 million tons of methane each year, compared with a total global input of 350-1,120 million tons. The level of atmospheric methane is increasing at about 2% per year; most of the increase appears to originate from tropical areas where land use is changing (Zimmerman and Greenberg 1983). Conversion of swamp forests to wet-rice fields and of forests to grazing areas for cattle also leads to an increase in global methane production. Farting cows are a major source of methane, thus contributing to global warming.

Development and industrialisation have played a major role in atmospheric pollution. Forest destruction, burning of fossil fuels and the release of a host of man-made compounds into the atmosphere are seriously affecting global climate. In addition to causing global warming and acid rain, atmospheric pollution helps to destroy the ozone layer which protects the Earth from ultraviolet radiation (box 13.3). Fluctuations in ultraviolet levels are known to have deleterious effects on phytoplankton, at the base of the ocean food chain, so that greater radiation could have serious ecological effects in the southern oceans. Increased levels of ultraviolet radiation also cause skin cancer, so that the incidence of this disease can be expected to increase, especially among agricultural and other outdoor workers. Increasing concern over these issues has led to international initiatives to reduce atmospheric pollution and to protect the ozone layer. At the Rio Earth Summit in 1992 these issues were on the agenda.

Rehabilitating Critical Lands

In Kalimantan the main focus of land management, now and for the foreseeable future, will be management of forests. At the same time, as population pressure increases and less new land is available or suitable for agriculture, more attention will have to be paid to the large areas of "critical lands" that have been degraded by shifting cultivation and forestry activities but could still be put to productive use. These areas, often characterised by alang-alang *Imperata cylindrica* or other invasive grasses, are being extended more rapidly by forest depletion than they are being reduced by conversion to more successful forms of land management (Dove and Martopo 1987). Rehabilitation of the critical alang-alang lands (table 13.6) is one of the greatest challenges in Kalimantan today.

Tropical grasslands are a spreading habitat type in Kalimantan, now estimated to cover at least 1.4 million hectares (RePPProT 1985, 1987). The areas of nonproductive land, under grass or *belukar* (scrub and regrowth), are much greater than this, estimated at 29% of South Kalimantan (RePPProT 1987) and 21% of West Kalimantan (RePPProT 1988), the two most densely populated provinces. Extensive grasslands in the wet tropics are usually considered to be the result of human activity, especially shifting cultivation. Large areas of grasslands were first recorded in the upper Mahakam in 1870 (Eichelberger 1924) and the extensive areas of grassland in the Kapuas foothills and Kelabit highlands are a consequence of decades of agricultural activities (Blower et al. 1981). Similarly the "pepper hills" of the Hulu Sungai, South Kalimantan, were converted to grassland by the activities of pepper farmers in the past (Potter pers. com.).

Inappropriate logging techniques, and the shifting cultivators who follow the loggers, have created extensive new grasslands in Kalimantan (Kartawinata and Vayda 1984). Where land has been cleared of forest, has been burned, and now is under alang-alang or shrubs, the fertility of the land is often low, with low levels of available nutrients, high acidity, low cation exchange capacity, poor structure and high susceptibility to erosion (Nye and Greenberg 1960). The process by which invasive grasses replace forest is closely linked to fire, which kills young trees and seedlings but allows grasses to survive. *Imperata cylindrica* is among the most fire-resistant of all tropical plants; alang-alang grasslands are now estimated to cover a tenth of the Outer Islands, areas formerly covered by forests (Kartawinata 1980b; RePPProT 1990).

Imperata cylindrica thrives on poor soils, producing a strong system of

Table 13.6. Causes of degradation of forest land and possible rehabilitation measures.

Type of degraded forest land	Cause of degradation	Rehabilitation measures
Barren land	Surface compaction; toxic excess minerals, (e.g., salinisation); major pH change; nutrient deficiencies; air pollution; and animal damage.	Stabilisation crops (exotic species may be necessary); soil treatment, protection from fire, grazing, harvest; subsequent establishment of useful tree species may be possible.
Grassland	Forest clearing by agriculture or by grazing and fire.	Establish forest plantations, agroforestry (desirable species characteristics include nitrogen-fixing and deep tap roots); exclusion of fire; grazing controlled.
Low secondary forest	Destruction of primary forest cover; fire, forest clearing, earth-moving, and overgrazing.	Planting of (preferably mixed) species to improve soil and also to give economic yields, e.g., fruit, coppice; agroforestry with shade-tolerant species, exclusion of fire and heavy browsing; research is needed.
Degraded logged forest	Damage to soil and biota from heavy machinery, inappropriate logging sites (on steep slopes), non-directional felling, skidding, and poorly designed access roads.	Enrichment planting and rapid re-establishment of vegetative cover on cleared areas.

Source: Lovejoy 1985

underground rhizomes, and is very easily ignited. The roots also produce an exudate that inhibits the growth of other grass crops such as rice and maize (Donner 1987). The flat-bladed coarse grass grows to two metres high, but the leaves are only palatable for cattle for a few weeks after sprouting. The rhizomes protect the soil against erosion and speed the surface runoff of rainwater. Alang-alang is a fire-climax vegetation which burns easily without being destroyed; indeed, fire helps *Imperata* to maintain its dominance over competing vegetation. Fire spreads rapidly through the alang-alang but does not affect the rhizomes which lie well below the soil surface. Fire used in clearing forest also encourages the invasion of *Imperata*. Escaping fires, especially into areas that are already grassland, extend the area of this invasive and fire-resistant plant. Since alang-alang has no leaf drop, it does not renew soil fertility, and nitrogen taken up by the grass is volatilised during burning.

A certain amount of *Imperata* grassland is useful. Young growth is palatable and attracts wild deer and mousedeer, which can be hunted (Dove and Martopo 1987). For this reason some Dayak villages maintained alang-alang areas by burning close to their fields (Seavoy 1975). Alang-alang can also be used for thatch, and new growth feeds domestic cattle. The sugar-rich rhizomes can be used medicinally as a diuretic or to make a wine, although the resulting beverage is an acquired taste. Most areas of alang-alang, however, are not maintained deliberately but are a consequence of poor land use.

Alang-alang land is difficult to manage. The grass can be eradicated for cultivation only by heavy hoe-work or by ploughing (Potter 1987) unless it can be shaded out. Some farmers in South and East Kalimantan cultivate alang-alang lands, but most prefer to clear forest (fig. 13.8). Hence while natural forest remains, little use is made of the grassland for agriculture. Shifting cultivators leave it behind and work a hollow frontier of surviving forest (Blaikie and Brookfield 1987).

Recently the extensive areas of alang-alang grassland in Kalimantan have been considered as possible sites for transmigration settlements (Burbridge et al. 1981). These grasslands, and those created by the great fires of 1983, can become productive agricultural and forestry lands if appropriate management systems are introduced.

Clearing land by burning and hoeing to eradicate *Imperata* and *glagah* (*Saccharum spontaneum*) is difficult and impracticable on a large scale. Trials using chemical weedkillers have proved expensive and rather unsatisfactory for treating large areas (Donner 1987). However, the Finnish Reforestry Project finds this a useful way to prepare the land prior to planting out young trees at Riam Kiwa in South Kalimantan (Vakko pers. com.). Shallow ploughing with harrows is more effective in eradicating the grasses, especially when combined with fertilising or manuring and the planting of leguminous plants (Donner 1987). The legumes can compete

Box 13.4. Riam Kiwa – agriculture in a degraded valley.

The upper Riam Kiwa Valley in South Kalimantan occupies an area of 1,420 km^2 on hilly lands. It is among the most heavily deforested areas in Indonesia. In 1938 the Dutch sent Madurese colonists to part of the Riam Kiwa Valley for an experiment in dry upland agriculture. These grassland farmers have been joined by fugitives from transmigration schemes in the difficult tidal swamps to the west. The Javanese and Madurese are adapting their agriculture to the grasslands, using cattle for the heavy ploughing needed to make these lands cultivatable.

The valley supports four different groups of people with different agricultural traditions: Dayak shifting cultivators, Banjarese and Javanese (traditionally cultivators of wet sawah) and Madurese accustomed to dryland agriculture. Within the valley there are some interesting contrasts; areas of deforestation as well as reclamation of deforested land for agriculture, some sustained production but also evident failure of management, leading to degraded lands and creation of *alang-alang*. Active creation of further grassland is now confined to the upper part of the basin; the lower valley was already deforested as a result of shifting cultivation and fires by the late 1800s.

Belimbing is an old Banjarese settlement with coffee, *kemiri* (candlenuts), cloves, kapok and fruit trees grown around the houses, and agricultural fields on the moderate slopes of scrub and secondary forest about 4-6 km away. Between the gardens and the fields the lower hillsides are covered in *alang-alang* and the grass *Saccharum spontaneum*. Rice, peanuts and bananas are grown in the hillside fields. Village shifting cultivation is restricted to this scrub zone since the government banned the cutting of new swiddens in primary forest. Farmers in Belimbing have three choices:

- to work present lands more frequently. At present land is worked for 18 months to give one crop of rice and two of peanuts and one year's bananas. The land is then returned to scrub for five years. Any increase in the frequency of cropping will result in invasion by *alang-alang*. Cultivation with digging sticks creates minimal disturbance and minimises erosion.
- to work the *alang-alang*, a difficult task, without access to cattle for ploughing. Initial land opening takes 30 person-days/ha by hoe or 6 person-days/ha with a plough; these activities must be repeated twice, with intervals between to dry out and break up the grass sods and destroy the rhizomes. Teams may be hired but ploughing is expensive, about $100/ha.
- to travel even further from the village to open new areas of secondary forest. Clearing secondary forest takes 32 person-days/ha, and clearing *Eupatorium* scrub only eight person-days, in both cases a once-only operation.

Kupang Rejo is a mainly Javanese village. Both the gently rolling slopes near the village and the much steeper country beyond are mainly under *alang-alang*. In the valley bottoms there is some sawah but no terracing. Crops of rice and peanuts can be grown in the flatlands but are dependent on fertilisers. Hillsides are worked on a rotation of one to three years of crops, and one year fallow with the regular burning of *alang-alang*. The cow-and-plough technology makes it possible to work the grassland but leaves large stretches of ground bare for several months at the start of the rainy season. Erosion is occurring, even though trees are sometimes planted along the contours, and deep gulleys on the steepest slopes speed runoff.

> **Box 13.4.** Riam Kiwa – agriculture in a degraded valley *(continued)*.
>
> **Baliangin** has more intensive land management and farms a larger area. The site of this village of Madurese transmigrants was originally chosen because of fairly good soils, and some soil conservation measures such as terracing were implemented. The area was already *alang-alang* when the village was established in 1938. Baliangin is a typical Madurese farming settlement with each household surrounded by its own lands. Although early rice yields were high and the second crop produced 80 times the seed planted, fertility has declined dramatically and present yields are much lower, 10-15 times the seed planted (360-540 kg/ha). The best rice crops come from the valley bottom sawah. Hill lands are ploughed, no longer to remove *alang-alang* but to eliminate weed seedlings, which increase with greater use of fertiliser. Although the Madurese have persisted with a mixed farming system and an emphasis on tree crops which is more protective of the soil, more intensive farming has led to degradation, land shortage and the need for additional fertilisers for sustainable crop yields.
>
> The different farming systems involve different inputs of labour and capital, including fertilisers and working livestock. They also lead to differing perceptions of land resources. The shifting cultivators of Belimbing regard the growth of *Eupatorium* scrub as a sign of restored soil fertility, whereas *alang-alang* is a pest to be eradicated or an indicator that land must soon be abandoned. In cattle-owning villages *alang-alang* is seen as both potential arable land and essential grazing for livestock. The cattle will not eat *Eupatorium* which is regarded as a useless weed. Such differences in perception are widespread in Indonesia (Sherman 1980; Dove 1984, 1985) and affect how people can be persuaded to change their land management systems. Hill rice is grown by all three villages, but yields are highest from the shifting cultivation system of Belimbing and at hill sites in Kupang Rejo, that is the areas with some period of fallow, however short. Valley bottom sawah is important in the cattle-owning villages. Although the *alang-alang* areas are being cultivated using cattle and ploughs, soil fertility seems to be declining and crops require substantial fertiliser inputs. While the ADB cattle scheme will enable more land to be worked per farmer, over-intensification obviously leads to land degradation. Inevitably the people who suffer most from land degradation are the poorest farmers with fewest available resources, whether these be cattle for ploughing or fertilisers. For the larger farmer, land degradation matters less, at least at Riam Kiwa, where there are still opportunities to open new land.
>
> *Source: Potter 1987*

successfully with re-invading *Imperata,* and restore fertility to the leached soil. At various sites in Nusa Tenggara the legumes *Stylosanthes guianensis* and *Leucaena leucocephala* have proved successful. The legumes *Acacia* and *Albizia* are being used to rehabilitate degraded alang-alang lands in the Riam Kiwa watershed, South Kalimantan. These tree crops shade out alang-alang, restore soil fertility and produce useful wood for pulp after only a few years. Although these rehabilitation techniques are neither simple nor cheap, they do increase the amount of land available for agriculture. Whether sustainable dry land agriculture is possible on many areas in Kali-

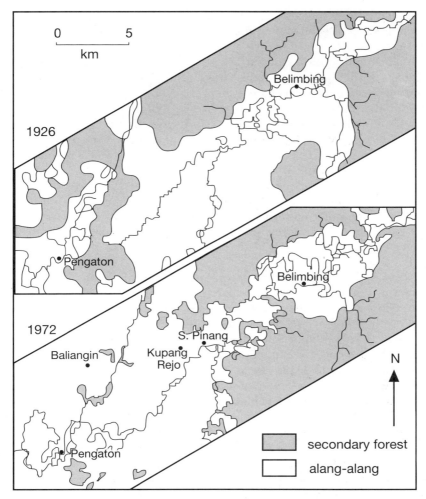

Figure 13.8. Changing vegetation cover in the Riam Kiwa Valley. (After Potter 1987.)

mantan now covered by alang-alang, even with large amounts of chemical fertilisers, is debatable (Daroesman 1981). Choice of crops will be crucial.

The conversion of alang-alang grasslands into pastures of arable and digestible grasses and legumes could contribute greatly to the improvement of livestock farming in Kalimantan. Plants suitable as fodder crops include *Panicum maximum, Stylosanthes guianensis, Centrosema pubescens, Pueraria phaseolodes* and *Desmodium ovalifolium* (Donner 1987). Some success with introducing these species to poor-quality watershed grasslands has given encouraging results on cattle ranches in Sumatra and Sulawesi (Donner 1987).

Kalimantan's critical land surface is increasing. In 1982 26% of South Kalimantan (one million hectares) was classified as critical land, with an annual increase of 180,000 ha since then, caused by illegal logging and agricultural encroachment (Donner 1987). The Riam Kanan watershed has been identified as one of twenty-eight critical watersheds countrywide and is the site for several rehabilitation and community forestry projects (Schweithelm 1987). The provincial administration suggested a regreening project in cooperation with local people, using cashew nuts, coconuts, cloves and the legumes *Acacia* and *Albizia* (Donner 1987). To date the planned regreening around the Riam Kanan Lake has met with only limited success because many of the newly planted areas of *Acacia mangium* have been destroyed by fire sweeping through alang-alang lands. The danger of fire is particularly acute in drought years like 1983, 1987 and 1991.

Many of the species most appropriate for rehabilitating alang-alang and other critical lands are secondary forest species (Jacobs 1988). Secondary forest tree species are fast-growing, making them useful for suppressing light-loving weeds such as alang-alang and for reforestation schemes. Moreover their soft wood has commercial potential for use as pulp and chipwood, for extraction of chemicals, and for local use as firewood. Those that can be planted as "sticks" have a head start in shading out alang-alang. Well-known genera of secondary forest trees that are useful for reforestation projects and rehabilitation of degraded lands include: *Albizia* and *Leucaena, Anthocephalus, Cecropia, Eucalyptus, Gmelina, Macaranga, Ochroma, Pinus, Terminalia* and *Trema*. Bamboos also have considerable potential for rehabilitation of degraded lands (Jacobs 1988). Several of these genera are already being planted on a large scale, and further research may reveal the usefulness of more native species for plantation and reforestation projects.

Agroforestry on Critical Soils

Kalimantan's poor soils limit development options. Only half of West Kalimantan is regarded as suitable for agriculture or plantation development; half of Central Kalimantan consists of infertile sand terraces, deep peat or tidal swamps; only two-fifths of East Kalimantan is regarded as suitable for agriculture; and more than one-third of South Kalimantan is unsuitable for development because of soils and topography (RePPProT 1985, 1987, 1988). Most of Kalimantan's best land is already being used for agriculture.

Of the rest, large areas of the potentially suitable land have already been degraded by inappropriate agricultural practices. Tree crops and agroforestry are the most appropriate forms of land use for newly opened lands, such as in transmigration settlements (RePPProT 1985) and for rehabilitating many degraded areas.

In agroforestry the key feature is the combination of tree and food crops. All agroforestry systems have certain common characteristics. Perennial (tree and shrub) crops rather than annual crops are dominant, so that there is a higher nutrient value in the vegetation than the soil. If undisturbed the perennials protect the soil from exhaustion, leaching and erosion. Agroforestry systems usually have two or more layers characterised by different crops. Optimum results in production and soil protection depend on the right combination of crops, since different crops have different requirements in terms of light, water, space and so on. Many basic food crops, including rice, have higher light requirements than vegetables and fruit, so are inappropriate for agroforestry systems. Systems that combine forestry and agriculture may have a number of advantages when compared with pure forestry or pure agriculture. Diversity of species, and spacing between individuals of the same species, minimise the spread of diseases and pests while providing a regular supply of food crops and other necessities such as timber and firewood. Agroforestry systems are usually ecologically and economically sound if managed properly, and feature in the traditional land-use systems of Kalimantan, such as Dayak home gardens (Seibert 1988; Lahjie and Seibert 1988; see box 8.3 and chapter 12).

Land-use planning in Kalimantan now stresses the appropriateness of cropping systems and management systems (RePPProT 1990). Traditional agroforestry systems make a useful base model for agricultural developments on the island's poor soils. Combinations of native and/or introduced plants enrich the island's agriculture and increase productivity. Cocoa, for instance, is an important crop in Sabah and is becoming increasingly important in Indonesia, with the first plantation in East Kalimantan started in 1978. The HASFARM cocoa plantation is now the largest cocoa plantation in the world, covering an area of 5,000 ha. Here cocoa is being planted with the introduced legume *Gliricidia sepium* as a shade tree. *Gliricidia* is a multipurpose tree, a nitrogen-fixer with wood useful for fencing and as supports for cash crops like pepper and vanilla. It also provides firewood and forage for cattle and goats (Lahjie and Sirait 1988).

While agroforestry can be expected to play an increasing role in industrial plantations and government-sponsored transmigration settlements, there is also some potential for improving productivity and income from the fields of traditional shifting cultivators. In Sabah, the company Sabah Softwoods donates *Acacia mangium* seedlings to shifting cultivators to be planted with the rice crop, and agrees to purchase the wood for pulp when the ladang is cleared eight to ten years later, after the fallow. Reforestation schemes could also employ the techniques of agroforestry as well as pure silviculture. Certain native forest species including *Shorea*, *Dipterocarpus* and *Dryobalanops*, which are relatively fast-growing, yield commercial timber and are tolerant of growing in relatively pure communities, could be incorporated into shifting cultivation systems with long fallows (Wyatt-Smith 1987).

The Urban Environment

There are very few densely populated areas in Borneo. Of the population of Kalimantan, 80% is rural, with very low population densities: only nine people/km^2 in Central and East Kalimantan. Major urban centres include Banjarmasin (with a population of 450,000), Pontianak (400,000), Balikpapan (400,000), Samarinda (over 300,000), Tarakan (50,000) and Palangkaraya (100,000). Outside Kalimantan the other Bornean towns are also fairly small. Kuching, provincial capital of Sarawak, had a population of only 300,000 in 1980 (Avé and King 1986).

Although urban centres in Kalimantan are small, they provide unique man-modified habitats for plants and animals. Urban ecology in Indonesia is poorly studied, yet it is a good way of introducing students to simple ecological projects (Collins 1984; Whitten et al. 1987a, 1987b). Generally towns have much less diverse plant and animal life than the surrounding countryside, since most natural habitats are destroyed or simplified in the creation of large settlements. In addition human activities, and the wastes they produce, alter the environment so that many species simply cannot survive in close proximity to people. Nevertheless, towns and large settlements provide numerous habitats and opportunities for adaptable plant and animal species:

- **roadside drainage ditches** provide a habitat for aquatic plants, frogs, toads, fish and insects. Some, like mosquitoes, are disease vectors;
- **garden ponds and town fountains** are exploited by amphibians, fish, dragonflies and other small invertebrates and plants;
- **rubbish tips** provide food and shelter for rats and flies (vectors of disease) and food for scavenging monitor lizards and nocturnal civets;
- **temples**, like the Chinese temple on Pulau Kembang, Banjarmasin, provide security and supplementary food for long-tailed macaques, which reach high population levels because of the abundant food supply provided by visitors;
- **human dwellings** provide shelter for commensals such as rats and mice;
- **eaves of roofs and watertowers** provide nesting and roosting sites for birds and bats;
- **high watertowers** provide a secure site for honey-bee nests, as on the

> **Box 13.5.** Simplified comparison of some system properties between a natural ecosystem and a man-made structure.
>
> | Natural ecosystem: pond, marsh, grassland, forest | Man-made system: house, (non-solar) factory, parking lot |
> |---|---|
> | Captures, converts, and stores energy from the sun | Consumes energy from fossil or nuclear fuels |
> | Produces oxygen and consumes carbon dioxide | Consumes oxygen and produces carbon dioxide |
> | Produces carbohydrates and proteins; accomplishes organic synthesis | Cannot accomplish organic synthesis; produces only chemical degradation |
> | Filters and detoxifies pollutants and waste products | Produces waste materials that must be treated elsewhere |
> | Is capable of self-maintenance and renewal | Is not capable of self-maintenance and renewal |
> | Maintains beauty if not disturbed excessively | Usually causes unsightly deterioration if not properly engineered and maintained |
> | Creates rich soil | Destroys soil |
> | Stores and purifies water | Often contributes to water pollution and loss |
> | Provides wildlife habitat | Destroys wildlife habitat |
>
> Source: Southwick 1985

watertower at Banjarbaru;
- **grasslands between airport runways** provide secure feeding and breeding sites for insect-eating birds and small mammals;
- **bare plots of rubble** in building plots are colonised by pioneer species with distinctive life histories;
- **house walls** provide specialised niches for some pioneer plants (e.g., lichens) and animals (e.g., geckos);
- **gardens and parks** are planted with ornamental plants, many of them introduced rather than native species. Some of these plants provide new food sources for native wildlife (e.g., for nectar-sipping butterflies);
- **gardens, parks and golfcourses** provide pockets of wildlife habitat. Small fruit bats feed on pawpaws and mango trees in urban gardens;
- **electricity wires** are a hazard to roosting fruit bats but provide roosts for migrating swallows.

Many of the plants found in towns, especially on newly disturbed land, are pioneer species. Many of the animals are "opportunist" species with broad niche requirements. It is not surprising, therefore, that man-made and disturbed habitats not only provide a niche for more adaptable native

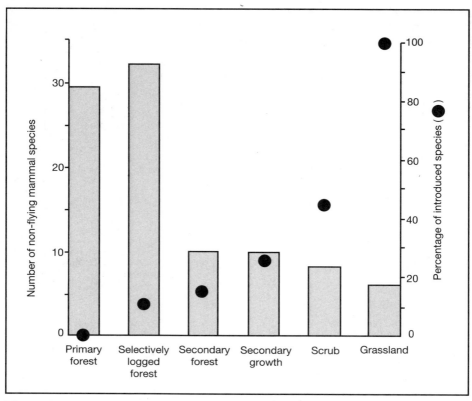

Figure 13.9. The number of non-flying mammal species in different habitat types and the percentage of introduced species. (After Harrison 1968.)

species but are also often colonised by introduced exotics (fig. 13.9). Water hyacinth and waterlilies thrive in urban ditches. The alien soft-shelled turtle *Trionyx sinensis* and the American bullfrog *Rana catesiana* are now common residents in Kuching ditches (Chan et al. 1985). Even with the addition of such exotics, species diversity remains low, and few species are abundant. Unfortunately the less desirable pest species and carriers of disease are an exception to this rule and thrive in urban environments. More thoughtful town planning, with the planting and cultivation of native plant species and the creation of microhabitats such as ponds and lakes, could lead to enrichment of urban wildlife and make towns more attractive for both people and animals.

Urban trees provide both shade and beauty for people as well as food and shelter for many animals, including butterflies and birds. Trees planted

Table 13.7. Birds of Bornean towns and villages.

	Feeding guilds	Mangrove, estuaries, rivers	Buildings, dockyards, streets	Woodland, forest edge/ gardens	Open country, scrub airfields	Swamps, ricefield	Migrant
Haliastur indus	C	x					
Haliaeetus leucogaster	C	x					
Amaurornis phoenicurus	I					x	
Pluvialis dominica	IC				x		
Glareola maldivarum	I				x		x
Charadrius dubius	IC				x		x
Streptopelia chinensis	G		x			x	
Geopelia striata	G		x				
Columba livia	G		x	x	x		
Psittacula alexandri*	F			x			
Cacomantis merulinus	I		x				
Centropus bengalensis	IC				x		
Otus bakkamoena	C		x	x			
Caprimulgus affinis	I				x		
Apus affinis	I	x	x		x		
Cypsiurus balasiensis	I					x	
Collocalia esculenta	I		x				
Collocalia fuciphaga	I		x				
Alcedo atthis	C	x					x
Halcyon chloris	IC	x		x			
Merops philippinus	I				x		x
Hirundo rustica	I	x	x				x
Hirundo tahitica	I	x					
Lalage nigra	I	x			x		
Pycnonotus aurigaster	IF			x			
Pycnonotus goiavier	IF			x	x		
Aegithina tiphia	IF			x			
Copsychus saularis	IF			x			
Gerygone sulphurea	I			x			
Orthotomus ruficeps	I			x			
Prinia flaviventris	I			x	x		
Rhipidura javanica	I	x		x			
Anthus novaeseelandiae	I				x		
Motacilla cinerea	I		x				
Artamus leucorhynchus	I		x				
Lanius schach	I			x	x		
Aplonis panayensis	IF	x		x			
Acridotheres tristis	IFC			x			
Sturnus contra	IFC			x			
Anthreptes malacensis	IN			x			
Nectarinia jugularis	IFN			x			
Dicaeum trochileum	IF			x			
Dicaeum cruentatum	IF			x			
Passer montanus	G				x		
Lonchura fuscans	G				x		
Lonchura malacca	G				x		

* common in coastal areas of South Kalimantan only
C = carnivore; F = frugivore; G = granivore; I = insectivore; N = nectarivore

Box 13.6. Habitat manipulations that may be ecologically beneficial.

Not all development-induced alterations need be damaging to the environment. Sometimes it is possible to incorporate features into the development to increase the ecological value of an area. The following ideas suggest just a few of the possibilities.

(1) It is often possible to increase the diversity of plant and animal species present in an environment. Carefully designed clearings in forests, or excavated ponds or waterways in marshes or mangrove swamps can result in markedly higher fish and wildlife production as well as making the area more interesting to visitors.

(2) Leaving strips of riverine forest along waterways provides a habitat for local wildlife (e.g., proboscis monkeys and native birds), even in highly developed areas. Creation and enlargement of ecotones between ecosystems can result in increased fish and wildlife populations. Rows of shrubs between agricultural fields and marsh, or mangrove strips along canal banks, enrich habitat for wildlife.

(3) Desirable and beneficial plant species can be encouraged and undesirable plants discouraged with the aid of certain management practices (e.g., selective cutting, controlled burning, liberation thinning and discriminate use of herbicides). Careful management and planting for species diversity (e.g., agroforestry systems) can lead to better economic return from increased plant and animal production.

(4) Pest and disease-carrying organisms can be controlled effectively in certain cases either by protecting the predators which feed upon them or through environmental manipulation. Mosquitoes can be controlled by draining stagnant pools. The introduction of small fish which consume mosquito larvae to drainage ditches can also be effective. Bothersome insect pests can be reduced in the vicinity of dwellings by trimming and thinning bushy vegetation.

(5) Land that has been devastated by strip mining and other destructive practices can be reclaimed through soil preparation and revegetation to yield useful recreational, agricultural and urban landscapes. Creation of waterholes and pockets of vegetation encourage the return of wildlife.

(6) Watershed runoff can be controlled or reduced through water resource management and ecologically sound land use to provide flood control, soil erosion control and more dependable and stable freshwater supplies. Terracing of agricultural fields, small size of hillside ladangs, long fallows, planting of tree crops and mixed cover crops are all sound soil conservation practices.

(7) Spoil material from dredging operations can be used to provide recreational land near the water in towns or to construct vegetated marshes in areas of low ecological value.

(8) Artificial reefs can be constructed from low-cost materials such as used automobile tyres, concrete pipes or old shipwrecks. They attract concentrations of fish and other marine organisms which are of direct benefit to both subsistence and commercial fishermen.

(Adapted from Odum 1976.)

in town centres, gardens and along roadside verges are often exotic species, but many of these can be attractive to insectivorous, frugivorous and nectar-eating birds (Whitten et al. 1987b). Compared with old tree-filled towns like Bogor on Java, Kalimantan towns are noteworthy for their lack of birdlife. The one common bird seen regularly in Banjarbaru is the long-tailed shrike, a common bird of open country. The paucity of birds seen in a new town like Banjarbaru may be due partly to lack of attractive mature trees. New plantings increasingly are of leguminous species such as *Acacia mangium*, chosen because of their rapid growth on poor soils and their ability to fix nitrogen.

As one might expect, the bird fauna of towns tends to have a lower species richness and diversity than that of natural forests. In addition, the major guild shifts from insect-eaters, associated with bark and canopy, to ground feeders (Ward 1968; Beisinger and Osborne 1982; Yorke 1984). These changes are not surprising, since very few town trees are above 10 m tall, and they are often widely dispersed and far from remaining areas of forest. The low number of birds in Kalimantan's urban centres can be attributed to ecological factors and lack of suitable niches, lack of bird-fruit trees, and the few insects associated with introduced trees (so that less food is available to insectivorous and partially insectivorous species). In Sulawesi towns, over half of the bird fauna originated from coastal habitats, about 5% from lowland forests and a similar percentage were cliff or cave-mouth nesters (Whitten et al. 1987b). Swifts and swiftlets take advantage of the similarity between cliffs and buildings, including watertowers, and are some of the commonest urban birds. Since most sizeable Kalimantan towns are on the coast or close to it, coastal species are likely to account for a large proportion of the urban list (table 13.7). Predatory fish-eaters such as the Brahminy kite are a common sight in all the main Kalimantan towns, and ospreys also swoop above the river in Pontianak. Introduced species such as the Java sparrow, an escaped cagebird, have now also become part of the native Kalimantan fauna (Smythies 1970).

Urban Ditches and Ponds

Kalimantan towns, as elsewhere in Asia, are drained by deep storm drains and ditches. A ditch is a simple, small river. Ditches in towns and large villages often have a high organic content derived from human waste. Study of the aquatic life in ditches can give indications of water quality.

Ditches are commonly inhabited by two species of small fishes, the mosquito fish *Aplocheilus panchax* and the guppy *Poecilia reticulata* (fig. 13.10). Both species are native to South America but were introduced to Kalimantan, and other Indonesian islands, to control mosquito larvae. The rice fish *Oryzias javanicus* recorded from Pontianak streams was presumably also introduced for mosquito control (Roberts 1989). Mosquito

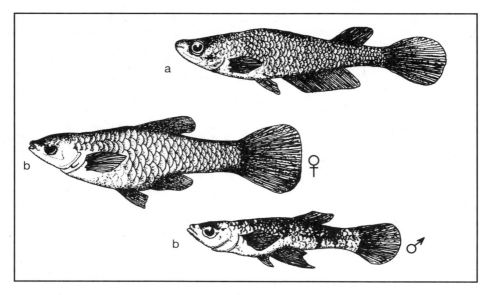

Figure 13.10. Fishes in urban ditches: a) mosquito fish *Aplocheilus panchax;* b) guppy *Poecilia reticulata*.

fish feed primarily on small aquatic larvae whereas guppies feed mainly on algae but will also take insects. These fish are often seen mouthing at the surface, apparently gulping in air. In fact they are taking in water from the air-water interface where the concentration of dissolved oxygen is greatest (Kramer and Mehegan 1981). Such "aquatic surface respiration" is an adaptation to enable the fish to survive in waters where the oxygen level is low; in water that is better oxygenated they respire normally. The ability to breathe in this fashion allows these species to survive in areas where other species cannot; therefore they can be used as indicators of water quality.

The fish fauna of urban ditches may repay careful study. Several endemic fish species are known only from Kalimantan suburban ditches (Roberts 1989). These fish are probably native to swamp habitats. Generally these fish are outcompeted in ditches by the introduced guppy.

Houses

Towns are full of people, who create their own microhabitats in houses. Human dwellings also provide a habitat for several other animals that have learned to live in close proximity to people: rats, mice and a host of other animals, including parasitic lice and blood-sucking bedbugs. Spiders are common predators in and around human dwellings. One of the largest garden spiders, *Nephila maculata*, constructs huge webs of tough silk between trees, telephone wires and bushes. Some of the most common and attractive of the house spiders are the brightly coloured jumping spiders Salticidae. These amazing spiders can leap up to forty times their body length and have very acute vision. Jumping spiders do not make webs but hunt on vertical walls or rockfaces. They trail a strand of silk, which is attached to the surface at intervals and acts as a safety rope, or dragline, when they pounce on insect prey (Preston-Martin and Preston-Martin 1984). Spider silk is a fibrous protein (fibroin) almost as strong as high tensile nylon but much more elastic, so that it can stretch by as much as one-third of its length before breaking (O'Toole 1987). All spiders use silk, for webs, egg sacs, nursery tents, or to wrap insect prey, or for the draglines used by the jumping spiders.

Geckos are some of the most familiar of animal commensals in Kalimantan. There are several gecko species, big-eyed nocturnal visitors, patrolling houses. Surprisingly little study has been made of geckos. The three smaller geckos *Hemidactylus frenatus*, *H. platurus* and *Gehyra mutilata* are similar in size and diet, although the last seems to shun brightly lit areas. Geckos run up and down the walls and across ceilings, preying on insects, including mosquitoes, thereby benefitting their human hosts. Although primarily insect eaters, house geckos will also feed on all kinds of household scraps, including rice, crumbs of bread, honey, meat and fish (Church 1962).

Geckos are remarkable for their ability to climb vertical surfaces and run upside-down on ceilings. Their feet do not have suckers but bear small overlapping flaps of skin. These flaps are covered with minute hairs which make contact with the slight irregularities of the surface and enable the geckos to cling. Geckos have relatively short tails which they can shed if caught, leaving the surprised predator with the cast tail while the gecko escapes (Vitt 1983). Tail-shedding is also known among some other lizards and snakes. Losing a tail may save the gecko's life, but it is a costly survival mechanism. Fat is usually stored in the tail and may be used at times of food shortage or, in females, for yolk production (Halliday and Adler 1987). Regrowing a new tail also uses up a considerable amount of energy.

Geckos are habitat specialists, relatively cryptic with a "sit-and-wait" hunting strategy. They feed on walls and also lay their eggs there. The female lays one or two hard white eggs in a crack on the wall, behind a picture or cornice, and three months later young geckos hatch out.

Houses, and kitchens in particular, harbour several species of ants. The very small yellow Pharaoh's ant *Monomorium pharaonis* has been carried unintentionally all over the world. This human commensal is quick to locate sugar or other sweet foodstuffs in open containers and in spilt sugary liquids. Once they have found a food source, the tiny ants may keep other species away with a repellent odour secreted from a poison gland. Ants are social insects, and several species commonly communicate the location of food sources with scent trails. Yet other species (e.g., *Leptothorax*) avoid attracting more aggressive ants by not emitting odours (O'Toole 1987).

Several other animals have become commensals of people, sharing dwellings and feeding directly or indirectly off their human hosts. The list includes bedbugs, rats, mice and cockroaches. Cockroaches are found just about everywhere and eat virtually anything. They are among the most ancient of all insect groups, with fossil forms known from the Carboniferous period, 345-280 million years ago (O'Toole 1987). Cockroaches can be described as the insect equivalent of rats and mice, and are extremely successful at living in association with people. Because of their flattened bodies, they are able to crawl into narrow crevices, under floorboards, and into drains and sewers. They are active by night when they feed on household scraps as well as papers and even book bindings. Cockroaches can transmit disease and often foul household foods with their excreta and dirty feet. They are extremely difficult to eradicate, long-lived and produce large numbers of eggs. One female may produce a thousand eggs in a lifetime. No matter how unattractive they may be, cockroaches are extremely successful as commensals with people.

INDONESIA'S EXPANDING POPULATION

Kalimantan's changing ecology can be attributed to people. Development and loss of natural habitats are a direct result of increasing population pressure, not just in Kalimantan itself but at the national level. Faced with an increasing population (fig. 13.11) and the desire to improve standards of living, what can the government do? Already Indonesia has implemented a successful birth control programme, Keluarga Berencana (KB), a fact that was recognised by a United Nations award to President Soeharto in 1989. Nevertheless, population stabilisation is not expected to happen until well into the next century (fig. 13.12). In the short term there is increasing pressure on land and natural resources to feed a growing population and to fuel national economic development. Kalimantan is at the forefront of these developments. With increasing exploitation, industrialisation and urbanisation the pattern is changing, and Kalimantan's population is now growing at a faster rate than that of Java (Buku Statistik 1991). It may therefore be pertinent to look at some of the ecological factors affecting human populations.

Population density in Kalimantan is low, which seems surprising when one considers the island's closeness to overcrowded Java and the fact that people have been sailing between islands in the archipelago for thousands of years. Kalimantan would seem to be an obvious destination for immigrants, especially if we consider the island's huge "empty" spaces. Cer-

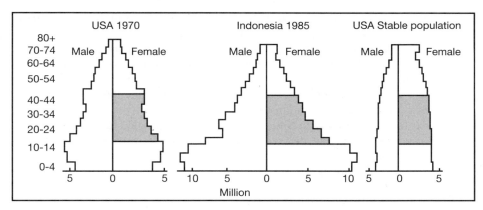

Figure 13.11. Population pyramids showing age structure of human populations in the U.S.A. and Indonesia. The population in the U.S.A. is approaching stability whereas the Indonesian population is growing. The shaded area represents women of child-bearing age.

Source: Buku Statistik 1988

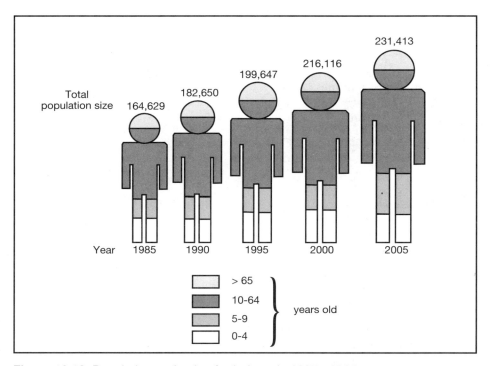

Figure 13.12. Population projection for Indonesia 1985 - 2005.

Source: Buku Statistik 1988

tainly this argument has appealed to planners and regional developers in the last few decades. Why has Kalimantan's population remained so low?

Human population density can be related to factors such as climate, availability of food and the prevalence of diseases. Even though rainforests are among the most productive of natural ecosystems, they provide little food suitable for people, and densities of foragers are low (see table 8.3). Manipulation of ecosystems, especially via agriculture, allows the possibility of supporting higher human densities. Kalimantan, with its poor, non-volcanic soils, has never experienced the wet-rice revolution which characterised Javan agricultural history and allows that island to support such a large population. Java, with less than a quarter of the land area of Kalimantan, supports twelve times as many people.

Population density alone tells little about population trends. Demographic data and age structure tell more about population growth and whether populations are increasing or declining. The age structure of the Indonesian population reveals that the population is growing and will continue to grow far into the future, because of the large number of young girls who will join the reproductive sector (fig. 13.11). Human populations may always have had high birth rates, but in the past these were balanced by high mortality, which maintained populations at low levels. Today, with better health care, more of the population survive to reproductive age, and mortality is declining, yet fertility rates are still high in many developing countries, including Indonesia. Programmes such as KB seek to redress the balance by substantially reducing the birth rate of the population. Birth control programmes are not a new phenomenon. Many Indonesian societies, including Dayaks, had cultural taboos which restricted population growth (Whittier 1973).

Ecology of Diseases

Disease is an important factor in tropical populations, causing mortality and debilitation. Some of the common parasitic diseases of tropical human populations are shown in table 13.8. Two groups of animals are the main contributors: protozoa, and helminths (roundworms and flatworms, such as tapeworms and flukes). Parasitic diseases may be endemic to certain populations and are easily transmitted through communities, especially where conditions are overcrowded and sanitation is poor. Such diseases are always present in the population at roughly constant levels of infection. Epidemic diseases such as cholera arise suddenly, sweep through a population, then disappear again.

Understanding the ecology of a disease can help combat its spread. Intestinal nematodes and dysentery, for instance, are transmitted directly from one human host to the next. The incidence of these diseases can be reduced simply by better hygiene. Other diseases, such as malaria and filariasis, have two or more hosts, one of which is a vector transmitting the disease from one person to the next. The vector is also infected but by a different growth stage of the parasite. Malaria and filariasis are spread by blood-sucking flies, while schistosomiasis is transmitted by snails. These vectors all have aquatic stages in their life cycles (see table 3.8), and the disease can be controlled or reduced by ensuring that there are no suitable habitats for these larvae or snails near to human dwellings. With one vector stage missing the disease cycle is then disrupted. Eliminating the larval habitats is not easy in a stilted town in a tidal swamp, with little water flow,

Table 13.8. Diseases related to deficiencies in water supply or sanitation.

Group	Disease	Route leaving human*	Route entering human*
Waterborne disease	Cholera	F	O
	Typhoid	F,U	O
	Leptospirosis	U,F	P,O
	Giardiasis	F	O
	Amoebiasis (b)	F	O
	Infectious hepatitis (c)	F	O
Water-washed disease	Scabies	C	C
	Skin sepsis	C	C
	Leprosy	N(?)	?
	Lice carrying typhus	B	B
	Trachoma	C	C
	Conjunctivitis	C	C
	Bacillary dysentery	F	O
	Salmonellosis	F	O
	Enterovirus diarrheas	F	O
	Paratyphoid fever	F	O
	Ascariasis	F	O
	Trichuriasis	F	O
	Whipworm (*Enterobius*)	F	O
	Hookworm (*Ankylostoma*)	F	O,P
Water-based disease	Urinary schistosomiasis	U	P
	Rectal schistosomiasis	F	P
Water-related vectors	Arbovirus encephalitis	B	B mosquito
	Dengue/haemorrhagic fever	B	B mosquito
	Bancroftian filariasis	B	B mosquito
	Malaria (c)	B	B mosquito
Faecal disposal disease	Hookworm (*Necator*)	F	P
	Clonorchiasis	F	Fish
	Diphyllobothriasis	F	Fish
	Fasciolopciatis	F	Edible plant

* F = faeces; O = oral; U = urine; P = through the skin; B = bite; N = nose; S = saliva; C = contact
(b) = Though sometimes waterborne, more often water-washed.
(c) = Unusual for domestic water to affect these much.

the conditions prevailing in many Kalimantan settlements.

Many parasites have co-evolved with their hosts to a state where the two live in balance. Successful parasites usually do not kill their hosts, though heavy parasite loads may be seriously debilitating and may even lead to death. Most parasites are highly specialised, with specific adaptations for life on, or within, their host and for dispersal from one host to another. Since parasites are detrimental to their hosts either by causing illness or by diverting energy from growth or reproduction, natural selection favours host individuals that can evade or reduce the effects of parasites. Immunological responses are often powerful means of suppressing parasites. Co-evolution may lead to reduced host mortality so that there is a long-term home for the parasite. Trypanosomiasis, for instance, has insignificant effects on wild African ungulates with which the parasites evolved, whereas it decimates introduced domestic cattle. The case of rabies is rather different (Deshmuk 1986). The virus is spread in the saliva of an infected host. The final stages of the disease induce biting behaviour; before dying the rabid animal runs wild, biting other victims and thus transmitting the disease. Sometimes genetic traits have evolved that confer enhanced resistance to disease to some population members, for example, the sickle-cell gene found among Africans in areas where high levels of malaria are endemic (Ford 1976).

The effects of ecological changes in human environments can increase the incidence of many diseases. Similarly the ecological changes that occur when people switch from being mobile forest foragers to settled farmers or urban dwellers may increase the prevalence of some diseases. For successful transmission, all diseases need a threshold level of host population density. Epidemics, in particular, require dense populations. Overcrowding in towns or new settlements, poor hygiene, and lack of clean water due to contamination by human, animal or industrial waste, all encourage the spread of disease. In these conditions the infective stages of vectors are in constant contact with their human hosts.

Changing land use and the establishment of agroecosystems mean that many species are eliminated locally (chapter 12), but those that survive, including pests and diseases, may become more successful. Diseases that were formerly restricted to nonhuman populations may infect workers involved in forest clearance. In natural forests, for instance, there are relatively few habitats with standing water, the habitat required by the aquatic larvae of the *Anopheles* mosquito, the females of which spread malaria. Where such habitats do occur (e.g., in tree holes or water-holding epiphytes) they are found in the canopy, and the adult mosquitoes feed off canopy mammals such as monkeys. Felling forest for timber, agriculture or mining creates many transient and permanent areas of standing water at ground level: wheel ruts, fish ponds, drainage ditches, abandoned pits and mine-workings. All these are potential breeding sites for malarial mosquitoes

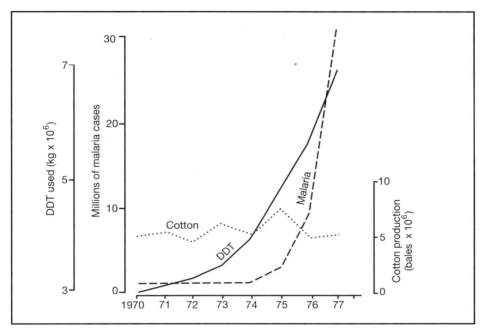

Figure 13.13. The incidence of malaria in India increased dramatically as the application of DDT to agroecosystems increased. Heavier pesticide applications were considered necessary to maintain cotton crop production. (After Chapin and Wasserstrom 1981.)

and other disease-carrying vectors. In Kenya, when irrigated ricefields replaced a diverse environment of scattered maize fields, seasonal swamps and overgrown waterholes, the ratio of malarial to nonmalarial mosquitoes changed dramatically from 1 : 99 to 2 : 1 (Desowitz 1980).

Large-scale applications of pesticides to agroecosystems may also affect the ecology of diseases. Thus the incidence of malaria in India, once close to eradication, has risen as the application of DDT to agroecosystems has increased. Heavier pesticide applications, needed to maintain crop production, have had the undesirable side-effect of promoting evolutionary selection for mosquito vectors resistant to the DDT used in disease control (fig. 13.13).

As with pests (chapter 12) integrated control is the best way of attacking diseases. Some of the control methods may be the same as for pest control, such as the use of pesticides, and the management of areas to make them less suitable for vectors. Unlike pest control, which leads to higher crop yields, disease control offers no immediate economic benefits; indeed, it may lead to greater demands from a burgeoning population on resources that are already limited. In the long term, however, disease control and better health care not only improve human welfare but ensure that less time is lost to illness; a healthier workforce can contribute to increased economic activity.

Integrated programmes against disease involve better hygiene, improved environmental conditions, good quality water, education, chemical control of vectors, and drug therapy. Careful consideration of ecological and environmental factors prior to development, together with public health programmes, can reduce the effects of many diseases (Deshmuk 1986).

The Future

Indonesia, like many other developing countries, is said to be experiencing a population explosion. In fact, consideration of human history shows that world population has grown steadily, influenced by different cultural and economic changes. The major steps in development are from subsistence economies (shifting agriculture) through settled agriculture to industrial agriculture (plantations) and industrial development. These stages of development can be seen in Kalimantan and elsewhere in Indonesia today (table 13.9). Indonesia's population is expected to increase by 40% over the next 15 years (fig. 13.12). Economic development will be fuelled by increased exploitation of natural resources and a movement towards more intensive forms of agriculture and industrialisation.

Local and global carrying-capacities have increased dramatically during human history, but recent increases in food production depend heavily on fossil fuels and a few staple crops (Deshmuk 1986; Wilson 1988). Continuing high population growth could have catastrophic consequences, with eventual shortages of non-renewable resources and increasing pollution leading to a rapid decline in carrying capacity followed by large-scale human mortality (figs. 13.14 and 13.15). This pessimistic picture can be applied to Indonesia as well as on a global scale. New discoveries of non-renewable resources, new technologies, wiser use of renewable resources, and population control could all change this projection, and hopefully that is what the future holds.

Indonesia is luckier than many countries because of her great wealth of natural resources, many of them on the Outer Islands including Kalimantan. Wise development, with sound management to minimise environmental degradation, will lead to healthy economic growth and improved social conditions. The management of Kalimantan's forest resources will be crucial in this context (see chapters 9 and 14). Box 13.7

Table 13.9. Ecosystem types classified by energy flow.

Ecosystem type	Annual energy flow (Kcal/m^2/yr)	
	Range	Average (estimated)
1. Unsubsidised natural solar-powered ecosystems (e.g., open oceans, upland forests). Human role: hunter-gatherer, shifting cultivation.	1,000-10,000	2,000
2. Naturally subsidised solar-powered ecosystems (e.g., tidal estuary, lowland forests, coral reef). Natural processes aid solar energy input (e.g., tides, waves bring in organic matter or do recycling of nutrients so most energy from sun goes into production of organic matter). These are the most productive natural ecosystems on the earth. Human role: fisherman, hunter-gatherer.	10,000-50,000	20,000
3. Human-subsidised solar-powered ecosystems. Food and fibre producing ecosystems subsidised by human energy as in simple farming systems or by fossil fuel energy as in advanced mechanized farming systems. (e.g., Green Revolution crops are bred to use not only solar energy but fossil energy as fertilisers, pesticides and often pumped water). Applied to some forms of aquaculture also.	10,000-50,000	20,000
4. Fuel-powered urban-industrial systems. Fuel has replaced the sun as the most important source of immediate energy. These are the wealth-generating systems of the economy and also the generators of environmental contamination: in cities, suburbs and industrial areas. They are parasitic upon types 1-3 for life support (e.g., oxygen supply) and for food; also for wood and charcoal where these are still important domestic fuels.	100,000-3,000,000	2,000,000

The most productive natural ecosystems and the most productive agriculture seem to have upper limits of 50,000 Kcal/m^2/yr.

Source: Odum 1975

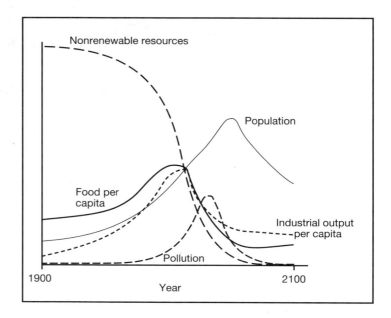

Figure 13.14. A model to show interactions between population and factors such as food, consumption of non-renewable resources, and industrial output and pollution. (After Deshmukh 1986.)

presents two views of Indonesia in the future (MacKinnon 1981; Whitten et al. 1987b). It needs little understanding of ecology to see which is preferable. Implementation of ecological principles could lead to better resource management.

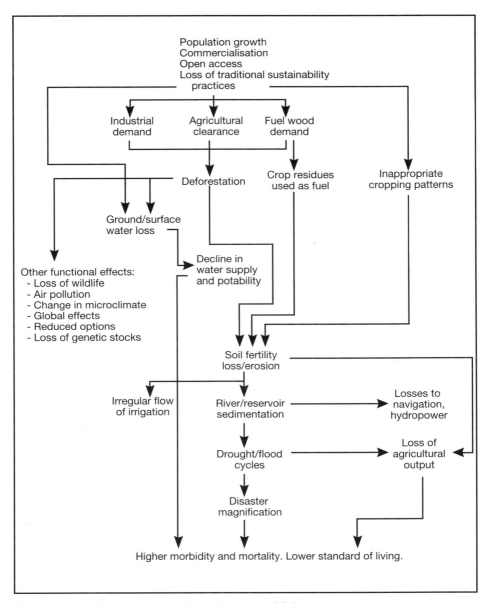

Figure 13.15. General economic-environmental linkages.

Source: Pearce and Markandaya 1986

Box 13.7. Two visions for the future of Indonesia.

	VISION 1 *Unbridled resource exploitation*	VISION 2 *Sound resource management*
Population	500 million and growing	250 million and stable. Effective birth control programme.
Lowlands	All lowland forest converted to agricultural land, much of it very marginal with poor yields. Lowland frequently damaged by floods, with economic and human losses; area of critical lands increasing.	Suitable lowlands under intensive, high-yielding agriculture based on a wide variety of crops and systems, including agroforestry. Former critical lands being rehabilitated for productive agriculture or social forestry.
Land use	Little land under forest; reserves and protection forests exist on paper only. Widespread shifting cultivation and poor logging practices create more degraded lands. Loss of genetic resources.	30% of land under natural forest (reserves and protection forests), 20% forest plantations, 10% under fruit trees; no boundary encroachment of reserves, many with utilised buffer zones.
Mountains	Montane forests being stripped for firewood with consequent erosion and serious hydrological impacts in lowlands.	Major watersheds actively protected and managed, thereby sustaining agriculture in the lowlands.
Transmigration	Programme halted because of serious regional unrest and poverty, and because marginal soils had begun to be settled.	Programme halted because aims achieved and no longer regarded as necessary.
Economy	Government unable to raise enough taxes, borrow enough or sell enough natural capital to pay off interest on development loans (much less the loans themselves) or to tackle increasingly wretched social and economic problems. Increasingly dependent on "soft" loans. Living standards eroding, dwindling foreign currency reserves.	Government reaping benefits of sustainable development and setting an example to other countries. Living standards comparable to, or above, those of today and well above tropical average. Healthy economy and substantial foreign currency reserves from sustainable harvesting of forests and from manufactured goods.
Tourism	Few tourists, generating little income.	Tourism an important source of income in many areas.
Defence	Military defences weak due to enormous budget reductions. Attempted schisms in resource-rich, outlying provinces.	Strong military defence, providing regional stability.
Agriculture	Poor land use and inappropriate crops and cropping systems degrading land. Heavy reliance on fossil-based fuels; agricultural sector collapses as oil reserves run out.	Better land use with diversification of crops. On poor soils emphasis on tree crops and agroforestry. Agricultural systems favoured that do not rely heavily on fertilisers.

Box 13.7. Two visions for the future of Indonesia *(continued)*.

	VISION 1	VISION 2
Pollution	Frequent pollution of waterways as cost-cutting increases and supervision decreases. Inland and coastal fisheries seriously affected.	Environmental regulations enforced and improved.
Industry	Industrial sector declining, relying increasingly on imported goods, purchased by sales of dwindling natural resources.	Manufacture of industrial goods thriving.
Research	Moribund.	Important centre for research in tropical forestry, agriculture, ecology and environmental problems.

(Adapted from MacKinnon 1981; Whitten et al. 1987b.)

Chapter Fourteen

Conservation: Protecting Natural Resources for the Future

Economists distinguish between non-renewable natural resources, such as oil, coal, gold and other minerals (chapter 13), and renewable resources, such as forests and wildlife. In theory the latter can be harvested in perpetuity, provided that harvesting is at sustainable levels (chapters 8 and 9). Natural habitats are also environmental resources, protecting watershed functions, ensuring clean air and conserving biological diversity. Biological resources can be considered in terms of three functions: production, protection and potential for future development. Effective systems of management of biological resources allow both utilisation and sustainable development (McNeely 1988). Conservation is the protection, management and wise utilisation of biological resources to ensure that benefits not only accrue today but will also be available tomorrow.

BIODIVERSITY

Biological diversity is the variety and variability found among living organisms and the ecosystems where they occur. Biodiversity is usually considered at three different levels: genetic diversity (within species), species diversity and ecological or habitat diversity (Wilson 1988; McNeely et al. 1990). Species diversity is the easiest to measure, both in terms of species richness and uniqueness or endemism. The biological diversity of an area depends not just on the total number of species found there but also on the uniqueness of those species, whether they are endemic to certain habitats or localities. The number of species found in a disturbed habitat may be equal or greater to that found in a pristine wildland. However, the species composition is different, and some original rare or endemic species may be replaced by common and widely distributed species. Overall this leads to reduced biological diversity.

The number of species of animals and plants known to science is about 1.4 million, but there are believed to be at least three times this number still to be discovered and possibly 30 million species more (Wilson 1988;

Erwin 1982, 1988; Myers 1983, 1984). A few of these new species may be large mammals like the recently discovered ox *Pseudoryx nghetinhensis* in Vietnam (Vu Van Dung et al. 1993), but many more will be small, obscure invertebrates. Indeed, there could be as many as 30 million species of insects alone (Erwin 1983).

Indonesia covers just 1.3% of the earth's land surface, yet 10% of all known plants are found here, as well as one-eighth of all mammals and one-sixth of all birds, reptiles and amphibians (BAPPENAS 1991; KLH 1992). Because of its wealth of plant and animal species, Indonesia has been identified as one of the world's "mega-diversity" countries, second only to Brazil for biodiversity (Mittermeier 1988). Within Indonesia the two major centres for species richness and biological diversity are Kalimantan (Borneo) and Irian Jaya (New Guinea).

THE BIOLOGICAL IMPORTANCE OF BORNEO

Two-thirds of all species occur in tropical regions, and probably half are confined to rainforests (Myers 1986; Raven 1988). Borneo supports the largest expanse of tropical rainforest in the Indomalayan Realm. It is a main centre of distribution for many genera of the Malesian flora and the Indomalayan fauna. Forest types include mangrove forests, large areas of peat swamp and freshwater nonpeaty swamp, the most extensive heath (kerangas) forests in the realm, lowland dipterocarp forest, forests on limestone and various montane formations. This wealth of habitats supports high species diversity (table 14.1).

Borneo, rich both in flora and fauna, is the richest unit of the Sundaic subregion, with small-plot tree diversity as high as found anywhere in New Guinea or South America. The great species richness of the island's lowland rainforests is well illustrated by a count of more than 700 different species of trees from ten selected one-hectare plots in Sarawak (Wilson 1988). With 267 species (155 endemic) of the family Dipterocarpaceae (Ashton 1982), Borneo is a centre of diversity for dipterocarps, the most important group of commercial timber trees in Southeast Asia, and source of most of the valuable timber exports from Kalimantan, Sabah and Sarawak.

Among the Indonesian islands Kalimantan is second only to Irian Jaya in terms of species richness for plants, mammals, birds and reptiles (table 14.2). The whole island of Borneo is a major centre for biodiversity and a priority area for conservation. Borneo covers less than 0.2% of the earth's land surface, yet one in twenty-five of all known plants are found here as well as one-twentieth of all known birds and mammals. The island is important both for species richness and endemism. There is a high level of

Table 14.1. Biological diversity in Borneo and selected reserves.

	World	Malesia	Indonesia	Borneo	Kinabalu N.P.	Kutai N.P.	Gunung Palung N.P.
Bacteria	3,600						
Blue-green algae	1,700		300				
Fungi	46,983		12,000				
Mosses	17,000		1,500	649	323		
Ferns	13,000		1,250		450		
Gymnosperms	750						
Flowering plants	250,000	30,000	25,000	10,000			
Trees			10,000	3,000		523	
Dipterocarps		386		267			
Figs		500		137	78		
Rhododendrons		300	287	50	25		
Orchids	20,000	3,500		2,000	1,000		
Pitcher plants	70			28			
Rattans	575	500		137			
Animals							
Sponges	5,000						
Corals and jellyfish	9,000						
Worms	36,000						
Crustaceans	38,000						
Molluscs	50,000		20,000				
Echinoderms	6,100						
Insects	750,000 *		250,000				
Butterflies				1,000	290		
Beetles				30,000			
Fish	19,056		8,500	394	32		
Amphibians	4,184		1,000	100	45		
Reptiles	6,300		2,000	254	100 +		
Birds	9,036		1,531	502 **	289	274	191
Mammals	4,010		530	221	102	74	66

* conservative estimate.
** 478 recorded in Kalimantan, of which 358 are resident land birds.

Table 14.2. A comparison of biotic richness and endemism throughout Indonesia.

Island	Plants (revised)		Mammals		Resident birds		Reptiles	
	species richness	% endemism	species	% endemism	species	% endemism	species	% endemism
Sumatra	820	11	221	10	465	2	217	11
Java and Bali	630	5	113	12	362	7	173	8
Borneo	900	34	221	19	358	10	254	24
Sulawesi	520	7	127	62	289	32	117	26
Lesser Sundas	150	3	41	12	242	30	77	22
Moluccas	380	6	69	17	210	33	98	18
Irian Jaya	1,030	55	125	58	602	52	223	35

Revised plant groups are those families which had been revised in Flora Malesiana by 1982.

Source: MacKinnon 1982

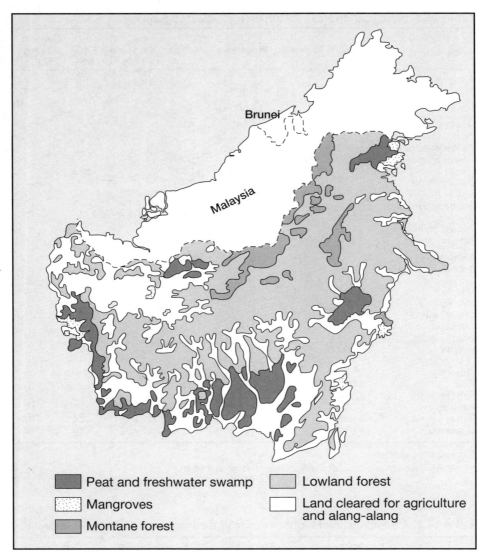

Figure 14.1. The distribution of cleared land and remaining natural habitats in Kalimantan. (After RePPProT 1990.)

floral endemism, with about 34% of all plant species, and 59 genera, unique to the island (MacKinnon and MacKinnon 1986). Borneo has 37 species of endemic birds (MacKinnon and Phillipps 1993) and 44 endemic land mammals (MacKinnon 1990a).

Borneo's biological diversity is now threatened by poorly regulated development and the nonsustainable harvesting of the island's rich natural resources, especially timber. About 60% of the island still remains under forest cover, but deforestation, due to poor agricultural and logging practices, is proceeding at an alarming rate (fig. 14.1). The habitats most threatened by these developments are the more accessible lowland forests, where species richness is greatest (MacKinnon 1990b). Yet these are also the least well represented habitats in the protected area system (MacKinnon and MacKinnon 1986).

In tropical forests the ambient conditions ideal for organic life, the struggle for nutrients, competition between predators and prey, and competition between herbivores and plants, have led to astonishing levels of species diversity and biological specialisation. Competition has led to many species having a very narrow set of ecological requirements. The number of species is high, but individual species tend to be rare and to occur at low densities. Once a tract of forest is reduced to isolated pockets, the fragments of remaining forest begin to lose species (Terborgh 1974; Lovejoy et al. 1983).

Box 14.1. Within- and between-habitat diversity. High diversity in the tropics could result from (a) more species occurring in each habitat; (b) more habitats, each containing the same number of species; or (c) a combination of both.

	Tropical	Temperate
(a) High within-habitat diversity:		
Average number of species per habitat	50	10
Number of different habitats	10	10
Total number of species	500	100
(b) High between-habitat diversity:		
Average number of species per habitat	10	10
Number of different habitats	50	10
Total number of species	500	100
(c) Combination of both types of diversity:		
Average number of species per habitat	20	10
Number of different habitats	25	10
Total number of species	500	100

Source: Deshmukh 1986

Species Extinctions

Evolution is a continuous process. New characters and new species are evolving constantly in response to changes in the environment. All species are eventually heading towards extinction, to be replaced by better-adapted forms. The present few million species are the survivors and descendants of the estimated half billion species that have ever existed. The best current estimates suggest that on average 900,000 species have become extinct every one million years during the last 200 million years, about 90 species each century (Raup 1986).

In the distant past, extinctions occurred as a result of natural processes. In the last 10,000 years, humans have played an increasingly active role in species extinctions, both by overhunting and by destroying habitats. It has been estimated that the rate of mammal extinction worldwide increased from 0.1 species per century during most of the Pleistocene to 17 species per century during the period 1600-1980, and the rate of loss is still rising (Wolf 1985).

Major extinctions of large mammals and birds can often be attributed to human settlement; this is illustrated clearly in the New World, Madagascar and Oceania (Martin and Klein 1984). Overhunting by early humans probably eliminated the tapir from Borneo, where it occurred in Pleistocene times (Medway 1977b). Today the main cause of species extinctions is habitat destruction.

Box 14.2. Characteristics of species affecting their survival.

Endangered species	Species that are "safer"
Large size	Small size
Predator	Grazer, scavenger, insectivore
Narrow habitat tolerance	Wide habitat tolerance
Valuable fur, oil, hide, etc.	Not a source of natural products
Restricted distribution	Broad distribution
Lives mainly in international waters or migrates across international boundaries	Lives mainly in one country
Reproduction in one or two vast aggregates	Reproduction by solitary parts or in many small aggregates
Long gestation period	Short gestation period
Small litters	Big litters and quick maturation
Behavioural idiosyncrasies that are nonadaptive today	Adaptive behaviour
Intolerant of the presence of humans	Tolerant of humans

Source: Goudie 1981

The destruction of rainforests and other major tropical ecosystems is sure to lead to significant loss of species. International conservation agencies and eminent scientists warn that the earth may be losing one species per day at present, and this could rise to a hundred species per day by the end of the century. If the present rate of habitat destruction continues, possibly 35% to 50% of all species could be lost by the year 2000 (Ehrlich and Ehrlich 1981; Myers 1983). This would be a mass extinction equivalent to the loss of the dinosaurs at the end of the Cretaceous period, 67 million years ago. During prehistoric extinctions, however, species did not disappear so quickly. The dinosaurs, for instance, died out gradually over a period of two million years and were replaced by the better-adapted ancestral birds and mammals. Moreover, during most mass extinction phases in prehistory, terrestrial plants survived with very few losses (Knoll 1984). Even when tropical forests were severely depleted and contracted during the Ice Ages, substantial forest refuges still remained, harbouring reservoirs of species that recolonised when warmer, moister conditions prevailed again (Jacobs 1988). The new mass extinction phase which the world faces will be more sudden, and there will be few natural refuges to provide colonisers for restocking plant and animal populations.

Tropical forest species are especially prone to extinctions. Species richness is linked to species rareness (Elton 1975; Whitmore 1984a). Most species of tropical trees, birds and mammals, and even invertebrates, occur at low densities. Many species of rainforest animals are limited to specific forest types. Collections of rainforest beetles from the canopy in the Amazon basin revealed that 83% of the beetles were found in only one forest type, 14% were shared between two types and only 1% were found in all forest types (Erwin 1983). Thus degradation of a forest type may lead to loss of some species dependent on that habitat.

Some animals are even more host-specific, associated with just one or a few species of plants. Thus a single leguminous tree in Tambopata Reserve, Peru, was found to be host to 43 species of ants belonging to 26 genera, equivalent to the entire ant fauna of the British Isles (Wilson 1987). Although many of those species will not be totally dependent on that tree, it can be seen that loss of even a relatively small block of forest will affect a large number of species and perhaps lead to local extinctions of some species.

Plant-animal relationships are complex and are little understood. Although animals usually comprise only a small proportion of the total biomass of an ecosystem, they are very important in regulating that ecosystem through their roles as pollinators, predators and seed dispersers (chapter 4). In rare cases a plant may be totally dependent on just one species of animal. In Mauritius, for instance, the formerly common tree *Calvaria major* has not produced seedlings since its obligate seed disperser, the dodo bird, was hunted to extinction three hundred years ago. The seeds needed to pass through the dodo's gut to germinate (Temple 1977). There are other examples of the "living dead" (Janzen 1986) where mature trees survive but cannot reproduce, because their pollinators or fruit dispersers have been exterminated locally; or where a population of plants or animals is so small and so isolated from other populations that there are no opportunities for outbreeding.

The greatest richness of mammal and bird species (and other animal and plant groups) is concentrated in lowland rainforests. Thus the Taman Negara (an area of 4,343 km^2 in West Malaysia), which encompasses a large block of lowland rainforests, contains 60% of the endemic mammal species of the entire Sunda Shelf region, with 142 of the 198 mammals recorded being dependent on rainforest for their existence (Medway 1971b). Of the 241 lowland bird species in Peninsular Malaysia, 172 (71%) have been recorded in the Taman Negara (Wells 1971). Similarly in Borneo 78% of all resident birds depend on some form of closed woodland, and 244 species (61% of resident birds) are confined to mixed lowland rainforests (Wells 1985). Of these, 146 species (60%) are Sunda endemics. Mammal and bird lists for Kalimantan reserves confirm that lowland forest habitats are especially rich in species. Thus Kutai National Park, a lowland rainforest reserve, supports at least 74 mammals (excluding bats) and 274 species of resident and visiting birds, about half the total listed for Borneo (MacKinnon 1990c).

What does habitat loss mean in terms of species conservation? It has been estimated that clearance of 1,000 km^2 of lowland forest in West Malaysia would lead to the loss of 1,190 siamangs, 10,900 gibbons, 60,000 banded langurs, 61,400 dusky langurs and 15,800 long-tailed macaques (Marsh and Wilson 1981). For Sumatra it has been suggested that clearance of 1,000 km^2 of forest would lead to loss of 50,000 monkeys, 9,000 siamang, 6,000 gibbons, 30,000 squirrels, 15,000 hornbills, 200 tigers and 100 elephants (Whitten et al. 1987b). Of course these figures are "guesstimates," but the numbers of animals involved are considerable.

The loss of even one individual before reproduction leads to loss of genetic diversity within the species. The loss of many individuals and the fragmentation of habitats associated with forest clearance can lead to such reduced and dispersed populations that there are few or no opportunities for outbreeding, and small isolated populations are doomed. Inevitably, destruction of lowland rainforests by human activities will lead to mass extinctions of species, perhaps in the order of several million species (Erwin 1988).

Endangered Species in Borneo

Several Bornean plants and animals are already considered to be threatened, the animals listed in the *IUCN Red List of Threatened Animals* (IUCN 1990). Bornean plants threatened by forest clearance include ironwood *Eusideroxylon zwageri* and five, possibly six, species of *Rafflesia* (fig. 14.2a and b; table 14.3). Threatened mammal species include Borneo's largest cat, the clouded leopard *Neofelis nebulosa*, the bay cat *Felis badia* (very rare), the marbled cat *F. marmorata* and the flat-headed cat *F. planiceps*. Rare and threatened primates include the orangutan *Pongo pygmaeus*, the endemic proboscis monkey *Nasalis larvatus* and the western tarsier *Tarsius bancanus*, all threatened by destruction of their forest habitats. The elephant *Elephas maximus* and banteng *Bos javanicus* are also listed. In the last few decades the Sumatran rhinoceros *Dicerorhinus sumatrensis* has disappeared from Tanjung Puting and Kutai national parks, and now there may be no surviving rhino population in Kalimantan (fig. 14.3).

The status of several bird species such as hornbills and pheasants also gives cause for concern. The only Bornean bird listed as endangered in the Red Data Book (IUCN 1990) is the Malay peacock pheasant *Polyplectron m. schleiermacheri*, but several other birds are rare and may be threatened, including the great argus *Argusianus argus* (common but easily snared), the helmeted hornbill *Rhinoplax vigil* (hunted for its feathers and valuable ivory casque), Bulwer's pheasant *Lophura bulweri*, the swiftlets *Collocalia fuciphaga* and *C. maxima* (locally threatened by overcollection of edible nests), the black-browed jungle babbler *Trichastoma perspicillatum* (very rare and threatened by habitat loss), Everett's ground thrush *Zoothera everetti* (rare on a few mountains) and the bald-headed woodshrike *Pityriasis gymnocephala* (a rare lowland bird threatened by habitat loss). The latest field guide to Bornean birds lists the black-nest swiftlet and another 13 species as endangered (MacKinnon and Phillipps 1993).

Table 14.3. Distribution of Rafflesia species in Borneo.

	West Kalimantan	Gunung Niut	East Kalimantan	Kutai	Sangkulirang	Sabah			
						Kinabalu	Crocker Range	Gunung Lotung	Mount Gading
R. borneensis			•	•					
R. tengku-adlinii							•	•	
R. pricei						•	•		
R. arnoldii	•	?							•
R. keithii				?	?	•	•		
R. witkampii				?	?				

Borneo has five (or possibly six) species of *Rafflesia*.
Rafflesia tuan-mudae described by Beccari (1896) is *R. arnoldii*.
Source: Meijer in litt.

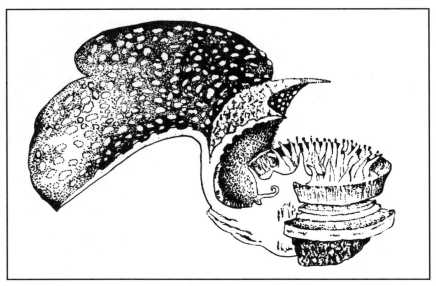

Figure 14.2a. Section through *Rafflesia* bloom.

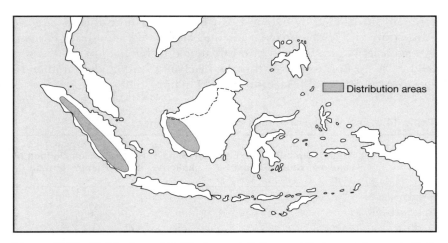

Figure 14.2b. Distribution of *Rafflesia arnoldii*, the world's largest flower, in Kalimantan and Sumatra.

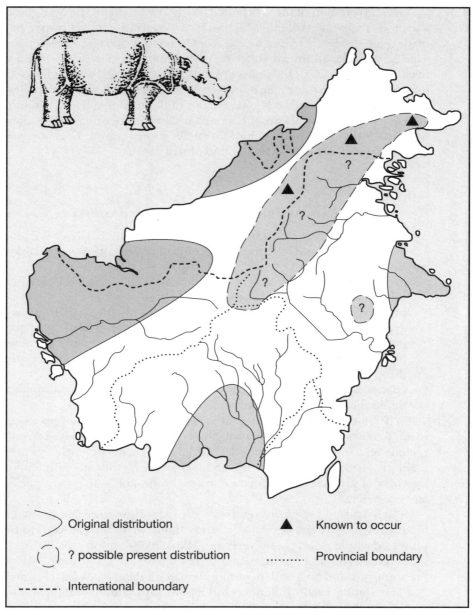

Figure 14.3. Distribution of Sumatran rhinoceros in Borneo: a species heading for extinction?

Overexploitation of the crocodile *Crocodylus porosus* and the false gavial *Tomistoma schlegeli*, and of other reptiles such as marine and river turtles, has made these species increasingly rare. In addition to these large, conspicuous species, many small invertebrates may be slipping into extinction unnoticed and unrecorded. One insect group which is relatively well known and for which critical habitats have been studied is the swallowtail butterflies. Borneo has 40 species of swallowtails, four of them endemic to the island. Three of these butterflies are threatened because of their limited distributions and loss of their lowland forest habitats: *Papilio acheron*, *Graphium procles* and *Troides andromache* (Collins and Morris 1985).

The Need to Conserve Biological Resources

The World Conservation Strategy, launched jointly by IUCN, WWF and UNEP in 1980, advocated conservation of living resources as essential for sustaining development by:
- maintaining the essential ecological processes and life-support systems on which human survival and development depend;
- preserving genetic diversity on which depend the breeding programmes necessary for the protection and improvement of cultivated plants and domesticated animals, as well as much scientific advance, technical innovation and the security of the many industries that use living resources; and
- ensuring the sustainable use of species and ecosystems, which support millions of human communities as well as major industries.

In 1991 the same international agencies released a follow-up document, *Caring for the Earth*, emphasising that conservation is an essential component of a strategy for sustainable living. The cornerstone of any programme attempting to achieve the above objectives is the establishment of networks of protected areas for *in situ* conservation of gene pools, species and ecosystems.

A well-managed system of reserves is the best way to maintain Borneo's biodiversity. IUCN has suggested a target figure of 10% of all land to be protected within reserves (McNeely and Miller 1983). Among the Bornean states, Brunei already has an extensive reserve network, covering 34% of the country's land area and including all major habitat types (MacKinnon and MacKinnon 1986). Indonesia and East Malaysia still have some way to go to meet these figures, but Kalimantan could reach the 10% target if all proposed reserves were gazetted. The National Conservation Plan for Indonesia outlines conservation priorities to protect biological diversity in all provinces including Kalimantan (MacKinnon and Artha 1981). Its recommendations are endorsed by the National Biodiversity Strategy (KLH 1989), the Forestry Action Plan (MoF/FAO 1991) and the Biodiversity Action Plan (BAPPENAS 1991).

HABITAT PROTECTION

Protecting habitats, and the individuals and species within them, can provide social and aesthetic benefits as well as many economic benefits, both directly via utilisation of biological resources and indirectly through protection of the environment. Most protected areas can be justified on the basis of traditional economic cost-benefit criteria (MacKinnon et al. 1986; McNeely 1988).

Wild species, and the genetic variation within them, make contributions to Indonesian agriculture, medicine and industry worth many billions of dollars per year (BAPPENAS 1991; KLH 1992). Perhaps even more important are the essential life processes that depend on natural habitats, including stabilisation of climate, protection of watersheds, conservation of soil and protection of nurseries and breeding grounds. Conserving these processes cannot be divorced from conserving the individual species that make up natural ecosystems. Some of the main benefits supplied by natural habitats are summarised below; together they make an excellent argument for the conservation and protection of at least some areas of natural habitat.

Box 14.3. Nonconsumptive benefits of conserving biological resources.
The benefits accruing to society in return for investments in conserving biological resources will vary considerably from area to area and from resource to resource. Most such benefits will fall into one or another of the following categories:

- photosynthetic fixation of solar energy, transferring this energy through green plants into natural food chains, and thereby providing the support system for species that are harvested;
- ecosystem functions involving reproduction, including pollination, gene flow, cross-fertilisation; protection of genetic reservoirs, and maintenance of environmental forces and species communities which influence and maintain evolutionary processes;
- maintaining water cycles, including recharging ground water, protecting watersheds and buffering extreme water conditions (such as flood and drought);
- regulation of climate, at both macro-climatic and micro-climatic levels (including influences on temperature, precipitation and air turbulence);
- soil production, and protection of soil from erosion, including protecting coastlines from erosion by the sea;
- storage and cycling of essential nutrients, e.g., carbon, nitrogen and oxygen; and maintenance of the oxygen-carbon dioxide balance;
- absorption and breakdown of pollutants, including the decomposition of organic wastes, pesticides and air and water pollutants; and
- maintenance of the recreational, cultural, scientific, educational, spiritual, and historical values of natural environments.

Source: McNeely 1988

Stabilisation of Hydrological Functions

Natural vegetation cover on water catchments regulates and stabilises water runoff. Deep penetration by tree roots or other vegetation makes the soil more permeable to rainwater so that runoff is slower and more uniform than on cleared land. As a consequence, streams in forested regions continue to flow in dry weather, and floods are minimised in rainy weather. Studies in Malaysia show that the peak runoff per unit area of forested catchments is about half that of rubber and oil palm plantations, while during dry periods the flow is roughly double (Daniel and Kulasingham 1974). Watershed protection of hydrological functions can be of enormous value, worth many millions of dollars per year (Garcia 1984; McNeely 1987). Often this means that the costs of establishing and managing reserves that protect water catchment areas can be met and justified as part of the hydrological investment (MacKinnon 1983b). With this in mind, the Dumoga-Bone National Park in North Sulawesi was established with part of a multimillion dollar loan from the World Bank, to protect a major irrigation project in the adjacent lowlands (Sumardja et al. 1984). The role of Kutai National Park, East Kalimantan, in protecting the aquifer on which urban and industrial water supplies depend has been recognised by local industry (Petocz et al. 1990).

Soil Protection

Exposed tropical soils degrade quickly due to leaching of nutrients, burning of humus, laterisation of minerals and accelerated erosion of topsoil. Good soil protection by natural vegetation cover and plant litter can preserve productive capacity, prevent dangerous landslides, safeguard coastlines and river banks and prevent the destruction of coral reefs and freshwater and coastal fisheries by siltation. Maintaining natural forests in the watershed can prevent costly and damaging siltation of fields, irrigation canals and hydroelectric dams. The forests of Pleihari-Martapura Reserve in South Kalimantan help to reduce siltation levels in the Riam Kanan dam and irrigation scheme in South Kalimantan.

Stabilisation of Climate

There is growing evidence that undisturbed forest helps to maintain local rainfall patterns by recycling water vapour back into the atmosphere and by the canopy's effect in promoting atmospheric turbulence (Dickinson 1981; Henderson-Sellers 1981; Myers 1988). This function may be particularly important in the drier parts of East Kalimantan, and in unusual drought years when extended dry periods are often more critical to agriculture than the heavier monsoon rains. Forest cover also helps to moderate local ambient temperatures, benefiting the surrounding agricultural areas by lowering transpiration levels and water stress.

Conservation of Renewable Harvestable Resources

Biological productivity under natural conditions is "cost free" and often higher than from any form of artificially planted alternative (Odum 1971). Forest products, both hardwood timbers and nontimber products such as rattan, damar, birds' nests, wild honey, medicinal plants and wild game, play an important role in local and national economies in Borneo (chapter 8). Harvesting these products at low levels from natural forests can be sustainable as just another form of "predation" on the resource. The value of these harvests to local and surrounding communities may be as great as, or greater than, the value of any other form of land use. The economic returns accruing from protecting natural habitats can be substantial. In India, for example, a partly protected mangrove forest yielded some 110 kg of prawns/ha/year, while in a nearby unprotected estuary, with damaged mangrove habitat, prawn production was only 20 kg/ha/year (Krishnamurthy and Jeyaseelan 1980).

Protection of Genetic Resources

Only a very small proportion of species are used directly by people in agroecosystems, fisheries, forests and other industries or for medicinal purposes. Probably many more species have potential for human use, either in their wild forms or after domestication. It has been estimated that worldwide, about 80,000 species of plants are likely to be edible, yet only 3,000 are used as human food, and a mere 150 are cultivated to any extent (Myers 1979). All domestic plants and animals were originally derived from the wild, and many can only be maintained and improved by crossbreeding with their wild relatives. The potential values of such genetic resources are enormous, and most future improvements in tropical agriculture and silviculture will depend on their preservation. Protected areas can be extremely valuable as *in situ* gene banks. Moreover the gene pool value of reserves will increase, as remaining areas of natural habitat decline in the face of development (MacKinnon et al. 1986).

Preservation of Breeding Stocks, Population Reservoirs and Biological Diversity

Natural habitats, including protected forests and designated reserves, protect crucial life stages or elements of populations that are widely and profitably used outside those habitats. They are sources of wildlife and of seed for dispersal, and fish spawning areas. The maintenance of parent populations of economically important plants and animals is a valuable role of protected habitats.

Protected areas also act as refuges to maintain biological diversity. This is particularly important for keeping open future options, since species of actual or possible value may be dependent on other species that have no known economic value. The disruption of complex tropical food webs may lead to unpredicted extinctions, especially when mutualism is involved (Gilbert 1980).

Maintenance of the Natural Balance of the Environment

The existence of a protected area may help to maintain a more natural balance of the ecosystem over a much wider area. Protected areas afford sanctuary to breeding populations of birds that control insect and mammal pests in agricultural areas. Studies in Sabah suggest that high densities of forest birds in commercial *Albizia* plantations limit the damage caused by leaf-eating caterpillars. The birds require natural forest for nesting (Sheldon 1992). Bats, birds and bees that nest, roost and breed in reserves may range far outside their boundaries and may pollinate fruit trees in the surrounding areas. The pollination of the durian tree, for example, is dependent on wild nectar-feeding bats for its pollination. When durians are not in flower the bats, especially *Eonycteris spelaea* and *Macroglossus minimus*, depend on other wild trees for nectar, so loss of forest can lead to reduced bat populations and failure of the valuable durian crop (Start and Marshall 1976). Monkeys, apes, many ungulates and frugivorous birds, such as hornbills, are important seed dispersers, and play an important role in regulating regeneration in Kalimantan forests. Conservation of these species is central to conservation of the ecosystems in which they live.

Economic Benefits from Tourism

Species conservation is a major source of revenue in some tropical countries such as Kenya, where wildlife-related tourism is the largest major earner of foreign revenue. In 1977 tourism returns from Amboseli National Park, Kenya, were equivalent to $40/ha/year, 50 times the potential agricultural value of the area. In Amboseli it is estimated that one lion is worth $25,000 and an elephant herd has a value of $600,000 as tourist attractions (Western and Henry 1979). While many Indonesian parks and reserves have only limited potential for tourism, because of the difficulties of viewing wildlife in rainforest conditions, some have potential for specialist tours, such as birdwatching or trekking. In Kalimantan several conservation areas such as Kutai, Kayan-Mentarang, Gunung Niut and Tanjung Puting could be developed for wildlife tourism. Local communities would benefit economically from providing services such as guest houses, restaurants, transport, local handicraft souvenir outlets, and guide facilities. Tourism can provide employment and help to preserve traditional cultures and handicrafts.

Environmental Monitors

Natural habitats have a proven role in maintaining a healthy environment. Forests act as "carbon sinks" and also remove pollutants from the atmosphere. Freshwater habitats and their component species play a key role in water purification processes. Many toxic chemicals and dangerous pollutants are gradually broken down and rendered harmless by natural processes. Several plants are valuable in fixing heavy metals in soil (Brooks et al. 1988). Animals may also be valuable indicators of ecosystem responses to air pollution (Newman and Schreiber 1984). Seabirds are useful indicators of pollution in coastal waters; sampling of eggs and tissues allows monitoring of organochlorines and heavy metals in marine environments (Diamond and Filion 1987). Dragonflies also may be good indicators of conservation needs. They have complex life cycles spanning both land and water, and often depend on complicated food chains. As such they are sensitive to disturbance and give a reasonable indication of environmental health (O'Toole 1987).

The economic and environmental benefits of conserving species and natural ecosystems are many and various. Only a few species are recognised as being of commercial importance at present, so the greatest economic contribution of protected areas may come from environmental protection functions. Other benefits that can come from conservation and protection of habitats include: opportunities for recreation; creation of employment opportunities; and the provision of facilities for research and education.

Future Benefits

Popular conservation literature assigns great importance to the "serendipity" value of species, that is, the potential that each species (especially those not yet discovered) may have for human use in the future (Pearsall 1984; Myers 1984). Certainly species can have unexpected uses. The horseshoe crab, found on Kalimantan's beaches, has bright blue blood that congeals rapidly when exposed to even minute concentrations of bacterial endotoxins. This property is exploited in tests of the purity of fluids intended for injections and also in the identification of a serious form of meningitis. The structure of the skull and vertebrae of woodpeckers has given insights into the design of crash helmets. The wing movements of chalcid wasps have assisted the design of improved helicopter rotors (Myers 1983). Here observations of simple natural phenomena have led to new technologies of immense economic benefit.

Not all plants and animals have economic values, however, nor is it wise to try to justify conservation of species in these terms alone. Even if species can be assigned economic values, it will probably be more profitable for the individual to exploit rather than to conserve resources (Clark 1973; Ehrenfeld 1988).

Box 14.4. Simplified scheme for assessing suitable protection category for protected habitats.

			Recommended Status	IUCN Category
Protection of nature highest priority	visitor use disturbing, or of low priority	primarily for preservation	Strict Nature Reserve	I*
		primarily for research	Scientific Reserve	I
		biologically valuable	Wildlife/Game Reserve	IV*
	zoned visitor use and/or some management desirable	geophysically or biologically spectacular	Natural Monument	III
	visitor use high priority	global priority	World Heritage Site	X*
	not for consumptive use	national priority	National Park	II*
		local priority	Provincial Park	II
	consumptive uses for local people	global interest	Biosphere Reserve	IX*
		regional interest	Anthropological Reserve	VII
Protection of nature secondary priority	water catchment vital	high visitor potential	Protective Recreation Forest	VIII*
		low visitor potential	Hydrological Protection Forest	VIII*
	water catchment not vital	hunting or harvesting value high	hunting a priority	Hunting Reserve	VIII*
			traditional use a priority	Wildlife Management Zone	VIII*
	hunting or harvesting value low	essentially natural	Agro-forestry Reserve	VIII*
		essentially agricultural	Protected Landscape	V

* – categories of protected areas in Indonesia

Source: MacKinnon et al. 1986

Apart from economic benefits, there are many social and moral arguments for conserving natural habitats. Conservation areas can improve quality of life, preserve traditional and cultural values and encourage national and regional pride. In addition, this generation has moral and social obligations to protect resources and a healthy living environment for future generations. No monetary value can be placed on any of these benefits, yet many people would agree that protecting the nation's natural heritage is as important as maintaining its cultural, religious and artistic traditions. In many traditional Kalimantan cultures wildlife plays an important symbolic role; maintenance of these traditions is just as important for the community's spiritual health as economic development is for improving living standards.

Genetic Resources

Conservation of genetic resources is particularly important because genetic variability is the raw material for evolution. All living organisms are made up of many genes, from 1,000 in a bacteria to 10,000 in a fungus and as many as 400,000 in a flowering plant (Wilson 1988). A small mammal such as a mouse has in the order of 100,000 genes. Almost every individual organism on this planet, apart from members of clones, possesses some unique genetic combination, so that even a fairly small local population of plants or animals represents a large pool of genetic variability.

In a sense, genetic erosion (the loss of genetic diversity) starts when an individual plant or animal dies without reproducing. Genetic erosion is a gradual process, but there are critical events in the process, where a whole "package" of genetic qualities are lost. Such danger points, in ascending order of seriousness, are:
- local loss of a breeding population
- loss of a distinct subspecies
- loss of a species
- loss of a genus
- loss of a block of habitat
- loss of a whole habitat type.

The establishment and protection of conservation areas is aimed at guarding against the loss of the larger units such as habitat types (MacKinnon et al. 1986). Conservation of gene pools of species of actual or potential value to humankind is a primary objective of management in many categories of protected areas. These are *in situ* gene banks. In addition, measures may also be taken to establish gene banks *ex situ* in botanic and zoological gardens and as germplasm banks. Concern about the loss of genetic diversity in many crop species has led to the establishment of gene banks such as the Indonesian Germplasm Bank to preserve varieties that are not widely used.

Destruction of tropical habitats, especially lowland rainforests, leads to the irreversible loss of biological diversity and genetic resources. The genetic variation in wild populations is the raw material not only for evolution but also for the improvement of domesticated and cultivated forms of animals and plants, and for the development of medicinal and industrial products. Conservation of genetic resources is one of the highest environmental priorities (Frankel and Soulé 1981; Prescott-Allen and Prescott-Allen 1983; Myers 1983, 1984, 1988). The need to protect biological diversity and genetic resources in Indonesia was recognised in the Indonesian Biodiversity Strategy (KLH 1989), the Biodiversity Action Plan (BAPPENAS 1991) and the Indonesian Country Study on Biological Diversity (KLH 1992).

A few species of plants and animals have already made major contributions to human welfare, providing the means whereby humans could progress from a nomadic hunter-gatherer lifestyle to settled agriculture. Of several thousand plant species known to be edible, fewer than twenty species provide 85% of the world's food, with two-thirds coming from just three crops: rice, maize and wheat (Vietmeyer 1986; Plotkin 1988). The species which were brought into cultivation in the past were chosen because they were easy to grow with primitive farming methods, not because of their ability to contribute to a modern industrial society (Raven 1988). Tropical forests and other tropical habitats may harbour many new species of potential food value, as well as many relatives of existing cultivars, with potential for improving yields and disease resistance. Many crop relatives are annuals or weeds, found in abandoned fields, such as more than twenty varieties of wild rice found in the Apo Kayan (Jessup pers. com.). These species are not only of botanical interest. One wild species of rice, *Oryza nivara* from India, has been used in breeding programmes to produce crops resistant to grassy stunt virus, a serious disease of cultivated Asian rice (Chang 1984).

To date, few of the native species of Kalimantan have been domesticated or have made much contribution to agricultural productivity. Among plant species, rattans and several native fruit trees including durian, mangoes and wild *Citrus* all have potential for more intensive cultivation. Animal species with known potential include the banteng *Bos javanicus*, the wild relative of the domesticated Bali cattle, and perhaps the wild pig *Sus barbatus*. Crocodile farms have been established at Banjarmasin, Pontianak and Balikpapan to produce skins and meat for local consumption and export (Luxmoore et al. 1985). At present these farms rely on wild populations for stocks of young animals and eggs; this could create incentives to protect the mature breeding populations and their wetland habitats, thus linking economic development to conservation.

Knowledge of the potential of many wild species is scanty. The quest for new medicines to cure or alleviate modern diseases, such as AIDS and cancer, has led to a resurgence of interest in rainforest plants. One rainforest tree genus *Calophyllum* is believed to have potential for treating AIDS. Recent surveys in West Kalimantan (Burley 1991) and the Apo Kayan (Leaman et al. 1991) aim to inventory the plants used by local communities, particularly medicinal plants.

About 40% of all drug prescriptions in the U.S.A. (valued at $8 billion per year) are compounds of natural plant origin. Worldwide, 119 pure chemical substances extracted from fewer than ninety species of higher plants are used in medicine (Farnsworth and Soedjarto 1985). The importance of traditional knowledge and medical folklore is well illustrated by the fact that of these 119 natural substances, 74% have the same, or related, current medical use as the plants from which they are derived (Farnsworth 1988). One plant alone, the rosy periwinkle *Catharanthus roseus*, native to Madagascar but a common garden plant throughout Indonesia, has yielded two major drugs: vincristine, effective against childhood leukaemia, and vinblastine, a secondary drug for Hodgkin's disease. By 1985 these two drugs were responsible for international and domestic sales in the U.S.A. worth more than $100 million annually, a sum equivalent to 80% of Indonesia's annual revenue from exports of nontimber forest products. In Indonesia and many other tropical Asian countries, natural (rather than synthetic) plant compounds are even more important. Many Kalimantan societies rely on herbal medicines (Pearce et al. 1987), at least some of which could have important pharmaceutical properties.

Surprisingly, very few of the major drug companies are interested in monitoring wild plants, or in the conservation of lowland rainforest areas (Elliott and Brimacombe 1986). In part, this is due to the substantial costs of collecting, testing and developing drugs from natural plant products. The National Cancer Institute, for instance, has tested 35,000 species of higher plants; many show anticancer activity, but none has yet provided safe and effective compounds for human use (Farnsworth 1988). The potential value of certain genetic resources, and to whom these benefits should accrue, are also matters still to be resolved. Many of the most valuable genetic resources, including food and industrial crops, originate in the world's tropics, and it is likely that there are many other species of potential benefit in tropical forests. When commercial firms exploit and develop these resources, they recover their development costs by charging high prices for their products; at present none of these profits return to the country of origin. The question of "property rights", and the granting of royalties to indigenous communities for ethnobotanical knowledge, are being discussed under the international Convention on Biological Diversity.

The Protected Areas System within Borneo

Destruction of tropical habitats, especially lowland rainforests, leads to the irreversible loss of biological diversity and genetic resources. As forest clearance continues, the total area of natural habitats is reduced and becomes fragmented. This has important implications for the survival of plant and animal species. For some lowland habitats, the only undisturbed areas of forest remaining in Borneo are within the protected area system.

Box 14.5. Different categories of protected areas within Kalimantan.

Taman Nasional (National Park)
Large, relatively undisturbed areas of outstanding natural value with high conservation importance, high recreation potential, of easy access to visitors and clearly of benefit to the region.
Examples: Tanjung Puting, Kutai, Gunung Palung, Bukit Baka

Cagar Alam (Nature Reserve)
Areas requiring strict protection, containing undisturbed fragile habitats of high conservation importance, unique natural sites, and homes of particular rare species.
Examples: Kayan Mentarang, Bentuang and Karimun

Suaka Margasatwa, Suaka Alam (Game or Wildlife Reserve)
Generally medium or large areas of relatively undisturbed habitats of moderate to high conservation importance, where management may be appropriate.
Examples: Pleihari-Martapura, Danau Sentarum

Taman Wisata (Recreation Park)
Small natural areas or landscaped sites with attractive or interesting features, of easy access for visitors, where conservation value is low or not threatened by visitor activities and recreation-orientated management.
Examples: Pulau Kembang, Mandor, Batu Hapu.

Taman Buru (Hunting Reserve)
Medium or large, natural or seminatural habitats with potential for hunting permitted game species (pigs, deer, wild cattle, fish, etc.). Such reserves should be of low conservation importance or have conservation values that are not threatened by the hunting/fishing activities.
Example: Apar Besar (proposed)

Hutan Lindung (Protection Forest)
Medium to large areas of natural or planted forested land on steep, high, erodible, rainwashed lands, where forest cover is essential to protect important catchment areas and prevent landslips and erosion, but where conservation priorities are not so high or urgent as to justify reserve status.
Examples: Batikap I, II and III, Bukit Batu Tenobang

Cagar MAB (Biosphere Reserve)
Sites of natural habitats with high species richness and of known conservation value, where resource extraction by indigenous people is allowed in a traditional manner.
Example: Apo Kayan (proposed)

Various attempts have been made to estimate the minimum size of protected habitat required to include viable populations of all essential component species in each ecotype (Frankel and Soulé 1981; Soulé et al. 1979; Diamond 1975; Lovejoy and Oren 1981; Wilcox 1984). To avoid extinction a species population must be large enough to maintain a sufficient proportion of its genetic variation to avoid the deleterious effects of inbreeding. This is the minimum viable population. Two hundred and fifty *reproducing* individuals is usually regarded as the minimum effective population size to maintain genetic variability (Wilcox 1984). This is not the same as a total population of 250 individuals. Many individuals may not breed at all; for example, among polygynous species such as orangutans or long-tailed macaques, only the dominant males breed with receptive females. Since only the individuals that breed successfully contribute to the gene pool of the next generation, the effective population size may be much less than the total population size.

Thus the maintainence of genetic diversity in an isolated population normally requires a breeding population several times larger than 250 individuals. Estimates for viable populations vary widely between 500 and 10,000. If we take a population of 5,000 as a conservative estimate, it is possible to calculate the area of habitat needed to protect that number of individuals of any species. Obviously the size of area required will be very different for species that live at high densities and for species that are dispersed. Lowland rainforests are characterised by high species richness and low species density. Most trees are present at densities of less than one tree per hectare, and many species at only one tree per 10 hectares. To include a viable population of most tree species would therefore require a reserve of at least 50,000 ha (500 km^2) for the richest habitat. Slightly smaller areas should suffice for less rich habitats.

Although the Bornean forests are species rich, individuals of any one species are rare. The area required to protect a minimum viable population of any species will, therefore, be large. Davies and Payne (1982) estimated the minimum continuous area required to conserve populations of 200 adults of various mammal and bird species (table 14.4). For large animals such as elephant and rhino, an area of 6,000 km^2 was estimated; for honey bears and clouded leopard, 800 km^2. Even a viable breeding population of hornbills is likely to require 150-400 km^2. These estimates give some idea of the amount of land needed to protect most of the Bornean flora and fauna.

From the above discussion it can be seen that protected areas must be large to conserve species, preferably large enough to include thousands of individuals of even the least abundant species. As a general rule reserves are of little long-term conservation value if smaller than 5,000 ha (50 km^2) unless they are protecting small sites of special interest such as turtle beaches or limestone caves, or areas where a rare species survives because of impoverishment of other species and reduced rates of competition. The smaller the reserve and the more isolated it is, the greater the degree of management required to maintain the features and species that the reserve was established to protect (Diamond 1984). The implications of habitat fragmentation, and the isolation of small reserves, for species survival are considered later.

Species losses can be countered by active management and artificially maintained levels of immigration and outbreeding (translocation), but these options are expensive in terms of money, time and effort. The best and most cost-effective way of conserving species is to establish land-use practices that allow many species to maintain themselves. Complete habitat protection in conservation areas is the best way of conserving species-rich communities.

Because almost all habitat types contain some unique species, an adequate protected area network must protect representative samples of all ecosystems to protect the widest possible range of species. The island of Borneo and its offshore islands can be divided into several distinct biogeographic units, and it is important to protect representative habitat types within all of them (MacKinnon and MacKinnon 1986). Biological diversity is not distributed uniformly; centres of biological richness can be recognised. These are areas or habitats of high species richness or endemism. All of these factors were taken into consideration in the preparation of the National Conservation Plan for Kalimantan (MacKinnon and Artha 1981).

Table 14.4. Estimated minimum continuous areas for conservation of 200 adult individuals of some mammal and bird species.

Species	Preferred habitat	Minimum area (km^2)
Sumatran rhinoceros	LD, UD	6,000
Elephant	L	6,000
Orangutan	LD	150
	UD	200
Honey bear	LD, UD, (HD?)	800
Clouded leopard	LD, UD, (HD?)	800
White-crested hornbill	LD, UD	150
Bushy-crested hornbill	LD, UD, HD	150
Wrinkled hornbill	(LD, FS?)	320
Wreathed hornbill	LD, UD, HD	300
Black hornbill	LD	150
Pied hornbill	R	50
Rhinoceros hornbill	LD, UD, HD	250
Helmeted hornbill	LD, UD, HD	400

FS : Freshwater swamp forest
HD : Primary highland dipterocarp forest
L : Logged lowland dipterocarp forest
LD : Primary lowland dipterocarp forest
R : Riparian (riverine) forest
UD : Primary upland dipterocarp forest

Source: Davies and Payne 1982

The gazetted and proposed reserves in Kalimantan account for 7.1% of the total land area (MacKinnon and Artha 1981) and cover most major habitat types and areas of high biological diversity (table 14.5). The amount of original habitat protected in Kalimantan varies from 5% in the extensively deforested province of South Kalimantan to 12% in East Kalimantan. Much of the remaining forest occurs on steep slopes or above 600 m altitude, and helps to protect the watersheds. While montane habitats are well represented in the Kalimantan protected area network, threatened habitats like lowland rainforest and coastal wetlands are poorly protected. The most important additions to the Kalimantan protected area network would be the declaration of the proposed Sangkulirang limestone reserve and reserve extensions to include more areas of lowland dipterocarp forests.

On the whole island of Borneo, only 3% of the original natural habitat has been protected within reserves, but another 8.7% is proposed for gazetting, much of this within Kalimantan (table 14.6). Many Bornean

Table 14.5. Areas of different habitat types in Kalimantan, and the proportion of habitats protected.

	Original area (km^2)	Remaining (km^2)	Remaining %	Protected* (km^2)	Protected* %
Montane and hill forest	98,220	94,390	96	29,210	29
Lowland rainforest	249,440	127,980	55	13,280	5.3
Limestone forest	2,270	1,450	63	–	–
Ironwood forest	2,100	730	34	200	9.5
Heath forest	80,760	24,750	30	2,100	2.6
Freshwater swamps	38,950	17,170	44	3,620	9.0
Peat swamps	44,130	35,310	80	2,570	5.8
Mangrove	18,010	9,800	54	780	4.3

* includes areas gazetted, designated and proposed as conservation areas.
Data from National Conservation Plan V: Kalimantan. (MacKinnon and Artha 1981)

Table 14.6. Areas of different habitat types in Borneo, and the proportion of habitats protected (actual or proposed).

Habitat	Original area (km^2)	Remaining (%)	Protected area (km^2)	Protected (%)	Proposed area (km^2)	Proposed (%)
Limestone forest	3,752	80	140	2.3	1,280	33
Freshwater swamp forest	31,644	82	1,710	5.4	3,370	11
Heath forest	66,882	48	1,440	2.1	2,975	4.4
Ironwood forest	1,440	41	200	13.8	0	0
Lowland dipterocarp forest	511,119	69	15,088	2.9	47,482	9.2
Montane forest	22,700	76	2,490	10.8	2,730	12
Mangrove	24,492	69	225	0.9	3,678	15
Peatswamp forest	58,102	92	1,500	2.5	2,680	4.6
Totals	720,131	68	22,793	3.0	64,195	8.7

Source: MacKinnon and MacKinnon 1986

Table 14.7. Areas of special conservation importance in Borneo.

Conservation area	Plants	Primates	Mammals	Birds	Insects	Fish	Reptiles	Caves	Utilisable plants	Research	Tourism
Kayan Mentarang	*		*	*	**				*	**	*
Kutai	*	**	**	*					*	*	**
Sangkulirang	**	*	*	*				**			
Apo Kayan	*								*	*	*
Tanjung Puting	*	**	*	*		*	*		*	*	**
Gunung Palung	**	**	*	*					*	**	
Bukit Raya/Bukit Baka	**	*	*	*					*	*	
Gunung Niut	*	*	*	*					*		
Danau Sentarum	*	*	*	*		**			*	*	
Pleihari Martapura	*	*	*								*
Gunung Mulu	**	*	*	**	*	*	**	**	*	*	**
Kinabalu	***	**	*	**	*	*	*		*	**	**
Danum Valley	**	**	**	**	*	*	*		*	**	*

reserves support high biological diversity and are of regional and global significance (table 14.7), in particular Gunung Palung National Park, Bukit Raya/Bukit Baka National Park, Gunung Bentuang and Karimun Natural Reserve, Danau Sentarum Wildlife Sanctuary, Tanjung Puting National Park, Kayan-Mentarang Wildlife Sanctuary, Kayan Mutlak and the proposed Sangkulirang area. Others, such as Kutai National Park and Gunung Niut Becapa, have suffered badly from agricultural encroachment, timber felling and other destructive activities officially prohibited in conservation areas (table 14.8). Kutai was also further damaged by the fires which raged through East Kalimantan in 1982-83.

This highlights the main problem facing Kalimantan parks and reserves: the lack of any effective protection and management. Many reserves have no staff, no budgets and even no properly marked boundaries. These constraints, together with the fact that many timber and mining concession boundaries overlap with conservation areas, make it imperative that action is taken soon to improve reserve protection. Table 14.9 sets out reserves likely to be destroyed within the next few years unless effective action is taken. Loss of these reserves will mean loss of much of Kalimantan's rich biological diversity. Recognising this need, the Conservation Department (PHPA) has approached several overseas agencies for help in upgrading the Kalimantan protected area network. USAID is

Table 14.8. Activities permitted (•) and prohibited (-) in different categories of protected area in Indonesia.

	National Parks	Nature Reserves	Wildlife Refuges	Tourist Parks	Hunting Parks	Protection Forests
Growing food crops	–	–	–	–	–	–
Growing tree crops	–	–	–	•	•	•
Human settlement	–	–	–	–	–	–
Commercial logging	–	–	–	–	–	–
Collecting herbs and firewood	–	–	–	–	•	–
Hunting	–	–	–	–	•	•
Fishing	•	–	–	–	•	•
Camping	•	–	•	•	•	•
Scientific collecting (under permit)	•	•	•	•	•	•
Active habitat management	•	–	•	•	•	•
Introduction of nonexotic species	•	–	•	•	•	•
Collecting rattan (under permit)	–	–	–	–	•	•
Mineral exploration	•	–	•	–	•	•
Wildlife control	•	–	•	•	•	•
Visitor use	•	–	•	•	•	•
Introduction of exotic species	–	–	–	•	–	•

Source: MacKinnon 1982

Table 14.9. Maximum estimated survival time of larger Kalimantan reserves (over 5,000 ha) without improved protection and management.

Reserve	Status	Area (ha)	D	0-5	5-10	10-20
West Kalimantan						
Gunung Niut	C.A.	140,000		**		
Muara Kendawangan	C.A.	150,000		**		
Gunung Palung	T.N.	30,000			**	
extension	T.N.	60,000			**	
Bentuang/Karimun	C.A.	600,000				**
Danau Sentarum	S.M.	80,000			**	
Hutan Sambas	pr	120,000	*	**		
Bukit Perai	H.L.	162,000				**
Gunung Tunggal	pr	61,000		**		
Bukit Ronga	H.L.	260,000				**
Bukit Batutenobang	H.L.	883,000				**
Bukit Baka	T.N.	100,000			**	
Central Kalimantan						
Bukit Raya	C.A.	110,000			**	
extension	pr	590,000		*	**	
Tanjung Puting	T.N.	305,000	*			**
extension	pr	70,000	*	**		
Parawen I & II	C.A.	6,200		**		
Kuala Kayan	pr	73,500		**		
Batikap	H.L.	740,375				**
Parawen Baru	pr	81,500		?		
Tanjung Penghujan	T.W.	40,000	*	**		
Hutan Kahayan	C.A.	150,000		**		
South Kalimantan						
Pleihari Tanah Laut	S.M.	35,000	**	*		
Pleihari Martapura	S.M.	30,000	*	*	*	
extension	T.W.	70,400	**			
Pamukan	pr	10,000		**		
Kelumpang Baai	pr	13,750		**		
Tanjung Dewa Barat	pr	16,250		**		
Pulau Sebuku	H.L.	14,400		**		
Meratus Hulu Barabai	pr	200,000	**		**	
Meratus Hulu Tanjung	pr	46,250			**	*
Muara Uya	pr	25,000		**	**	

Table 14.9. Maximum estimated survival time of larger Kalimantan reserves (over 5,000 ha) without improved protection and management *(continued)*.

Reserve	Status	Area (ha)	D	0-5	5-10	10-20
East Kalimantan						
Kutai	T.N.	200,000	**	**	**	
Muara Kaman	S.M.	62,500		*	*	
Kayan Mentarang	C.A.	1,600,000			**	
Muara Sebuku	pr	110,000			**	
Muara Kayan	pr	80,000		**		
Ulu Kayan	pr	800,000			**	
Gunung Berau	pr	110,000			**	
Sangkulirang	pr	200,000			**	
Sungai Barambai	pr	110,000		**		
Gunung Beratus	pr	130,000		**		
Apar Besar	pr	90,000	**			
Gunung Lumut	pr	30,000				**
Mahakam Lakes	pr	200,000	**	**		
Pantai Samarinda	pr	95,000	**			
Ulu Sembakung	pr	500,000			**	*
Apo Kayan	pr	100,000			*	*
Batu Kristal	pr	10,000				*

D – already partially destroyed; Pleihari Tanah Laut degazetted 1992.

C.A. = Cagar Alam (Nature Reserve)
H.L. = Hutan Lindung (Protection Forest)
S.M.= Suaka Margasatwa (Game/Wildlife Reserve)
T.N. = Taman Nasional (National Park)
T.W. = Taman Wisata (Recreation Park)
pr = proposed protected area

Source: MacKinnon 1988

Box 14.6. Linking conservation with development.

Kutai National Park in East Kalimantan was originally established as a game reserve of 360,000 hectares in 1936 by the Sultan of Kutai to protect species such as Sumatran rhino, banteng and orangutan. Subsequently 100,000 ha were excised for logging and oil exploration. In 1971 the logging stopped, and the logged-over forest was reinstated in the reserve. Shortly afterwards another 106,000 ha of pristine rain forest and some previously logged coastal forest were excised from the southern part of the reserve and allocated to the P.T. Kayu Mas timber concession and for site development to the two natural gas industries, P.T. Badak and P.T. Pupuk Kaltim.

By 1980 the eastern third of the reserve was already badly disturbed by logging, oil exploration and production facilities, the expansion of Bontang and Sangatta townships and uncontrolled forest clearance for agriculture. Nevertheless, in 1982 Kutai was declared a national park of 200,000 ha. In 1982-1983 fire swept through Kutai during abnormal drought, and destroyed large areas of forest; areas that had been logged suffered more damage, since the fire fed on dead trees and other debris left behind after the logging. About half the park, some 100,000 ha of lowland forest in the west, escaped the fire, but even here the drought killed many emergent trees. In 1984 Kutai National Park was included in IUCN's shortlist of the world's most threatened protected areas; since then encroachment and agricultural clearance have continued to degrade the park.

In spite of this unfortunate history, Kutai is still an important conservation area, since it encompasses the largest remaining tracts of the lowland rainforests and mixed ironwood forests that were once typical of East Kalimantan. Kutai includes a wide spectrum of lowland habitats and is a centre of biodiversity and high species richness. Half of all Bornean mammals and birds are recorded from the park, as well as 500 species of trees and a rich variety of other flora. Several rare, endangered and endemic species occur here. In addition to these biological values, the park protects the natural aquifer that supplies fresh water to the industrial towns of Bontang and Sangatta.

In the late 1970s the government of Indonesia adopted a policy to substitute coal for oil as fuel for electricity generation. South and East Kalimantan were known to have large Tertiary coal deposits, and exploration agreements were offered to foreign companies. Kaltim Primacoal, an Indonesian registered company owned by CRA Ltd. of Australia, and the British Petroleum Company (BP) of the U.K. were granted a 790,000 ha concession between Samarinda and Sangkulirang (see fig. 13.1). KPC discovered high-quality thermal coal in an area just north of Kutai National Park, and the Sangatta coal mine began production in early 1989. Mining is by conventional open-pit methods with production scheduled to rise from 2 million tonnes in 1991 to 7 million tonnes per year from 1994 onwards. Coal reserves exist for a 30-year mine life. A new town has been created for the workforce of 2,300 and their families, a total of some 10,000 persons. In addition, many other newcomers and spontaneous settlers have moved into the area, drawn by the new industrial developments. The mine and the increased settlement place new pressures on Kutai.

The opening of the mine, and the building of a new access road which cuts through the park, could have been the "last straw" for Kutai. Instead, the industrial development has opened up new opportunities for strengthening the protection and management of the conservation area. In an exciting new conservation initiative, Kaltim Primacoal (KPC) determined to enlist local industries to work with the national conservation agency (PHPA) and local government, to help to conserve Kutai. Under the sponsorship of KPC, a development

Box 14.6. Linking conservation with development *(continued).*

plan has been prepared with PHPA to identify key projects for the park. Other local industries have agreed to help, initially by providing recreational facilities. KPC is also seeking to establish a consortium of potential international donors. One potential source to fund restoration and reforestation of degraded areas in the park is the Dutch Association of Electricity-producing Companies (SEP) through its Forests Absorbing Carbon dioxide Emissions foundation (FACE). FACE is funded by a tax on some industrial carbon dioxide producers in the Netherlands. It seems appropriate that some of these funds should be used to restore degraded forests in a coal-producing province of Indonesia.

Kutai National Park has been recognised, both nationally and internationally, as a site of high conservation value. Nevertheless, with limited resources and increasing pressure on park land, the future of the park has not always seemed secure. Now the park's neighbours, the firms responsible for much of East Kalimantan's economic development, are turning their attention to conservation. With friends in industry and local government, Kutai could become one of the best-managed parks in Indonesia.

Source: MacKinnon et al. 1994

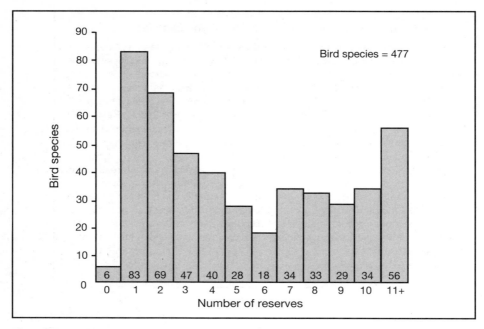

Figure 14.4. Incidence of Bornean bird species recorded in 15 reserves. (About half of all birds occur in five or more reserves.)

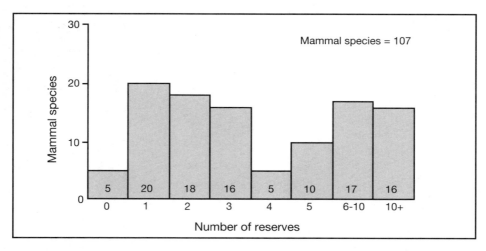

Figure 14.5. Incidence of selected mammals in Bornean reserves. (45% of all mammals selected are recorded for four or more reserves.)

assisting in the management of Gunung Palung and Bukit Baka, ODA in Danau Sentarum and WWF in Kayan Mentarang. Local industries are providing support to Kutai National Park, an exciting innovation that could lead to greater involvement of the private sector in conservation in Indonesia.

Species Protection in Reserves

The rationale for protecting large blocks of habitat is the assumption that if adequate habitat is protected then all the constituent species will also be protected. This assumption can be checked by examining the distribution of various indicator species of birds and mammals in the protected areas of Kalimantan and other Bornean states.

All of the Bornean birds (Smythies 1960; MacKinnon and Wind 1980; MacKinnon and Phillipps 1993) are recorded from at least one Bornean reserve, and many are represented in several (fig. 14.4). The same is true for mammals, certainly the most conspicuous species (table 14.10; fig. 14.5). All Bornean primates (including five, possibly six, endemic species) are protected in at least one major reserve (table 14.11). All endemic birds and mammals (MacKinnon and Artha 1981; Francis 1984; Payne et al. 1985) are represented in one or more reserves (tables 14.10 and 14.12). It should be emphasised, however, that while these species are recorded in reserves at present, their continued survival there will depend upon effective protection and management of those reserves.

Will the Reserve System Alone Be Adequate to Protect Borneo's Biodiversity?

Even with an extensive system of well-protected and well-managed protected areas, some species extinctions can be expected on Borneo. Even a large block of habitat is likely to lose some species after isolation; this can be thought of as decay of the ecosystem.

Several classic studies of the biogeography of islands (Diamond 1975; Simberloff 1974) have shown that small islands are unable to support as many species as larger islands of similar habitat, and that isolated islands support fewer species than islands close to the mainland (see chapter 1). As a rough guide, a tenfold increase in land area results in a near doubling of the number of species that an island can support at equilibrium. Where conservation areas become habitat "islands" (Diamond 1975; Willis 1974), they will lose some of their original species until a new equilibrium is reached, dependent on the size, richness and diversity of the area and its degree of isolation from other similar habitats. Larger patches of habitat or protected areas will lose species more slowly, but any loss of habitat will lead to some loss of species. As a rough generalisation, a single reserve con-

Table 14.10. Distribution of selected mammals in Bornean reserves.

Reserve	KM	Ku	S	PM	TP	BR	GN	GP	DS	DV	K	T	Mu	LE	Sm
Echinosorex gymnurus				X	X					X		X			
Hylomys suillus											X	X			
Suncus etruscus												X			
Suncus ater										X					
Suncus murinus										?					
Crocidura fuliginosa										X					X
Crocidura monticola										X		X			
Chimarrogale himalayica										X					
Tupaia gracilis							X			X				X	
Tupaia minor											X	X			X
Tupaia glis							X			X	X	X		X	
Tupaia montana							X				X		X		
Tupaia dorsalis														X	
Tupaia tana								X			X	X		X	X
Tupaia picta	?											X			
Tupaia splendidula		X					X								
Ptilocercus lowii										X					
Dendrogale melanura										X	X				
Manis javanica	X	X	X	X	X	X	X	X	X	X	X			X	X
Cynocephalus variegatus										X	X		X		X
Nycticebus coucang	X	X	X		X		X	X		X	X	X			
Tarsius bancanus	X	X			X	X	X	X		X	X	X	X	X	X
Presbytis rubicunda	X	X		X	X	X	X	X		X	X	X	X		
Presbytis hosei		X								X	X	X	X		
Presbytis frontata	X	X												X	
Presbytis cristata		X		X	X					X		X	X		X
Presbytis melalophos							X	X							X
Nasalis larvatus		X	X	?	X			X	X	X					X
Macaca fascicularis	X	X	X	X	X	X	X	X	X	X	X	X	X	X	X
Macaca nemestrina	X	X	X		X	X	X	X		X	X	X	X	X	X
Hylobates agilis					X	X		X							
Hylobates muelleri	X	X	X	X			X		X	X	X	X	X	X	X
Pongo pygmaeus		X	X		X	X	X	X	X	X	X		X		
Ratufa affinis		X			X		X	X		X	X	X	X		
Callosciurus baluensis											X		X		
Callosciurus prevostii		X	X	X	X	X	X	X		X	X	X	X	X	X
Callosciurus orestes		X									X		X		
Callosciurus notatus		X	X	X		X	X	X		X	X	X	X	X	X
Callosciurus adamsi	?									X	X				
Sundasciurus hippurus		X	X	X			X	X			X	X	X	X	X
Sundasciurus lowii		X									X	X		X	X
Sundasciurus tenuis		X					X				X			X	X
Sundasciurus brookei	X	X					X			X					
Sundasciurus jentinki										X	X				
Dremomys everetti										X	X				
Lariscus hosei			X							X					
Lariscus insignis		X			X		X			X		X	X		
Rhinosciurus laticaudatus										X					
Exilisciurus whiteheadi										X		X	X		X
Exilisciurus exilis		X								X	X	X	X		X
Nannosciurus melanotis		X				X	X	X					X		X
Glyphotes simus										X					
Rheithrosciurus macrotis							X			X	X	X	X		
Petaurillus hosei*															
Petaurillus emiliae*															
Petinomys setosus										X					
Petinomys vordermanni*															
Petinomys genibarbis		X								X		X			

Table 14.10. Distribution of selected mammals in Bornean reserves *(continued)*.

Reserve	KM	Ku	S	PM	TP	BR	GN	GP	DS	DV	K	T	Mu	LE	Sm
*Petinomys hageni**															
Iomys horsfieldi	?									X					
Hylopetes spadiceus	X									X					
Hylopetes lepidus										X					
Aeromys tephromelas		X								X		X			
Aeromys thomasi	X	X								X					
Pteromyscus pulverulentus										X					
Petaurista elegans										X					
Petaurista petaurista		X	X	X			X			X	X				
Hystrix brachyura	X	X			X	X	X	X	X	X	X		X	X	X
Trichys fasciculata	X						X					X			
Thecurus crassispinis	X	X	X							X		X			
Dugong dugon					X										X
Helarctos malayanus	X	X	X		X	X	X	X	X	X	X	X			X
Martes flavigula		X			X		X	X			X	X	X	X	X
Mustela nudipes		X								X	X				
Melogale personata										X					
Mydaus javanensis				X				X			X				
Lutra sumatrana				X		X				X					
Lutra lutra	?														
Lutra perspicillata	X									X					
Aonyx cinerea		X	X					X			X	X	X	X	X
Viverra tangalunga	X			X	X	X	X				X				
Cynogale bennettii		X		X											
Arctictis binturong	X	X			X		X			X	X	X	X		X
Arctogalidia trivirgata	X		X			X	X		X	X	X				
Paguma larvata	X					X				X					
Paradoxurus hermaphroditus		X								X	X	X			X
Hemigalus hosei										X					
Hemigalus derbyanus		X	X				X			X	X	X			
Prionodon linsang					X					X					
Herpestes semitorquatus		X	X							X	X				X
Herpestes brachyurus		X													
Herpestes hosei													X		
Neofelis nebulosa	X				X	X	X	X	X		X			X	X
Felis marmorata										X					
Felis planiceps								?					?		
Felis bengalensis		X	X		X					X	X				X
Felis badia										?			?		
Elephas maximus	X									X	X				
Dicerorhinus sumatrensis		e			e					X	X				
Sus barbatus	X	X	X	X	X	X	X	X	X	X	X	X	X	X	X
Tragulus javanicus	X	X			X	X	X	X		X	X	X	X	X	X
Tragulus napu	X	X			X	X	X	X		X	X	X		X	X
Muntiacus muntjak	X	X	X	X		X	X			X	X		X		X
Muntiacus atherodes		X		X		X		X		X		X			
Cervus unicolor	X	X	X		X		X	X	X	X		X	X	X	
Cervus timorensis (i)					X										
Bos javanicus	X	X	X		X					X	X	X			

Kalimantan
KM – Kayan Mentarang
Ku – Kutai
S – Sangkulirang
PM – Pleihari Martapura
TP – Tanjung Puting
BR – Bukit Raya/Bukit Baka
GN – Gunung Niut
GP – Gunung Palung
DS – Danau Sentarum

Sabah
DV – Danum Valley
K – Kinabalu
T – Tabin

Sarawak
Mu – Gunung Mulu
LE – Lanjak Entimau
Sm – Samunsam

e – extinct in this reserve
i – introduced species
* – known from one or few specimens; **Petaurillus hosei** recorded at Niah; **Petinomys vordermanni** probably occurs in Apo Kayan Reserve.

underlined = endemic species (e.g., *Muntiacus atherodes*)

Table 14.11. Areas of original, remaining and protected habitat for primates in Borneo.

Species	Indonesian distribution	Habitat type	Orig. habitat (km²)	Habitat remaining (km²)	% habitat loss	Prot. area (km²)	% of orig. protected	Reserves
Orangutan *Pongo pygmaeus*	Borneo N. Sumatra	Lowland & hill forest swamp and heath forest	464,000	184,000	63	8,185	2.1	Kutai, Tanjung Puting, Gn. Palung, Bukit Raya, Bt. Baka, Gn. Niut (Danum, Lanjak Entimau)
Agile gibbon *Hylobates agilis*	C & S. Sumatra S.W. Kalimantan	Lowland forest, excluding mangrove	115,134	75,988	34	1,328	1.1	Gn. Palung, Tj. Puting, Kendawangan, Bt. Raya, Bt. Baka, Batikap
Bornean gibbon *Hylobates muelleri*	Borneo	Lowland forest, excluding mangrove and kerangas	395,000	253,000	36	20,385	5.1	D. Sentarum, Kayan-Mentarang, Kutai, Pleihari-Martapura, (Danum, Kinabalu, Mulu)
Long-tailed macaque *Macaca fascicularis*	Sumatra Java and Bali Borneo	Coastal and riverine forest up to 1,000 m, disturbed forest	100,380	52,660	48	3,837	3.3	Kutai, Tanjung Puting, Meratus Hulu Barabai, Sangkulirang, Bukit Raya, D. Sentarum, (Tabin, Samunsam)
Pig-tailed macaque *Macaca nemestrina*	Sumatra Borneo	Primary and secondary lowland and hill forest; often raid farmland	113,179	82,347	28	4,516	4.0	Kutai, Tanjung Puting, Bukit Raya, Kayan-Mentarang, Sangkulirang, (Danum, Gn. Mulu)
Proboscis monkey *Nasalis larvatus*	Borneo	Estuarine and riverine forest, often far inland	29,496	17,750	40	1,225	4.1	Tanjung Puting, Kutai, Gn. Palung, D. Sentarum, Sangkulirang, P. Kaget, (Samunsam, Sepilok)

Table 14.11. Areas of original, remaining and protected habitat for primates in Borneo (continued).

Species	Indonesian distribution	Habitat type	Orig. habitat (km²)	Habitat remaining (km²)	% habitat loss	Prot. area (km²)	% of orig. protected	Reserves
Silvered langur *Prescytis cristata*	Sumatra Borneo	Coastal and riverine forest	153,600	89,230	42	7,610	5.0	Kutai, Gn. Palung, Pleihari-Martapura, (Samunsam, Mulu)
White-fronted langur *Presbytis frontata*	Borneo	Primary forest up to 300 m excluding swamp forest	125,000	61,500	51	7,260	5.8	Kutai, Pleihari? Kayan Mentarang, (Lanjak Entimau)
Bornean or grey langur *Presbytis hosei*	Borneo	Primary lowland & lower montane forest up to 1,000 m.	104,000	54,000	48	7,120	6.8	Ulu Kayan, Kutai, Kayan-Mentarang, (Danum, Kinabalu, Mulu)
Maroon langur *Presbytis rubicunda*	Borneo	Dry lowland & montane forest up to 1,500 m, secondary forest	415,000	266,000	36	19,670	4.7	Kutai, Tanjung Puting, Gn. Palung, Gn. Niut, Pleihari Martapura, (Mulu, Lanjak Entimau)
Banded langur *Presbytis melalophos**	Sumatra Borneo	Swamp forest & lowland forest	95,289	45,267	53	1,908	2.0	Gn. Palung, Gn. Niut, (Samunsam)
Western tarsier *Tarsius bancanus*	Sumatra Borneo	Lowland primary & secondary rainforests	450,513	313,903	31	14,828	3.3	Kutai, Gunung Palung, Kayan Mentarang, Bukit Raya, Gn. Niut, (Danum, Tabin, Mulu)
Slow loris *Nycticebus coucang*	Sumatra, Java and Borneo	Primary & secondary rainforest	450,513	313,903	31	14,828	3.3	Kutai, Tanjung Puting, Gn. Palung

( ) reserves outside Kalimantan
* *P. melalophos* in Borneo = *P. femoralis*
Data on habitat areas: from *National Conservation Plan for Indonesia*: V
Source: MacKinnon 1986

taining 10% of the original habitat will retain only 50% of the original species present (Diamond 1975).

The problem of attrition of species from habitat fragments is being researched in northern Brazil, where large areas of forest are being cleared for cattle ranches. Plots of forest ranging in size from 1 ha to 10,000 ha are being studied for rates of species loss (Lovejoy et al. 1983). Very small forest fragments, with areas of 10 ha, show a rapid loss of species, particularly of larger mammals and predators at the head of the food chain. Apart from species loss and genetic drift, isolated blocks of forest also suffer change in forest structure, caused by increased wind damage to small remaining plots of forest, more tree falls and the replacement of mature forest species with colonising species. The reproductive behaviour of some trees and plant-animal interactions have also changed in some

Table 14.12. Occurrence of endemic birds in Bornean reserves.

	A	B	C	D	E	F	G	H	I	J	K	L	M	N	O
Arachnothera juliae								•	•						
Arborophila hyperythra								•	•						
Batrachostomus harterti														•	
Bradypterus accentor								•							
Calyptomena hosei								•	•						
Calyptomena whiteheadi								•	•						
Cettia whiteheadi								•	•						
Chlamydochaera jefferyi							•	•	•			•			
Chlorocharis emiliae								•	•						
Cyornis superba		•	•		•		•			•					
Haematortyx sanguiniceps							•	•							
Harpactes whiteheadi								•	•						•
Lonchura fuscans	•	•	•	•	•	•	•	•			•	•	•		•
Lophura bulweri					•	•	•		•	•			•	•	
Megalaima eximia						•		•	•	•					
Megalaima monticola							•	•	•						•
Megalaima pulcherrima								•	•						
Microhierax latifrons		•	•					•							
Napothera atrigularis		•						•							
Napothera crassa								•	•						
Oculocincta squamifrons						•		•							
Oriolus hosei [1]															
Pachycephala hypoxantha								•	•						
Pitta arquata							•		•						
Pitta baudi		•	•	•				•							
Pityriasis gymnocephala	•	•	•	•	•							•			
Ptilocichla leucogrammica		•	•						•			•			
Spilornis kinabaluensis								•	•						
Zoothera everetti								•	•						

A = Ulu Barito
B = Danum Valley
C = Tabin
D = Samunsam
E = Gunung Lotung/Maliau
F = Danau Sentarum
G = Gunung Niut
H = Kinabalu
I = Gunung Mulu
J = Lanjak Entimau
K = Kutai
L = Pleihari-Martapura
M = Bukit Raya/Baka
N = Tanjung Puting
O = Gunung Palung

[1] occurs on Mts. Dulit and Murud

small forest blocks. These ecological changes also apply to larger blocks of forest over a longer time span. However, species losses are slowed considerably if protected areas are not totally isolated, but are connected together or are adjacent to exploited but still forested lands. This has important implications for reserve design (fig. 14.7).

Box 14.7. Habitat fragmentation and loss of species: an Indonesian example.

The loss of bird species on Java is an indicator of what can be expected in Borneo and the other islands of the Greater Sundas as forests are cut and fragmented. Studies of distribution of forest birds in remaining blocks of forest on Java show an altitudinal distribution of species that is atypical when compared with the avifauna of Borneo, Sumatra or other large islands such as New Guinea. In Borneo and Sumatra maximum species richness is found in lowland forests, and species number decreases with altitude (fig. 14.6). The Javan graph is peculiar in having a dip in the hill zone between 300 m and 1,500 m. This is interpreted as the result of long term deforestation. Lowland species have failed to survive in remaining hill forests because they have been cut off from lower altitude populations which were a source of colonisers.

Families of large birds such as malkohas have lost proportionally more species than families of small birds such as flowerpeckers. Moreover Java has lost more birds which are exclusive to lowland rainforests than species which can occupy secondary habitats, forest edge or open habitats. Small forest patches have lost more species than larger blocks, with extinction rates as high as 80% in small 10-40 ha plots compared to rates of 25% for areas over 10,000 ha. These results have important implications for reserve design and management. They emphasise that reserves must be large, cover a wide spectrum of habitat types and be connected by corridors of natural habitats if they are to retain the maximum number of species in the longterm.

Almost 60% of Borneo is still forested, and 37 large reserves cover a total of 35,000 km^2 or 5% of the island. In parts of West and South Kalimantan, Sarawak and Sabah much of the lowland forests have been lost, but there are still huge tracts of forest in the south and centre of the island. Most endemic birds are montane species which are well-protected in reserves and protection forests. As yet only 13 bird species are listed as endangered in Borneo, compared to 73 birds on Java.

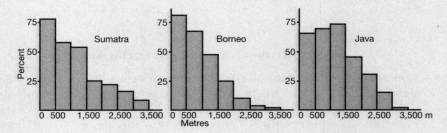

Figure 14.6. Proportions of island bird fauna at different altitudes on Borneo, Sumatra and Java.

Source: MacKinnon and Phillipps 1993

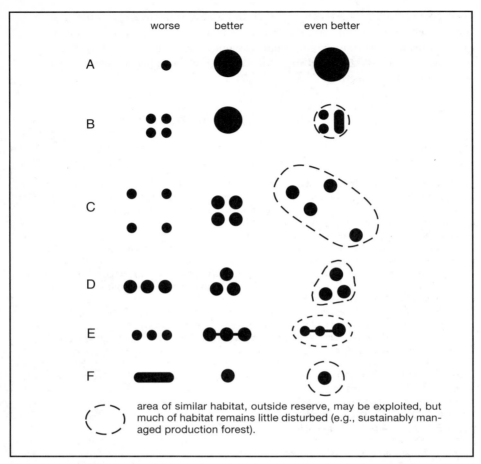

Figure 14.7. The use of biogeographical principles in reserve design. (After MacKinnon et al. 1986.)

Conservation Outside Protected Areas

Conservation of natural habitats makes good ecological and economic sense, but less than a twentieth of the whole island of Borneo currently lies within protected areas (MacKinnon and MacKinnon 1986). Moreover it is unrealistic to expect provincial governments to allocate much more land for total protection. Not all the land outside protected areas, however, will be converted to permanent agriculture, grazing, plantation or urban use. There will always be large areas of selectively logged forests, fallow land, secondary habitats and unused areas, which may constitute suitable habitat for much of Borneo's wildlife.

The future survival of many Bornean species will depend upon such "unprotected" habitats. The Bornean avifauna, for instance, is enriched by large numbers of migrant birds that breed in the Palearctic and migrate south during the northern winter. In Borneo these visitors depend on vital stop-over points and dispersed feeding grounds, often outside the protected area system. These include coastal mudflats, wetlands, forests, mountains and even agricultural lands and gardens. Even in heavily utilised areas, species can be conserved and encouraged if care is taken to employ the least disruptive and most ecologically sound technologies.

Wetlands are a good example of habitats facing multiple pressures. Kalimantan's rivers, lakes and swamps are critical habitats for many of the island's fish, amphibians, reptiles, birds and invertebrates. Yet these wetlands are also crucial fishing grounds, transport routes and water sources for people and domestic livestock, so that few wetland sites can be protected totally as conservation areas. The viability of protected wetlands is dependent, more than for most other habitats, upon appropriate land-use practices outside the wetland site, which maintain the quantity and quality of water flow. Accordingly, establishment of wetland protected areas is only one element in a more comprehensive approach to water resource and landuse planning. Figure 10.10 shows some major wetland sites in Kalimantan and elsewhere in Borneo. Other sites, such as the waterbodies of the Mahakam Lakes and the Sungai Negara wetlands in South Kalimantan (see chapter 10) are so heavily utilised by people that they cannot be afforded full legal protection as conservation areas. Nevertheless, they can be managed in a way that is beneficial to both people and wildlife.

Caves, too, are critical habitats. Many ranges of limestone hills and limestone outcrops in Borneo are riddled by interesting caves and potholes, including some of the most extensive cave systems in the world. Some of these caves are well protected in reserves such as Gunung Mulu National Park in Sarawak. However, many caves are outside protected areas.

Cave habitats are threatened by several factors: vandalism and excessive visitor use; overharvesting of guano and birds' nests; and direct destruction of the limestone to make lime, cement and marble chips. At Batu Hapu in South Kalimantan, for instance, guano collecting has stopped, and the caves have become a popular tourist site, but vandalism by tourists is destroying the very resource which visitors want to see (KPSL-UNLAM 1989c). Caves are interesting and valuable resources that require better protection and special management.

As forest felling and land opening continues in Kalimantan, much greater attention will need to be paid to the needs of wildlife on lands outside protected areas in the next decade. It will be necessary to examine the policies and practices of forestry, animal husbandry and fisheries, as well as establishing guidelines for the use and protection of rivers and waterways and transfrontier resources. With sensible management, lands outside protected areas can provide useful habitat for wildlife as well as livelihoods for local people.

Buffer Zones

Buffer zones are legally defined as areas adjacent to, or surrounding, national parks or other protected areas, where land-use is restricted to give extra protection to the conservation area while providing some socio-economic benefits to local communities. These zones are usually accessible to both people and animals, but act as a buffer between the intensively used lands outside the park and the protected habitats within. Some buffer zones extend around the total area of protected habitats, thus providing for a larger breeding population of plant and animal species than could survive in the protected area alone. Selectively logged production forests, natural forests harvested by villagers, rubber, rattan and firewood plantations, and even golf courses, can all be effective buffer zones.

Buffer-zone management requires careful planning. The first priority is protection, and the second is to meet local people's needs for harvestable products; the establishment of cash crops is usually the lowest priority (MacKinnon et al. 1986). Buffer zone crops should be unattractive to wildlife but still should be selected to give good soil cover and to maintain soil fertility (see table 12.6). For instance, elephants love young oil palms, so these are obviously not a suitable crop in a buffer zone near to an elephant reserve.

No permanent settlement or damaging activities, such as burning, can be allowed in buffer zones. Shifting agriculture is not a suitable buffer-zone activity. Where crops are planted, care should be taken not to introduce any animal or plant species that is likely to invade the protected area or threaten the species in it. Many of the traditional forest resources collected in Kalimantan are suitable crops for buffer zones; these include rattans and medicinal herbs grown beneath tree crops or regenerating forest.

The Value of Secondary Forests

In fallow agricultural fields and abandoned logged areas, secondary forest soon becomes established. As more areas of Bornean forests are cleared and logged, secondary forests will become increasingly important for maintaining species diversity. Therefore it is important in land use planning to save plots of mature primary and mature secondary forests as seed sources for recolonisation. The management of these secondary forests and the sil-

> **Box 14.8.** Benefits of buffer zones.
>
> **Biological benefits:**
>
> These zones:
> - provide extra protection, from human activities, for the strictly protected core zone;
> - protect the core reserve from biological changes;
> - provide extra protection from storm damage;
> - provide a larger forest unit for conservation, with less species loss through edge effects;
> - extend habitat and thus population size of large, wide-ranging species;
> - allow for a more natural boundary, relating to movements of species;
> - provide a replenishment zone for core area species.
>
> **Social benefits:**
> - Local people have access to traditionally utilised species.
> - People are compensated for loss of access to the strictly protected core zone.
> - Local people participate in conservation of the protected area.
> - More land is available for education, recreation and tourism.
> - Wildlife conservation becomes a part of local and regional rural development planning.
> - Traditional land rights of local people are safeguarded.
> - Conservation-related employment is increased.
>
> **To achieve these benefits, the following basic criteria must be observed:**
> - Tree cover and habitats should be maintained as far as possible in their natural state.
> - The vegetation of buffer zones should resemble that of the protected area, both in species composition and physiognomy.
> - Buffer zones should be biologically diverse.
> - The physiognomy of the vegetation should be heterogeneous and stratified.
> - It is important to maintain an ecosystem in the buffer zone that can retain and recycle soil nutrients. Buffer zone activities should not have negative impacts on the physical structure of the soil or on its water-regulating capacity.
> - Exploitation of buffer zones should, whenever possible, make use of traditional, locally adapted lifestyles and resource management practices.
>
> *Source: Oldfield 1988*

viculture and utilisation of their species will be one of the main tasks for tropical forestry in the future (Jacobs 1988).

Production Forests

Traditionally, land-use classifications made a clear distinction between areas allocated for conservation and those to be exploited, but any long-

Box 14.9. Selection of crops for buffer zones.

1. Natural forest, disturbed forest and secondary forest provide additional habitat for wildlife species and excellent protection of the soil.
2. Fast-growing firewood plantations are of some use to wildlife and can provide good soil protection. Firewood plantations of coppiceable species are effective buffer zones in areas where land is limited and where the demand for firewood is pressing.
3. Plantations of timber trees (e.g., *Shorea*) give good soil protection and are a useful way of reforesting steep slopes.
4. Mixed plantations providing firewood, poles and timber are of some use to wildlife and give good soil protection.
5. Plantations of fruit trees protect the soil and provide cash and food, but problems may arise with some animals (e.g., monkeys or bats) raiding the crops.
6. Cinnamon provides cash and firewood but is of no use to wildlife; soil protection is good, although the trees have to be cut every few years. Cinnamon plantations are suitable on less steep, lower mountain slopes.
7. Cloves provide cash but require suitable climate and soil. Clove plantations are of no use to wildlife and provide only moderate soil protection. They can be undercropped for fodder or grazing.
8. Coconuts provide some cash, some food and some wood, but are of little use to wildlife and provide poor soil protection. Undercropping is possible. Raiding sun-bears can be a serious problem.
9. Rubber and oil palm provide cash crops and some firewood, but are of little use to wildlife. Both species need moderately level ground and give good protection to the soil. Oil palm is unsuitable when the conservation area contains elephants.
10. *Eucalyptus* and *Melaleuca* provide wood and valuable resins; they are of little value to wildlife but give good soil protection. Both species are susceptible to accidental fires.
11. Tea is a profitable cash crop in wet climates at higher altitudes, but is unattractive to wildlife and soil protection is poor.
12. Coffee requires good soils, is of little use to wildlife and gives poor to moderate soil protection. It can be mixed with other crops but is not suitable on steep slopes.
13. *Leucaena* provides fodder (edible leaves and pods) as well as fuelwood, but gives poor soil protection. *Leucaena* is a fast-growing coppiceable tree, which fixes nitrogen in soil. It can be seeded by air and is a good species to cover grassland quickly.
14. Bamboos provide cash, building material and some fuel. Bamboos are fast-growing and coppiceable and give moderate soil protection.
15. Rattans provide cash and are useful to wildlife; the degree of soil protection varies with conditions. Rattans can be mixed with other tree crops or planted among secondary regrowth. Cultivated rattans will become more valuable as wild stocks dwindle. Good silvicultural techniques have been developed in Central Kalimantan.
16. Sago provides food and fodder and gives an excellent cover in damp or swampy areas.
17. Mangrove plantations provide tannin, timber, seafood and firewood. They are good for wildlife and soil protection and are an excellent use of saline swamp areas. Plantations can also be combined with fishponds.
18. Herbs and grasses provide fodder, cash and medicine. Numerous herbs and grasses are suitable for planting under tree cover. Many of the ritual and medicinal herbs used traditionally in Kalimantan can be planted in buffer zone areas.

Source: MacKinnon 1981

term conservation strategy needs to recognise the role of production forests. Worldwide, the area of selectively logged rainforest already exceeds that of unlogged forest in a ratio of four to one (Brown and Lugo 1984). The ratio in lowland forests in Kalimantan and other Bornean states is probably much higher. Indeed, for some forest habitats, such as peat swamp forests in Sarawak, there are no remaining unexploited areas. The area of logged forest will continue to increase, even though some logged forests are clear-felled for small-scale or plantation agriculture. Selectively logged areas, particularly when large, are able to support a high proportion of mature forest species, including many that are unable to survive in isolated primary forest reserves (Johns 1983, 1988).

While unlogged forests may support a greater variety of mammals than logged areas (Davies and Payne 1982), most species persist to at least some extent in logged-over forests (Johns 1988). Habitat disturbance usually causes changes in numbers of some species rather than species deletions. Only a few species, such as the endemic large tufted ground squirrel *Rheithrosciurus macrotis*, disappear from logged forest. Several species, such as primates and bats, are less abundant in logged forests, but large browsing and grazing herbivores, such as deer, elephant and banteng, may become more numerous. Predators such as the clouded leopard may be more abundant in logged-over forests where they feed largely on rats (abundant in disturbed forests); sunbears also seem common in some logged forests (Johns 1982). Logged-over forests are also useful to those animals that show seasonal migrations, including bearded pigs, orangutans, bats and the Sabah elephant herds that cross the Kalimantan border to venture south into Ulu Sembakung.

Although many species are able to adapt to the changed conditions of logged forest, the effects of logging on rainforest communities depend on the degree of disturbance, itself a reflection of forest management and logging practices. Clearly the value of logged forests for conservation depends on the ways in which they are managed to minimise forest disturbance. Moreover, while many vertebrate species may be able to survive in patches of less-damaged forest if it is logged selectively once, repeat-logging for small-sized timber stocks and short rotation cycles, now practised in Kalimantan concessions, have severe and permanent effects on wildlife. Natural forest recovery is dependent to some extent on the presence of natural pollination and dispersal agents. Many nondipterocarp timber trees rely on animals for pollination or seed dispersal. Therefore it is not only ecologically preferable but often to the foresters' advantage to let forests regenerate naturally and to maintain as far as possible their natural complement of animal life (Johns 1988).

Logged forests can play an important role in maintaining wildlife populations, but their long-term value depends on long rotation cycles (perhaps fifty years or more) to allow forests to regenerate naturally. Shorter logging cycles will cause progressive changes in forest regeneration and forest composition, with associated loss of floral and faunal diversity and abundance. Replanting of logged areas with fast-growing alien species may deflect regenerative processes and lead to gradual breakdown of the food web. It is therefore vital to conserve some primary forests in an unlogged state. Logged forests can be an important supplement to primary forest reserves, but they are not an alternative. Logged forests surrounding fully protected areas will increase the effective conservation estate, to help to preserve much of Kalimantan's lowland wildlife.

Captive Breeding and Reintroductions

The best, cheapest and easiest way of protecting species is usually the protection and management of their natural habitats. Sometimes, however, wild populations have become so small or isolated that they may need active management, and introduction of "new blood". Captive breeding schemes may then be a useful supplement to habitat protection.

Many species are now housed in botanical and zoological gardens, and captive breeding programmes have emerged as a popular argument for the existence of zoos (Tudge 1991). Zoos involved in such programmes aim to breed sufficient individuals of endangered species for eventual reintroduction into the wild. Often these programmes involve capturing wild individuals as breeding stock.

Indonesia and Malaysia are presently collaborating with British and American zoos on a captive breeding programme for the Sumatran rhinoceros, a species which is probably now extinct in Kalimantan, though still recorded in protected areas in Sarawak and Sabah. Here the aim is to remove "doomed" rhinos from pockets of habitat threatened by development schemes and to establish zoo populations for captive breeding. At first sight this seems commendable. In practice the success of this programme is questionable; there have been no breeding successes and several unfortunate fatalities among captured animals (Santiapillai and MacKinnon 1991). Rather than remove animals to scattered zoos, it may be far more effective to invest increased funding on *in situ* protection and to translocate "doomed" rhinos to large, well-protected reserves within their known range, a practice that has been employed with considerable success for African rhinos.

Even if captive breeding programmes do produce large numbers of suitable progeny, there can be considerable problems with reintroducing captive animals to the wild (MacKinnon et al. 1986). With a few notable exceptions such as the Arabian oryx (Stanley Price 1989), rehabilitation projects generally play an insignificant role in species conservation. The difficulties of reintroductions are well exemplified by attempts to rehabilitate captive orangutans to wild living (MacKinnon and MacKinnon 1991). Numerous

Box 14.10. Habitat conservation within a large timber concession: Danum Valley, Sabah.

Over the last thirty years logging has been the most rapid single agent of forest modification and degradation in Southeast Asia. This effect has been compounded by the opening of remote areas to other forms of encroachment, such as cultivation, harvesting of forest products and hunting. On the other hand, moderately logged forests seem able to support almost all of the species complement of primary forest, and many of its ecological functions (Johns 1992). Since the area of production forests is increasing while areas of undisturbed natural forest decline, the challenge is to combine sustainable forest exploitation and conservation. In Sabah timber interests and conservationists have embarked on a joint venture that should benefit both.

Yayasan Sabah is a parastatal charitable organisation which was allocated a large timber concession to generate revenue from log sales. The concession covers 972,804 ha in six contiguous forest reserves. Six categories of land, including steep slopes, water catchments and riparian reserves and amounting to 20% of the concession, are excluded from timber production. The most important are two designated conservation areas: Danum Valley (43,800 ha) and the Maliau Basin (39,000 ha).

Danum Valley Conservation Area (DCVA) lies in the upper catchment of the river Segama in eastern Sabah. The area is predominantly lowland dipterocarp forest, uninhabited, and supports a fauna characteristic of this region of Borneo, including species such as orangutan, proboscis monkey, endemic langurs, Sumatran rhino, elephant, banteng, clouded leopard and sunbear. The conservation area lies within the working timber concession and is recognised in the concession management plan. Nevertheless, the conservation area has no formal legal status. Its protection relies on the goodwill of Yayasan Sabah and the Forestry Department; in effect it is a privately established reserve, managed by Yayasan Sabah.

In 1984 Yayasan Sabah signed a collaborative agreement with the local campus of the Universiti Kebangsaan Malaysia and the Royal Society of the U.K. to establish a permanent research centre and training programme at Danum Valley. To date, more than 70 research projects have been completed or are in progress. Research topics include forest regeneration studies and the effects of logging on watersheds, forest dynamics and animal communities (Marshall and Swaine 1992). Although most of this research is basically academic, useful applied results are also beginnning to emerge, for example, on the management of logging skid trails and the propagation and silviculture of dipterocarp tree seedlings.

The longterm scientific programme and the international attention it has drawn have convinced the Malaysian authorities of the value of maintaining the conservation area, which may now be gazetted as a national reserve. In addition to supporting research, Yayasan Sabah are exploring the possibilities of non-logging forest activities within the concession and are establishing a forest lodge for wildlife tourism. The success of the tourism venture will depend on maintenance of substantial tracts of lowland forests and healthy populations of interesting wildlife. This in turn should encourage better forest management to minimise disturbance to forests and wildlife during logging operations. Allowing visitors to see forest management practices encourages improved management, leads to a better informed public attitude to forestry and generates income. In the case of Danum Valley it is hoped to underwrite the running costs of the field centre with profits from the lodge.

The Danum Valley model is an interesting and innovative solution to conserving biodiversity in production forests. It demonstrates the value of collaboration between the private sector and the conservation community, and that different kinds of conservation activity can generate synergies, particularly between research, education and recreational/tourism activities.

Source: Marsh and Sinun 1992

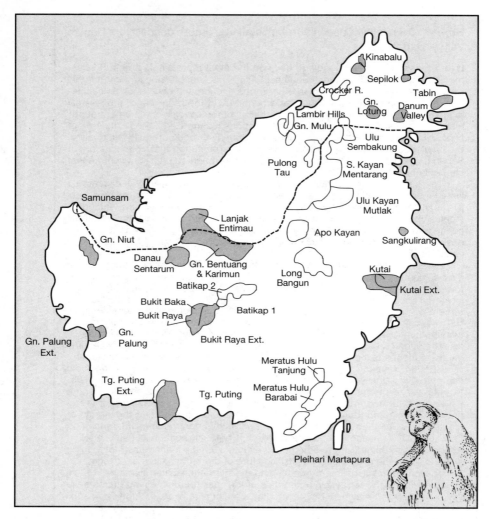

Figure 14.8. Major Bornean reserves, showing those with important populations of wild orangutans *Pongo pygmaeus*.

attempts have been made to return confiscated pet orangutans to the wild. Rehabilitation stations have been established at Tanjung Puting and Kutai in Kalimantan, Bohorok in Sumatra, Semoko in Sarawak and Sepilok in Sabah. Rehabilitating apes is difficult because of their social systems, the danger of introducing human diseases into the wild, and the fact that suitable habitats are probably already supporting their full carrying capacity of these species (Konstant and Mittermeier 1982; MacKinnon 1977). Reha-

Table 14.13. Some rare and endemic species protected in orangutan reserves.

	P.p.	H.a.	H.m.	N.l.	P.r.	P.h.	P.f.	E.m.	D.s.	N.n.	B.j.	He.m.	C.p.	T.s.	A.a.	L.b.
Kalimantan																
Gunung Palung	•	•		•	•							•			•	•
Danau Sentarum	•		•	•	•							•			•	
Gunung Niut/Becapa	•	•										•			•	•
Bukit Baka	•		•	•	•							•			•	
Karimun/Bentuang	•	•		•	•							•			•	•
Bukit Raya	•	•			•							•			•	
Batikap	•				•							•				
Tanjung Puting	•	•		•	•	•	•		ex	•		•		•	•	•
Kutai	•	•	•	•	•	•	•	•	ex	•	•	•	•		•	•
Sangkulirang (pr)	•		•	•	•	•	•	•	•	•	•	•	•		•	
Sabah																
Danum Valley	•		•	•	•	•	•		•	•	•	•	•		•	
Tabin	•		•	•	•	•	•		•	•	•		•		•	
Kinabalu	•		•	•	•	•				•		•			•	
Gunung Lotung	•		•	•	•					•					•	
Sarawak																
Lanjak Entimau	•		•	•	•	•				•		•			•	
	15	4	10	10	15	6	4	2	3	11	5	15	3	1	13	7

P.p. = *Pongo pygmaeus* — orangutan
H.a. = *Hylobates agilis* — agile gibbon
+H.m. = *Hylobates muelleri* — Bornean gibbon
+N.l. = *Nasalis larvatus* — proboscis monkey
+P.r. = *Presbytis rubicunda* — red langur
+P.h. = *Presbytis hosei* — grey langur
+P.f. = *Presbytis frontata* — white-fronted langur
E.m. = *Elephas maximus* — Asian elephant
D.s. = *Dicerorhinus sumatrensis* — Sumatran rhinoceros
N.n. = *Neofelis nebulosa* — clouded leopard
B.j. = *Bos javanicus* — banteng
He.m. = *Helarctos malayanus* — sunbear
C.p. = *Crocodylus porosus* — estuarine crocodile
T.s. = *Tomistoma schlegeli* — false gavial
A.a. = *Argusianus argus* — Argus pheasant
+L.b. = *Lophura bulwerii* — Bulwer's pheasant

+ = endemic species   ex = extinct at this locality

Source: MacKinnon and MacKinnon 1991

bilitation programmes are expensive in terms of time and resources, yet few captive orangutans have been returned successfully to the wild. The main value of these stations is their role in promoting conservation and interest in wildlife among the visitors attracted to the centres (Aveling and Mitchell 1980).

Both orangutans and rhinos are known to occur in several Bornean reserves. Such large, spectacular and well-known species can play a useful role as "flagship" species, attracting support for reserves that protect many thousands of other species of plants and animals as well. There still may be as many as 50,000 wild orangutans in Borneo (MacKinnon 1992). Orangutans occur throughout Borneo, except for the southeast between the Barito and Mahakam rivers and parts of the northwest of the island (Rijksen 1978). They are found in both fruit-rich forests and logged-over areas, and there seems to be little in floral composition to distinguish suitable orangutan habitat (Davies and Payne 1982). Orangutans are found in several major gazetted and proposed reserves in Kalimantan and East Malaysia (fig. 14.8). Most of these "orangutan reserves" score very highly for biodiversity and are considered to be conservation areas of global significance (MacKinnon and MacKinnon 1986). Between them they protect all the major habitat types on Borneo, as well as most of that island's flora and fauna, including numerous endemics. Good protection and management of these reserves is the best and most effective way of conserving orangutans and thousands of other species (table 14.13).

Research Needs

If future developments in Kalimantan are to be carried out in a way least damaging to biodiversity and the environment, planners and regional developers will need access to a broad base of ecological information. Kalimantan universities, environmental study centres and research institutes have a key role to play in this research. Long-term study sites investigating forest regeneration and plant-animal interactions have already been established at Gunung Palung, Tanjung Puting, Wanariset and Barito Ulu in Kalimantan, and Danum Valley in Sabah (fig. 14.9).

Topics that need further investigation are outlined below; work is already in progress on several of these themes.
1. Regeneration of natural forests.
2. Regeneration and propagation of dipterocarps (e.g., the Tropenbos programme).
3. Effects of selective logging and logging intensities on wild populations of plants and animals.
4. Effects of logging on important animal pollinators and seed dis-

persers (e.g., bats).
5. Effects of relogging on plant and animal communities and species diversity in logged forests.
6. Regeneration after fires; fire ecology.
7. Effects of watershed disturbance and soil erosion on aquatic communities.
8. Basic inventory and monitoring to determine species distribution and abundance in natural habitats, degraded lands, logged forests, protected forests and reserves.
9. Appropriate agriculture and tree crops for Kalimantan soils.
10. Surveys and inventory of limestone habitats and cave ecosystems, such as the Sangkulirang Peninsula.
11. Rehabilitation of critical lands and the return of wildlife.
12. Importance of minor forest products and native wildlife to local communities and economies.
13. The ecology of commercially important species; the establishment of commercial plantations and captive propagation programmes to reduce the drain on wild resources (e.g., rattans, orchids, crocodile farms, arowana fish).
14. Ecological studies, including plant-animal interactions, to attempt to understand how forest disturbance is likely to disrupt natural ecological processes.
15. Sustainable harvesting levels for forest products, including timber and rattan.
16. Fisheries research and sustainable levels of exploitation in inland fisheries in Kalimantan; the effects of deforestation and industrialisation on these resources.
17. Mangrove ecology.
18. Better utilisation of swamplands with potential acid sulphate soils.
19. Surveys of distribution, ecology, and breeding patterns of, and threats to, endangered species. Species databases have already been set up at KPSL-UNLAM and PSL-UNTAN to handle this information for the whole of Borneo (see appendix 8).

The Future

Since the early 1960s Kalimantan and the northern Bornean states of Sarawak, Sabah and Brunei have seen increasingly rapid economic development, fuelled by profits from harvesting forest resources. East Kalimantan, Indonesia's second largest province, supports only 1.6 million people (1% of the nation's population) but provides more than 25% of Indonesia's export revenues from exploitation of its natural resources. The future for Borneo's forests is clear; exploitation will continue.

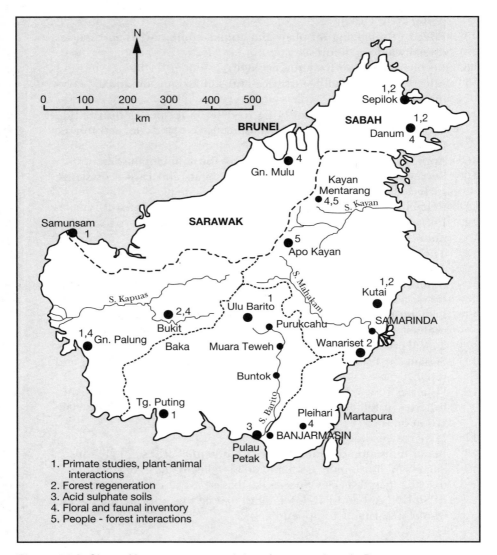

Figure 14.9. Sites of long-term research into forest ecology in Borneo.

Deforestation, caused by logging activities, agricultural encroachment and changing land-use patterns, is fast reducing the areas of natural habitat remaining on the island. Careful management could minimise the environmental degradation caused by resource exploitation and agricultural developments, but inevitably there will be some species lost as areas of natural habitat contract. Protection of large areas of natural forests must be the priority for conservation, but less than a twentieth of Borneo lies within protected areas. Increased protection of lowland forests and better management of existing conservation areas are essential for conserving the island's biodiversity and endangered species. The Kalimantan conservation areas should become prime targets for government and international assistance.

Even with a target of 10% of land designated as reserves, the area of protected habitats will be insufficient to maintain all species. Not all lands outside protected areas will be converted to agriculture or plantations. There will always be large areas of selectively logged forest, wetlands, fallow lands and secondary habitats, which are valuable habitat for wildlife (Johns 1992a, 1992b). Since much of the island's biological estate will remain outside reserve borders, the fate of many species of plants and animals depends on efforts to slow down deforestation, to log in a more sustainable and less damaging manner, to find alternative, more ecologically sound agricultural options and to rehabilitate degraded lands. There are exciting new possibilities for cooperation and collaboration between the Bornean states, in research and in management of the transnational reserves spanning their common borders. Genuine commitment to wise utilisation and resource management can lead to conservation and development in Kalimantan, and can make the foresters' new slogan of "Forests are Forever" a reality rather than an optimistic dream.

Appendices

Appendix 1. Plant genera and families, mentioned in text.

Genus	Family
Acacia	Leguminosae/Mimosoideae
Acanthus	Acanthaceae
Achasma	Zingiberaceae
Acrostichum	Pteridaceae
Aegialitis	Plumbaginaceae
Aegiceras	Myrsinaceae
Aeschynanthus	Gesneriaceae
Aetoxylon	Thymelaceae
Agathis	Araucariaceae
Aglaia	Meliaceae
Agrostistachys	Euphorbiaceae
Alangium	Alangiaceae
Albertisia	Menispermataceae
Albizia	Leguminosae/Mimosoideae
Aleurites	Euphorbiaceae
Alocasia	Araceae
Alseodaphne	Lauraceae
Alstonia	Apocynaceae
Amorphophallus	Araceae
Ananas	Bromeliaceae
Anisoptera	Dipterocarpaceae
Annona	Annonaceae
Antiaris	Moraceae
Antidesma	Euphorbiaceae
Antrophyum	Adiantaceae
Aquilaria	Thymelaceae
Araucaria	Araucariaceae
Areca	Arecaceae, Palmae
Arenga	Palmae
Aromadendron	Magnoliaceae
Artocarpus	Moraceae
Artrophyllum	Adiantaceae
Asplenium	Aspleniaceae
Auricularia	Fungae
Austrobuxus	Euphorbiaceae
Avicennia	Avicenniaceae
Azadirachta	Meliaceae
Azolla	Azollaceae
Baccaurea	Euphorbiaceae
Bambusa	Gramineae, Bambusaceae

Appendix 1. Plant genera and families, mentioned in text *(continued)*.

Genus	Family
Barringtonia	Lecythidaceae
Bauhinia	Leguminosae/Caesalpinoideae
Begonia	Begoniaceae
Blechnum	Acanthaceae
Blumea	Compositae
Blumeodendron	Euphorbiaceae
Boea	Gesneriaceae
Borassodendron	Palmae
Bouea	Anacardiaceae
Brownlowia	Tiliaceae
Bruguiera	Rhizophoraceae
Buchanania	Anacardiaceae
Bulbophyllum	Orchidaceae
Caesalpinia	Leguminosae/Caesalpinoideae
Calamus	Palmae, Arecaceae
Callicarpa	Verbenaceae
Calophyllum	Guttiferae
Campnosperma	Anacardiaceae
Camptostemon	Bombacaceae
Cananga	Annonaceae
Canarium	Burseraceae
Canavalia	Leguminosae/Papilionatae
Canthium	Rugiaceae
Cardamine	Cruciferae
Caryota	Palmae
Cassia	Leguminosae/Caesalpinoideae
Castanopsis	Fagaceae
Casuarina	Casuarinaceae
Cecropia	Urticaceae
Ceiba	Bombacaceae
Cephalomappa	Euphorbiaceae
Ceratolobus	Palmae
Cerbera	Apocynaceae
Ceriops	Rhizophoraceae
Chisochekon	Meliaceae
Chloranthus	Chloranthaceae
Cinnamomum	Lauraceae
Citrus	Rutaceae
Clerodendrum	Verbenaceae
Cocos	Palmae
Codiaeum	Euphorbiaceae
Coelogyne	Orchidaceae
Coelostegia	Bombacaceae
Coffea	Rugiaceae

Appendix 1. Plant genera and families, mentioned in text *(continued)*.

Genus	Family
Colubrina	Rhamnaceae
Colocasia	Araceae
Combretocarpus	Rhizophoraceae
Combretum	Combretaceae
Copaifera	Leguminosae
Cordyline	Liliaceae
Corybas	Orchidaceae
Coscinium	Menispermaceae
Cotylelobium	Dipterocarpaceae
Cratoxylum	Hypericaceae
Crinum	Liliaceae
Croton	Euphorbiaceae
Crytandra	Gesneriaceae
Ctenolophon	Linaceae
Cucumis	Cucurbitaceae
Curculigo	Amaryllidaceae
Curcuma	Zingiberaceae
Cyathea	Cyatheaceae
Cyathocalyx	Annonaceae
Cycas	Cycadaceae
Cymodocea	Petamogetonaceae
Cynometra	Leguminosae/Caesalpinoideae
Cyperus	Cyperaceae
Cyrtandra	Gesneriaceae
Cyrtosperma	Araceae
Cyrtostachys	Palmae, Arecaceae
Dacrydium	Podocarpaceae
Dacryodes	Burseraceae
Dactylocladus	Crypteroniaceae
Daemonorops	Palmae, Arecaceae
Dehaasia	Lauraceae
Dendrobium	Orchidaceae
Dendrocalamus	Gramineae, Bambusaceae
Derris	Leguminosae/Papilionatae
Dialium	Leguminosae/Caesalpinoideae
Dianella	Liliaceae
Dicranopteris	Gleicheniaceae
Dillenia	Dilleniaceae
Dimocarpus	Sapindaceae
Dinochloa	Gramineae
Dioscorea	Dioscoreaceae
Diospyros	Ebenaceae
Diplazium	Athyriaceae
Dipteris	Polypodiaceae

Appendix 1. Plant genera and families, mentioned in text *(continued)*.

Genus	Family
Dipterocarpus	Dipterocarpaceae
Dischidia	Asclepiadaceae
Disepalum	Annonaceae
Donax	Marantaceae
Dolichandrone	Bignoniaceae
Dracontomelum	Anacardiaceae
Drimycarpus	Anacardiaceae
Drosera	Droseraceae
Dryobalanops	Dipterocarpaceae
Duabanga	Sonneratiaceae
Durio	Bombacaceae
Dyera	Apocynaceae
Dysoxylum	Meliaceae
Eichhornia	Pontederiaceae
Elaeis	Palmae
Elaeocarpus	Elaeocarpaceae
Elateriospermum	Euphorbiaceae
Eleiodoxa	Palmae
Enhalus	Hydrocharitiaceae
Erechtites	Compositae
Eria	Orchidaceae
Erica	Ericaceae
Eryngium	Umbelliferae
Erythrina	Leguminosae/Papilionatae
Eucalyptus	Myrtaceae
Eugeissona	Palmae, Arecaceae
Eugenia	Myrtaceae
Eupatorium	Compositae
Euphorbia	Euphorbiaceae
Euphrasia	Schrophulariaceae
Eusideroxylon	Lauraceae
Evodia	Rutaceae
Excoecaria	Euphorbiaceae
Fagraea	Loganiaceae
Fagus	Fagaceae
Fibraura	Menispermaceae
Ficus	Moraceae
Fimbristylis	Cyperaceae
Flagellaria	Flagellariaceae
Flaucourtia	Flacourtiaceae
Fordia	Leguminosae/Papilionatae
Galium	Rubiaceae

Appendix 1. Plant genera and families, mentioned in text *(continued).*

Genus	Family
Ganua	Sapotaceae
Garcinia	Guttiferae
Gaultheria	Ericaceae
Geanthus	Zingiberaceae
Gendarusa	Acanthaceae
Gentiana	Gentianaceae
Gigantochloa	Gramineae, Bambusaceae
Gliricidia	Leguminosae
Glochidion	Euphorbiaceae
Gluta	Anacardiaceae
Gmelina	Verbenaceae
Gnetum	Gnetaceae
Goniothalamus	Annonaceae
Gonystylus	Thymelaeaceae
Gossypium	Malvaceae
Grewia	Tiliaceae
Guettarda	Rubiaceae
Gunnera	Haloragidaceae
Halodule	Cymodoceaceae
Halophila	Hydrocharitaceae
Haloragis	Haloragaceae
Helicia	Polygalaceae
Heliciopsis	Protaceae
Heritiera	Sterculiaceae
Hernandia	Hernandiaceae
Hevea	Euphorbiaceae
Hibiscus	Malvaceae
Hopea	Dipterocarpaceae
Hornstedtia	Zingiberaceae
Horsfieldia	Myristicaceae
Hydnophytum	Rubiaceae
Iguanura	Palmae
Ilex	Aquifoliaceae
Impatiens	Balsaminaceae
Imperata	Gramineae
Intsia	Leguminosae/Caesalpinoideae
Ipomoea	Convolvulaceae
Irvingia	Simaroubaceae
Ischaemum	Gramineae
Ixonanthus	Linaceae
Kandelia	Rhizophoraceae
Koompassia	Leguminosae/Caesalpinoideae

Appendix 1. Plant genera and families, mentioned in text *(continued)*.

Genus	Family
Korthalsia	Palmae, Arecaceae
Lagerstroemia	Lythraceae
Lansium	Meliaceae
Lantana	Verbenaceae
Lecea	Leeaceae
Leea	Leeaceae
Lemna	Lemnaceae
Lepironia	Cyperaceae
Lepisanthes	Sapindaceae
Leptospermum	Myrtaceae
Lepturus	Gramineae
Leucaena	Leguminosae/Mimosoideae
Licuala	Palmae
Linariantha	Acanthaceae
Lindera	Lauraceae
Lithocarpus	Fagaceae
Litsea	Lauraceae
Livistona	Palmae
Lophopetalum	Celastraceae
Lumnitzera	Combretaceae
Lycopodium	Lycopodiaceae
Macaranga	Euphorbiaceae
Mammea	Guttiferae
Mangifera	Anacardiaceae
Manihot	Euphorbiaceae
Maranthes	Rosaceae
Medinilla	Melastomataceae
Melaleuca	Myrtaceae
Melanorrhoea	Anacardiaceae
Melastoma	Melastomataceae
Meliosma	Sabiaceae
Memecylon	Melastomataceae
Mesua	Guttiferae
Metroxylon	Palmae
Mezoneuron	Leguminosae/Caesalpinoideae
Mezzettia	Annonaceae
Microcos	Tiliaceae
Mitrella	Annonaceae
Monophyllaea	Gesneriaceae
Morinda	Rubiaceae
Musa	Musaceae
Mussaendopsis	Rubiaceae
Myrica	Myricaceae

Appendix 1. Plant genera and families, mentioned in text *(continued)*.

Genus	Family
Myristica	Myristicaceae
Myrmecodia	Rubiaceae
Myosotis	Boraginaceae
Nelumbo	Nymphaeaceae
Neoscortechinia	Euphorbiaceae
Nepenthes	Nepenthaceae
Nephelium	Sapindaceae
Nephrolepis	Nephrolepidaceae
Nertera	Rubiaceae
Nicolaia	Zingiberaceae
Norrisia	Loganiaceae
Nothofagus	Fagaceae
Nymphoides	Gentianaceae
Nypa	Palmae
Ochanostachys	Olacaceae
Ochroma	Bombacaceae
Ocimum	Labiatae
Octomeles	Datiscaceae
Oncosperma	Palmae
Oroxylum	Bignoniaceae
Orthosiphon	Labiatae
Osbornia	Myrtaceae
Palaquium	Sapotaceae
Pandanus	Pandanaceae
Pangium	Flacourtiaceae
Paphiopedilum	Orchidaceae
Paraboea	Gesneriaceae
Paramignya	Rutaceae
Paranephelium	Sapindaceae
Paraserianthes	Leguminosae
Parashorea	Dipterocarpaceae
Parastemon	Rosaceae
Paratocarpus	Moraceae
Parishia	Anacardiaceae
Parkia	Leguminosae
Passiflora	Passifloraceae
Pemphis	Lythraceae
Pentaspadon	Anacardiaceae
Pericopsis	Leguminosae
Phalaenopsis	Orchidaceae
Phanerosus	Matoniaceae
Phoebe	Lauraceae

Appendix 1. Plant genera and families, mentioned in text *(continued)*.

Genus	Family
Pholidocarpus	Palmae
Phrynium	Podocarpaceae
Phyllocladus	Podocarpaceae
Pinanga	Palmae, Arecaceae
Pinus	Pinaceae
Piper	Piperaceae
Pisonia	Nyctaginaceae
Pithecellobium	Leguminosae/Mimosoideae
Pityrogramma	Hemionitidaceae
Platycerium	Polypodiaceae
Plectocomia	Palmae
Plectranthus	Labiatae
Ploiarium	Theaceae
Pluchea	Compositae
Polyalthia	Annonaceae
Pometia	Sapindaceae
Pongamia	Leguminosae/Papilionatae
Potentilla	Rosaceae
Prainea	Moraceae
Premna	Verbenaceae
Prunus	Rosaceae
Psychotria	Rubiaceae
Pterocarpus	Leguminosae/Papilionatae
Ptychopyxis	Euphorbiaceae
Quassia	Simaroubaceae
Quercus	Fagaceae
Rafflesia	Rafflesiaceae
Randia	Rubiaceae
Ranunculus	Ranunculaceae
Rhododenron	Ericaceae
Rhizophora	Rhizophoraceae
Rubus	Rosaceae
Saccharum	Gramineae
Salacca	Palmae
Samadera	Simaroubaceae
Sandoricum	Meliaceae
Santiria	Burseraceae
Sapium	Euphorbiaceae
Saraca	Leguminosae/Caesalpinoideae
Sarcotheca	Oxalidaceae
Scaevola	Goodeniaceae
Schima	Theaceae

Appendix 1. Plant genera and families, mentioned in text *(continued).*

Genus	Family
Schizostachyum	Gramineae, Bambusaceae
Scolopia	Flacourtiaceae
Scorodocarpus	Olacaceae
Scyphiphora	Rubiaceae
Sebastiana	Euphorbiaceae
Setaria	Gramineae
Shorea	Dipterocarpaceae
Sida	Malvaceae
Sindora	Leguminosae/Caesalpinoideae
Solichandrone	Bignoniaceae
Sonneratia	Sonneratiaceae
Sophora	Leguminosae/Papilionatae
Spathodea	Bignoniaceae
Spathoglottis	Orchidaceae
Sphagnum	Bryophyta
Spinifex	Gramineae
Spondias	Anacardiaceae
Stachyprynium	Maranthaceae
Stemonurus	Icacinaceae
Stenochlaena	Blechnaceae
Sterculia	Sterculiaceae
Strombosia	Olacaceae
Syringodium	Cymodoceaceae
Syzigium	Myrtaceae
Tacca	Taccaceae
Teijsmanniodendron	Verbenaceae
Terminalia	Combretaceae
Ternstroemia	Theaceae
Tetracera	Dilleniaceae
Tetractoma	Rutaceae
Tetrameles	Datiscaceae
Tetramerista	Theaceae
Tetrastigma	Vitaceae
Thalassia	Hydrocharitaceae
Thalassodendron	Cymodoceaceae
Thelymitra	Orchidaceae
Thespesia	Malvaceae
Thottea	Aristolochiaceae
Timonius	Rubiaceae
Tournefortia	Boraginaceae
Trema	Ulmaceae
Trigoniastrum	Trigoniaceae
Trigonobalanus	Fagaceae
Tristaniopsis	Myrtaceae

Appendix 1. Plant genera and families, mentioned in text *(continued).*

Genus	Family
Triumfetta	Tiliaceae
Uncaria	Rubiaceae
Urena	Malvaceae
Utricularia	Lentibulariaceae
Vaccinium	Ericaceae
Vanda	Orchidaceae
Vatica	Dipterocarpaceae
Ventilago	Rhamnaceae
Vigna	Leguminosae/Papilionatae
Viola	Violaceae
Viscum	Loranthaceae
Vitex	Verbenaceae
Vitis	Vitaceae
Willughbeia	Apocynaceae
Wolffia	Lemnaceae
Xanthomyrtus	Myrtaceae
Xanthophyllum	Polygalaceae
Xerospermum	Sapindaceae
Ximenia	Olacaceae
Xylocarpus	Meliaceae
Xylopia	Annonaceae
Zea	Gramineae
Zingiber	Zingiberaceae
Zizyphus	Rhamnaceae

Appendix 2. Land birds found on offshore islands of Borneo.

Species	K	M	T	B	A
Ardea sumatrana		•			
Bubulcus ibis		•			
Egretta sacra	•	•	•	•	
Pandion haliaetus				•	
Haliaeetus leucogaster		•		•	
Accipiter soloensis			•		
Megapodius cumingii/reinwardt		•		•	
Treron vernans		•	•		
Ptilinopus melanospila		•		•	
Ducula aenea			•		
Ducula bicolor	•	•	•	•	
Ducula pickeringi		•			
Columba vitiensis		•		•	
Columba argentina	•	•			
Streptopelia chinensis				•	
Goepelia striata		•			
Chalcophaps indica		•			
Caloenas nicobarica		•	•	•	
Tanygnathus lucionensis				•	
Cuculus fugax				•	
Chrysococcyx minutillus		•			
Phaenicophaeus chlorophaeus				•	
Phaenicophaeus curvirostris				•	
Otus lempiji			•		
Ninox scutulata				•	
Batrachostomus javensis				•	
Caprimulgus indicus				•	
Collocalia fuciphaga				•	
Collocalia esculenta				•	
Cypsiurus balasiensis			•		
Todirhamphus chloris		•	•		
Merops viridis	•				
Anthracoceros albirostris		•			
Dinopium javanense		•			
Meiglyptes tristis				•	
Dryocopus javensis				•	
Dendrocopus moluccensis		•			
Hirundo rustica				•	•
Hirundo tahitica	•	•	•	•	•
Hemipus hirundinaceus				•	
Coracina striata				•	•
Lalage nigra		•		•	
Aegithina tiphia		•		•	
Pycnonotus atriceps		•			
Pycnonotus goiavier		•			
Pycnonotus plumosus				•	•
Pycnonotus simplex					•
Pycnonotus brunneus	•			•	•

Appendix 2. Land birds found on offshore islands of Borneo *(continued)*.

Species	K	M	T	B	A
Iole olivacea				•	•
Dicrurus sumatranus/hottentottus		•			
Dicrurus paradiseus				•	
Oriolus xanthonotus				•	
Oriolus xanthornus		•			
Sitta frontalis		•			
Pellorneum capistratum				•	
Malacocincla malaccensis					•
Trichastoma rostratum				•	
Malacopteron magnirostre					•
Macronous gularis				•	•
Copsychus saularis		•		?	
Copsychus malabaricus	•	•			
Gerygone sulphurea		•			
Acrocephalus orientalis					•
Orthotomus atrogularis					•
Rhinomyias olivacea				•	
Cyornis rufigaster	•	•		•	•
Rhipidura javanica		•		•	
Hypothymis azurea	•	•	•	•	•
Pachycephala grisola	•	•			
Motacilla flava			•		
Artamus leucorhynchus		•		•	
Lanius cristatus					•
Lanius tigrinus					•
Aplonis panayensis	•	•	•	•	•
Gracula religiosa	•		•		•
Anthreptes malacensis	•	•		•	
Anthreptes singalensis				•	
Nectarinia sperata		•			•
Nectarinia calcostetha	•	•			
Nectarinia jugularis	•			•	
Aethopyga siparaja	•	•		•	•
Dicaeum trigonostigma	•			•	
Dicaeum cruentatum	•			•	
Lonchura fuscans				•	

K = Karimata Islands
M = Maratuas
T = Tambelan
B = Banggi (Sabah)
A = Anambas

Source: MacKinnon and Phillipps 1993

Appendix 3. Distribution of Bornean montane birds.

Name	K	T	M	M	K	B	U	B	D	H	L	L	M	P	P	S	P	K
Spilornis kinabaluensis	•		•	•														
Spizaetus alboniger	•		•		•								•				•	•
Falco peregrinus	•		•		•		•											
Arborophila hyperythra	•		•			•					•							
Caloperdix oculea					•	•	•		•									
Haematortyx sanguiniceps	•		•	•	•	•	•	•	•	•								
Lophura bulweri					•	•	•					•			•			
Ducula badia	•	•	•	•	•	•		?	•	•	•	•	•	•	•			•
Macropygia emiliana	•	•	•		•			•	•	•	•	•	•	•				
Macropygia ruficeps	•	•	•		•			•	•	•	•	•	•	•	•			•
Cuculus sparverioides	•	•	•	•	•			•	•	•	•	•	•	•	•		•	
Cuculus saturatus	•			•	?													
Otus rufescens	•		•	•							•		•				•	•
Otus brookii			•				•											
Glaucidium brodiei	•		•	•		•					•							
Batrachostomus harterti	?		?	•		•		•										
Batrachostomus poliolophus				•			•											
Harpactes whiteheadi	•		•		•		•											
Harpactes orrhophaeus	•		•						•		•							
Harpactes oreskios	•		•		•		•				•	•						
Aceros undulatus	•	•	•	•	•	•	•	•	•	•	•	•	•	•	•		•	•
Megalaima monticola	•		•	•	•		•		•	•								
Megalaima pulcherrima	•	•	•															
Megalaima eximia	•	•	•	•	•	•	•	•	•	•	•	•	•	•	•			•
Psarisomus dalhousiae	•		•	•					•									
Calyptomena hosii	•		•	•	•		•	•										
Calyptomena whiteheadi	•		•	•		•		•										
Pitta arquata	•	•	•	•	•	•	•	•	•	•	•	•					•	•
Hemipus picatus	•	•	•	•	•	•	•	•	•	•	•	•	•					
Coracina larvata	•	•	•	•		•		•			•							
Pericrocotus solaris	•	•	•	•	•	•	•	•	•	•								
Pericrocotus flammeus	•		•	•	•	•	•	•	•	•	•	•	•				•	•
Chloropsis cochinchinensis	•	•	•	•	•	•	•	•	•	•	•	•	•	•			•	•
Pycnonotus melanicterus	•		•	•		•			•			•						
Pycnonotus flavescens	•	•	•	•	•													
Alophoixus ochraceus	•	•	•	•	•	•	•	•	•	•	•	•	•	•			•	•
Hypsipetes flavala	•	•	•	•	•	•	•	•	•	•	•	•	•	•				
Dicrurus leucophaeus	•	•	•										•					
Oriolus hosei			•				•		•									
Oriolus cruentus	•	•	•	•	•		•											
Cissa thalassina	•		•	•			•											
Dendrocitta cinerascens	•	•	•	•	•		•		•									
Pellorneum pyrrogenys	•	•	•	•	•	•	•	•	•			•	•					
Napothera crassa	•			•	•								•					
Napothera epilepidota	•		•	•	•								•					
Stachyris leucotis					•						•	•					•	
Garrulax palliatus	•		•	•	•	•		•										
Garrulax lugubris	•		•	•	•	•	•		•									
Garrulax mitratus	•	•	•	•	•	•	•		•	•							•	

Appendix 3. Distribution of Bornean montane birds *(continued)*.

Name	K	T	M	M	K	B	U	B	D	H	L	L	M	P	P	S	P	K
Pteruthius flaviscapis	•	•	•	•	•	•	•	•	•	•	•	•	•	•	•	•	•	
Alcippe brunneicauda	•	•		•	•	•						•		•	•		•	•
Yuhina everetti	•			•	•	•						•	•					•
Brachypteryx montana	•	•		•	•													
Enicurus leschenaulti	•			•		•						•	•			•	•	
Chlamydochaera jefferyi	•							•					•	•				
Myiophonus glaucinus	•	•	•	•	•	•	•	•	•	•	•	•	•	•	•	•		
Zoothera citrina	•	•													•			
Zoothera everetti	•	•	•	•				•				•						
Turdus poliocephalus	•	•																
Seicercus montis	•	•	•	•	•	•		•		•		•	•	•	•	•		•
Abroscopus superciliaris	•			•		•						•				•	•	•
Phylloscopus trivirgatus	•	•	•	•	•	•								•	•		•	
Orthotomus cuculatus	•			•	•	•							•					
Urosphena whiteheadi	•	•	•	•				•	•	•			•	•				
Cettia vulcania	•			•	•													
Bradypterus accentor	•	•																
Rhinomyias ruficauda	•		•	•	•			•					•					
Rhinomyias gularis	•	•	•	•	•													
Eumyias indigo	•		•	•	•													
Ficedula hyperythra	•	•	•	•	•									•	•			
Ficedula westermanni	•	•	•	•	•											•		
Cyornis superbus	•			•		•						•		•			•	•
Muscicapella hodgsoni	•			•		•			•			•						
Rhipidura albicollis	•	•	•	•	•									•	•			
Pachycephala hypoxantha	•	•	•	•	•	•	•	•	•	•		•	•	•	•			
Aethopyga mystacalis	•	•	•	•	•	•	•	•	•	•	•	•	•	•	•	•		•
Arachnothera everetti	•		?					•										
Arachnothera juliae	•		•	•	•			•		•								
Dicaeum monticolum	•		•	•	•	•	•	•		•			•	?				
Zosterops atricapilla	•			•		•												
Zosterops everetti	•	•		•	•	•	•	•	•	•	•		•	•		•	•	
Oculocincta squamifrons	•	•		•	•	•		•				•	•					
Chlorocharis emiliae	•	•		•	•								•	•				
Erythrura hyperythra	•		•															

Key to mountain ranges in order (read horizontally):

K = Kinabalu,	T = Trus Madi,	M = Murud,
M = Mulu,	K = Kelabits,	B = Brassey Range**
U = Usun Apau,	B = Batu Song,	D = Dulit,
H = Hose Mts.,	L = Lanjak Entimau,	L = Liang Kubung,
M = Müller Range,	P = Niut/Penrissen,	P = Gn. Pueh (Poi Mts.),
S = Schwaner Range,	P = Gn. Palung,	K = Kayan-Mentarang.

** (including Magdalena, Maliau and Danum/Segama)

Source: MacKinnon and Phillipps 1993

Appendix 4. Distribution of Bornean snakes.

	K	Sab	Sw
Family Typhlopidae			
Ramphotyphlops braminus	•	•	•
Ramphotyphlops lineatus	•		•
Ramphotyphlops lorenzi		•	
Ramphotyphlops olivaceus	•		•
Typhlops koekkoeki	•		
Typhlops muelleri	•	•	•
Family Uropeltidae			
Subfamily Cylindrophinae			
Cylindrophis lineatus			•
Cylindrophis rufus	•		•
Family Boidae			
Python curtus	•	•	•
Python reticulatus	•	•	•
Family Xenopeltidae			
Xenopeltis unicolor	•	•	•
Family Acrochordidae			
Acrochordus granulatus		•	
Acrochordus javanicus	•		•
Family Colubridae			
Subfamily Xenodermatinae			
Stoliczkaia borneensis		•	
Xenelaphis ellipsifer		•	•
Xenelaphis hexagonotus	•		•
Xenodermus javanicus		•	•
Subfamily Pareatinae			
Aplopeltura boa		•	•
Pareas carinatus	•	•	
Pareas laevis	•	•	•
Pareas malaccanus	•	•	•
Pareas nuchalis	•	•	•
Pareas vertebralis		•	
Subfamily Homalopsinae			
Cerberus rynchops	•	•	•
Enhydris alternans			•
Enhydris doriae	•	•	•
Enhydris enhydris	•		•
Enhydris plumbea	•	•	

Appendix 4. Distribution of Bornean snakes *(continued).*

	K	Sab	Sw
Enhydris punctata	•		
Fordonia leucobalia	•		•
Homalopsis buccata	•	•	•
Subfamily Lycodontinae			
Lycodon albofuscus		•	•
Lycodon effraensis	•		•
Lycodon subcinctus	•	•	•
Oligodon annulifer		•	
Oligodon cinereus		•	
Oligodon everetti	•	•	
Oligodon octolineatus	•	•	•
Oligodon purpurescens	•	•	•
Oligodon subcarinatus		•	•
Oligodon vertebralis	•	•	
Psammodynastes pictus	•	•	•
Psammodynastes pulverulentus	•	•	•
Subfamily Colubrinae			
Ahaetulla fasciolata	•		•
Ahaetulla prasina	•	•	•
Boiga cynodon	•	•	•
Boiga dendrophila	•	•	•
Boiga drapiezii		•	•
Boiga jaspidea	•	•	•
Boiga nigriceps		•	•
Calamaria battersbyi[+]	•		
Calamaria bicolor	•	•	•
Calamaria borneensis	•	•	•
Calamaria everetti	•	•	•
Calamaria gervaisii		•	
Calamaria grabowskyi	•	•	•
Calamaria gracillima			•
Calamaria griswoldi		•	
Calamaria hilleniusi	•	•	
Calamaria lateralis		•	
Calamaria leucogaster	•	•	•
Calamaria lovi	•	•	•
Calamaria lumbricoidea	•	•	•
Calamaria lumholtzi	•		
Calamaria melanota	•		•
Calamaria modesta		•	
Calamaria prakkei[+]		•	
Calamaria rebentischi	•		
Calamaria schlegeli		•	•
Calamaria schmidti		•	
Calamaria suluensis		•	•
Calamaria virgulata	•	•	•

Appendix 4. Distribution of Bornean snakes *(continued)*.

	K	Sab	Sw
Chrysopelea paradisi		•	•
Chrysopelea pelias		•	•
Dendrelaphis caudolineatus	•	•	•
Dendrelaphis formosus	•	•	•
Dendrelaphis pictus	•	•	•
Dryocalamus subannulatus		•	
Dryocalamus tristigatus		•	•
Dryophiops rubescens		•	•
Elaphe erythrurus		•	
Elaphe flavolineata	•	•	•
Elaphe radiata	•		
Elaphe taeniura		•	•
Gonyophis margaritatus		•	•
Gonyosoma oxycephala	•	•	•
Lepturophis borneensis		•	•
Liopeltis baliodeirus	•	•	•
Liopeltis longicaudus		•	•
Liopeltis tricolor			•
Pseudorabdion albonuchalis		•	
Pseudorabdion collaris	•	•	
Pseudorabdion longiceps	•		
Pseudorabdion saravacensis			•
Ptyas carrinatus	•		•
Ptyas korros	•		
Sibynophis melanocephalus		•	•
Stegonotus borneensis		•	•
Zaocys fuscus	•	•	•

Subfamily Natricinae

	K	Sab	Sw
Amphiesma flavifrons		•	•
Amphiesma frenata			•
Amphiesma sarawacensis	•	•	•
Amphiesma stolata		•	
Hydrablabes periops	•	•	•
Hydrablabes praefrontalis		•	
Macropisthodon flaviceps			•
Macropisthodon rhodomelas		•	•
Opisthotropis typicus		•	
Oreocalamus hanitschi		•	•
Pseudoxenodon baramensis			•
Rhabdophis chrysarga	•	•	•
Rhabdophis conspicillatus	•	•	•
Rhabdophis subminiatus	•		
Xenochrophis maculatus	•	•	•
Xenochrophis piscator	•	•	•
Xenochrophis trianguligera	•	•	•

Appendix 4. Distribution of Bornean snakes *(continued)*.

	K	Sab	Sw
Family Elapidae Subfamily Elapinae			
Bungarus fasciatus	•	•	•
Bungarus flaviceps	•	•	•
Maticora bivirgata	•	•	•
Maticora intestinalis	•	•	•
Naja sumatrana	•	•	•
Ophiophagus hannah	•	•	•
Subfamily Laticaudinae			
Laticauda colubrina		•	
Laticauda laticaudata		•	
Subfamily Hydrophiinae			
Aiepysurus eydouxii		•	
Enhydrina schistosa		•	•
Hydrophis brookii			•
Hydrophis caerulescens		•	•
Hydrophis fasciatus		•	•
Hydrophis klossi		•	
Hydrophis melanosomna		•	
Hydrophis ornatus		•	
Hydrophis torquatus	•	•	•
Kerilia jerdoni		•	
Lapemis curtus		•	•
Leioselasma cyanocincta	•	•	•
Leioselasma spiralis		•	•
Microcephalophis gracilis	•	•	•
Pelamis platurus		•	•
Praescutata viperina			•
Thalassophis anomalus		•	•
Family Viperidae Subfamily Crotalinae			
Ovophis chaseni		•	
Trimeresurus popeorum		•	•
Trimeresurus puniceus	•	•	•
Trimeresurus sumatranaus	•	•	•
Tropidolaemus wagleri	•	•	•

+ = endemic
K = Kalimantan
Sab = Sabah
Sw = Sarawak

Source: Stuebing 1991

Appendix 5. Simple methods for inventory and monitoring of species.

Floral inventory

A simple method for estimating the floral diversity of a site involves identifying all trees greater than 50 cm in girth within 5 m either side of a 500 m survey trail. This is equivalent to sampling a half-hectare forest plot.

Survey trails can be forest rentices or transects, narrow forest paths or specially cut transects. They should not be along major forest trails which show signs of human disturbance (e.g., stumps, or pole collecting), beside roads or at the edge of the forest, nor follow stream beds or the tops of narrow ridges. The trail does not necessarily need to follow a straight compass bearing but should be a representative section through the general forest types of the locality.

The marker pegs are placed at 100 m intervals for a distance of 500 m along the survey trail. With the aid of a good local guide, who can give reliable vernacular names, identify all trees within 5 m of either side of the trail. Divide the lists into 100 m units. Girth is measured with a tape, or even knotted string, at breast height. Tree height can be measured with the use of a reflex camera and telefocus lens; measurements can be made on the top branches of occasional trees to keep estimates within acceptable limits of accuracy.

It is important to distinguish between different tree species. Try to find a local man who can give local names. Test his reliability by checking the same trees twice. Tease out the qualifying names ("perempuan", "laki-laki", "halus", "lebar", "keras", etc.) that are used to separate taxa given the same general name.

Try to collect twig and leaf specimens (flowers and fruits if available) of the commonest species in the inventory, by sending up a climber or throwing sticks to knock down specimens. Preserve floral specimens in a polythene bag with some alcohol or dry between newspaper in sunshine. It is only necessary to collect the few commonest species which characterise the forest type. Other trees can usually be identified, to genus at least, from the Forestry Department lists of vernacular names available for most areas of Indonesia.

Record the locality and approximate altitude of each half-hectare sample, the general topography and the mean land slope. Note this is the typical slope of the area covered rather than the mean slope of the survey trail which may be considerably shallower if the path cuts across the slopes.

Each sample will take about one-and-a-half hours to complete, depending on the number of helpers available and their experience. Four or five half-hectare inventories can be completed in a day. Each sample should be made at a similar interval of height (e.g., at altitudes of 500 m, 1,000 m, 1,500 m, etc.), time, or distance from the previous one, to achieve a picture of habitat variability. No two samples should be contiguous.

Results from the inventory can be represented as a series of graphs and histograms which characterise the richness and structure of the forest. A diversity histogram shows whether the forest is dominated by a few common species (fig. 1). Histograms of height and girth (fig. 2) show the structure of the forest, are a guide to the degree of disturbance and a gauge to the forest age. Species discovery curves give a measure of richness (fig. 3), and also indicate where the sampling has crossed into different floral communities.

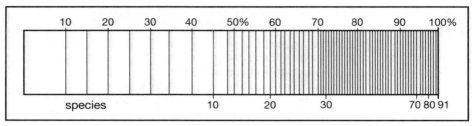

Figure 1. Diversity histogram for trees.

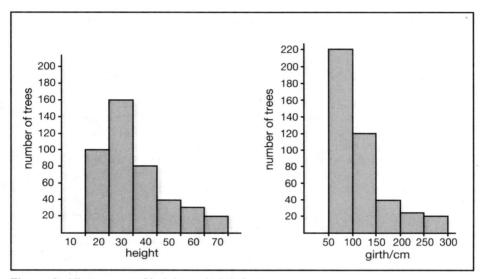

Figure 2. Histograms of height and girth for trees.

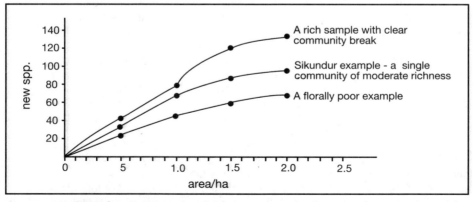

Figure 3. Species discovery curve for tree species richness. Total spp. for Sikundur = 91 (40-29-15-7).

Density estimates for flora: point centre quarter method.

The inventory method gives an estimate of general floral richness, but the point centre quarter (PCQ) method gives a more accurate picture of a species' abundance:
1) Make a series of transects (avoiding long-standing footpaths) which traverse different habitat types or randomly traverse the whole reserve area.
2) Make measurements at sample points at regular intervals (10 m, 100 m, or whatever is suitable for the size of the area to be surveyed).
3) At each sample point draw a line at 90° to the path, forming four quadrats.
4) Note the nearest tree within the size-range sample to the sample point in each of the quadrats, record its girth, estimated height and local name of specimen for identification, and also its distance from the sample point.

When all sample points have been covered, the abundance of each tree species can be estimated as a percentage of the total trees sampled, or can be calculated as a density by the formula:

$$D_a = \frac{t_a \times 1000}{N \times \sim d^2}$$

where:
- D_a = density of species a per hectare
- t_a = total trees of species a sampled
- N = total number of sample points
- $\sim d$ = mean distance of all sampled trees from point centres

A simple method for estimating bird richness.

This method can be used by anyone with a basic knowledge of the main bird groups. Recognition or identification of all species is not necessary, but the reseacher must be able to distinguish between all species seen. The method is in effect a species/area curve method with area equivalent to bird-finding success.

The method involves making a series of bird lists. Each new species is recorded until 20 species have been encountered, then a new list is begun. No species is entered twice in the same list but each new list starts from scratch so the same species may occur on several or all lists.

A sample of 10 or more lists for a given area will give a good picture of bird diversity, and even an estimate of the approximate total number of bird species present in the area at the time of the sample. After 10 sample lists, a species discovery curve is plotted.

If the total number of species seen is plotted graphically against the total number of species recorded on the 10 lists, this will give a species discovery curve whose steepness reflects species richness and indicates how many more species are likely still to be found.

Species which occur on a high proportion of lists are clearly the most abundant or conspicuous species of the local avifauna. The graphs below show examples of field data of this sort for three reserves in Java and one in Borneo (fig. 4).

The great advantage of this method is that it is relatively independent of observer experience and expertise, and independent of birdwatching intensity, weather or other factors.

The only rules to keep in mind are that:
a. All species seen should be included in the list even if they are not recognised; give a name or code to the unknown species so that they can be recognised as the same species again.
b. The results reflect where the data were collected and will show any bias in the survey pattern. To reflect accurately the bird fauna of the whole area of a reserve or park, all habitat types should be surveyed in similar proportions to their abundance.

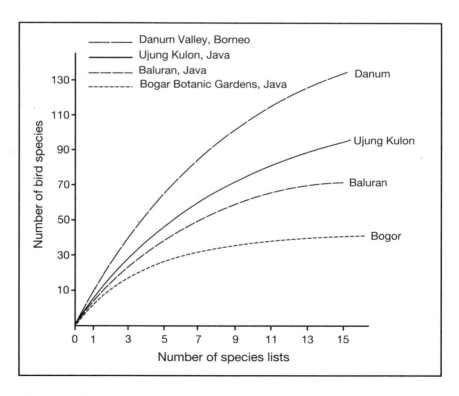

Figure 4. Species discovery curves for birds at four Sundaic localities.

A simple method for estimating animal densities.

The density of any mammal, or large bird, species can be estimated roughly by the formula:

$$D = \frac{f \times 100}{w \times c}$$

- D = density/km²
- f = frequency of encounter per km² surveyed
- w = effective search width in m
- c = crypsis factor (proportion of animals that the observer can expect to see)

The frequency of encounter (f) is the mean number of animals of a given species seen per kilometre of survey trail in a given area or habitat type. The effective search width (w) is an estimate of the width of habitat in the survey. On a forest trail this is approximately 10 m either side of the survey path.

The crypsis factor (c) is the most difficult to determine. For large animals (e.g., elephants in open terrain), it can be taken to be 1.0. For smaller animals in open terrain it approximates the probability that a given individual will be active, (i.e., the proportion of the population active at any one time). In thick cover for animals which give alarm calls (such as hornbills and some monkeys) the crypsis factor is also approximately 1.0. For small and/or quiet, inactive animals, the crypsis factor may be very small. For strictly nocturnal animals it is almost zero.

The crypsis factor in a given set of circumstances can be estimated only if there is an independent density figure for mammal populations from mark/release/recapture studies. In the absence of such data, the following table of effective search widths and crypsis factors can be applied. These figures are estimates based on average rates of encounter by quiet observers travelling at about 2 km per hour along good trails through rainforest.

	w (m)	c
Orangutan	80	0.2
Siamang *	80	0.7
Gibbon	80	0.4
Macaques	120	1.0
Langurs	80	0.7
Large hornbills	100	0.9
Small hornbills	80	0.5
Large squirrels	50	0.1
Small squirrels	40	0.04
Bear cuscus *	60	0.02
Tarsiers at night	30	0.03
Elephant	100	0.5
Banteng	60	0.8
Deer	60	0.8
Pigs	60	0.3
Pheasants	50	0.9
Peacock	50	0.3
Jungle fowl *	50	0.7
Monitor lizard	50	0.5
Civets (at night)	60	0.5

(* not found in Borneo)

Record actual rates of encounter as well as the indices used to estimate density, the former are more useful for comparison with other field studies.

Other methods for estimating population size.

A. **Mark/release/recapture method** (suitable for use only by scientists). For sampling resident populations;
   1) Catch a sample of the population;
   2) Mark the animals with a permanent or semipermanent mark, (e.g., hair clipping, paint, leg rings, ear tags);
   3) Release the animals into the population;
   4) Catch another sample (time interval between samples must be negligible in relation to life span of species);
   5) Estimate resident population in area sampled by the formula:

$$N = \frac{n_1 \times n_2}{m_2}$$

where: N = total number of residents in the area sampled
n_1 = number caught in sample (1)
n_2 = number caught in sample (2)
m_2 = number of marked animals in sample (2)

B. **Total count methods.**

e.g., by:
   1) aerial survey over open terrain (specialist techniques and equipment required);
   2) recording known groups or herds in known ranges (ecological studies needed);
   3) counting all nests of a breeding bird population at communal nest sites;
   4) counting all holes/dens of animals x average number of occupants;
   5) several observers regularly surveying a defined area for long enough to be sure to see all animals (monkeys and other large primates);
   6) surrounding a known sample area with observers close enough together to see wildlife passing between them, then observers converge, flushing out and counting all animals within the sample site.

C. **Minimal count methods.**

In cases where total counts are impossible, it is sometimes useful to make minimal counts of the number of animals known to exist.

A count of all banteng feeding at several feeding areas at one time (with different observers at each feeding area) would give such a result. Another method is to count individually-recognised individuals or their tracks (only for big animals). Large mammals, such as elephants and rhinoceros can also be recorded by photographic surveys using pressure pads and remote telephoto lenses.

D. Density estimates using indirect signs.

Indirect signs include nest counts (e.g., for orangutans), footprints, feeding signs, elephant faeces, etc.

Density (D) is estimated as :

$$D = \frac{S}{A \times f \times d}$$

where:
- S = number of signs found
- A = area searched
- f = mean signs made per individual animal per day
- d = mean duration signs remain recognisable in days

In each case:
i) number of signs must be measured per unit area; searching must be thorough to get a total count;
ii) the number of signs (e.g., nests) made by each animal per day must be known;
iii) the duration of recognisable signs must be known;
iv) sampled areas should extend over a wide range of habitats to avoid bias caused by patchy habitat use. Knowledge is needed from long-term observations of the animals in the wild, or even in captivity, to estimate appropriate values for ii) and iii).

Calling frequency is generally a poor guide to density, since it can vary with season, social relations and other factors. However, data on calling can provide useful information for populations of some animals which call regularly each day.

1) Most gibbon groups call once or more each day and can be plotted.
2) Some langurs (e.g., *Presbytis melalophos*) call from their sleeping positions each dusk and dawn. Troops can be plotted.
3) Groups of tarsiers call each dawn. The numbers of groups calling, multiplied by the mean group size, gives a population estimate.
4) Cock Argus pheasants call from individual dancing rings each morning. The number calling in a valley can be counted, but this does not provide information on the number of females.

Source: MacKinnon 1981

Appendix 6. Useful forest plants used by Iban people.

species	plant type	local name	use
Pityrogramma calomelanos	fern	pantang	medicinal
Cordyline fruticosa	shrub	sabang tilan; sabang amat	ritual, ornamental plant
Tetrastigma pedunculare	climber	akar kelalau	string, dyeing
Campnosperma squamatum	tree	terentang paya	building, padi bins
Mangifera havilandii	tree	raba	fruit, building, fire wood
Goniothalamus dolichocarpus	tree/bark	lukai bukit	insect repellent, ritual
Goniothalamus macrophyllus	tree	semeliok	ritual
Goniothalamus umbrosus	tree	lukai	ritual
Goniothalamus ridleyi	tree/bark	lukai	ritual
Polyalthia glauca	tree	pendok	string, adornment
Xylopia lanceolata	tree	suloh tulang	baskets, chicken-coops, firewood, building
Xylopia malayana	tree	sengkajang	kindling, honey bees
Disepalum	tree	selali padang	chicken-coops, building
-------	tree	semukau	medicinal
Alstonia macrophylla	tree	jelutong	latex, substrate for edible fungus
Alstonia spatulata	tree	pelai paya	icons, shields, masks corks, medicinal
Willughbeia coriacea	climber	akar kubal	fruit
Cyrtosperma lasioides	marsh plant	turak	food, flavouring
Cyrtosperma merkusii	marsh plant	semelaong	wrapping food
Aralidium pinnatifidum	tree	pawa	medicinal
Blechnum orientale	fern	paku kelindang	medicinal
Blumea blasamifera	herb	mambong	medicinal, insect repellent
Nephrolepis biserrata	fern	kubok; paku kero	food, medicinal
Dillenia suffruticosa	tree	buan	medicinal, food, ritual, wrapping
Tetracera macrophylla	climber	akar empelas	sandpaper
Cotylelobium burckii	tree	resak sabang	building
Dipterocarpus borneensis	tree	resak kulat	kindling, building
Hopea kerangasensis	tree	temang luis	building
Hopea semicuneata	tree	kayu besi; mang	building, boat
Shorea albida	tree	empenit	building
Shorea seminis	tree	tegelam; ajol; engkabang bung-kus; e. terendak	timber, fruit
Vatica albiramis	tree	resak duduk	building
Diospyros piscicapa	tree	tuba buah	fish poison
Diospyros sp.	tree	kayu entegeram; kayu malam	firewood

Appendix 6. Useful forest plants used by Iban people *(continued)*.

species	plant type	local name	use
Agrostistachys longifolia	tree	singka	medicinal
Antidesma sp.	tree	jirak	dyeing
Baccaurea lanceolata	tree	empaong; lempaong	fruits
Codiaeum variegatum	shrub	sabang bunga	ornamental plant
Cephalomappa	tree	mereberas	building, firewood
Macaranga gigantea	tree	merakubong	medicinal, ritual, kindling fire
-------	tree	lia padang	firewood, medicinal
-------	tree	gambih	ingredient
Lithocarpus dasystachyus	tree	empili	fruits, firewood
Pangium edule	tree	kepayang	perfumes, brine pickle
Flagellaria indica	climber	wee buntak	fishing baskets
Dicranopteris linearis	fern	demam	decorative lashing, bracelets, bangles
Gigantochloa hasskarliana	bamboo	buluh munti	food (shoots), building
Gigantochloa sp.	bamboo	buluh baloi	hats
Gigantochloa sp.	bamboo	buluh prien	food (shoots), building
Hymenachne aglutigluma	grass	kumpai ai	adornment
Schizostachyum latifolium	bamboo	buluh engkalat	blow-pipes, baskets
Schizostachyum sp.	bamboo	buluh amat	cooking glutinous rice (container), ritual
Calophyllum sclerophyllum	tree	chenaga	rice mills
Cratoxylum arborescens	tree	geronggang	building, boat-building, scabbards, shingles
Cratoxylum glaucum	tree	ketemau	building
Curculigo villosa	shrub	lemba	tie-dyeing, fruit
Gigantochloa latifolia	bamboo	buluh betong	building, container, fish trap, food (shoots), tools, cooking glutinous rice
Stemonorus scorpioides	tree	entaburok	building, substrate for edible fungus
Ocimum basilicum	shrub	bangkit tual	adornment
Ocimum tenuiflorum	shrub	bangkit simbang sembawang	medicinal
Plectranthus scutellarioides	herb	daun nurun darah tinggi	medicinal
Dehaasia brachybotrys	tree	kayu antu	building, boat-building
Eusideroxylon melangangai	tree	belian kebuau	building, carving
Lindera	tree	pawas	building
Litsea	tree	libas	building
Leea aculeata	shrub	kemali	ritual

Appendix 6. Useful forest plants used by Iban people *(continued)*.

species	plant type	local name	use
Cassia alata	shrub	rugan	medicinal
Derris elliptica	shrub	tubai akar	fish poison, insecticide
Fordia coriacea	tree	meragantong	tool handle, for carrying poles, fish poison
Pithecellobium kunstleri	tree	kemudak	shampoo
Spatholobus	climber	akar kemedu	binding and fastening, source of water
Fagraea racemosa	tree	sukong	ritual
Lycopodium cernuum	club moss	selap padi	baskets, decoration
Gossypium barbadense	tree	taya	thread for weaving
Sida acuta	shrub	siar uras	broom
Urena lobata	shrub	kejumpang ukoi	medicinal
Donax arundastrum	tree	bemban ai	mats
Donax grandis	tree	bemban batu	mats and baskets
Phacelophrynium maximum	tree	daun ririk	roofing, plates
Phrynium	tree	mulong itik	food
Melastoma malabathricum	shrub	kemunting	medicinal, fruit
Memecylon myrsinoides	tree	nipis kulit	spear handles, fishing spear, climbing spikes
Ochthocharis	shrub	riang tekura	cooking ingredients
Sandoricum koetjape	tree	kelampo	fruit, building, firewood
Albertisia ? papuana	climber	sengkubai	food flavouring
Artocarpus elasticus	tree	tekalong	mats, pack straps, loincloth, coats, blankets, catching birds, fruits
Artocarpus integer	tree	temedak	fruits, knife hilts
Artocarpus kemando	tree	pudu; pudau	padi bins, walling, fruits
Ficus grossularioides	tree	lengkan	fruits, medicinal
Parartocarpus venenosus	tree	kati	poison (for blowpipe darts), fish poison
Prainea frutescens	tree	selangking padi	posts, drums. fruit
Horsfieldia polyspherula	tree	kumpang	carved charms (ritual)
Myristica elliptica	tree	balau	building
	tree	kumpang	carvings
Eugenia acuminatissima	tree	bungkang bukit	food (fruits)
Melaleuca leucadendron	tree	gelam	caulking boats, ornamental plant
Tristania beccarii	tree	melaban	firewood, spear shafts, tool handles, building
Nepenthes gracilis	climber	akar entuyut, entuyut melaban	lashing
Nepenthes rafflesiana	climber	pok yok	containers, tying
Areca catechu	palm	pinang	ritual, chewing betel, carrying food, ornamental plant

Appendix 6. Useful forest plants used by Iban people *(continued)*.

species	plant type	local name	use
Calamus caesius	climber	wee sega	rattan girdles, baskets, mats
Calamus convallium	climber	wee sero	baskets and mats
Calamus javensis	climber	wee anak	fishing baskets
Calamus pseudoulur	climber	wee tapah	rattan girdles, lashing, threading palm leaves, string
Calamus scipionum	climber	semambu	baskets, chairs, tables
Ceratolobus concolor	climber	danan tai manok	baskets
Korthalsia flagellaris	climber	danan	baskets
Licuala bintuluensis	palm	gerenis	baskets
Licuala petiolulata	palm	gerenis	baskets
Metroxylon sagu	palm	mulong	fruits, sago, breeding-place for edible larvae, dart, blowpipe, gum, baskets, hats, buckets, roofing, building
Pinanga malaiana var. *barramensis*	palm	daun mera-tudong	
Salacca vermicularis	palm	ridan	fishing rods, fruits, food (apical shoot), baskets
-------	palm	belibih	baskets, mats
Pandanus borneensis	screwpine	repo	mats
Pandanus kami	tree	kerupok	mats, hats, matting
Pandanus odoratissimus	screwpine	pandan	cooking, making ketupat
Pandanus vinaceus	screwpine	akas	mats
Pandanus sp.	screwpine	meredang	mats and hats
Piper betel	climber	sireh	betel liquid, ritual, medicinal
Piper caninum	climber	pantong	medicinal
Xanthophyllum amoenum	tree	langgir	washing, fruits
Xanthophyllum scortechinii	tree	langgir bagok	bark-cloth beater
Ventilago malaccensis	climber	akar rarak	fish trap, source of water in the jungle
Combretocarpus rotudantus	tree	keruntum; beluan paya	building, posts, source of income, padi bins, building
Prunus arborea	tree	enteli	
Morinda citrifolia	tree	engkudu	dyeing
Psychotria woodii	tree	engkerebai	brown dyeing
Randia	tree	enchepong	building, fruit
Uncaria sp.	climber	akar klauh	drinking water
Lepisanthes divaricata	tree	kayu rias	decorative sticks
Ploiarium alternifolium	tree	jinggau mudai; jinggau kampong	posts for temporary construction, food (young leaves)

Appendix 6. Useful forest plants used by Iban people *(continued)*.

species	plant type	local name	use
Aquilaria malaccensis	tree	engkaras	blankets, loincloths, string
Eryngium foetidum	herb	tepus libang	flavouring food
Clerodendrum adenophysum	tree	kempait ukoi	medicinal
Achasma megalocheilos	tree	tepus	spice or flavouring, perfume
Curcuma domestica	shrub	kunyit, entemu	spice, dyeing
Hornstedtia scyphifera	shrub	senggang	coarse mats, baskets, fruits
Languas galanga	shrub	engkuas	spice, food, medicinal
Phaeomeria speciosa	shrub	kecala	fruits, flavouring food
Zingiber cassumunar	shrub	lia betong; tepus merah	medicinal
Family unknown			
------	moss	lumut	stuffing a pillow
------	tree	ruan	building
------	shrub	daun sabong	vegetable

Source: Pearce et al. 1987

Appendix 7. Field-sketch landscape profile, based on Field Rapid Rural Appraisal, for reforestation planning in South Kalimantan.

	A	B	C	D	E	F	G	H
LANDSCAPE UNIT								
Age								
Altitude								
Slope								
Soils & Hydrology								
Zone width								
Main crops Vegetation								
Animal husbandry								
Wildlife/ Resources Fuelwood NTFP								
Land status								
Wildlife, pest problem								
Pressure on wildlife								
Management input								
Biodiversity								
Typical, common mammals								
Typical, common birds								
Rare, endangered species								

Source: DHV Environmental Impact Analysis 1992

APPENDIX

Appendix 8. Species Databases - sample sheet.

STATUS SUMMARY FOR Pongo pygmaeus   Orangutan
DISTRIBUTED IN BIOUNITS : – , 25b, 25e, 25f, 25g, 25h, Sum,
OCCUPIES HABITATS : – , PSV, LRF, IFV, HFV,

EVALUATION OF HABITAT STATUS (Areas given in square kms.)

Unit;	Country;	Admin. Div.	Vegtn. type	Origin; area	Remain; area	Rem. %;	Protect; area	Prot. %;	Propos; area	Prop. %
TOTALS				495986	352446	71	29513	6.0	32505	6.6

% original habitat lost =   28.9
% original habitat protected =   6.0
% original habitat proposed =   6.6
% remaining habitat protected =   8.4
Expected protected population =   16067
Expected total population =   60018
% age population protected =   27
Red Data Book Status =   Endangered
Species listed on CITES appendix   1
Species protected in countries :   ,ID, MA, BR,

Orangutan is expected/confirmed in the following areas

Name	Country	Admin.	Total Area;	Confirmed;	Status
Apo Kayan	ID	KTI	1,000.0		5
Batikap 1	ID	KTE	720.0		5
Batikap 2,3	ID	KTE	6,784.0		5
Batu Kristal	ID	KTI	100.0		5
Bentuang-Karimun	ID	KBA	6,000.0	•	4
Bukit Baka	ID	KBA	705.5	•	5
Bukit Perai	ID	KBA	1,680.0		5
Bukit Raya	ID	KTE	1,100.0	•	4
Bukit Raya extn.	ID	KTE	5,900.0	•	5
Danau Sentarum	ID	KBA	800.0	•	4
Danum Valley	MA		427.0	•	5
Gn. Lotong	MA		390.0		5
Gn. Niut	ID	KBA	1,400		5
Gn. Palung	ID	KBA	300.0	•	5
Gn. Palung extn.	ID	KBA	700.0	•	5
Hutan Kahayan	ID	KTE	1,500.0		5
Kayan Mentarang	ID	KTI	16,000		5
Kinabalu	MA		754.0	•	4
Kuala Kayan	ID	KTE	735.0		5
Kutai	ID	KTI	2,000	•	4
Lanjak Entimau	MA		1,700.0	•	4
Long Bangun	ID	KTI	3,500.0		5
Muara Kendawangan	ID	KBA	1,245		5
Muara Sebuku	ID	KTI	1,327.0		5
Parawen Baru	ID	KTE	815.0		5
Pulong Tau	MA		1,645.0		5
Sangkulirang	ID	KTI	2,000	•	4
Tanjung Puting	ID	KTE	3,050.0	•	5
Tj. Puting extn.	ID	KTE	700.0	•	5
Ulu Kayan Mutlak	ID	KTI	9,800.0		5
Ulu Sembakung	ID	KTI	5,000.0		5

Bibliography

Abdulhadi, R. Kartawinata, K. and Sukardjo, S. 1981. Effects of mechanized logging in the lowland dipterocarp forest at Lempake, East Kalimantan. *Malay Forester* 44: 407-418.

Abubakar, M.L. and Seibert, B. (eds.) 1988. *Agroforestry untuk pengembangan daerah pedesaan di Kalimantan Timur.* Prosiding Seminar Kehutanan, Samarinda.

Achmad, S. 1986. *Tambak atau hutan mangrove.* Diskusi panel dayaguna dan lebar jalur hijau mangrove, Ciloto, Indonesia.

Adiwilaga, E.M., Djokosetiyanto, D., Harris, E., Sumantadinata, K. and Sukimin, S. 1987. Survai potensi ikan siluk (*Scleropages formosus*) di Kabupaten Daerah Tingkat II Sintang dan Kapuas Hulu, Kalimantan Barat. Kerjasama Dinas Perikanan Propinsi Daerah Tingkat I Kalimantan Barat dengan Fakultas Perikanan Institut Pertanian Bogor.

Ahmad, A. 1978. Klasifikasi rawa-rawa pesisir (coastal swamps) di Pasaman Barat. *Terubuk* 10: 30-36.

Aldrich-Blake, F.P.G. 1980. Long-tailed macaques. In *Malayan forest primates. Ten years study in tropical rain forest* (ed. D.J. Chivers), pp. 147-165. Plenum, New York.

Alexander, I., Ahmad, N. and Lee, S.S. 1992. The role of mycorrhizas in the regeneration of some Malaysian forest trees. In *Tropical rain forest: disturbance and recovery* (eds. A.G. Marshall and M.D. Swaine), pp. 379-388. The Royal Society, London.

Alfred, E.R. 1966. The freshwater fishes of Singapore. *Zool. Verh.* 78: 1-68.

Allan, J. and Muller, K. 1988. *Borneo adventurers in East Kalimantan* (ed. P. Zach). Times Editions, Singapore.

Alrasjid, 1978. Tending of a residual stand in a logged tropical rainforest in East Kalimantan. In *Proceedings of the Symposium on long-term effects of logging in Southeast Asia*, Bogor, pp 127 - 136. BIOTROP.

Anderson, A.J.V. 1977. Sago and nutrition in Sarawak. *Sarawak Mus. J.* 25: 71-80.

Anderson, A.J.V. 1982. *Subsistence of the Penan in the Mulu area of Sarawak.* Unpubl. ms.

Anderson, J.A.R. 1958. Observations on the ecology of the peat- swamp forests of Sarawak and Brunei. In *Proceedings of the symposium on humid tropics vegetation.* pp. 141-148. Council for Sciences of Indonesia/UNESCO, Ciawi, Indonesia.

Anderson, J.A.R. 1961a. *The ecology and forest types of the peat swamp forests of Sarawak and Brunei in relation to their silviculture.* Ph.D thesis, University of Edinburgh.

Anderson, J.A.R. 1961b. The destruction of *Shorea albida* forest by an unidentified insect. *Empire For. Rev.* 40(1): 19-29.

Anderson, J.A.R. 1963. The flora of the peat swamp forests of Sarawak and Brunei, including a catalogue of all recorded species of flowering plants, ferns and fern allies. *Garden's Bull. Singapore* 29: 131-228.

Anderson, J.A.R. 1964. The structure and development of the peat swamps of Sarawak and Brunei. *J. Trop. Geog.* 18: 7-16.

Anderson, J.A.R. 1965. The limestone habitat in Sarawak. In *Proceedings of the symposium on ecological research in humid tropics vegetation*, pp. 49-57. UNESCO, Kuching.

Anderson, J.A.R. 1972. *Trees of peat swamp forests of Sarawak.* Forest Dept., Sarawak.

Anderson, J.A.R. 1976. Observations on the ecology of five peat swamps in Sumatra and Kalimantan. *Soil Res. Inst. Bogor. Bull.* 3: 45-55.

Anderson, J.A.R. 1975. Illipe nuts *Shorea* spp. as potential agricultural crops. In *Proceedings of symposium on South East Asian plant genetic resources* (eds. J.T. Williams, G.H. Lamoureux and N. Wulijarni-Soetjipto, pp. 217-230, Bogor.

Anderson, J.A.R. 1983. The tropical peat swamps of western Malesia. In *Mires, swamp, bog, fen and moor* (ed. A.J.P. Gore). Ecosystems of the World 14b. Elsevier, Amsterdam.

Anderson, J.A.R. and Chai, P.P.K. 1982. Vegetation. Gunung Mulu National Park, Sarawak. *Sarawak Mus. J. Special Issue* No. 2, Vol.30 (51): 195-223.

Anderson, J.A.R., Jermy, A.C. and Cranbrook, Earl of. 1982. *Gunung Mulu National Park: a management and development plan.* Royal Geographic Society, London.

Anderson, J.A.R. and Muller, J. 1975. Palynological study of a Holocene peat and a Miocene coal deposit from NW Borneo. *Review of Palaeobotany and Paly-*

nology 19: 291-351.
Anderson, J.M. and Swift, M.J. 1983. Decomposition in tropical forests. In *Tropical rain forest: ecology and management* (eds. S.L. Sutton, T.C. Whitmore and A.C. Chadwick), pp. 287-309. Blackwell, Oxford.
Anderson, P.K. 1981. The behaviour of the dugong (*Dugong dugon*) in relation to conservation and management. *Bull. Mar. Sci.* 31: 640-647.
Andriesse, J.P. 1974. Nature and management of tropical peat soils. *FAO Soils Bulletin* 59. FAO, Rome.
Andriesse, J.P. 1974. *Tropical lowland peats in Southeast Asia.* Koninklijk Instituut voor de Tropen, Amsterdam.
Anon. 1976. *Pest control in rice.* Centre for Overseas Pest Research, London.
Anon. 1976. Peat and podzolic soils and their potential for agriculture in Indonesia. *Soil Res. Inst. Bogor Bull.* 3.
Anon. 1985. *A review of policies affecting the sustainable development of forest lands in Indonesia.* Vol. I. GOI Department of Forestry, State Ministry of Population, Environment and Development, Department of the Interior and International Institute for Environment and Development, Jakarta.
Antaran, B. 1985. Omen birds: their influence on the life of the Dusuns with special reference to the Merimbun Dusuns. *Brunei Mus. J.* 6(1): 105-109.
Anwar, A., Wibowo, R. and Rachman, A.M.A. 1986. *Konsepsi pembangunan wilayah pantai dalam hubungan dengan penentuan jalur hijau mangrove.* Diskusi panel dayaguna dan lebar jalur hijau mangrove, Ciloto, Indonesia.
Appanah, S. 1981. Pollination in Malaysian primary forests. *Malay. Forester* (44)1: 37-42.
Appanah, S. 1985. General flowering in the climax rain forests of Southeast Asia. *J. Trop. Ecol.* 1: 225-240.
Appanah, S. 1990. Plant-pollinator interactions in Malaysian rainforests. In *Reproductive ecology of tropical forest plants* (eds. K.S. Bawa and M. Hadley), pp. 85-101. Man and the Biosphere Series vol. 7. UNESCO and The Parthenon Publishing Group.
Appanah, S. and Chan H.T. 1981. Thrips: the pollinators of some dipterocarps. *Malay Forester* 4: 234-252.
Appell, G.N. 1976 (ed). *Studies in Borneo societies.* Social process and anthropological explanation. Special Report No. 12, Center for Southeast Asian Studies, Univ. of Northern Illinois.
Argent, G., Lamb. A., Phillipps, A. and Collenette, S. 1988. *Rhododendrons of Sabah.* Sabah Parks Publication No. 8, Sabah.
Ashton, P.S. 1964. Ecological studies in the mixed dipterocarp forests of Brunei State. *Oxford Forestry Memoirs No.* 25.
Ashton, P.S. 1965. Some problems arising in the sampling of mixed rain forest communities for floristic studies. In *Proceedings symposium on ecological research in humid tropics vegetation.* pp. 235-240. UNESCO. Kuching, Sarawak.
Ashton, P.S. 1969. Speciation among tropical forest trees: some deductions in the light of recent evidence. *Biol. J. Linn. Soc. Lond.* 1: 155-162.
Ashton, P.S. 1971. The plants and vegetation of Bako National Park. *Malay. Nat. J.* 24: 151-162.
Ashton, P.S. 1972. The Quaternary geomorphological history of Western Malesia and lowland rainforest phytogeography. In *The Quaternary era in Malesia* (eds. P. and M. Ashton), Geog. Dept. Univ. of Hull. Misc. Series 13.
Ashton, P.S. 1978. Crown characteristics of tropical trees. In *Tropical trees as living systems* (eds P.B. Tomlinson and M.H. Zimmermann), Cambridge University Press.
Ashton, P.S. 1982. Dipterocarpaceae. *Fl. Mal. Ser. I.* 9: 237-552.
Ashton, P.S. 1981. The need for information regarding tree age and growth in tropical trees. In *Age and growth rate of tropical trees: new directions for research* (eds. F.H. Bourman and G. Berlyn), Yale University School of Forestry and Environmental Studies, Bull. No. 94: 3.
Ashton, P.S. 1984. The biological significance of complexity in lowland tropical rain forest. *J. Indian Bot. Soc.* 50A: 530-537.
Ashton, P.S. 1984. Techniques for the identification and conservation of threatened species in tropical rain forests. In *The biological aspects of rare plant conservation,* pp. 155-163. John Wiley and Sons Ltd.
Ashton, P.S. 1988. Dipterocarp biology as a window to the understanding of tropical forest structure. *Ann. Rev. Ecol. Syst.* 19: 347-70.
Ashton, P.S. 1989. Sundaland. In *Floristic inventory of tropical countries.* (eds D.G. Campbell and H.D. Hammond), pp. 91-

99, New York Botanic Garden.
Ashton, P.S., Givnish, T.J. and Appanah, S. 1988. Staggered flowering in the Dipterocarpaceae: new insights into floral induction and the evolution of mast fruiting in the aseasonal tropics. *Am. Nat.* 132(1): 44-66.
Audley-Charles, M.G. 1981. Geological history of the region of Wallace's Line. In *Wallace's Line and plate tectonics* (ed. T.C. Whitmore), pp. 24-35. Oxford University Press, Oxford.
Audley-Charles, M.G. 1987. Dispersal of Gondwanaland: relevance to the evolution of the angiosperms. In *Biogeographical evolution of the Malay Archipelago* (ed. T.C. Whitmore), pp. 5-25. Oxford Science Publications.
Augspurger, C.K. 1984. Seedling survival of tropical tree species: interactions of dispersal distance, light gaps and pathogens. *Ecology* 65: 1705-1712.
Avé, J.B. and King, V.T. 1986. *Borneo: the people of the weeping forest; tradition and change in Borneo*. National Museum of Ethnology, Leiden.
Avé, J.B., King, V.T. and de Wit, J.G.W. 1983. *West Kalimantan: a bibliography*. Dordecht: Foris.
Avé, W. 1985. *Small-scale utilization of rotan by Semai in West Malaysia*. M.Sc thesis, Leiden.
Avé, W. 1986. The use of rattan by a Semai community in West Malaysia. *Principes* 30(4): 143-150.
Aveling R. and Mitchell, A. 1980. Is rehabilitating orangutans worthwhile ? *Oryx* 16: 263-271.
AWB/KPSL-UNLAM. 1989. *Conservation of Sungai Negara wetlands, Barito Basin, South Kalimantan*. Papers presented to the Sungai Negara workshop, Banjarbaru, Indonesia.
Bachruddin, M.A., Irving A. and MacKinnon, K. 1992. *Local industry support for Kutai National Park*. Paper presented to 4th World Congress on National Parks, Caracas, Venezuela.
Baillie, I.C. 1972. *Further studies on the occurrence of drought in Sarawak*. Forest Department, Sarawak: Soil Survey Research Section.
Baillie, I.C., and Ashton, P.S. 1983. Some soil aspects of the nutrient cycle in mixed dipterocarp forests in Sarawak. In *Tropical rain forest: ecology and management* (eds. S.L. Sutton, T.C. Whitmore and A.C. Chadwick), pp. 347-358. Blackwell Scientific Publications, Oxford.
Baillie, I.C., Tie, Y.L., Lim , C.P. and Phang C.M.S. 1982. Soils of the Gunung Mulu National Park. *Sarawak Mus. J.* 30: 183-193.
Baker, H.G. 1978. Chemical aspects of the pollination biology of woody plants in the tropics. In *Tropical trees as living systems* (eds. P.B. Tomlinson and M.H. Zimmermann), Cambridge University Press.
Baker, J.M. 1982. Mangrove swamps and the oil industry. *Oil Petrochem. Poll.* 1: 5-22.
Baker, J.M. 1983. Impact of oil pollution on living resources. *Commission on Ecology Papers No.* 4, IUCN.
Baker, R.R. 1982. *Migration: paths through time*. Hodder and Stoughton, London.
Balgooy, M.M.J. van 1987. A plant geographical analysis of Sulawesi. In *Biogeographical evolution of the Malay archipelago* (ed. T.C. Whitmore), pp. 94-102. Oxford Science Publications.
Banks, E. 1981. Mountain birds and plate tectonics. *Brunei Mus. J.* 5(1): 78-91.
Banks, E. 1983. A note on Iban omen birds. *Brunei Mus. J.* 5(3): 104-107.
Banister, K. and Campbell, A. (eds.) 1985. *The encyclopedia of underwater life*. Allen and Unwin Ltd., London.
BAPEDAL and EMDI, 1992. AMDAL. *A guide to environmental assessment in Indonesia*. EMDI, Jakarta.
BAPPENAS, 1991. *Biodiversity action plan for Indonesia. Final draft.* BAPPENAS, Jakarta.
Barlow, B.A. (ed.) 1986. *Flora and fauna of alpine Australasia, ages and origins*. C.S.I.R.O.
Barnes, R.D. 1968. *Invertebrate zoology.* Saunders, Philadelphia.
Barnes, R.S.K. 1984. *A synoptic classification of living organisms*. Blackwell, Oxford.
Barrett, E. 1984. *Ecology of nocturnal arboreal mammals in Peninsular Malaysia*. Ph.D. thesis, University of Cambridge.
Basri, A., Husodo, T. and Firmansyah, A. 1987. Setelah hutan jadi lautan api. *TEMPO* 32(17): 48.
Bawa, K.S., Ashton, P.S., and S.M. Nor, 1990. Reproductive ecology of tropical forest plants: management issues. In *Reproductive ecology of tropical forest plants* (eds. K.S. Bawa and M. Hadley). Man and the Biosphere Series, vol. 7: 3-12. UNESCO and The Parthenon Publishing Group.
Beadle, N.C.W. 1966. Soil phosphate and its role in moulding segments of the Aus-

tralian flora and vegetation with special reference to xeromorphy and sclerophylly. *Ecology* 47: 992-1007.

Beaman, R.S., Beaman, J.H., Marsh, C.W. and Woods, P.V. 1985. Drought and forest fires in Sabah in 1983. *Sabah Soc. J.* 8(1): 10-30.

Beaman, R.S., Decker, P., and Beaman, J.H. 1988. Pollination of *Rafflesia* (Rafflesiaceae). *Am. J. Bot.* 75(8): 1148-1162.

Beaman, J.H., Parris, B.S and Beaman, R.S. 1991. Pteridophytes of Mount Kinabalu, Borneo. *Am. J. Bot. 78* (6 Suppl.): 166.

Bearder, S.K. 1984. Lorises, bushbabies, and tarsiers: diverse societies in solitary foragers. In *Adaptations for foraging in nonhuman primates* (eds. P.S. Rodman and J.G.H. Cant), pp. 11-23. Columbia Univ. Press, New York.

Beattie, A.J. 1985. *The evolutionary ecology of ant-plant mutualisms.* Cambridge Studies in Ecology, Cambridge University Press.

Beattie, A. 1989. Myrmecotrophy: plants fed by ants. *TREE* 4(6): 172-176.

Beaver, R.A. 1979a. Fauna and foodwebs of pitcher plants in West Malaysia. *Malay. Nat. J.* 33(1): 1-9.

Beaver, R.A. 1979b. Biological studies of the fauna of pitcher plants (*Nepenthes*) in West Malaysia. *Ann. Soc. Ent. Fr. (N.S.)* 15(1): 3-17.

Beaver, R.A. 1983. The communities living in *Nepenthes* pitcher plants: fauna and food webs. Plexus Publishing, Inc. pp. 129-159.

Beccari, O. 1904/1986. *Wanderings in the great forests of Borneo.* Oxford in Asia Hardback Reprints.

Becker, P., Leighton, M. and Payne, J.B. 1985. Why tropical squirrels carry seeds out of source crowns. *J. Trop. Ecol.* 1: 183-186.

Beekman, H.A.J.M. 1949. *Houttelt in Indonesia.* H. Veenan and Zonnen, Wageningen.

Beer, J.H. de and McDermott, M.J. 1989. *The economic value of non-timber forest products in Southeast Asia.* Netherlands Committee for IUCN, Amsterdam.

Bellwood, P.S. 1978. *Man's conquest of the Pacific. The prehistory of Southeast Asia and Oceania.* Collins, Auckland, London.

Bellwood, P.S. 1980. The peopling of the Pacific. *Sci. Am.* 243(5): 174-185.

Bellwood, P. 1985. *Prehistory of the Indo-Malaysian Archipelago.* Academic Press, London, Sydney.

Bellwood, P.(ed.) 1988. Archaeological research in south-eastern Sabah. *Sabah Museum Monograph 2,* Sabah.

Bemmelen, R.W. van. 1942/1970. *The geology of Indonesia.* Martinus Nijhoff, The Hague.

Bennett, E.L. 1984. *The banded langur: ecology of a colobine in a West Malaysian rain forest.* Ph.D thesis, University of Cambridge.

Bennett, E.L. 1986. *Proboscis monkeys in Sarawak: their ecology, status, conservation and management.* World Wildlife Fund Malaysia and New York Zoological Society.

Bennett, E.L. 1987. The value of mangroves in Sarawak. *Sarawak Gazette* 63: 12-21.

Bennett, E.L. 1988a. Cyrano of the swamps. *BBC Wildlife* (February): 71-75.

Bennett, E.L. 1988b. Proboscis monkeys and their swamp forests in Sarawak. *Oryx* 22(2): 69-74.

Bennett, E.L., Caldecott, J.O. and Davidson, G.W.H. 1984. *A wildlife survey of Ulu Temburong, Brunei.* Report on the biological findings of Exercise Temburong Ringer, WWF, Malaysia.

Bennett, E.L., Caldecott, J., Sebastian, A. and Kavanagh, M. 1981. Conservation status of Sarawak primates. *Primate Conservation* 8:37-41.

Bennett, E. and Gombek F. 1993. *Proboscis monkeys of Borneo.* Natural History Publications (Borneo) Sdn. Bhd. and KOKTAS Sabah Berhad.

Bennett, E.L. and Sebastian, A.C. 1988. Social organization and ecology of proboscis monkeys (*Nasalis larvatus*) in mixed coastal forest in Sarawak. *Int. J. Primatol* 9: 233-256.

Berry, A.J. 1972. The natural history of West Malaysian mangrove faunas. *Malay. Nat. J.* 5: 135-162.

Bessinger, S.R. and Osborne, D.R. 1982. Effects of urbanization on avian community organization. *Condor* 84: 75-83.

Beveridge, A.E. 1953. The Menchali forest reserve. *Malay. Forester* 16: 87-93.

Biro Pusat Statistik. 1988. *Statistik Indonesia (Statistical Yearbook of Indonesia).* Central Bureau of Statistics, Jakarta.

Biro Pusat Statistik. 1991. *Statistik Indonesia (Statistical Yearbook of Indonesia).* Central Bureau of Statistics, Jakarta.

Birowo, A.T. 1979. Landwirtschaft. In *Indonesien* (eds. H. Kotter et al.), pp. 382-447.

Bishop, J.E. 1973. *Limnology of a small Malayan river, Sungai Gombak.* Dr. W. Junk Publishers, The Hague.

Bishop, W.F. 1980. Structure, stratigraphy, and hydrocarbons offshore southern

Kalimantan, Indonesia. *Am. Assoc. Petrol. Geol. Bull.* 64(1): 7-58.

Blaikie, P. and Brookfield, H. 1987. *Land degradation and society.* Methuen and Co. Ltd., London.

Blasco, F. 1977. Outlines of ecology, botany and forestry of the mangals of the Indian subcontinent. In *Wet coastal ecosystems.* (ed. V.J. Chapman), pp. 241-260, Elsevier, Amsterdam.

Blockhus, J.M., Dillenbeck, M., Sayer, J.A. and Wegge, P. (eds.) 1992. *Conserving biological diversity in managed tropical forests.* IUCN/ITTO.

Blower, J.H., Wirawan, N., and Watling, R. 1981. *Preliminary survey of Sungai Kayan-Sungai Mentarang Nature Reserve in East Kalimantan.* WWF, Bogor.

Bock, C. 1881/1985. *The head-hunters of Borneo.* Oxford University Press.

Bodmer, R.E., Mather, R.J., and Chivers, D.J. 1991. Rainforests of Central Borneo: habitats threatened by modern development. *Oryx* 25(1): 21-26.

Boerboom, J.H.A. and Wiersum, K.F. 1983. Human impact on tropical moist forests. In *Man's impact on vegetation* (eds. W. Holzner, M.J.A. Werger and I. Ikusuma), pp. 83-106. Dr. W. Junk Publishers, The Hague.

Bohap bin Jalan and Galdikas B.M.F. 1987. Birds of Tanjung Puting National Park, Kalimantan Tengah, a preliminary list. *Kukila* 3: 33-37.

Bolin, B. Doos, B.R., Jager, J and Warrick, R.A. (eds).1986. *The greenhouse effect: climate change and ecosystems.* John Wiley and Sons, Chichester, U.K.

Bompard, J.M. 1988. Wild *Mangifera* species in Kalimantan (Indonesia) and in Malaysia. Report prepared for IBPGR and IUCN-WWF.

Bompard, J.M. and Kostermans, A.J.G.H. 1992. The genus *Mangifera* in Borneo: results of a IUCN-WWF/IBPGR project. In *Forest biology and conservation in Borneo* (eds. Ghazally, Murtedza and Siraj) pp 61-71. Yayasan Sabah, Kota Kinabalu, Sabah.

Borner, M. 1979. *A field study of the Sumatran rhinoceros, Dicerorhinus sumatrensis (Fischer, 1914). Ecology, behaviour and conservation situation in Sumatra.* Juris Druck, Zurich.

Bouman, M.A. 1924. Ethnographische aanteekeningen omtrent de Gouvernementslanden in de boven-Kapoeas, Westerafdeeling van Borneo. *Tijdschrift voor Indische Taal-, Landen Volkenkunde* 64: 173-195

Boutin, M. and Boutin, A. 1984. Indigenous groups of Sabah: an annotated bibliography. *Sabah Mus. Monograph* No. 1.

Bovbjerg, R.V. 1970. Ecological isolation and competition exclusion in two crayfish (*Oronectes virilis* and *Oronectes immunis*). *Ecology* 51: 225-236.

Boyce, D. 1986. *Kutai, East Kalimantan. A journal of past and present glory.* Kota Bangun, East Kalimantan.

Braam, J. van 1914. Verslag omtrent een reis naar de z. en o. Afdeeling van Borneo. *Tectona* XLIII: 614-635.

Brassel, H.M. and Sinclair, D.F. 1983. Mineral elements returned to the forest floor in two rainforest and three plantation plots in tropical Australia. *Ecology.* 1: 367-378.

Bratawinata, A.A. 1984. *BentandesGliederung eines Bergregenwaldes in Ostkalimantan/Indonesien nach floristischen und strukturellen Merkmalen.* Ph.D. thesis, Georg-August-Universitat, Gottingen.

Brinkman, W.J. and Vo Tong Xuan. 1986. *Melaleuca leucadendron,* a useful and versatile tree for acid sulphate soils and some other poor environments. M.Sc. thesis, Wageningen Agricultural University.

Brockelman, W.Y. and Gittins, S.P. 1984. Natural hybridization in the *Hylobates lar* species group: implications for speciation in gibbons. In *The lesser apes: evolutionary and behavioural biology* (eds. H. Preuschoft, D.J. Chivers, W.Y. Brockelman & N. Creel), pp. 498-532, Edinburgh University Press.

Brook, D.B. and Waltham, A.C. 1978. The underworld of Mulu, part 1. *Caving Internat. Mag.* 1: 3-6.

Brook, D.B. and Waltham, A.C. (eds.) 1979. *Caves of Mulu. The limestone caves of Gunong Mulu National Park, Sarawak,* (2nd. ed.). Royal Geographic Society, London.

Brook, D.B., Eavis, A.J., Lyon, M.K. and Waltham, A.C. 1982. Caves of limestone. Gunung Mulu National Park, Sarawak. *Sarawak Mus. J. Special Issue* No. 2: 95-120.

Brooke, C. 1866. *Ten years in Sarawak* (2 vols.). Tinsley, London.

Brooks, K.N., Gregerson H.N., Berglund E.R. and Tayaa M. 1982. Economic evaluation of watershed projects: an overview, methodology and application. *Water Resource Bull.* 18(2): 245-249

Brooks, W.R. 1988. Are hermit crabs placing anemones on the crab's gastropod shell randomly, for balance, or for maximum protection? *Am. Zool.* 28(4).

Brooks, W.R. and Mariscal, R.N. 1984. The acclimation of anemone fishes to sea-anemones, protection by changes in the fish's mucous coat. *Exper. Mar. Biol. and Ecol.* 80(3): 277-285.

Brosius, J.P. 1986. River, forest, and mountain: the Penan gang landscape. Development versus traditional and wise use of forest resource managers. *Sarawak Mus. J.* (36)57: 173-184.

Brosius, P. 1988. A separate reality: comments on Hoffman's The Punan: hunters and gatherers of Borneo. *Borneo Res. Bull.* 20(2): 81-105.

Brown, S. and Lugo, A.E. 1982. The storage and production of organic matter in tropical forests and their role in the global carbon cycle. *Biotropica* 14(3): 161-187.

Brown, S. and Lugo, A.E., 1984. Biomass of tropical forests: a new estimate based on forest volumes. *Science* 223: 1290-1293.

Brown, N.D. and Whitmore, T.C. 1992. Do dipterocarp seedlings really partition tropical rain forest gaps? In *Tropical rain forest: disturbance and recovery.* (eds. A.G. Marshall and M.D. Swaine), pp. 369-378. The Royal Society, London.

Bruijn, G.H. de and Veltkamp, H.J. 1989. *Manihot esculenta* Crantz. In *Plant resources of South-East Asia* (eds. E. Westphal and P.C.M. Jansen), pp. 175-180. Pudoc, Wageningen.

Bruijnzeel, L.A. 1982. *Hydrological and biogeochemical aspects of man-made forest in south Central Java, Indonesia.* Academisch Proefschrift, Vrije Universiteit te Amsterdam.

Bruijnzeel, L.A. 1984. Elemental content of litterfall in a lower montane rainforest in Central Java, Indonesia. *Malay. Nat. J.* 37: 199-208.

Bruijnzeel, L.A. 1990. *Hydrology of moist tropical forests and effects of conversion: a state of knowledge review.* IHP-UNESCO, Paris.

Bruijnzeel, L.A., Waterloo, M.J., Proctor, J., Kiuters, A.T. and Kotterink, B. 1993. Hydrological observations in montane rainforests of Gn. Silam, Sabah, Malaysia with special reference to the "Massenerhebung" effect. *J. Ecology* 81: 145-167.

Brünig, E.F. 1965. A guide and introduction to the vegetation of the kerangas forests and the padang of the Bako National Park. In *Proceedings of the symposium of recent advances in humid tropics vegetation.* UNESCO, Kuching.

Brünig, E.F. 1969. The classification of forest types in Sarawak. *Malay Forester* 32: 143-149.

Brünig, E.F. 1970. Stand structure, physiognomy and environmental factors in some lowland forests in Sarawak. *Trop. Ecol.* 11: 26-43.

Brünig, E.F. 1971. On the ecological significance of drought in the equatorial wet evergreen (rain) forest of Sarawak (Borneo). *Erdkunde* 23: 127-133.

Brünig, E.F. 1973. Species richness and stand diversity in relation to site and succession of forests in Sarawak and Brunei (Borneo). *Amazoniana* 4: 293-320.

Brünig, E.F. 1974. *Ecological studies in the kerangas forests of Sarawak and Brunei.* Borneo Literature Bureau, Kuching.

Brünig, E.F. 1976. Comparison of the phytomass structure of equatorial "rainforest" in central Amazones, Brazil and in Sarawak, Borneo. *Garden's Bull. Singapore* 30: 81-101.

Brünig, E.F. 1977. The tropical rain forest - a wasted asset or an essential biospheric resource?. *Ambio* 6(449): 187-191.

Budiman, A. 1985. The molluscan fauna in reef-associated mangrove forests in Elpaputih and Wailale, Ceram, Indonesia. In *Coastal and tidal wetlands of the Australian monsoon region.* (eds. K.N. Bardsley, J.S. Davie and D. Woodroffe), pp251-256. Australian National Univ. Mangrove Monograph No.1, Darwin.

Budiman, A. and Darnaedi, D. 1982. Struktur komunitas moluska di hutan mangrove, Morowali, Sulawesi Tengah. In *Prosiding seminar II: Ekosistem mangrove.* Lembaga Oseanologi Nasional, Jakarta.

Bullock, J.A. 1966. The ecology of Malaysian caves. *Malay Nat. J.* 19: 57-63.

Burbridge, P.R. 1983. *Coastal resource management.* Government of Indonesia-United Nations Development Program, Environmental Sector Reviews. UNDP, Jakarta, Indonesia.

Burbridge, P., Dixon, J. and Soewardi, B. 1981. Forestry and agriculture: options for resource allocation in choosing lands for transmigration development. *Appl. Geog.* 1: 237-258.

Burbridge, P. and Koesoebiono. 1980. *Mangrove exploitation.*Proceedings of Asian mangrove symposium. Kuala Lumpur.

Burbridge, P. and Maragos, J. 1985. *Coastal resources management and environmental assessment needs for aquatic resources development in Indonesia.* International Institute for Environment and Development, Washington, USA.

Burbridge, P., Koesoebiono and Dahuri R.

1988. Problems and issues in coastal resources management and planning in East Sumatra, and the Strait of Malacca. In *Coastal zone management in the Strait of Malacca* (eds. P.R. Burbridge, Koesoebiono, H. Dirschl and B. Patton), pp. 8-117, SRES, Dalhousie.

Burgess, P.F. 1961. The structure and composition of lowland tropical rain forest in North Borneo. *Malay. Forester* 24: 66-80.

Burgess, P.F. 1966. *Timbers of Sabah*. Sabah Forest Record No. 6, Forest Dept. Sabah.

Burgess, P.F. 1971. The effect of logging on hill dipterocarp forests. *Malay. Nat. J.* 24: 231-237.

Burgess, P.F. 1972. Studies on the regeneration of the hill forests of the Malay Peninsula. The phenology of dipterocarps. *Malay. Forester* 35: 103-123.

Burgess, P.F. 1973. The impact of commercial forestry on the hill forests of the Malay Peninsula. In *Proceedings of the symposium on biology, resources and national development.* (eds. E. Soepadmo and K.G. Singh). Malay. Nat. Soc., Kuala Lumpur.

Burgess, P.F. 1975. *Silviculture in the hill forests of the Malay Peninsula.* Malay For. Dept. Research Pamphlet 66.

Burgess, P.F. 1988. *Natural forest management for sustainable timber production in the Asia/Pacific region.* Report to ITTO.

Burghouts, T., Ernsting, G., Korthals, G. and de Vries, T. 1992. Litterfall, leaf decomposition, and litter invertebrates in primary and selectively logged dipterocarp forest in Sabah, Malaysia. In *Tropical rain forest: disturbance and recovery* (eds. A.G. Marshall and M.D. Swaine), pp. 407-416. The Royal Society, London.

Burhannudin, 1980. Pengamatan terhadap ikan gelodok *Periophthalmodon schlosseri* di Muara Sungai Banyuasin. In *Sumber daya hayati bahari* (ed. Burha-nuddin, M.K. Moosa, and H. Razak), pp. 117-124. Lembaga Oseanologi Nasional, Jakarta.

Burhannudin and Martosewojo, S. 1978. Pengamatan terhadap ikan gelodok. *Periophthalmus koelreuteri* (Pallas) di Pulau Pari. In *Prosiding seminar ekosistem hutan mangrove* (ed. S. Soemodihardjo, A. Nontji, A. Djamali, pp. 86-92. Lembaga Oseanologi Nasional, Jakarta.

Burkholder, P.R. 1973. The ecology of marine antibiotics and coral reefs. In *Biology and geology of coral reefs. Vol. 2, Biology 1.* (eds. O.A. Jones and R. Endean), Academic Press, New York, San Francisco, London.

Burkill, I.H. 1935/1966. *A dictionary of the economic products of the Malay Peninsula.* Government Printing Office, Singapore.

Burley, J. 1991. A floral inventory of West Kalimantan. Project proposal.

Burnham, C.P. 1984. Soils. In *Tropical rain forests of the Far East* (ed. T.C. Whitmore), pp. 137-152. Oxford University Press.

Burrett, C., Duhig, N., Berry, R. and Varne, R. 1991. Asian and south-western Pacific continental terranes derived from Gondwana and their biological significance. *Aust. Syst. Bot.* 4: 13-24.

Burrett, C., Long, J. and Strait, B. 1990. Early-Middle Palaeozoic biogeography of Asian terranes derived from Gondwana. In *Palaeozoic palaeogeography and biogeography.* (eds. W.S. McKerrow and C.R. Scotese) pp. 163-174, Geol. Soc. Memoir No 12.

Caldecott, J.O. 1980. Habitat quality and populations of two sympatric gibbons (Hylobatidae) on a mountain in Malaya. *Folia Primatol.* 33: 291-309.

Caldecott, J. 1988a. *Hunting and wildlife management in Sarawak.* IUCN Tropical Forest Programme.

Caldecott, J. 1988b. Climbing towards extinction. *New Scientist* (9 June): 62-66.

Caldecott, J.O. and Nyaoi, A. 1985. Sarawak's wildlife: a resource to be taken seriously. *Sarawak Gazette* 111: 31-32.

Carcasson, R.H. 1977. *A field guide to the coral reef fishes of the Indian and West Pacific Oceans.* Collins, London.

Carpenter, R.A. 1981. *Assessing tropical forest lands, their suitability for sustainable uses.* Tycooly International Publishing Ltd, Dublin.

Carter, S. 1985. Comparison of bird numbers in primary and selectively-felled tropical rain forest in Brunei. *Brunei Mus. J.* 6(1): 125-130.

Cassells, D., Hamilton L. and Saplaco, S.R. 1983. Understanding the role of forests in watershed protection. In *Natural systems for development - what planners need to know* (ed. R.A. Carpenter), pp. 52-98, Macmillan Publishing Co.

Caufield, C. 1985. *In the rainforest.* W. Heinemann Ltd.

Caughley, G. 1977. *Analysis of vertebrate populations.* J.Wiley and Sons, New York.

Chadwick, A.C. and Sutton. S.L. (eds.) 1984. *Tropical rain forest: the Leeds symposium.*

Leeds Philosophical and Literary Society.
Chaeruddin, G. 1990. Fisheries of the Sungai Negara wetlands. In *Conservation of Sungai Negara wetlands, South Kalimantan.* pp. 173-200. PHPA/AWB-Indonesia and KPSL-UNLAM.
Chaeruddin, G, Rusmayadi, G., Djasmani, H. 1990. *Trade in python and other reptile skins in South Kalimantan, Indonesia.* Report prepared for the World Conservation Monitoring Centre and CITES. KPSL-UNLAM, Banjarbaru.
Chai, P.P.K. 1975a. Mangrove forests in Sarawak. *Malay. Forester* 38(2): 108-134.
Chai, P.P.K. 1975b. The mangrove trees and shrubs of Sarawak. *Malay. Forester* 38(3): 187-207.
Chai, P.P.K. 1978. Ethnobotany, Part II. *Sarawak Mus. J.* 26(47): 243-270.
Chai, P.P.K. 1982. *Ecological studies of mangrove forest in Malaysia.* Ph.D. thesis, Universiti Malaya, Kuala Lumpur.
Chai, P.K. and Lai, K.K. 1980. Management and utilization of the mangrove forest of Sarawak. In *Asean symposium on mangrove environment: research and management.* Kuala Lumpur, Malaysia.
Chai, P.K., Lee, M.H. and Ismawi, O.Hj. 1986 *Sarawak plants of economic importance.* Dun Special Select Commitee on Flora and Fauna, Sarawak.
Chambers, M.J. 1980. The environment and geomorphology of deltaic sedimentation. In *Tropical ecology and development.* (ed. J. Furtado), pp. 1091-1095, Univ. of Malaya, Kuala Lumpur.
Chan, H.T., Ujang, R. and Putz, F.E. 1982. A preliminary study on planting of *Rhizophora* species in an *Avicennia* forest at the Matang mangroves. In *Proceedings Seminar II Ekosistem Mangrove,* pp. 340-345. Lembaga Oseanologi Nasional, Jakarta.
Chan, H.T. 1984. *The Malaysian mangrove ecosystem.* Report presented at the second regional task force meeting of the UNDP/UNESCO mangrove regional project Bogor, Indonesia.
Chan, L., Kavanagh, M., Cranbrook, Earl of, Langub, J. and Wells, D.R. 1985. *Proposals for a conservation strategy for Sarawak.* WWF Malaysia/State Planning Unit of Sarawak, Kuching.
Chan, N. W. 1986. Drought trends in northwestern Peninsular Malaysia: is less rain falling? *Wallaceana* 44:8-9.
Chang T.T. 1984. Conservation of rice genetic resources: luxury or necessity? *Science* 224: 251-256.
Chapin, G. and Wasserstrom R. 1981. Agricultural production and malaria resurgence in Central America and India. *Nature* 293: 181-185.
Chapman, E.C. 1975. Shifting agriculture in tropical forest areas of Southeast Asia. In *The use of ecological guidelines for development in tropical areas of Southeast Asia.* pp. 120-135. IUCN, Gland.
Chapman, P. 1980. *Studies of the invertebrate cave fauna of the Gunung Mulu National Park, Sarawak, with a discussion of the possible mechanisms involved in the evolution of tropical cave faunas.* M.Sc. thesis, Univ. of Bristol.
Chapman, P. 1981. The biology of caves in the Gunung Mulu National Park, Sarawak. *Trans. Brit. Cave Res. Assoc.* 7(3): 141-149.
Chapman, P. 1982. The ecology of caves in the Gunung Mulu National Park, Sarawak. *Trans. Brit. Caves Res. Assoc.* 9(2): 142-162.
Chapman, P. 1983. Species diversity in a tropical cave ecosystem. *Proc. Univ. Bristol Spelaeol. Soc.* 16(3): 201-213.
Chapman, P. 1984. The invertebrate fauna of the caves of Gunung Mulu National Park. *Sarawak Mus. J.* 30(51): 18
Chapman, P. 1985. Cave-frequenting vertebrates in the Gunung Mulu National Park, Sarawak. *Sarawak Mus. J.* 34(55): 101-113.
Chapman, V.J. (ed.) 1977. *Ecosystems of the world. I. Wet coastal ecosystems.* Elsevier, Amsterdam.
Cheke, A.S., Nanakorn, W. and Yankoses, C. 1979. Dormancy and dispersal of secondary forest species under the canopy of a primary tropical rain forest in Northern Thailand. *Biotropica* 11(2): 88-95.
Cheng, L. (ed.) 1976. *Marine insects.* North-Holland, Amsterdam.
Chin, F.H. 1981. Edible and poisonous fungi from the forests of Sarawak. Part 1. *Sarawak Mus. J.* 29 : 211-225.
Chin, F.H. 1988. Edible and poisonous fungi from the forests of Sarawak. Part II. *Sarawak Mus. J.* 39(60): 195-202.
Chin, L. 1980. *Cultural heritage of Sarawak.* Sarawak Museum, Kuching.
Chin, S.C. 1977. The limestone hill flora of Malaya I. *Garden's Bull. Singapore* 30: 166-219.
Chin, S.C. 1979. The limestone hill flora of Malaya II. *Garden's Bull. Singapore* 32: 64-

203.

Chin, S.C. 1982. The significance of rubber as a cash crop in a Kenyah swidden village in Sarawak. *Federn Mus. J.* 27:23-38.

Chin, S.C. 1983a. The limestone hill flora of Malaya III. *Garden's Bull. Singapore* 35(2): 137-190.

Chin, S.C. 1983b. The limestone hill flora of Malaya IV. *Garden's Bull. Singapore* 36(1): 31-91.

Chin, S.C. 1984. Kenyah tops and top playing, an integral part of the agricultural cycle. *Sarawak Mus. J.* 33: 33-53.

Chin, S.C. 1985. Agriculture and resource utilization in a lowland rainforest Kenyah community. *Sarawak Mus. J. Special Monograph* No. 4.

Chin, S.C. 1987. Deforestation and environmental degradation in Sarawak. *Wallaceana* 48/49: 6-8.

Chin, S.C. 1988. Shifting cultivation, a tropical landuse system. *Wallaceana* 49: 3-7.

Chin, S.C. and Chua, T.H. 1984. The impact of man on a Southeast Asian tropical forest. *Malay. Nat. J.* 36: 255-269.

Chiu, S.C. 1979. Biological control of the brown planthopper. In *Brown planthopper: a threat to rice production in Asia.* pp. 335-355. International Rice Research Institute, Los Banos.

Chivers, D.J. (ed.) 1980. *Malayan forest primates. Ten years study in tropical rain forest.* Plenum Press, New York and London.

Chivers, D.J., Burton, K.M. and Marshall, J.T. 1987. *Some observations on the primates of Kalimantan Tengah, Indonesian Borneo.* Unpublished expedition report.

Chou, L.M. and Leong, C.F. 1984. Activity cycles of the house geckos, *Cosymbotus platyurus* and *Hemidactylus frenatus*. *Malay. Nat. J.* 36: 247-252.

Christensen, B. 1979. *Mangroves - what are they worth?* FAO, Rome.

Christensen, B. and Wium-Anderson S. 1977. Seasonal growth of mangrove trees in southern Thailand I. The phenology of *Rhizophora apiculata* Bl. *Aquat. Bot.* 3: 281-286.

Church, G. 1962. The reproductive cycles of the Javanese house geckos, *Cosymbotus platurus*, *Hemidactylus frenatus* and *Peropus mutilatus*. *Copeia* 2: 262-269.

Church, G. and Lim, C.S. 1961. The distribution of three species of house geckos in Bandung (Java). *Herpetologica* 17: 119-201.

Clark, C.W. 1973. Profit maximization and the extinction of animal species. *J. Pol. Econ.* 81: 950-961.

Cochrane, G.R. 1969. Problems of vegetation change in western Viti Levu, Fiji. In *Settlement and encounter: geographical studies presented to Sir Grenfell Price* (eds. F. Gale and G.H. Lawton), pp. 115-147, Oxford Univ. Press, Melbourne.

Cockburn, P.F, 1978. Flora. In *Kinabalu, summit of Borneo* (eds. M. Luping, N. Chin and E.R. Dingley), pp. 179-198. Sabah Society Monograph.

Cockburn, P.F. 1976 *Trees of Sabah* (vol. 1). Sabah Forest Record No. 10. Borneo Literature Bureau, Jabatan Hutan, Sabah.

Cole, R. 1959. Temiar Senoi agriculture I and II. *Malay. For.* 22: 191-207, 260-271.

Coley, P.D. 1983. Herbivory and defensive characteristics of tree species in a lowland tropical forest. *Ecological Monographs* 53(2): 209-233.

Colfer, C.J.P. 1981. Women, men and time in the forests of East Kalimantan. *Borneo Res. Bull.* 13(2): 75-85.

Colfer, C.J.P. 1983. Change and indigenous agroforestry in East Kalimantan. *Borneo Res. Bull.* 15(2): 3-21.

Colfer, C.J.P. 1983. Change and indigenous agroforestry in East Kalimantan (continued). *Borneo Res. Bull.* 15(2): 70-87.

Colfer, C.J.P. and Soedjito H. 1988. On resettlement: from the bottom up. In *Some ecological aspects of tropical forest of East Kalimantan* (ed. S. Soemodihardjo), pp. 87-105. Indonesian Institute of Sciences (LIPI), Jakarta.

Collette, B.B. and Talbot, F.H. 1972. Activity of coral reef fishes with emphasis on nocturnal-diurnal changeover. *Bull. Nat. Hist. Mus., Los Angeles County* 14.

Collier, W.L. 1977. Income, employment and food systems in Javanese coastal villages. South East Asia Series, No. 44. Athens.

Collier, W.L. 1979. *Social and economic aspects of tidal swampland development in Indonesia.* Occasional Paper No. 15. Development Studies Centre, Australian National University.

Collier, W.L. 1980. Resource use in the tidal swamps of Central Kalimantan: a case study of Banjarese and Javanese rice and coconut producers. In *Tropical ecology and development* (ed. J.I. Furtado), pp. 1047-1064. International Society of Tropical Ecology, Kuala Lumpur.

Collier, W.L., Hadikoesworo and Malingreau, M. 1979. Coastal resources and seafishing in Kalimantan. In *Economics of agriculture, seafishing and coastal resource use in Asia.* (eds. A. Librero and W.L. Collier) Philippines Council for Agriculture and Resource Research, Manila.

Collins, M. 1984. *Urban ecology: a teacher's resource book.* Cambridge University Press, Cambridge.

Collins, N.M. 1979a. Observations on the foraging activity of *Hospitalitermes umbrinus* (Haviland), (Isoptera: Termidae) in the Gunung Mulu National Park, Sarawak. *Ecological Entomology* 4: 231-238.

Collins, N.M. 1979b. A comparison of the soil macrofauna of three lowland forest types in Sarawak. *Sarawak Mus. J.* 27: 267-282.

Collins, N.M. 1980a. The distribution of soil macrofauna on the west ridge of Gunung (Mount) Mulu, Sarawak. *Oecologia* 44: 263-275.

Collins, N.M. 1980b. The habits and populations of terrestrial crabs (Brachyura: Gacarcinucoidea and Grapsoidea) in the Gunung Mulu National Park, Sarawak. *Zool. Meded. Leiden* 55(7): 81-85.

Collins, N.M. 1980c. The effect of logging on termite (Isoptera) diversity and decomposition processes in lowland dipterocarp forests. In *Tropical ecology and development* (ed. J.I. Furtado), pp. 114-121. International Society of Tropical Ecology, Kuala Lumpur.

Collins, N.M. 1983. Termite populations and their role in litter removal in Malaysian rain forests. In *Tropical rain forest: ecology and management* (eds. S.L. Sutton, T.C. Whitmore and A.C. Chadwick), pp. 311-325. Blackwell Scientific Publication, Oxford.

Collins, N.M. 1984. The termites (Isoptera) of the Gunung Mulu National Park with a key to the genera known from Sarawak. *Sarawak Mus. J.* 30: 65-87.

Collins, N.M. 1989. Termites. In *Tropical rain forest ecosystems.* Ecosystems of the World 14B. Elsevier, Amsterdam.

Collins, N.M., Anderson, J.M., Vallack, H.W. 1984. Studies on the soil invertebrates of lowland and montane rain forests in the Gunung Mulu National Park. *Sarawak Mus. J.* 30(51): 19-33.

Collins, N.M., Holloway, J.D. and Proctor, J. 1984. Notes on the ascent and natural history of Gunung Api, a limestone mountain in Sarawak. *Sarawak Mus. J.* 33(54): 219-235.

Collins, N.M. and Morris, M.G. 1985. *Threatened swallowtail butterflies of the world.* The IUCN Red Data Book. IUCN. Gland.

Collins, N.M., Sayer, J.A. and Whitmore, T.C. (eds.) 1991. *The conservation atlas of tropical forests: Asia and the Pacific.* Macmillan Press Ltd., London and Basingstoke.

Collis, M. 1966. *Raffles.* Faber and Faber, London.

Combaz, A. and M. de Matharel. 1978. Organic sedimentation and genesis of petroleum in the Mahakam Delta, Borneo. *Am. Assoc. Petrol. Geol.* 62(9): 1684-1695.

Conley, W. 1974. Kenyah cultural themes and inter-relationships. In *The peoples of central Borneo* (ed. J. Rousseau), pp. 303-309. *Sarawak Mus. J.* vol. 22. Special issue.

Conley, W. 1975. *The Kalimantan Kenyah. A study of tribal conversion in terms of dynamic cultural themes.* World Mission.

Connell, J.H. 1972. Community interactions on marine rocky intertidal shores. *Ann. Rev. Ecol. Syst.* 3: 169-132.

Connell, J.H. 1978. Diversity in tropical rain forests and coral reefs. *Science* 199: 1302-1309.

Connell, D.W. and Miller, G.J. 1984. *Chemistry and ecotoxicology of pollution.* John Wiley, New York.

Conway, G.R. (ed.) 1982. *Pesticide resistance and world food production.* Imperial College, University of London, London.

Conway, G.R. 1985. Agroecosystem analysis. *Agric. Admin.* 20: 31-55.

Cooke, F.P., Brown, J.P. and Mole, S. 1984. Herbivory, foliar enzyme inhibitors, nitrogen and leaf structure of young and mature leaves in a tropical forest. *Biotropica* 16(4): 257-263.

Corbet, G.B. and Hill, J.E. 1980. *A world list of mammalian species.* British Museum (Natural History), Comstock Publishing, London.

Corner, E.J.H. 1940. *Wayside trees of Malaya* (2 vols). Government Printing Office, Singapore.

Corner, E.J.H. 1946. Suggestions for botanical progress. *New Phytol.* 45: 185-192.

Corner, E.J.H. 1949. The durian theory on the origin of the modern tree. *Ann. Bot.* 52: 367-414

Corner, E.J.H. 1966. *The natural history of palms.* Weidenfeld and Nicolson, London. pp. 201-224.

Corner, E.J.H. 1978a. The freshwater swamp-forest of South Johore and Singapore. *Garden's Bull. Singapore Suppl.* 1.

Corner, E.J.H. 1978b. Plant life. In *Kinabalu, summit of Borneo*. (eds. M. Luping, N. Chin and E.R. Dingley), pp. 112-178. The Sabah Society, Kota Kinabalu.

Cousens, J.E. 1974. *An introduction to woodland ecology*. Oliver and Boyd, Edinburgh.

Crain, J.B. 1973. Mengalong Lun Dayeh agricultural organisation. *Brunei Mus. J.* 3: 1-25.

Crain, J.B. 1978. The Lun Dayeh. In *Essays on Borneo societies* (ed. V.T. King), pp. 123-142. Oxford University Press. Hull Monographs on South-East Asia No. 7.

Cramb, R. A. 1989 Shifting cultivation and resource degradation in Sarawak: perception and policies. *Borneo Res. Bull.* 21(1): 22-49.

Cranbrook, Earl of, 1981. The vertebrate faunas. In *Wallace's Line and plate tectonics* (ed. T.C. Whitmore), pp. 57-69, Oxford University Press.

Cranbrook, Earl of, 1982. The significance of the vertebrate fauna. In *Gunung Mulu National Park a management and development plan* (eds. J.A.R. Anderson, A.C. Jermy, and Earl of Cranbrook). Royal Geographical Society, London.

Cranbrook, Earl of, 1984. Report on the birds' nest industry in the Baram District and Niah, Sarawak. *Sarawak Mus. J.* 34: 145-170.

Crawfurd, J. 1820. *History of the Indian archipelago; containing an account of the manners, arts, languages, religions, institutions and commerce of its inhabitants*. Constable, Edinburgh.

Crisswell, C.N. 1978. *Rajah Charles Brooke, monarch of all he surveyed*. Oxford University Press.

Crome, F.H. J. and Brown, H. E. 1979. Notes on the social organization and breeding of the orange-footed scrubfowl *Megapodius reinwardt. Emu* 79: 111-119.

Crompton, R.H. 1984. Foraging, habitat structure and locomotion in two species of *Galago*. In *Adaptations for foraging in nonhuman primates*. (eds. P.S. Rodman and J.G.H. Cant), pp. 73-111. Columbia University Press, New York.

Crowther, J. 1982a. The thermal characteristics of some West Malaysian rivers. *Malay. Nat. J.* 35: 99-109.

Crowther, J. 1982b. Ecological observations in a tropical karst terrain, West Malaysia. I. Variations in topography, soils and vegetation. *J. Biogeogr.* 9: 65-78.

Crowther, J. 1986. Karst environments and ecosystems in Peninsular Malaysia. *Malay. Nat. J.* 39: 231-257.

Crowther, J. 1987a. Ecological observations in a tropical karst terrain, West Malaysia. II. Rainfall interception, litterfall and nutrient cycling. *J. Biogeogr.* 14: 145-155.

Crowther, J. 1987b. Ecological observations in a tropical karst terrain, West Malaysia. III. Dynamics of the vegetation-soil-bedrock system. *J. Biogeogr.* 14: 157-164.

Croxall, J.P. 1979. *The montane birds of Gunung Mulu*. Paper presented at Gunung Mulu symposium. Royal Geographical Society, London.

Croxall, J.P., Evans, P.G.H. and Schreiber, R.W. 1982. *Status and conservation of the world's seabirds*. ICBP Technical Publication No. 2., ICBP, Cambridge.

Curtin, S.H. 1976. Niche separation in sympatric Malaysian leaf-monkeys (*Presbytis obscura* and *Presbytis melalophos*). *Yearbook of Physical Anthropology* 20: 421-439.

Curtin, S.H. 1980 Dusky and banded leaf monkeys. In *Malayan forest primates: ten years study in tropical rain forest*. (ed. D.J. Chivers), pp. 107-145. Plenum Press, New York and London.

Dahl, A.L. 1981. Monitoring coral reefs for urban impact. *Bull. Mar. Sci.* 31: 544-551.

Dahuri 1990 Development activities and coastal environment in East Kalimantan. Unpubl. ms.

Daniel, J.G. and Kulasingham, A. 1974. Problems arising from large scale forest clearing for agricultural use: the Malaysian experience. *Malay. Forester* 37: 152-160.

Danser, B.H. 1928. The Nepenthaceae of the Netherlands Indies. *Bull. Jard. Bot. Buitenzorg*, Serie III. 9(3-4): 250-438.

Daroesman, R. 1979. An economic survey of East Kalimantan. *Bull. Indonesian Economic Studies* 15(3): 43-82.

Daroesman, R. 1981. Vegetative elimination of alang-alang. *Bull. Indonesian Economic Studies* 17(1): 83-107.

Darus, A. and Stuebing, R. 1986. Vertebrate exotics in Sabah. *Sabah Soc. J.* 8(2): 296-307.

Dasmann, R.F. and Poore, D. 1979. *Ecological guidelines for balanced land use, conservation and development in high mountains*. UNEP/IUCN/WWF.

Davidson, J. 1985. *Economic use of tropical moist forests*. Commission on Ecology Papers No.9. IUCN.

Davidson, J. 1987. *Conservation planning in Indonesia's transmigration programme. Case studies from Kalimantan*. The IUCN Tropical Forest Programme, IUCN. Gland.

Davidson, J., Tho, Y.P. and Bijleveld, M. 1985. *The future of tropical rain forests in South East Asia*. Commission on Ecology Papers No.10. IUCN.

Davies, G. 1982. Distribution, abundance and conservation of simian primates in Borneo. In *Primate conservation in the wild* (ed. D. Harper), pp. 122-148, Univ. of Leicester.

Davies, G. and Payne, J. 1982. *A faunal survey of Sabah*. WWF, Malaysia, Kuala Lumpur.

Davis, D.D. 1962. Mammals of the lowland rain forest of North Borneo. *Bull. Nat. Mus. Singapore* 31: 1-129.

Davis, G.E. 1977. Anchor damage to a coral reef on the east coast of Florida. *Biol. Cons.* 11: 29-34.

Davis, T.A., Sudarsip, H. and Darwis, S.N. 1985. *Coconut research in North Sulawesi*. Coconut Research Institute, Manado, Indonesia.

Davison, G.W.H. 1980. Territorial fighting by lesser mouse-deer. *Malay. Nat. J.* 34(1): 1-6.

Davison, G.W.H. 1981. Sexual selection and the mating system of *Argusianus argus* (Aves: Phasianidae). *Biol. J. Linn. Soc.* 15: 91-104.

Day, M.J. 1981. Rock hardness and landform development in the Gunong Mulu National Park, Sarawak, E. Malaysia. *Earth surface processes and landforms.* 6: 165-172.

Deeleman-Reinhold, C.L. 1983. Leaf-dwelling Pholcidae in Indo-Australian rain forests. In *Proceedings of the 9th International Congress of Arachnology*. pp. 45-48. Panama.

Delsman, H.C. 1929. The distribution of freshwater eels in Sumatra and Borneo. *Treubia* 11: 287-292.

Deneven, W.M. and Padoch, C. 1988. *Swidden-fallow agroforestry in the Peruvian Amazon.* Advances in Economic Botany 5. New York Botanical Garden, New York.

Dent, D. 1986. *Acid sulphate soils: a baseline for research and development*. International Institute for Land Reclamation and Improvement/ILRI Publication No. 39.

Dentan, R.K. 1968. *The Semai. A nonviolent people of Malaya*. Holt, Rinehart and Winston, New York.

Deshmukh, I 1986. *Ecology and tropical biology.* Blackwell Scientific Publications, Oxford.

Desowitz, R.S. 1980. Epidemiological-ecological interactions in savanna environments. In *Human ecology in savanna environments* (ed. D.R. Harris), pp. 457-477. Academic Press, London.

De Wulf, R. 1982. *A field guide to common Sumatran trees*. FAO Special Report FO/INS/78/ 061. Bogor.

Diamond, A.W. and Filion, F.L.(eds). 1987. *The value of birds*. ICBP Technical Publication No. 6. Cambridge.

Diamond, A.W. and Lovejoy, T.E. (eds.) 1985. *Conservation of tropical forest birds.* ICBP Technical Publication No. 4. Cambridge.

Diamond, J.M. 1972. *Avifauna of the eastern highlands of New Guinea.* Nuttall Ornithological Club, Publications 12. Cambridge, Mass.

Diamond, J.M. 1973. Distributional ecology of New Guinea birds. *Science* 179: 759-765.

Diamond, J.M. 1975. The island dilemma: lessons of modern biogeographic studies for the design of natural reserves. *Biol. Cons.* 7: 129-146.

Diamond, J.M. 1978. Niche shifts and the rediscovery of interspecific competition. *Am. Scient.* 66(3): 322-331.

Diamond, J.M. 1984. Biological principles relevant to protected areas design in the New Guinea region. In *National parks, conservation, development: the role of protected areas in sustaining society* (eds. J.A. McNeely and K.R. Miller), pp. 330-332, IUCN/ Smithsonian Institution Press, Washington D.C.

Dickinson, R.E. 1981. Effects of tropical deforestation on climate. In *Blowing in the wind: deforestation and long-range implications.* Studies in the Third World Societies, No. 14. College of William and Mary, Williamburg.

Dieterlen, F. 1982. *Fruiting seasons in the rain forest of Eastern Zaire and their effect upon reproduction.* Poster presentation at tropical rain forest symposium. Leeds University, Leeds.

Dilmy, A. 1965. Ecological data from the Sampit area. In *Proceedings of the symposium on humid tropics vegetation, Ciawi, Indonesia.* UNESCO.

Dilmy, A. 1971. *Natural products of the lowland tropical forests (mainly of Kalimantan).* Paper presented to pre-congress conference in Indonesia: Planned utilization of the lowland tropical forests, Jakarta.

Dinerstein, E. 1986. Reproductive ecology of fruit bats and the seasonality of fruit pro-

duction in a Costa Rican cloud forest. *Biotropica* 18(4): 307-318.
Ding Hou 1958. Rhizophoraceae. *Flora Malesiana Ser.I.* 5:429-493.
Ditlev, H. 1980. *A field guide to the reef-building corals of the Indo Pacific.* Backhuys, Rotterdam.
Dixon, J.A. and Sherman, P.B. 1990. *Economics of protected areas: a new look at benefits and costs.* Earthscan, London.
Dixon, J.A., Carpenter, R.A., Fallon, L.A., Sherman, P.B. and Manipomoke, S. 1988. *Economic analysis of the environmental impacts of development projects.* Earthscan, London.
Djasmani, and Rifani. 1988. *Monitor lizard populations in Kalimantan.* Report prepared by KPSL UNLAM and PSL UNTAN for CITES.
Dobias, R.J., Tech, T., Wangwacharakul, V. and Sangswang, N. 1988. *Beneficial use quantifications of Huai Kha Khaeng/Thung Yai Naresuan Wildlife Sanctuary complex: executive summary and main report.* WWF, Thailand.
Doi, T. 1988. Present status of the large mammals in the Kutai National Park, after a large scale fire in East Kalimantan, Indonesia. In *A research on the process of earlier recovery of tropical rain forest after a large scale fire in Kalimantan Timur, Indonesia* (eds. H.Tagawa and N. Wirawan), pp. 2-11, Occ. papers No. 14, Kagoshima Univ.
Donner, W. 1987. *Land use and environment in Indonesia.* University of Hawaii Press, Honolulu.
Doty, M.S., Soeriaatmadja, R.E., and Soegiarto, A. 1963. Penelitian laut di Indonesia. *Mar. Res. Indonesia* 5: 201-219.
Douglas, I. 1992. Hydrological and geomorphic changes following rainforest disturbance with special references to studies in Borneo. In *Forest biology and conservation in Borneo* (eds. G. Ismail, M. Mohamed and S. Omar), pp. 261-303. Center for Borneo Studies Publication No. 2.
Douglas, I., Spencer, T., Greer, T., Bidin, K., Sinun, W. and Wong, W.M. 1992. The impact of selective commercial logging on stream hydrology, chemistry and sediment loads in the Ulu Segama rain forest, Sabah, Malaysia. *Phil. Trans. R. Soc. Lond. B* 335: 397-406.
Dove, M.R. 1980a. Development of tribal land-rights in Borneo: the role of ecological factors. *Borneo Res. Bull.* 12(1): 3-19.
Dove, M.R. 1980b. The swamp rice swiddens of the Kantu's of west Kalimantan, Indonesia. In *Tropical ecology and development.* (ed.J.I. Furtado), pp. 953-956. Kuala Lumpur, Malaysia.
Dove, M.R. 1981. *Subsistence strategies in rain forest agriculture, the Kantu' at Tikul Batu.* Ph.D. thesis, Stanford University.
Dove, M.R. 1983. Theories of swidden agriculture, and the political economy of ignorance. *Agroforestry Systems* 1: 85-99.
Dove, M.R. 1983. Forest preference in swidden agriculture. *Trop. Ecol.* 24(1): 122-142.
Dove, M.R. 1985. *Swidden agriculture in Indonesia. The subsistence strategies of the Kalimantan Kantu.* Mouton Publishers, Berlin, New York, Amsterdam.
Dove, M.R. 1986a. Peasant versus government, perception and use of the environment: a case study of Banjarese ecology and river basin development in south Kalimantan. *J. Southeast Asian Studies* 17(1): 113-136.
Dove, M.R. 1986b. The practical reason of weeds in Indonesia: peasant vs. state views of *Imperata* and *Chromolaena*. *Human Ecology* 14(2): 163-190.
Dove, M.R. 1986c. Plantation development in West Kalimantan II: the perceptions of the indigenous population. *Borneo Res. Bull.* 18(1): 3-27.
Dove, M.R. and Martopos, 1987. *Manusia dan alang-alang di Indonesia.* Gadjah Mada University Press, Yogyakarta.
Dransfield, J. 1971. *Report of a field trip to Kalimantan Selatan, report on the palms.* Regional Centre for Tropical Biology (BIOTROP), Bogor.
Dransfield, J. 1974. *A short guide to rattans.* Regional Centre for Tropical Biology (BIOTROP), Bogor.
Dransfield, J. 1979. A manual of the rattans of the Malay peninsula. *Mal. For. Rec.* 29: 7-36.
Dransfield, J. 1981. The biology of Asiatic rattans in relation to the rattan trade and conservation. In *The biological aspects of rare plant conservation* (ed. H. Synge), 179-186. John Wiley and Sons Ltd.
Dransfield, J. 1984. *The rattans of Sabah.* Forest Department, Sabah.
Dransfield, J. 1988. Prospects for rattan cultivation. *Advances in Economic Botany* 6: 190-200.
Dransfield, J. 1992. Rattans in Borneo: botany and utilisation. In *Forest biology and conservation in Borneo* (eds. G. Ismail,

M. Mohamed and S. Omar), pp. 22-31. Center for Borneo Studies Publication No. 2.
Dransfield, S. 1992. Bamboos of Borneo and their uses. In *Forest biology and conservation in Borneo* (eds. G. Ismail, M. Mohamed and S. Omar), pp. 14-21. Center for Borneo Studies Publication No. 2.
Dransfield, S. 1992. *The Bamboos of Sabah*. Sabah For. Rec. 14.
Driessen, P.M. 1978. Peat soils. In *Soils and rice*, pp. 763-779, IRRI, Los Banos, Philippines.
Driessen, P.M., Buurman, P., and Permadhy, S. 1976. The influence of shifting cultivation on a "podzolic" soil from Central Kalimantan. *Soils Res. Inst., Bogor, Bull.* 3: 95-115.
Dring, J. 1980. *Preliminary report on the frogs of Gunung Mulu National Park, Sarawak.* Unpubl. ms.
Dring, J. 1987. Bornean tree frogs of the genus *Philautus* (Rhacophoridae). *Amphibia-Reptilia* 8: 19-47.
Dubinsky, Z. (ed.) 1990. *Coral reefs. Ecosystems of the world 25*. Elsevier, Amsterdam.
DuBois, R., Berry, L. and Ford, R. 1984. Catchment land use and its implications for coastal resources conservation. In *A casebook of coastal management*. Research Planning Institute, Columbia, S.C.
Duckett, J.E. 1976. Plantations as a habitat for wildlife in Peninsular Malaysia with particular reference to the oil palm (*Elaeis guineensis*). *Malay. Nat. J.* 29: 176-182.
Duckett, J.E. 1982. The plantain squirrel in oil palm plantations. *Malay. Nat. J.* 36: 87-98.
Duckham, A.N. and Masefield G.B. 1970. *Farming systems of the world*. Chatto and Windus, London.
Dunn, I.G. and Otte, G. 1983. *Fisheries development in the middle Mahakam area*. East Kalimantan Transmigration Area Development (TAD) Project. TAD report No. 13, for the Ministry of Transmigration, Jakarta.
Eavis, A.J. (ed.) 1981. *Caves of Mulu '80. The limestone caves of the Gunung Mulu National Park Sarawak*. The Royal Geographical Society, London.
Edwards, P.J. 1982. Studies of mineral cycling in a montane rainforest in New Guinea. V. Rates of cycling in throughfall and litterfall. *J. Ecol.* 70: 807-827.
Edwards, P.J. and Grubb, P.J. 1977. Studies of mineral cycling in a montane rain forest in New Guinea. I. The distribution of organic matter in the vegetation and soil. *J. Ecol.* 65: 943-969.
Edwards, P.J. and Grubb, P.J. 1982. Studies of mineral cycling in a montane rainforest in New Guinea. IV. Soil characteristics and the division of mineral elements between the vegetation and soil. *J. Ecol.* 70: 649-666.
Ehrenfeld, D. 1986. Thirty million cheers for diversity. *New Scientist* 110: 38-43.
Ehrenfeld, D. 1988. Why put a value on biodiversity. In *Biodiversity* (ed. E.O. Wilson), pp. 212-216. National Academy Press, Washington, D.C.
Ehrlich, P. and Ehrlich, A.H. 1970. *Population, resources, environment, issues in human ecology*. W.H. Freeman and Co., San Francisco.
Ehrlich, P. and Erhlich, A. 1981. *Extinction, the causes and consequences of the disappearance of species*. St. Edmundsbury Press.
Eichelberger, R. 1924. Regenverteilung, Pflanzendecke und Kulturentwicklung in der ostindischen inselwelt. *Geographische Zeitschrift* 30: 103-116.
Eisenberg, J.F. and Seidensticker, J. 1976. Ungulates in southern Asia: a consideration of biomass estimates for selected habitats. *Biol. Conserv.* 10: 283-308.
Elliott, S. and Brimacombe, J. 1986. *The medicinal plants of Gunung Leuser National Park, Indonesia*. WWF, Gland.
Elton, C.S. 1966. *The pattern of animal communities*. Methuen, London.
Elton, C.S. 1973. The structure of invertebrate populations inside neotropical rain forest. *J. Anim. Ecol.* 42: 55-104.
Elton, C.S. 1975. Conservation and the low population density of invertebrates inside neotropical rain forest. *Biol. Cons.* 7:3-15.
Emmons, L.H. 1992. The roles of small mammals in tropical rainforest. In *Forest biology and conservation in Borneo* (ed. G. Ismail, M. Mohamed and S. Omar). pp. 512-513, Center for Borneo Studies Publication No. 2.
Emmons, L.H., Nias, J. and Biun, A. 1991. The fruit and consumers of *Rafflesia keithii* (Rafflesiaceae). *Biotropica* 23: 197-199.
Endert, F.H. 1925. *Verslag van de Midden-Oost-Borneo-expeditie, 1925*. Batavia.
Endert, F.H. (ed.) 1927. *Midden-Oost Borneo expeditie*. Weltevreden: Kolff.

Erwin, T.L. 1982. Tropical forests: their richness in Coleoptera and other arthropod species. *Coleopt. Bull.* 34: 305-322.
Erwin, T.L 1983. Tropical forest canopies, the last biotic frontier. *Bull. Entomol. Soc. Am.* 29(1) 14-19.
Erwin, T.L. 1988. The tropical forest canopy, the heart of biotic diversity. In *Biodiversity* (ed. E.O. Wilson), pp. 123-129. National Academy Press, Washington, D.C.
Erwin, T.L. and Scott, J.C. 1980. Seasonal and size patterns, trophic structure and richness of Coleoptera in tropical arboreal ecosystems: the fauna of the tree *Luehea seemanii* in the canal zone of Panama. *Coleopt. Bull.* 34: 305-322.
ESFIK, 1983. *The caves of the Sangkulirang peninsula, East Kalimantan.* Report of the French Speleological Expedition, Kalimantan 1982.
Euroconsult, 1984. *A preliminary assessment of peat development potential.* Final report to Ministry of Mines and Energy, Jakarta.
Euroconsult, 1986. *Nationwide survey of coastal and near-coastal swamplands.* Executive Report. Euroconsult, Arnhem, The Netherlands/BIEC, Bandung, Indonesia.
Eussen, J.H.H. and Soerjani, M. 1976. Problems and control of "Alang-alang" *Imperata cylindrica* (L.) Beauv. in Indonesia. In *Proceedings of the 5th Asian-Pacific weed science society conference.* pp. 58-65. Tokyo, Japan.
Eve, R. and Guigue, A.M. 1989. *Survey of the Mahakam river delta, East Kalimantan, with special reference to its waterbirds.* Asian Wetland Bureau Publication No. 45. Kuala Lumpur.
Ewel, J.J. 1986. Designing agricultural ecosystems for the humid tropics. *Ann. Rev. Ecol. Syst.* 17: 245-271.
FAO. 1978. *Forestry for rural communities.* FAO, Rome.
FAO. 1981. *The global assessment of tropical forest resources.* GEMS PAC Information series No. 3.
Faegri, K. and Pijl, L. van der. 1979. *The principles of pollination ecology.* Pergamon International Press, Oxford.
Farnsworth, N.R. 1988. Screening plants for medicines. In *Biodiversity* (ed. E.O. Wilson), pp. 83-97. National Academy Press, Washington, D.C.
Farnsworth, N.R. and Soedjarto, D.D. 1985. Potential consequences of plant extinction in the United States on the current and future availability of prescription drugs. *Econ. Bot.* 39(3): 231-240.
Farnsworth, N.R. and Soedjarto, D.D. 1988. *The global importance of medicinal plants.* Paper presented at the international consultation on the conservation of medicinal plants. Chiang Mai, Thailand.
Fasseur, C. 1979. Een koloniale paradox; De Nederlandse expansie in de Indonesische archipel in het midden van de negentiende eeuw (1830-1870). *Tijdschrift voor Geschiedenis* 92: 162-87.
Fenton, M.B. 1975. Acuity of echolocation in *Collocalia hirundinacea* (Aves: Apodidae), with comments on the distribution of echolocating swiftlets and molossid bats. *Biotropica* 7: 1-7.
Fenton, M.B. and Flemming, T.H. 1976. Ecological interactions between bats and nocturnal birds. *Biotropica* 8(2): 104-110.
Fernando, C.H. 1977. Investigations on the aquatic fauna of ricefields with special reference to Southeast Asia. *Geo. Eco. Trop.* 3: 169-188.
Fernando, C.H., Furtado, J.I. and Lim, R.P. 1980. The ecology of ricefields with special reference to the aquatic fauna. In *Tropical ecology and development* (ed. J.I. Furtado), pp. 943-951. Universiti Malaya Press, Kuala Lumpur.
Ffolliott, P. and Thames, J.L. 1983. *Environmentally sound small-scale forestry projects. Guidelines for planning.* CODEL/VITA Publications.
Flach, M. 1983. *The sago palm.* FAO Plant Production and Protection Paper No. 47. FAO, Rome.
Flach, M. 1987. Cropping systems for development of tidal lowlands. In *Proceedings of symposium on lowland development in Indonesia.* pp. 84-92. ILRI, The Netherlands.
Flach, M. and Schuiling, D.L. 1989. Revival of an ancient starch crop: a review of the sago palm. *Agroforestry Systems* 7: 259-281.
Flenley, J.R. 1979. *The equatorial rain forest: a geological history.* Butterworth, London.
Flenley, J.R. 1980. The quaternary history of the tropical rain forest and other vegetation of tropical mountains. In *Proceedings of the 4th Int. Palyn. Conf.* pp. 21-27, Lucknow.
Flenley, J.R. 1985. Quaternary vegetational and climatic history of Island South-East Asia. *Mod. Quat. Res. in Southeast Asia* 9: 55-63.
Flenley, J.R. and Morley, P.J. 1978. A minimum age for the deglaciation of Mount Kinabalu, East Malaysia. *Mod. Quat. Res.*

in Southeast Asia 9: 55-63.
Fogden, M.P.L. 1972. The seasonality and population dynamics of equatorial forest birds in Sarawak. *Ibis* 114(3): 307-342.
Fogden, M.P.L. 1974. A preliminary field study of the western tarsier, *Tarsius bancanus* Horsfield. In *Prosimian biology* (eds. R.D. Martin, G.A. Doyle and A.C. Walker), pp. 151-165. Duckworth and Co., London.
Fogden, M.P.L. 1976. A census of a bird community in tropical rainforest in Sarawak. *Sarawak Mus. J.* 24: 251-267.
Fogden, M. and Fogden, P. 1974. *Animals and their colours*. Eurobook Ltd.
Fogden, S.C.L. and Proctor, J. 1985. Notes on the feeding of land leeches (*Haemadipsa zeylanica* Moore and *H. picta* Moore) in Gunung Mulu National Park, Sarawak. *Biotropica* 17(2):172-174.
Fong, F.W. 1988. The apung palm: traditional techniques of sugar tapping and alcohol extraction in Sarawak. *Principes* 33(1):
Ford, E.B. 1976. *Genetics and adaptation*. Studies in Biology No. 69. Edward Arnold, Southampton.
Foster, R.B. 1982. Famine on Barro Colorado Island. In *The ecology of a tropical forest* (eds. E.G. Leigh, A.S. Rand and D.M. Windsor), pp. 201-212, Smithsonian Institution, Washington.
Fox, J.E.D. 1972. *The natural vegetation of Sabah and natural regeneration of the dipterocarp forests*. Ph.D. thesis, University of Wales.
Fox, J.E.D. 1973. Dipterocarp seedling behaviour in Sabah. *Malay. Forester* 36: 205-214.
Fox, J.E.D. and Tan, T.H. 1971. Soils and forest on an ultrabasic hill north east of Ranau, Sabah. *J. Trop. Geog.* 32: 38-48.
Francis C.M. 1984. *Pocket guide to the birds of Borneo*. Sabah Society/ WWF Malaysia, Kuala Lumpur.
Francis, C.M. 1987. *The management of edible birds nest caves in Sabah*. Wildlife Section, Sabah Forestry Department.
Frankel, O.H. and Soulé, M. 1981. *Conservation and evolution*. Cambridge University Press, Cambridge.
Freeman, J.D. 1955. *Iban agriculture: a report on the shifting cultivation of hill rice by the Iban of Sarawak*. University of London.
Freeman, J.D. 1960. Iban augury. In *Birds of Borneo* (ed. B.E. Smythies), pp. 73-97. Oliver and Boyd, Edinburgh.
Freeman, J.D. 1970. *Report on the Iban*. The Athlone Press.
Furtado, J.I. 1980. Freshwater swamp and lake resources: a synthesis. In *Tropical ecology and development*. (ed. J.I. Furtado), pp. 797-798, International Society of Tropical Ecology, Kuala Lumpur.
Gagné, W.C. 1982. Staple crops in subsistence agriculture, their major insect pests, with emphasis on biogeographical and ecological aspects. *Monographiae Biologicae* 42: 229-259. Dr W. Junk Publishers, The Hague.
Galdikas, B.M.F. 1978. *Orangutan adaptation at Tanjung Puting Reserve, Central Borneo*. Ph.D. thesis, Univ. of California, Los Angeles.
Galdikas, B.M.F. 1979. Orangutan adaptation at Tanjung Puting Reserve: mating and ecology. In *The great apes*. (eds. D.Hamburg and E.R. McCown), pp. 194-233. Benjamin/Cummings, Menlo Park.
Galdikas, B.M.F. 1982. Orangutans as seed dispersers at Tanjung Puting, Central Kalimantan: implications for conservation. In *The orangutan, its biology and conservation* (ed. L.E.M de Boer), pp. 285-298. Dr W. Junk Publishers, The Hague.
Galdikas, B.M.F. 1983. The orangutan long call and snag crashing at Tanjung Puting reserve. *Primates* 24(3): 371-384.
Galdikas, B.M.F. 1985a. Orangutan sociality at Tanjung Puting. *Am. J. Primatol.* 9: 101-119.
Galdikas, B.M.F. 1985b. Short communication: crocodile predation on a proboscis monkey in Borneo. *Primates* 26(4): 495-496.
Galil, J. 1973. Pollination in dioecious figs. Pollination of *Ficus fistulosa* by *Ceratosolen hewitti*. *Garden's Bull. Singapore* 26: 303-311.
Garcia, J.R. 1984. Waterfalls, hydro-power, and water for industry: contributions from Canaima National Park. In *National parks, conservation and development: the role of protected areas in sustaining society* (eds. J.A. McNeely and K.R. Miller), pp. 588-591. IUCN/Smithsonian Institution Press, Washington D.C.
Gauld, I.D. 1987. Some factors affecting the composition of tropical ichneumonid faunas. *Biol. J. Linn. Soc.* 30: 299-312.
Gautier-Hion, A. 1990. Interactions among fruit and vertebrate fruit-eaters in an African tropical rainforest. In *Reproductive ecology of tropical forest plants* (eds. K.S. Bawa and M. Hadley). Man and the Biosphere Series, vol. 7: 219-230. UNESCO and The Parthenon Publishing Group.

George, W. 1987. Complex origins. In *Biogeographical evolution of the Malay archipelago* (ed. T.C. Whitmore), pp. 119-130. Oxford Science Publications.

Ghani, M.N.A, Ong, S.H. and Wessel, M. 1989. *Hevea brasiliensis*. In *Plant resources of South-East Asia* (eds. E. Westphal and P.C.M. Jansen), pp. 152-161. Pudoc, Wageningen.

Ghazally, I. 1988. *Rafflesia* of Sabah, case study for conservation. *Sabah Soc. J.* 9: 437-456.

Ghosh, D. 1983. *Sewage treatment fisheries in East Calcutta wetlands: low cost, resource-conserving option in environment repair*. Project report, Government of West Bengal, Calcutta, India.

Giesen, W. 1986. The status of *Scleropages formosus* (Asian arowana) in Indonesia's West Kalimantan Province. WWF, Bogor.

Giesen, W. 1987. *Danau Sentarum wildlife reserve. Inventory, ecology and management guidelines*. WWF/ PHPA, Bogor, Indonesia.

Giesen, W. 1990. Vegetation of the Negara river basin. In *Conservation of Sungai Negara wetlands, South Kalimantan*. pp. 1-51. PHPA/AWB-Indonesia and KPSL-UNLAM.

Gilbert, L.E. 1980. Food web organisation and conservation of neo-tropical diversity. In *Conservation biology* (eds. M.E. Soulé and B.A. Wilcox). Sinauer, Sunderland, Mass.

Gillis, M. 1988a. Indonesia: public policies, resource management and the tropical forest. In *Public policies and the misuse of forest resources* (eds. R. Repetto and M. Gillis), pp. 43-114. Cambridge University Press, Cambridge.

Gillis, M. 1988b. Malaysia: public policies and the tropical forest. In *Public policies and the misuse of forest resources* (eds. R. Repetto and M. Gillis), pp. 115-164. Cambridge University Press, Cambridge.

Gittins, S.P. 1979. The behaviour and ecology of the agile gibbon *Hylobates agilis*. Ph.D. thesis, University of Cambridge.

Gittins, S.P. 1980. Territorial behaviour in the agile gibbon. *Int. J. Primatol.* 1 (4): 381-399.

Gittins, S.P. 1982. Feeding and ranging in the agile gibbon. *Folia Primatol.* 38: 39-71.

Gittins, S.P. and Raemaekers, J.J. 1980. Siamang, lar and agile gibbons. In *Malayan forest primates: ten years study in tropical rain forest* (ed. D.J. Chivers), pp. 63-105. Plenum, New York.

Godoy, R.A. and Tan, C. F. 1988. *Agricultural diversification among smallholder rattan cultivators in southern Borneo, Indonesia*. Harvard Institute for International Development, Cambridge, U.S.A.

Glover, I.C. 1979. The late prehistoric period in Indonesia. In *Early South East Asia* (eds. R.B. Smith and W. Watson), pp. 167-184. Oxford University Press, Oxford.

Goeltenboth, F. (ed.) 1985. *Subsistence agriculture, improvement manual*. Wau Ecology Institute Handbook No. 10.

Goldammer, J.G. and Seibert, B. (in press). The impact of droughts and forest fires on tropical lowland rainforest of East Kalimantan. In *Fire in the tropical biota* (ed. J.D. Goldammer). Ecological Studies, Springer, Berlin.

Golley, F.B. 1983. Decomposition. In *Tropical rain forest ecosystems. Structure and function* (ed. E.B. Golley), pp. 157-165. Elsevier Scientific Publishing Company, Amsterdam.

Gomez-Pompa, A. and Burley, F.M. 1991. The management of natural tropical forests. In *Rain forest regeneration and management* (eds. Gomez-Pompa, T.C. Whitmore and M. Hadley), pp. 3-18. UNESCO and The Parthenon Publishing Group, Paris.

Goodland, R.J.A., Watson, C. and Ledec, G. 1984. *Environmental management in tropical agriculture*. Westview Press, Boulder.

Goor, C.P. van, and Kartasubrata, J. 1982. *Indonesian forestry abstracts*. Centre for Agricultural Publishing and Documentation, Wageningen.

Gorman, M. 1979. *Island ecology. Outline studies in ecology*. Chapman and Hall.

Goudie, A. 1981. *The human impact on the natural environment*. Basil Blackwell Ltd., Oxford

Gould, E. 1978. Foraging behaviour of Malaysian nectar-feeding bats. *Biotropica* 10(3): 184-193.

Gould, E., Andau, M. and Easton, E.G. 1987. Observations of earthworms in Sepilok forest, Sabah, Malaysia. *Biotropica* 19 (4): 370-372

Graaf, J. de. 1986. The economics of coffee. In *Economics of crops in developing countries No. 1*. Pudoc, Wageningen.

Graaf, N.R. de. 1987 A silvicultural system for natural regeneration of tropical rain forest in Suriname. *Series: Ecology and management of tropical rain forest in Suriname*. Wageningen.

Grandison, A.G.C. 1972. The Gunong Benom expedition 1967. Reptiles and amphibians of Gunong Benom, with a description of a new species of *Macroca*. *Bull. Brit. Mus. Nat. Hist. (Zool.)* 23: 43-107.

Green, J., Corbet, S.A., Watts, E. andOey, B.L. 1976. Ecological studies on Indonesian lakes. Overturn and restratification of Ranu Lamongan. *J. Zool. London.* 180: 315-354

Green, J. and Trett, M.W. 1989. *The fate and effects of oil in freshwater.* Elsevier, London and New York.

Greenland, D.J. 1975. Bringing the green revolution to shifting cultivation. *Science* 190: 841-844.

Greenpeace, 1989. *Sea turtles and Indonesia.* Paper prepared for the 7th conference of the parties to the Convention on International Trade in Endangered Species of Wild Flora and Fauna (CITES), Lausanne, Switzerland.

Greig-Smith, P. 1965. Notes on the quantitative description of humid tropical forest. In *Proceedings of symposium on ecological research in humid tropic vegetation.* pp. 227-234. Kuching, Sarawak.

Grist, D.H. 1953 *Rice.* Longman, London.

Groombridge, B. 1982. *The IUCN Amphibia-Reptilia red data book.* IUCN. Gland.

Groves, C.P. 1984. Mammal faunas and the palaeogeography of the Indo-Australian region. *Cour. Forsch. Inst. Seckenberg* 69: 267-273.

Groves, C.P. 1985. Plio-Pleistocene mammals in island Southeast Asia. *Mod. Quaternary Res. S.E. Asia* 9: 43-54.

Groves, C.P. 1992. Endemism in Bornean mammals. In *Forest biology and conservation in Borneo* (eds. G. Ismail, M. Mohamed and S. Omar), pp. 152-168. Center for Borneo Studies Publication No. 2.

Grubb, P.J. 1974. Factors controlling the distribution of forest types on tropical mountains: new factors and a new perspective. In *Altitudinal zonation in Malesia* (ed. J.R. Flenley), pp. 13-46. University of Hull, Hull.

Guppy, N. 1984. Tropical deforestation: a global view. *Foreign Affairs* 62(4): 929-965.

Haeruman, 1985. Future of tropical forests in Indonesia, resolving land resource conflicts. In *The future of tropical rain forests in South East Asia.* Commission on Ecology Papers No. 10: 111-114.

Haeruman, H. 1988. Conservation in Indonesia. *Ambio* 17(3): 218-222.

Hahude, A.G. Ongkosongo, O.S.R. and Praseno, D.P. 1979. *The oceanographic features of the coastal region between Jakarta and Cirebon.* LIPI workshop, Lembaga Oseanologi Nasional, Jakarta.

Haile, N.S. 1958. The snakes of Borneo, with a key to the species. *Sarawak Mus. J.* 8: 743-771.

Haile, N.S. 1975. Postulated late Cainozoic high sea level in the Malay peninsula. *J. Malay. Brit. Roy. Asiatic Soc.* 48: 78-88.

Hails, C.J. and Amirrudin, A. 1981. Food samples and selectivity of white-bellied swiftlets *Collocalia esculenta*. *Ibis* 123: 328-333.

Hallé, F. and Oldeman, R.A.A. 1970. Essai sur l'architecture et la dynamique du croissance des arbres tropicaux. *Collect. Monogr. Bot. Biol. Veget.* 6, Paris. [Translated into English by B.C.Stone], Univ.of Malaya Press, Kuala Lumpur (1975).

Hallé, F., Oldeman, R.A.A. and Tomlinson, P.B. 1978. *Tropical trees and forests: an architectural analysis.* Springer, Berlin.

Hallé, F. and Ng, F.S.P. 1981. Crown construction in mature dipterocarp trees. *Malay. Forester* 44: 222-233.

Halliday, T. and Adler, K. (eds.) 1987. *The encyclopedia of reptiles and amphibians.* Guild Publishing, London.

Halteren, P. van. 1979. The insect pest complex and related problems of lowland rice cultivation in South Sulawesi. *Indonesia. Meded. Landbouwhogeschool, Wageningen* 79:1-112.

Hamer, W.I. 1981. *Soil conservation.* Consultant's report. INS/78/006. Technical Note No. 7 and No. 10. UNDP.

Hamilton, A. 1930. *A new account of the East Indies.* London (2 vols).

Hamilton, L.S. and King, P.N. 1983. *Tropical forested watersheds: hydrologic and soils response to major uses or conversions.* Westview Press, Boulder, Colorado.

Hamilton, L.S. and Snedaker, S.C. (eds.) 1984. *Handbook for mangrove areas management.* UNEP and East-West Center.

Hamilton, W. 1989. Tectonics of the Indonesian region (with 1:5 million scale map). In *Geologi Indonesia J.A. Katili commemorative volume,* Ikatan Ahli Geologi Indonesia. pp. 35-88.

Hamzah, Z. 1978. Some observations on the effects of mechanical logging on regeneration, soil and hydrological conditions in East Kalimantan. *BIOTROP Spec. Publ.* 3: 73-78.

Hanbury-Tenison, A.R. and Jermy, A.C. 1979.

The RGS expedition to Gunung Mulu, Sarawak 1977-78. *Geog. J.* 145(2): 175-191.
Hanks, J. 1984. *Traditional lifestyles, conservation and rural development.* Commission on Ecology Papers No.7, IUCN.
Hanski, I. 1983. Distributional ecology and abundance of dung and carrion-feeding beetles (Scarabaeidae) in tropical rain forests in Sarawak, Borneo. *Acta Zool. Fenn.* 167: 1-45.
Hanski, I. 1989. Dung beetles. In *Tropical rain forest ecosystems. Ecosystems of the World 14B* (eds. H. Leith and M.J.A. Werger). Elsevier, Amsterdam.
Hanson, A.J. and Koesoebiono 1977. *Settling coastal swamplands in Sumatra: a case study for integrated resource management.* PSL Research Report 004, Institut Pertanian Bogor.
Hardenberg, J.D.F. 1936. On a collection of fishes from the estuary and the lower and middle course of the River Kapuas (W.Borneo). *Treubia* 15: 225-254.
Hardenberg, J.D.F. 1937. Hydrological and ichthyological observations in the mouth of the Kumai River (SW Borneo). *Treubia* 16: 1-14.
Hardjono, J.M. 1977. *Transmigration in Indonesia.* Oxford University Press, Kuala Lumpur.
Hardjosuwarno, S. 1980. *The impact of Cilacap refinery to the mangrove vegetation.* Mangrove Symposium, BIOTROP. Bogor.
Hardon, J.J. 1989. *Elaeis guineensis* Jacq. In *Plant resources of South-East Asia* (eds. E. Westphal and P.C.M. Jansen), pp. 118-123. Pudoc, Wageningen.
Harrison, C.J.O. (ed.) 1978. *Bird families of the world.* Peerage Books.
Harrison, J.L. 1956. Survival rates of Malayan rats. *Bull. Raffles Mus.* 27: 5-26.
Harrison, J.L. 1962a. The natural food of some Malayan mammals. *Bull. Nat. Mus. Singapore* 30: 5-18.
Harrison, J.L. 1962b. The distribution and feeding habits among animals in tropical rainforest. *J. Anim. Ecol.* 31: 53-64.
Harrison, J.L. 1965. *The effect of forest clearance on small mammals.* Bangkok Conference IUCN (Mimeo ms.).
Harrison, J.L. 1968. The effect of forest clearance on small mammals. In *Conservation in tropical Southeast Asia.* IUCN, Gland.
Harrison, J.L. 1969. The abundance and population density of mammals in Malayan lowland forests. *Malay. Nat. J.* 22: 174-178.

Harrisson, T. 1949. Explorations in central Borneo. *Geog. J.* 54: 140-150.
Harrisson, T. 1956. Rhinoceros in Borneo: and traded to China. *Sarawak Mus. J.* 7: 263-274
Harrisson, T. 1958. The caves of Niah: a history of prehistory. *Sarawak Mus. J.* 8: 549-595.
Harrisson, T. 1959a. New archaelogical and ethnological results from Niah caves, Sarawak. *Man* 1: 1-8.
Harrisson, T. 1959b/1984. *World within. A Borneo story.* Oxford University Press.
Harrisson, T. (ed.) 1959c. *The peoples of Sarawak.* Sarawak Museum, Kuching.
Harrisson, T. 1960. Birds and men in Borneo. In *Birds of Borneo* (ed. B.E. Smythies), pp. 21-30. The Sabah Society, Kota Kinabalu.
Harrisson, T. 1961. The threat to rare animals in Borneo. *Oryx* 6: 126-128.
Harrisson, T. 1962. Megaliths of central Borneo and Western Malaya, compared. *Sarawak Mus. J.* 10: 376-382
Harrisson, T. 1964. The "palang", its history and proto-history in West Borneo and the Philippines. *J. Malay. Br. R. Asiatic Soc.* 37: 162-174.
Harrisson, T. 1965. Six specialized stone tools from upland and southwest Borneo. *Sarawak Mus. J.* 12: 133-142
Harrisson, T. 1966. The "palang": II. Three further notes. *J. Malay. Br. R. Asiatic Soc.* 39(1): 172-174.
Harrisson, T. 1966. The gibbon in West Borneo, folklore and augury. *Sarawak Mus. J.* 14: 132-145.
Harrisson, T. 1968. Prehistoric fauna changes and losses in Borneo. In *Nature conservation in western Malesia* (ed. J. Wyatt-Smith and P.R. Wycherley), Malayan Nature Society, Kuala Lumpur.
Harrisson, T. 1970. Birds from the rest house verandah, Brunei. *Brunei Mus. J.* 2(1): 269-278.
Harrisson, T. 1972. The Borneo Stone Age in the light of recent research. *Sarawak Mus. J.* 20: 40-41.
Harrisson, T. 1974. The food of *Collocalia* swiftlets (Aves: Apodidae) at Niah Great Cave in Borneo. *J. Bomb. Nat. Hist. Soc.* 71: 376-393.
Harrisson, T. 1975. The upper palaeolithic in Borneo and adjacent areas: gateways to the Pacific?. *Brunei Mus. J.* 3(3): 175-185.
Harrisson, T. and Harrisson, B. 1971. *The prehistory of Sabah.* Sabah Society, Kota Kinabalu.

Harrisson, T., Hooijer D. and Medway, Lord. 1961. An extinct giant pangolin and associated mammals from Niah cave, Sarawak. *Nature* 189: 166.

Hart, T.B. 1990. Monospecific dominance in tropical rain forests. *TREE* 5(1): 6-11.

Hartog, C. den 1970. *Sea grasses of the world.* North Holland, Amsterdam.

Hatch, T. 1980. Shifting cultivation in Sarawak. In *Tropical ecology and development* (ed. J.I. Futardo). International Society of Tropical Ecology, Kuala Lumpur.

Hatch, T. 1983. Soil erosion and shifting cultivation in Sarawak. In *Proceedings of the regional workshop on hydrological impacts of forestry practices and reafforestation* (eds. A. Kamis, F.S. Lai, S.S. Lee and A.R.M. Derus), pp. 51-60. Universiti Pertanian Malaysia.

Hatch, T. and Lim, C.P. 1978. *Shifting cultivation in Sarawak.* Department of Agriculture, Sarawak.

Heaney, L.R. 1984. Mammalian species richness on islands on the Sunda shelf, Southeast Asia. *Oecologia* 61: 11-17.

Heaney, L.R. 1986. Biogeography of mammals in southeast Asia: estimates of rates of colonization, extinction and speciation. *Biol. J. Linn. Soc.* 28: 127-165.

Heckman, C.W. 1979. *Ricefield ecology of northeastern Thailand. The effect of wet and dry seasons on a cultivated aquatic ecosystem.* Junk, The Hague.

Heekeren, H.R. van, 1972. *The stone age in Indonesia.* (2nd ed.) Martinus Nijhoff, The Hague.

Hegerl, E.J. and Davie, J.D.S. (eds.) 1983. *Global status of mangrove ecosystems.* Commission on ecology papers No. 3. IUCN, Gland, Switzerland.

Helbig, K.M. 1955. Die Insel Borneo in Forschung und Schrifttum. *Mitteilungen der Geographischen Gesellschaft in Hamburg* 52: 105-395.

Henderson, M.R. 1939. The flora of the limestone hills of the Malay Peninsula. *J. Malay Br. R. Asiat. Soc.* 17: 13-87.

Henderson, M.R. 1961. *Common Malayan wildflowers.* Longmans, London.

Henderson-Sellers, A. 1981. The effects of land clearance and agricultural practices on climate. In *Blowing in the wind: deforestation and long-range implications.* pp. 443-486. Studies in Third World Societies, No. 14. College of William and Mary, Williamsburg.

Henrey, L. 1982. *Coral reefs of Malaysia and Singapore.* Longman, Singapore.

Herrera, C.M. 1982. Defence of ripe fruit from pests: its significance to plant-disperser interactions. *Am. Nat.* 12: 218-241.

Herzog, T. 1950. Hepaticae Borneenses (Oxford University expedition to Sarawak, 1932). *Trans. Br. Bryol Soc.* 1: 275-326.

Heywood, V.H. 1985. *Flowering plants of the world.* Croom Helm Publishers Ltd., Beckenham and Sydney.

Hill, J.E. and Smith J.D. 1984. *Bats: a natural history.* British Mus. Nat. Hist. London.

Hinckley, A.D. 1973. Ecology of the coconut rhinoceros beetle, *Oryctes rhinoceros* (L.) (Coleoptera: Dynastidae). *Biotropica* 5(2): 111-116.

Hladik, M. 1979. Diet and ecology of prosimians. In *The study of promisian behaviour* (eds. G.A. Doyle and R.D. Martin), pp. 307-358. Academic Press, New York.

Hobbs, R.J., 1992. The role of corridors in conservation: solution or bandwagon? *TREE* 7(11): 389-392.

Hoffman, C.L. 1981. Some notes on the origins of the "Punan", Borneo. *Borneo Res. Bull.* 13: 71-75.

Hoffman, C. 1983. *The Punan: hunters and gatherers of Borneo.* Ph.D. thesis, Univ. of Pennsylvania.

Hoffman, C.L. 1986. *The Punan; hunters and gatherers of Borneo.* Studies in Cultural Anthropology, No. 12. UMI Research Press, Ann Arbor.

Hoffman, C.L. 1988. The "Wild Punan" of Kalimantan: a matter of economics. In *Function and change in traditional Indonesian cultures* (ed. M.R. Dove), pp. 1-46. The Obor Foundation, Jakarta.

Holdgate, M.W. 1987. Changing habitats of the world. *Oryx* 21(3): 149-159.

Holdridge, L.R. 1959. Ecological indications of the need for a new approach to tropical land-use. *Econ. Bot.* 13: 271-280.

Holdridge, L.R. 1967. *Life zone ecology.* Tropical Science Centre, San José, Costa Rica.

Holloway, J.D. 1970. The biogeographical analysis of a transect sample of the moth fauna of Mt. Kinabalu, Sabah, using numerical methods. *Biol. J. Linn. Soc.* 2: 259-286.

Holloway, J.D. 1978. Butterflies and moths. In *Kinabalu, summit of Borneo* (eds. M. Luping, N. Chin and E.R. Dingley), pp. 255-265. The Sabah Society, Kota Kinabalu.

Holloway, J.D. 1983. Insect surveys - an approach to environmental monitoring. *Atti XII Congr. Naz. Ital. Entomol.*, Roma, 1980. 1: 239-261.

Holloway, J.D. 1984. Notes on the butterflies of the Gunung Mulu National Park. *Sarawak Mus. J.* 30 (51): 89-131.

Holloway, J.D. 1986. The moths of Borneo: key to families Cossidae, Metarbelidae, Ratardidae, Dudgeonidae, Epipyropidae and Limacodidae. *Malay. Nat. J.* 40: 1-166.

Holloway, J.D. 1986. Origins of lepidopteran faunas in high mountains of the Indo-Australian Tropics. In *High altitude tropical biogeography* (eds. M. Monasterio and F. Vuilleumier). Oxford University Press, Oxford.

Holloway, J.D. 1987. Macrolepidoptera diversity in the Indo-Australian tropics: geographic, biotopic and taxonomic variations. *Biol. J. Linn. Soc.* 30: 35-341.

Holloway, J.D., Kirk-Spriggs, A.H. and Chey, V.K. 1992. The response of some rain forest insect groups to logging and conversion to plantation. In *Tropical rain forest: disturbance and recovery* (eds. A.G. Marshall and M.D. Swaine), pp. 425-436. The Royal Society, London.

Holmes, D.A. and Burton, K. 1987. Recent notes on the avifauna of Kalimantan. *Kukila* 3: 2-32.

Holthuis, L.B. 1979. Caverniculous and terrestrial decapod Crustacea from northern Sarawak, Borneo. *Zool. Verdhandelingen* 171: 1-47.

Holttum, R.E. 1974. The tree-ferns of the genus *Cyathea* in Borneo. *Garden's Bull. Singapore* 27: 167-181.

Holttum, R.E. 1978. The ferns of Kinabalu national park. In *Kinabalu, summit of Borneo* (eds. M. Luping, N. Chin and E.R. Dingley), pp. 199-210. The Sabah Society, Kota Kinabalu.

Hong, E. 1987. *Natives of Sarawak, survival in Borneo's vanishing forests*. Institut Masyarakat, Malaysia.

Hong, L.T., Thillainathan, P. and Omar, A. 1984. Observations on the fruiting and growth of some agarics in a dipterocarp stand. *Malay. Nat. J.* 38: 81-88.

Hooijer, D.A. 1972. Prehistoric evidence for *Elephas maximus* L. in Borneo. *Nature* 239: 228.

Hooijer, D.A. 1975. Quaternary mammals west and east of Wallace's line. *Netherlands J. Zool.* 25(1): 45-56.

Hose, C. 1929/1985. *The fieldbook of a jungle wallah. Shore, river and forest life in Sarawak*. Oxford University Press.

Hose, C. 1926/1988. *Natural man, a record from Borneo*. Oxford University Press, Singapore.

Hose, C. and McDougall, W. 1912. *The pagan tribes of Borneo*. 2 vols. Macmillan, London.

Howe, C.P., Claridge, G.F., Hughes, R. and Zuwendra 1991. *Manual of guidelines for scoping EIA in tropical wetlands*. Asian Wetland Bureau, Bogor.

Howe, H.F. 1984. Implications of seed dispersal by animals for tropical reserve management. *Biol. Conserv.* 30: 261-281.

Huc, R. and Rosalina, U. 1981. *Aspects of secondary forest succession in logged over lowland forests in East Kalimantan*. BIOTROP report, Bogor.

Hudson, A.B. 1972. *Padju Epat. The Ma'anyan of Indonesian Borneo*. Holt, Rinehart and Winston, New York.

Hunter, L. 1983. Tropical forest plantations and natural stand management: a national lesson from East Kalimantan. *Bull. Indonesian Economic Studies* 20 (1): 98-116.

Husin, A. and Suradi, T. 1973. *Some notes on the regional geology of East Kalimantan*. Geological Survey of Indonesia.

Hutchinson, G.E. 1975. *A treatise on limnology. III Limnological botany*. Wiley, New York.

Hutterer, K.L., Rambo, A.T. and Lovelace, G. (eds.) 1985. *Cultural values in human ecology in South East Asia*. Michigan Papers on South and South-East Asia No.27. University of Michigan.

Hutton, A.F. 1985. Butterfly farming in Papua New Guinea. *Oryx* 19(3): 158-162.

Huxley, C.R. 1978. The ant-plants *Myrmecodia* and *Hydnophytum* (Rubiaceae) and the relationships between their morphology, ant occupants, physiology and ecology. *New Phytol.* 80: 231-268.

Inger, R.F., 1966. The systematics and geography of the amphibians of Borneo. *Fieldiana Zool.* 52: 1-402.

Inger, R.F. 1978. The frogs and toads. In *Kinabalu, summit of Borneo* (eds. M. Luping, N. Chin and E.R. Dingley), pp. 311-320. The Sabah Society, Kota Kinabalu.

Inger, R.F. 1979. Abundance of amphibians and reptiles. In *Tropical forests in Southeast Asia* (ed. A.G. Marshall). Trans. Aberdeen-Hull symposium on Malesian ecology. Univ. of Hull Misc. Ser. 22: 93-112.

Inger, R.F. 1980a. Densities of floor-dwelling

frogs and lizards in lowland forests of Southeast Asia and Central America. *Am. Nat.* 115(6): 761-770.

Inger, R.F. 1980b. Relative abundances of frogs and lizards in forests of Southeast Asia. *Biotropica* 12(1): 14-22.

Inger, R.F. 1985. Tadpoles of the forested regions of Borneo. Field Museum of Natural History. *Fieldiana Zool.* 26: 1-89.

Inger, R.F. 1986. Diets of tadpoles living in a Bornean rain forest. *Alytes* 5(4): 153-164.

Inger, R.F. and Chin, P.K. 1962. The freshwater fishes of North Borneo. *Fieldiana Zool.* 45: 1-268.

Inger, R.F. and Greenberg, B. 1966. Ecological and competitive relations among three species of frogs (genus *Rana*). *Ecology* 45: 746-759.

Inger, R.F., Voris, K. and Walker, P. 1985. A key to the frogs of Sarawak. *Sarawak Mus. J.* 34: 161-182.

Inger, R.F., Voris, H.K. and Frogner, K.J. 1986. Organization of a community of tadpoles in rain forest streams in Borneo. *J. Trop. Ecol.* 2: 193-205.

Inger, R.F., Ean, Y.M. and Lee, Y.K. 1987. Key to the frogs of Danum Valley. *Sabah Soc. J.* 7(3): 373-379.

Ismail, G., Mohamed, M. and Omar, S. 1992. *Forest biology and conservation in Borneo.* Center for Borneo Studies Publication No. 2. Kota Kinabalu, Sabah.

IUCN. 1985. *Viet Nam: national conservation strategy.* Prepared by the Committee for Rational Utilisation of Natural Resources and Environmental Protection, Vietnam, with assistance from IUCN. WWF-India, New Delhi.

IUCN. 1990. *IUCN red list of threatened animals of the world.* IUCN, Gland.

IUCN/UNEP/WWF. 1991. *Caring for the Earth. A strategy for sustainable living.* IUCN, Gland, Switzerland.

Jackson, J.C. 1970. *Chinese in the West Borneo goldfields. A study in cultural geography.* Occasional Papers in Geography 15, University of Hull.

Jacobs, M. 1976. The study of lianes. *Fl. Males. Bull.* 29: 2610-2618.

Jacobs, M. 1982. The study of minor forest products. *Fl. Males. Bull.* 35: 3768-3782.

Jacobs, M. 1984. The study of non-timber products. *Environmentalist* 4. Suppl. No. 7.

Jacobs, M. 1988. *The tropical rainforest. A first encounter.* Springer-Verlag, Berlin Heidelberg.

Jacobson, S.K. 1986. *Kinabalu Park.* Sabah Parks Publication No. 7.

Janos, D.P. 1980. Mycorrhizae influence tropical succession. *Biotropica* 12: 56-64.

Janzen, D.H. 1970. Herbivores and the number of tree species in tropical rain forests. *Am. Nat.* 104: 501-528.

Janzen, D.H. 1971. Seed predation by animals. *Ann. Rev. Ecol. Syst.* 2: 465-92.

Janzen, D.H. 1973. Rate of regeneration after a tropical high elevation fire. *Biotropica* 5(2): 117-122.

Janzen, D.H. 1974a. Tropical blackwater rivers, animals and mast fruiting by the Dipterocarpaceae. *Biotropica* 6(2): 69-103.

Janzen, D.H. 1974b. Epiphytic myrmecophytes in Sarawak: mutualism through the feeding of plants by ants. *Biotropica* 6(4): 237-359.

Janzen, D.H. 1976. Why tropical trees have rotten cores. *Biotropica* 6: 69-103.

Janzen, D.H. 1977. Promising directions of study in tropical animal-plant interactions. *Ann. Missouri Bot. Garden.* 64: 706-736.

Janzen, D.H. 1979. How to be a fig. *Ann. Rev. Ecol. Syst.* 10: 13-51.

Janzen, D.H. 1983. Seed and pollen dispersal by animals: convergence in the ecology of contamination and sloppy harvest. *Biol. J. Linn. Soc.* 20(1): 103-113.

Janzen D.H. 1985. Mangroves, where's the understorey? *J. Trop. Ecol.* 1: 89-92.

Janzen, D.H. 1986. The future of tropical ecology. *Ann. Rev. of Ecol. and Syst.* 17: 305-324.

Janzen, D.H. 1987. Insect diversity of a Costa Rican dry forest: why keep it, and how? *Biol. J. Linn. Soc.* 30: 343-356.

Jeffrey, S.M. 1982. Threats to the proboscis monkey. *Oryx* 16: 337-339.

Jermy, A. C. 1983. Gunung Mulu National Park, Sarawak. *Oryx* 17(1): 6-14.

Jermy, A.C. and Kavanagh, K.P. (eds.) 1982. Gunung Mulu National Park. *Sarawak Mus. J.* 51. Special issue No. 2 Part I.

Jermy, A.C. and Kavanagh, K.P. (eds.) 1984. Gunung Mulu National Park, Sarawak. *Sarawak Mus. J.* 30(51), Special issue No. 2. Part II.

Jessup, T.C. 1981. Why do Apo Kayan shifting cultivators move? *Borneo Res. Bull.* 13(1): 16-32.

Jessup, T.C. and Peluso, N.L. 1985. Minor forest products as common property resources in East Kalimantan, Indonesia. In *Proceedings of the conference on common property resource management.* pp. 505-531. National Academy Press, Washington,

D.C.

Jessup, T.C. and Vayda, A.P. 1988. Dayaks and forests of interior Borneo. The University Museum Magazine of Archeology/ Anthropology University of Pennsylvania. *Expedition* 30(1): 5-17.

Jett, S.C. 1970. The development and distribution of the blowgun. *Ann. Ass. Am. Geog.* 60: 662-688.

Jhamtani 1989. Mangroves for chipwood and pulp in East Kalimantan. *Jakarta Post*, June.

Jimenez, J.A., Lugo, A.E. and Cintron, G. 1985. Tree mortality in mangrove forests. *Biotropica* 17: 177-185.

Jin, E.O. 1982. Mangroves and aquaculture in Malaysia. *Ambio* 11(5): 252-257.

Jin, E.O., Wooi, K.G., Chee H.W. and Bin, H.J.W. 1985. *Productivity of the mangrove ecosystem: a manual of methods*. Unit Pencetakan Pusat, Universiti Sains Malaysia.

Johannes, R.E. 1975. Pollution and degradation of coral reef communities. In *Tropical marine pollution* (eds. E.J.F. Wood and R.E. Johannes), Elsevier, Amsterdam.

Johannes, R.E. and Betzer, S.S. 1975. Marine communities respond differently to pollution in the tropics than at higher latitudes. In *Tropical marine pollution* (eds. E.J.F. Wood and R.E. Johannes) pp 1-12, Elsevier, Amsterdam.

Johns, A.D. 1981. The effects of selective logging on the social structure of resident primates. *Malay. Appl. Biol.* 10(2): 221-226.

Johns, A.D. 1983a. *Ecological effects of selective logging in a West Malaysian rain forest*. Ph.D. thesis, University of Cambridge.

Johns, A.D. 1983b. Tropical forest primates and logging - can they co-exist?. *Oryx* 17(3): 114-118.

Johns, A.D. 1986. Effects of selective logging on the ecological organization of a Peninsular Malaysian rainforest avifauna. *Forktail* 1: 65-79.

Johns, A.D. 1987. The use of primary and selectively logged rainforest by Malaysian hornbills (Bucerotidae) and implications for their conservation. *Biol. Cons.* 40: 179-190.

Johns, A.D. 1988. Effects of "selective" timber extraction on rainforest structure and composition and some consequences for frugivores and folivores. *Biotropica* 20(1): 31-37.

Johns, A.D. 1992a. Vertebrate responses to selective logging: implications for the design of logging systems. In *Tropical rain forest: disturbance and recovery* (eds. A.G. Marshall and M.D. Swaine), pp. 437-442. The Royal Society, London.

Johns, A.D. 1992b. Species conservation in managed tropical forests. In *Realistic strategies for tropical forest conservation* (eds. J.A. Sayer and T.C. Whitmore). IUCN, Gland, Switzerland.

Johns, A.D. and Marshall, A.G. 1992. Wildlife population parameters as indicators of the sustainability of timber logging operations. In *Forest biology and conservation in Borneo* (eds. G. Ismail, M. Mohamed, and S. Omar), pp. 366-374. Center for Borneo Studies Publication No. 2.

Johns, A.D. and Skorupa, J.P. 1987. Responses of rain forest primates to habitat disturbance: a review. *Int. J. Primatol.* 8(2): 157-191.

Johnson, B. 1984. *The great fire of Borneo. Report of a visit to Kalimantan Timur a year later*. WWF.

Johnson, D.S. 1967. Distributional patterns in Malayan freshwater fish. *Ecology* 48: 722-730.

Johnson, D.S. 1968. Malayan black waters. *Proc. Symp. Rec. Adv. Trop. Ecol.* 1: 303-310.

Johnstone, I.M. 1981. Consumption of leaves by herbivores in mixed mangrove stands. *Biotropica* 13: 252-259.

Johnstone, I.M. and Hudson, B.E.T. 1981. The dugong diet: mouth sample analysis. *Bull. Mar. Sci.* 31: 681-690.

Jordan, C.F. and Farnworth, E.F. 1982. Natural vs. plantation forests: a case study of land reclamation strategies for the humid tropics. *Env. Managt.* 6: 485-492.

Jordan, C.F. 1985. *Nutrient cycling in tropical forest ecosystems*. J. Wiley and Sons.

Jordano, P. 1983. Fig-seed predation and dispersal by birds. *Biotropica* 15: 138-41.

Joy, D. and Wibberley, E.J. 1979. *A tropical agriculture handbook*. Cassell, London.

JP-Energy Oy. 1987a. *Peat production experiments in Central Kalimantan*. FINNIDA and Ministry of Mines and Energy, Jakarta.

JP-Energy Oy. 1987b. *Socio-economic analysis: different impacts of peat utilisation in Pontianak and Palang-karaya*. FINNIDA and Ministry of Mines and Energy, Jakarta.

Junk, W.J. 1975. The bottom fauna and its distribution in Bung Borapet, a reservoir in central Thailand. *Verh. Internat. Verein Limnol.* 19: 1935-1946.

Kam S.P. and Leong, Y.K. 1985. *Mangrove and corals, an evaluation for the national coastal erosion study*. Environmental Research Group, Industrial Research and Consultancy Services, Universiti Sains Malaysia.

Kamminga, C.H., Wiersma, H. and Heel, W.H.D. van 1983. Investigations on cetacean sonar. VI. Sonar sounds in *Orcaella brevirostris* of the Mahakam river, East Kalimantan, Indonesia. *Aquatic Mammals* 10: 83-94.

Kampen, P.N. van. 1923. *The Amphibia of the Indo-Australian archipelago*. E.J. Brill, Ltd.

Karpowicz, Z. 1985. *Wetlands in East Asia - a preliminary review and inventory*. ICBP Study Report No.6. Cambridge.

Kartawinata, K. 1977. A report on a study of floristic, faunistic and other ecological changes in the lowland rainforest after destruction by man in East Kalimantan. In *Transactions of the international MAB-IUFRO workshop on tropical rainforest ecosystems research* (ed. E.F. Brunig), pp. 233-237, Hamburg.

Kartawinata, K. 1978. The "kerangas" heath forest in Indonesia. In *Glimpses of ecology* (eds. J.S. Singh and B. Gopal), pp. 145-152. International Scientific Publications, Jaipur.

Kartawinata, K. 1980a. Note on a kerangas (heath) forest at Sebulu, East Kalimantan. *Reinwardtia* 9(4): 429-447.

Kartawinata, K. 1980b. The environmental consequences of tree removal from the forest in Indonesia. In *Where have all the flowers gone? Deforestation in the Third World*. Studies in Third World Societies Publication No. 13.

Kartawinata, K. 1980c. Classification and utilization of Indonesian forests. *BioIndonesia* 7: 95-106.

Kartawinata, K. 1984. *Environmental effects of different kinds of land use*. LIPI-Indonesian MAB Committee. Doc. No. 39.

Kartawinata, K. (in prep). Vegetation and environment of East Kalimantan. In *People and forests in East Kalimantan* (eds. T.C. Jessup, K. Kartawinata, and A.P. Vayda).

Kartawinata, K., Abdulhadi, R. and Partomihardjo, T. 1981. Composition and structure of a lowland dipterocarp forest at Wanariset, East Kalimantan. *Malay. Forester* 44(2): 397-406.

Kartawinata, K., Adisoemarto, S., Riswan, S. and Vayda, A.P. 1981. The impact of man on tropical forest in Indonesia. *Ambio* 10(2/3): 114-119.

Kartawinata, K., Adisoemarto, S., Soemodihardjo, S. and Tantra, I.G.H. 1979. Status pengetahuan hutan bakau di Indonesia. In *Prosiding seminar ekosistem hutan mangrove* (eds S. Soemodihardjo, A. Nontji and A. Djamali), pp. 21-39, Lembaga Oseanologi Nasional, Jakarta.

Kartawinata, K., Jessup, T.C., and Vayda, A.P. 1989. Exploitation in Southeast Asia. In *Tropical rain forest ecosystems* (eds. H. Lieth and M.J. Werger), pp. 591-610 Elsevier, Amsterdam.

Kartawinata, K., Soedjito, H., Jessup, T., Vayda, A.P. and Colfer, C.J.P. 1984. The impacts of development on interactions between people and forests in East Kalimantan: a comparison of two areas of Kenyah Dayak settlement. *Environmentalist* 4. Suppl. No. 7: 87-95.

Kartawinata, K. and Vayda, A.P. 1984. Forest conversion in East Kalimantan, Indonesia: the activities and impact of timber companies, shifting cultivators, migrant pepper-farmers, and others. In *Ecology in practice. Part I: ecosystem management* (eds. F. di Castri, F.W.G. Baker and M. Hadley), pp. 98-125. Tycooly International Publishing Ltd, Dublin/UNESCO Paris.

Kartawinata, K., Vayda, A. P. and Wirakusumah, R.S. 1977 East Kalimantan and the Man and Biosphere Program. *Berita Ilmu Pengetahuan dan Teknologi* 21(2).

Katili, J.A. 1989. Evolution of the Southeast Asian arc complex. In *Geologi Indonesia. J.A. Katili commemorative volume*. Ikatan Ahli Geologi Indonesia, Jakarta.

Kato, R., Tadiki, Y. and Ogawa, H. 1978. Plant biomass and growth increment studies in the Pasoh forest. *Malay Nat. J.* 30: 211-224.

Kaul, R.B. 1982. Floral and fruit morphology of *Nepenthes lowii* and *N. villosa*, montane carnivores of Borneo. *Am. J. Bot.* 69 (5): 793-803.

Kedit, P.M. 1978. Gunong Mulu report: a human-ecological survey of nomadic/settled Penan within the Gunong Mulu National Park area, Fourth/Fifth Divisions. *Sarawak Museum Field Report* No. 1, Kuching.

Kedit, P.M. 1982. An ecological survey of the Penan. *Sarawak Mus. J.* 30(51): 225-279.

Kelly, M. 1988. *Mining and the freshwater environment*. Elsevier Science Publ., New York.

Kennedy, D.N. 1991. *The role of colonising species in the regeneration of dipterocarp rain forest*. Ph.D. thesis, University of Aberdeen.

Kennedy, D.N. and Swaine, M.D. 1992. Germination and growth of colonising species in artificial gaps of different sizes in dipterocarp rain forest. In *Tropical rain forest: disturbance and recovery* (eds. A.G. Marshall and M.D. Swaine), pp. 357-367. The Royal Society, London.

Kent, G.A. 1986. Assessment of mangrove regeneration after logging for woodchips in Sabah. Forest Research Centre, Sabah. FRC Publication No. 29.

KEPAS, 1985. *Tidal swamp agro-ecosystems of southern Kalimantan.* The research group on Agro-ecosystems. Kelompok penelitian Agro-eksosistem (KEPAS). Ford Foundation and Agency for Agricultural Research and Development, Ministry of Agriculture, Indonesia.

Keuning, S.J. 1984. Farm size, land use and profitability of food crops in Indonesia. *Bull. Indon. Econ. Studies* 20(1): 58-82.

Khoo, S.G. 1974. Scale insects and mealy bugs: their biology and control. *Malay. Nat.J.* 27: 124-30.

Kiew, B.H. 1978. *A preliminary report on the lowland fishes, amphibians and reptiles of Gunung Mulu National Park, Sarawak.* Unpubl. ms.

Kiew, R. 1978. Floristic components of the ground flora of a tropical lowland rainforest at Gunung Mulu National Park, Sarawak. *Pertanika* 1(2): 112-119.

Kikkawa, J. and Williams, W.T. 1971. Altitudinal distribution of land birds in New Guinea. *Search* 2:64-65.

King, B., Woodcock, M. and Dickinson, E.C. 1975. *A field guide to the birds of South-East Asia.* Collins, London.

King, V.T. 1974. Notes on Penan and Bukit in West Kalimantan. *Borneo Res. Bull.* 6(2): 39-42.

King, V.T. 1978. *Essays on Borneo societies.* Oxford University Press. Hull Monographs on Southeast Asia No. 7.

King, V.T. 1985. *The Maloh of West Kalimantan.* Foris, Dordrecht.

King, V.T. 1986. Land settlement schemes and the alleviation of rural poverty in Sarawak, East Malaysia: a critical commentary. *Southeast Asian J. Soc. Sci.* 14(1): 71-100.

Kitching and Schofield, 1986. Every pitcher tells a story. *New Scientist* 109: 48-50.

Klepper, O. 1990. An overview of the hydrology of the Sungai Negara wetlands in relation to their ecology. In *Conservation of Sungai Negara wetlands, South Kalimantan.* pp. 229-254. PHPA/AWB-Indonesia and KPSL-UNLAM.

Klepper, O. 1992. Model study of the Negara river basin to assess the regulating role of its wetlands. *Regulated Rivers: research and management,* 7: 311-325.

Klepper, O. and Asfihani, 1990. Legend to the reconnaissance soil map of the Sungai Negara basin, South Kalimantan. In *Conservation of Sungai Negara wetlands, South Kalimantan.* pp. 95-125. PHPA/AWB-Indonesia and KPSL-UNLAM.

Klepper, O., Hatta, G., Chairuddin, G, Sunardi and Iriansyah, 1990. *Acid sulphate soils in the humid tropics. Ecology component.* First interim report. RIN/KPSL-UNLAM.

Klepper, O., Chairuddin, G.T., Iriansyah, and Rijksen, H.D. 1992. Water quality and the distribution of some fishes in an area of acid sulphate soils, Kalimantan, Indonesia. *Hydrobiol. Bull.* 25(3): 217-224.

KLH, 1989. *National strategy for the management of biodiversity.* KLH (Ministry of Population and Environment), Jakarta.

KLH, 1992. *Indonesian country study on biological diversity.* Prepared for the United Nations Environment Programme (UNEP), KLH (Ministry of Population and Environment). Jakarta.

Knoll, A.H. 1984. Patterns of extinction in the fossil record of vascular plants. In *Extinction* (ed. M.H. Nitecki), pp. 21-68. University of Chicago Press, Chicago.

Knox, G.A. and Miyabara, T. 1984. *Coastal zone resource development and conservation in Southeast Asia.* UNESCO/East-West Center.

Ko, R.K.T. 1986. *Conservation and environmental management of subterranean biota.* Paper presented at the BIOTROP Symposium on the Conservation and management of endangered plants and animals.

Koesoebiono, Collier, W.L. and Burbridge, P.R. 1982. Indonesia: resource use and management in the coastal zone. In *Man, land and sea* (eds. Soysa et al), pp. 115-134. Bangkok.

Koesmawadi, N. 1984. Analisa konsumsi masyarakat dalam kaitannya dengan pendapatan penggunaan gelam *Melaleuca* sp. sebagai bahan kayu bakar di sekitar desa Pandahan kecamatan Bai-Bati, Kabupaten Tanah Laut. Fakultas Kehutanan, UNLAM, Banjarbaru.

KOMPAS BORNEO 1988. *Laporan singkat perjalanan di Gunung Aurbunak, Suaka Mar-*

gasatwa Pleihari Martapura. Universitas Lambung Mangkurat, Banjarbaru.

Konstant, W.R. and Mittermeier, R.A. 1982. Introduction, reintroduction and translocation of neotropical primates: past experiences and future possibilities. *Int. Zoo Yearbook* 22: 69-77.

Konsten, C.J.M. and Klepper, O. 1992. *Pyrite in coastal wetlands: a natural chemical time bomb.* Paper presented at the European state-of-the-art conference of delayed effects of chemicals in soils and sediments (Chemical Time Bombs), Veldhoven, the Netherlands.

Koopman, M.J.F. and Verhoef, L. 1938. *Eusideroxylon zwageri* Het ijzerhout van Borneo en Sumatra. *Tectona* 31: 381-399.

Kostermans, A.J.G.H. 1953. Notes on durian (*Durio*) species of east Borneo. *De Tropische Natuur* 33: 31-36.

Kostermans, A.J.G.H. 1958. Secondary growth on areas of former peat swamp-forest. In *Proceedings of the symposium on humid tropics vegetation, Tjiawi.* pp. 155-169, UNESCO, Jakarta.

Kottelat, M. 1982. A small collection of freshwater fishes from Kalimantan, Borneo, with descriptions of one new genus and three new species of Cyprinidae. *Revue Suisse Zool.* 89 (2): 419-437.

Kottelat, M., Whitten, A.J., Kartikasari, S.N. and Wiroatmodjo, S. 1993. *The freshwater fishes of western Indonesia and Sulawesi.* Periplus, Singapore.

KPSL-UNLAM 1989a. *Longitudinal variation of the ecological condition of the Kala'an river in Pleihari-Martapura forest reserve.* Report to the Biodiversity Support Program. KPSL-UNLAM, Banjarbaru.

KPSL-UNLAM 1989b. *A study of tree composition and flowering and fruiting in Pleihari-Martapura forest reserve, South Kalimantan.* KPSL-UNLAM, Banjarbaru.

KPSL-UNLAM, 1989c. *Informasi lingkungan bukit kapur dan gua kapur Batu Hapu.* Kelompok Program Studi Lingkungan, Universitas Lambung Mang-kurat, Banjarbaru.

Kramer, D.L. and Mehegan, J.P. 1981. Aquatic surface respiration, an adaptive response to hypoxia in the guppy *Poecilia reticulata* (Pisces, Poecilidae). *Env. Biol. Fish.* 6: 299-313.

Kramer, F. 1926. Onderzoek naar de natuurlijke verjonging in den uitkap in Praenger gebergtebosch. *Med. Proefst. Boschw. Bogor* 14.

Kramer, F. 1933. De natuurlijke verjonging in het Goenoeng Gedeh complex. *Tectona* 26: 156-185.

Krishnamurthy, K. and Jeyaseelan, M.J.P. 1980. *The impact of the Pichavaran mangrove ecosystem upon coastal natural resources: a case study from southern India.* Asian symposium on mangrove environments: research and management. Kuala Lumpur.

Kurata, S. 1976. *Nepenthes of Mount Kinabalu.* Sabah National Parks Publications No.2. Sabah.

Laan, E. van der 1925. De boschen van de Zuider en Oosterafdeeling van Borneo. *Tectona* 18: 925-952.

Lahjie A.M. and Seibert, B. (eds). 1988. *Prosiding agroforestry untuk pengembangan daerah pedesaan di Kalimantan Timur.* Fakultas Kehutanan, Universitas Mulawarman, Samarinda.

Lahjie, A.M. and Sirait, T.M. 1988. Potensi pagar hidup sebagai kayu bakar dan makanan ternak. In *Prosiding agroforestry untuk pengembangan daerah pedesaan di Kalimantan Timur* (eds. A.M. Lahjie and B. Seibert), pp. 223-232, Fakultas Kehutanan, UNMUL, Samarinda.

Lam, H.J. 1927. *Een plantengeografisch Dorado.* Handelingen IV Nederlands-Indisch natuurwetenschappenlijk congres, pp. 386-397, Welteneden.

Lam, T.M. 1983. Reproduction in the rice field rat *Rattus argentiventer. Malay. Nat. J.* 36: 249-282.

Lamb, A. and Chan, C.L. 1978. The orchids. In *Kinabalu, summit of Borneo* (eds. M. Luping, W. Chin and R.E. Dingley), pp. 209-252. The Sabah Society, Kota Kinabalu.

Lambert, F.R. 1989. Fig-eating by birds in a Malaysian tropical lowland rain forest. *J. Trop. Ecol.* 5: 401-12.

Lambert, F. 1991. The conservation of fig-eating birds in Malaysia. *Biol. Cons.* 58: 31-40.

Lambert, F.R. 1992. The consequences of selective logging for Bornean lowland forest birds. In *Tropical rain forest: disturbance and recovery* (eds. A.G. Marshall and M.D. Swaine), pp. 443-457. The Royal Society, London.

Langenhoff, R. 1986. *Distribution, mapping, classification and use of acid sulphate soils in the tropics.* A literature study. Soil Survey Institute (STIBOKA) Wageningen, The Netherlands.

Langham, N. 1980. Breeding biology of the edible-nest swiftlet *Aerodramus fuciph-*

agus. Ibis 122: 461-477.

Langham, N. 1982. The ecology of the common tree shrew *Tupaia glis* in Peninsular Malaysia. *J. Zool.* 197: 323-344.

Langham, N. 1983. Distribution and ecology of small mammals in three rain forest localities of Peninsular Malaysia with particular reference to Kedah Peak. *Biotropica* 15(3): 19-206.

Langub, J. 1974. Adaption to a settled life by the Punan and Penan of the Belaga subdistrict. In *The peoples of central Borneo* (ed. J. Rousseau), pp. 295-302. *Sarawak Mus. J. Special Issue* vol. 22.

Langub, J. 1988. Some aspects of life of the Penan. Paper presented at the Orang Ulu culture heritage seminar, Miri.

Lau, D. 1987. *Penans: the vanishing nomads of Borneo*. Inter-State Publishing Company Sdn. Bhd, Sabah, Malaysia.

Lau, B.T. 1979. The effects of shifting cultivation on sustained yield management for Sarawak national forests. *Malay. Forester* 42(4): 418-422.

Lau, B.T. and Chung, K.S. 1978. Forest protection against shifting cultivation. In *Shifting cultivation in Sarawak*. Appendix 3. Report based upon the Workshop on shifting cultivation, Department of Agriculture, Kuching.

Laurie, E.M.O. and Hill, J.E. 1954. *List of land mammals of New Guinea, Celebes and adjacent islands*. British Museum, London.

Lavieren, L.P. van. 1982. *Wildlife management in the tropics, with special emphasis on South East Asia: a guide book for the warden*. School of Environmental Conservation Management, Bogor. 2 vols.

Lawlor, T.E. 1986. Comparative biogeography of mammals on islands. *Biol. J. Linn. Soc.* 28: 99-125.

Layne, E.N. 1976. *The natural environment: a dimension of development*. National Audubon Society.

Leach, G.J. and Burgin, S. 1985. Litter production and seasonality of mangroves in Papua New Guinea. *Aquatic Botany* 23: 215-224.

Leakey, R.A. and Proctor, J. 1987. Invertebrates in litter and soil at a range of altitudes on Gunung Silam, a small ultrabasic mountain in Sabah. *J. Trop. Ecol.* 3: 119-129.

Leaman, D.J., Yusuf, R. and Sangat-Roemantyo, H. 1991. *Kenyah Dayak forest medicines*. WWF Indonesia Programme, Jakarta.

Lee, D. 1969. Some soils of the British Solomon Islands Protectorate. *Phil. Trans. Roy. Soc. B.* 225: 211-258.

Lee, D.W. and Lowry, J.B. 1980. Solar ultraviolet on tropical mountains. Can it affect plant speciation? *Am. Nat.* 115: 880-883.

Lee, H.S. 1979. Natural regeneration and reforestation in the peat swamp forests of Sarawak. *Trop. Agric. Res. Soc.* 12: 51-60.

Lee, H.S. 1981. *Silvicultural management options in the mixed dipterocarp forests of Sarawak*. M.Sc. thesis, Austalian National University.

Lee, H.S. and Lai, K.K. 1977. *A manual of silviculture for the permanent forest estate of Sarawak*. Sarawak Forest Department Silviculture Pamphlet SR/177.

Lee, J., Brooks, R.R., Reeves, R.D., Boswell, C.R. and Jaffre, T. 1977. Plant-soil relationships in a new Caledonian serpentine flora. *Plant and Soil* 46: 675-650.

Lee, J.A. 1985. *The environment, public health, and human ecology. Considerations for economic development*. Johns Hopkins University Press.

Leigh, E.G. 1975. Structure and climate in tropical rain forest. *Ann. Rev. Ecol. Syst.* 6: 67-86.

Leigh, E.G., Rand, A.S. and Windsor, D.M. 1982. *The ecology of a tropical forest. Seasonal rhythms and long-term changes*. Smithsonian Inst.

Leighton, D.S.R. and Whitten, A.J. 1984. Management of free-ranging gibbons. In *The lesser apes, evolutionary and behavioural biology* (eds H. Preuschoft, D.J. Chivers, W.Y. Brockelman and N. Creel), pp. 32-43, Edinburgh University Press.

Leighton, M. 1982. *Fruit sources and patterns of feeding, spacing and grouping among sympatric Bornean hornbills* (Bucerotidae). Ph.D. thesis, University of California, Davis.

Leighton, M. and Leighton, D. 1983. Vertebrate responses to fruiting seasonality within a Bornean rain forest. In *Tropical rain forest ecology and management* (eds. S.L. Sutton, T.C. Whitmore, and A.C. Chadwick), pp. 181-196. Blackwell Scientific Publications, Oxford.

Leighton, M. and Wirawan, N. 1986. Catastrophic drought and fire in Borneo tropical rainforest associated with the 1982-1983 El-Nino Southern Oscillation event. In *Tropical rain forests and the world atmosphere*. AAAS Symposium 101, Westbury Press, Boulder.

Lembaga Penelitian Tanah. 1973. *Soil survey of the delta Pulau Petak (South and Central Kalimantan)*. Lembaga Penelitian Tanah, Bogor.

Lennertz, R. and Panzer, K.F. 1983. *Preliminary assessment of the drought and forest fire damage in Kalimantan Timur*. Transmigration Areas Development Project (TAD), German Agency for International Cooperation (GTZ).

Lenton, G.M. 1983. Wise owls flourish among the oil palms. *New Scientist* 97: 436-437.

Levine, N.D. 1975. *Human ecology*. Duxbury Press.

Lewis, J.G.E. 1981. The relative abundance of myriapods in the Gunung Mulu National Park, Sarawak. *Entomologist's Monthly Magazine* 116: 219-220.

Li, H. 1970. The origin of cultivated plants in Southeast Asia. *Economic Botany* 24(1): 3-19.

Lieftinck, M.A. 1950. Further studies of S.E. Asiatic species of *Macromia* Rambur. *Treubia* 20: 657-716.

Liew, T.C. 1974. A note on soil erosion study at Tawau hills forest reserve. *Malay Nat. J.* 27: 20-6.

Liew, T.C. and Wong, F.O. 1973. Density, recruitment, mortality and growth of dipterocarp seedlings in virgin and logged-over forests in Sabah. *Malay. Forester* 36: 3-15.

Lim, B.L. 1973. Breeding patterns, food habits and parasitic infestations of bats in Gunong Brinchang. *Malay. Nat. J.* 26: 6-13.

Lim, B.L. and Heyneman, D. 1968. A collection of small mammals from Tuaran and the southwest face of Mt Kinabalu, Sabah. *Sarawak Mus. J.* 16: 257-276.

Lim, B.L. and Muul, I. 1978. Small mammals. In *Kinabalu, summit of Borneo* (eds. M. Luping, W. Chin and R.E. Dingley), pp. 403-457. The Sabah Society, Kota Kinabalu.

Lim, R.P., Furtado, J.I. and Liew, K.S. 1978. *The physical and chemical environment, and invertebrate fauna of some rivers in the Gunong Mulu National Park, Sarawak*. Unpubl. ms.

Lindblad, J.T. 1988. *Between Dayak and Dutch. The economic history of Southeast Kalimantan 1880-1942*. Foris, Dordrecht, Holland.

Ling Roth, H. 1896. *The people of Sarawak and British North Borneo*. British Museum, London.

Lincoln, R.J., Boxshall, G.A. and Clark, P.F. 1982. *A dictionary of ecology, evolution and systematics*. Cambridge University Press.

Lloyd, M., Inger, R.F. and King, F.W. 1968. On the diversity of reptile and amphibian species in a Bornean rainforest. *Am. Nat.* 102: 497-516.

Lovejoy, T. E. 1985. Rehabilitation of degraded tropical forest lands. *The Environmentalist* 5: 1-8.

Lovejoy, T.E., Bierregaard, R.O., Rankin, J. and Schubart, H.O.R. 1983. Ecological dynamics of tropical forest fragments. In *Tropical rain forest ecology and management* (eds. S.L. Sutton, T.C. Whitmore and A.C. Chadwick). pp. 377-386. Blackwell Scientific Publications, Oxford.

Lovejoy, T.E. and Oren, D.C. 1981. The minimum critical size of ecosystems. In *Forest island dynamics in man-dominated landscapes* (eds. R.L. Burgess and D.M. Sharpe), pp. 7-12. Springer-Verlag, New York.

Lovelock, J.E. 1979. *Gaia. A new look at life on Earth*. Oxford University Press.

Lowe-McConnell, R.H. 1977. *Ecology of fishes in tropical waters*. Studies in Biology No.76. Edward Arnold.

Lowry, J.B. 1971. Conserving the forest, a phytochemical view. *Malay. Nat. J.* 24: 225-230.

Lowry, J.B., Lee, D.W. and Stone, B.C 1973. Effects of drought on Mount Kinabalu. *Malay Nat. J.* 26: 178-178.

Lugo, A.E. and Snedaker, S.C. 1974. The ecology of mangroves. *Ann. Rev. Syst. Ecol.* 5: 39-64.

Lugo, A.E., Cintron, G. and Geonaga, C. 1978. Mangrove ecosystems under stress. In *Stress effects on natural ecosystems*. (eds. G.W. Barret and R. Rosenberg), pp 1-32, UNESCO, Montevideo.

Lumholtz, C. 1920. *Through Central Borneo*. T.F. Unwin, London. 2 vols.

Luping, M., Chin, N. and Dingley, E.R. (eds.) 1978. *Kinabalu, summit of Borneo*. The Sabah Society, Kota Kinabalu.

Luxmoore, R.A., Barzdo, J.G., Broad, S.R. and Jones, D.A. 1985. *A directory of crocodilian farming operations*. CITES-IUCN.

Luxmoore, R. and Groombridge, B. 1989. *Asian monitor lizards*. Report to CITES. World Conservation Monitoring Centre, Cambridge.

Maas, E.F., Tie, Y.L. and Lim, C.P. 1979. *Sarawak land capability classification and evaluation for agricultural crops*. Technical

Paper 5, Soils Division. Sarawak Dept. of Agriculture, Kuching.

Mabberley, D.J. 1983. *Tropical rain forest ecology*. Blackie.

MacArthur, R.H. and Wilson, E.O. 1967. *The theory of island biogeography*. Princeton University Press.

Macdonald, D. 1984. *The encyclopedia of mammals*. (2 vols). George Allen and Unwin.

MacDonald, M. 1956/1985. *Borneo people*. Oxford University Press, Oxford.

MacKenzie, N. 1981. *A faunal inventory of Samunsam wildlife sanctuary*. National Parks and Wildlife Office (Forest Department), Kuching, Malaysia.

Mackie, C. 1984. The lessons behind East Kalimantan's forest fires. *Borneo Res. Bull.* 16: 63-74.

Mackie, C. 1986. *The landscape ecology of traditional shifting cultivation in an upland Bornean rainforest*. Regional workshop on impact of man's activities on tropical upland forest ecosystems. Universiti Pertanian Malaysia, Serdang, Selangor, Malaysia.

Mackie, C., Jessup, T.C., Vayda, A.P. and Kartawinata, K. 1987b. Shifting cultivation and patch dynamics in an upland forest in East Kalimantan, Indonesia. In *Proceedings of regional workshop on impact of man's activities on tropical upland forest ecosystems* (eds. Y. Hadi, K. Awang, N.M. Majid and S. Mohamed), pp. 465-518. Faculty of Forestry, Universiti Pertanian Malaysia.

MacKinnon, J. 1971. The orangutan in Sabah today. A study of a wild population in the Ulu Segama reserve. *Oryx* 11(2/3): 141-191.

MacKinnon, J. 1974a. The behaviour and ecology of wild orangutans (*Pongo pygmaeus*). *Anim. Behav.* 22: 3-74.

MacKinnon, J. 1974b. *In search of the red ape*. Collins, London.

MacKinnon, J. 1975. *Borneo*. Time-Life International (Netherlands), Amsterdam.

MacKinnon, J. 1976. A comparative ecology of Asian apes. *Primates* 18(4): 749-771.

MacKinnon, J. 1977. Pet orangutans: should they return to the forest? *New Scientist* 74: 697-699.

MacKinnon, J. 1981a. *National conservation plan for Indonesia Vol. 8 national park development and general topics*. FAO, Bogor.

MacKinnon, J. 1981b. *Guidelines for the development of conservation buffer zones and enclaves*. Nature conservation workshop PPA/WWF/FAO. Bogor.

MacKinnon, J. 1982a. *National conservation plan for Indonesia Vol. 1: introduction, evaluation methods and overview of national nature richness*. FAO, Bogor.

MacKinnon, J. 1982b. *Methods and criteria for selecting and evaluating reserves for conservation in Indonesia*. Paper presented to World National Parks Congress, Bali, Indonesia.

MacKinnon, J. (ed.) 1983a. *Tanjung Puting National Park. Management plan for development*. WWF/PPA, Bogor.

MacKinnon, J. 1983b. *Irrigation and watershed protection in Indonesia*. Report to the World Bank.

MacKinnon, J. 1988. *Field guide to the birds of Java and Bali*. Gadjah Mada Univ. Press, Yogyakarta.

MacKinnon, J. 1992. Species survival plan for the orangutan. In *Forest biology and conservation in Borneo* (eds. G. Ismail, M. Mohamed and S. Omar), pp. 209-219. Center for Borneo Studies Publication No. 2.

MacKinnon, J. and Artha, M.B. 1981. *National conservation plan for Indonesia Vol. 5: Kalimantan*. FAO, Bogor.

MacKinnon, J. et al. 1981-1982. *National conservation plan for Indonesia*. 8 vols. FAO, Bogor.

MacKinnon, J. and MacKinnon, K. 1974. *Animals of Asia - the ecology of the Oriental region*. Eurobook, London.

MacKinnon, J. and MacKinnon, K. 1977. The formation of a new gibbon group. *Primates* 18(3): 701-708.

MacKinnon, J. and MacKinnon, K. 1978. Comparative feeding ecology of six sympatric primates in West Malaysia. In *Recent advances in primatology, vol.1. Behaviour* (eds. D.J. Chivers and J. Herbert), pp. 305-321. Academic Press, London.

MacKinnon, J. and MacKinnon, K. 1980a. Niche differentiation in a primate community. In *Malayan forest primates: ten years study in tropical rain forest* (ed. D.J. Chivers), pp. 167-190. Plenum, New York.

MacKinnon, J. and MacKinnon, K. 1980b. The behaviour of wild spectral tarsiers. *Int. J. Primatol.* 1(4): 361-379.

MacKinnon, J. and MacKinnon, K. 1984. Territoriality, monogamy and song in gibbons and tarsiers. In *The lesser apes. Evolutionary and behavioural biology* (eds. H. Preuschoft, D.J. Chivers, W.Y. Brockelman and N. Creel), pp. 291-297. Edin-

burgh University Press.
MacKinnon, J. and MacKinnon, K. 1986. *Review of the protected areas system in the Indo-Malayan Realm.* IUCN, Gland.
MacKinnon. J., MacKinnon K., Child, G. and Thorsell, J. 1986. *Managing protected areas in the tropics.* IUCN, Gland.
MacKinnon, J. and Phillipps, K. 1993. *A field guide to the birds of Borneo, Sumatra, Java and Bali.* Oxford University Press, Oxford, New York and Tokyo.
MacKinnon, J. and Warsito. 1982. *Gunung Palung Reserve Kalimantan Barat. Preliminary management plan.* FAO, Bogor.
MacKinnon, J. and Wind, J. 1980. *Birds of Indonesia (draft).* FAO, Bogor.
MacKinnon, K. 1978. Stratification and feeding differences among Malayan squirrels. *Malay. Nat. J.* 30 (3/4): 593-608.
MacKinnon, K. 1983. Report of a WHO consultancy to Indonesia to determine population estimates of the cynomolgus or long-tailed macaque *Macaca fascicularis* (and other primates) and the feasibility of semi-wild breeding projects of this species. WHO Primate Resources Programme Feasibility Study: Phase II.
MacKinnon, K. 1986. The conservation status of nonhuman primates in Indonesia. In *Primates, the road to self sustaining populations* (ed. K. Benirschke), pp. 99-126, Springer-Verlag.
MacKinnon, K. 1986. *Alam asli Indonesia. Flora, fauna dan keserasian.* Gramedia, Jakarta.
MacKinnon, K. 1988. *Natural resources management in Kalimantan.* Consultant's report of biodiversity specialist. USAID, Jakarta.
MacKinnon, K. 1990a. *Biological diversity in Indonesia: a resource inventory.* WWF, Bogor.
MacKinnon, K. 1990b. *Conserving biodiversity and endangered species in Borneo.* Keynote paper to Conference on forest biology and conservation in Borneo. Kota Kinabalu, Sabah.
MacKinnon, K. 1990c. Kutai National Park. *Voice of Nature* 80: 51-55.
MacKinnon, K. 1992. *Nature's treasurehouse. The wildlife of Indonesia.* Gramedia, Jakarta.
MacKinnon, K., Irving, A. and Bachruddin, M.A. 1994. A last chance for Kutai National Park: local industry support for conservation. *Oryx* 28: 191-198.
MacKinnnon, K. and MacKinnon, J. 1991. Habitat protection and re-introduction programmes. In *Beyond captive breeding. Re-introducing endangered mammals to the wild.* (ed. J. Gipps), pp. 173-198, Oxford University Press.
MacNae, W. 1968. A general account of the fauna and flora of mangrove swamps and forests in the Indo-Pacific region. *Adv. Mar. Biol.* 6: 673-720.
Madiapura, T., Amir and Zulfahmi. 1977. *Batugamping dan dolomit di Indonesia.* Publikasi teknik - Seri Geologi Ekonomi No. 8, Direktorat Geologi, Dirjen Pertambangan Umum, Departemen Pertambangan.
Magnusson, W.E. 1980. Hatching and creche formation by *Crocodylus porosus. Copeia* 2: 359-362.
Majid, Z. 1982. The West Mouth, Niah, in the prehistory of Southeast Asia. *Sarawak Mus. J.* 31. Special Monograph 3.
Malingreau, J.P., Stephens, G., and Fellows, L., 1985. The 1982-83 forest fires of Kalimantan and North Borneo: Satellite observations for detection and monitoring. *Ambio* 14(6): 314-321.
Malley, D.F. 1977. Adaptations of decapod crustaceans to life in mangrove swamps. *Mar. Res. Indonesia* 18: 63-72.
Mallinckrodt 1928. *Het adatrecht van Borneo,* Dubbeldeman. Leiden.
Maloney, B.K. 1985. Man's impact on the rainforests of West Malesia: the palynological record. *J. Biog.* 12: 537-558.
Maloney, B.K. 1987. The pollen and pteridophyte spore record. *Report of the 1982-1983 Bukit Raya Expedition* (ed. H.P. Nooteboom), Rijksherbarium, Leiden.
Maltby, E. 1986. *Waterlogged wealth. Why waste the world's wet places?.* Earthscan, London and Washington.
Manaputty, D.N. 1955. Keluarga *Agathis* di Indonesia. *Rimba Indonesia* 4: 132-188.
Mann, K.H. 1975. Patterns of energy flow. In *River ecology* (ed. B.A. Whitton), pp. 248-263. Blackwell, Oxford.
Mann, K.H. 1982. *Ecology of coastal waters. A systems approach.* Blackwell Scientific Publications, Oxford. Studies in Ecology vol. 8.
Manokaran, N. 1979. Stemflow, throughflow and rainfall interception in a lowland tropical rainforest in Peninsular Malaysia. *Malay. Forester* 42: 174-201.
Manokaran, N. 1980. The nutrient contents of precipitation, throughfall and stemflow in a lowland tropical rainforest in Peninsular Malaysia. *Malay. Forester* 43: 266-289.
Manokaran, N. and Wong, K.M. 1983. The

silviculture of rattans–an overview with emphasis on experiences from Malaysia. Paper presented at the Bangladesh small and cottage industries corporation training course on rattan (cane) furniture manufacturing, Dhaka, Bangladesh.

Mapala Sylva, 1989. *Laporan perjalanan ekspedisi penelusuran gua dan pemanjatan tebing Kotabaru.* Fakultas Kehutanan, UNLAM, Banjarbaru.

Marlier, G. 1973. Limnology of the Congo and Amazon rivers. In *Tropical forest ecosystems in Africa and South America; a comparative review.* (eds B.J. Meggers, E.S. Ayensu and D.W. Duckworth), pp. 223-238, Smithsonian Institution Press, Washington.

Marquis, R.J., Young, H.J. and Braker, H.E. 1986. The influence of understory vegetation cover on germination and seedling establishment in a tropical lowland wet forest. *Biotropica* 18(4): 273-278.

Marsh, C.W. and Wilson, W.L. 1981. *A survey of primates in Peninsular Malaysian forests.* Universiti Kebangsaan Malaysia and University of Cambridge, U.K.

Marsh, C.W., Johns, A.D. and Ayres, J.M. 1987. Effects of habitat disturbance on rain forest primates. In *Primate conservation in the tropical rain forest* (eds. C.W. Marsh and R.A. Mittermeier), pp. 83-107. Alan R. Liss, New York.

Marsh, C. and Sinun, W. 1992. *Pragmatic approaches to habitat conservation within a large timber concession in Sabah, Malaysia.* Paper presented to the Fourth World Congress on National Parks. Caracas, Venezuela.

Marshall, A.G. 1971. The ecology of *Basilla hispida* (Diptera: Nycteribiidae) in Malaysia. *J. Anim. Ecol.* 40: 141-154.

Marshall, A.G. 1980. The comparative ecology of insects ectoparasitic upon bats. In *Ecology of bats* (ed. T.H. Kunz), pp. 369-397. Plenum, New York.

Marshall, A.G. 1985. Old World phytophagous bats (Megachiroptera) and their food plants: a survey. *Zool. J. Linn. Soc.* 83: 351-369.

Marshall, J.T. and Sugardjito, J. 1978. The mystery of the colonization of Kalimantan by gibbons. Unpubl. ms.

Martin, R.D., Doyle, G.A. and Walker, A.C. (eds.) 1976. *Prosimian behaviour.* Gerald Duckworth and Co. Ltd.

Martin, E.B. 1983. *Rhino exploitation.* WWF, Hong Kong.

Martin, P.S. and Klein, R.G. (eds.) 1984. *Quaternary extinctions: a prehistoric revolution.* University of Arizona Press, Tucson.

Martosewojo, S., Burhannudin and Sutomo, A.B. 1982. Makanan ikan gelodok *Boleophthalmus boddaerti* dari muara sungai Banyuasin dan sungai Jenerbang. In *Prosiding seminar II ekosistem mangrove* pp 259-269, Lembaga Oseanologi Nasional, Jakarta.

Martosubroto, P. and Naamin, N. 1977. Relationships between tidal forests (mangroves) and commercial shrimp production in Indonesia. *Mar. Res. Indonesia* 18: 81-86.

Masefield, G.B. (ed.) 1969. *The Oxford book of food plants.* Oxford University Press.

Mash, K. 1975. *How invertebrates live.* Elsevier Phaidon, Oxford.

Massing, A.W. 1981. The journey to paradise; funerary rites of the Benuaq Dayak of East Kalimantan. *Borneo Res. Bull.* 14: 56-84.

Massing, A.W. 1982. Where medicine fails: belian disease prevention and curing rituals among the Lawangan Dayak of East Kalimantan. *Borneo Res. Bull.* 14: 56-84.

Massing, A.W. 1986. The Central Mahakam basin in East Kalimantan. A socio-economic survey. *Borneo Res. Bull.* 18-1: 64-100.

Mathias, J.A. 1977. The effect of oil on seedlings of the pioneer *Avicennia intermedia* in Malaysia. *Mar. Res. Indonesia* 18: 17.

Matsumoto, T. and Abe, T. 1979. The role of termites in an equatorial rain forest ecosytem of West Malaysia. II. Leaf litter consumption on the forest floor. *Oecologica* 38: 261-274.

Matteson, P.C., Altieri, M.A. and Gagné. 1984. Modification of small farmer practices for better pest management. *Ann. Rev. Entomol* 29: 383-402.

Mayer, J. 1988. Letter from East Kalimantan. *Wallaceana* 52-53: 19-23.

McClure, H.E. 1966. Flowering, fruiting and animals in the canopy of a tropical rain forest. *Malay. Forester* 29: 182-203.

McClure, H.E. 1978. Some arthropods of the dipterocarp forest in Malaya. *Malay. Nat. J.* 32: 31-51.

McClure, H.E., Lim, B.L. and Winn, S.E. 1967. Fauna of the Dark Cave, Batu Caves, Kuala Lumpur, Malaysia. *Pac. Insects.* 9: 399-428.

McKey, D. 1975. The ecology of coevolving seed dispersal systems. In *Coevolution of animals and plants* (eds. L.E. Gilbert and P. Raven), pp. 159-191. University of Texas Press, Texas.

McKey, D. 1978. Soils, vegetation and seed eating by black colobus monkeys. In *The ecology of arboreal folivores* (ed. S. Montgomery), pp. 423-437. Smithsonian Institution Press, Washington.

McKey, D., Waterman, P.G., Mbi, C.N., Gartlan, J.S. and Struhsaker, T.T. 1978. Phenolic content of vegetation in two African rain forests; ecological implications. *Science* 202: 61-64.

McLay, C.L. 1970. A theory concerning the distance travelled by animals entering the drift of a stream. *J. Fish. Res. Board Canada* 27: 359-370.

McNeely, J.A. 1987. How dams and wildlife can coexist: natural habitats, agriculture, and major water resource development projects in tropical Asia. *Cons. Biol.* 1(3): 228-237.

McNeely, J.A. 1988. *Economics and biological diversity. Developing and using economic incentives to conserve biological resources.* IUCN, Gland, Switzerland.

McNeely, J.A. and Miller, K. R. 1983. *IUCN, national parks and protected areas.* UN Economic and Social Commission for Asia and the Pacific, Bangkok.

McNeely, J.A., Miller, K.R., Reid, W.V., Mittermeier, R.A. and Werner, T.B. 1990. *Conserving the world's biological diversity.* IUCN, World Resources Institute, Conservation International, WWF-US and the World Bank.

Medway, Lord 1957. Birds nest collecting. *Sarawak Mus. J.* 8(10): 252-260.

Medway, Lord. 1958. 300,000 bats. *Sarawak Mus. J.* 8(12): 667-668.

Medway, Lord, 1959. Niah animal bones II. *Sarawak Mus. J.* 9(13-14): 151-163.

Medway, Lord 1960a. The Malay tapir in Quaternary Borneo. *Sarawak Mus. J.* 9: 364-367.

Medway, Lord 1960b. Cave swiftlets. In *The birds of Borneo* (ed. B.E. Smythies), Oliver and Boyd, Edinburgh.

Medway, Lord. 1962. The swiftlets (*Collocalia*) of Niah Cave, Sarawak. Part 1, Breeding biology. *Ibis* 104: 45-66.

Medway, Lord. 1963. The antiquity of trade in edible birds' nests. *Fed. Mus. J.* 8: 36-47.

Medway, Lord. 1969. Studies on the biology of the edible-nest swiftlets of South-East Asia. *Malay Nat. J.* 22(2): 57-63.

Medway, Lord. 1971a. Observations of social and reproductive biology of the bent-winged bat *Miniopterus australis* in northern Borneo. *J. Zool. Lond.* 165: 261-273.

Medway, Lord. 1971b. The importance of Taman Negara in conservation of mammals. *Malay. Nat. J.* : 24: 212-214.

Medway, Lord 1972a. The Quaternary mammals of Malesia: a review. *Univ. of Hull Dept. Geog. Misc. Ser.* 13: 63-83.

Medway, Lord. 1972b. Phenology of tropical rainforest in Malaya. *Biol. J. Linn. Soc.* 4: 117-146.

Medway, Lord 1972c. The Gunong Benom Expedition 1967. The distribution and altitudinal zonation of birds and mammals on Gunong Benom. *Bull. Brit. Mus. Nat. Hist. (Zool.)* 23: 105-154.

Medway, Lord. 1976. Hunting pressure on orangutans in Sarawak. *Oryx* 13(4): 332-333.

Medway, Lord 1977a. *Mammals of Borneo. Field keys and an annotated checklist.* Monograph of the Royal Asiatic Society, Kuala Lumpur.

Medway, Lord. 1977b. The Niah excavations and an assessment of the impact of Early Man on mammals in Borneo. *Asian Perspectives* 20(1): 51-69.

Medway, Lord. 1983. *The wild mammals of Malaya (Peninsular Malaysia) and Singapore.* Oxford University Press, Kuala Lumpur.

Medway, Lord and Wells, D.R. 1971. Diversity and density of birds and mammals at Kuala Lompat, Pahang. *Malay. Nat. J.* 24: 238-247.

Medway, Lord and Wells, D.R. 1976. *The birds of the Malay Peninsula.* Witherby, London.

Meene, E.A. van de 1982. Geological aspects of peat formation in the Indonesian - Malaysian lowlands. In *Seminar on peat for energy use.* Bandung, Indonesia.

Meher-Homji, V.M. 1988. Effects of forests on precipitation in India. In *Forests, climate and hydrology* (eds E.R.C. Reynolds and F.B. Thompson), pp. 131-143, United Nations University, Tokyo.

Meijer, W. 1965. Forest types in North Borneo and their economic aspects. In *Proceedings of the symposium on humid tropics vegetation.* UNESCO, Kuching.

Meijer, W. 1969. Fruit trees in Sabah (North Borneo). *Malay. Forester* 32(3): 252-265.

Meijer, W. 1970. Regeneration of tropical lowland forest in Sabah, Malaysia, forty years after logging. *Malay. Forester* 33:

204-229.
Meijer, W. 1971. Plant life in Kinabalu National Park. *Malay. Nat. J.* 24: 184-189.
Meijer, W. 1985. Saving the world's largest flower. *Nat. Geog.* (July): 136-140.
Meijer, W. 1982. Plant refuges in the Indo-Malesian region. In *Biological diversification in the tropics* (ed. G.T. Prance), pp. 576-584. Columbia Univ. Press, New York.
Meijer, W. and Wood, G.H.S. 1964. *Dipterocarps of Sabah.* Sabah Forest Record 5. Forest Department, Sandakan, Sabah.
Metcalf, P. 1976. Birds and deities in Borneo. *Bijd. Tot. de Taal. Land en Volkenkunde* 132: 96-123.
Miller, T.B. 1981. Growth and yield of logged-over mixed dipterocarp forest in East Kalimantan. *Malay. Forester* 44: 419-424.
Mirmanto, E., Kartawinata, K. and Suriadarma, A. 1989. Mangrove and associated plant communities in the Barito river estuary and its vicinity, South Kalimantan. *Ekologi Indonesia* 1(2): 42-56.
Mitani, J.C. 1984. The behavioral regulation of monogamy in gibbons (*Hylobates muelleri*). *Behav. Ecol. Sociobiol.*15: 225-229.
Mitani, J.C. 1985. Mating behavior of male orangutans in the Kutai Game Reserve, Indonesia. *Anim. Behav.* 33: 392-402.
Mitani, J.C. 1987. Territoriality and monogamy among agile gibbons (*Hylobates agilis*). *Behav. Ecol. Sociobiol.* 20: 265-290.
Mitchell, B.A. 1963. Forestry and tanah beris. *Malay. Forester* 26: 160-170.
Mittermeier, R.A. 1988. Primate diversity and the tropical forest: case studies from Brazil and Madagascar and the importance of the megadiversity countries. In *Biodiversity* (ed. E.O. Wilson), pp. 145-156. National Academy Press, Washington, D.C.
MOF/FAO, 1991. *Indonesian tropical forestry action plan.* 3 vols. (second draft). Ministry of Forestry, Jakarta.
Moffet, M.W. 1986. Observations on *Lophomyrmex* ants from Kalimantan, Java and Malaysia. *Malay. Nat. J.* 39: 207-211.
Mogea, J. 1984. Struktur dan komposisi hutan primer dan sekunder di Tanah Grogot, Kalimantan Timur. In *Laporan teknik 1982-1983, Penelitian peningkatan pendayagunaan sumberdaya hayati* (ed. S. Wirjoatmodjo), pp. 179-181. Lembaga Biologi Nasional-LIPI, Bogor.
Mogea, J. 1987. Notes on the Palmae collected during the expedition.In *Report of the 1982-1983 Bukit Raya Expedition* (ed. H.P. Nooteboom). Rijksherbarium, Leiden.
Mogea, J.P. and Wilde de, W.J.J.O. 1982. *Short report on the visit to the forest area at the upper Sambas river and Bukit Raya mountain, Central Kalimantan, Indonesia. November - December 1982.*
Mohr, E.C.J. 1945. The relation between soil and population density in the Netherlands Indies. In *Science and scientists in the Netherlands Indies* (eds P. Honig and F. Verdoorn), pp. 254-262, New York.
Molengraaff, G.A.F 1900. *Geologische Verkenningstochten in Centraal Borneo. 1893-94.* Brill, Leiden.
Molengraaff, G.A.F., 1902. *Borneo Expedition - Geological Explorations in Central Borneo (1893-94).* Kegan Paul, Trench, Trubner and Co., London.
Moll, E.O. 1980. Natural history of the river terrapin, *Batagur baska* (Gray) in Malaysia (Testudines: Emydidae). *Malay. J. Sci.* 6(A): 23-62.
Mooney, H.A., Vithousek, P.M. and Matson, P.A. 1987. Exchange of materials between terrestrial ecosystems and the atmosphere. *Science* 238: 926-992.
Moore, M. 1982. *Economic concepts and methods for valuing protected areas.* Paper presented to 3rd World Congress on National Parks, Bali.
Morat, P., Veillon, J.M. and Mackee, H.S. 1984. Floristic relationships of New Caledonia rain forest phanerograms. In *Biogeography of the tropical Pacific* (eds. R. Radovsky, P.H. Raven and S.H. Sohmer), pp. 71-128. Association of Systematics Collections and Bernice P. Bishop Museum Special Publications No. 72. Honolulu.
Morgan, S. 1968. Iban aggressive expansion: some background factors. *Sarawak Mus. J.* 16: 141-185.
Morgan, R.P.C. 1986. *Soil erosion and conservation* (ed. D.A. Davidson). Longman Scientific and Technical, Essex, England.
Mori, T., Rahman, Z. H. Abd. and Tan, C.H. 1980. Germination and storage of rotan manau (*Calamus manan*) seeds. *Malay. Forester* 43(1): 44-55.
Mori, T. 1980. Growth of rotan manau (*Calamus manan*). *Malay. Forester* 43(2): 187-192.
Morley, R.J. 1981. Development and vegetation dynamics of a lowland ombroge-

nous peat swamp in Kalimantan Tengah, Indonesia. *J. Biog.* 8: 383-404.
Morley, R.J. 1982. A palaeoecological interpretation of a 10,000 year pollen record from Danau Padang, Central Sumatra, Indonesia. *J. Biogeogr.* 9: 151-90.
Morley, R.J. and Flenley, J.R. 1987. Late Cainozoic vegetational and environmental changes in the Malay archipelago. In *Biogeographical evolution of the Malay Archipelago*. (ed. T.C. Whitmore), pp. 50-59. Clarendon Press, Oxford.
Morris, H.S. 1953. *Report on the Melanau sago producing community in Sarawak*. HMSO, London.
Morris, H.S. 1978. The coastal Melanau. In *Essays on Borneo societies* (ed V.T. King), pp. 37-58, Oxford University Press.
Morrison, H. 1957/1976. *Sarawak*. Federal Publ., Singapore.
Morton, J.F. 1976. Craft industries from coastal wetland vegetation. In *Estuarine products* (ed. M.L. Wiley), pp. 254-266, Academic Press, New York.
Morton, J. 1990. *The shore ecology of the tropical Pacific*. UNESCO, Jakarta.
Mosby, H.S. (ed.) 1963. *Wildlife investigational techniques*. 2nd edition. The Wildlife Society, Washington, D.C.
Moss, B. 1980. *Ecology of fresh waters*. Blackwell Scientific Publications, Oxford.
Moyle, P.B. and Senanayake, F.R. 1984. Resource partitioning among the fishes of rainforest streams in Sri Lanka. *J. Zool.* 202: 195-223.
Muller, J. 1965. Palynological study of Holocene peat in Sarawak. In *Proceedings of the symposium on humid tropics vegetation*. UNESCO, Kuching.
Muller, J. 1970. Palynological evidence on early differentiation of angiosperms. *Biol. Rev.* 45: 417-450.
Muller, J. 1972. Palynological evidence for change in geomorphology, climate and vegetation in the Mio-Pliocene of Malesia. In *The Quaternary era in Malesia* (eds. P. Ashton and M. Ashton). Department of Geography, University of Hull. Misc. Series 13.
Murphy, D.H. 1973. Animals in the forest ecosystem. In *Animal life in Singapore* (ed. C.H. Chuang). University of Singapore Press.
Muscatine, L. 1973. Nutrition of corals. In *Biology and geology of coral reefs*. (eds. O.A. Jones and R. Endean), pp. 77-115, Academic Press, New York, San Francisco and London.

Musser, G.G. 1987. The mammals of Sulawesi. In *Biogeographical evolution of the Malay archipelago* (ed. T.C. Whitmore), pp. 73-91, Oxford Science Publications.
Myers, L.C. 1978. Geomorphology. In *Kinabalu, summit of Borneo* (eds. M. Luping, W. Chin and R.E. Dingley). The Sabah Society, Kota Kinabalu.
Myers, N. 1979. *The sinking ark*. Pergamon Press, Oxford.
Myers, N. 1983. *A wealth of wild species: storehouse for human welfare*. Westview, Boulder, Colorado.
Myers, N. 1984. *The primary source: tropical forests and our future*. W.W. Norton, New York.
Myers, N. 1986. *Tackling mass extinction of species: A great creative challenge*. Albright Lecture, University of California, Berkeley.
Myers, N. 1988a. Tropical forests: much more than stocks of wood. *J. Trop. Ecol.* 4: 209-221.
Myers, N. 1988b.Tropical forest and their species: going, going...? In *Biodiversity* (ed. E.O. Wilson), pp. 28-37. National Academy Press, Washington, D.C.
Myers, N. 1988c. Tropical deforestation and climatic change. *Environmental Conservation* 15(4): 293-297.
Nagendran, J. 1991. *PROKASIH: a river cleanup program in Indonesia*. EMDI, Jakarta.
Nash, S.V. and Nash, A.D. 1986. *The ecology and natural history of birds in the Tanjung Puting National Park, Central Kalimantan, Indonesia*. WWF/IUCN Project 1687, Bogor.
National Research Council 1976. *Making aquatic weeds useful: some perspectives for developing countries*. National Academy of Sciences.
National Research Council 1983. *Little-known Asian animals with a promising economic future*. National Academy Press, Washington, D.C.
Nayoan, G.S. 1981. Offshore hydrocarbon potential of Indonesia. *Energy* 6(11): 1225-1246.
Nectoux, F. and Kuroda, Y., 1989. *Timber from the south seas: an analysis of Japan's tropical timber trade and its environmental impact*. WWF International.
Needham, R. 1953. *The social organisation of Penan, a Southeast Asian people*. Ph.D. thesis, Oxford.
Newbery, D.McC., Campbell, E.J.F., Lee, Y.F.,

Ridsdale, C.E. and Still, M.J. 1992. Primary lowland dipterocarp forest at Danum Valley, Sabah, Malaysia: structure, relative abundance and family composition. In *Tropical rain forest: disturbance and recovery* (eds. A.G. Marshall and M.D. Swaine), pp. 341-356. The Royal Society, London.

Newbery, D. McC. and Proctor, J. 1984. Ecological studies in four contrasting lowland rain forests in Gunung Mulu National Park, Sarawak. IV. Associations between tree distribution and soil factors. *J. Ecol.* 72: 475-493.

Newbery, D. McC., Renshaw, E. and Brünig, E.F. 1986. Spatial patterns of trees in kerangas forest, Sarawak. *Vegetatio* 65:77-89.

Newbold, J.D., Erman D.C. and Roby K.B. 1980. Effects of logging on macroinvertebrates in streams with and without buffer strips. *Can. J. Fish. aquat. Sci.* 37: 1076-1085.

Newman, J.R. and Schreiber, R.K. 1984. Animals as indicators of ecosystem responses to air emissions. *Environmental Management* 8(4): 309-324.

Newton, K.1960. Shifting cultivation and crop rotations in the tropics. *Papua New Guinea Agric. J.* 13(3).

Ng, F.S.P. 1983. Ecological principles of tropical lowland rain forest conservation. In *Tropical rain forest ecology and management* (eds. S. L. Sutton, T.C. Whitmore and A.C. Chadwick), pp. 359-376, Blackwell, Oxford.

Ng, K.L. and Kang, N. 1988. The mud lobster *Thalassina. Nat. Malay.* 13(4): 28-31.

Ng, R. and Lee, S.S. 1982. The vertical distribution of insects in a tropical primary lowland dipterocarp forest in Malaysia. *Malay. J. Sci.* 7: 37-52.

Niemitz, C. 1979. Outline of the behaviour of *Tarsius bancanus*. In *The study of prosimian behaviour* (eds. G.A. Doyle and R.D. Martin). Academic Press, New York.

Niemitz, C. (ed.) 1984. *Biology of tarsiers.* Gustav Fischer Verlag.

Nieuwenhuis, A.W. 1904, 1907. *Quer durch Borneo.* 2 vols. Brill, Leiden.

Nirarita, C.E. 1992. A preliminary study on litterfall, litter turnover, time and nutrient status in Gunung Palung National Park, West Kalimantan, Indonesia. In *Forest biology and conservation in Borneo* (eds. G. Ismail, M. Mohamed and S.Omar), pp. 331-338. Center for Borneo Studies Publication No. 2.

Nisbet, I.C.T. 1968. The utilization of mangrove by Malayan birds. *Ibis* 110: 345-352.

Norman, M.J.T., Pearson, C.J. and Searle, P.G.E. 1984. *The ecology of tropical food crops.* Cambridge University Press.

Nontji, A. 1988. Coral reef pollution and degradation by LNG plant in south Bontang Bay (East Kalimantan), Indonesia. In *Coastal zone management in the Strait of Malacca* (eds. P.R. Burbridge, Koesoebiono, H. Dirschl and B. Patton), pp. 98-117, SRES, Dalhousie.

Nooteboom, H.P.(ed.) 1987. *Report of the 1982-1983 Bukit Raya expedition.* Rijksherbarium, Leiden, The Netherlands.

NPWO/Interwader 1985. *Survey of the Sarawak coast to evaluate the status of coastal wetlands and to identify key sites for migrating waterbirds.* Interwader, Kuala Lumpur.

Nursall, J.R. 1981. Behaviour and habitat affecting the distribution of five species of sympatric mudskippers in Queensland. *Bull. Mar. Sci.* 3(3): 730-735.

Nye, P.H. and Greenland, D.J. 1960. The soils under shifting cultivation. *Tech. Commun. Commonw. Bur. Soil Sci.* 51.

O'Brien, W.J. and Dixon, P. 1976. The effects of oil and oil components on algae: a review. *Br. Phycol. J.* 11: 115-142.

Odum, H.T. 1971. *Environment, power and society.* John Wiley and Sons, New York.

Odum, W.E. 1976. *Ecological guidelines for tropical coastal development.* IUCN Publications New Series No. 42.

Odum, W.E. 1984. The relationship between protected coastal areas and marine fisheries genetic resources. In *National parks, conservation and development* (eds. J.A. McNeely and K.S. Miller), pp. 648-655, Smithsonian Institution Press, Washington.

Ohler, J.G. 1984. *Coconut, tree of life.* FAO plant production and protection paper no. 57. FAO, Rome.

Ohler, J.G. 1989. *Cocos nucifera* L. In *Plant resources of South-East Asia* (eds. E. Westphal and P.C.M. Jansen), pp. 91-95 Pudoc, Wageningen.

Oldeman, L.R., Las, I. and Muladi. 1980. The agroclimatic maps of Kalimantan, Maluku, Irian Jaya and Bali, West and East Nusa Tenggara. *Contr. Centr. Res. Inst. Agric. Bogor.* No. 60.

Oldfield, S. 1988a. *Rare tropical timbers.* The IUCN Tropical Forest Programme, IUCN, Cambridge.

Oldfield, S. 1988b. *Buffer zone management in*

tropical moist forests. Case studies and guidelines. The IUCN Tropical Forest Programme, IUCN, Cambridge.

Olivier, R. 1978. Distribution and status of the Asian elephant. *Oryx* 14(4): 379-424.

Ong, J.E. 1982. Mangroves and aquaculture in Malaysia. *Ambio* 11: 252-257.

Ong, J.E. 1985. Seven years of productivity studies in a Malaysian managed mangrove forest. Then what? In *Coasts and tidal wetlands of the Australian monsoon region* (eds. K.N. Bardsley, J.D.S. Davie and C.D. Woodroffe), pp. 213-223. Mangrove Monograph No. 1, Australian National University North Australia Research Unit, Darwin.

Ong, J.E., Gong, W.K. and Wong, C.H. 1980. *Ecological survey of the Sungai Merbok estuarine mangrove ecosystem.* School of Biological Sciences, Universiti Sains Malaysia, Penang.

Orians, G.H. 1982. The influence of tree falls in tropical forests on tree species richness. *J. Trop. Ecol.* 23: 68-84.

Osmaston, H.A. and Sweeting, M.M. 1982. Geomorphology, Gunung Mulu National Park, Sarawak. *Sarawak Mus. J. Special Issue* 51: 75-94.

O'Loughlin, C.L. 1974. The effect of timber removal on the stability of forest soil. *Hydrology* 13: 121-134.

O'Toole, C. (ed.) 1987. *The encyclopedia of insects.* Guild Publishing, London.

Ovington, J.D. 1984. Ecological processes and national park management. In *National parks, conservation, and development: the role of protected areas in sustaining society* (eds. J.A. McNeely and K.R. Miller), pp. 60-64, Smithsonian Institution Press, Washington, D.C.

Padoch, C. 1982. *Migration and its alternatives among the Iban of Sarawak.* Martinus Nijhoff, The Hague.

Padoch, C. 1985. Labor efficiency and intensity of land use in rice production: an example from Kalimantan. *Human Ecol.* 13(3): 271-289.

Padoch, C. 1988. Agriculture in interior Borneo: shifting cultivation and alternatives. The University Magazine of Archaeology/Anthropology, University of Pennsylvania. *Expedition* (30)1: 18-28.

Padoch, C., Jessup, T.C., Soedjito, H. and Kartawinata, K. 1991. Complexity and conservation of medicinal plants: anthropological cases from Peru and Indonesia. In *Conservation of medicinal plants* (eds. O. Akerele, V. Heywood and H. Synge), pp. 321-328, Cambridge University Press.

Padoch, C. and Vayda, A.P. 1983. Patterns of resource use and human settlements in tropical forests. In *Tropical rain forest ecosystems: structure and function* (ed. F.B. Golley), pp. 301-313. Ecosystems of the World, Vol. 14A. Elsevier, Amsterdam.

Paijmans, K. 1976. Vegetation. In *New Guinea vegetation* (ed. K. Paijmans). Elsevier, Amsterdam.

Palmieri, J.R., Purnowo, Dennis, D., Marwoto and Harijono, A. 1980. Filarid parasites of South Kalimantan (Borneo) Indonesia *Wuchereria Kalimantani* sp. (Nematoda: Filarioidea) from silvered leaf monkey, *Presbytis cristatus* Escholtz 1921. *J. Parasitol* 66(4): 645-651.

Partomihardjo, T., Yusuf, R., Sunarti, S., Purwaningsih, Abdulhadi, R. and Kartawinata, K. 1987. A preliminary note on gaps in a lowland dipterocarp forest in Wanariset, East Kalimantan. In *Proceedings of the Third round table conference on dipterocarps* (ed. A.J. G.H. Kostermans), pp. 241-253. UNESCO/ROSTSEA. Jakarta.

Paterson, S. 1982. Short notes: observations on ant associations with rainforest ferns in Borneo. *Fern Gazette* 12(4): 243-245.

Payne, A.J. 1986. *The ecology of tropical lakes and rivers.* J. Wiley and Sons. Chichester.

Payne, J.B. 1979. Synecology of Malayan tree squirrels with special reference to the genus *Ratufa.* Ph.D. thesis, University of Cambridge.

Payne, J.B. 1980. Competitors. In *Malayan forest primates: ten years study in tropical rain forest* (ed. D.J. Chivers), pp. 261-277. Plenum, New York.

Payne, J. 1992. Why are rhinoceroses rare in Borneo forests? In *Forest biology and conservation in Borneo* (eds. G. Ismail, M. Mohamed and S. Omar), pp. 169-174. Center for Borneo Studies Publication No. 2.

Payne, J. Francis, C.M. and Phillipps, K. 1985. *A field guide to the mammals of Borneo.* The Sabah Society and WWF Malaysia.

Peace, W.J.H. and Macdonald, F.D. 1981. An investigation of the leaf anatomy, foliar mineral levels, and water relations of trees of a Sarawak forest. *Bio-tropica* 13(2): 100-109.

Pearce. D. and Markandaya, A. 1986. *The costs of natural resource depletion in low income developing countries.* Discussion Paper No. 85-23, University College, London.

Pearce, K.G., Aman, V.L. and Jok, S. 1987. An ethnobotanical study of an Iban community of the Pantu subdistrict, Sri Aman, Division 2, Sarawak. *Sarawak Mus. J.* 37(58): 193-270.

Pearsall, S. 1984. In absentia benefits of natural resources: a review. *Environmental Conservation* 11(1): 3-10.

Pearson, D.L. 1975. A preliminary survey of the birds of the Kutai Reserve, Kalimantan Timur. *Treubia* 28(4): 157-162.

Pearson, D.L. 1977. A pantropical comparison of bird community structure on six lowland forest sites. *Condor* 79: 231-244.

Pearson, D.L. 1980. Patterns of limiting similarity in tropical forest tiger beetles (Coleoptera: Cicindelidae). *Biotropica* 12(3): 195-204.

Peluso, N.L. 1983a. *Markets and merchants: the forest products trade of East Kalimantan in historical perspective.* Ph.D. thesis, Cornell University.

Peluso, N.L. 1983b. Networking in the commons: a tragedy for rattan?. *Indonesia* 35: 95-108.

Peluso, N.L. 1986. *Rattan industries in East Kalimantan, Indonesia.* FAO.

Peluso, N.L. 1989. *The role of non-timber forest products in shifting cultivation communities and households: current knowledge and prospects for development.* Ministry of Forestry and FAO, Jakarta.

Peluso, N.L. 1992. The ironwood problem: (mis)management and development of an extractive rainforest product. *Cons. Biol.* 6(2): 210-219.

Peluso, N.L. 1993. *The impacts of social and environmental change on indigenous peoples' forest management in West Kalimantan, Indonesia.* Case studies in community forestry. FAO, Rome.

Perry, L.M. 1980. *Medicinal plants of East and Southeast Asia: attributed properties and uses.* MIT Press, Cambridge.

Peters, C.M., Gentry, A.H. and Mendelsohn, O. 1989. Valuation of an Amazonian rainforest. *Nature* 339: 655-656.

Petocz, R., Wirawan, N. and MacKinnon, K. 1990. *The Kutai National Park: planning for action.* WWF, Bogor.

Pfeffer, 1959. Biologie et migrations du sanglier de Borneo (*Sus barbatus* Muller 1869). *Mammalia* 23: 277-303.

Pfeffer, P. and Caldecott, J. 1986. The bearded pig in East Kalimantan and Sarawak. *J. Malay. Br. Royal Asiat. Soc.* 59(2): 81-100.

Phillipps, A. and Lamb, A. 1988. Pitcher plants of East Malaysia and Brunei. *Nat. Malay.* 13(4): 8-27

Phillipson, J. 1966. *Ecological energetics.* The Institute of Biology's Studies in Biology No. 1. Edward Arnold, Great Britain.

PHPA, AWB and KPSL-UNLAM. 1990. *Conservation of the Sungai Negara wetlands, Barito Basin, South Kalimantan.* PHPA/AWB-Indonesia and KPSL-UNLAM.

Pieters, P.E. and Supriatna, S. 1990. Late Cretaceous-Early Tertiary continent collision in Borneo. In *Terrane analysis of China and the Pacific rim* (eds. T.J. Wiley, D.G. Howell and F.L. Wong), pp. 193-194, Circum-Pacific Council for Energy and Mineral resources, Houston, Texas.

Piggott, A.G. 1988. *Ferns of Malaysia in colour.* Tropical Press Sdn. Bhd., Kuala Lumpur.

Pijl, L. van der, 1957. The dispersal of plants by bats (chiropterochory). *Acta Bot. Nederl.* 6: 291-315.

Pijl, L. van der 1982. *The principles of dispersal in higher plants.* Third edition. Springer, Berlin.

Plotkin, M.J. 1988. The outlook for new agricultural and industrial products from the tropics. In *Biodiversity* (ed. E.O. Wilson), pp. 106-118, National Academy Press, Washington.

Plowman, K. 1981. Resource utilisation by two New Guinea rain forest ants. *J. Anim. Ecol.* 50: 903-916.

Pocs, T. 1982. Tropical forest bryophytes. In *Bryophyte ecology* (ed. A.F.J. Smith), pp. 59-104, Chapman and Hall, London.

Polak, E. 1949. *Verslag einer tournee naar Kapuas en Kapuasmeren van 18 September - 39 October 1949.* Verslag van het Bodemkundig Instituut van het Algemeen Proefstation voor de Landbouw, Bogor, Indonesia.

Polak, B. 1950. Occurrence and fertility of tropical peat soils in Indonesia. Transactions IVth International Congress of Soil Science, vol. 2, pp. 183-185, Amsterdam.

Polunin, I. 1988. *Plants and flowers of Malaysia.* Times Editions, Singapore.

Polunin, N.V.C. 1983. The marine resources of Indonesia. *Oceanogr. Mar. Biol. Ann. Rev.* 21: 455-531.

Poore, D. 1989. *No timber without trees: sustainability in the tropical forest.* Earthscan, London.

Poore, D. and Sayer, J. 1988. *The management of tropical moist forest lands. Ecological guidelines.* The IUCN Tropical Forest

Programme, IUCN, Gland.
Poore, M.E.D. 1968. Studies in Malaysian rain forest I. The forest on the Triassic sediments in Jengka Forest Reserve. *J. Ecol.* 56: 143-196.
Popham, P. 1987. A rumble in the jungle. *Sunday Times Magazine* (December 6): 38-44.
Potter, L. 1987. Degradation, innovation and social welfare in the Riam Kiwa valley, Kalimantan, Indonesia. In *Land degradation and society* (eds. P. Blaikie and H. Brookfield), pp. 164-176. Methuen, London and New York.
Potter, L. 1988a. Eating the forest in gulps and nibbles. Concessionaires, transmigrants and 'free loggers' in South Kalimantan. *Inside Indonesia* October: 19-21.
Potter, L. 1988b. *Indigenes and colonisers: Dutch forest policy in south and east Borneo (Kalimantan), 1900 to 1950.* Paper presented for IUFRO. Tropical Forests Working Group, Conference on Tropical Forest History in South and Southeast Asia, ANU, Canberra.
Potter, L. and Hasymi, A. 1988 Indonesian regional economicsurveys: South Kalimantan, the Banjarese heartland. Draft, to be published in *Indonesian regional economics surveys* (ed. H. Hill). Oxford University Press, Singapore.
Prance, G.T. (ed.) 1986. *Tropical forests and the world atmosphere.* Westview Press, Boulder, Colorado.
Pratt, T.K. and Stiles, E.W. 1985. The influence of fruit size and structure on composition of frugivore assemblages in New Guinea. *Biotropica* 17(4): 314-321.
Prescott-Allen, R. and Prescott-Allen, C. 1982. *What's wildlife worth? Economic contributions of wild plants and animals to developing countries.* Earthscan, London.
Prescott-Allen, R. and Prescott-Allen, C. 1983. *Genes from the wild.* Earthscan, London.
Preston-Martin, R. and Preston-Martin, K. 1984. *Spiders of the world.* Blandford Press, Poole.
Prieme, A. and Heegard, M. 1988. A visit to Gunung Nyiut in West Kalimantan. *Kukila* 4: 138-140.
Primack, R.B. 1985. Comparative studies of fruits in wild and cultivated trees of chempedak (*Artocarpus integer*) and terap (*Artocarpus odoratissimus*) in Sarawak, East Malaysia with additional information on the reproductive biology of the Moraceae in Southeast Asia. *Malay. Nat. J.* 39: 1-39.
Primack, R.B. and Tomlinson, P.B. 1978. Sugar secretions from the buds of *Rhizophora. Biotropica* 10: 74-75.
Proctor, J. (ed.). 1989. *Mineral nutrients in tropical forest and savanna ecosystems.* Special Publication No. 9, British Ecological Society, Blackwell, Oxford.
Proctor, J. 1992. The vegetation over ultramafic rocks in the tropical Far East. In *The ecology of areas with serpentized rocks* (eds. B.A. Roberts and J. Proctor), pp. 249-270. Kluwer Academic Publishers, The Netherlands.
Proctor, J., Anderson, J.M., Chai, P. and Vallack, H.W. 1983. Ecological studies in four contrasting lowland rainforests in Gunung Mulu National Park, Sarawak. I. Forest environment, structure and floristics. *J. Ecol.* 71:237-260.
Proctor, J., Anderson, J.M., Fogden, S.C.L. and Vallack, H.W. 1983. Ecological studies in four contrasting lowland rain forests in Gunung Mulu National Park, Sarawak. II. Litterfall, litter standing crop and preliminary observations on herbivory. *J. Ecol.* 71: 261-283.
Proctor, J., Anderson, J.M. and Vallack, H.W. 1982. Ecological studies in four forest types. *Sarawak Mus. J.* 51: 207-224.
Proctor, J., Anderson, J.M. and Vallack, H.W. 1983. Comparative studies on forests, soils and litterfall at four altitudes on Gunung Mulu, Sarawak. *Malay. Forester* 46(1):60-76.
Proctor, J., Baker, A.J.M. and Reeves, R.D. (eds.) 1992. *The vegetation of ultramafic (serpentine soils).* Lavoisier Publishing Inc., Newark, N.J.
Proctor, J., Lee, Y.F., Langley, A.M., Munro, W.R.C. and Nelson, T. 1988. Ecological studies on Gunung Silam, a small ultrabasic mountain in Sabah, Malaysia. I. Environment, forest structure and floristics. *J. Ecol.* 76: 320-340.
Proctor, J., Phillipps, C., Duff, G.K., Heaney, A. and Robertson, F.M. 1989. Ecological studies on Gunung Silam, a small ultrabasic mountain in Sabah, Malaysia. II. Some forest processes. *J. Ecol.* 77: 317-331.
Proctor, J. and Woodell, S.R.J. 1975. The ecology of serpentine soils. *Adv. Ecol. Res.* 9: 225-366.
Proctor, M.C. and Yeo, P.F. 1973. *The pollination of flowers.* Collins, London.
Proud, K.R.S. 1978. Some notes on a captive earless monitor lizard *Lanthanotus borneensis. Sarawak Mus. J.* 24: 47:235-242.
Proud, K.R.S. and Hutchinson, I.D. 1980.

Management of natural reserves to maintain faunal diversity: a potential use for forest silviculture. In *Tropical ecology and development* (ed. J.I. Furtado), pp. 247-255. Institute of Tropical Ecology, Kuala Lumpur.
Prowse, G.A. 1958. Fish and food chains. *Malay. Nat. J.* 12: 66-71.
Prowse, G.A. 1968. Pollution in Malayan waters. *Malay. Nat. J.* 21: 149-158.
Pupipat, U. 1983. The ecology of pests: pathogens. *BIOTROP Spec. Publ.* 18: 65-78.
Purchon, R.D. and Enoch, I. 1954. Zonation of the marine fauna and flora on a rocky shore near Singapore. *Bull. Raffles Mus. S'pore* 25: 47-65.
Purseglove, J.W. 1974. *Tropical crops: dicotyledons* (2nd edition). Longman Group Ltd., London.
Purseglove, J.W. 1975. *Tropical crops: monocotyledons* (2nd edition). Longman Group Ltd., London.
Putman, R.J. 1983. *Carrion and dung: the decomposition of animal wastes.* Edward Arnold, Great Britain.
Putz, F.E. 1979. A seasonality in Malaysian tree phenology. *Malay. Forester* 42: 1-24.
Putz, F.E. and Appanah, S. 1987. Buried seeds, newly dispersed seeds, and the dynamics of a lowland forest in Malaysia. *Biotropica* 19(4): 326-333.
Putz, F.E. and Chan, H.T. 1986. Tree growth, dynamics, and productivity in a mature mangrove forest in Malaysia. *Forest Ecol. and Manag.* 17: 211-230.
Rabinowitz, A., Andau, P. and Chai, P.P.K. 1987. The clouded leopard in Malaysian Borneo. *Oryx* 21(2): 107-111.
Rachman, A.R. and Johari, R. 1987. Prospects for integrating timber production into agricultural plantations and of timber trees as the next major agricultural crop. In *Proceedings of seminar on the future role of forest plantations in the national economy and incentives required to encourage investments in forest plantation development.* pp. 39-52. Tropenbos.
Raffles, T.S. 1817/1988. *The history of Java.* Oxford University Press, Singapore.
Raich, J.W. 1983. Effects of forest conversion on the carbon budget of a tropical soil. *Biotropica* 15(3): 177-184.
Rambo, A.T. 1979. Primitive man's impact on genetic resources of the Malaysian tropical rain forest. *Malay. Appl. Biol.* 8(1): 59-65.
Rambo, A.T. 1980. Of stones and stars: Malaysian orang asli environmental knowledge in relation to their adaptation to the tropical rain forest ecosystem. *Fed. Mus. J.* 25: 77-89.
Rambo, A.T. 1982. Human ecology research on tropical agroecosystems in Southeast Asia. *Singapore J. Trop. Geog.* 3(1): 86-99.
Rambo, A.T. and Sajise, P.E. 1985. *An introduction to human ecology research on agricultural systems in Southeast Asia.* University of the Philippines, Los Banos.
Rappaport, R.A. 1967. *Pigs for the ancestors.* Yale University Press, New Haven.
Raup, D.M. 1986. Biological extinction in Earth history. *Science* 231: 1528-1533.
Raven, P.H. 1988. Our diminishing tropical forests. In *Biodiversity* (ed. E.O. Wilson), pp. 119-122. National Academy Press, Washington, D.C.
Repetto, R. 1987. Creating incentives for sustainable forest development. *Ambio* 16(2-3): 94-99.
Repetto, R. 1988. *The forest for the trees? Government policies and misuse of forest resources.* World Resources Institute, Washington.
Repetto, R. and Gillis, M. (eds.) 1988. *Public policies and the misuse of forest resources.* Cambridge University Press, Cambridge, UK.
RePPProT, 1985. *Review of phase I results: Central Kalimantan.* Regional Physical Planning Programme for Transmigration, Direktorat Bina Program, Indonesia.
RePPProT, 1987. *Review of phase I results, East and South Kalimantan.* 2 vols. Regional Physical Planning Programme for Transmigration, Direktorat Bina Program, Indonesia.
RePPProT, 1988. *Review of phase I results, West Kalimantan.* Regional Physical Planning Programme for Transmigration, Direktorat Bina Program, Indonesia.
RePPProT, 1990. *The land resources of Indonesia.* ODA/Ministry of Transmigration, Jakarta.
Reuler, H. van. 1987. Introduction. *Report of the 1982-1983 Bukit Raya Expedition* (ed. H.P. Nooteboom). Rijksherbarium, Leiden.
Reynolds, E.R.C. and Thompson, F.B. (eds). 1987. *Forests, climate and hydrology, regional impacts.* United Nations University, Tokyo, Japan.
Rice, C.G. 1989. A further range extension of the black-breasted thrush *Chlamydochaera jefferyi* in Kalimantan. *Kukila* 4: 47-48.

Richards, P.W. 1936. Ecological observations on the rain forest of Mount Dulit, Sarawak. Parts I, II. *J. Ecol.* 24: 1-37, 340-363.

Richards, P.W. 1952. *The tropical rain forest. An ecological study.* Cambridge University Press.

Richards, P.W. 1970. *The life of the jungle.* McGraw-Hill, New York and London.

Richards, P.W. 1973. The tropical rain forest. *Sci. Am.* 229(6): 59-67.

Richards, P.W. 1983. The three-dimensional structure of tropical rain forest. In *Tropical rain forest ecology and management* (eds. S.L. Sutton, T.C. Whitmore and A.C. Chadwick), pp. 3-10. Blackwell Scientific Publications, Oxford.

Rijksen, H.D. 1978. *A field study on Sumatran orangutans (Pongo pygmaeus abelii Lesson 1827). Ecology, behaviour and conservation.* H. Veenman and Zonen B.V., Wageningen.

Riswan, S. 1982. *Ecological studies on primary, secondary and experimentally cleared mixed dipterocarp forest and kerangas forest in East Kalimantan, Indonesia.* Ph.D. thesis, Aberdeen University.

Riswan, S. 1985a. The estimation of temporal processes in tropical rain forest: a study of primary mixed dipterocarp forest in Indonesia. *J. Trop. Ecol.* 1: 171-182.

Riswan, S. 1985b. *Kerangas forest at Gunung Pasir, Samboja, East Kalimantan: its structural and floristic composition.* Paper presented to the Third Round Table Conference of the International Working Group on Dipterocarps, Samarinda.

Riswan, S. 1985c. Nitrogen content of topsoil in lowland tropical forest in East Kalimantan (before and after clear-cutting and burning) *Reinwardtia* 10(2): 131-138.

Riswan, S. and Kartawinata, K. 1986. *Regeneration after disturbance in a kerangas (heath) forest in East Kalimantan, Indonesia.* Paper for International workshop on rain forest regeneration and management.

Riswan, S. and Kartawinata, K. 1987. *Natural regeneration in primary and secondary mixed dipterocarp forest in East Kalimantan, Indonesia.* Paper presented to The international workshop on reproductive ecology of tropical forest plants, Bangi, Malaysia.

Riswan, S. and Kartawinata, K. 1988. *A lowland dipterocarp forest 35 years after pepper plantation in East Kalimantan, Indonesia.* Unpubl. ms.

Riswan, S., Kenworthy, J.B., and Kartawinata, K. 1985. The estimation of temporal processes in the tropical rain forest: a study of primary mixed dipterocarp forest in Indonesia. *J. Trop. Ecol.* 1: 171-182.

Riswan, S. and Yusuf, R. 1986. Effects of forest fire on trees in the lowland dipterocarp forest, East Kalimantan, Indonesia. In *Forest regeneration in Southeast Asia,* BIOTROP.

Robbins, R.G. 1968. The biogeography of tropical rain forest in Southeast Asia. In *Proceedings of the symposium on recent advances in tropical ecology* (eds. R. Misra and B. Gopal). International Society for Tropical Ecology, Varanasi.

Robbins, R.G. 1969. A prerequisite to understanding tropical rain forest. *Malay. Forester* 32: 361-367.

Roberts, T.R. 1973. Ecology of fishes in the Amazon and Congo basins. In *Tropical forest ecosystems in Africa and South America: a comparative review* (eds. B.J. Meggers, E.S. Ayensu and W.D. Duckworth), pp 239-254, Smithsonian Institution Press, Washington, D.C.

Roberts, T.R. 1989. *The freshwater fishes of western Borneo (Kalimantan Barat, Indonesia).* California Academy of Sciences. San Francisco.

Robertson, J.M.Y. and Soetrisno, B.R. 1982. Logging on slopes kills. *Oryx* 16: 229-230.

Robinson, G.S. 1980. Cave dwelling tineid moths: a taxonomic review of the world species (Lepidoptera: Tineidae). *Trans. British Cave Res. Assoc.* 7(2): 83-120.

Rodelli, M.R., Gearing, J.N., Gearing, P.J., Marshall, N. and Sasekumar, A. 1984. Stable isotope ratio as a tracer of mangrove carbon in Malaysian ecosystems. *Oecologia* 61: 326-333.

Rodgers, W.A. and Homewood, K.M. 1982. Species richness and endemism in the Usambara mountain forests, Tanzania. *Biol. Proc. Linn. Soc.* 18: 197-242.

Rodman, P.S. 1973. Population composition and organisation among orangutans of the Kutai reserve. In *Comparative ecology and behaviour of primates* (eds. R.P. Michael and J.H. Crook), pp. 172-209. Academic Press.

Rodman, P.S. 1973. Synecology of Bornean primates I. A test for interspecific interactions in spatial distribution of five species. *Amer. J. Phys. Anthropol.* 38: 655-660.

Rodman, P.S. 1979. Individual activity patterns and the solitary nature of orangutans. In *The great apes* (ed. D. Hamburg and E.R. McCown), pp. 234-255. Benjamin Cummings, Menlo Park.

Rooij, N. de. 1915. *The reptiles of the Indo-Australian Archipelago.*. 2 vols., Brill. Leiden.

Rookmaaker, L.C. 1977. The distribution and status of the rhinoceros, *Dicerorhinus sumatrensis*, in Borneo - a review. *Bijdragen tot de Dierkunde* 47(2): 197-204.

Ross, M.S. 1982. *The southsea log market in relation to the Indonesian Transmigration Program.* East Kalimantan Transmigration Development Project.

Ross, M.S. 1984. *Forestry in land use policy for Indonesia.* Ph.D. thesis, University of Oxford.

Ross, M.S. 1985. *A review of policies affecting the sustainable development of forest lands in Indonesia.* International Institute for Environment and Development, London.

Ross, M.S. and Donovan, D.G. 1986. *Land clearing in the humid tropics.* IUCN/IIED Tropical Forest Policy Paper No. 1.

Roth, H.L. 1968. *The natives of Sarawak and British North Borneo.* 2 vols. University of Malaya Press, Kuala Lumpur.

Roth, L. M. 1980. Cave-dwelling cockroaches from Sarawak, with one new species. *System. Ent.* 5: 97-104.

Rousseau, J. (ed.) 1974. The peoples of Central Borneo. *Sarawak Mus. J. Special Issue* 22.

Rousseau, J. 1977. Kayan agriculture. *Sarawak Mus. J.* 25: 129-156.

Rousseau, J. 1978. The Kayan. In *Essays on Borneo societies* (ed. V.T. King), pp. 78-91. Oxford University Press, Oxford.

Rousseau, J. 1988. Central Borneo: a bibliography. *Sarawak Mus. J. Special Monograph* No. 5.

Rousseau, J. 1990. *Central Borneo, ethnic identity and social life in a stratified society.* Clarendon Press, Oxford.

Royen, van, P. and Kores, P. 1982. The Ericaceae of the high mountains of New Guinea. In *Vegetation of New Guinea* (ed. J. Cramer), pp. 1485-1911.

Rubeli, K. 1986. *Tropical rain forest in South-East Asia. A pictorial journey.* Tropical Press Sdn. Bhd. Kuala Lumpur, Malaysia.

Ruddle, K., Johnson, D., Townsend, P.K. and Rees, J. 1978. *Palm sago, a tropical starch from marginal lands.* Univ. Press of Hawaii, Honolulu.

Ruggieri, C.D. 1976. Drugs from the sea. *Science* 194: 491-497.

Rutter, O. 1929/1985. *The pagans of North Borneo.* Oxford University Press.

Rutter, O. 1930. *The pirate wind. Tales of the sea-robbers of Malaya.* Oxford University Press.

Sabar, F., Djajasasmita, M. and Budiman, A. 1979. Susunan dan penyebaran moluska dari krustacea pada beberapa hutan rawa payau: suatu studi pendahuluan. In *Prosiding seminar ekosistem hutan mangrove* (eds. S. Soemodihardjo, A. Nontji and A. Djamali), pp. 120-125, Lembaga Biologi Nasional, Jakarta.

Saenger, P., Hegerl, E.J. and Davie, J.D.S. (eds.) 1983. *Global status of mangrove ecosystems.* IUCN Commission on Ecology Papers No. 3.

Salati, E., Dall'Olio, A., Matsui, E. and Gat, J.R. 1979. Recycling of water in the Amazon basin: an isotopic study. *Water Resources Res.* 15: 1250-1258.

Salati, E., Lovejoy, T.E. and Vose, P.B. 1981. Precipitation and water recycling in tropical rain forests with special reference to the Amazon basin. *The Environmentalist* 3(1): 67-72.

Salati, E. and Vose, P.B. 1984. Amazon Basin: a system in equilibrium. *Science* 225: 129-138.

Salick, J. 1983. Natural history of crop-related wild species: uses in pest habitat management. *Env. Manag.* 7: 85-90.

Salm, R.V. 1981. Fried rice without shrimp? *Conservation Indonesia* 5(3,4): 4-6.

Salm, R.V. 1984a. *Conservation of marine and littoral habitats.* IUCN/WWF, Bogor.

Salm, R.V. 1984b. *Sea turtle trade.* IUCN/WWF, Bogor.

Salm, R.V. 1984c. *Conservation of marine species in Indonesia.* IUCN/WWF, Bogor.

Salm, R.V. and Clark, J.R. 1984. *Marine and coastal protected areas: a guide for planners and managers.* IUCN, Gland, Switzerland.

Salm, R.V. and Halim, M. 1984. *Marine conservation data atlas.* IUCN/WWF, Bogor.

Salter, R.E. and MacKenzie, N.A. 1985. Conservation status of the proboscis monkey in Sarawak. *Biol. Cons.* 33: 119-132.

Salter, R.E., MacKenzie, N.A., Aken, K.M. and Chai, P.K. 1985. Habitat use, ranging behaviour, and food habits of the proboscis monkey, *Nasalis larvatus* (van Wurmb), in Sarawak. *Primates* 26(4): 436-451.

Sammy, N. 1980. Lichens from Gunong Mulu National Park, Sarawak, East Malaysia. Part 2. Folicolous lichens.

Malay. Nat. J. 34(2): 65-72.
Santiapillai, C. and Suprahaman, H. 1984. *An ecological study of the riverine habitats in the Way Kambas Game Reserve.* WWF, Bogor.
Santiapillai, C. and MacKinnon, K. 1991. Conservation and management of Sumatran rhino (*Dicerorhinus sumatrensis*) in Indonesia. Paper presented to Conference on biology and conservation of rhinoceros, San Diego.
Sasekumar, A. 1974. Distribution of macrofauna on a Malayan mangrove shore. *Anim. Ecol.* 3: 51-69.
Sasekumar, A. and Loi, J.J. 1983. Litter production in three mangrove forest zones in the Malay peninsula. *Aquat. Bot.* 17: 283-290.
Sastrapradja, D.S., Adisoemarta, S., Kartawinata, K., Sastrapradja, S. and Mien Rifai, A. 1989. *Keanekaragaman hayati Indonesia untuk kelangsungan hidup bangsa.* LIPI Bogor.
Sather, C. 1978. Iban folk mycology. *Sarawak Mus.J.*26(47): 81-102.
Sawyer, R.T., Taylor, A. and Sahat, Hj. M.J. bin. 1982. The leeches of Brunei (Annelida: Hirudinea) with a checklist and key to the known and expected freshwater, terrestrial and marine leeches of Borneo. *Brunei Mus. J.* 5(2): 168-197.
Sayer, J. 1991. *Rain forest buffer zones, guidelines for protected area managers.* IUCN, Gland.
Schaik, C.P. van 1985. *The socio-ecology of Sumatran long-tailed macaques (Macaca fascicularis) I. Costs and benefits of group living.* Elinkwijk BV, Utrecht.
Scharer, H. 1946/1963. *Ngaju religion - the concept of God among a South Borneo people.* Nijhoff, The Hague.
Schemske, D.W. and Brokaw, N. 1981. Treefalls and the distribution of the understory birds in a tropical forest. *Ecology* 62(4): 938-945.
Schneeburger, W.F.1979. *Contributions to the ethnology of central northeast Borneo.* Institute of Ecology, University of Berne.
Schophuys, H.J. 1936. *Het stroomgebied van de Barito; Landbouwkundige kenschets en landbouwvoorlichting,* Ph.D. thesis,Wageningen.
Schroeder, R.E. 1980. *Philippine shore fishes of the western Sulu sea.* NMPC Books, Manila.
Schulz, J.P. 1984. *Turtle conservation strategy in Indonesia.* IUCN/ WWF, Bogor.

Schultz, J.P. 1987. *Status of Chelonia mydas and Eretmochelys imbricata in Indonesia.* Consultancy report prepared for the Conservation Monitoring Centre, Cambridge.
Schuster, W.H. 1950. Comments on the importation and transplantation of different species of fish into Indonesia. *Contr. Gen. Agr. Res. Stn. Bogor* 111: 1-31.
Schuster, W.H. and Djajadiredja, R.R. 1952. *Common names of Indonesian fishes.* Min. of Agric. and Lab. for Inland Fisheries, Indonesia. Bandung/The Hague.
Schwaner, C.A.L.M. 1853. *Borneo.* Van Kampen, Amsterdam.
Schweithelm, J. 1987. The need for a method of land evaluation for watershed land use planning in the Outer Islands of Indonesia: a case study of Riam Kanan, Kalimantan. In *Proceedings of the international workshop on quantified land evaluation procedures* (eds. K.J. Beek, P.A. Burrough and D.E. McCormack). ITC. Pub. No. 6.Washington.
Scott, D.A. 1989. *A directory of Asian wetlands.* Compiled for WWF, IUCN, ICBP and IWRB. IUCN, Cambridge.
Scott, N.J. 1976. The abundance and diversity of the herpetofaunas of tropical forest litter. *Biotropica* 8(1): 41-58.
Seavoy, R.E. 1973. The transition to continuous rice cultivation in Kalimantan. *Ann. Assoc. Am. Geog.* 63(2): 218-225.
Seavoy, R.E. 1975. The origin of tropical grasslands in Kalimantan Indonesia. *J. Trop. Geog.* 40: 48-51.
Secrett, C. 1986. The environmental impact of transmigration. *Ecologist* 16:77-88.
Seibert, B. 1988a. Agroforestry for the conservation of genetic resources. In *Agroforestry* (eds. A.B. Lahjie and B. Seibert), pp. 235-251, Fakultas Kehutanan, UNMUL, Samarinda.
Seibert, B. 1988b. *Forest fires in East Kalimantan 1982-83 and 1987: the press coverage.* Faculty of Forestry, Mulawarman University, Samarinda, East Kalimantan. Mimeograph.
Sellato, B. 1986a. *Les nomades forestiers de Borneo et la sédentarisation. Essai de l'histoire economique et sociale.* Ph.D. thesis, Paris.
Sellato, B. 1986b. An ethnic sketch of the Melawi area, West Kalimantan. *Borneo Res. Bull.* 18(1): 46-45.
Sellato, B. 1989a. *Hornbill and dragon.* Elf Aquitaine Indonesie-Elf Aquitaine Malaysia.
Sellato, B. 1989b. *Nomades et sédentarisation à*

Borneo. *Histoire économique et sociale.* Maison des Sciences de l'Homme-Archipel, Paris.

Seow, R.C.W. 1976. *The effect of a mixed organic effluent on the distribution of pelagic macrofauna at Sungai Puloh with special reference to water quality.* B.Sc. thesis. University of Malaya, Kuala Lumpur.

Shanmugasundarum, S. and Sumarno. *Glycine max* (L.). In *Plant resources of South East Asia* (eds. E. Westphal and P.C.M. Jansen) pp. 139-143, Pudoc, Wageningen.

Sheldon, F.H., Mitra, S. and Kennard, J. 1992. The birds of Sabah softwoods exotic tree plantation. In *Forest biology and conservation in Borneo* (eds. G. Ismail, M. Mohamed, and S. Omar), pp. 498-499, Center for Borneo Studies, Publication No. 2.

Shelford, R.W.C. 1916. *A naturalist in Borneo.* Fischer Unwin, London.

Shimokawa, E. 1988. Effects of a fire of tropical rain forest on soil erosion. In *A research on the process of earlier recovery of tropical rain forest after a large scale fire in Kalimantan Timur, Indonesia* (eds. H.Tagawa and N. Wirawan), pp. 2-11, Occ. papers No. 14, Kagoshima Univ.

Short, L.L. 1978. Sympatry in the woodpeckers of lowland Malayan forest. *Biotropica* 10(2): 122-133.

Silvius, M.J., Djuharsa, E., Taufik, A., Steeman, A. and Berczy, E. (comps.) 1987. *The Indonesian wetland inventory.* 2 vols. PHPA-AWB/Interwader and Edwin.

Sim, S.L. 1978. *Illipe nuts and illipe nut research in Sarawak.* Unpubl. ms.

Sim, E.S. 1988. A preliminary phytochemical survey of peat swamps in Sarawak. *Sarawak Mus. J.* 39(60): 203-208.

Simberloff, D.S. 1974. Permo-Triassic extinctions: effects of area on biotic equilibrium. *J. Geol.* 82: 267-274.

Simberloff, D. 1992. Species-area relationships, fragmentation and extinction in tropical forests. In *In harmony with nature* (eds. S.K. Yap and S.W. Lee), pp. 398-413. Proc. Int. Conf. on Conserv. of Trop. Biodiversity.

Simberloff, D.S. and Abele, L.G. 1976. Island biogeography theory and conservation practice. *Science* 191: 285-286.

Simmermon, F. 1967. *The effect of shifting cultivation on the agricultural economy of the peoples of Sarawak, Malaysia.* M.A. thesis. Miami University.

Simmons, I.G. 1981. The ecology of natural resources. Butler and Tanner, Frome and London.

Simons, H. 1987. *Gunung Niut Nature Reserve. Proposed management plan.* WWF/PHPA, Bogor.

Sinclair, A.R.E. and Fryxell, J.M. 1985. The Sahel of Africa: ecology of a disaster. *Can. J. Zool.* 63: 987-994.

Slobodkin, L.B. and Fishelson, L. 1974. The effect of the cleaner fish *Labroides dimidiatus* on the point diversity of fishes on the reef front at Eilat. *Am. Nat.* 108: 369-376.

Smith, C.L. and Tyler J.C. 1972. Space resource sharing in a coral reef fish community. *Bull. Nat. Hist. Mus.* 14. Los Angeles County.

Smith, J.M.B. 1970. Herbaceous plant communities in the summit zone of Mount Kinabalu. *Malay. Nat. J.* 24: 16-29.

Smith, J.M.B. 1977. An ecological comparison of two tropical high mountains. *J. Trop. Geog.* 44: 71-80.

Smith, J.M.B. 1979. Vegetation recovery from drought on Mount Kinabalu. *Malay Nat. J.* 32: 341-342.

Smith, J.M.B. 1980. The vegetation of the summit zone of Mount Kinabalu. *New Phytol.* 84: 547-573.

Smith. M.A. 1931. The herpetology of Mt. Kinabalu, North Borneo, 13,455 ft. *Bull. Raffles Mus.* 5: 3-35.

Smith T.J. 1986. *The influence of seed predators on the structure of tropical tidal forests.* Paper presented at Symposium on Australia's wet tropics, Brisbane.

Smits, W. 1989. *Shorea johorensis.* In *Plant resources of South- East Asia* (eds. E. Westphal and P.C.M. Jansen), pp. 253-254. Pudoc, Wageningen.

Smits, W. 1990. *Forest resource partition through participation with local people.* Voluntary Paper, ASEAN seminar on the management of tropical forests for sustainable development, Jakarta.

Smits, W.T.M., Leppe, D., Yasman, I. and Noor, M. 1992. Ecological approaches to commercial dipterocarp forestry. In *Forest biology and conservation in Borneo* (eds. G. Ismail, M. Mohamed and S. Omar), pp. 432-435. Center for Borneo Studies Publication No. 2.

Smits, W.T.M., Oldeman, R.A.A. and Limonard, T. 1987. Mycorrhizae and Dipterocarpaceae in East Kalimantan rain forests. Reports on some of the research projects. *Tropenbos Newsletter.* 67-77.

Smythies, B.E. 1965. *Common Sarawak trees.* Borneo Literature Bureau.
Smythies, B.E. 1960/1981. *The birds of Borneo.* The Sabah Society and The Malayan Nature Society, Kuala Lumpur.
Smythies, B.E. 1978. Some interesting birds of Kinabalu National Park. In *Kinabalu, summit of Borneo* (eds. M. Luping, N. Chin and R.E. Dingley), pp. 321-346. The Sabah Society, Kota Kinabalu.
Snow, D.W. 1981. Tropical frugivorous birds and their food plants: a world survey. *Biotropica* 13(1): 1-14.
Soedjito, H. 1988. *Spatial patterns, biomass, and nutrient concentrations of root systems in primary and secondary forest trees of a tropical rainforest in Kalimantan, Indonesia.* Unpubl. report. MAB Program.
Soegiarto, A. 1985. The mangrove ecosystem in Indonesia: its problems and management. In *Coasts and tidal wetlands of the Australian monsoon region* (eds. K.N. Bardsley, J.D.S. Davie and C.D. Woodroffee), pp. 313-326. Mangrove Monograph No. 1, Australian National University North Australia Research Unit, Darwin.
Soegiarto, A. and Polunin, N. 1980. *The marine environment of Indonesia.* IUCN/WWF, Bogor.
Soermarwoto, O. 1977. Nitrogen in tropical agriculture: Indonesia as a case study. *Ambio* 6(2-3): 162-165.
Soemodihardjo, S. (ed.) 1988. *Some ecological aspects of tropical forests of East Kalimantan.* Final report, LIPI - Indonesian MAB Committee, UNESCO.
Soepadmo, E. and Eow, B.K. 1976. The reproductive biology of *Durio zibethinus* Murr. *Gardens' Bull. Singapore* 29: 25-33.
Soepraptohardjo, M. 1972. *Generalized soil map, Indonesia.* Scale 1:2,500,000. 3rd ed. Soil Research Institute, Bogor.
Soepraptohardjo, M. and Driessen, P.M. 1976. The lowland peats of Indonesia, a challenge for the future. *Soil Res. Inst. Bogor Bull.* 3: 11-19.
Soerjani, M. 1980. Aquatic plant management in Indonesia. In *Tropical ecology and development* (ed. J.I. Furtado), pp. 725-737. Kuala Lumpur, Malaysia.
Soil Research Institute. 1973a. *Report on soil investigations of the Delta Pulau Petak, South and Central Kalimantan.* Soil Research Institute, Bogor.
Soil Research Institute. 1973b. *Soil types and suitability for agricultural crops.* Soil Research Institute, Bogor.

Sopher, D.E. 1965. *The sea nomads. A study of the maritime boat people of Southeast Asia.* National Museum Publication, Singapore.
Soulé, M.E. (ed.) 1987. *Viable populations for conservation.* Cambridge University Press, Cambridge.
Soulé M.E., Wilcox B.A and Holtby C. 1979. Benign neglect: a model of faunal collapse in the game reserves of East Africa. *Biol. Cons.* 15: 259-272
Specht, R.L. and Womersley, J.S. 1979. Heathlands and related shrublands of Malesia (with particular reference to Borneo and New Guinea). In *Heathlands and related shrublands* (ed. R.L. Specht), pp. 321-378. Elsevier, Amsterdam.
Spellerberg, I.F. 1982. *The biology of reptiles: an ecological approach.* Blackie, Glasgow and London.
Spragg W.T. and Paton R. 1980. Tracing, trophallaxis and population measurement of colonies of subterranean termites (Isoptera) using a radioactive tracer. *Ann. Am. Entomol. Soc.* 73: 708-714.
St. John, S. 1974. *Life in the forests of the Far East.* Oxford in Asia Historical Reprints. Oxford Univ. Press, Kuala Lumpur.
Stanley Price, M.R. 1989. *Animal reintroduction: the Arabian oryx in Oman.* Cambridge University Press.
Stanton, W.R. and Flach, M. (eds.) 1980. *Sago: the equatorial swamp as a natural resource.* Martinus Nijhoff, The Hague.
Stark, N. 1978. Man, tropical forest and the biological life of a soil. *Biotropica* 10(1): 1-10.
Start, A.N and Marshall, A.G. 1976. Nectarivorous bats as pollinators of trees in West Malaysia. In *Tropical trees: variation breeding and conservation* (eds. J. Burley and B.T. Styles), pp. 141-150. Academic Press, London.
Stebbings, R.C. and Kalk, M. 1961. Observations on the natural history of the mudskipper *Periopthalmus sobrinus. Copeia* 1961: 18-27.
Stebbings, R.E. 1984. Bats. In *The encyclopedia of mammals vol.2* (ed. D. Macdonald), pp. 786-817, George Allen and Unwin, London.
Steenis, C.G.G.J. van. 1950. The delimitation of Malaysia and its main plant geographical subdivisions. *Fl. Mal.*1:70-75.
Steenis C.G.G.J. van, 1957. Outline of vegetation types in Indonesia and some adjacent regions. *Proc. Pac. Sci. Cong.* 8: 61-

Steenis, C.G.G.J. van, 1958a. Rhizophoraceae. *Fl. Mal. Ser I* 5: 431-436.
Steenis, C.G.G.J. van, 1958b. Discrimination of tropical shore formations. In *Proceedings of the symposium on humid tropics vegetation.* UNESCO, Tjiawi.
Steenis, C.G.G.J. van, 1958c. Rejuvenation as a factor for judging the status of vegetation types. The biological nomad theory. In *Proceedings of the symposium on humid tropics vegetation,* Kandy, UNES-CO.
Steenis, C.G.G.J. van, 1962. The mountain flora of the Malaysian tropics. *Endeavour* 21: 183-193.
Steenis, C.G.G.J. van, 1964. Plant geography of the mountain flora of Mount Kinabalu. *Proc. Roy. Soc.* (B)161: 7-38.
Steenis, C.G.G.J. van. 1971. Plant conservation in Malaysia. *Bull. Jard. Bot. Nat. Belg.* 41: 189-202.
Steenis, C.G.G.J. van 1972. *The mountain flora of Java.* E. J., Brill, Leiden.
Steenis, C.G.G.J. van. 1977. Autonomous evolution in plants. Differences in plant and animal evolution. *Garden's Bull. Singapore* 29: 103-126.
Steenis, C.G.G.J. van. 1981. *Rheophytes of the world. An account of the flood-resistant flowering plants and ferns and the theory of autonomous evolution.* Sijthoff and Noordhoff, Alpen a/d Rijn.
Stern V.J., Smith R.F., Bosch R. van den, and Hagen K.S. 1959. The integrated control concept. *Hilgardia* 29: 81-101.
Stevens, W.E. 1968. *The conservation of wildlife in West Malaysia.* Office of the Warden, Federal Game Department, Ministry of Lands and Mines Serembang.
Stevenson, D.K. and Marshall, N. 1974. Generalizations on the fisheries potential of coral reefs and adjacent shallow water environments. In *Proceedings of the Second international coral reef symposium,* pp. 147-156. Great Barrier Reef Committee.
Stocker, G.C. 1981. Regeneration of a North Queensland rain forest following felling and burning. *Biotropica* 13(2): 86-92.
Stork, N.E. 1986. An annotated checklist of the Carabidae (including Cicindelinae, Rhysodinae and Paussinae) recorded from Borneo. *Occ. Papers Syst. Entomol.* No 2.
Strickland, S.S. 1986. Long term development of Kejaman subsistence, an ecological study. *Sarawak. Mus. J.* 36(57): 117-178.
Strien, N. J. van. 1985. *The Sumatran rhinoceros Dicerorhinus sumatrensis (Fisher, 1814) in Gunung Leuser National Park, Sumatra, Indonesia: its distribution, ecology and conservation.* Parey, Berlin.
Stuebing, R. 1984. Creeping frogs of Sabah. *Sabah Soc. J.* 7(4): 299-302.
Stuebing, R.B. 1991. A checklist of the snakes of Borneo. *Raffles Bull. Zool.* 39(2): 323-362.
Suchtelen, B.C.C. van. 1933. *Memorie van overgave van de residentie zuideren ooster afdeeling van Borneo.* Unpubl. ms. KIT series, The Hague.
Sudagung, H.S. 1984. *Migrasi swakarsa orang Madura ke Kalimantan Barat.* Universitas Gadjah Mada, Yogyakarta.
Sudjadi, 1988. *Daftar peta tanah (list of soil maps).* Soil Research Institute, Bogor.
Sukardjo, S. 1987. *The biological resources of the mangrove swamp forest in the Apar deltaic system, Tanah Grogot, East Kalimantan.* Paper for conference on wetland and waterfowl conservation, Malacca.
Sukardjo, S. 1988. *Tumpang sari pond as a multiple use concept to save the mangrove forest in Java.* Symposium on mangrove management: its ecological and economic considerations. Bogor.
Sukardjo, S and Toro, A.V. 1987. *Guidelines for the protection and management of mangrove and estuarine wetland in Indonesia.* Paper for Conference on wetland and waterfowl conservation, Malacca.
Sumardja, E.A., Tarmudji and Wind, J. 1984. Nature conservation and rice production in the Dumoga area, North Sulawesi, Indonesia. In *National parks, conservation and development: the role of protected areas in sustaining society* (eds. J.A. McNeely and K.R. Miller), pp. 224-227. Smithsonian Institution Press, Washington, D.C.
Supardi. 1980. The tidal swamp resources as a food supplier in South Kalimantan. In *Proceedings of the 5th international symposium of tropical ecology* (ed. J.I. Furtado), pp. 1065-1069. Institute Trop. Ecology, Kuala Lumpur, Malaysia.
Supriatna, J., Manullang, B.A and Soekara, E. 1984. Diets and densities of maroon leaf-eating monkey (*Presbytis rubicunda*) in Tanjung Puting Reserve, Central Borneo. In *Wildlife ecology in Southeast Asia.* Biotrop Special Publication No. 21.
Suselo, T.B. and Riswan, S. 1985. *Compositional and structural patterns of lowland mixed dipterocarp forest in Kutai National Park, East Kalimantan.* Paper presented to the Third Round Table Conference I.W.G. on Dipterocarps, Samarinda.

Sutisna, U. 1985. Analisa komposisi jenis pohon hutan rawa gambut di Sei Mandor, Kalimantan Barat. *Bull. Penelitian Hutan* 469: 39-66.

Sutlive, V.H. 1978. *The Iban of Sarawak.* AHM Publishing Corporation, Illinois.

Sutton, S.L. 1983. The spatial distribution of flying insects in tropical rain forests. In *Tropical rain forest: ecology and management* (eds. S.L. Sutton, T.C. Whitmore and A.C. Chadwick), pp. 71-97. Blackwell Scientific Publications, Oxford.

Sutton, S.L. and Hudson, P.J. 1980. The vertical distribution of small flying insects in the lowland rain forest of Zaire. *Zool. J. Linn. Soc.* 68: 111-123.

Suzuki, A. 1984. *Socio-ecological studies on the cercopithecoid primates in Kalimantan.* Primate Research Institute, Kyoto University, Japan.

Suzuki, A. 1992. The population of orangutans and other non-human primates and the forest conditions after the 1982-83 fires and droughts in Kutai National Park, East Kalimantan, Indonesia. In *Forest biology and conservation in Borneo* (eds. G. Ismail, M. Mohamed and S. Omar), pp. 190-205. Center for Borneo Studies Publication No. 2.

Swaine, M.D. and Whitmore, T.C. 1988. On the definition of ecological species groups in tropical rain forests. *Vegetatio* 75: 81-86.

Swindler, D.R. and Erwin J. 1986. *Comparative biology Vol. 1. Systematics, evolution and anatomy.* Alan Ness, New York.

Syukur, S.A. 1982. *Penyelamatan hutan mangrove di Kalimantan Barat.* Unpubl. ms. PSL-UNTAN.

TAD 1981. *Forest for food.* Phase I. East Kalimantan Transmigration Area Development Project. Report 11.

Tagawa, H. and Wirawan, N. 1988. *A research on the process of earlier recovery of tropical rainforest after a large scale fire in Kalimantan Timur, Indonesia.* Kagosyma University Research Centre for the South Pacific. Occasional paper No. 14.

Talling, J.F. 1957. Diurnal changes of stratification and photosynthesis in some tropical African waters. *Proc. Roy. Soc. Lond.* B. 17: 57-83.

Tan, E.S.P. 1980. Ecological aspects of some Malaysian riverine cyprinids in relation to their aquaculture potential. In *Proceedings of the Fifth international symposium of tropical ecology* (ed. J.I. Furtado), pp. 757-762. Kuala Lumpur, Malaysia.

Tan, K. 1980. Logging the swamp for food. In *Sago, the equatorial swamp as a natural resource* (eds. W.R. Stanton and M. Fiach), pp. 13-34. Martinus Nijhoff, The Hague.

Tang, H.T. 1986. *Problems and strategies in regenerating dipterocarp forests in Malaysia.* Paper prepared for symposium on natural management of tropical moist forests, Yale School of Forestry and Environmental Studies, U.S.A.

Tang, H.T., Pinso, C. and Marsh, C. 1988. *Proceedings of seminar on the future role of forest plantations in the national economy and incentives required to encourage investments in forest plantation development.* Sabah World Bank Forestry Technical Assistance Project.

Taufik, A.W., Somantri, A. and Utadi, T. 1986. *Monitoring dan evaluasi satwa dan habitat biota laut langka (penyu) di Kalimantan Barat.* PHPA, Bogor.

Taylor, M. and Soekarsono, S.H. 1991. *Development of water quality standards: an Indonesian study.* EMDI, Jakarta.

Temple, S.A. 1977. Plant-animal mutualism: coevolution with dodo leads to near extinction of plant. *Science* 197: 885-886.

Tempo, 1987. Setelah hutan jadi arang. No. 32(10): 49-71.

Teng, S.K. 1970. A preliminary report on the species of planktonic marine diatoms found in Brunei waters. *Brunei Mus. J.* 2(1): 279-315.

Terborgh, J. 1974. Preservation of natural diversity: the problem of extinction prone species. *BioScience* 24: 715-722.

Terborgh, J. 1990. Seed and fruit dispersal. In *Reproductive ecology of tropical forest plants* (eds. K.S. Bawa and M. Hadley). Man and the Biosphere Series, vol. 7: 181-190. UNESCO and The Parthenon Publishing Group.

Thaithong, O. 1984. Bryophytes of the mangrove forest. *J. Hattori. Bot. lab.* 56: 85-87.

Thapa, R.S. 1981. *Termites of Sabah.* Sabah Forest Record No. 12. Sabah.

Tho, Y.P. 1982. Gap formation by the termite *Microcerotermes dubius* in lowland forests of Peninsular Malaysia. *Malay. Forester* 45(2): 184-192.

Tie, Y.L., Baillie, I.C., Phang, C.M.S. and Lim, C.P. 1979. *Soils of Gunung Mulu National Park.* Department of Agriculture, Kuching.

Tillema, H.F. 1938. *Apo Kajan - Een filmreis naar en door Central Borneo.* Van Muster, Amsterdam.

Tinal, U. and Palanewen, J.L. 1978. Mechanical logging damage after selective cutting in the lowland dipterocarp forest at Beloro, East Kalimantan. *BIO-TROP Spec. Publ.* 3: 91-96.
Tjia, H.D. 1980. The Sunda Shelf, Southeast Asia. *Z. Geomorph.* 24: 405-427.
Tjia, H.D., Sujitno, S., Suklija, Y., Harsono, R.A.F., Rachmat, A., Hainim, J. and Djunaedi. 1984. Holocene shorelines in the Indonesian islands. *Mod. Quat. Res. S.E.Asia* 8: 103-117.
Tomlinson, P.B., Primack, R.B. and Bunt, J.S. 1979. Preliminary observations on floral biology in mangrove Rhizophoraceae. *Biotropica* 11(4): 256-277.
Torquebiau, E. 1984. Man-made dipterocarp forest in Sumatra. *Agroforestry Systems* 2: 103-127.
Townsend, P.K.W. 1974. Sago production in a New Guinea economy. *Human Ecology* 2: 217-236.
Townsend, C.R. 1980. *The ecology of streams and rivers.* Edward Arnold, London.
Townsend, C.R. and Hildrew, A.G. 1976. Field experiments on the drifting colonization and continuous redistribution of stream benthos. *J. Anim. Ecol.* 45: 759-772.
Tsing, A.L. 1984. *Politics and culture in the Meratus mountains.* Ph.D. thesis, Stanford University.
Tudge, C. 1991. *Last animals at the zoo. How mass extinction can be stopped.* Hutchinson Radius.
Turner, R.E. 1975. *The Segara Anakan reclamation project; the impact on commercial fisheries.* Report to Engineering Consultants, Denver, Colorado, USA.
Turner, R.E. 1977. Intertidal vegetation and commercial yield of penaeid shrimp. *Trans. Amer. Fish. Soc.* 106(5): 411-416.
Turner, R.E. 1980. Protein yield from wetlands. In *Wetlands ecology and management* (eds. B. Gopal, R.E. Turner, R.G. Wetzel and D.F. Whigham), Nat. Inst. of Ecol. and Int. Science Publishers, Lucknow, India.
Tweedie, M.W.F. 1961. On certain Mollusca of the Malay limestone hills. *Bull. Raffles Mus.* 26: 49-65.
Uhl, C. and Buschbacher, R. 1985. A disturbing synergism between cattle ranch burning practices and selective tree harvesting in the eastern Amazon. *Biotropica* 17(4): 265-268.
Uhl, C., Clark H., Clark K. and Maquirino P. 1982. Successional patterns associated with slash-and-burn agriculture in the upper Rio Negro region of the Amazon basin. *Biotropica* 14: 249-254.
Uhl, C., Clark, K., Dezzeo N. and Maquirino P. 1988. Vegetation dynamics in Amazonian treefall gaps. *Ecology* 69: 751-763.
Uhlig, H. 1980. Man and tropical karst in Southeast Asia: geoecological differentiation, land use and rural development potentials in Indonesia and other regions. *Geo. J.* 4: 31-44.
USAID. 1987. *Natural resources and environmental management in Indonesia: an overview.* USAID, Jakarta.
Usop, M. 1984. *Comprehensive investment profile Central Kalimantan.* UNIDO and Palangka Raya University.
Vayda, A. 1979. Human ecology and economic development in Kalimantan and Sumatra. *Borneo Res. Bull* 2-1: 23-32.
Vayda, A.P. 1980. Buginese colonization of Sumatra's coastal swamplands and its significance for development planning. In *Proceedings of the Jakarta workshop on coastal resources management* (eds. E.C.F. Bird and A. Soegiarto), pp. 80-87. United Nations University, Tokyo.
Vayda, A.P., Colfer, C.J.P. and Brotokusumo, M. 1980. *Interactions between people and forests in East Kalimantan.* East-West Environment and Policy Institute. Reprint No. 13.
Vayda, A.P. and Sahur, A. 1985. *Forest clearing and pepper farming by Bugis migrants in East Kalimantan: antecedents and impact.* East-West Environment and Policy Institute. Reprint No. 83.
Veevers, J.J. 1988. Morphotectonics of Australia's northwestern margin - a review. In *The North West shelf, Australia,* pp. 19-27, Petroleum Exploration Society, Perth.
Vergara, B.S. and De Datta, S.K. 1989. *Oryza sativa.* In *Plant resources of South-East Asia* (eds. E. Westphal and P.C.M. Jansen), pp. 206-213. Pudoc, Wageningen.
Verstappen, H.T. 1960. Some observations on karst development in the Malay archipelago. *J. Trop. Geogr.* 14: 1-10.
Verstappen, H.T. 1973. *A geomorphical reconnaissance of Sumatra and adjacent islands.* International Training Centre, Netherlands.
Vietmeyer, N. 1986b. Lesser-known plants of potential use in agriculture and forestry. *Science* 232 1379-1384.
Vitt L.J. 1983. Tail loss in lizards: the significance of foraging and escape modes. *Herpetol.* 39: 151-162.
Vogel, E.F. de. (ed.) 1987. *Manual of*

herbarium taxonomy. Theory and practice. UNESCO - MAB.

Voous, K.H. 1961. Birds collected by Carl Lumholtz in eastern and central Borneo. *Nytt Magasin for Zoologi* 10: 127-180.

Voss, F. 1979. *Natural resources inventory, East Kalimantan.* TAD Report No. 9.Transmigration Area Development Project, Samarinda.

Voss, F. 1983. *Kalimantan Timur.* TAD Atlas. Transmigration Area Development Project (TAD), Samarinda.

Vossen, H.A.M. van der and Soenaryo, 1989. *Coffea* L. In *Plant resources of South-East Asia* (eds. E. Westphal and P.C.M. Jansen), pp. 95-102. Pudoc, Wageningen.

Vu Van Dung, Pham Mong Giao, Nguyen Chinh, Do Tuoc, Arctander, P. and MacKinnon, J. 1993. A new species of living bovid from Vietnam. *Nature* 363(June): 443-445.

Waard, P.W.F. de. 1989. *Piper nigrum.* In *Plant resources of South-East Asia* (eds. E. Westphal and P.C.M. Jansen), pp. 225-230. Pudoc, Wageningen.

Wade, P. 1958. Breeding seasons among mammals in the lowland rainforest of North Borneo. *J. Mammal.* 39(3): 429-433.

Walker, D. 1982. Speculations on the origin and evolution of Sunda-Sahul rain forests. In *Biological diversification in the tropics* (ed. G. Prance), pp. 554-575. Columbia University Press, New York.

Wallace, A.R. 1869/1986. *The Malay archipelago.* Oxford University Press.

Wallace, B.J. 1989. Vascular epiphytism in Australo-Asia. In *Tropical rain forest ecosystems* (ed. H.Leith and M.J.A. Werger), pp. 261-282. Elsevier, Amsterdam.

Walsh, R.P.D. 1978. Solute concentrations of rain, soil and river water in two very wet tropical environments: Dominica and Sarawak. *Swansea Geographer* 16: 26-32.

Walsh, R.P.D. 1981. River basin planning and conservation in a newly-created national park in a tropical rain forest region. The Gunung Mulu National Park, Sarawak. In *River basin planning: theory and practice* (eds. S.K.Saha and C.J. Barrow), pp. 79-92. Wiley, London.

Walsh, R.P.D. 1982. Hydrology and water chemistry. *Sarawak Mus. J. Special Issue* No. 2: 121-182.

Waltham, A.C. and Brook, D.B. 1979. Caves in Mulu hills. *Geogr. Mag.* 51: 486-491.

Waltham, A.C. and Brook D.B. 1980. Geomorphological observations in the limestone caves of the Gunung Mulu National Park, Sarawak. *Trans. British Cave Res. Assoc.* 7(3): 123-139.

Walton, O.E. 1978. Substrate attachment by drifting aquatic insect larvae. *Ecology* 59: 1027-1030.

Ward, P. 1968. Origin of the avifauna of urban and suburban Singapore. *Ibis* 110: 239-255.

Watanabe, H., Ruaysongern S. and Takeda H. 1983. Soil animals. In *Shifting cultivation*, pp. 110-126. Min. Sciences Technology and Energy, Bangkok.

Waterman, P.G. 1983. Distribution of secondary metabolites in rain forest plants: towards an understanding of cause and effect. In *Tropical rain forest:. ecology and management* (eds. S.L. Sutton, T.C. Whitmore and A.C. Chadwick), pp. 167-179. Blackwell, Oxford.

Waterman, P.G. and Mole, S. 1989. Soil nutrients and plant secondary compounds. In *Mineral nutrients in tropical forest and savanna ecosystems* (ed. J.Proctor), pp. 241-252. Blackwell Scientific Publications, Oxford.

Waterman, P.G., Ross, J.A.M., Bennett, E.L. and Davies A.G. 1988. A comparison of the floristics and leaf chemistry of the tree flora in two Malaysian rain forests and the influence of leaf chemistry on populations of colobine monkeys in the Old World. *Biol. J. Linn. Soc.* 34: 1-32.

Watson, D.J. 1982. Subsistence fish exploitation and implications for management in the Baram river system, Sarawak. *Malay Fisheries Res* 1: 299-310.

Watson, D.J. and Balon, E.K. 1984. Structure and production of fish communities in tropical rain forest streams of northern Borneo. *Can. J. Zool.* 62: 927-940.

Watson, D.J. and Balon, E.K. 1984. Ecomorphological analysis of fish taxocenes in rainforest streams of northern Borneo. *J. Fish Biol.* 25: 371-384.

Watson, H. 1985. *An overview of the conservation of reptiles and amphibians in Sarawak.* Report prepared for Special Select Committee of Dewan Undangan Negeri, Sarawak.

Watt, A.S. 1947. Pattern and process in the plant community. *J. Ecol.* 35:1-22.

Weinstock, J.A. 1979. *Land tenure practices of the swidden cultivators of Borneo.* Ph.D. thesis, Cornell University.

Weinstock, J.A. 1983. Rattan ecological bal-

ance in a Borneo rainforest swidden. *Economic Botany* 37(1): 58-68.
Weinstock, J.A. 1987. Land or plants: agricultural tenure in agroforestry systems. *Economic Botany* 41(2): 312-322.
Weinstock, J.A. 1989. *Study on shifting cultivation in Indonesia*. Phase I Report. FAO, Jakarta.
Welcomme, R.L. 1979. *Fisheries ecology of floodplain rivers*. Long-man, London.
Wells, D.R. 1971. Survival of the Malaysian bird fauna. *Malay. Nat. J.* 24: 248-56.
Wells, D.R. 1976a. Resident birds. In *Birds of the Malay Peninsula* (eds. Lord Medway and D.R. Wells) chapter 1. Witherby, London.
Wells, D.R. 1976b. Some bird communities in western Sabah, with distributional records. *Sarawak Mus. J.* 14: 277- 286.
Wells, D.R. 1978. Numbers and biomass of insectivorous birds in the understorey of rain forest at Pasoh. *Malay. Nat. J.* 30: 353-362.
Wells, D.R. 1985. Forest avifauna of West Malesia and its conservation. In *Conservation of tropical forest birds* (eds. A.W. Diamond and T.E. Lovejoy), pp. 213-232. ICBP Technical Publication No. 4, Cambridge.
Wells, D.R., Hails, C.J. and Hails, A.J. 1979. *A study of the birds of Gunung Mulu National Park, Sarawak, with special emphasis on those of lowland forests*. Royal Geography Society Gunung Mulu Expedition.
Wells F.E. 1984. Comparative distribution of macromolluscs and macrocrustaceans in a northwestern Australian mangrove system. *Aust. J. Mar. Freshwater Res.* 35: 591-596.
Wells, K.D. 1979. Reproductive behaviour and male mating success in a neotropical toad *Bufo typhonius*. *Biotropica* 11(4): 301-307.
Wells, M., and Brandon, K. with Hannah, L. 1992. *People and parks. Linking protected area management with local communities*. The World Bank, Washington, D.C.
Wessel, M. and Toxopeus, H. 1989. *Theobroma cacao*. In *Plant resources of South-East Asia* (eds. E. Westphal and P.C.M. Jansen), pp. 265-270. Pudoc, Wageningen.
Western, D. and Henry, W. 1979. Economics and conservation in third world national parks. *BioScience* 29(7): 414-418.
Westphal, E. and Jansen, P.C.M. (eds.) 1989. *Plant resources of South-East Asia. A selection*. Pudoc, Wageningen.
Wheelwright, N.T., Haber, A.W., Murray, K.G. and Guindon, C. 1984. Tropical fruit-eating birds and their food plants: a survey of a Costa Rican lower montane forest. *Biotropica* 16(3): 173-192.
Wharton, C. 1968. Man, fire and wild cattle in South East Asia. *Proceedings of the Tall timber fire ecology conference*. 8: 107-167.
White, A. 1987. *Philippine coral reefs, a natural history guide*. New Day Publishers, Quezon City, Philippines.
Whitmore, T.C. (ed.) 1972. *Tree flora of Malaya*, Vol.1, Longman, Kuala Lumpur.
Whitmore, T.C. 1973. *Palms of Malaya*. Oxford University Press, Kuala Lumpur.
Whitmore, T.C. 1977. A first look at *Agathis*. *Trop. For. Papers* 2. Commonwealth Forestry Institute, Oxford.
Whitmore, T.C. 1978. Gaps in the forest canopy. In *Tropical trees as living systems* (eds. P.B. Tomlinson and M.H. Zimmermann), pp. 639-655. Cambridge University Press.
Whitmore, T.C. 1983. Secondary succession from seed in tropical rain forests. *Forestry Abstracts* 44(12): 767-779.
Whitmore, T.C. 1984a. *Tropical rain forests of the Far East*. (2nd ed.). Clarendon Press, Oxford.
Whitmore, T.C. 1984b. Vegetation map of Malesia, at scale 1:5 million. *J. Biogeogr.* 11: 461-471.
Whitmore, T.C. (ed.) 1987. *Biogeographical evolution of the Malay archipelago*. Oxford Science Publications.
Whitmore, T.C. 1990. *An introduction to tropical rain forests*. Clarendon Press, Oxford, UK.
Whitmore, T.C., Peralta, R. and Brown, K. 1985. Short communication: total species count in a Costa Rican tropical rain forest. *J. Trop. Ecol.* 1: 375-378.
Whitmore, T.C. and Tantra, I.G.M. (eds.). 1987. *The flora of Indonesia. Draft check list for Borneo*. Forest Research and Development Centre, Bogor.
Whitten, A.J., Damanik, S.J., Anwar, J. and Nazaruddin, H. 1987a. *The Ecology of Sumatra*. Second edition. Gadjah Mada University Press, Yogyakarta.
Whitten, A.J., Mustafa, M. and Henderson, G.S. 1987b. *The Ecology of Sulawesi*. Gadjah Mada University Press, Yogyakarta.
Whitten, A.J., Haeruman, H., Alikodra, H.S. and Thohari, M. 1987c. *Transmigration and the environment in Indonesia. The past, present and future*. IUCN Forest Programme, IUCN.
Whitten, A.J. and Whitten, J.E.J. 1981. The sago palm and its exploitation on Siberut island, Indonesia. *Principes* 25(3): 91-100.

Whitten, J.E.J. 1981. Ecological separation of three diurnal squirrels in tropical rainforest on Siberut island, Indonesia. *J. Zool. Lond.* 193: 405-420.

Whitten, J.E.J. and Whitten, A.J. 1987. Analysis of bark eating in a tropical squirrel. *Biotropica* 19(2): 107-115.

Whittier, H.L. 1973. *Social organization and symbols of social differentiation; an ethnographic study of the Kenyah Dayaks of East Kalimantan (Borneo)*. Ph.D. thesis, Michigan State University.

Whittier, H.L. 1978. The Kenyah. In *Essays on Borneo societies* (ed. V.T. King), pp. 92-122. Oxford University Press, Oxford.

Whittier, H.L. and Whittier, P.R. 1974. The Apo Kayan of East Kalimantan. In *The peoples of central Borneo* (ed. J. Rousseau), pp. 8-15. Sarawak Mus. J. Special Issue Vol. 22.

Widen, K. and Setiaven, N. 1990. Aspek-aspek sosial ekonomi de daerah lahan basah. In *Conservation of Sungai Negara wetlands, Barito Basin, South Kalimantan*, pp. 143-171. PHPA/AWB, Bogor.

Wiebe W.J., Johannes, R.E. and Webb, K.L. 1975. Nitrogen fixation in a coral reef community. *Science* 188: 257-259.

Wiersum, K.F. 1979. *Introduction to principles of forest hydrology and erosion, with special reference to Indonesia*. Lembaga Ekologi Universitas Padjadjaran, Bandung.

Wilcox, B.A. 1984. In situ conservation and genetic resources: determinants of minimum area requirements. In *National parks, conservation and development: the role of protected areas in sustaining society* (eds. J.A. McNeely and K.R. Miller), pp. 639-647. IUCN/ Smithsonian Institution Press, Washington, D.C.

Wilcox, B.A. 1986. Extinction models and conservation. *Trends in Ecol. Evol.* 1: 46-48.

Wilford, G.E. 1951. Cave phosphate deposits in Sarawak and North Borneo. *Brit. Borneo Geol. Survey Ann. Report* 1951: 32-41, 55-62.

Wilford, G.E. 1960. Radiocarbon age determinations of Quaternary sediments in Brunei and North east Sarawak. *British Borneo Geol. Survey Ann. Report*, 1959.

Wilken, G.A. 1893. *Handleiding voor de Vergelijkende volkenkunde van Nederlandsch-Indie*. Brill, Leiden.

Wilkinson, R., Dutson, G. and Sheldon, B. 1991. *The avifauna of Barito Ulu, Central Borneo*. Report of the Barito Bird Project, 1989. ICBP Study Report No. 48.

William, W., Wong, W. and Lamb, A. 1992. Species diversity of wild fruit trees in the forests of Sabah as illustrated by the genera *Artocarpus, Durio*, and *Mangifera*. In *Forest biology and conservation in Borneo* (eds. G. Ismail, M. Mohamed and S. Omar), pp. 41-60. Center for Borneo Studies Publication No. 2.

Williams, D.D. 1981. Migrations and distributions of stream benthos. In *Perspectives in running water ecology* (eds. M.A. Lock and D.D. Williams), pp. 155-207. Plenum, New York.

Williams, P.R. and Harahap, B.H. 1986. Geochemistry, age and origin of post-subduction intrusive rocks in West Kalimantan and Sarawak. *Bull. Geol. Res. and Dev. Centre* 12: 43-53.

Williams, P.R. 1989. A late Cretaceous to early Tertiary accretionary complex in West Kalimantan. *Geol. Res. and Development Centre Bull.* 13, Bandung.

Williams, T.R. 1965. *The Dusun. A North Borneo society*. Holt, Rinehart and Winston, Inc., New York.

Willis, E.O. 1974. Population and local extinction of birds on Barro Colorado island, Panama. *Ecol. Monogr.* 44: 153-159.

Wilson, C.C. and Wilson, W.L. 1975. The influence of selective logging on primates and some other animals in East Kalimantan. *Fol. Primatol.* 23: 245-274.

Wilson, D.S. 1974. Prey capture and competition in the ant lion. *Biotropica* 6(3): 187-193.

Wilson, E.O. 1987. The arboreal ant fauna of the Peruvian Amazon forests: a first assessment. *Biotropica* 2: 245-251.

Wilson, E.O. (ed.) 1988. *Biodiversity*. National Academy Press. Washington, D.C.

Wilson, W.L. and Johns, A.D. 1982. Diversity and abundance of selected animal species in undisturbed forest, selectively logged forest and plantations in East Kalimantan Indonesia. *Biol. Cons.* 24: 205-218.

Winser, S. and Jermy, C. 1985. *Expedition and survey of Gunung Mulu National Park, Sarawak: a bibliography*. Royal Geographical Society.

Wint, G.R.W. 1983. Leaf damage in tropical rain forest canopies. In *Tropical rain forest: ecology and management* (eds. S.L. Sutton, T.C. Whitmore, and A.C. Chadwick), pp. 229-239. Blackwell, Oxford.

Wirakusumah, N. 1977. *From Kutai to Dayak*. Pemerintah Daerah Kabupaten Tingkat II Kutai, Kalimantan Timur.

Wirawan, N. 1984. *Good forests within the burned*

forest area in East Kalimantan. WWF, Bogor.
Wirawan, N. 1985. *Kutai National Park management plan 1985- 1990.* WWF/IUCN.
Wirawan, N. 1986. Protecting the pesut (freshwater dolphin) in the Mahakam river of Kalimantan, Borneo. *Wallaceana* 44: 3-6.
Witkamp, H. 1925. Een en ander over krokodillen in Koetai. *Tropische Natuur* 14: 178-183.
Witkamp, H. 1932. Het voorkomen van eenige diersoorten in het landschap Koetai. *Tropische Natuur* 21: 169-177.
Wium-Anderson, S. 1981. Seasonal growth of mangrove trees in southern Thailand III. Phenology of *Rhizophora micronata* Lamk and *Scyphiphora hydrophyllacea* Gaertn. *Aquat. Bot.* 10: 371-376.
Wolf, E.C. 1985. Conserving biological diversity. In *State of the World 1985* (ed. L.R. Brown), pp. 124-146. Norton, New York.
Wong, K.M. 1990. *In Brunei forests. An introduction to the plant life of Brunei Darussalam.* Borneo Natural Heritage Series, Forestry Dept, Brunei.
Wong, M. 1983. Understory phenology of the virgin and regenerating habitats in Pasoh Forest reserve, Negeri Sembilan, West Malaysia. *Malay. Forester* 46(2): 197-223.
Wood, G.H.S. 1956. The dipterocarp flowering season in North Borneo, 1955. *Malay. Forester* 19: 193-201.
Wood, J.J. 1984. New orchids from Gunung Mulu National Park, Sarawak. *Kew Bulletin* 39(1): 73-98.
Woodland, D.J. and Hopper, J.N.A. 1977. The effect of trampling on coral reefs. *Biol. Conserv.* 11: 1-4.
Woodwell G.M., Hobbie J.E., Houghton R.A., Melillo J.M., Moore B., Peterson B.J. and Shaver G.R. 1983. Global deforestation: contribution to atmospheric carbon dioxide. *Science* 222: 1081-1086.
World Bank, 1988. *Forests, land and water: issues in sustainable development.* World Bank, Jakarta.
WWF, 1980. *Saving Siberut. A conservation masterplan.* WWF, Bogor.
WWF, 1982. *Lanjak Entimau orangutan sanctuary. A management plan.* IUCN/WWF Malaysia, Kuala Lumpur.
WWF, 1985. *Tabin wildlife reserve, Sabah. A preliminary management plan.* WWF Malaysia, Kuala Lumpur.
World Resources Institute. 1987. *Tropical forestry action plan: recent developments.* World Resources Institute, Washington.
Wyatt-Smith, J. 1954. Storm forest in Kelantan. *Malay. Forester* 17: 5-11.
Wyatt-Smith, J. 1963. Manual of Malayan silviculture for inland forests (2 vols). *Malay. For. Rec.* 23.
Wyatt-Smith, J. 1987. *The management of tropical moist forest for the sustained production of timber: some issues.* IUCN/IIED Tropical Forest Policy Paper No. 4.
Yajima, T. 1988. Change in the terrestrial invertebrate community structure in relation to large fires at the Kutai National Park, East Kalimantan (Borneo), Indonesia. In *A research on the process of earlier recovery of tropical rain forest after a large scale fire in Kalimantan Timur, Indonesia* (eds. H.Tagawa and N. Wirawan), pp. 2-11, Occ. papers No. 14, Kagoshima University.
Yamakura, T., Hagihara, A., Sukardjo, S. and Ogawa, H. 1986. Tree size in a mature dipterocarp forest stand in Sebulu, East Kalimantan, Indonesia. *Tonan Ajia Kenkyu (Southeast Asian Studies)* 23(4): 451-477.
Yamakura, T., Hagihara, A., Sukardjo, S. and Ogawa, H. 1986. Aboveground biomass of tropical rain forest stands in Indonesian Borneo. *Vegetatio* 68: 71-82.
Yap, S.K. 1982. The phenology of some fruit tree species in a lowland dipterocarp forest. *Malay. Forester* 45(1): 21-35.
Yap, S.K. and Lee, S.W. (eds.) 1992. *In harmony with nature.* Proceedings International Conference on Conservation of Tropical Biodiversity.
Yeager, C.P. 1989. Proboscis monkey (*Nasalis larvatus*) social organization and ecology. Ph.D. thesis, University of California.
Yeager, C. and Blondal, T.K. 1992. Conservation status of proboscis monkeys *Nasalis larvatus* at Tanjung Puting National Park, Kalimantan Tengah, Indonesia. In *Forest biology and conservation in Borneo* (eds. G. Ismail, M. Mohamed and S. Omar), pp. 220-228, Center for Borneo Studies Publication No. 2.
Yorke, C.D. 1984. Avian community structure in two modified Malaysian habitats. *Biol. Cons.* 29: 345-362.
Yunus, A. and Lim, G.S. 1971. A problem in the use of insecticides in paddy fields in West Malaysia. A case study. *Malay Agric. J.* 48: 168-178.
Zeven, A.C. and Koopmans, A. 1989. Ceiba pentandra In *Plant resources of South East Asia* (eds. E.Westphal and P.C.M. Jansen), pp.79-83, Pudoc, Wageningen.
Zieman, J.C. 1975. Tropical sea grass ecosys-

tems and pollution. In *Tropical marine pollution* (eds. S.J.F. Wood and R.E. Johannes), pp. 68-74, Elsevier, Amsterdam.

Zimmerman, P.R., Greenberg, J.P., Wandiga, S.O. and Crutzen, P.J. 1982. Termites: a potentially large source of atmospheric methane, carbon dioxide and molecular hydrogen. *Science* 218: 563-565.

Zimmerman, P.R. and Greenberg, J.P. 1983. Termites and methane. *Nature* 302: 354-355.

Zon, A.P.M. van der, 1980. *Mammals of Indonesia*. Special Report of Project FO/INS/78/061, Bogor.

Bibliography Addendum

Since this book was finalized for publication in 1993 the following useful references have been published.

Abdulhadi, R. and Suhardjono. 1994. The remnant mangroves of Sei Kecil, Simpang Hilir, West Kalimantan, Indonesia. *Hydrobiologia* 285 (1-3): 249-255.

Andau, P.M., Hiong, L.K. and Sale, J.B. 1994. Translocation of pocketed orangutans in Sabah. *Oryx* 28 (4): 263-268.

Ashton, P.S. and Hall, P. 1992. Comparisons of structure among mixed dipterocarp forests of northwestern Borneo. *J. Ecol.* 80 (3): 459-481.

Baki, B.B. 1993. Spatial pattern analysis of weeds in selected rice fields of Samarahan, Sarawak. *MARDI Research J.* 21 (2): 121-128.

Balen, B. van. 1992. Distribution status and conservation of the forest partridges in the Greater Sundas, Indonesia, with special reference to the chestnut-bellied partridge *Arborophila javanica*. *Gibier Faune Sauvage* 9 (Dec.): 561-569.

Beaman, J.H. and Beaman, R.S. 1993. The gymnosperms of Mount Kinabalu. *Contrib. Univ. Mich. Herb.* 19 (1): 307-340.

Bennett, E.L., Nyaoi, A. and Sompud, J. 1995. *A conservation management study of wildlife hunting in Sabah and Sarawak:* Report on the completion of fieldwork. Wildlife Conservation Society.

Bennett, J. 1992. A glut of gibbons in Sarawak, is rehabilitation the answer? *Oryx* 26 (3): 157-164.

Bossel, H. and Krieger, H. 1994. Simulation of multi-species tropical forest dynamics using a vertically and horizontally structured model. *Forest Ecology and Management* 69 (1-3): 123-144.

Brooks, S.M., Richards, K.S. and Spencer, T. 1993. Tropical rain forest logging, modeling slope processes and soil erosion in Sabah, East Malaysia. *Singapore J. Trop. Geog.* 14 (1): 15-27.

Burghouts, T.B.A., Campbell, E.J.F. and Kolderman, P.J. 1994. Effects of tree species heterogeneity on leaf fall in primary and logged dipterocarp forest in the Ulu Segama Forest Reserve, Sabah, Malaysia. *J. Trop. Ecol.* 10 (1): 1-26.

Campbell, E.J.F. 1994. *A walk through the lowland rain forest of Sabah*. Natural History Publications (Borneo) Sdn. Bhd., Kota Kinabalu, in association with Borneo Rainforest Lodge.

Campbell, E.J.F. and Newbery, D.Mcc. 1993. Ecological relationships between lianas and trees in lowland rain forest in Sabah, East Malaysia. *J. Trop. Ecol.* 9 (4): 469-490.

Cannon, C.H., Peart, D.R., Leighton, M. and Kartawinata, K. 1994. The structure of lowland rain forest after selective logging in West Kalimantan, Indonesia. *Forest Ecology and Management* 67 (1-3): 49-68.

Choy, S.C. and Booth, W.E. 1994. Prolonged inundation and ecological changes in an Avicennia mangrove: implications for conservation and management. *Hydrobiologia* 285 (1-3): 237-247.

Christensen, M.S. 1993a. The artisanal fishery of the Mahakam River floodplain in East Kalimantan, Indonesia: I. Composition and prices of landings, and catch rates of various gear types including trends in ownership. *J. Applied Ichthyology* 9 (3-4): 185-192.

Christensen, M.S. 1993b. The artisanal fishery of the Mahakam River floodplain in East Kalimantan, Indonesia: II. Catch, income and labour requirements of fisher households. *J. Applied Ichthyology* 9 (3-4): 193-201.

Christensen, M.S. 1993c. The artisanal fishery of the Mahakam River floodplain in East Kalimantan, Indonesia: III. Actual and estimated catches, their relationship to water levels and management options. *J. Applied Ichthyology* 9 (3-4): 202-209.

Clarke, C.M. and Kitching, R.L. 1993. The metazoan food webs from six Bornean *Nepenthes* species. *Ecol. Entomol.* 18 (1): 7-16.

Cox, J.H., Frazier, R.S. and Maturbongs, R.A. 1993. Freshwater crocodiles of Kalimantan, Indonesian Borneo. *Copeia* 1993 (2): 564-566.

Cranbrook, Earl of and Curran, L.M. 1994. Reproductive ecology of mast-fruiting Dipterocarpaceae in Kalimantan, Indonesia: An experimental test of the predator-satiation hypothesis. *Bull. Ecol. Soc. Am.* 75 (2 part 2): 47.

Cranbrooke, Earl of and Edwards, D. 1995. *Belalong: A tropical rainforest.* Royal Geographical Society, U.K. and Sun Tree, Singapore.

Dahuri, R. 1992. Dynamic interactions between regional development and Kutai National Park East Kalimantan Indonesia. In *'Developments in landscape management and urban planning 7: Science and the management of protected areas'* (ed. J.H.M. Willison) pp 55-58, Elsevier Science Publishers, Amsterdam.

Deeleman-Reinhold, C.L. 1993. An inventory of the spiders in two primary tropical forests in Sabah, North Borneo. *Memoirs Queensland Museum* 33 (2): 491-495.

Dove, M.R. 1994a. Smallholder rubber and swidden agriculture in Borneo: A sustainable adaptation to the ecology and economy of the tropical forest. *Econ. Bot.* 47 (2): 136-147.

Dove, M.R. 1994b. Transition from native forest rubbers to *Hevea brasiliensis* (Euphorbiaceae) among tribal smallholders in Borneo. *Econ. Bot.* 48 (4): 382-396.

Dransfield, J. 1992. Morphological considerations: The structure of rattans. *Malayan Forest Record* 35: 11-26.

Dutrieux, E. 1992. Experimental study of the impact of hydrocarbons on the inter-tidal benthic community of the Mahakam Delta East Kalimantan Indonesia. *Oceanol Acta* 15 (2): 197-209.

Emmons, L.H. and Biun, A. 1991. Maternal behavior of a wild tree shrew *Tupaia tana* in Sabah. *Res. Explor.* 7 (1) 70, 72-81.

Floren, A. and Linsenmair, K.E. 1994. About the diversity and re-colonization dynamics of arthropods on tree species in a lowland rain forest in Sabah, Malaysia. *Andrias* 13 (1): 23-28.

Halenda, C.J. 1993. Aboveground biomass production and nutrient accumulation of a *Gmelina arborea* plantation in Sarawak, Malaysia. *J. Trop. For. Sci.* 5 (4): 429-439.

Heley, C. 1994. Tribes, states, and the exploitation of birds: some comparisons of Borneo and New Guinea. *J. Ethnobiology* 14 (1): 59-73.

Hoffman, P. 1994. Record of a rare *Rana palavanensis* at Mt. Kinabalu, Borneo. *Salamandra* 30 (3): 223-224.

Holloway, J.D. 1993. The moths of Borneo: Family Geometridae, subfamily Ennominae. *Malay. Nat. J.* 47 (1-2): 1-307.

Holloway, J.D. 1994. The relative vulnerabilities of moth higher taxa to habitat change in Borneo. In *'Systematics and conservation evaluation'* (eds. P.L. Forey, C.J. Humphries and R.I. Vane-Wright) pp 197-205, Oxford University Press.

Inger, R.F. and Stuebing, R.B. 1992. The montane amphibian fauna of northwestern Borneo. *Malay. Nat. J.* 46 (1):41-51.

Inger, R.F. and Voris, H.K. 1993. A comparison of amphibian communities through time and from place to place in Bornean forests. *J. Trop. Ecol.* 9 (4): 409-433.

Kitamura, K., Rahman, M.Y.B.A., Ochiai, Y. and Yoshimaru, H. 1994. Estimation of the outcropping rate on *Dryobalanops aromatica* Gaertn. f. in primary and secondary forests in Brunei, Borneo, Southeast Asia. *Plant Species Biology* 9 (1): 37-41.

Kitayama, K. 1992. An altitudinal transect study of the vegetation on Mount Kinabalu Borneo. *Vegetatio* 102 (2): 149-171.

Kleine, M. and Heuveldop, J. 1993. A management planning concept for sustained yield of tropical forests in Sabah, Malaysia. *Forest Ecology and Management* 61 (3-4): 277-297.

Kofron, C.P. 1994. Bamboo-roosting of the thick-thumbed pipistrelle bat (*Glischropus tylopus*) in Borneo. *Mammalia* 58 (2): 306-309.

Lambert, F.R. 1991. Fruit-eating by purple-naped sunbirds *Hypogramma Hypogrammicum* in Borneo. *Ibis* 133 (4): 425-426.

Lambert, F.R. 1994. Some key sites and significant records of birds in the Philippines and Sabah. *Bird Conservation International* 3 (4): 281-297.

Lawrence, D.C., Leighton, M. and Peart, D.R. 1995. Availability and extraction of forest products in managed and primary forest around Dayak village in West Kalimantan, Indonesia. *Conserv. Biol.* 9 (1): 76-88.

Lee, Y.H., Stuebing, R.B. and Ahmad, A.H. 1993. The mineral content of food plants of the Sumatran Rhinoceros (*Dicerorhinus sumatrensis*) in Danum Valley, Sabah, Malaysia. *Biotropica* 25 (3): 352-355.

Leh, C.M.U. 1994. Hatch rates of green turtle eggs in Sarawak. *Hydrobiologia* 285 (1-3): 171-175.

Malmer, A. 1992. Water-yield changes after clear-felling tropical rain forest and establishment of forest plantation in Sabah Malaysia. *J Hydrol.* 134 (1-4); 77-94.

Marsh, C.W. 1995. Danum Valley Conservation Area: management plan 1995-2000. Yayasan Sabah, Kota Kinabalu.

Mitra, S.S. and Sheldon, F.H. 1993. Use of an exotic tree plantation by Bornean lowland forest birds. *Auk* 110 (3): 529-540.

Moffett, M.W. 1993. The high frontier. Harvard U.P.

Mohamedsaid, M.S. 1994. The Chrysomelidae (Coleoptera) of Danum Valley, Sabah, Malaysia: II. Subfamily Cassidinae. *Malay. Nat. J.* 47 (3): 369-371.

Mohammed, W.R., Dransfield, J. and Manokoran, N. (eds.) 1992. A guide to the cultivation of rattans. *Malayan Forest Record* 35: 11-26.

Mori, A. and Hikida, T. 1994. Field observations on the social behavior of the flying lizard, *Draco volans sumatranus*, in Borneo. *Copeia* 1994 (1): 124-130.

Nasi, R. 1993. Analysis of the spatial structure of a rattan population in a mixed dipterocarp forest of Sabah Malaysia. *Acta. Oecol.* 14 (1): 73-85.

Newbery, D.Mcc. 1991. Floristic variation within kerangas heath forest re-evaluation of data from Sarawak and Brunei. *Vegetatio* 96 (1): 43-86.

NRMP, 1993. *Effective protection and natural resource management in Indonesia*. Natural Resources Management Project, Jakarta, Report No. 26, October 1993.

Oelze, K.A and Heinrich, M. 1994. Documentation, effectivity and importance in forestry of medicinal plants from Sabah, Malaysia. *Angewandte Botanik* 68 (5-6): 177-186.

Ohta, S., Effendi, S., Tanaka, N. and Miura, S. 1993. Ultisols of lowland dipterocarp forest in East Kalimantan Indonesia III. Clay minerals, free oxides and exchangeable cations. *Soil Science Plant Nutrition* 39 (1): 1-12.

Orr, A.G. 1994. Life histories and ecology of Odonata breeding in phytotelmata in Bornean rain forest. *Odonatologica* 23 (4): 365-377.

Padoch, C. and Peters, C. 1993. Managed forest gardens in West Kalimantan Indonesia. In '*Perspectives on Biodiversity: Case studies of genetic resource conservation*' (eds. C.S. Potter, J.I. Cohen and D. Janczewski) pp 167-176, American Association for the Advancement of Science, Washington.

Padoch, C. and Peluso, N.L. 1996. *Borneo in transition: People, forest, conservation, and development*. Oxford University Press.

Pearce, K.G. 1994. The palms of Kubah National Park, Kuching Division, Sarawak. *Malay. Nat. J.* 48 (1): 1-36.

Phillipps, A. and Lamb, A. 1996. *Pitcher-plants of Borneo*. Natural History Publications (Borneo) Sdn. Bhd., Kota Kinabalu, in association with Royal Botanic Garden Kew and Malaysian Nature Society.

Pinard, M.A and Putz, F.E. 1994. Vine infestation of large remnant trees in logged forest in Sabah, Malaysia: Biomechanical facilitation in vine succession. *J. Trop. For. Sci.* 6 (3): 302-309.

Primack, R.B and Hall, P. 1992. Biodiversity and forest change in Malaysian Borneo-Long-term studies of trees provide insight on the conservation and management of Asian rain forests. *Bioscience* 42 (11): 829-837

Primack, R.B. and Lee, H.S. 1991. Population dynamics of pioneer *Macaranga* trees and understory *Mallotus* trees Euphorbiaceae in primary and selectively logged Bornean rain forests. *J. Trop. Ecol.* 7 (4): 439-457.

Primack, R.B., Hall, P. and Ashton, P.S. 1991. Maintenance of rare tree species in the mixed dipterocarp forests of Sarawak, East Malaysia, with implications for conservation biology. *Malay. Nat.* 45 (1-4): 55-68.

Putz, F.E. and Susilo, A. 1994. Figs and fire. *Biotropica* 26 (4): 468-469.

Rieley, J.O., Sieffermann, R.G. and Page, S.E. 1992. The origin, development, present status and importance of the lowland peat swamp forests of Borneo. *Suo (Helsinki)* 43 (4-5): 241-244.

Robinson, G.S. and Tuck, K.R. 1993. Diversity and faunistics of small moths (Microlepidoptera) in Bornean rain forest. *Ecol. Entomol.* 18 (4): 385-393.

Salafsky, N. 1993. Mammalian use of a buffer zone agroforestry system bordering Gunung Palung National Park, West Kalimantan, Indonesia. *Conserv. Biol.* 7 (4): 928-933.

Salafsky, N. 1995. Forest gardens in the Gunung Palung region of West Kalimantan, Indonesia: Defining a locally-developed, market-oriented agroforestry system. *Agroforestry Systems* 28 (3): 237-268.

Salafsky, N., Dugelby, B.L. and Terborgh, J.W. 1993. Can extractive reserves save the rain forest? An ecological look and socioeconomic comparison of non-timber forest product extraction systems in Peten Guatemala and West Kalimantan Indonesia. *Conserv. Biol.* 7 (1): 39-52.

Schairer, G. and Zeiss, A. 1992. First record of callovian ammonites from West Kalimantan Middle Jurassic, Kalimantan Barat, Borneo, Indonesia. *J. Aust. Geol. Geophys.* 13 (3): 229-236.

Schintlmeister, A. 1994. Check-list of the Notodontidae of Sundaland (excluding Java) with description of new species (Lepidoptera, Notodontidae). *Heterocera Sumatrana* 7 (2): 207-251.

Stone, B.C. 1991. Myrsinaceae as an example of plant diversity in Malesia with special reference to the species in Borneo. *Malay. Nat.* 45 (1-4); 230-237.

Strigel, G., Ruhiyat, D., Prayitno, D. and Sarmina, A. 1994. Nutrient input by rainfall into secondary forests in East Kalimantan, Indonesia. *J. Trop. Ecol.* 10 (2): 285-288.

Stuebing, R.B., Ismail, G. and Ching, L.H. 1994. The distribution and abundance of the Indo-Pacific crocodile *Crocodylus porosus* Schneider in the Klias River, Sabah, East Malaysia. *Biol. Cons.* 69 (1): 1-7.

Sunquist, M., Leh, C., Sunquist, F., Hills, D.M. and Rajaratnam, R. 1994. Rediscovery of the Bornean bay cat. *Oryx* 28 (1): 67-70.

Vermeulen, J.J. 1991. Notes on the non-marine molluscs of the island of Borneo 2. The genus *Opisthostoma* (Gastropoda Prosobranchia Diplommatinidae). *Basteria* 55 (1-4): 139-164.

Vermeulen, J.J. 1992. Notes on the non-marine molluscs of the island of Borneo 4. The genus *Eostobilops* (Gastropoda Pulmonata Strobilopsidae). *Basteria* 56 (1-3): 65-68.

Vermeulen, J.J. 1993. Notes on the non-marine molluscs of the island of Borneo 5. The genus *Diplommatina* (Gastropoda Prosobranchia: Diplommatinidae). *Basteria* 57 (1-3): 3-69.

Vermeulen, J.J. 1994. Notes on the non-marine molluscs of the island of Borneo: 6. The genus *Opisthostoma* (Gastropoda Prosobranchia: Diplommatinidae): Part 2. *Basteria* 58 (3-4): 75-191.

Voris, H.K. and Inger, R.F. 1995. Frog abundance along streams in Bornean forests. *Conserv. Biol.* 9 (3): 679-683.

Weaver, J.S. and Huisman, J. 1992. A review of the *Lepidostomatidae Trichoptera* of Borneo. *Zool. Meded.* 66 (16-40): 529-560.

Yeager, C.P. 1992. Proboscis monkey *Nasalis larvatus*, social organization, nature and possible functions of intergroup patterns of association. *Am. J. Primatol.* 26 (2): 133-137.

Index

Bold type denotes main reference or definition.

Acacia, 434, 556-58, 606-9, 615
Acanthuridae, 496
Acanthus, 96, 100, 102, 509
Aceraceae, 323
Acetes, 105, 504
Achatina, 554
Acid drainage, 582
Acid sulphate soils, 29
  management of, 448, 467-68, 484, 516, 518-19, 536, 681
Acridiidae, 381
Acrocercops, 556
Acropora, 86
Acrostichum, 96, 100, 509, 516
Adat (customary tribal laws), 358, 365
Aedes, 170
Aegialitis, 97-98
Aegiceras, 97-98
Aeredes, 267
Aeromys, 45, 423
Agaonidae, 206
Agar-agar, 494, 496
Agathis, 245, 256-57, 414
Aglaia, 150
Agolonidae, 216
Agricultural clearance, 117, 127, 130, 144, 271, 401
  threats to wildlife, 350, 372, 390-92
Agricultural conversion
  effects on ecosystem, 447, 465-72, 483-84, 514-16, 531-38, 601
  extent, 403, 445, 455, 569
Agricultural pests, *see* pests
Agriculture
  benefits of wildlife to, 111, 292, 309, 312, 645
  intensive, 30, 366, 374, 392, 538-40, 624
  suitability of sites, 28-30, 63, 65, 67, 256-57, 465-72, 514, 516, 531-38, 569, 608-10
  traditional practices, 28, 58, 63, 271, 363-75, 384, 386, 412-13, 469-71, 480-81, 540-43, 565, 604-6, 650, *see also* ladang
Agrochemicals, *see* fertilisers; herbicides; pesticides
Agroclimate, 31, 33-34
Agroecosystems, 474-81, 536, **538-50**, 566
Agroforestry, 392, 543, 569, 608-10
Alang-alang grasslands, 390-92, 547, 602-4
  extent, 63, 65, 418, 603, 607

fire in, 434, 439, 554, 568, 603, 608
  replanting of, 554, 557, 604-8
  use by tribal peoples, 372-74, 604-6
Albizia, 557, 558, 560, 562, 606, 608, 646
Alcohol production, 451, 510, 513, 540
Algae, 92, 104, 587
  blue-green, 88, 474-75, 494, 539, 589
  commercial use, 84, 494, 496, 588, 595
  epiphyllic, 185
  freshwater, 139, 141, 145, 153-54, 161, 171, 172, 473
  macroscopic, 84, 86
  nitrogen-fixing, 84, 474-75, 494, 539
  symbiotic, 87, 90
Algal blooms, 145, 153-54, 163, 448, 456-57, 592
Alien species, *see* introduced species
Allochthonous material, **141-42**, 148, 161, 171
Aloe wood, *see* incense wood (gaharu)
Alphaeus, 109
Alpine zones, 317-20, 352
Alseodaphne, 452
Alstonia, 125, 129
Altitudinal zonation
  Massenerhebung effect, 266, 320-21
  of mountain animals, 329-39, 343-48
  of mountain plants, 19, 266, 267, 269, 282, 317, 320-29, 339-43, 349, 410
Aluminium, 67, 277, 410, 448, 467, 518, 533
  toxicity, 28, 29, 467, 468
Amorphophallus, 204
Amphibians, 44, 632, 671
  endemic, 352
  habitats, 148-49, 204, 213, 232, 258, 337-38, 348, 610
  threats to, 422, 438
Amphiprionidae, 93
Ampullaria, 458, 480
Amyciaea, 110
Anabantids, 160, 462
Anabas, 110, 157, 475, 477
Anacardiaceae, 162, 323
Anadara, 105
Ananas, 383
Anhinga, 112, 151
Animal adaptations
  against predation, 93, 109, 111, 140, 149, 151, 203, 215, 223, 229, 235, 236, 292, 553, 617
  for arboreal life, 232, 235, 347
  for commensalism, 617-18

775

776 INDEX

for herbivory, 219, 230, 235
for marine life, 75, 78, 86, 92, 158
for parasitism, 622
for predation, 75, 92, 106, 109, 150-51, 157, 229, 235-37, 299, 617
to aquatic life, 139-40, 147, 151
to dry habitats or water stress, 75, 78, 106, 232, 302, 318, 329, 477
to habitat disturbance, 131, 424-26, 439, 561, 611, 675
to island habitats, 73-74
to low light intensity, 140, 147, 292-95
to low oxygen concentrations, 107-10, 157, 477, 616
to nutrient deficiency or imbalance, 226
to pollution, 596
Animal-plant interactions, *see* plant-animal interactions
Annona, 209
Annonaceae, 180, 183
Anopheles, 169-70, 622
Anous, 73
Ant plants, *see* myrmecophytes
Anthocephalus, 440, 608
Antiaris, 212, 213, 382
Antibiosis, 500
Antimony, 575
Antrophyum, 281
Ants, 110, 111, 216, 224, 227, 228, 332, 617-18, *see also* myrmecophytes (ant plants)
Apis, 102, 381
Aplocheilus, 157, 615-16
Aquaculture, 169, 447, 464, 474, 484, 514, 516
Aquatic macrophytes, 140, 141, 151, 155, 466, 480
Aquatic plants, *see* algae; aquatic macrophytes; seagrass
Aquilaria, 415
Arabs, 357, 363
Arachis (groundnut), 549
Arachnis (orchid), 340
Arachnothera, 46, 54
Araucaria, 558
Araucariaceae, 323
Arboreal animals, 214, 232, 345, 439
Arborophila, 46, 54
Archaeological studies, 21, 55, 56, 59, 303, 305-8, 309, 310, 355
Arctictis, 210, 237
Ardea, 112
Ardisia, 77
Areca, 349
Arenga, 41, 375
Argus pheasant (*Argusianus*), 229, 639
Arius, 166
Arixenia, 303
Aroids, 282

Arowana, 127, 457, 681
Arsenic, 172, 595
Artocarpus, 41, 209, 375, 467, 546
Aspidontus, 92
Asplenium, 340
Augury, *see* omens and augury; religious and ritual practices
Auricularia, 379
Autochthonous material, 141
Avicennia, 96-103, 506, 508, 510
Azadirachta, 280
Azolla, 474

Babblers, 215, 229, 639
Baccaurea, 41, 180, 196, 209, 280, 375
Bacteria, 104, 141, 154, 172, 204, 219, 223, 225, 327, 592, 595
nitrogen-fixing, 185, 247, 474, 475
pathogenic, 448, 450, 592
Badgers, *see* ferret badger; stink badger
Baeckea, 244, 246
Bagworms, 555
Bajau, 362
Balaenoptera, 497
Baleh, 372
Balitoridae, 46
Balsams, 284
Bamboos, 41, 230, 267, 342, 355, 379, 608
Bananas, 182, 230, 379, 543, 552
Banjarese, 361, 387, 389, 450, 465, 467, 469, 565
Banteng, 43, 148, 230, 284, 311, 428-30, 438, 443, 447, 565, 639, 650, 675
Barbets, 46, 215, 439
Barnacles, 75, 106
Barramundi, 105, 504, 517
Barringtonia, 75, 77, 97, 244
Batrachostomus, 46, 54
Bats, 44-45, 199, 214, 216, 232, 345, 347, 610
as pollinators, 77, 102, 151, 204, 205, 206, 292, 309, 422, 426, 562, 580, 646
as seed dispersers, 209, 232
fruit-eating, 73, 151, 209, 213, 222, 229, 232, 294, 379, 426, 432, 611
in caves, 290-94, 297-303, 309
in forest, 426, 675, 681
in pest control, 292, 309
nectar-eating, 232, 292, 294, 309, 422, 646
Bauhinia, 183
Bauxite, 67
Beaches, 75, 76, 77, 78, 81, 83
Beans, 552
Bears, 210, 236, 311, 357, 653
Bedbugs, 298, 616, 618
Bedstraw, 323
Beermaking, 510, 540
Bees, 160, 179, 180, 228, 236, 357, 381, 554,

INDEX

610
  as pollinators, 102, 160, 204, 205, 411, 646
Beetles, 111, 213, 259, 264, 285, 336, 337, 381, 421, 422, 439, 637
  aquatic, 172, 474
  as crop pests, 552, 553, 555, 564
  as detritivores, 224, 227, 228, 260, 332
  as pollinators, 204, 205, 411, 555
  in caves, 298, 299, 301, 302
Begonias, 182, 282, 284
Belangiran, 480
Belida (*Notopterus*), 461, 462, 463
Belontiidae, 46
Benthic animals, 134, 139-40, 144-46, 151, 168, 171, 473, 595
Benuaq, 60
Berawan, 359
Betta, 161
Bezoar stones, 59, 357, 382
Bidayuh, 359, 384
Bindweed, 76
Binturong, 210, 379
Bioconversion, 451
Biodiversity, 284, 310-12, 543, 631-35, 642-83, *see also* genetic diversity; species richness
Biogeography, 12, 15, 18-22, 35-54, 247, 323-24, 343-44, 654
Biological control, *see under* pest control
Biological indicators, *see* indicator species
Biomass, 108, 190-93, 213-14, 227, 259-61, 278, 439, *see also* productivity
Birds
  as pollinators, 102, 184, 204-5, 327, 422, 562, 646
  as rice-field pests, 472, 478, 563
  as seed dispersers, 73, 188, 208-13, 411, 646
  breeding behaviour, 73, 199, 229, 295-97, 384
  coastal, 73, 78-79, 82, 94, 588, 647, appendix 2
  effects of logging on, 222, 423, 426, 428-31
  endemic, 44, 46, 52, 54, 315, 345, 348, 638-39, 663, 668
  fish-eating, 141, 145, 147, 150-51, 162, 476
  fruit-eating, 209-16, 218, 221, 438-41, 646
  in caves, 294-97
  in pest control, 111, 309, 478, 518, 562, 646
  insect-eating, 199-200, 214-15, 294-97, 332, 345, 562, 611
  migratory, 157, 160, 384, 518, 611, 671
  montane, 311, 316, 327, 330-32, 343-45, 348, appendix 3
  of protected areas, 348, 662, 663
  rainforest, 126, 213-16, 229, 258, 559, 638
  urban, 610, 613, 615

  wetland, 82, 112-3, 127, 147, 150-51, 157, 172, 423, 429, 445, 457-59, 478, 480
  see also egg collecting
Birds of prey, 150, 298, 361, 553, 561-63, 615
Birds'-nests, edible, 59, 75, 294-98, 309-11, 357, 580, 645, 671
Bisaya, 451
Bitterns, 150
Bivalves, 92, 105, 108, 493
Blackfish, 462
Blackwater rivers, *see under* rivers
Bladderworts, 249, 323
Blechum, 41, 382
Blennies, 75, 92
Blowpipes, 57-59, 60, 357, 379-80, 382
Blythipicus, 384
Boatbuilding, 414, 506
Boleopthalmus, 109-10
Borassodendron, 41, 375
Borneodendron, 266
Bos, *see* banteng
Botia, 167
Bovidae, 55
Bradypterus, 46, 54, 345
Brahminy kite, 150, 361, 384, 615
Breadfruits, 41, 375, 546
Brittle stars, 90
Brontispa, 553
Brownlowia, 97, 281
Bruguiera, 97-103, 509-10
Bryophytes, 184-85, 245, 282, 321, 323
Bubalus, 565
Buceros, 382, 384
Buchaninia, 266
Buffer zones, 171, 413, 562, 594-95, 672-74
Bugau, 358
Bugis, 59, 362, 403, 413
Bugs, 172, 299
Building, *see* boatbuilding; construction industry; roadbuilding
Building materials, 84, 265, 451, 501, 510, 513, 525, 540, 550, 554, 604, 671
  traditional, 41, 130, 398, 443, 452, 453, 506, 552
Bukit Raya/Bukit Baka National Park, 44, 202, 244, 315-16, 345, 348-52, 657, 663
Bukitan, *see* Penan
Bulbophyllum, 245, 248, 267, 340
Bulbuls, 215
Bulweria, 73
Burseraceae, 180, 209, 211, 323
Butorides, 112
Buttercup, 323, 343
Butterflies, 102, 216, 228-29, 259, 344, 440, 611-12
  as pollinators, 205
  endemic, 46

montane, 285, 332, 335, 344
  threatened, 642
Buttresses, *see under* root systems

Cactoblastis, 564
Caddis flies, 299
Cadmium, 448, 595
Caesalpinia, 77
Calamus, 41, 281, 349, 410, 411, 412
Calcicoles (chalk-loving plants), 277, 280-82
Calcifuges, 282
Calcium, 29, 267, 271, 275, 277, 280
  deficiency, 120, 265, 467
Callosciurus, 45, 53, 264, 439, 553, 561
Caloenas, 73
Calophyllum, 75, 76, 77, 129, 245, 247, 651
Calyptomena, 46, 54
Cambrian, 15
Camphor, 357
Campnosperma, 125, 129
Camptostemon, 97, 506
Canals and ditches, 157, 387, 471, 517, 581, 610, 612, 614-16, 622, *see also* irrigation canals
Canarium, 129
Canavalia, 77
Candlemaking, 544, 552
Candlenuts, 379
Canopy structure, 123-25, 128-30, 243-55, 262-66, 321-23, 346-48, 438, 550, 569, 609
Capparidaceae, 323
Capricornis, 284
Captive breeding and species reintroductions, 458, 497, **676, 678-81**
Captotermes, 554
Carabidae, 228
Carbon dioxide
  atmospheric concentrations, 3, 17, 193, 443, 599
  fixation, 3, 87, 193, 224, 599, 601, 647
Carbon-14 dating, 55, 58
Carboniferous, 15, 578, 618
Cardamine, 323
Caretta, 79, 80, 81
Carnivores, 44, 162, 210, 235-37
Carnivorous plants, 35
Carp, 172, 450, 464
Caryota, 281
Cash crops, 271, 362, 365, 369, 372-75, 391-92, 403, 441, 543-58, 609, 672
Cashew-nuts, 608
Cassava, 543, 544
Cassia vera, *see Cinnamomum*
Cassis, 494
Castanopsis, 320, 341
Casturi, 375, 544
Casuarina, 71, 75, 76, 77, 243-48, 266, 282, 327

Catfish, 160, 161, 299, 462
Catharanthus, 651
Cats, 235, 236, 447, 563, 639
Cattle
  domesticated, 230
  wild, 55, 230, 565, 650, *see also* banteng
Caulerpa, 496
Cauliflory, 229, 323
Caves
  archaeological remains in, 273, 289, 303-2, 354
  as sacred sites, 308, 311, 312
  coastal, 73, 75
  conservation, 303, 308-12, 653, 671
  fauna, 274, 290-91, 299-304, 681
  formation and structure, 274, 287-90, 300
  threats to, 310, 312, 580, 671
Cecropia, 608
Ceiba, 467, 550
Celrodendrum, 41, 382
Cenozoic, 17, 19
Centipedes, 299
Centriscidae, 93
Centrosema, 607
Ceratolobus, 349
Ceratopogonidae, 216
Cerbera, 97, 130
Cercyon, 299
Cereals, 57
Ceriops, 97, 98, 99, 102, 506, 510
Cervus, 55, 148, 230, 284, 379, 382, 447
Cettia, 345
Chaerilus, 299
Channa, 160-61, 462, 464, 475
Chanos, 105, 494
Charaxes, 228
Charcoal, 510, 552
Charidae, 161
Cheiromeles, 292, 294, 298
Chela, 167
Chelonia, 79-81, 86, 497
Chestnuts, 320, 341, 342, 347, 349
Chickens, 565
Chinese, 59, 61, 296, 309, 357, 361-62, 450, 497, 575-76
Chipwood, *see under* timber products
Chironomidae, 172
Chiropodomys, 45
Chitons, 587
Chlamydochaera, 44, 46, 54
Chloranthus, 280
Chlorocharis, 46, 54, 345
Chromium, 29, 265, 448
Chrysopelea, 223
Cicadas, 236, 259
Cimicidae, 298
Cinnamomum, 379
Cipangopaludina, 381

Cissus, 383
Citrus, 540, 650
Civets, 44, 147, 210, 214, 232-33, 236, 298, 347, 379, 447, 610
Clams, 78, 84, 90, 494, 500, 504, 587, 595
Clamydochaera, 345
Clarias, 150, 157, 166
Clay, 576
Clethraceae, 323
Climate
  agroclimate, 31, 33-34
  ancient, 12, 17, 21, 121, 310, 317
  global, 443, 602
  influence of, 30, 290, 308, 317, 435, 531, 586
  local, 31, 418, 443
Climbing palms, *see* rattans
Climbing perch, 109, 110
Climbing plants, 96, 179, 182-83, 190, 246, 282, 323, 406, 438, 480, 547, *see also* pitcher plants (*Nepenthes*); rattans
Cloud forest, 282, 317-18, 323, 341-43, 418, 432
Clouded leopard, 44, 235-36, 343, 386, 447, 639, 653, 675
Cloves, 550, 552, 608
Clownfish, 93
Clubmosses, 121, 343, 578
Clupea, 493
Cnidarians, 86, 93
Coal, 17, 25, 67, 311, 436, 573-82
Coastal ecosystems, 71-94, *see also* mangrove ecosystems
  fauna, 73-75, 78-84
  flora, 75-78, 84
  management, 495, 523-25
  threats to, 79, 81, 491, 500, 520-26, 586-89
  value, 79, 491, 498, 526
Coastal peoples, 361-63
Coastal protection, 75, 84, 94, 101, 114, 448, 491, 500, 504, 513-14, 525-26, 589, 644
Coastline changes, 81, 101, 121, 510, 513
Cobalt, 29, 265
Cobitidae, 139
Cobras, 308
Coccus, 204
Cockles, 105, 493-94, 504, 587
Cockroaches, 298-302, 336, 618
Cocoa, 552, 555-56, 609
Cocoa moth, 292
Coconut (*Cocos*), 77, 467, 516, 538, 550-52, 608
  plantations, 73, 78, 471, 550, 552, 553
Codium, 496
Coelenterates, 86, 93
Coelogyne, 245, 267, 340
Coffee, 549, 550, 552

Collocalia, 73, 75, 294, 300, 639
Colocasia, 548
Colubrina, 77
Combretaceae, 123, 323
Combretocarpus, 456
Commelina, 466
Commensalism, 92, 93, 562, 610, 616, 617, 618
Competition, 90, 102, 144-46, 500, 604, 608, 635
  between crops and weeds, 469, 543, 549
  between people and wildlife, 554, 562
  between plants, 257, 277, 319, 340, 455, 604
  in lowland rainforest, **214-22**
  with introduced species, 479, 484, 616
  see also niche separation
Cone shells, 92
Conifers, 244, 256, 257, 282, 342, 343, 349
Connaraceae, 323
Conservation
  funding, 350, 390, 465, 536, 657, 676
  of coral reefs, 500, 526, 644
  of forests, 129, 382, 396, 409, 430, 432, 442, 443, 562
  of habitats, 643-70
  of living resources, 4, 352, 631-49, 670, 670-83
  of soil, *see* soil conservation
  of wetlands, 144, 483-88, 528
Conservation areas, *see* protected areas
Construction industry, 79, 262, 398
Continental drift, 12-17
Conversion forest, 401
Copaifera, 454
Copepods, 78
Copper, 410, 448, 575
Coppicing, 256, 265, 356, 557
Coprophages (dung-eaters), 298-303
Copsychus, 46, 384
Coral reefs
  ancient, 273
  distribution, 73, 85-88
  fish *see* fish, reef
  invertebrates, 89-90
  stabilisation of, 75, 496
  threats to, 501-3, 513, 520, 526, 584, 588-89, 644
  value, 93-94, 494, 500
Corals, 85-90, 92, 500
Corbicula, 168
Cormorants, 151, 157
Corybas, 323
Corypha, 349
Cotton, 58, 565
Cotylelobium, 245
Cover crops
  for soil protection, 387, 555, 558

value in disease control, 554
Cowries, 92
Crabs
  freshwater, 381, 474, 477
  in caves, 300, 310
  land, 224
  marine, 75, 78, 82, 84, 93, 107, 131, 500
  of mangrove, 102-7, 494, 504, 517, 520
  see also horseshoe crabs
Crassostrea, 105
Cratoxylum, 245, 248, 455
Crayfish, 137, 464, 494
Crematogaster, 204
Cretaceous, 15, 637
Crickets, 298, 299, 302, 303
Crinum, 77
Critical lands, 65, 434, 598, 602, 608, 681
Crocidura, 55
Crocodiles (*Crocodylus*), 55, 95, 127, 147, 157-59, 162, 311, 447, 483, 504, 642
  farming, 457-58, 650, 681
Crop raiding, 562
Crops
  choice of, 389, 465, 546, 552-55, 561, 568-69, 607-9, 672-74, 681
  resistance to pests or diseases, 564, 568, 650
  varietal diversity, 543-4, 568, 642, 649-50
  wetland, 451, 465-68, 536-38, 548
Crustaceans
  freshwater, 137, 141, 474
  in caves, 299-300
  in forest, 236
  marine, 78, 82-83, 493, 584
  nursery sites, 82, 95, 504
  of mangroves, 95, 106, 504
Culex, 169-70
Cuon, 43, 308
Curculionidae, 208
Cuttlefish, 493
Cyanide, 172, 578
Cyathea, 379
Cycas, 77
Cylocheilichthys, 167
Cymodocea, 85
Cynocephalus, 223
Cynogale, 147, 236, 447
Cynoglossus, 167
Cynometra, 97
Cyornis, 46, 95, 112
Cyperus, 77
Cyprinids, 144, 160, 172, 461-62
Cyrtodactylus, 299, 303
Cyrtostachys, 123, 148

Dacrydium, 244-45, 266, 282, 329, 343
Dactylocladus, 125, 245
Daemonorops, 411

Dam construction, threats to fisheries, 163, 168-69, 447, 484, 526
Damar, 349, 357, 414, 645
Danaidae, 216
Danau Sentarum Wildlife Reserve, 159, 160, 162, 163, 453, 657, 663
Danum Valley, 677
Daphnia, 141
Dayaks, 356, **358-61**, 386, 391
  agriculture and horticulture, 58, 179, 180, 349, 358, 363, 370-82, 391, 403, 412-16, 450, 474, 540-46, 604, 609
  culture and lifestyle, 59, 60, 150, 308, 358-59, 384-86, 458, 540, 575, 620
  migrations, 60, 356, 360
Decomposers, 223, 224, 318, 559, 594, 595, 601
Decomposition, 104, 121, 135, 141, 153-54, 171-72, 222-28, 350, 439, 558, 594
Deer, 204, 229, 428, 438, 458, 675
  as human food, 374, 379-80, 565, 604
  sambar, 148, 230, 447, 560
Deforestation
  by displaced or migrant people, 169, 271, 360, 372, 390-91, 516
  by shifting agriculture, 58, 171, 369-70, 390, 403, 435, 604
  degradation of habitat by, 163, 177, 251, 256, 350, 372, 392, 398, 418, 443, 448, 471, 472, 514, 526, 603-4, 635, 638, 668-69
  effects on climate, 417-19, 598, 601-2
  effects on local people, 358, 379, 380, 382, 622, 646
  effects on wildlife, 136, 143, 171, 230, 271, 386, 421-30, 560, 637-38, 668-69
  extent of, 34, 63, 395, 401, 598, 635, 682
  see also logging; shifting cultivation
Delias, 344
Deltas, 17, 73, 80, 94
Dendrobium, 197, 245, 267, 340
Dendrocitta, 46
Dendrogale, 45
Dendronephthya, 86
Dermochelys, 79, 80
Derris, 41, 96, 281, 382
Desa, 358
Desmids, 172
Desmodium, 77, 607
Detritivores, 84, 90, 104-5, 141, 162, 172, 222-28, 260, 291, 298-99, 336, 595, 601
Detritus, *see* litterfall
Development
  sustainable, 4-5, 63-68, 614, 624-29, 642, 660-61
  threats of, 2, 63, 447, 520, 593, 602, 635
Dialium, 179
Diamonds, 60, 61, 67, 573, 575-76, 580-81

Diaphera, 286
Dicaeum, 46, 54
Dicerorhinus, 43, 230-31, 639, 641
Dillenia, 130
Dilleniaceae, 323
Dioecism, 206
Dioscorea, 378
Diospyros, 41, 180, 209, 280, 382
Diplazium, 379
Dipseudopsis, 168
Diptera, 141, 216
Dipterocarp forest, *see* heath forest; hill dipterocarp forest; lowland dipterocarp forest; montane forest
Dipterocarps, 162, 175, 222, 414
  as timber trees, 35, 177, 396, 399, 403, 454
  characteristics, 177-80, 213
  distribution, 20, 120, 175, 202, 244-47, 265-66, 278-84, 323, 349, 632
  flowering and fruiting, 34, 182, 189, 196-99, 263, 406, 409, 416, 435, 557
  mycorrhizal associations, 181, 441, 558
  regeneration, 188-89, 209, 263, 409, 435, 557, 680
Dipterocarpus, 130, 150, 179, 190, 265, 399, 454, 610
Dischidia, 184, 247
Disease
  control, 169, 517, 543, 609, 621-24; *see also* medicinal products; pharmaceutical products
  ecology, 479, 620-23
  from pollutants, 468, 471, 578, 584, 592, 594, 596, 602
  plant, 472, 478, 540, 547, 554, 650
  vectors, 169-70, 309, 472, 517, 581, 610, 612, 618, 620-24
Disoxylum, 150, 211
Ditches, *see* canals and ditches
Diversity, *see* biodiversity; genetic diversity; species richness
Dog, wild, 43, 308
Dolichandrone, 97
Dolphins, 82, 147, 447
Domestic livestock, 230, 458, 480, 484, 538, 565-68, 604, 622, 642, 650, 671, *see also* live stock farming
Domestication
  animals, 443, 538, 565, 645, 650
  fruit trees, 41, 211, 443, 544-46, 568
  plants, 538, 568, 645, 650
Donax, 78
Doryichthys, 161
Dowitchers, 82
Draco, 223
Dracontomelum, 130, 209
Dragonflies, 172, 474, 610, 647
Drainage, 280, 284, 447-48, 471, 516, 581,
  see also canals and ditches
Drainage patterns, 79, 120, 133, 532
Drainage systems, effects on habitats, 447, 465, 468, 471, 473, 484, 536
Dredging, 84, 132, 501
Dremomys, 45
Drimys, 323
Drosera, 248
Drought, 34, 125, 130, 160, 171, 198, 251, 255, 282, 319, 350, 386, 435-38, 455, 463, 477, 608, 644
Drynaria, 183
Dryobalanops, 179, 222, 248, 262, 399, 403, 454, 455, 610
Ducks, 157, 458-59, 476
Duckweed, 140, 448
Ducula, 73
Dugongs (*Dugon*), 82, 84, 86-87, 493, 497, 520
Durians, 180, 196
  as human food, 39, 57, 211, 375, 453, 546, 569, 646, 650
  pollination by bats, 206, 292, 309, 422, 562, 580, 646
  value to wildlife, 39, 211, 221, 347, 422
Dusun, 308, 359-60, 374
Dwarf forest, *see* elfin forest
Dyera (jelutong), 123-25, 413, 453

Earthworms, 224, 235, 236, 332, 336, 350, 439
Earwigs, 298, 303, 310
Eatogenia, 168
Ebenaceae, 162
Echinoderms, 78, 79, 84, 500, 584
Echinosorex, 235
Echinothrix, 93
Echolocation
  aquatic mammals, 147
  bats, 292, 294
  swiftlets, 294, 295
Ecosystems
  capacity for self-maintenance, 594, 663
  comparison of productivity and species richness between, 75
  interactions between, 80-82, 104-5, 129, 147-48, 157, 171, 332, 339, 465, 484-85, 491, 501-3, 513-15, 521-26, 592, 646
Education, *see* scientific research and education
Eels, 157, 477, 517
Effluents, *see under* pollution
Egg collecting, 73, 79, 458, 497
Egrets (*Egretta*), 82, 111, 157, 476, 478, 518
Eichhornia, 168, 479-80
Elaeis, 451, 555
Elaphe, 297-98
Eleocharis, 466

Eleoctris, 151
Elephantiasis, 169
Elephants (*Elephas*), 43, 211, 230, 428, 639, 646, 653
  as pests, 230, 553-54
  behaviour, 230, 672
  distribution, 52, 230-31, 675
  migration, 430, 675
  threats to, 638, 639
Elfin forest, 282, 322-23, 340-43
Ellobium, 108
Emballonura, 292, 426
Embaloh, 359
Emilia, 466
Empran (seasonal swamp forest), 129, 262
Endangered species, *see under* species
Endemism
  birds, 44, 46, 49, 52, 54, 95, 315, 345, 348, 633-39, 663, 668, 679
  fish, 44, 143, 447, 616
  invertebrates, 284-86, 310, 344-45, 352, 446-47, 642
  mammals, 43-45, 49, 52-53, 215, 271, 315, 348, 352, 429, 447, 633-34, 638, 663-64, 675, 679
  plants, 35, 38, 40-43, 49, 123, 177, 248, 266-67, 277-80, 284, 315, 318, 327, 375, 453, 544, 631-32, 634, 639
  primates, 52-53, 95, 112, 126, 447, 639, 663-64, 679
  reptiles, 43-44, 47, 49, 95, 259, 633, 679
Energy flow in ecosystems, 82, 92, 104-5, 115, 141-42, 147, 153, 161, 191, 291, 300, 625
Enhalus, 85
Enrichment planting, 433
Eocene, 15, 583
Eonycteris, 102, 292, 309, 422, 646
Epalzeorhynchus, 167
Ephemeroptera, 172
Epiphylls, 185
Epiphytes, 128, 179, **183-85**, 209, 216, 243, 245-46, 248, 256, 281-82, 321-23, 342, 622
  algae, 84, 141
  ferns, 183, 282, 559-60
  figs, 183, 281, 424
  orchids, 183, 282, 340
Epithema, 281
Eretmochelys, 79, 80, 81
Eria, 244, 267, 340
Ericaceae, 320, 323, 349
Erosion, *see* coastal ecosystems, threats to; soil, erosion
Erythrina, 77
Estuaries, 80-82, 99, 120, 147, 491, 505, 519, 526
Ethnic groups, distribution in Borneo, 355-392

Ethnobotany, *see under* tribal peoples
Eucalyptus, 434, 557, 608
Euchema, 496
Eugeissona, 41, 57, 357-59, 375
Eugenia, 129, 148, 245, 247, 262, 329
Eumenes, 228
Euphorbia, 77
Euphorbiaceae, 41, 150, 162, 180
Euphrasia, 343
Eurycoma, 383
Eusideroxylon, see ironwood
Eutrophication, 136, 154, 448, 484
Evolution, *see* species, evolution
Excoecaria, 97-98, 100, 506, 510-f12
Exilisciurus, 45
Export
  agricultural products, 544, 547
  forest products (non-timber), 309, 382, 410-417, 651
  marine products, 79, 162, 447, 456, 461, 494, 496-97, 504, 516
  minerals and metals, 667, 575, 579, 581
  petroleum products, 67, 575, 583
  pharmaceutical products, 651
  plantation products, 554
  timber, 395, 398-99, 401, 506, 573-74, 632
  wetland products, 458, 650
Extinction, of lake ecosystems, 154
Extinctions
  local, 44, 231, 265, 308, 421-22, 426, 428-30, 438-41, 447, 483, 636-37, 646, 653, 676
  mass, 21, 636-38
  of unique species, 55, 310, 352, 355, 637, 642, 663
Eyebrights, 343

Fagaceae, 323, 349, 441
Fagraea, 281
Fagus, 342
Falco, 298
Fallow periods, importance, 28, 58, 63, 365-68, 371-72, 412, 421, 469, 538
False gavial (*Tomistoma*), 157-58, 447, 642
Fanworms, 90, 92
Featherback, *see Notopterus*
Feeding behaviour and patterns
  birds, 150, 216, 218
  fish, 161-62, 164-65, 462
  insects, 618
  mammals, 148, 209, 230-37
  primates, 131, 199, 219, 220-22, 232-34, 271, 425, 439, 563
Feeding guilds
  birds, 216, 428, 615
  fish, 91, 144, 146, 162, 164-65
  termites, 259
Feeding sites

INDEX   783

for marine turtles, 81
for migratory birds, 79, 82, 95, 671
Felis, 45, 236, 447, 639
Ferns
  diversity, 35, 96, 123, 341, 348, 509, 516
  epiphytic, 183, 282, 340, 559, 560
  filmy, 321, 323
  terrestrial, 77, 182, 256, 266, 281, 284, 290, 578
  use, 41, 375, 379
Ferret badger, 21, 343
Fertilisers, 448, 467, 476, 494, 516-18, 556, 584, 589, 604
  natural, 450, 565-66
Fibres, from plants, 550-52, 570
Fig wasps, 206-7, 295
Figs (*Ficus*), 75, 99, 150, 201, 211, 212, 281, 341, 347, 348
  geocarpic, 180
  predation on, 73, 206-9, 212-13, 216, 219, 221, 424, 430-31, 441
  strangling, 183-84, 206, 219, 221
Filariasis, 169, 621
Filter feeders, 92
Fimbristylis, 77
Fire
  catastrophic, 3, 34, 311, 401, 434-41, 456, 608, 657
  degradation of vegetation, 245, 256, 282, 435-41, 456, 480, 568, 604
  economic loss from, 435, 441, 554
  effects on ecosystem, 130, 230, 418, 435-40, 463, 568, 599
  factors increasing risk, 284, 350, 435, 438, 441-42, 455, 472
  for agricultural clearance, 2, 117, 193, 355, 363, 435, 441, 455, 469, 484
  for nutrient enrichment, 256, 480, 543
  regeneration after, 130, 256, 284, 350, 440-41, 681
  to promote grass growth, 374, 538, 565, 568, 604
Fire-adapted species, 130, 372, 466, 603
Fire-climax vegetation, 127, 604
Fire-susceptible species, 435, 436, 441, 558
Fireflies, 111
Firewood, *see* fuelwood
Fish
  air-breathing, 109-10, 157, 261, 477
  as seed dispersers, 146, 150, 161
  breeding behaviour, 110, 145, 462, 584
  cave, 299-300
  commercial, 144, 457, 493, 500, 504, 516-17, 525
  endemic, 44, 143, 447, 616
  estuarine and marine, 78, 82-84, 88, 90-94, 105, 107, 109-10, 143-44, 493
  feeding guilds of, 91, 144, 146

freshwater, 20, 44, 127, 137-46, 155-57, 160-69, 171-72, 271, 379, 456-57, 461, 474-75, 479, 578, 590, 610
  introduced, 475, 479, 615-16
  kills of, 154, 157, 171, 589, 592
  nursery sites, 84, 90, 95, 157, 447, 456, 461, 469, 492, 504, 520
  mangrove, 95, 106, 109-10, 504, 516-19
  rice-field, 475-6
  reef, 88, 90-93
Fish eaters, 75, 111, 145-47, 150-51, 157, 232, 236, 615
Fish fry collecting, 464, 494, 517
Fish poisons, 41, 357, 382
Fish ponds, freshwater, 146, 163, 450, 464, 474, 479, 592, 622, *see also* tambak culture
Fisheries
  estuarine and marine, 84, 114, 445, 491, 493, 500, 504-5, 519, 589, 644
  freshwater, 144-48, 153, 160-69, 171, 379, 438, 447, 456-64, 481-83, 644, 671
  threats of development to, 163, 169, 463-64, 469, 473, 484, 526, 582, 584, 589, 592, 594-96, 681
  see also fish ponds, freshwater; tambak culture
Flacourtiaceae, 323
Flagellates, 185
Flatworms, 78, 139, 227, 620
Fleas, 78
Flickingaria, 267
Flies, 139, 204, 216, 227, 228, 298, 302, 336, 610, 621
Flood control, 160, 163, 417, 432, 445, 448, 518, 644
Flooding
  nutrient enrichment after, 539
  seasonal, 117, 128, 144, 151, 157, 159, 160, 161, 162, 445, 447, 456, 460, 463, 477, 479, 480, 540
  unexpected, 133, 169, 171, 369, 370, 418, 432, 438, 471
  tidal, 96, 121
  see also sea level changes
Flowering and fruiting patterns, 34, 102, 160, 180, 182, 188, 196, 197, 206, 219, 263, 265, 267, 270, 409, 416, 422, 427, 557, 668
Flowerpeckers, 210
Flukes, 620
Flycatchers, 95, 332
Flying foxes, *see* bats, fruit-eating
Flying lemur, 74, 223, 345
Fodder and animal feed, 451, 474, 494, 513, 540, 548, 550, 604, 607, *see also* grazing land
Food resources, for forest animals, 199-204, 209-37, 332, 424-31, 438-39, 560-61

Food webs
　disruption of, 646, 668, 676
　in agroecosystems, 538, 566
　in caves, 290-91, 300-302
　in forest, 228, 235, 254, 259, 432
　in rivers, 141-42, 147, 157, 161
　in wetland ecosystems, 171-72, 447, 456-58, 482
　marine, 79, 84, 92-93, 104-6, 513, 525, 588, 602
　toxin concentration in, 448, 563, 578, 582, 594-97
Forest
　concessions, *see* timber, concessions
　hydrological functions, 67, 133, 271, 284, 396, 398, 417-21, 438, 441
　phases, 186-89, 223, 368, 559, 601
　primary, 130, 191, 365, 369-70, 390, 421-23, 438, 441, 443, 448, 559, 672
　secondary, 182, 226, 256, 366, 369-73, 423, 426-28, 438, 440-41, 456, 466, 558-59, 672-75
　total area covered by, 395-402
　value, 67, 420, 442-43, 562, 568, 598, 644, 647, 650, 651
　see also forest types
Forest conversion, *see* deforestation
Forest products, non-timber, 41, 375, 443, 452-53, 510-13, 672
　export of, 309, 382, **410-17**, 651
　sustainability of, 536, 645, 672, 681
　trade in, 57, 59, 357-59, 382, 410-17, 441-42
　use by local people, 41, 55, 349, 357, **375-83**, 385, **410-17**, 441-43, 568, appendix 6
Forest types
　comparative area and distribution, 35-37, 66, 397-98
　differences and similarities between, 96, 111, 125, 127, 148, 243-48, 251, 255, 258-60, 265-67, 278-80, 321, 323, 326, 329, *see also under* species richness
　see also under individual types: elfin forest; freshwater swamp forest; heath forest; limestone forest; lowland dipterocarp forest; mangrove; montane forest; peat swamp forest
Forktails, 423
Fossil remains, 354
Fregata, 73
Freshwater ecosystems, *see* lake ecosystems; lakes; sawah rice fields; wetland habitats
Freshwater swamp forest, 117, **127-131**, 160, 163, 177, 398, 480
　area and distribution, 29, 127
　characteristics and structure, 127-28, 455, 467
　conservation, 129

　fauna, 130-31, 157
　flora, 127-30
　threats to, 127, 129, 472, 473
Frogs, 136-37, 214, 232, 303, 381, 474, 610
　distribution, 127, 162, 259, 337-39
　introduced, 479, 612
Frugivores, 211, 229, 237, 441
　birds, 73, 111, **208-14**, 216, 218, 332, 424, 429, 438, 441, 646
　mammals, 214, 347, 438
　primates, 201, 210-211, 216, 271, 424
　threats to, 439
　see also bats, fruit-eating
Fruit trees, 375-78, 431
　for primates, 126, 131, 199, 201, 209-11, 219-22, 271
　seasonality, 196-201, *see also* flowering and fruiting patterns
　threats to, 403
　value to humans, 39, 41, 57, 211
　see also frugivores
Fuelwood, 130, 451, 452, 456, 510-13, 519, 525, 540, 608, 609, 672, *see also* charcoal; peat, as fuel
Fungi, 199, 379, 595
　as decomposers, 104, 139, 141, 172, 182, 185, 204, 222-28, 290, 340, 592
　mycorrhizal, 181-82, 327, 340, 441, 557-58
　pathogenic, 186, 263, 547, 554

Gaharu, *see* incense wood
Galium, 323
Gallus, 565
Gambier, 361
Ganua, 125
Garcinia, 196, 211, 375, 453, 544
Gardens, 611-15, *see also* home gardens
Garrulax, 54
Gas, *see* natural gas (LNG)
Gastromyzon, 139
Gastropods, 57, 82-84, 108, 307, 494
Gaultheria, 323
Gavial, *see* false gavial (*Tomistoma*)
Geckos, 223, 299, 303, 611, 617
Gehyra, 617
Gelam, 320, 343
Gelidiopsis, 496
Gene banks, 569, 645, 649
Genetic diversity, 4, 131, 631, 638, 642-43, 645, 649, 653, 668
Genetic resources, 68, 230, 443, 543, 568-70, 643, 645, 649-53
Gentiana, 323, 343
Geography, 2, 9-12, 61
Geology, 12-17, 22-25, 35, 41, 51, 274, 343, 533, 536
Germination, *see* seed germination
Gesneriaceae, 182, 281-84

Gibbons, 52, 126, 211-13, 218-22, 271, 284, 311, 332-34, 343, 347, 386, 421, 423-24, 638
Gingers, 77, 180, 182, 379, 552
Glaciation, 17-21, 120, 317
Glagah, *see Saccharum*
Gliders, 74, 223, 232-33, 235, 345-47, 423
Gliricidia, 556, 609
Global warming, *see* greenhouse effect
Glucosides, 544
Glycine, 549
Glyphotes, 45
Gmelina, 556, 608
Gneiss, 22
Gnetum, 211
Gold, 25, 60, 61, 67, 575-76
  mining, 172, 361, 390, 576-78
Gondwanaland, 14-16, 344
Goniothalamus, 41, 382
Gonystylus, 123, 125, 150, 276, 398-99, 443, 454
Goodyera, 340
Gourami, 157, 464, 479
Gracilaria, 496
Granite, 15, 22
Granodiorite, 25
Graphium, 642
Grapsus, 75
Grasses, 76-77, 96, 121, 230, 343, 379, 421, 428, 439, 472
Grasshoppers, 228, 381, 477
Grasslands, 256, 372, 456, 565, *see also* alang-alang grasslands
Grazing, 445, 447, 458, 484, 538, 566, 602, 609
Greenhouse effect, 3, 17, 193, 599, 601
Gross primary productivity, 190
Ground flora, 182, 183, 248-49, 262
Groundnuts, 549
Groundwater contamination, 582, 589
Growth inhibitors, 604
Guano, 291, 294, 298-304, 309-10, 312, 671
Guava, 309
Guettarda, 77
Gunnera, 323
Gunung Mulu National Park, 191, 251, 285, 312
  cave ecosystems, 274, 289-90, 299-303, 310, 312, 671
  fauna, 127, 216, 221, 258, 275, 285, 290-91, 300-304, 310, 330, 332, 336-37
  flora, 245, 279-81, 320, 329, 349
  limestone habitats, 274-85
Gunung Palung National Park, 352, 657, 663, 680
  fauna, 199, 222
  flora, 127-28, 202, 320
Gunung Silam, 266-67, 271, 336

Guppies, 157, 615
Gutta percha, 414
Guttiferae, 162
Gymnure, 21
Gypsum, 575

Habitat fragmentation, 635, 638, 652, 668, 669
Haemadapsa, 226, 227
Haematortyx, 44, 46, 54
Haeromys, 45
Halcyon, 111, 112
Haliastur, 112, 150, 298, 361, 384
Halimeda, 84, 494
Halodule, 84, 85
Halophila, 84, 85
Haloragis, 323
Hampala, 167
Hampung (horticulture on floating vegetation mats), 480
Handicrafts, 41, 55, 382, 452, 458, 494, 568, 646
Harpactes, 46, 54, 384
Headhunting, 60, 234, 358, 360, 370, 386
Heath forest, 125, 160, 226, 240-61, 266, 282, 319, 329, 479
  agricultural potential, 29, 256, 387, 536
  characteristics and structure, 240-48, 256-57, 398, 410
  fauna, 257-61, 336
  plant adaptations, 248-55
Heavy metals, 172, 265, 448, 594-96, 647
Hedyotis, 349
Helarctos, 44, 236, 382
Helastoma, 479
Helminths, 620
Helostoma, 167
Hemidactylus, 617
Hemigalus, 45, 236
Hemiptera, 172
Hemiramphodon, 161
Herbicides, 472, 476, 538, 559, 604
Herbivores, 86, 92, 141, 148, 161, 202-4, 214, 216-17, 219-21, 222, 230, 231, 235, 258, 635, 675
Heritiera, 97, 100, 130, 201, 506
Hernandia, 77
Herons, 150, 157, 476, 478, 518
Herpestes, 45, 52
Hesperidae, 216, 344
Heteropoda, 299
Hevea, 414, 554, *see also* rubber
Hibiscus, 76, 77, 97
Hill dipterocarp forest, 160, 342, 398
Hill rice, *see under* rice farming systems
Hill sago, *see* sago (*Eugeissona*)
Hindus, 59, 575
Hipposideros, 45, 292, 426

Histosols, 28
Holocene, 17, 57, 121, 308
Holothuroideae, 497
Home gardens, 41, 375, 378, 541, 543, 546, 548, 549, 550, 577
Homoptera, 216
Honey, 410, 510, 519
Hopea (merawan), 179, 180, 190, 246, 281, 282, 399
Hornbills, 44, 213, 215, 216, 379, 382, 384, 430, 438, 638, 639, 646, 653
  feeding behaviour, 199, 209, 211, 221, 424, 429, 439, 441
  "ivory" from, 59, 357, 382, 384, 639
Horseshoe crabs, 107-8, 647
Horsfieldia, 329
Horticulture, *see* hampung; home gardens
Hoya, 256
Hulu, 381
Human food
  commercial use, 84, 86, 105, 375-79, 416, 451, 494-96, 500, 504-5, 543-50, 552-54, 556, 588, 650
  local use, 41, 75, 78, 82, 84, 447, 450-53, 464, 480, 546, 550, 568-70, 609, 650
  protein sources, 86, 169, 379-81, 450-52, 456, 464, 473, 493, 516. 519
  staple, 357, 374-75, 445, 450-51, 516-18, 540, 538-44, 548-49, 568-69, 650
  subsistence use, 57, 84, 363-68, 374-82, 410, 443, 493-96, 500, 504, 549
Human health, *see* disease; *see under* local people
Human history, 55-62, 303-8
Humidity, 317, 323, 340, 422
Humus, 29-30, 194-95, 256, 282, 319, 644
Hunting, 55, 57, 58, 354, 357-58, 374, 375, 378, 379, 380, 382, 565, 604
Hybrids, 52
Hydnophytum, 125, 248-51
Hydrochous, 294
Hydroelectric projects, 163, 169, 644
Hydrological regimes, 163, 171, 266, 417, 436, 465, 471, 484, 510, 514, 526, 536
Hydrozoans, 162
Hylobates, 45, 52-53, 126, 219, 271, 284, 332
Hylomys, 21
Hymantridae, 203
Hymenoptera, 225
Hypnea, 496
Hystrix, 45

Iban, 60, 150, 229, 234, 241, 358, 360, 372, 379, 384, 386, 540, 543
Ice Age, *see* glaciation
Iguanura, 185
Ilex, 244
Illipe, 199, 222, 379, 415, 546, *see also* Shorea

Imperata, 65, 374, 391, 439, 558, 602-4, *see also* alang-alang
Incense wood (gaharu), 59, 311, 349, 357, 411-15, 417
Inceptisols, 28
Indians, 356, 363
Indicator species, 126, 172, 267, 440, 466, 615, 616, 647, 663
Industrial developments, 312, 491, 501, 514, 573, 584, 589-594, 624
Industrial processes, 169, 398, 451, 500, 506, 510, 540, 554, 555, 582, 594
Insecticides, 41, 382, 476-78, 555, 559, 564
Insectivores, 259, 299, 617
  bats as, 199, 214, 232, 292-94, 309, 426
  birds as, 150, 199-200, 214-215, 229, 295, 309, 332, 345, 429, 478, 562, 611, 615
  fish as, 106, 110
  mammals as, 201, 214, 229, 232-36, 347-48, 426, 561
  spiders as, 228, 299, 617
Insectivorous plants, 248-52, 327, 341
Insects, 111, 214-16, 259, 298, 381, 384, 421, 610
  aquatic, 139, 143, 172, 457, 477
  as pests, 309, 474-77, 549, 552-56, 564, 618
  as pollinators, 102, 182, 199, 203-4, 206-8, 327, 340, 411, 422
  breeding behaviour, 111, 143, 207, 564
  leaf-eating, 202-4
  social, *see* ants; bees; termites
Intercropping, 363, 366, 372, 374, 421, 452, 474, 540, 549-50, 552, 556, 558-59, 609
Introduced species, 168, 434, 475, 478, 479, 554-55, 557, 562, 564, 565, 612, 615-16
  as threats to native species, 432, 479, 484, 557, 564, 616, 672, 676
  crop and tree, 41, 57, 129, 144, 406, 450-51, 543-44, 547, 549, 554-55, 562, 568, 570, 609, 611, 615
Intsia, 99, 130, 262, 308, 399
Invertebrate drift, 143
Invertebrates
  biomass, 108, 227, 259, 260, 439
  endemic, 46, 310, 352
  freshwater, 139, 141, 143, 145, 163, 172, 457, 473, 477, 595, 610
  in caves, 298-304, 310
  in forest, 202-4, 215, 216, 258, 259, 332, 422
  in soil, 194, 199, 224-29, 259, 260, 285, 332, 336, 439, 456
  marine, 78, 84, 89, 90, 108, 493
  richness in Borneo, 44, 213, 637
  threats to, 439, 473, 586, 642, 671
  see also insects
Ipomoea (kangkong, water spinach), 76, 77, 451, 474, 578

Iron
  compounds, 28-29, 58, 171, 255, 265, 448, 452, 467-68, 533, 576, 582
  pyrites, 29, 467, 469, 480, 516, 518
  technology, 58, 60, 363, 573, 574, 575
  toxicity, 467, 518
Ironwood
  forest, 241, 262-65, 398, 639
  products, 58, 262, 265, 357, 391, 547
  trees, 130, 262-65, 276, 280, 382, 396, 399
Irrawaddy dolphin, 147
Irrigation, 169, 374, 450, 465-69, 474-79, 484, 644
Irrigation canals, 163, 168, 406, 447-48, 471
Ischaemum, 77
Islam, 59
Island biogeography, 41, 47-49, 73-74, 315, 663, 670
Islands, offshore, 51-52, 73, 75, 79, 85, 391, 494, 497-503, 506, 526, 654
Isoptera, 225
Ivory, *see* hornbills, "ivory" from

Jackfruits, 41, 375
Javanese, 59, 362, 387, 390, 391
Jays, 384
Jellyfish, 92, 493
Jelutong, *see Dyera*
Jurassic, 15

K-selection, *see* reproductive strategies
Kadazan, 359, 360
Kajang, 359
Kalis, 359
Kalotermitidae, 225
Kandelia, 97
Kangkong, *see Ipomoea*
Kantu, 358, 359, 367, 540, 543
Kaolin, 575
Kapok, 467, 550
Kapuas lakes, 12, 117, 119, 127, 128, 145, 151, 155, **159-67**
Kapur, *see Dryobalanops*
Karahan, *see Lepidochelys*
Karst landscape, 30, 273, 276, 284, 302-3, 311
Kayan, 57-58, 359-360, 381, 384, 575
Kecapi, 467
Kelabits, 60, 359-60, 374, 384
Kendayan, 391
Kendia, *see Thynnichthys*
Kenyah, 359-61, 370, 375, 379, 384, 391, 416, 575
Kerangas, *see* heath forest
Kerapah (waterlogged heath forest), 120, 241, 247, 479
Keruing, *see Dipterocarpus*
King crab, *see* horseshoe crab

Kingfishers, 111, 150, 160, 384, 423
Kites, 150, 615
Koompassia, 129, 179
Korthalsia, 203, 349, 410, 412
Kryptopterus, 166
Kumpai (floating mats of vegetation), 480
Kutai National Park, 199, 214, 216, 221-22, 231, 262, 438-41, 458, 584, 638-39, 644, 646, 657-61, 663, 678
Kutai people, 361, 395, 413

Labroides, 92
Lacedo, 384
Ladang (traditional agricultural fields), 30, 144, 179, 375, 401, 403, 412, 543, 546-49, 610
Lagerstroemia, 130, 271
Lake ecosystems
  fish fauna, 144-46, 151, 153-57, 162
  productivity, 153, 155, 157, 162-63, 456, 460, 480-81
  swamp and marsh, 155, 479-82
  threats to, 169-72, 482-83
Lakes
  area and distribution, 152
  eutrophication, 153-54, 484, 592
  formation and characteristics, 151, 163, 480
  oxbow, 117, 131, 151, 160, 447, 460-61
Land Dayaks, *see* Bidayuh
Land use, 62-68, 271, 373, 442-43, 447, 454-56, 469-88, 503-26, 602-10, 654, 670-76
Landslips, 28, 281, 418, 440, 535, 644
Langsats, 196, 375
Langurs, 59, 111, 204, 219, 347, 423
  banded, 52, 332
  grey, 382
  maroon, 126, 271, 447, 560
  silvered, 95, 126, 149, 169, 426
  threats to, 421, 423, 439, 638
Lanius, 384
Lansium, 196, 209, 211, 375
Lanthonotus, 290
Lariscus, 45, 429
Laterisation, 644
Lates, 105
Latex, 125, 413, 453, *see also* rubber
Latosols, 276
Launea, 77
Lauraceae, 180, 209, 211, 262, 323
Laurasia, 14
Lawangan, 358
Leaching, **27**, 135, 172, 193, 257, 349, 609
  of acids from soil, 448, 465, 468, 518
  see also under nutrient loss
Lead, 575, 595
Leaf predation
  by invertebrates, 202-4, 258

by vertebrates, 112, 234, 235, 271, 425
Leafhoppers, 477
Leaves
  adaptive modifications, 125, 183-84, 188, 203, 219, 243-44, 253, 258, 265, 267, 321, 323, 342
  production, 196-97, 204, 427
Leeches, 139, 226
Legal controls and regulations, 79, 434, 458, 464, 526, 582, 586, 671-72
Legumes
  as fertility restorers, 549, 555, 604, 606-8, 615
  tree, 35, 129-30, 179, 399, 557, 606, 609, 637
Leiocassis, 166
Lemna, 140, 448
Lempam carp, *see* Puntius
Lemurs, *see* flying lemur
Leopard (*Neofelis*), *see* clouded leopard
Leopard (*Panthera*), 41
Lepidocaryoidae, 410
Lepidochelys, 80, 497
Lepidodendron, 578
Lepironia, 480
Leptobarbus, 166, 461
Leptomyrmex, 227
Leptoptilus, 82, 112
Leptospermum, 266, 282, 320, 342-43, 349
Leptothorax, 618
Lepturus, 77
Leucaena, 606, 608
Lianas, 96, 128, 182-83
Lice, 616
Lichens, 185, 317, 329, 611
Limacodidae, 553, 555
Limestone, 22, 25, 29, 311-12, 576, 580
Limestone cliffs, 74-75, 279, 281, 288, 290, 311
Limestone forest, 241, 247, 280-84, 311, 398
Limestone formations, 90, 273-77, 284, 311
  conservation, 280, 285, 310-12, 653, 655, 671
  flora and fauna, 274-86, 290-95, 310-12, 410, 681
  origins of caves in, 287-89
  soils, 275-77, 280
  threats to, 311-12, 580, 671
Liming, 467, 516, 518
Limnodromus, 82
Limpets, 75, 587
Liparis, 245
Lithocarpus, 320, 341-42
Lithothamnion, 496
Litsea, 383, 455
Litter communities, 222-30, 259-60
Litterfall
  as component of production, 104, 191-92,
    368, 558
  decomposition, 181, 193-94, 224-26, 318-19
  into rivers and ponds, 139-41, 517
  minerals in, 193-96, 248, 319
  polyphenol levels in, 257-58
  rates of, 251, 255, 267, 319
  role in fires, 435
Littoraria, 108
Liverworts, 139, 185, 248, 323
Livestock farming, 362, 388-89, 392, 538-39, 541, 565-68, 604, 607-8, 609, 668
Lizards, 213, 214, 259, 300, 337, 338, 617
Loaches, 139, 171
Lobsters, 90, 109, 494, 500, 516
Local people
  agricultural practices, *see* agriculture, traditional practices
  health, 381, 468, 472, 589, 594
  livelihoods, 68, 169, 308-10, 386, 441-43, 464-65, 493-500, 504, 552, 554, 589, 608-9, 645, 646, 647, 672
  see also human food, local use
Locusts, 381
Logging, 73, 177, 284, 369-70, 398-410
  biological impacts, 117, 163, 195, 350, 392, 398, 429-30, 491, 501, 675
  concessions, *see* timber concessions
  effects on forest structure, 78, 180, 183, 186-90, 256, 400-401, 403-9, 430
  effects on landscape, 403-6, 417-18, 603
  methods, *see* rotation systems, in forestry
  selective, 403-9, 421-23, 430, 670, 672, 680
  with burning, 256, 387, 455-56
Lonchura, 46, 478
Lophura, 46, 382, 639
Loranthaceae, 184, 210
Loris, slow, 219, 232-34, 423
Lowland dipterocarp forest, 73, 160, 174-238, 342
  characteristics and structure, 177-90, 263-64, 349-50, 638, 655
  fauna, 213-38, 258-59, 421, 638
  threats to, *see* deforestation; *see under* logging
Lucinidae, 494
Luciocephalidae, 161
Luciosoma, 166
Lugats, 360
Lumnitzera, 97, 98, 506
Lun Bawang, 359
Lun Dayeh, 359, 374, 540, 542
Lutjanus, 517
Lycaenidae, 216, 344
Lycopodium, 121
Lythraceae, 123

Ma'anyan, 358, 359

Macaques (*Macaca*), 213, 219, 258, 421, 638
  long-tailed, 111, 126, 130, 131, 149, 201, 332, 426, 563, 610, 653
  pig-tailed, 126, 311, 332, 563
Macaranga, 188, 203, 440, 472, 608
Machaeramphus, 298
Mackerel, 493
Macrobrachium, 381
Macroglossus, 422, 646
Macrophytes, *see* aquatic macrophytes
Madurese, 362, 387
Magnesium, 29, 265, 266, 267, 271, 275-77, 467
Mahogany, 557
Maize, 543, 549, 552
Malaria, 169, 387, 621-22
Malays, 357, 359, 361
Mallotus, 129
Malocincla, 46
Maloh, 359, 384, 575
Mammals
  arboreal, 214, 234, 235, 345
  as pests, 478, 554, 555, 560, 561, 563
  breeding behaviour, 112, 131, 200, 229, 294, 421, 563, 653
  carnivorous, 236, 237
  endangered, 147, 349, 639
  endemic, 43, 44, 52, 271, 315, 348, 352, 633, 634, 638, 663, 675
  extinctions, 21, 354, 636
  in protected areas, 662-67
  insect-eating, 214, 235, 348
  island races, 73-74
  leaf-eating, 204
  marine, 84, 86, 497, 588
  nocturnal, 210, 213, 232-37
  population densities, 258, 271
  riverine, 141, 147
  seasonal movements, 160, 199-200
  species richness, 41, 213, 329, 343, 345, 612, 632, 638
  threats to, 222, 386, 421-23, 438, 559, 588, 668
Mammea, 77, 209, 276
Manatees, 86
Mandor Nature Reserve, 245, 246
Manganese, 267, 518, 576
Mangoes (*Mangifera*), 39, 209, 375, 467, 544-46, 611, 650
Mangosteens, 57, 196, 211, 375, 453, 544
Mangrove ecosystems, 94-115, 503-19
  conservation, 510, 526, 681
  distribution and extent, 81, 94-96, 128, 506-7
  energy flow in, 82, 105-6, 115
  fauna, 82, 96, **106-15**, 158, 504, 518
  management, 467, 514, 519, 528
  productivity, 103, 104, 504, 510-13, 515, 519, 645
  regeneration, 508-10, 514, 519
  threats to, 73, 114, 398, 484, 504-6, 514-516, 518-26, 586
  value, 95, 101, 105, 114, 469, 503-5, 510-14, 521, 526, 589
  vegetational structure, 78, 94-96, 101-2
  woody species, 96-98, 102-3, 506-9
  zonation, 96-101, 108
Manihot, 543-44
Manis, 55, 235, 382
Mansonia, 169
Manuring, *see* fertilisers, natural
Maranthes, 211
Marble, 22
Martes, 236
Masonia, 170
Massenerhebung effect, 266, 320, 321
Mast-fruiting, 34, 189-90, 196-99, 211, 222, 263
Mastocembelus, 161, 166
Maxomys, 45
Mayflies, 139, 172, 295
Medicinal products
  from animals, 227, 231, 458, 500
  from plants, 41, 271, 349, 379, 382, 383, 415, 416, 443, 451, 452, 453, 510, 544, 548, 550, 552, 568, 569, 570, 604, 645, 650-51, 672
Megachiroptera, 294
Megaderma, 292
Megalaima, 46, 54, 345
Megapodius, 73
Melaleuca, 100, 130, 415, 453, 456, 466, 472, 479, 519
Melanau, 57, 357, 359, 450, 568
Melange, 22
Melanorrhoea, 129
Melastoma, 209, 466
Melastomataceae, 150, 182, 247
Melawi, 359
Meliaceae, 150, 211
Melolonthis, 554
Meranti, *see Shorea*
Merawan, *see Hopea*
Mercury, 578, 594-95
Mesozoic, 15, 22, 25
Messerschmidia, 77
Mesua, 125
Metalworking, 58, 574-75
Metamorphism, 22
Methane, 601-2
Metroxylon, 41, 57, 129, 450-51
Mezoneuron, 183
Mice, 610, 618
Micro-organisms, 84, 185, 204, 219, 225, 451, 595, *see also* soil micro-organisms
Microchiroptera, 199, 292

Microclimate, 186, 188, 226, 228, 301, 302, 303, 317, 422, 438, 483
Microhabitats, 137, 185, 478, 612, 616
Microhierax, 46
Midges, 172, 299
Migration
  animal, after forest disturbance, 421, 422, 423, 426, 429, 430, 432
  animal, during Pleistocene, 18, 344
  fish, 161, 169
  human, 55, 57, 60, 355-57, 360-63, 372, 387, 391, 401, 531, 576, *see also* trans-migration
  seasonal, 199, 222, 430, 432, 477, 675
Migratory species, 79, 82, 95, 111, 131, 157, 160, 518, 671
Milkfish, 105, 464, 494, 504, 517, 519
Millipedes, 224, 228, 229, 301
Mimicry, 92, 110, 228, 340
Mineral licks, 231, 432
Mineral resources, 25, 67, 573, 574-84
Mining, environmental impacts, 67, 163, 581-82, 594, 596, 657
  see also under coal; gold; peat
Miniopterus, 292, 294
Miocene, 14, 582, 583
Mistletoe, 184, 210
Mites, 298, 554, 564
Modang, 359, 360
Molluscs
  as human food, 306, 307, 381, 494, 504
  endemic, on limestone, 284-86, 310
  for handicrafts, 494, 500
  forest, 224
  freshwater, 57, 139, 141, 162, 355, 381, 474
  mangrove, 95, 106, 108
  marine, 75, 82, 92, 307, 494, 504
  threats to, 587
Moluccana, 379
Mongoloids, 57, 356
Mongooses, 52, 210
Monitor lizards, 55, 95, 127, 147, 157, 290, 298, 381, 457, 458, 480, 504, 610
Monkeys, 111, 126, 211, 214, 216, 236, 258, 379, 439, 622, 646
Monoculture, 435, 540, 558, 559, 560, 561
Monomorium, 617
Monophyllaea, 281
Monopterus, 157, 166, 477, 517
Monospecific dominance, 244, 262, 264
Monsoons, 31-33
Montane forest, 120, 248, 282, 318-352
  fauna, 318, 327, 329-39, 343-48
  flora, 248, 320-29
  on limestone, 282, 284
  threats to, 350-52
  value as protection forest, 398, 418
  vegetational structure, 249, 319-29, 347-48, 349
  zonation, 19, 320-48
  see also mountain habitats
Moonrat, 235
Morinda, 77
Mosquitoes, 169-70, 472, 474, 581, 610, 615, 617, 622, 623
Moss forest, 28, 247-48, 317, 320, 323, 340, 342, 349
Mosses, 185, 248, 290, 340
  aquatic, 139, 140
  montane, 317, 323, 342, 348
Motacilla, 384
Moths, 207-8, 216, 259, 343, 344-45, 422
  as pests, 203, 292, 477, 556
  as pollinators, 102, 204-5
  in biological control, 564
  in caves, 298, 299, 302
Moundbuilders, 73
Mount Kinabalu National Park, 339-48
  fauna, 22, 343, 345-46, 348
  flora, 320, 340-43
Mountain formations, 9, 11, 17, 315, 317
Mountain habitats
  conservation, 351-52
  endemism in, 44, 52, 54, 315-16, 318, 343-45, 348, 350-52
  soils, 319
  threats to, 351-52
  see also montane forest
Mousedeer, 55, 230, 236, 284, 343, 345, 379, 386, 604
Mualang, 358
Mudflats, 80, 82, 106
Mudskippers, 109-10
Mullets (*Mugil*), 105, 504, 517
Multiple cropping, 464, 543, 552, 569
Muntjak (*Muntiacus*), 45, 148, 230, 284, 379, 382
Murut, 359, 360
Musa, 182, 209
Muslims, 59, 60
Mussels, 92, 504, 587
Mustela, 236
Mustelids, 44
Mutualism, 646
Mycorrhiza, 181, 195, 327, 340, 441, 557
Mydaus, 236
Myotis, 292
Myrica, 349
Myristicaceae, 180, 209, 323
Myrmecodia, 125, 247, 327
Myrmecophytes (ant plants), 125, 203-4, 248, 249, 250-51, 282, 319, 327, 411, 637
Myrmeleontidae, 299
Myrtaceae, 150, 162, 247, 266, 320, 323, 349, 415
Mystus, 166

INDEX 791

Naja, 561
Napothera, 46, 54
Nasalis, 45, 95, 111, 113, 126, 219, 447, 639
National parks, *see* protected areas
Natural gas (LNG), 67, 501, 520, 526, 573
Nectar eaters, 204
  bats, 214, 232, 292, 294, 309, 422, 646
  birds, 214, 327, 441, 615
  insects, 102, 204, 340, 611
  other mammals, 236, 347
Nectarinia, 112
Negritos, 57
Nekton (free-swimming community), 143, 151
Nelumbo, 480
Nematodes, 78, 169, 620
Neofelis, 44, 235, 447, 639
Neomeris, 84
Nepenthes, *see* pitcher plants
Nephelium, 41, 196, 375, 467, 544
Nephila, 616
Nephothera, 345
Nephrolepis, 41, 379, 382
Nerita, 75
Neritina, 108
Nertera, 323
Nesting sites, 157, 216, 445, 560, 610
  turtles, 79, 81, 497, 587, 653
Net primary productivity, 84, 103, 125, 191, 192, *see also* productivity
Ngaju, 358, 359, 384, 581
Ngorek settlements, 59
Niche separation, 93, 295, 296, 301, 478
  by food type, 86, 214-22, 227, 233, 236, 298, 332, 346-48
  in rivers, 236, 137-39, 144-46, 155-56, 161, 162
  temporal, 90, 216, 231-37
Niches, 90, 216, 260-61, 422, 479, 559, 562, 563, 611
Nickel, 29, 265, 267, 448, 575
Nilaparvata, 478
Nipa palm, 94, 99, 513
Nitrogen and its compounds, 86, 104, 135, 195, 226, 319, 327, 448, 450, 544, 594, 601
  deficiency, 196, 251, 277, 282, 319
  fixation, 76, 84, 88, 185, 248, 266, 327, 474, 494, 539, 549, 555, 609, 615
  uptake, 182, 248
Nocturnal animals, 214, **231-38**
  birds, 111, 151, 561
  fish, 90, 157
  invertebrates, 86, 109, 204, 228, 294, 565
  mammals, 148, 151, 210, 213, 219, 223, 232, 234-37, 610, *see also* bats
  reptiles, 235, 291, 617
Non-timber forest products, *see* forest products, non-timber
Nothofagus, 342
Notopterus, 167, 461, 462
Nutrient cycling
  between ecosystems, 80, 87, 504, 517
  decomposers in, 224, 226, 228, 558
  disruption, 365, 471, 516, 518, 525
  in rainforest, 182, 184, 192, **193-96**, 202, 247, 251, 319, 558
  in wetland habitats, 82, 104-5, 117, 127, 141, 147, 149, 157, 161, 447, 456, 465, 469, 480, 484, 491, 525
Nutrient deficiency
  see calcium, deficiency; nitrogen, deficiency; phosphorus, deficiency; potassium, deficiency; soils, nutrient-poor; *see under* plant adaptations
Nutrient enrichment, *see* eutrophication; *see under* fire; flooding
Nutrient loss
  by leaching, 27, 365, 372, 438, 467, 533, 534, 644
  through erosion, 30, 365-66, 372
Nycteris, 426
Nycticebus, 232
Nycticorax, 112
Nymphaea, 452
Nymphalidae, 216
Nymphoides, 140
Nypa, 41, 94, 97, 98, 510

Oaks, 320, 330, 341, 342, 343, 347, 349, 350, 430, *see also* trig-oak
Ochroma, 608
Octopus, 90, 92, 500
Oculocincta, 44, 46, 54
Ocypode, 78, 107, 108
Odonata, 172
Oecophylla, 110
Oil (petroleum)
  formation, 17, 25, 582
  pollution, 484, 518, 520, 526, 584, 586-89
  production and revenue, 67, 395, 401, 573, 574, 575, 577, 583-85
  revenue,
Oil palm
  importance and productivity, 451, 552, 555, 555
  plantations, 192, 451, 540, 550, 555, 558, 560, 561, 562, 644, 672
  processing, 589, 594
Oils
  edible, 379, 416, 451, 549, 550, 552, 555
  useful, 125, 415, 451, 453, 497, 550, 552
Oligocene, 15, 17
Ombrogenous (rain-fed) peats, 28, 120-21
Omens and augury, 234, 361, 363, 384, 386
Oncosperma, 97

Ophiocephalus, 166
Ophiolite, 22
Ophisternon, 157
Opuntia, 564
Orang Ot, *see* traditional people; Penan
Orangutans
  as pests, 386, 553
  as prey, 55, 236, 306, 565
  breeding behaviour, 422, 653
  conservation, 676, 678-80
  distribution, 52, 231, 332, 680
  feeding behaviour, 201, 211, 219, 220, 221, 222, 347, 439, 544
  habitats, 126, 219, 258, 284, 311, 343, 349, 447, 680
  in protected areas, 126, 222, 343, 347, 349, 678
  seasonal ranging, 199, 432, 675
  threats to, 386, 422, 423, 554, 639
Orcaella, 147, 148, 447
Orchids, 182, 197, 245, 246, 248, 267, 279, 323, 340, 342, 348, 350
  diversity, 35, 52, 175, 340, 341
  endemic, 248, 267, 279
  epiphytic, 183, 282, 340
  propagation, 681
  seed dispersal, 209, 340
Organic matter, *see* allochthonous material; litterfall; soil, organic matter
Oriolus, 46, 54
Orthotomus, 112, 258, 345
Oryctes, 552, 555
Oryza, 540, 650
Oryzias, 157, 615
Osbornia, 98
Osphronemus, 167
Ospreys, 150, 615
Osteochilus, 166
Ot, *see* Penan
Ot Danum, 358
Otters, 127, 147, 162
Otus, 54
Overexploitation of natural resources, 1-5
  by hunting, 21, 271, 355, 636
  forest, 63, 177, 284, 409-10, 413, 434, 506, 508, 514
  in caves, 297, 309-10, 312
  marine, 79, 84, 494, 497, 503, 504, 642
  prevention, 680-82
  wetland, 163, 447, 458, 462-64, 482-83, 642
Overgrazing, 418, 567-68
Owls, 151, 234, 553, 561-62
Oxycephala, 297
Oxygen depletion, 135-36, 153-54, 155, 157, 171, 448, 471, 473, 484, 520, 592, *see also* under animal adaptations; plant adaptations
Oxyleotris, 167

Oxytelinae, 259
Oysters, 75, 92, 105, 494, 504, 520, 587
Ozone layer, 600, 602

Pachycephala, 46, 54, 112
Padang (open scrubby vegetation), 245, 247, 256, 257-58
Padda, 478
Padi, *see* rice farming systems, hill rice cultivation
Palaeocene, 15
Palaeozoic, 107
Palaquium, 125, 148, 245, 263, 281, 414
Palinuridae, 494
Palms, 41, 230, 375, 379, 568
  climbing, *see* rattans
  distribution, 123, 148, 244-45, 267, 281, 349
  epiphylls on, 185
  see also coconuts; nipa palm; oil palm; sago palm; sugar palm
Pandans (*Pandanus*), 58, 75, 77, 127, 148, 282
Pandion, 150
Pangasius, 166, 461-62
Pangolin, 21, 55, 235, 379, 382
Panicum, 607
Panthera, 41, 44
Papermaking, 452, 513, 526, 589
Paphiopedilum, 279, 340
Papilio, 642
Papilionidae, 216
Paraboea, 281
Paradoxurus, 237
Parakeets, 222
Paramignya, 97
Paranephelium, 280
Parartocarpus, 41, 382
Paraserianthes, 556
Parashorea, 190
Parasites
  adaptations of, 226, 227, 622
  control of, 544, 559, *see also* pest control
  human, 169, 592, 616, 620, 621
  of cave animals, 291, 297-98, 310
  of fish, 92, 517
  plant, 179, 182, 183, 184
  wasps as, 206-7
Parkia, 379
Parrotfish, 496
Parrots, 199, 208, 210, 213
Parus, 112
Pastoralism, *see* livestock farming
Patin, 461-63
Pawpaws, 611
PCBs (polychlorinated biphenyls), 581
Peanuts, 271
Pearlfish, 93

Pearls, 500
Peat
  as fuel, 445, 472, 473
  degradation, 469, 471
  deposits, 25, 120, 460, 472, 480, 536
  formation, 28, 120-21, 193, 318, 349
  mining, 447, 448, 473, 484
  ombrogenous, 28, 120-21
  role in forest fires, 436, 455
  see also soil types, peaty
Peat swamp forest, 119-127
  characteristics and structure, 120, 122, 123, 124, 128
  effects of fire on, 436, 438, 456, 480
  extent and distribution, 117-20, 160, 163, 479
  fauna, 126, 127, 257
  flora, 123, 125, 198, 247-48, 329, 410, 453
  hydrological functions of, 438, 448, 455
  threats to, 117, 203, 390, 398, 435, 447, 455, 471-72, 484, 675
  value, 438, 443, 445, 448, 454, 455
  zonation, 121-24, 245
Pectinidae, 494
Pemphis, 77
Penaeus, 504, 517
Penan, 57, 349, **356-58**, 367, 375, 378
Penihing, 360
Penthetor, 294
Pepper, 61, 374, 410, **548**, 550
  plantations, 361, 362, 372, 403, 441, 540, 547, 603
Perch, 105
Pericopsis, 399
Periopthalmodon, 109
Periopthalmus, 109
Permian, 15
Pes-caprae communities, 76, 77
Pest control
  biological, 111, 292, 309, 372, 382, 458, 474, 478, 518, 553, 562, 564, 615, 616, 646
  by land management, 368, 543, 563, 609
  chemical, 472, 478, 555, 556, 564
  integrated, 549, 564, 565, 623
Pesticides, 172, 448, 476, 484, 518, 538, 563, 564, 594, 623
Pests
  agricultural, 230, 292, 390, 478, 516, 538, 562-65
  increase after forest clearance, 421, 472, 622
  insects as, 186, 203, 292, 474, 477, 549, 552-54, 556, 562-63, 564, 618
  mammals as, 230, 478, 553, 555, 561, 563
  urban, 612, 618
Pesut, *see Orcaella*
Petai (locust beans), 309, 378

Petaurillus, 45, 52
Petaurista, 423
Petroleum, *see* oil (petroleum); natural gas (LNG)
Pets, wild species as, 386, 457, 478, 615
Phalaenopsis, 340
Phallostethidae, 161
Phanera, 281
Phanerosus, 281
Pharmaceutical products, 41, 415, 500, 510, 540, 556, 569, 594, 647, 651
Phascolosoma, 109
Pheasants, 215, 229, 382, 639
Phenolic compounds, 125, 203, 251, 257-58
Pholidota, 267
Phosphates, 86-87, 182, 309-10, 467, 482, 518, 575, 594
Phosphorus
  compounds, 82, 104, 135, 195, 319, 448, 450, 472, 575
  deficiency, 125, 196, 251, 265, 277, 319
Photosynthesis, 190, 586, 589, 599
Phragmites, 452
Phyllocladus, 282, 342, 349
Physeter, 497
Phytophthora, 547
Picoides, 112
Piculet, 384
Pieridae, 216, 344
Pigeons, 73, 199, 208, 210, 215, 439
Pigs
  as pests, 390, 553, 563
  domesticated, 458, 565, 577
  seasonal migrations, 199, 222, 432, 675
  wild, 148, 157, 183, 211, 229, 236, 378, 379, 380, 565, 650
Pinctada, 494
Pineapple, 552
Pinus, 434, 608
Pioneer species, 188, 203, 455, 608, 611
  after disturbance, 188, 350, 368, 406, 428, 440, 611
  as fire risk, 441
  coastal, 76-77, 96, 101
Piper, see pepper
Pipistrellus, 45
PIR schemes, 554
Pisonia, 77
Pitcher plants (*Nepenthes*), 35, 125, 245, 249, 252-54, 256, 267, 282, 319, 327
  altitudinal zonation, 325, 326, 329, 341
  threats to, 350
Pithecanthropus (Homo) erectus, 55
Pits, *see* canals and ditches
Pittas (*Pitta*), 44, 46, 95, 112, 215, 229, 332, 430
Pityogramma, 41, 382
Pityriasis, 46, 639

Plankton, 82, 86, 90, 471, 474
  phytoplankton, 141, 143, 153, 161, 163, 448, 588, 589, 602
  zooplankton, 104, 108, 141, 143, 145-46, 162, 588
Plant adaptations
  against fire damage, 130, 372, 415, 466, 603
  against predation by herbivores, 103, 125, 199, 202, 211, 219, 257, 263, 265, 411, 544
  to animal pollination or seed dispersal, 204, 207, 209, 211, 216, 327, 340
  to aquatic habitats, 139, 140, 446
  to disturbed habitats, 372, 411, 604, 611
  to dry habitats or water stress, 102, 123, 183, 184, 188, 196, 247, 251, 255, 281, 340, 477, 552
  to exposure, 76, 179, 180, 181, 318, 327, 552
  to low light intensity, 140, 182, 183, 188, 209, 340
  to low oxygen concentrations, 99, 102, 123, 128, 129, 140, 445
  to nutrient deficiency or imbalance, 76, 125, 183, 184, 195, 246, 248-255, 266, 267, 282, 319, 327
  to salinity, 76, 102, 544
Plant breeding, 451, 564, 568, 642, 650
Plant diversity, *see* species richness
Plant propagation, 139, 435, 557
Plant-animal interactions, 93, 201, 216, 327, 411, 422, 637, 668
Plantations
  agroforestry in, 609
  as buffer zones, 672
  crops, **550-58**
  ecological costs and benefits, 193, 538, 558, 560-62, 601, 681
  for timber, 265, 391, 410, 421-22, 556-58
  productivity, *see* productivity, of plantations
  smallholder, 546, 550
  see also under coconut; oil palm; rattans; rubber; sago palm
Planthoppers, 478
Plate tectonics, 12-17, 22
Platycerium, 183, 340
Platylophus, 384
Pleihari-Martapura Wildlife Reserve, 267, 269-71, 549, 644
Pleistocene
  climate, 18-22, 175
  forest, area of, 175, 308
  glacial activity, 339, 343
  humans, 55, 57, 306, 354
  mammal extinctions, 21, 636
  refugia, 343, 352

  sea levels, 17, 19, 51, 52, 55, 240, 344
  species evolution, 318
Pliocene, 47, 583
Pluchea, 77
Plywood, *see under* timber products
Pneumatophores, *see under* root systems
Podocarpaceae, 323
Podocarpus, 244
Podzols, 28, 257, 536, 538
Poecilia, 157, 615-16
Poisons, *see under* toxicity
Polder systems, 471
Pollen analysis, 121
Pollination
  by animals, 151, 202, 204-7, 216, 406, 559, 638, 675
  by bats, 77, 102, 206, 292, 309, 422, 426, 562, 580, 646, 680
  by birds, 102, 184, 327, 562, 646
  by insects, 102, 161, 182, 199, 204, 207, 216, 327, 340, 411, 422, 555
  by wind, 204, 406, 422
Pollution
  atmospheric, 443, 584, 600-602, 647
  by fertilisers, 153-54, 172, 448, 482-84, 592, 594
  by heavy metals, 172, 578-81, 582, 594-95
  by industrial installations, 172, 484, 501, 525-26, 584, 586, 589-90, 592-96, 622
  by oil spills, 484, 518, 526, 584, 586-89
  by pesticides or herbicides, 172, 484, 518, 592, 594-96
  by human waste, 148, 154, 172, 482, 484, 525, 526, 592, 615, 622
  coastal, 71, 84, 85, 526, 589
  effects on freshwater fauna, 135, 138-39, 172, 483, 578, 590
  treatment by natural organisms, 172, 443, 448, 450, 578, 594, 595, 647
Polyplectron, 46, 639
Polyps, 86
Pometia, 280
Ponds, *see* canals and ditches; fish ponds
Pongamia, 77
Pongo, 52, 126, 284, 447, 553, 639
Population densities, 219, 392, 438, 560, 610, appendix 5
  after fire, 439
  in disturbed habitats, 421-24, 439
  in plantations, 559-60
  in forest, 126, 213-15, 221, 226, 258, 271, 635, 637
  of humans, 62-65, 372, 531, 573, 610, 618-20, 622, 624, 626-8
  of pests, 562-64
Porcupines, 236, 264, 379
Porolithon, 496
Porphyrodesme, 267

Portunus, 494
Potamonidae, 381
Potassium, 135, 193, 450
  deficiency, 29, 125, 265, 277, 467, 472, 544
Power stations, 473, 484, 526
Prawns
  brackish-water, 493, 504, 516, 517
  freshwater, 137, 147, 300, 381, 474
  marine, 95, 104, 493
  production, 82, 494, 504, 505, 516-19, 589, 645
Predation
  adaptations against, *see under* animal adaptations; plant adaptations
  adaptations for, *see under* animal adaptations
Predators
  in caves, 291, 292, 298, 299, 302, 303
  in forest, 228-29, 236, 259, 426, 668, 675
  in freshwater ecosystems, 143-48, 151, 157, 162, 172
  in pest control, 372, 478, 562-65
  insects as, 110, 224, 259, 332
  on reefs, 90, 92
  see also seed predators; birds of prey
Prehistoric people, 55-59, 305, 306, 355, 379, 565
Premna, 77
Presbytis, 45, 52, 53, 95, 111, 126, 219, 271, 332, 382, 447, 560
Primary forest, *see* forest, primary
Primates
  competition with squirrels, 214, 216
  densities in forest, 126, 130, 213, 258, 271, 332, 334, 423, 424, 439, 560, 675
  endemic, 45, 52, 126, 639, 663
  feeding behaviour and patterns, 199, 209, 211, 219-222, 271, 424-25, 441, 646
  niche separation, 219-22
  nocturnal, 232-34
  surveys of, 126, 216, 218-19
  "Primitive" species, 211, 227, 236, 340, 342, 343, 349, 578
Prinodon, 236
Prionochilus, 46
Pristolepis, 167
Proboscis monkeys, 95, 111, 113, 126, 149, 160, 219, 311, 439, 447, 639
Production
  see gross primary productivity; net primary productivity; productivity
Production forests, 395, 401, 432, 510, 672, 673, 675, 676
Productivity
  comparison between ecosystems, 445, 451, 492, 519, 539
  of coastal ecosystems, 75, 82, 84, 86-87, 103-5, 114, 491-94, 500, 504, 519, 645

of crops, 366-67, 369, 374, 467, 469, 540, 547, 550, 609
of fish ponds, 464, 515-19
of forests, 125, 127, 130, 192-93, 262, 319, 398, 507-8, 645
of lake ecosystems, 153, 155, 162-63, 168, 456, 460-61, 480
of plantations, 192, 391, 451, 552, 557
of wetlands, 145, 445, 447, 484, 513
Prosimians, 439
Proteaceae, 123
Protected areas, 348, 351
  design, 129, 332, 432, 652-55, 663, 668-70
  extent, 37, 94, 114, 177, 312, 443, 485-88, 526, 642, 655, 657
  fauna, 662-68, 678-79
  management strategies, 163, 443, 642, 654, 657-59, 663
  proposed, 85, 163, 311, 312, 351, 352, 386, 485, 510, 526-28, 642, 655, 683
  threats to, 657-59
  values, 163, 349, 642, 643, 645, 646, 647, 648, 649, 652-54, 663, 679, 680
  see also individual sites named in Table 14.7, page 656
Protection forest, 311, 350, 395, 398, 586
Protozoa, 141, 204, 225, 474, 592, 595, 620
Psocoptera, 216
Psychidea, 555
Pteridophytes, *see* ferns
Pterocarpus, 399, 557
Pteropus, 73, 151, 209, 232, 294, 426
Ptilocichla, 46
Ptychozoon, 223
Pueraria, 607
Pulmonates, 108
Pulpwood, *see under* timber products
Punan, *see* Penan
Punan Gang, 384
Puntius, 166-67, 461
Pycnonotus, 54
Pynoscelus, 298
Pyralidae, 208
Pyralis, 298
Pyrites, *see under* iron
Python, 381, 458

Quail, 430
Quarrying, 311, 312, 580
Quartz, 575
Quaternary, 17-21, 25
Quercus, 320, 341, 342
Quotas, *see* legal controls and regulations

R-selection, *see* reproductive strategies
Rabbitfish, 496
Rabies, 622
Radioactive pollutants, 172

Rafflesia, 182-83, 204, 639, 640
Rails, 478
Rainfall
  after drought, 438, 440, 544
  amount and distribution, 30-34, 160, 196, 198, 348-49, 531-32, 599
  capture by forests, 133, 135, 183, 417, 644
  crop requirements, 547, 552, 555
  effects on organisms, 199, 225, 232, 317
  effects on soil and rocks, 27, 121, 274, 288, 536
  enrichment with nutrients, 135, 193, 196, 319, 517
  heavy, 133, 143, 157, 160, 343, 366, 518
  local regeneration by forest, 418, 432, 598, 644
  see also acid rain; drought
Rambai, 196, 375
Rambutans, 41, 57, 196, 211, 375, 467, 544, 569
Ramin, *see Gonystylus*
Rana, 136, 303, 381, 479, 612
Randia, 349
Ranunculus, 323, 343
Raptors, *see* birds of prey
Rasbora, 167
Rastrelliger, 493
Rats, 45, 199
  as pests, 390, 421, 472, 478, 553, 554, 561, 563, 610, 616, 618
  control of, 458, 553, 561-62, 563
  in caves, 290, 298
  in forest, 229, 235, 236, 347, 426, 675
Rattans, 39, 41, 183, 203, **410-13**
  collection and processing, 284, 311, 413, 453, 589
  in forest, 243, 281, 410
  montane, 342, 349
  plantations, 412, 413, 550, 551, 650, 681
  uses, 349, 355, 357, 368, 382, 410, 441, 645, 672
Rattus, see rats
Ratufa, 439
Rays, 143
Recolonisation
  after coastal oil pollution, 588
  after Ice Age, 637
  by animals after forest fires, 438-41
  by aquatic organisms after flooding, 477
  of headwaters by invertebrates, 143
  see also regeneration
Reduviidae, 299
Reeds, 141, 452
Reefs, *see* coral reefs
Reforestation, 182, 256, 432-35, 601, 608-10, appendix 7
Refugia, 151, 343, 352, 430, 637, 645, 654
Regeneration
  after clearance, 189-90, 366, 370-72, 406-9, 432
  factors affecting, 182, 188, 406, 422, 432, 455, 472, 508-9, 568, 675-76
  of forest gaps, 186-90, 281, 408, 410, 646
  of heath forest, 256, 257
  of mangroves, 96, 507-10, 514,
  of montane forest, 350
Reintroductions, *see* captive breeding and species reintroductions
Religious and ritual practices, 58-60, 150, 308, 361, 363, 382, 384, 386, 548, 565, 575
Repelita programmes, 573
Reproductive strategies, 264, 563, 564, 618
Reptiles, 44, 84, 141, 149, 204, 213, 223, 258, 337, 352, 422, 438, 671
Research, *see* scientific research and education
Reservoirs, 271, 406, 474, 514
Resettlement, *see* transmigration
Resins, 125, 413-15, 436, 453, 552
Rhacophorus, 223
Rhapidophora, 298, 299
Rheithrosciurus, 45, 271, 429, 675
Rheophytes, 150
Rhinoceros, 43, 55, 386, 680
  conservation, 343, 639, 641, 653, 676, 680
  feeding behaviour, 230, 231, 237, 428
  horn from, 59, 231, 343, 357
  Javan, 308
Rhinoceros beetle, 228, 552, 555, 564
Rhinolophidae, 426
Rhinolophus, 292
Rhinomyias, 54
Rhinoplax, 382, 639
Rhinotermitidae, 225
Rhiomyias, 332
Rhipidura, 112
Rhizophora, 97-99, 100, 102, 103-4, 112, 506, 508, 510, 519
Rhodamnia, 247
Rhododendrons, 282, 319, 325, 327-28, 329, 341, 342, 348, 349
Rhyncophorus, 381, 553
Rice
  importance, 568, 650
  non-food uses, 540
  origins, 57-58, 356, 540
  pests and diseases, 563, 650
  varietal diversity, 568, 650
Rice farming systems,
  hill rice, 58, 358, 363-69, 374, 381, 450, 543, 569, 609
  rice-coconut, 466, 467, 552
  rice-fish, 464, 475
  sawah, *see* sawah (wet-rice) cultivation
Riodinidae, 216
Riparian forest, 128, 129, 479

INDEX 797

Ritual, *see* religious and ritual practices
River systems, *see* wetland habitats, area and distribution
Riverine forest, 117, 128, 148-51, 182, 271, 411, 594, *see also* riparian forest
 fauna, 112, 126, 130, 148, 149, 150, 447
Rivers
 as transport routes, 11, 60, 131, 151, 169, 413, 480, 484, 671
 blackwater, 120, 127, 135, 160, 261, 460-61
 chemistry, *see* water chemistry
 ecology, 126-27, 137-51, 157-59, 172, 299, 447, 458, *see also* fish, freshwater
 flow rates, 132-34, 139-43, 144, 151-52, 160-63, 168-69, 418, 421, 432, 443, 509, 589, 644
 physical characteristics, 131-37
 pollution, 135, 138-39, 148, 171-72, 484, 578, 581-82, 589-90, 592, 594-96
 threats to wildlife in, 589, 590
Roadbuilding, 262, 311, 372, 580
 threats to habitat, 79, 163, 312, 400, 406, 418, 501, 581
Rock crevice plants, 277, 280, 281, 343
Rocks, *see* geology
Rocky shores, 74, 75, 77, 83, 587
Rodents, 199, 200, 204, 345, 348, 458, 472, 616
 endemic, 44, 45, 52
Roosting sites
 bats, 232, 291, 292, 294, 298, 302, 303, 309, 426, 610
 birds, 79, 95, 111, 209, 295-96, 303, 478, 610, 611
Root systems
 adventitious roots, 340
 buttresses, 128, 180, 181, 243, 281, 292, 323
 effect on runoff, 644
 feeding roots, 123, 247, 256
 in mangrove, 99, 102, 103
 in strangling figs, 183
 nutrient uptake by, 195
 pneumatophores, 123, 128-29, 586
 stabilisation of soil by, 101, 418
 stilt roots, 97, 102, 128, 243
Ropemaking, 77, 451
Rosy periwinkle, 651
Rotation systems
 in agriculture, 357, 363, 610
 in forestry, 398, 407, 408, 409, 421, 422, 432, 455, 506-8, 510, 557, 430, 435, 508, 675-76, 681
Rotifers, 141, 474
Roundworms, 595, 620
Rousettus, 294
Rubber, 368, 413, 554, 556, 558
 as cash crop, 374, 391, 550, 554, 589, 590, 594
 plantations, 192, 361, 390, 414, 417, 540, 550, 554, 558, 560, 644, 672
 see also jelutong (*Dyera*)
Rubiaceae, 150, 180, 182, 247, 248, 349
Rubus, 343

Saccharum, 604, *see also* sugar, cane
Sago (*Eugeissona*), 41, 57, 58, 60, 355, 357, 375
Sago palm (*Metroxylon*), 41, 57, 129, 359, 450-52, 471, 568
 plantations, 451, 452, 471
Salacca, 267, 281
Salinity, 75, 78, 81, 83, 96, 158, 465, 471, 514, 517
 plant adaptations to, 76, 102, 544
Salt production, 514
Salticidae, 617
Sambar, 284
Samunsam Wildlife Reserve, 488
Sand dollars, 78
Sand dunes, *see* beaches
Sandoricum, 467
Sandstone, 12, 22, 25, 28, 75, 241, 282, 285, 290, 329, 533
Sangkulirang, 273, 275, 277, 284, 290, 308, 309, 311, 312
Sapotaceae, 180
Saraca, 128, 150
Sarax, 299
Sassia, 384
Satyridae, 216
Savanna, 256
Sawah (wet-rice) cultivation
 associated flora and fauna, 111, 474-78, 518, 623
 by tribal peoples, 359, 374, 391, 450, 474, 541-43, 569
 distribution and extent, 450, 532, 536, 569
 origins, 361, 362, 450, 577
 pests, 472, 477, 478, 563, 565, 650
 plant nutrition in, 539, 543
 suitability, 5, 65, 388-89, 392, 465, 468, 469, 471, 532, 535-39, 540
 threats from, 127, 130, 445, 465, 601, 623
 wetland creation by, 163, 474
 see also rice; rice farming systems
Scaevola, 77
Scale insects, 204, 554
Scallops, 494, 504, 587
Scaphula, 168
Scarabaeidae, 227, 228, 259
Scaridae, 496
Scat association, **301, 302**
Schima, 343
Schistosomiasis, 621
Schists, 22

Schizaea, 256, 266
Scientific research and education, 93, 169, 271, 309, 310, 315, 349-50, 350, 443, 484, 510, 569, 608, 610, 647, 677, 680-82
Scleropages, 127, 161, 166, 457
Scolopia, 97
Scorodocarpus, 280
Scorpions, 228, 299
Scutigera, 299, 303
Scylla, 105, 109, 494, 504, 517
Scyphiphora, 97, 98
Scyphostegiaceae, 35
Sea anemones, 90, 92, 93, 500
Sea cucumbers (trepang), 84, 93, 493, 497, 500
Sea fans, 500
Sea hares, 92
Sea level changes, 16-21, 41, 55, 120, 599
Sea lilies, 92
Sea slugs, 90, 108
Sea snails, 92
Sea squirts, 92
Sea stars, 84
Sea urchins, 78, 79, 84, 92, 93, 500
Seabirds, *see* birds, coastal
Seagrass, 84-6, 88, 92, 526, 589
Seasonality, 17-18, 31, 121, 160, 196-200, 263, 532
Seaweeds, 84, 92, 587
Seberuang, 358
Second World War, 62
Secondary forest, 126
Sedges, 76, 96, 130, 343, 428, 452, 456, 472, 480
Sedimentary rocks, 17, 25, 273, 311, 339, 533
Sedimentation
  as threat to coastal ecosystems, 27, 84, 101, 114, 501, 513, 518, 523-26
  as threat to freshwater ecosystems, 169, 171, 438, 473, 484, 644
  dolphins' adaptation to, 147
  effect of deforestation on, 171, 350, 406, 429
  effect of reforestation on, 433
  in soil formation, 29, *see also* peat, formation
  in wetlands, 131, 141, 163, 168-69, 448, 523
Sediments
  fauna, 82, 139, 141, 168
  flora, 140-41, 155
  from rivers, 80, 131, 491, 517
  immobilisation of phosphates in, 594
  micro-organisms in, 88, 595
  nutrient-rich, 82, 481
  particle size in, 137
Seed banks, 129, 188, 263, 367, 370, 406, 409, 441, 510, 672
Seed dispersal, 208, **209-13**, 264, 378
  by animals, 151, 183, 188, 202, 209-13, 216, 228, 406, 422, 675, 681
  by bats, 44, 209, 232, 426
  by birds, 73, 188, 209, 212, 213, 411, 637-38, 646
  by carnivores, 210, 411
  by fish, 146, 150
  by large herbivores, 229, 646
  by primates, 213, 411, 646
  by water, 76-77, 99, 103, 150
  by wind, 188, 209, 327, 340
Seed dormancy, 99, 188
Seed germination, 102, 103, 104, 188, 199, 263, 340, 416, 435, 557
  after fire, 440
  in forest gaps, 189
  threats to, 406, 422, 441, 638, *see also* seed predators
Seed predators, 44, 103, 161, 199, 207, 210, 211, 213, 263, 474, 478
Seedling survival
  factors affecting, 186, 188, 209, 256, 263, 264, 406, 455
  in forests, 99, 102, 103, 188-89, 256, 435, 586
  in orchids, 340
  in rattans, 410
Selako, 359
Selenium, 595
Serow, 284
Sesarama, 109
Settlements, human, 111, 372, 465, 514, 610-18, 636
  Penan, 357, 358
  prehistoric, 55, 57, 355
  riverine, 356, 359, 360, 361, 375, 379, 391
  sites of, 12, 63, 355, 390, 391, 491, 573, 579, 622
  see also transmigration; urban centres
Sewage treatment, *see* water purification
Shade plants, 244, 368, 549, 550, 556, 606, 609, 612
Shading, 136, 141, 144, 171, 182, 188, 209, 517, 552, 557, 601, 609
Shale, 25, 28, 285, 533
Shama, 384
Shellfish, 75, 307, 354, 492
Shifting cultivation, 363-75, 540
  as a Dayak tradition, 358, 391
  comparison with sedentary agriculture, 451, 538, 544, 565
  ecological effects, 369-72, 421, 602
  environmental impacts, 171, 366, 369-72, 390, 441
  extent, 363, 370
  of hill rice, 58, 359, **363-69**, 543

origins, 356, 363
sustainability, 28, 368, 372, 374, 535
see also deforestation, by shifting agriculture; swidden agriculture
Shorea, 263
   as timber tree (meranti), 120, 262, 399, 403, 454-56, 557, 610
   for illipe nuts, 130, 199, 222, 379, 415, 546
   distribution in forests, 179, 245, 246-47, 265, 281
   in swamp forest, 120, 123, 125, 130, 203, 245, 398, 456, 480
   productivity, 557
   propagation, 557
   regeneration, 190, 209
Shores, *see* beaches; rocky shores
Shrews, 55, 235, 290, 348
Shrikes, 384, 615
Shrimpfish, 93
Shrimps, 84, 90, 92, 105, 139, 494, 504
Siamang, 332
Siamese, 59
Sigamidae, 496
Siltation, *see* sedimentation
Silurus, 299
Silviculture, 130, 188, 190, 265, 406, 645
Sindora, 179, 280
Sirenia, 86
Slash-and-burn cultivation, 363, 369, 401, 435, *see also* shifting agriculture
Slime moulds, 185
Slopes, *see* topography
Slow loris, 219, 423
Slugs, 90, 108, 554
Smallholders, *see under* plantations
Snails
   air-breathing, 108, 477
   as disease vectors, 621
   concentration of heavy metals in, 595
   freshwater, 172, 299, 458, 474, 475, 477
   mangrove, 108-9
   marine, 75, 92, 494, 500
   terrestrial, 235, 284, 285, 286, 440, 554
Snakes, 44, appendix 4
   forest, 213, 214, 235, 259, 561
   freshwater, 148
   in caves, 297, 298
   sea, 75
   montane, 337-39
   value, 57, 381, 457, 458
Soapmaking, 544, 550, 552
Social behaviour, vertebrate, 131, 149, 230, 296, 421, 425, 678
Social insects, *see* ants; bees; termites
Sodium, 158, 231, 277
Soil
   cation-exchange capacity, 195, 257, 275, 603

conservation, 28, 67, 76, 366, 369, 370, 387, 398, 418, 421, 432, 443, 517, 541, 543, 552, 555, 558, 568, 581, 604, 609, 643, 644, 672
degradation, 256, 368, 369, 387, 406, 432-33, 438, 465, 534, 603, 609
erosion, 134, 171, 350, 365, 368, 369, 370, 406, 418, 421, 433, 438, 440, 501, 513, 532, 536, 542, 547, 549, 557, 558, 568, 603, 644, 681, *see also under* nutrient loss
fauna, 222-26, 285, 332, 335-36, 368, 438, 439
fertility, 27-30, 130, 365, 367, 432, 438, 443, 448, 533-39, 540, 606
formation, 25-28, 318
micro-organisms, 199, 368, 438, 589
organic matter, 30, 102, 120-21, 194-95, 255, 256-57, 280, 365-68, 447, 467, 534
temperature, 422, 471, 557
Soil types, **25, 26, 27, 28, 29, 30**, 121, 533, 536
   acid, 28-29, 120, 240, 251, 256, 282, 318, 327, 465-69, 516, 533, 535, 546, 603
   alluvial, 27, 29, 127
   association with plant richness, 35, 40, 96
   base-rich, 29, 265, 275-77, 280
   infertile, 4, 28, 193, 257, 265, 271, 387, 392, 465, 469, 472, 533, 603
   nutrient-poor, 28, 117, 120, 125, 196, 219, 231, 241, 245, 246, 247, 251, 256, 265-66, 277, 282, 318, 349, 374, 534
   peaty, 27-29, 120, 127, 240, 276, 282, 319, 327, 336, 455, 471-72, 536
   sandy, 241, 245, 387
   toxic, 125, 251, 265, 465, 467-68, 516
   ultrabasic, 29, 265-67, 271
   waterlogged, *see* waterlogging
Sonneratia, 96-103, 112, 506, 510
Sophora, 77
Soya beans, 549, 552
Sparrows, 478, 615
Spathoglottis, 340
Spawning sites, 82, 95, 160, 168, 447, 456, 484, 493, 504, 506, 645
   threats to, 171, 469, 473
Species,
   composition, 323, 372, 424, 462, 631, 675
   conservation, 631-35, 642, 645-46, 652-56, 662-69, 676-80,
   density (plants), 202, 262, 263, 421, *see also* population densities (animal)
   diversity, *see* endemism; species richness
   endangered, 157, 265, 348, 349, 639, 676, 681, appendix 8
   evolution, 74, 131, 202, 203, 230, 318, 352, 636
   interactions between, 500, *see also* plant-animal interactions

loss, 663, 668, 682
replacement, 111, 125, 332, 429, 668
Species richness
  as component of biodiversity, 35, 175, 631-4, 650
  comparison, between forest types, 117, 123-7, 129, 130, 175-77, 208, 213, 245-47, 258-59, 266-68, 271, 278-80, 333-39, 341, 347, 638, 681
  distribution of categories, 39-49, 52
  figures for, 35, 631-34
  in coastal and marine habitats, 82, 86, 90, 493, 500
  in freshwater habitats, 123, 127, 143, 161, 447, 474
  in mountain habitats, 315, 352
  in plantations, 557-59
  in secondary forest, 407, 422, 424, 426, 430, 443, 675
  in urban centres, 610
  on limestone hills, 284
Sphaerichthys, 161
Sphagnum, 323
Spices and flavourings, 379, 568
Spiders, 110, 216, 228-29, 299, 301, 477, 564, 616-17
Spilornis, 46, 54, 561
Spinifex, 76, 77
Spirogyra, 141
Spondias, 209
Sponges, 90
Springtails, 298, 299
Squid, 493
Squirrels, 214, 236, 341, 345, 348
  as human food, 379
  as pests, 421, 553, 555, 561
  as seed dispersers, 183, 188
  competition with primates, 216
  endemic, 45, 52, 215, 271, 429, 675
  feeding behaviour, 211, 213, 232-233, 264, 347, 425
  montane, 348
  niche separation, 214-15, 217, 235
  pigmy, 215
  threats to, 423, 438, 439, 638
Staetornis, 295
Staphylinidae, 259
Starfish, 92
Stenochlaeana, 379
Sterna, 73
Stink badger, 236
Stonefish, 92
Stoneworking, 55, 58, 59
Storks, 82, 518
Strangling figs, *see* figs, strangling
Stratification
  of animal communities in forest, 214-16, 238, 425

of forest canopy, 35, 179, 180
of water in lakes, 152-53, 154
Stylosanthes, 606-7
Styraceae, 123
Succession
  forest, 128, 186-90, 373
  in mangroves, 99, 101, 121
  of micro-organisms on leaves, 185
Sucker fish, 139
Sugar cane, 543
Sugar palms, 41, 375
Sugar refining, 513, 589, 594
Sula, 73
Sulawesi, 20, 31, 41, 44, 47, 84, 169, 234, 265-66, 361, 362, 516, 573, 608, 615, 644
Sulphates, 157, *see also* acid sulphate soils
Sulphides, 121, 582
Sulphur, 575, 579
Sunbears, 44, 55, 379, 382, 423, 426, 675
Sunbirds, 102
Suncus, 45
Sundasciurus, 45
Sundews, 249, 319
Surgeonfish, 496
Sus, 148, 199, 222, 553, 565, *see also* pigs
Sustainable agriculture, 30, 65, 365, 368-69, 371-72, 374, 539-40, 541-43, 568, 569-70, *see also* swidden agriculture
Sustainable aquaculture, 463, 484, 519, 681
Sustainable forestry, 130, 256, 357, 392, 407-9, 430, 443, 454-55, 469, 507-8, 510, 540, 554, 557, 598
Sustainable harvesting, 296, 410-17, 469, 484, 536, 645, 681
Swallows, 150, 611
Swallowtails, 46, 642
Swamp forests, 82, 117, 398, 447, 528, *see also* freshwater swamp forest; mangrove ecosystems; peat swamp forest
Swamps, *see under* lake ecosystems
Swidden agriculture, **363-65**, 369, 370-72, 378, 543
Swiftlets, **294-97**, 298, 300
  as energy importers into caves, 290, 291, 300, 302
  coastal, 73, 75
  feeding behaviour, 150, 294-95
  in sandstone caves, 290
  pest control by, 309
  threats to, 296-98, 580, 639
  urban, 615
  see also birds'-nests, edible
Swifts, 615
Symbiosis, 87, 90, 93, 181, 225-26, 248, 327, 340
Symplocaceae, 323
Syringodium, 85

Tadarida, 294, 300
Taman (Maloh people), 359
Tambak (brackish-water fish pond) culture, 94, 464, 494, 504, 514-19, 525
Tanjung Puting National Park, 125, 126-27, 157, 390, 458, 639, 646, 657, 678, 679, 680
Tanning, 594
Tannins, 103, 125, 257, 453, 510
Tapeworms, 620
Taphozous, 292, 294
Tapir (*Tapirus*), 21, 55, 354, 636
Taro, 58, 543, 548
Tarsier (*Tarsius*), 219, 232, 234, 639
Tea, 410
Tegelan (unirrigated fields), 392
Teijsmanniodendron, 280
Telescopium, 108
Tembadau, *see* banteng
Temperate species, 323, 343
Temperature regimes
  global, 17, 599, 601
  in forests, 255, 257, 422, 557, 601
  in freshwater habitats, 136-37, 144,152, 155, 157, 171
  in lowland Borneo, 30
  in sex determination, 584
  on mountains, 136, 317-18, 319, 348
  Pleistocene, 21
Tengkawang, *see Shorea,* for illipe nuts
Terminalia, 76, 77, 608
Termites, 216, 225, 235, 236, 332, 350, 601,
  as detritivores, 204, 224-26, 228, 259-60, 318, 439
  as pests, 554
Terns, 73
Territorial species, 107, 110, 234, 271, 421, 439
Tertiary, 15-22, 25, 28, 74, 311, 348, 576, 583
Tethys Ocean, 14, 15
Tetramerista, 125
Tetraodon, 167
Tetrastichus, 553
Tetrastigma, 182
Tettigoniidae, 381
Textiles, 58, 451, 480, 506
Thalassia (seagrass), 85
Thalassina (mud lobster), 109
Thalassodendron, 84, 85
Theaceae, 323
Theobroma, 555, 556
Thespesia, 77, 97
Thoracostachyum, 480
Thrips, 554
Thrushes, 215, 229, 639
Thuarea, 77
Thymelacaceae, 415
Thynnichthys, 461

Ticks, 564
Tidal swamplands, 465-71, 475, *see also* wetland ecosystems
Tides, 82, 101
  effects of, 74-76, 78, 82, 96, 107, 121, 514, 517, 525
Tiger, 44, 211
Tilapia, 144, 450, 464, 479, 517
Timber
  commercial, 265, 391, 395, 445, 454, 469, 484, 506, 610
  export policy, 399-401, 409, 589
  from heath forest, 256
  from wetlands, 454, 455, 484
  non-forest, 544, 546, 552, 554, 555, 570, 609
  plantations, *see* plantations, for timber
  production and revenue, 369-70, 395, 399-401, 407-10, 441, 573, 645
  see also logging
Timber concessions, 94, 231, 358, 391, 401, 408-9, 434, 657, 677
  extent, 67, 114, 177, 395, 396, 398, 506, 528
Timber products
  chipwood, 398, 505-8, 510, 514, 519, 525, 608
  plywood, 400, 409, 556, 589, 594
  pulpwood, 506, 510, 589, 594, 606, 608, 609
  sawnwood, 409, 556
  veneer, 556
Timber trees, 248, 398, 399, 403, 406, 443, 480
  productivity, 130, 180, 262
  see also Acacia, Avicennia, Dipterocarpus, Dryobalanops, Eusideroxylon, Gonystylus, Heritiera, Intsia, Lumnitzera, Pericopsis, Pterocarpus, Rhizophora, Shorea, Xylocarpus
Tin, 595
Tineid moths, 298
Toads, 339, 458, 610
Tobacco, 410
Tomistoma, 147, 157-8, 447, 642
Tools
  modern, 372, 387, 604
  primitive, 55, 355, 363
  traditional, 58, 363, 365, 366, 575
Topography, 532
Tor, 167
Tortoises, 57, 157, 229
Tourism and recreation
  interaction with conservation, 303, 312, 342, 350, 386, 498, 500, 501, 646, 671
  value of wildlife habitats for, 79, 169, 309, 310, 312, 386, 443, 484, 500
Towns, *see* urban centres
Toxicity

of animal parts and products, 92, 235, 553, 617
of leachates from acid sulphate soils, 467, 518
of plant parts and products, 41, 103, 125, 203, 211, 213, 219, 251, 257, 327, 544, 604
of plant products exploited by humans, 58, 212, 357, 379, 382
Toxotes, 106, 107
Trade and commerce, 59-60, 357, 362, 363, 390
  by tribal people, 57, 382, 403, 413, 416, 575
  in animal products, 59, 231, 357, 382, 497
  see also under export
Tragulus, 55, 230, 284
Transmigration, 362, 387-92, 403, 455, 465, 525, 573, 620
  recommended crops, 549, 554, 555, 569, 609
  sites, 257, 284, 311, 312, 387, 451, 469, 533, 538, 604
Tree crops, 390, 391, 538, 550, 569, 609, *see also under* plantations
Tree form, 180-82, 243-46, 264
Tree shrews, 43, 45, 74, 183, 229, 345, 347
Trema, 209, 608
Trepang, *see* sea cucumbers; Holothuroideae
Treron, 208
Triassic, 14
Tribal peoples
  culture, 309, 357-58, 384, 386, 646, 649
  ecological knowledge, 365, 382, 386, 609, 649, 651
  ethnobotanical knowledge, 41, 357, 382-83, 651
  forest use, 57, 375-9, 381-383, 385, 410-16
  origins, 57, 355-58
  resettlement, *see* transmigration
  sources of income, 386, 416-17, 646, *see also* local people, livelihoods
Trichastoma, 639
Trichogaster, 167, 475
Trichoglottis, 340
Trichoptera, 139, 299
Tridacna, 90, 494
Trig-oak (*Trigonobalanus*), 342
Trionyx, 479, 612
Tristania, 148
Tristaniopsis, 180, 245, 247, 329
Triumfetta, 77
Trochus, 494
Troglobites, 290, 300-301, 303
Troglophiles, 290, 301-2
Trogloxenes, 290
Trogona, 102

Trogons, 46, 215, 384
Troides, 642
Trombiculidae, 298
Trypanosomiasis, 622
Tubeworms, 92
Tupaia, 45, 53, 74
Tupaiidae, 229
Turbo, 494
Turdus, 345
Turtles
  breeding and nesting behaviour, 79, 81, 497, 498, 653
  egg collection, 79, 497, 498, 499
  exploitation, 79, 457, 497, 500
  freshwater, 147, 157, 258, 457, 479, 480, 612
  marine, 79, 80, 81, 84, 86, 493, 496
  threats to, 79, 483, 497, 520, 642
Tyto, 562

Uca, 106, 109
Ukit, *see* Penan
Ulin, *see Eusideroxylon*; ironwood, as timber tree
Ultisols, 28
Ultrabasic formations, 29, 247, 265, 336
  forest characteristics and structure on, 265-69, 271
  soils, 29, 265, 336
Ultraviolet radiation, 318
Ungulates, 646
Urban centres, ecology, 610-618, *see also* settlements, human
Urena, 41, 382
Urophyllum, 349
Urosphena, 46, 54
Urostigma, 441
Urticaceae, 280, 282
Useful (non-food) plants, 41, 77, 271, 379, 414-16, 452-53, 494, 544, 550, 552, 555, 569-70, 609, 651, appendix 6
Useful wildlife, 82, 84, 111, 379, 382, 458, 498, 645, 646
  ornamental uses, 58, 84, 90, 92, 229, 382, 386, 457-58, 480, 494, 500, 503, 568, 650
  see also indicator species; pest control; pollination
Utricularia, 248, 323

Vaccinium, 245, 247, 329, 349, 350
Vanda, 340
Varanus, 95, 381, 457, 458
Vatica, 179, 180
Vegetables, 374, 379, 403, 451, 474, 481, 543, 544, 578
Vegetation mosaics, 128, 245, 370, 438
Vigna, 77
Vines, 182

A Bugis schooner loads timber in the Kumai estuary, Central Kalimantan. Trading ships have visited Borneo for centuries to collect cargoes of the island's natural resources.

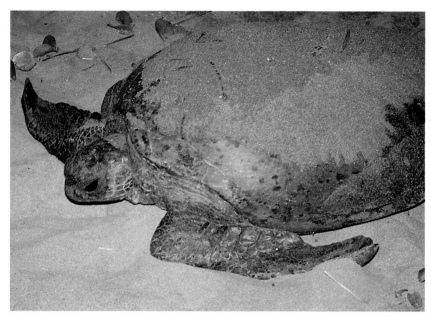

A female green turtle labours up the beach to lay her eggs above the tideline. Overcollection of eggs and harvesting of adult turtles have led to a serious decline in populations of marine turtles.

Colonial seasquirts and green sea anemones anchor on a skeleton of *Porites* coral. The many fine subdivisions of food and space on coral reefs support a high diversity of species.

A tangle of arched *Rhizophora* roots forms a coastal barrier between land and sea. The high productivity and detrital food chains of the mangrove forest support many marine animals, including several species of commercial value.

The dagger-like propagule of the *Rhizophora mucronata* is specially adapted for germination in the tide-washed mudflats.

The grapefruit-shaped fruits of the *Xylocarpus* are dispersed by river and tides to new areas of the mangrove swamp.

A proficient marksman, the archerfish, *Toxotes jaculator*, squirts a powerful jet of water to shoot down flies resting on mangrove roots above the water surface.

The mudskipper's ability to survive out of water allows it to forage for food on the exposed mud banks, a niche that no other fish exploits.

Head of the river food chain, the estuarine crocodile, *Crocodylus porosus,* feeds mainly on fish but sometimes takes larger prey such as migrating pigs, deer, monkeys and even people.

Weaver ants, fierce predators on fireflies and other insects, swarm over the mangrove bushes. Worker ants make compact nests by sewing leaves together with silk exuded by an ant larva, which is used as a living shuttle.

The proboscis monkey, *Nasalis larvatus,* is unique to Borneo. It is a common inhabitant of coastal mangrove and peat swamp forests and is found far inland along major rivers.

Upland streams tumble over rocky outcrops, carrying water and nutrients down to the great rivers of Kalimantan. The stream fauna is particularly sensitive to environmental disturbance which can lead to much reduced diversity of aquatic insects and fish species.

The wide lowland rivers of Kalimantan, winding through lowland dipterocarp forests, are the main highways into the island interior.

The false gavial, *Tomistoma schlegeli*, a specialist fish eater, is now rare in Kalimantan, but can still be seen in Tanjung Puting National Park.

Peat swamp forest and *Pandanus* fringe the tea-coloured waters of the Sikonyer River, Central Kalimantan.

The soft-shelled river turtle, *Trionyx cartilagineus* (*labi-labi*), preys on fish and frogs in muddy, slow-moving rivers and shallow, fast-flowing hill streams.

Lowland rainforest is the most species-rich habitat on earth, with 150 to 250 different species of trees in a one-hectare plot.

Winged fruits of *Dipterocarpus*. More than 260 species of dipterocarps are recorded in Borneo, many of them valuable timber trees.

The milkweed butterfly, *Idea,* alights on a lantana bush, the South American plant which has become a pantropical weed and is common on wastelands.

A wild ginger flower, Zingiberaceae.

The lantern flower of a *Bulbophyllum* orchid.

The fan palm, *Livistona rotundifolia*.

The *Rafflesia arnoldi*.

The green-crested wood partridge, *Rollulus roulroul,* scratches for insects and grubs among the leaf litter on the forest floor.

Lantern bugs, *Fulgora* sp. The 'lantern' at the tip of the insect's head has no known function.

The delicate maiden's veil fungus smells like rotten meat to attract flies to disperse its spores.

White parasol fungi grow on dead and dying wood, helping to speed up the process of decomposition.

Forest carnivores. The leopard cat, *Felis bengalensis*, smallest of the Kalimantan wild cats.

The green whip snake, *Ahaetulla prasina*.

Forest gliders. The flying lizard, *Draco volans,* extends a membrane supported by eight bony extensions of the ribs.

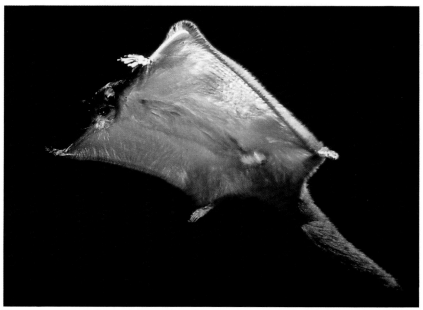
The nocturnal spotted giant flying squirrel, *Petaurista elegans*, found in hill dipterocarp forests, occupies a canopy niche exploited by fruit-eating squirrels and langurs during the day.

Nocturnal mammals. The big-eyed slow loris, *Nycticebus coucang,* forages for insects from the lower forest layers to the canopy.

The ant plants *Myrmecodia* and *Dischidia* accommodate *Iridiomyrmex* ants, whose faeces provide nutrients to the plant.

Kerapah forest at Mandor Reserve, West Kalimantan. Note the low canopy, air-breathing rootlets and the tea-coloured stream, draining white sandy soils.

A Rajah Brooke's birdswing butterfly, *Troides brookiana,* sips water at a mineral spring beside a lowland stream.

Nepenthes rafflesiana.

Nepenthes gracilis.

Pitcher plants derive nutrients by trapping and digesting insects in their liquid-filled pitchers. Twenty-eight species of *Nepenthes* are recorded for Borneo and are especially common on nutrient-poor soils in *kerangas*, peat swamps and mountain ridges.

Nepenthes ampullaria.

Nepenthes villosa.

The jagged limestone teeth of the Pinnacles rise 40 m among the high canopy of rainforest trees in Gunung Mulu N.P., Sarawak.

The forest-covered limestone massif of Batu Hapu, South Kalimantan, is a typical example of tower karst.

Stilted houses line the river at Pengandan, beneath a limestone escarpment in the Sangkulirang peninsula. Forest on these limestone hills burned during the Great Fire of 1982-1983.

The mouth of Deer Cave, Gunung Mulu N.P., where bat hawks and falcons hover at dusk to prey on bats exiting from the caves.

A Dusun coffin is hauled into limestone caves beside the Segama River in Sabah. Ancient cave burials tell much about human history in Borneo.

Shrouded in mist, the moss forest on upper mountain slopes is rich in ferns, orchids and mosses.

Moss festoons the branches and stems of trees and saplings in the damp forest above the cloud line.

Schima brevifolia, a relative of the tea bush, grows on the upper mountain slopes of Kinabalu up to about 3,800 m.

Begonias fringe the streams on the lower mountain slopes.

Rhododendron rugosum. Rhododendrons are characteristic plants of the acid soils of mountain forests. Fifty species occur in Borneo, with 25 recorded from Mount Kinabalu.

Rhododendron ericoides, named after its heather-like flowers.

The red berry of a *Vaccinium* plant provides food for montane birds.

Scarlet flowers of the epiphyte *Aeschynanthus*, on a moss-covered trunk, attract nectar-seeking sunbirds which act as pollinators.

Montane ferns, *Dipteris conjugata*, are close relatives of fossil ferns dating from Jurassic times.

The curiously-shaped crags of the Donkeys Ears rise above the granite pavement on Mount Kinabalu, Borneo's highest mountain.

Gnarled *Leptospermum recurvum* trees and rhododendrons on the upper mountain slopes of Kinabalu between 2,600 m and 3,200 m.

A Dayak climber pegs a ladder up a giant *Koompassia* tree to collect honey from the nests of wild bees.

A blazing fire sweeps through a newly cleared *ladang*, clearing the vegetation and providing nutrient-rich ash to fertilize the first year's crops.

Hill rice is the main staple of most shifting cultivators in Kalimantan.

The rhinoceros hornbill, *Buceros rhinoceros,* is an important omen bird for the Dayak tribes.

Whiskered catfish, fattened on falling figs, provide a rich food source for the villagers living along Kalimantan's great rivers.

A raft of logs on the Kapuas, West Kalimantan.

Bundles of rattan canes represent several days' work for collectors in the Sangkulirang peninsula. Almost half of the world's traded rattan comes from Kalimantan forests.

Even though only 10% of the forest trees may be extracted for timber, another 40% are damaged by falling trunks. Logging roads scar the forests and compact the soil.

The forested hills of the Pleihari-Martapura Reserve in the Meratus Mountains protect the watershed of the Riam Kanan Lake, a reservoir for a hydroelectric and irrigation scheme.

The orange patches on this computer-generated image show where fires blazed during the Great Fire of Borneo in 1982 and 1983, which affected more than four million hectares of forests in East Kalimantan and Sabah.

Fires set to clear agricultural fields during the exceptionally long dry season of 1982 spread into adjacent secondary logged forests and peat swamp forests.

Inland fisheries on the Kapuas lakes provide 60% of all fish eaten in West Kalimantan.

Clean water is vital for drinking, washing and domestic use.

The leaves of wetland plants such as reeds, sedges and pandans are used to weave baskets, mats and other household utensils.

Young rice plants ready for planting out on floating mats of vegetation in the Sungai Negara wetlands.

The leaves of the *Pandanus* are utilized by local communities for weaving and thatching materials.

Freshwater turtles, *Callagur borneensis*, are captured for food in the Barito basin.

The bark of the swamp tree gemor, *Alseodaphne coriacea*, is collected to manufacture mosquito repellent.

The traditional Banjarese system, growing rice interspersed with coconuts, is a successful agroecosystem in the wetlands of southern Kalimantan.

Troops of pig-tailed macaques, *Macaca nemestrina*, are destructive pests in farmers" fields. The pig-tail (beruk) is often kept as a pet and trained to collect coconuts.

Bali cattle, high-yielding milk producers, are the domesticated relatives of wild banteng, *Bos javanicus*, which roam Kalimantan forests.

Rubber is an important cash crop for shifting cultivators and other smallholders in Kalimantan.

An oil exploration platform in nipa swamps on the east coast is a potential hazard. Oil spills bring death and destruction to coastal habitats.

An oil pipeline runs beside the rainforest in East Kalimantan. Kalimantan provides 37% of Indonesia's timber and 20% of the country's oil and gas, all major export earners.

A miner pans for gold in a pit near Martapura. As gold prices have risen, large areas of Kalimantan have been allocated as gold-mining concessions to large companies.

A plywood mill at Banjarmasin. Effluent from the wood processing is discharged into the Barito River, while smoke and toxic waste are discharged into the atmosphere.

The green *Calotes* lizard is commonly seen in urban gardens.

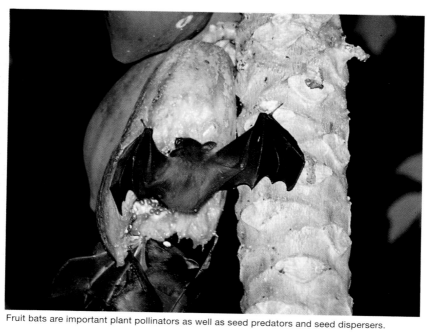
Fruit bats are important plant pollinators as well as seed predators and seed dispersers.

A citizen of tomorrow, this Dayak infant is unlikely to benefit from harvesting Kalimantan's rich natural resources unless measures for their conservation and sustainable utilization are implemented now.

Largest of the arboreal apes, the orangutan, *Pongo pygmaeus,* is a 'flagship' species for conservation. Orangutan reserves in Borneo protect a host of other plant and animal species. Habitat protection is the best, cheapest and most effective way to conserve biological diversity.

The elephant, *Elephas maximus,* occurs only in the northeast corner of Borneo. Like many other forest species, elephants can survive in logged forests, provided that those forests are carefully managed and allowed to regenerate after logging. Logged forests can be useful for conservation, by extending the total area of forest available to wide-ranging species such as the elephant.

Dayak artefacts, an orangutan skull and the casque of a rhinoceros hornbill, represent trophies from past Dayak hunting expeditions. The main threat to these species today comes from destruction of their lowland rainforest habitats.